Benchmark Papers
in Analytical Chemistry

Volume
1 ION-EXCHANGE CHROMATOGRAPHY / *Harold F. Walton*
2 THERMAL ANALYSIS / *W. W. Wendlandt and L. W. Collins*
3 DIGITAL COMPUTERS IN ANALYTICAL CHEMISTRY, PART I: 1950–1969 / *J. B. Justice, Jr. and T. L. Isenhour*
  DIGITAL COMPUTERS IN ANALYTICAL CHEMISTRY, PART II: 1970–1978 / *J. B. Justice, Jr. and T. L. Isenhour*

Benchmark Papers in Analytical Chemistry/3

A BENCHMARK® Books Series

DIGITAL COMPUTERS IN ANALYTICAL CHEMISTRY, PART II 1970–1978

Edited by

J. B. JUSTICE, Jr.
Emory University

and

T. L. ISENHOUR
University of North Carolina

Hutchinson Ross Publishing Company

Stroudsburg, Pennsylvania

Copyright © 1981 by **Hutchinson Ross Publishing Company**
Benchmark Papers in Analytical Chemistry, Volume 3
Library of Congress Catalog Card Number: 81-2933
ISBN: 0-87933-062-7

All rights reserved. No part of this book covered by the copyrights hereon may be reproduced or transmitted in any form or by any means—graphic, electronic, or mechanical, including photocopying, recording, taping, or information storage and retrieval systems—without written permission of the publisher.

83 82 81 1 2 3 4 5
Manufactured in the United States of America.

LIBRARY OF CONGRESS CATALOGING IN PUBLICATION DATA
Main entry under title:
Digital computers in analytical chemistry.
  (Benchmark papers in analytical chemistry; 3)
  Includes indexes.
  Contents: pt. 1. 1950–1969 — pt. 2. 1970–1978.
  1. Chemistry, Analytic—Data processing. I. Justice, J. B., 1946-   . II. Isenhour, Thomas L. III. Series.
QD75.4.E4D53   543'.0028'54   81-2933
ISBN 0-87933-061-9 (v. 1)   AACR2
ISBN 0-87933-062-7 (v. 2)

Distributed world wide by Academic Press,
a subsidiary of Harcourt Brace Jovanovich,
Publishers.

CONTENTS

Preface ix
Contents by Author xiii

Introduction 1

LABORATORY AUTOMATION

Editors' Comments on Papers 1 Through 5   10

1   MARGOSHES, M.: When the Computer Becomes a Part of the Instrument   13
      Anal. Chem. 43(4):101A, 103A, 105A–107A, 109A (1971)

2   FRAZER, J. W.: Computer Automation in Chemistry   17
      Chem. Instrum. 2:271–295 (1970)

3   BAUMANN, F., J. HENDRIKSON, and D. WALLACE: Microprocessors in Chemical Instrumentation   42
      Chromatographia 7:530–538 (1974)

4   GOEDERT, M., S. A. WISE, and R. S. JUVET, Jr.: Application of an Inexpensive General-Purpose Microcomputer in Analytical Chemistry   51
      Chromatographia 7:539–546 (1974)

5   WORDWARD, W. S., T. H. RIDGWAY, and C. N. REILLEY: An Instrumentation-oriented Micro-computer: An Extremely Inexpensive Data Acquisition Computer Optimised for the Automated Laboratory   59
      Analyst 99:838–852 (1974)

CHROMATOGRAPHY

Editors' Comments on Papers 6 Through 10   76

6   GLENN, T. H., and S. P. CRAM: A Digital Logic System for the Evaluation of Instrumental Contributions to Chromatographic Band Broadening   80
      J. Chromatogr. Sci. 8:46–56 (1970)

7   SWINGLE, R. S., and L. B. ROGERS: Computer-Controlled Gas Chromatograph Capable of Real-Time Readout of High-Precision Data   91
      Anal. Chem. 43:810–818 (1971)

Contents

8 DENTON, M. S., B. A. JEZL, W. R. HEINEMAN, and H. B. MARK, Jr.: Computer Control of Data Acquisition, Reduction, and Display in Rapid Scanning Liquid Chromatography — 100
J. Chem. Inf. Comput. Sci. **17**:238–242 (1977)

9 DROMEY, R. G., M. J. STEFIK, T. C. RINDFLEISCH, and A. M. DUFFIELD: Extraction of Mass Spectra Free of Background and Neighboring Component Contributions from Gas Chromatography/Mass Spectrometry Data — 105
Anal. Chem. **48**:1368–1375 (1976)

10 GATES, S. C., M. J. SMISKO, C. L. ASHENDEL, N. D. YOUNG, J. F. HOLLAND, and C. C. SWEELEY: Automated Simultaneous Qualitative and Quantitative Analysis of Complex Organic Mixtures with a Gas Chromatography-Mass Spectrometry-Computer System — 113
Anal. Chem. **50**:433–441 (1978)

SPECTROSCOPY

Editors' Comments on Papers 11 Through 15 — 124

11 HOLLAND, J. F., R. E. TEETS, and A. TIMNICK: A Unique Computer Centered Instrument for Simultaneous Absorbance and Fluorescence Measurements — 128
Anal Chem. **45**:145–153 (1973)

12 CUSHLEY, R. J., D. R. ANDERSON, and S. R. LIPSKY: Computer Controlled Fourier Transform Nuclear Magnetic Resonance System for Carbon-13 and Phosphorus-31 Spectrometry — 137
Anal. Chem. **43**:1281–1287 (1971)

13 PERRY, J. A., M. F. BRYANT, and H. V. MALMSTADT: Microprocessor-Controlled, Scanning Dye Laser for Spectrometric Analytical Systems — 144
Anal. Chem. **49**:1702–1710 (1977)

14 VENKATARAGHAVAN, R., R. J. KLIMOWSKI, and F. W. McLAFFERTY: On-Line Computers in Research. High-Resolution Mass Spectrometry — 153
Acc. Chem. Res. **3**:158–165 (1970)

15 SWEELEY, C. C., B. D. RAY, W. I. WOOD, J. F. HOLLAND, and M. I. KRICHEVSKY: On-Line Digital Computer System for High-Speed Single Focusing Mass Spectrometry — 161
Anal. Chem. **42**:1505–1516 (1970)

ELECTROCHEMISTRY

Editors' Comments on Papers 16, 17, and 18 — 174

16 CREASON, S. C., R. J. LLOYD, and D. E. SMITH: Evaluation of a Computerized Sampling Technique for Digital Data Acquisition of High-Speed Transient Waveforms: Application to Cyclic Voltammetry — 178
Anal. Chem. **44**:1159–1166 (1972)

17	SCHWALL, R. J., A. M. BOND, R. J. LLOYD, J. G. LARSEN, and D. E. SMITH: High Speed Synchronous Data Generation and Sampler System: Application to On-Line Fast Fourier Transform Faradaic Admittance Measurements *Anal. Chem.* **49**:1797–1805 (1977)	186
18	HANAFEY, M. K., R. L. SCOTT, T. H. RIDGWAY, and C. N. REILLEY: Analysis of Electrochemical Mechanisms by Finite Difference Simulation and Simplex Fitting of Double Potential Step Current, Charge, and Absorbance Responses *Anal. Chem.* **50**:116–137 (1978)	195

DATA ENHANCEMENT

	Editors' Comments on Papers 19 Through 24	218
19	HORLICK, G., and H. V. MALMSTADT: Basic and Practical Considerations for Sampling and Digitizing Interferograms Generated by a Fourier Transform Spectrometer *Anal. Chem.* **42**:1361–1369 (1970)	223
20	COOPER, J. A.: Errors in Computer Data Handling *Anal. Chem.* **50**:801A, 802A, 804A, 807A–808A, 810A, 812A (1978)	232
21	den HARDER, A., and L. de GALAN: Evaluation of a Method for Real-Time Deconvolution *Anal. Chem.* **46**:1464–1470 (1974)	239
22	SKOGERBOE, R. K., P. J. LAMOTHE, G. J. BASTIAANS, S. J. FREELAND, and G. N. COLEMAN: A Dynamic Background Correction System for Direct Reading Spectrometry *Appl. Spectrosc.* **30**:495–500 (1976)	246
23	NIEMCZYK, T. M., and D. G. ETTINGER: A Computer-controlled Photon Counting Spectrometer for Rapidly Scanning Low Light Level Spectra *Appl. Spectrosc.* **32**:450–453 (1978)	252
24	McLAFFERTY, F. W., J. A. MICHNOWICZ, R. VENKATARAGHAVAN, P. ROGERSON, and B. G. GIESSNER: Signal Enhancement in Real-Time for High-Resolution Mass Spectra *Anal. Chem.* **44**:2282–2287 (1972)	256

INFORMATION RETRIEVAL

	Editors' Comments on Papers 25 Through 29	264
25	LYTLE, F. E.: Computerized Searching of Inverted Files *Anal. Chem.* **42**:355–357 (1970)	267
26	GROTCH, S. L.: Matching of Mass Spectra When Peak Height Is Encoded to One Bit *Anal. Chem.* **42**:1214–1222 (1970)	270
27	HERTZ, H. S., R. A. HITES, and K. BIEMANN: Identification of Mass Spectra by Computer-Searching a File of Known Spectra *Anal. Chem.* **43**:681–691 (1971)	279

Contents

28   WANGEN, L. E., W. S. WOODWARD, and T. L. ISENHOUR: Small Computer, Magnetic Tape Oriented, Rapid Search System Applied to Mass Spectrometry   290
Anal. Chem. **43**:1605–1614 (1971)

29   FELDMANN, R. J., G. W. A. MILNE, S. R. HELLER, A. FEIN, J. A. MILLER, and B. KOCH: An Interactive Substructure Search System   300
J. Chem. Inf. Comput. Sci. **17**:157–163 (1977)

INTERPRETATION

Editors' Comments on Papers 30 Through 35   308

30   ISENHOUR, T. L., and P. C. JURS: Some Chemical Applications of Machine Intelligence   312
Anal. Chem. **43**(10):20A–21A, 23A–26A, 29A, 31A, 33A–35A (1971)

31   ACZEL, T., D. E. ALLAN, J. H. HARDING, and E. A. KNIPP: Computer Techniques for Quantitative High Resolution Mass Spectral Analyses of Complex Hydrocarbon Mixtures   321
Anal. Chem. **42**:341–347 (1970)

32   CRAWFORD, L. R., and J. D. MORRISON: Computer Methods in Analytical Mass Spectrometry: Development of Programs for Analysis of Low Resolution Mass Spectra   328
Anal. Chem. **43**:1790–1795 (1971)

33   BUCHS, A., A. M. DUFFIELD, G. SCHROLL, C. DJERASSI, A. B. DELFINO, B. G. BUCHANAN, G. L. SUTHERLAND, E. A. FEIGENBAUM, and J. LEDERBERG: Applications of Artificial Intelligence for Chemical Inference. IV. Saturated Amines Diagnosed by Their Low Resolution Mass Spectra and Nuclear Magnetic Resonance Spectra   334
Am. Chem. Soc. J. **92**:6831–6838 (1970)

34   DELFINO, A. B., and A. BUCHS: Heuristic Programming as an Ion Generator in Mass Spectrometry. I. Generation of Primary Ions with Charge Localization   342
Helv. Chim. Acta **55**:2017–2029 (1972)

35   COREY, E. J.: Computer-assisted Analysis of Complex Synthetic Problems   355
Q. Rev. Chem. Soc. (London) **25**:455–482 (1971)

Author Citation Index   383
Subject Index   391
About the Editors   397

PREFACE

The role of computers in analytical chemistry is continually expanding. Digital computers—from the early versions used for off-line analysis and data processing to the microprocessor-controlled instruments currently available—have been incorporated ever further into the everyday world of the analytical chemist. The papers in these volumes (Parts I and II) trace the evolution of computer use by analytical chemists. They are representative of the wide variety of ways in which digital computers have been applied to analytical chemistry. Other editors undoubtedly would have chosen different papers in many instances. And, indeed, we have certainly omitted many excellent articles. Our intention was to select papers that would be indicative of various developments and ideas, not to produce a comprehensive treatise. We believe that our selections are valuable contributions in their own areas and time periods and, certainly, many of them must be considered outstanding by any criterion.

Our initial search* revealed the typical literature explosion so often observed when technology and ideas come into phase. Although much work had been done in applying analog computers to certain areas of analytical chemistry (particularly electrochemistry), few papers were written on the applications of digital computers before 1960. A clearly exponential growth in the publication rate on this subject has occurred from about 1960 through the time of this publication.

It became immediately apparent that there were far more interest-

---

*The following journals have published most of the papers on computers relevant to analytical chemistry: *Analytical Chemistry, Analytica Chimica Acta, Angewandte Chemie* (International Edition), *Applied Spectroscopy, Biomedical Mass Spectrometry, Chemical Instrumentation* (N.Y.), *Chromatographia, Computers in Chemistry, Electrochimica Acta, Fresenius' Zeitschrift fur Analytische Chemie, Journal of the American Chemical Society, Journal of Chemical Education, Journal of Chemical Information and Computer Science, Journal of Chromatographic Science, Journal of Chromatography, Journal of Electroanalytical Chemistry and Interfacial Electrochemistry, Journal of Electrochemical Society, Journal of Magnetic Resonance, Journal of Organic Mass Spectrometry, Pure and Applied Chemistry, Review of Scientific Instruments, Spectrochimica Acta* (A & B), *Talanta,* and *Zhurnal Analiticheskoi Khimii.*

Preface

ing and innovative papers than could possibly be included in one or two volumes of reasonable size. Therefore, a comprehensive treatment of the subject seemed impractical. Furthermore, attempts to organize the subject matter along computer lines (such as hardware versus software, data acquisition versus data processing, and so forth) seemed inappropriate because the analytical chemistry applications did not develop along computer lines but rather developed with respect to various analytical techniques within the limitations of available equipment. For example, some of the earliest extensive applications of digital computers to analytical chemistry occurred in the subject area of activation analysis. This is understandable because the development of modern digital computers was at least partially stimulated by the needs of nuclear physics, which provided the basic techniques and nuclear equipment for activation analysis.

A historical approach to the application of digital computers to analytical chemistry, while it might be interesting in its own right, seemed somewhat far afield from our purpose as well as inordinately difficult. Incredible overlaps in sophistication of applications occurred from one technique to another—again, because of the availability of equipment and the need for the applications.

We chose, therefore, to use a combined chronological-technical organization. We felt that in this way the reader could gain some insight into the development of computer applications to analytical chemistry while simultaneously realizing the variety of possible applications of and approaches to the digital computer's utilization.

Part I includes papers up through 1969, when large-scale data processing facilities were developing throughout the United States. The minicomputer emerged late in this era. Hence, most of this work involved off-line numerical processing of data. Part II includes papers from 1969 through 1978, when the minicomputer was becoming ubiquitous and data acquisition, correspondingly, much more popular. Furthermore, microprocessors were starting to appear just as this volume was produced. Both volumes include individual papers that fit none of our specific categories, but we include them because of their interest, presentation, or vision.

In a preface such as this, one always feels called upon to play sage and prophet simultaneously. The temptation could be no greater than on the subject of computer applications, and we succumbed with only token resistance. Therefore, we speculate on the future development of digital computer applications to analytical chemistry by asking some questions.

Where will the computer go? Where will it lead us, and how far will we be willing to follow? Currently, many believe that modern analytical chemists are very weak on the use of classical techniques. If this is true it must result from the emergence of instrumental methods and the corresponding advantages and glamour of using such techniques. Will the computer bring about a comparable weakness in data-interpretation

Preface

skills? Will the probable emergence of advanced software systems bring about a total dependence on computers for the interpretation of experimental results? It certainly appears that the not too distant future will see analytical laboratories in which virtually every instrument is interfaced to some sort of computer. In the past computer interfaces were, for the most part, one-way paths—that is, the instrument produced information that the computer collected and processed. Today, however, many computer interfaces are two-way paths; the computer not only collects data, but often controls the instrument itself. Furthermore, systems are being developed in which the computer makes logical decisions as to how the instrument should be operated. The road now seems open for the emergence of Norbert Weiner's cybernetics. If this happens, will the computer replace analytical chemists' decision making, and hence their minds, as instruments have largely replaced their hands? In other words, are we looking toward a future in which the analytical chemist will be a complex organization of electronic components rather than the present complex organization of macromolecules?

The answer to these questions will, of course, become known in the future. But one thing is certain. Analytical chemists of the coming computer generation will be satisfied that things are as they should be and will continue to defend the nature of their discipline as they always have.

J. B. JUSTICE, Jr.
T. L. ISENHOUR

CONTENTS BY AUTHOR

Aczel, T., 321
Allan, D. E., 321
Anderson, D. R., 137
Ashendel, C. L., 113
Bastiaans, G. J., 246
Baumann, F., 42
Biemann, K., 279
Bond, A. M., 186
Bryant, M. F., 144
Buchanan, B. G., 334
Buchs, A., 334, 342
Coleman, G. N., 246
Cooper, J. A., 232
Corey, E. J., 355
Cram, S. P., 80
Crawford, L. R., 328
Creason, S. C., 178
Cushley, R. J., 137
de Galan, L., 239
Delfino, A. B., 334, 342
den Harder, A., 239
Denton, M. S., 100
Djerassi, C., 334
Dromey, R. G., 105
Duffield, A. M., 105, 334
Ettinger, D. G., 252
Feigenbaum, E. A., 334
Fein, A., 300
Feldmann, R. J., 300
Frazer, J. W., 17
Freeland, S. J., 246
Gates, S. C., 113
Giessner, B. G., 256
Glenn, T. H., 80
Goedert, M., 51
Grotch, S. L., 270

Hanafey, M. K., 195
Harding, J. H., 321
Heineman, W. R., 100
Heller, S. R., 300
Hendrikson, J., 42
Hertz, H. S., 279
Hites, R. H., 279
Holland, J. F., 113, 128, 161
Horlick, G., 223
Isenhour, T. L., 290, 312
Jezl, B. A., 100
Jurs, P. C., 312
Juvet, R. S., Jr., 51
Klimowski, R. J., 153
Knipp, E. A., 321
Koch, B., 300
Krichevsky, M. I., 161
Lamothe, P. J., 246
Larsen, J. G., 186
Lederberg, J., 334
Lipsky, S. R., 137
Lloyd, R. J., 178, 186
Lytle, F. E., 267
McLafferty, F. W., 153, 256
Malmstadt, H. V., 144, 223
Margoshes, M., 13
Mark, H. B., Jr., 100
Michnowicz, J. A., 256
Miller, J. A., 300
Milne, G. W. A., 300
Morrison, J. D., 328
Niemczyk, T. M., 252
Perry, J. A., 144
Ray, B. D., 161
Reilley, C. N., 59, 195
Ridgway, T. H., 59, 195

Contents by Author

Rindfleisch, T. C., 105
Rogers, L. B., 91
Rogerson, P., 256
Schroll, G., 334
Schwall, R. J., 186
Scott, R. L., 195
Skogerboe, R. K., 246
Smisko, M. J., 113
Smith, D. E., 178, 186
Stefik, M. J., 105
Sutherland, G. L., 334

Sweeley, C. C., 113, 161
Swingle, R. S., 91
Teets, R. E., 128
Timnick, A., 128
Venkataraghavan, R., 153, 256
Wallace, D., 42
Wangen, L. E., 290
Wise, S. A., 51
Wood, W. I., 161
Woodward, W. S., 59, 290
Young, N. D., 113

INTRODUCTION

The development of the computer from Babbage's appropriately titled "analytical engine" to today's microprocessors has been central to the development of automated analytical chemistry. Generally speaking, developments in the use of computers in analytical chemistry have occurred soon after the introduction of available technology. In particular, the growth in computer use accelerated rapidly with the advent of the minicomputer. When individual chemists no longer had to transfer data from the laboratory to a central facility, but could instead process the data in the laboratory, then new ways of thinking about the use of computers began. The computer in the lab meant the beginning of computer interfacing: first the collection of data, then control of the instrument. The initial interfaces were primarily one-way communication channels as data flowed from the instrument to the computer. Later, two-way communication developed. The computer told the instrument when to start, change conditions, and stop. This was still not a cybernetic situation in which the state of the instrument was monitored and modified by the computer via feedback, but it was and still is the most common form of interface.

As instrumentation was growing more complex—for example, the continued development of mass and nuclear magnetic resonance spectrometers—it was natural to interface it to computers. Since these instruments are capable of generating large quantities of data, the need for computer manipulation of the data was clearly established. While the computer's ability to acquire data and provide instrument control was the reason to interface, other factors were necessary to make interfacing possible. Two lines of development that have been important in allowing interfacing to be successful have been in operational amplifiers and in analog-to-digital and digital-to-analog converters. Operational

Introduction

amplifiers have become cheaper, faster, and more stable. They have also evolved into specific types of amplifiers; instrumentation amplifiers, isolation amplifiers, and logarithmic amplifiers all perform functions that satisfy specific needs of various measurement processes.

Analog-to-digital converters have become faster and more accurate, as have digital-to-analog converters. The continued incorporation of more functions into single integrated circuits has led to easily interfaced multiplexed converters that require relatively simple software control.

Perhaps the most significant development in hardware was the introduction of the four-bit, single chip microcomputer in 1971. This device was quickly followed by eight-bit devices that were soon applied to chemical problems. Microprocessors have become so inexpensive that they can be incorporated into almost any instrumentation. For example, entire home computer systems built around microprocessor chips can be purchased for a few hundred dollars. Such systems containing read-only memory can have entire operating systems and languages, such as Basic, built in. Undoubtedly most analytical instrumentation will be controlled by microprocessors in the future.

Although hardware developments were most significant, developments in software also expanded the role of computers in analytical chemistry. Software or algorithmic developments cannot be underestimated. Cooley and Tukey's fast Fourier transform algorithm has had enormous impact in spectroscopy as well as in other areas of analytical chemistry.[1] New uses are still being found for the fast Fourier transform. Several papers that make use of transform methods are included in this volume.

Cross-fertilization has played an important role in the development of analytical methodology and instrumentation. In several papers in this book, the authors have recognized that a conceptual or technological development in another field is a solution to a problem in their area of research. In transferring ideas or technology from one field to another, a new research instrument may be produced or a new method of analyzing data developed. If the new instrument is sufficiently useful, commercial development will follow. The resulting product is then available to others in the scientific community.

Chromatography is one area in which the developments in digital devices were applied to control problems in gas chromatography. Glenn and Cram in 1970 (Paper 6) and Swingle and Rogers in 1971 (Paper 7) illustrated what could be accomplished with the technology then available. The dedication of minicomputers to chromatographic detectors started the evolution of multichannel chromatographic detectors. One application of multichannel detectors is in the area of high-pressure liquid chromatography (HPLC). Denton et al. (Paper 8) present a rapid scanning detector for HPLC that would not be possible without computer control of data acquisition, reduction, and display. Perhaps one of the most significant impacts of computers has been in the development of gas chromatography/mass spectrometry (GC/MS) instruments. Two approaches to the processing of GC/MS data are presented in the Chromatography section. Dromey et al. (Paper 9) discuss extraction of spectra from background interferences, while Gates et al. (Paper 10) describe an automated system for qualitative and quantitative analysis. In the Spectroscopy section several other papers concerning computerization of mass spectrometry are presented. Venkataraghavan, Klimowski, and McLafferty (Paper 14) discuss on-line computers in high resolution mass spectrometry, while Sweeley et al. (Paper 15) also present a computerized approach to mass spectrometry. Improvements in fluorescence spectroscopy in the form of improved reproducibility and accuracy have been made through computer control of the instrument, as demonstrated in the paper by Holland, Teets, and Timnick (Paper 11). NMR has had a tremendous increase in sensitivity and resolution as a result of computer approaches to NMR spectrometry. Pulse triggering and control is an area that has received attention recently.[2,3,4]

Electrochemistry has been advanced over the years by improvements in electronics. The operational amplifier has been used for some time for potentiostatic control and current amplification in electrochemical experiments. As computers became more available, they were also applied to electrochemistry. Papers 16 and 17 in the Electrochemistry section describe how computerized sampling can be used to study transient electrochemical phenomena. A much different approach to studying electrochemistry is presented in the paper by Hanafey et al. (Paper 18). Here simulation is used to analyze electrochemical mechanisms. Simulation has also been used to examine the behavior of surface reactions.[5-9]

One of the growth areas of analytical chemistry has been the computerized handling of information. Data enhancement and information retrieval have both been revolutionized by the computer. The application of statistical methods to chemical data has proved especially productive. Particularly useful for analytical chemists are deconvolution procedures to recover undistorted spectra, such as the one described by den Harder and de Galan (Paper 21). Background correction is another problem that is of almost universal concern in analytical measurements. Skogerboe et al. (Paper 22) describe a dynamic background correction method that they have applied to emission spectroscopy.

Related to the above problems is the concept of signal-to-noise ratio. Niemczyk and Ettinger (Paper 23) describe how the signal-to-noise ratio can be used to control the measurement process via feed-

Introduction

back. In the final paper of the Data Enhancement section, McLafferty (Paper 24) demonstrates further how computer control of the data collection process can enhance the quality of the data.

No discussion of the computer in analytical chemistry would be complete without a mention of the research done in information storage and retrieval. Several search systems are presented, as well as an interesting paper by Grotch (Paper 26) on the remarkable degree of spectral compression possible in storing and searching mass spectra.

The last section of the book contains papers concerned with interpretation of chemical data. Several different approaches are represented. The interpretation of data adds a new dimension to chemical research. Interpretation has historically been the one area that man has had to do completely by himself. Before the emergence of the computer, all instrumentation was fundamentally designed to produce data by experimentation. Computers add a new dimension because with sophisticated programming they can actually participate in the interpretation process.

It has been said that man's two greatest inventions are language and writing—language because it allows the communication of ideas and writing because it allows the communication of ideas over time. Computers may be the third great invention because they are capable of reorganizing information to generate new ideas. No other inanimate device has had that ability.

In its use in chemical research, the computer is a solution to many problems and also a problem in itself. As demonstrated by the papers in this collection, the computer has provided solutions to a number of problems in the areas of spectroscopy, chromatography, and electrochemistry in regard to experiment control, data collection, and data analysis. However, the computer is not a panacea for analytical problems. It is itself a problem in several respects. The use of a computer in an experiment frequently results in the collection of more data than can be easily processed. Data reduction programs must be written and debugged. In developing programs for data collection, one must often write in assembly language to obtain the requisite speed for high data rates. This takes the chemist away from the main task of performing the actual experiment. Microcomputers in particular have been a problem in regard to software and debugging facilities. Although the cost of microcomputer hardware is low, the overhead in software development time can be high. A researcher must decide if the time required to set up a microcomputer controller and write the necessary software is justified in terms of the expected benefit.

Of the recent trends in digital computers, a significant difference at present is in word length between sixteen-bit minicomputers and eight-bit microcomputers. The enormous increase in possibilities with each additional bit of word length (see, for example, how word length effects signal processing in NMR spectroscopy in the Data Enhancement section of this book) means that a sixteen-bit machine is a more flexible and therefore a more powerful machine than an eight-bit machine. However, the level of flexibility required is directly related to the complexity of the task. One can solve quite complex control problems with eight-bit machines. In fact, most microprocessor chips sold are four-bit devices used in games and other low-level control situations. A complex task can frequently be broken down into simpler tasks, each of which can be under individual microprocessor control. An example from the microcomputer itself is the microprocessor disk controller used to control information flow to and from disks. However, sixteen-bit microcomputers call into question the appropriateness of the prefix "micro," since these devices are equivalent to minicomputers.

Another development among the seemingly endless advances should be noted. This is the decreasing cost of mass storage, either in the form of relatively slow memories such as bubble memories or in the form of newer types of disks. Bubble memories offer the promise of low-cost, large memories. Although slow by solid state and core memory standards, they could easily take the place of mechanical mass storage devices such as floppy disks. This would accomplish two things: It would lower the cost of mass storage and it would replace a mechanical device by something with no moving parts. One potential use would be for built-in spectral libraries in spectrometers.

Another trend in computer technology is the proliferation of languages. Fortran has been the common scientific language for some time, although Basic has replaced Fortran in many applications. A newer language gaining in popularity is Pascal, a simple but powerful language. Other languages used by chemists for various problems include APL, a very concise language, and LISP, a language useful in artificial intelligence applications. For many applications the high-level languages are ideal. However, for data collection at high rates of information flow, it is still usually necessary to write in the assembly language of the particular computer. For example, for most versions of Basic, it is not possible to collect data at a rate faster than one point per millisecond.

The area of chemometrics[10] is a growing one. Biology has had several journals devoted to statistical questions for some time, of which *Biometrika* and *Biometrics* are well known. Because chemists have perhaps been better able to control their variables than have biologists, the questions raised in these journals have not affected most chemists. But now many more multivariable chemical experiments are being done that require more sophisticated statistical examination of the results.

When one wonders just how great will be the impact of computers in chemistry, one place to look is the marketplace. A sign of the suc-

Introduction

cessful application or usefulness of an idea or process is its commercial development. In this regard, we note a *C & E News* article that described a business based on computer assisted molecular design.[11] Wipke and Marson have combined a great deal of information about organic structures and a substantial programming effort to generate plausible structures of organic compounds that may exhibit certain types of activity. Descriptions of similar processes can be found in the Interpretation section.

Another way to assess the impact of computers in analytical chemistry is to attend a large analytical meeting such as the Pittsburgh conference where the latest in instrumentation is displayed. Major instruments have computer hardware as a matter of course, but they also have increasingly capable software, in particular, programs to get the data volume down to manageable size and make it easier to interpret. At the other extreme of the cost of instrumentation, even the pipet is under microprocessor control.

An often overlooked benefit of the decrease in cost of instrumentation is a new phenomenon. Prices are such that individuals can now own computers. Because advances are made by those with the tools, even greater advances in the use of computers should be made in the future as people from many different backgrounds explore the computer's possibilities.

During the preparation of these volumes the editors read many excellent papers on the use of computers. It is unfortunate that they all could not be included. We have, however, cited a number of them in each section.

REFERENCES

1. J. W. Cooley and J. W. Tukey, "An Algorithm for the Machine Calculation of complex Fourier Series," *Math. Comput.*, **19**, 297 (1965).
2. A. O. Goedde, M. F. Froix, and D. J. Williams, "A Computer Assisted Method for Measuring NMR Relaxation Times," *Chem. Instrum.*, **7**, 179 (1976).
3. R. E. Adder, A. R. Lepley, and D. C. Songco, "Utilization of a Microprocessor in a Pulsed NMR Spectrometer," *J. Magn. Reson.*, **29**, 105 (1978).
4. B. L. Neff, J. L. Ackerman, and J. S. Waugh, "Fully Automatic Software Correction of Fourier Transform NMR Spectra," *J. Magn. Reson.*, **25**, 335 (1977).
5. M. Angerstein-Kozlowska, J. Klinger, and B. E. Conway, "Computer Simulation of the Kinetic Behavior of Surface Reactions Driven by a Linear Potential Sweep. Part 1. Model 1-Electron Reaction with a Single Adsorbed Species," *J. Electroanal. Chem*, **75**, 45 (1977).
6. M. Angerstein-Kozlowska, J. Klinger, and B. E. Conway, "Computer Simulation of the Kinetic Behavior of Surface Reactions Driven by a Linear Potential Sweep. Part II. Sequential Reactions of Adsorbed Species," *J. Electroanal. Chem.*, **75**, 61 (1977).

7. M. Angerstein-Kozlowska, B. E. Conway, and J. Klinger, "Computer Simulation of the Kinetic Behavior of Surface Reactions Driven by a Linear Potential Sweep. Part III. Monolayer Formation by a Nucleation and Growth Mechanism," *J. Electroanal. Chem.*, **87**, 301 (1978).
8. M. Angerstein-Kozlowska, B. E. Conway, and J. Klinger, "Computer Simulation of the Kinetic Behavior of Surface Reactions Driven by a Linear Potential Sweep. Part IV. Kinetic Behavior of a Nucleation and Growth Controlled Surface Process Under Potentiostatic Conditions and Comparison with Conclusions for Potentiodynamic Conditions," *J. Electroanal. Chem.*, **87**, 321 (1978).
9. R. G. Barradas, F. C. Benson, and S. Fletcher, "A Computer Simulation of the Voltammogram Corresponding to the Two-Dimension Progressive Nucleation and Growth of a Passivating Film," *Electrochim. Acta*, **22**, 1197 (1977).
10. B. R. Kowalski, ed., *Chemometrics: Theory and Application*, ACS Symposium Series No. 52, 1977.
11. *Chemical and Engineering News*, **57**, 29 (1979).

LABORATORY AUTOMATION

Editors' Comments
on Papers 1 Through 5

1   MARGOSHES
    When the Computer Becomes a Part of the Instrument

2   FRAZER
    Computer Automation in Chemistry

3   BAUMANN, HENDRIKSON, and WALLACE
    Microprocessors in Chemical Instrumentation

4   GOEDERT, WISE, and JUVET
    Application of an Inexpensive General-Purpose Microcomputer in Analytical Chemistry

5   WOODWARD, RIDGWAY, and REILLEY
    An Instrumentation-oriented Micro-computer: An Extremely Inexpensive Data Acquisition Computer Optimised for the Automated Laboratory

In the introductory article for this second volume on computers in analytical chemistry, Margoshes discusses the advantages and frequent necessity of including the computer as an integral component of the contemporary analytical instrument. The direct coupling of a computer to an analytical instrument means that the instrument should be designed with computer capability for instrument control and data handling in mind. Fourier transform infrared spectroscopy is used to illustrate the advantages of interfacing a computer to an analytical instrument.

Margoshes mentions two trends for the future of computerized analytical chemistry. The first is in methods of getting maximum information from the analytical signal, while the second, which will probably have the greatest impact, is towards more sophisticated control of the instrument by the computer. Both of these trends are evident in the papers in this collection.

Margoshes points out that it is always better to achieve a physical separation of variables than to have to resort to mathematical techniques to extract the desired information, but that frequently a mathematical separation is necessary. Fourier transform spectroscopy is a case in which variables are separated mathematically.

Paper 2 is a review by Frazer of the development of automation in chemistry, in which he discusses types of automation along with the associated advantages and disadvantages. Hardware and software considerations are discussed, with greater emphasis on software, and several approaches to laboratory automation are critically treated. Time-sharing, dedicated computers, assembly language programming, and high-level languages are evaluated from the practicing chemist's point of view.

In another paper on laboratory automation, Dulaney[1] has considered the requirements of computer interfacing to and processing of data from chemical instrumentation, with particular emphasis on cost effectiveness. The paper is a good introduction to computerized signal processing but, due to limitations of space, could not be included here.

Frazer (Paper 2) lists several reasons for automating an experiment. Among these are: (1) to perform data acquisition functions, (2) to control the experiment, (3) to provide on-line computation, and (4) to permit adaptive experimentation. He also states, and it is still true today, that "the largest single stumbling block to computer automation of the chemical laboratory appears to be the implementation of proper software programs." Hardware developments, such as microprocessors, have far outpaced the development of software.

The last three papers of this section (Papers 3, 4, and 5) indicate the future direction of digital electronics with respect to chemical instrumentation. The minicomputer increased the chemist's ability to control his experiments and analyze his data. The microprocessor does the same at a greatly reduced cost. Microprocessors are to the minicomputer what the minicomputer was to the maxi. They bring computers closer to the experiment. Their greatest impact will be in the control of analytical instruments, where their low cost will make them more available than minicomputers. One pays for the reduced cost, however, in loss of minicomputer flexibility. In the future most major chemical instrumentation will probably be equipped with microprocessors to aid in instrument control. Currently, a user needs more knowledge to implement a microprocessor than to implement a minicomputer to accomplish the same task; however, as microprocessor technology advances, this situation should change.

Baumann, Hendrikson, and Wallace provide an overview of microprocessors, in Paper 3, while Goedert, Wise, and Juvet (Paper 4) report on the specific application of microprocessors in analytical chemistry. In another microcomputer paper, Woodward, Ridgway, and Reilley (Paper 5) discuss design and realization of a digital computer system for an entire chemistry department, which uses microprocessors as "intelligent" data acquisition terminals of a central system. As stated in the article, it was the intention of the authors to "design a modular unit that would be sufficiently inexpensive and yet adaptable to various applications to possess the potential to become as ubiquitous as the research-quality oscilloscope or digital voltmeter."

Among other applications of microprocessors not included in this volume is the interesting paper of Eaton and Stuart,[2] in which an inexpensive microcomputer assisted single-beam photoacoustic spectrometer is described. Other papers that may be useful to chemists using or considering use of microprocessors in their work include those of Ritz et al. and O'Haver. Ritz et al.[3] have described a microprocessor controlled data acquisition system for pulsed laser spectroscopy that uses a signal processing system consisting of gated integration with pulse averaging and ratioing to a reference signal. O'Haver[4] has described a microprocessor-based, linear response time low-pass filter. The response characteristics relative to an RC filter are examined and several advantages of the microprocessor-based filter are pointed out, not the least of which is programmability. O'Haver mentions an important point in using microprocessors in chemical instrumentation—that is, the need to consider the manhours required of the investigator inexperienced in such matters to construct, program, and debug such a device. If the device will be of significant advantage in his work, or if the experience gained in the development will be of future value, then such an undertaking may be worthwhile.

REFERENCES

1. Dulaney, "Computerized Signal Processing," *Anal. Chem.*, **47**, 24A (1975).
2. H. E. Eaton and J. D. Stuart, "Microcomputer Assisted Single Beam Photoacoustic Spectrometer System for the Study of Solids," *Anal. Chem.*, **50**, 587 (1978).
3. G. P. Ritz, D. J. Wallen, and M. D. Morris, "Microprocessor-controlled Data Acquisition System for Pulsed Laser Spectroscopy," *Appl. Spectrosc.*, **32**, 493 (1978).
4. T. C. O'Haver, "Microprocessor-Based, Linear Response Time Low-Pass Filter," *Anal. Chem.*, **50**, 676 (1978).

When the Computer Becomes a *Part* of the Instrument

MARVIN MARGOSHES
6806 Delaware St.
Chevy Chase, Md. 20015

Computers have revolutionized analytical instrumentation and the problems of instrument-computer interfacing are continually confronting analytical chemists. Making the computer a component part of the instrument should provide the most feasible means of realizing the maximum information content of an analytical method

COMPUTERS ARE CHANGING analytical chemistry. Their influence on our instrumentation and methods will be as dramatic and pervasive as the earlier effect of electronics. Those who recognize and make intelligent use of these new developments will reap considerable benefit. The field of X-ray crystallography provides an example of the likely effects. With computerized techniques, the same group of crystallographers was able to increase its number of publications sixfold (1). The accuracy and precision of the data are also improved. As a result, entirely new types of crystallographic studies are being made.

In analytical chemistry, computers are used mainly to automate existing instruments and conventional computations. These applications have been useful, but the real benefits of computerizing will come with the development of new measurement methods that may be possible only with computerized instruments, and from techniques which take advantage of more of the information content of an analytical method. Kaiser (2) recently discussed, in this journal, the information content of an analytical method. Although Kaiser stated that he was not necessarily referring to computers, it is apparent that routine realization of his suggestions requires the use of these machines at least for the computations.

In this article, I will stress the instrumentation aspect, though one really should not isolate the measurement from how the data are used. In particular, I will stress types of measurements that, perhaps, can be accomplished without a computer, but which are most feasible when the computer is made a part of the instrument.

Attaching the Computer to an Instrument

Before going into the main theme of this article, it is worthwhile to take a brief look of the use of computers with conventional analytical instruments. In the first applications, the computer was used off-line only, and analog computers often were used because the digital machines were still expensive and unreliable. Despite these limitations in early work, one of the first uses of computers in analytical chemistry is a good illustration of a technique that is greatly improved by being able to carry out fairly extensive computations.

The application was the analysis of petroleum fractions by infrared spectrometry. These fractions frequently are mixtures of similar compounds, and it is nearly impossible to find an infrared frequency which allows measurement of the concentration of one component specifically, without interference from the other components. Quantitative concentration data were computed by equating the absorbances at each of several wavelengths with the sum of the absorbances of the individual compounds. For a ten-component mixture, measurements were made at ten wavelengths and ten simultaneous equations were solved on an analog computer. In principle, the analytical computations could have been made by hand or with a desk calculator, but the greater speed of the computer made the method much more attractive for routine use.

The gas chromatograph eventually took over this type of analysis. It has the advantage of giving a physical, rather than mathematical, separation of the compounds in a complex mixture. Here, too, the computer has become prominent. The original hardware integrating circuits have been replaced by computer attachments of varying capability, providing for one or many chromatographs and for different data reduction techniques. The instrument industry has developed these many types of systems in response to the widely differing needs of analytical laboratories. The chemist is faced with the task of selecting both the best chromatograph and the best data system for his present and projected needs. This cost-benefit analysis requires time, but there is an obvious advantage in having flexibility of choice.

Numerous types of computation devices also have been used in emission spectrometry over the past two decades. These include analog and digital devices; the digital computers were used in the dedicated, time-sharing, and batch-loading modes. Mostly, the computers were programmed to automate the usual types of computations. It has been proved (3) that improvements in analytical accuracy and precision are possible when more complex calculations are made and/or when more of the information in an emission spectrum is used. Until quite recently, there were no instruments which could make use of more than a small fraction of this information. Automatic, high-speed microphotometers now have been developed (4) to read all of the information on a photographic plate and convert it into machine-readable form. A spectrometer capable of measuring photoelectrically at up to 2048 wavelengths is under development (5). This spectrometer *requires* that a computer be designed in; it cannot be operated in any other way. The automatic microphotometer does not require a computer as a component, but it produces more data than can be handled manually. It is safe to predict that these instrumental developments, together with computational improvements, will have a dramatic impact on methods of emission spectrochemical analysis far beyond mere automation of existing methods.

Pulse-height analysis is done routinely for collection of data in such analytical methods as nuclear activation and X-ray spectrometry. Signal averagers are less common, but they are used in several methods where the signal measured in a short time is accompanied by excessive noise. It is most common to use special pulse-height analyzers and signal averagers, but, in fact, these are forms of computers. The data collected by these instruments are often fed to a larger computer, and it is logical to use a general-purpose computer to combine the tasks of data collection and analysis. The general-purpose computer is usually less expensive than the special-purpose device, though savings here must be balanced against programming cost. The great advantage of the general-purpose computer in these applications is flexibility in data collection and processing through changes in software. The major disadvantage is that general-purpose computers are normally slower in carrying out a function than are machines wired for the specific operation. It is still necessary to use special-purpose signal averagers and pulse-height analyzers when the data input rates are high.

Building the Computer In

When it is recognized that the instrument will always be used with a computer, it becomes logical to design the instrument with the computer as a component rather than as an add-on attachment. Two major advantages come from this approach: (1) It may be possible to make new types of measurements when the computer is used to operate the instrument and/or to convert raw instrument output into a useful form. (2) The computer can serve several purposes and can replace costly specialized circuits.

Fourier transform spectrometers illustrate both advantages. Both nmr and infrared spectra are now being measured by Fourier transform methods. The ir measurements are probably easier to explain, so they will be described first.

Infrared Fourier Transform Spectrometry. The essential components of an infrared Fourier transform spectrometer are the interferometer and the data system. The latter consists of a computer, interface circuits, a plotter or other device for displaying the spectrum, and a Teletype or similar unit for input and output of nonspectral information.

Figure 1 is a schematic representation of a Michelson interferometer. The beam splitter sends beams of the incoming light to the two mirrors. Light reflected from these mirrors is recombined at the beam splitter and emerges at right angles to the incident beam.

We may now consider what happens when the incoming light is monochromatic and one of the mirrors is moved toward the beam splitter at a constant velocity. At the starting point, the two mirrors are equidistant from the beam splitter, so that both light beams travel the same distance and arrive back at the beam splitter in phase. These beams interfere constructively and the intensity of the beam emerging from the interferometer is at a maximum. When the mirror has moved a distance equal to one-fourth of the wavelength of the incident light, the distance traveled by the beam in that path is one-half wavelength shorter than the path to and from the fixed mirror. The two beams then interfere destructively and give a minimum in the output intensity. When a detector senses the intensity of the beam from the interferometer and its signal is plotted against the distance moved by the mirror, the curve is a cosine function. This is the *interferogram* for monochromatic light.

When monochromatic light of another wavelength is used, the interferogram is identical except that it has a different period. The interferogram for polychromatic light is the sum of the individual cosine functions which result from all of the discrete wavelengths. Each cosine function is weighted by the intensity at that wavelength. This interferogram is simply the Fourier transform of the spectrum, and computing the Fourier transform of the interferogram gives the desired spectrum.

The basic functions of the data system in this case are to store measurements made at equal intervals of mirror

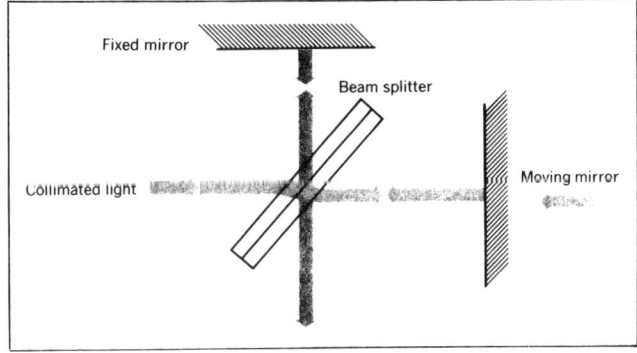

Figure 1. Michelson interferometer

movement, to compute the spectrum from the interferogram, and to plot the spectrum. The computation step includes phase correction and apodization as well as the Fourier transform. Phase correction compensates for the probability that the two mirrors are not initially at the same distance from the beam splitter. Apodization removes certain artifacts from the spectrum which arise from the fact that the measured interferogram is only of finite length.

Improved signal-to-noise in the spectrum can be obtained by signal-averaging repeated scans of the interferometer. This becomes another valuable function of the data system. In addition, the computer can be programmed to perform other useful functions. One is to control the resolution by changing the total distance moved by the mirror. Another is to control the form of the spectral plot. Depending on how the computer is programmed, the size of the plot may be varied and wavelength or wavenumber can be plotted vs. transmittance, absorbance, or log absorbance. In addition, the data system can continuously monitor certain instrument functions and alert the operator if certain key components are not operating properly. When the computer controls the instrument, less skilled operators are needed than with conventional instruments. The instrument operation is changed by typing commands and data into the keyboard of the Teletype, rather than by adjustment of knobs.

Automation of the measurement process is only a fringe benefit in this case. The key reason for computerizing is that it is the only practical way to do Fourier transform spectroscopy, and this method has important advantages compared to scanning through the spectrum with a grating or prism monochromator (6). In the Fourier transform method the detector senses signals for all wavelengths simultaneously instead of one at a time. This is called the *multiplex advantage* or *Fellgett's advantage*. Also, an interferometer is easily made with mirrors perhaps five centimeters in diameter and thus, the input aperture is much larger than the slit of a monochromator. This is called the *throughput advantage* or *Jacquinot's advantage*. The combination of the two advantages gives a Fourier transform spectrometer the ability to record a spectrum in a much shorter time than can a prism or grating spectrometer. Alternatively, signal-averaging can be used to give a much better signal-to-noise in the spectrum for a given measurement time. The throughput and multiplex advantages are more than large enough to overcome the extra time required in the Fourier transform spectrometer to carry out the computations and to plot the spectrum.

Pulsed Nmr Spectrometry. Pulsed nmr spectrometry is a more recent application of Fourier transforms. The sample in a magnetic field is excited by a radiofrequency pulse of short duration. After the pulse, the resonance signal emitted by the sample is recorded as a function of time. This signal is the Fourier transform of the nmr spectrum.

An nmr measurement made this way has the multiplex advantage, but not the aperture advantage. Also, the measurement time for a single pulse is quite short. As a result, the signal-to-noise in a spectrum from a single pulse is quite poor, and signal-averaging of repeated pulse signals is needed. Various pulse signal sequences may be used, depending on particular measurement requirements. One of the functions of the computer may be to control the pulse sequence. This allows more flexibility than hardware control of the pulses.

A typical application of pulsed nmr is to record spectra from ^{13}C nuclei in samples which were not enriched in this isotope. Such measurements are effectively impossible by the older nmr methods. Commercial computer systems for this purpose have only recently become available, but it is clear that the Fourier transform method will have a major effect on analysis by nmr spectrometry.

The Economics of Computerized Instruments

A change in the design of an instrument to improve one aspect of its performance nearly always involves some drawback. In the case of computerized instruments, the drawback is cost. Computers and interface components are declining in cost, but it is expensive to design the interface and program the computer. The design and programming costs are minimized when a manufacturer can spread them over many instruments, but computerized instruments are generally much more expensive than conventional instruments. For example, a grating infrared spectrometer can be purchased for one-half to less than one-tenth the price of a Fourier transform spectrometer.

The extra cost is justified if it allows new types of measurements which have high economic or scientific value, or if the usual measurements can be made for a total lower expense. It is difficult to make generalizations about the value of new types of measurements, but we can examine costs of making experimental measurements of the usual type with conventional and computerized instruments.

Computerized instruments can be more economical if they significantly reduce the time needed for measurement and computation. When a computer is attached to a conventional instrument, measurement times are not reduced, and only the time for computation is affected. When the instrument is designed for computer operation, significantly shorter measurement times may be achieved. This is true for the Fourier transform infrared spectrometer, where the multiplex and throughput advantages result in much shorter times for recording a spectrum.

Both advantages are greatest in recording high-resolution spectra. Then there are many resolution elements and in conventional instruments it is necessary to use very narrow slits. If a spectrum is measured from 400 to 3800 cm^{-1} at 0.5-cm^{-1} resolution, there are 6800 resolution elements, and the multiplex advantage is 6800:1. A typical value for the aperture advantage is 100:1; this is less than the ratio of the area of the interferometer aperture to the area of a slit, but the interferometer must be illuminated by parallel light and a grating or prism spectrometer gains back some of the throughput in the external optics.

A valid comparison of dispersive and interferometer instruments must take into account all time factors, not just the measurement time. In practice, a Fourier transform spectrometer will record a complete spectrum from 400 to 3800 cm^{-1} at better than 0.5-cm^{-1} resolution in less than one-half hour. Conservative estimates show that the best commercial grating infrared spectrometer will take more than nine months, at a minimum, to record this spectrum.

The cost of running this one spectrum with the Fourier transform spectrometer is only about $15, including amortization of the instrument, cost of the operator, laboratory space, etc. Amortization alone for the grating spectrometer is about $5000 for nine months. The economics clearly favor the Fourier transform spectrometer even if only a few high-resolution spectra are needed.

Most laboratories have only a limited need for high-resolution spectra. A typical infrared spectrum is recorded in 20 min on a $9000 spectrometer. At this rate, the spectrometer will produce 24 spectra in an eight-hour day, 120 per five-day week, and 6000 per 50-week year. The five-year amortized cost of the spectrometer is only $1800 per year, but the technician cost (with overhead) will be about $15,000 per year. The total cost for the 6000 spectra is about $16,800, neglecting such items as laboratory space and chart pa-er which are essentially the same for the two types of spectrometer being considered.

The amortized cost of the Fourier transform spectrometer is $13,000 per year, and adding the cost of the technician brings the cost of operation to $28,000 per year. When the work load is only 24 samples per day, it costs about $2.00 more per spectrum to do the work on the interferometer spectrometer.

These cost estimates favor the dispersive spectrometer, but it is necessary to keep in mind several items which can swing the balance. First of all, the Fourier transform spectrometer can record a spectrum in three minutes which is at least as good as that recorded by the dispersive spectrometer in 20 min. The interferometer instrument can run many more than 24 samples per day with almost no increase in cost. When the spectrometer is being used to control a process stream, the ability to record spectra more often can have considerable economic impact. When the spectrometer is being used to support a research scientist, the cost calculations should include the value of his time while he is waiting for the spectrum. Ideally, the scientist will be doing other useful work in this time, but in practice he may use the delay for an impromptu coffee break.

For the one extreme case of high-resolution spectra, it is easy to show that the computerized system is actually much less expensive than the nominally "cheaper" dispersive spectrometer. In the other extreme case, the economic advantage may go one way or the other, depending on individual circumstances. In this latter example, the cost of manpower becomes the key factor, and historically this has been increasing faster than instrument costs. Economic factors thus favor increasing use of computerized instruments in the future.

For the Future

These first steps toward fully computerized instruments will certainly be followed by the development of many new types of instruments, some of which will be more complex. This prediction is based on advantages of these instruments which have already been cited, and on other benefits of computerizing. For example, the following benefits come from having the analytical information in digital form in a computer:

(1) Higher accuracy and precision in the output compared to analog representation.

(2) Facile control of output size and form.

(3) Easy transfer of data to a larger computer.

(4) Computer correction of data for experimental error.

(5) Computer interpretation of data.

(6) Digital storage of data for later retrieval, interpretation, and comparison with new results.

At least two trends in computerized analytical chemistry are already discernible. New types of research studies are being directed to methods of making use of more information in the analytical signal. When such research was done earlier, it was chiefly of academic interest because the measurements and computations would have required too much time for most applications. New instruments and computer methods eliminate the time restriction (in fact, less time may be needed to measure more data and do more calculations). Acceptance of the new methods requires only that advantages be demonstrated to justify the extra costs. Such demonstrations already have been made in several cases.

A second trend is toward more sophisticated control of the instrument by the computer. Presently, the computer is programmed to operate the instrument by a fixed scheme. The computer potentially can be programmed to alter the experiment in response to analysis of the data being recorded. When this is done, the analytical chemist will specify the desired results to the instrument, rather than stating how the measurements are to be made. At the same time, the computer will be checking the instrument operation. It will monitor several components directly, as is now being done in a few instances, and also will analyze the output for other signs of malfunction. When these tests indicate an instrument malfunction, the computer can terminate the measurement process and indicate to the operator the nature of the disorder. The extra cost for these operations would be regained by more efficient instrument operation.

References

(1) H. Cole, *IBM J. Res. Develop.* **13**, 5 (1969).
(2) Heinrich Kaiser, ANAL. CHEM. **42** (2), 24A (1970).
(3) Marvin Margoshes, *Spectrochim. Acta* **25B**, 113 (1970).
(4) A. W. Helz, F. G. Walthall, and Sol Berman, *Appl. Spectrosc.* **23**, 508 (1969).
(5) Marvin Margoshes, Pittsburgh Conference on Analytical Chemistry and Applied Spectroscopy, Cleveland, Ohio, March 1970.
(6) M. J. D. Low, ANAL. CHEM. **41** (6), 97A (1969).

2

Copyright © 1970 by Marcel Dekker, Inc.

Reprinted from *Chem. Instrument.* 2:271-295 (1970)
by courtesy of Marcel Dekker, Inc.

Computer Automation in Chemistry

Jack W. Frazer

*Lawrence Radiation Laboratory,
University of California
Livermore, California 94550*

INTRODUCTION

The digital computer is a common instrument in many laboratories; it is no longer necessary to describe its operating characteristics in detail. However, it is not yet generally used in an entirely satisfactory manner nor are its full capabilities understood and exploited in laboratory automation. Some of the more pressing issues to be understood and resolved are optimal trade-offs between hardware and software, dedicated computer automation versus automation via time-shared systems, single-processor time-shared systems versus multiprocessor hierarchical time-shared systems, and the development of a real-time high-level control language.

Computer automation in the chemistry laboratory is a rapidly growing area of technology; most literature references are dated after 1965. However, despite the present rate of growth of laboratory automation, many leading scientists have refused to become entangled in the vulgarities of on-line computer automation. The availability of computational resources in a batch-processing mode has answered many of their more pressing needs. Hence, while almost all scientists recognize and use, to a greater or lesser extent, the computational facilities of a computer center, the value of a computer in their laboratory is often not readily apparent to them. Perhaps, however, a suspicion is arising that this reluctance to become involved in computer automation may be somewhat short-sighted. Scientists in

---

†This work was performed under the auspices of the U.S. Atomic Energy Commission.

increasing numbers are beginning to use small computers for the more obvious tasks of data acquisition and on-line computations.

First, this review briefly outlines the growth of chemistry automation, giving specific examples of the classes and types of automation and of the advantages and disadvantages of each approach with respect to evolving technology. Last, it presents the author's opinion as to the direction of effort over the next few years.

APPROACHES TO LABORATORY AUTOMATION

What the scientist wants to know is why he should automate his laboratory via a digital computer. Depending upon the activities of his laboratory, the answer could be any one or any combination of the following.

(1) To perform data acquisition functions. The digital computer gives more precise and accurate results than can be obtained by analog techniques. For example, in a digital computer system the signal-to-noise ratio can be greatly improved by means of ensemble-averaging. Spectral data can be smoothed and reproduced up to the inherent capabilities of the instrument, that is, the quality of the analysis is not limited by the accuracy of the output device (e.g., a strip chart recorder).

(2) To control the experiment. The computer can perform many of the routine tasks that are time-consuming and uninteresting.

(3) To provide on-line computations. Many benefits can be obtained from such a capability. First is an overall saving of manpower. Second is turnaround time, that is, the answers appear on the output device almost immediately after the experiment or analysis has been completed. Third, when coupled with instrument control, the user can have "closed-loop" control, that is, the computer can rapidly make any computations required for control of the instrument of process by means of a predetermined logarithm.

(4) To control an instrument or perform an experiment in a way that is either very difficult or impossible by more standard techniques. For example, several instrument parameters can be measured at very nearly the same instant and the output can be corrected according to the cross correlation of these functions. For mass spectrometry of gases, this might mean measurement of ionization current and isotron temperature during the time the ion current (for every ion fragment) is at a maximum; the maximum ion current could then be corrected for the interdependencies of these two parameters.

(5) To provide for more flexible, interactive experimentation and

experiment control. For example, this could mean that the course of an experiment could be followed by means of a CRT display. When the proper software-hardware is provided, the chemist can then interact with the data (mathematical manipulations), interactively change the course of the experiment, and observe the effect of his interaction.

(6) To permit adaptive experimentation. In the future the direction of the experiment (via the parameters under computer control) can be changed almost continuously in response to all of the information available at the instance of change.

It is important that the computer be introduced into the scientific laboratory in such a fashion that scientific experimentation and endeavor be enhanced, not degraded. Although much is to be gained by automation, digital automation is no panacea. Computer automation is not a cure for poorly conceived or executed experimentation; such an application merely spawns more useless data. However, when properly implemented, the computer becomes part of the overall experiment or process. In addition, it can provide maximum freedom and capability for the imaginative scientist.

Expertise in the programming or interfacing of a computer to the equipment is necessary but not sufficient to assure success in laboratory automation. More than any other laboratory instrumentation, automation via a digital computer is a systems problem; that is, automation is not only a problem of computer hardware and software, but must also include the necessary interfaces and instrumentation (with the required modifications). Furthermore, to develop an optimum automation system, careful attention must be paid to the many subtle interdependencies of its components. For example, the scientist must consider the interdependencies of computer hardware and software, that is, the subtle trade-offs of one for the other. Software is still usually more expensive than hardware although this phenomenon is beginning to change, as noted later. Therefore, as an alternative to sophisticated programming, it is often better and less expensive to expand the hardware in the forms of more sophisticated digital and analog interfaces between the instrument and the computer.

The computer system should be capable of responding to the imaginative scientist's needs. The computer should perform the tedious, mundane tasks and the predetermined mathematical analyses, thus providing the scientist with freedom and time for more creative work. Too often systems are built in such a way that the computer offers the scientist limited assistance instead of responding to his overall needs in the context of the experiment.

Hardware

New small computers are being developed and marketed at an ever-increasing rate. Hardly a month passes without the appearance of a new computer reputed to provide more computational power, a faster CPU, an expanded instruction set, greater I/O capability, and so on. Experience in the on-line use of the small computers, together with improved and less expensive integrated circuits, will undoubtedly result in even more rapid changes in hardware. Even now, some CPUs and core memories of the small 12- to 16-bit computers are being assembled on two or three large boards and thus can "disappear" into the standard electronic hardware of many laboratory instruments.

In some instances these computers may be of the type having read-only memories (ROM), that is, applications programs are wired into the computer memory and can be read, but not changed, by the CPU. Unlike the general-purpose computers, which can be easily reprogrammed by reading in a new binary listing from disk, paper tape, magnetic tape, cards, and so on, these computers can only be reprogrammed by rewiring the instructions. The programs for such a computer are designated "firmware."

Some of the new small computers now consist of ROM mixed with the standard read-write memory. These instruments often offer great versatility but are not necessarily for amateurs. The utilization of the full capability of these computers is not at all well understood. In addition, there is virtually no software available to support their implementation.

The trend in small computer hardware appears to be toward modularity. In a few more years it will be possible to purchase "parts" for the construction of a computer system tailored to the specific needs of the purchaser. In the meantime the newer computers are constantly being improved. Some of these units have been in continuous operation for a year or longer without any hardware breakdown. In some cases the CPU is more reliable than the I/O equipment, the interfaces, or the actual laboratory instrument being controlled.

At the present time we are entering a period that will see a great improvement in I/O devices. These peripheral devices will become less expensive, more reliable, and more versatile. Notable examples are some of the newer CRT display units. Also under development are new, inexpensive, small disks and drums that will greatly expand the capability of the small inexpensive computer system. Moreover, in the newer computers, the I/O functions are becoming more nearly integral parts of the computer.

Software

To date, the largest single stumbling block to computer automation of the chemistry laboratory appears to be the implementation of proper software programs. This is especially true for systems and applications programs written in assembly language. Software implementation is made more difficult because the specifications often are neither well planned nor well defined. The proper design of good software is an important concept that should command more resources than are generally allotted for its implementation.

Despite the present disagreement on time-shared systems versus dedicated systems, it is our belief that the time-shared system is the proper configuration and that it is in fact the configuration of the future. However, based on present computer designs, the future time-shared configurations will probably be different from the systems now in operation. Later, we shall propose a configuration for such a system.

In the future, dedicated computers will be used for the most part where an experiment or instrument requires all or nearly all of the computer capacity. Obvious exceptions are small laboratory facilities that can profitably use only one small computer that is operated intermittently. However, these computers will often be operated in a time-shared mode using a core-resident interpreter and a properly designed monitor (as discussed below).

For most laboratory instruments the very small computer is much too fast and has far more computing power than is required to service a single instrument. Instruments generally require very slow data rates and not much computational power or extensive I/O requirements. In fact, even when the scientist wishes to operate interactively with his experiments the CPU is used very little if the hardware has been properly designed. Therefore, even very small computers will be used in the future in a time-shared mode; however, as stated earlier, the time-shared systems developed to date will not necessarily be the preferred implementation of the future.

Up to the present time, time-shared systems generally have been developed around two different philosophies. The first is based on operation in a foreground-background mode, and the second, on the foreground mode only. The first philosophy is exemplified by the IBM 1800 (16-bit, 2-μsec CPU) laboratory system as implemented in a very dynamic way at the Procter and Gamble Laboratories and at the IBM Research Laboratories in San Jose, California. A similar system is installed in the Max Planck Institut für Kohlenforschung, Mülheim-Ruhr, Germany, where a 36-bit PDP-10 is used in a time-shared configuration. The second philosophy of time-shared implementation is exemplified by that developed at the Lawrence Radiation Laboratory (LRL), Livermore, California, and other laboratories.

In systems based on the first philosophy, operation in a foreground-background mode, data acquisition, instrument control, and on-line calculations are executed in a foreground mode. These applications programs are written in FORTRAN and executed from compiled programs. The user can compile and debug new programs in the background mode for use in real-time applications. In addition, the computer time not required to service the real-time operation can be used in the background mode for the usual batch-processing type of tasks. This means that the monitor not only must have good real-time response characteristics but also must incorporate simultaneously the usual time-shared scheduling algorithms to process properly the multiuser background mode. It is very difficult to perform these varied tasks properly with a single computer of the present design. Furthermore, any modifications of these monitors are difficult and time-consuming. It seems more reasonable to assign these two functions to two different computers connected in a hierarchical arrangement. (This point is developed further in the Discussion.)

The advantages of this approach are that the user can write his applications programs in a high-level language and that he can compile and debug new programs without shutting down the time-shared operations. It is often said that a further advantage is that all users have access to very expensive I/O devices, such as line printers, disks and so on, that they could not afford for each individual dedicated computer facility. This argument is no longer valid, however, when small computers are interconnected in a hierarchical arrangement.

For those applications, the overall systems are fairly large and expensive. Core memory requirements are usually 32K or greater. Large core memories are required to support the monitors (8K to 22K) and the applications programs compiled from FORTRAN code, and to provide adequate area for batch-processing in the background mode.

The second type of time-shared system, exemplified by the system developed at LRL, provides no background capabilities. Instead, the scientists have chosen either to perform the assemblies and debugging in another computer or to shut down the time-shared system on a regular basis to accommodate those tasks. The monitors are generally of three-level construction to provide the time-response characteristics necessary for data acquisition, instrument control, and computations.

The advantages of the second type of monitor are its fast time-response characteristics and its relatively small size. Its disadvantages include the necessity for shutting down the time-shared system to assemble, compile, and debug new programs. An additional disadvantage is that these monitors usually but not always require that the user write the application programs in assembly language.

Until recently, anyone who wished to use a small computer was forced to write programs in assembly language. Hence during the first years of laboratory automation, when many scientists employed small computers dedicated on a one-to-one basis to an instrument, most of their applications programs were written in assembly language. However, within the past year, scientists have begun to make increasing use of high-level languages for laboratory automation via dedicated computers. The compiler language FORTRAN has been the most extensively used, but the interpreter languages FOCAL and BASIC are beginning to come into their own.

One variation based on the intermingled use of two languages was developed at the University of Louisville under the direction of S. L. Cooke [1]. The dedicated computer for this project was a PDP-9 (18-bit). Software standardization was effected by subprograms that enabled graduate and undergraduate students to write all their programs in higher-level languages. Additional packages of assembly language subprograms were written to facilitate the use of the analog converters and the clock with both the compiler language FORTRAN and the interpreter language FOCAL. In practice, a student selects the best language for the particular job. A combination of the two languages has many advantages. The more efficient FORTRAN language, modified with real-time subroutines, is used to collect and punch either raw or slightly refined data on paper tape in an appropriate format. These data are then subsequently processed and analyzed with a program written in the more convenient FOCAL. In this manner the operator can interact with the computer, modifying the FOCAL program immediately after seeing numerical and/or graphical results from the last analysis. As a result, statistical refinement is greatly facilitated since only the necessary amount of analysis is performed.

At the University of Oregon, C. E. Klopfenstein [2] is using small (620i, 16-bit) computers in dedicated modes for teaching and chemical experimentation. He found in the initial stages of automation that over 95% of the small computer machine time was being used for the editing of paper tapes and assembly of programs from paper tapes. To improve the effective utilization of the small computers, an assembler program was developed that would function within the university-supported computer system but would produce a code that could be executed on the small computer. In addition, a simulator (written in FORTRAN) was developed that can use a large computer to debug certain kinds of errors in the programs of the small machines.

As a final example of more recent software developments, consider the use of a core-resident interpreter as part of a real-time operating system. Frederic Strange[†] has developed such a monitor for laboratory automation. Its major features are multilevel priority structure; simple but powerful real-time control, plus capability for data acquisition and use of a computational language (modified FOCAL); unified buffer and file structure, providing for efficient, uncomplicated management of data flow through the system; and provisions for multiprocessor configurations that use local slave processors to provide computing and data file management services.

A four-level priority structure is inherent in the monitor. The two highest levels are concerned mainly with the standard problems of device servicing, error-condition identification and recovery procedures, data transport, and the execution of system-level services. At the third priority level resides a highly modified version of DEC's FOCAL interpreter. This version of FOCAL has been linked to the real-time and data acquisition facilities of the monitor to provide a high-level user-oriented language that can invoke most of the system-level functions in addition to providing an arithmetic processing language. The lowest priority level is occupied by a general background job slot. In the 12K system, the background job can be assigned to field 2 on a core-resident basis or, if required and permitted by system timings, the background job can be swapped out in favor of foreground activity.

Let us now consider approaches to software development for laboratory automation. When the problems are better understood, and when large-scale integrated circuits become both more reliable and less expensive, the solutions are likely to differ considerably from those in present use. It seems that the more immediate solution of laboratory automation problems will involve more use of high-level languages and core-resident interpreters that are imbedded in good monitor systems. However, the final solution may well be the development of more sophisticated hardware. In that case monitors, high-level control macros, and a high-level language may be implemented entirely in hardware. Such systems would require a minimum of software support.

AUTOMATION SYSTEMS IN OPERATION

The several laboratory systems chosen for illustrative purposes are not intended to be completely representative of all of the systems in existence, but they do represent a fair cross section of the computer-automated systems

[†]Delphic Industries, Inc., P. O. Box 47, Livermore, California.

in the United States. The reasons for their selection are twofold. First, each one represents an approach to laboratory automation that needs to be discussed. Second, they are systems with which we are familiar to some degree.†

Time-Shared Systems

The first discussion deals with the IBM 1800 time-shared laboratory automation system. Probably the most outstanding example of this time-shared system is the one implemented in the IBM San Jose Research Laboratory, which employs approximately 150 to 200 personnel. The experiments for which the laboratory automation system was designed are listed in Table 1. The references in the table give detailed descriptions of the instruments and the techniques used to accomplish the automation. Each may be classified as belonging to one of three general categories: (a) slow scanning and time independent, (b) slow scanning and time dependent, (c) fast scanning.

The machine on which the system was implemented is a 32K, 16-bit, μsec, IBM 1800 with 12 levels of priority interrupt. It is equipped with three 512K-1810-B disks, a line printer, a card-read punch, an incremental plotter, a keyboard typewriter, three output typewriters, and three internal timers, one with a time base of 0.125 μsec and two with time bases of 1 μsec. In addition, the 1800 is equipped with digital I/O, both high- and low-speed analog I/O, and external or process-interrupt terminals. The laboratory automation system was written as a special-purpose submonitor of the TSX (time-shared executive) system control program.

The TSX monitor is a foreground-background monitor used for laboratory automation. It allows the individual scientist to program solely in FORTRAN if he so desires. Such a monitor system is valuable in a laboratory in which most or all of the scientists have some limited computer and programming background; this is the case at the IBM Research Laboratory. However, to implement this system properly, a single individual was assigned to develop and maintain systems programs for the laboratory monitor and applications programs.

This system presently allows a maximum of 6 simultaneous users among the 11 listed in Table 1. This limitation is reasonable, since the nature of the laboratory is such that most applications do not require continual use of the instrument over extended periods of time. The number of users

†The use of these examples in no way implies endorsement by the University of California, Lawrence Radiation Laboratory, of these computers or computer systems.

Table 1
Instruments On-Line to the IBM 1800 at the IBM
San Jose Research Laboratory

Instrument	Category	Reference
EPR spectrometer	a	[3]
Vacuum UV spectrophotometer	a	[4]
Visible-region monochromator	a	[5]
Gas-liquid chromatograph	b	[6]
Far-IR interferometer	a	[7]
Low-resolution mass spectrometer	c	
High-resolution Stark spectrometer	a	
NMR spectrometer	a	
Gel-permeation chromatograph	b	[8]
Photoconductive decay studies	c	
Differential scanning calorimeter	a	

has increased at a steady pace from the initial 2 in 1967 to the 11 presently on line today.

As mentioned earlier, this time-shared system has both foreground and background capabilities. It can be used for control as well as data acquisition, but its main use has been for data logging and on-line processing. It allows data rates up to 20 points per second per instrument. However, the users' experience with the TSX monitor system over the last few years has enabled them to design significant improvements in the new MPX system control program for the 1800. This system allows improved data rates, simultaneous overlap of fast and slow experiments, and more efficient use of hardware resources.

In summary, the advantages of this system are that the user can write his applications programs in FORTRAN and then have them compiled in a background mode and executed in the foreground of the monitor. Some disadvantages of this approach (i.e., the implementation of batch processing in conjunction with real-time data acquisition and control functions) are the large core-memory requirements and the very complicated scheduling algorithms for time-shared batch processing that often interfere with the required real-time response. However, such criticism should be tempered by the realization that most laboratory instruments require maximum data rates of from 1 to 10 points per second and that small discrepancies in the time-response factors are not often critical.

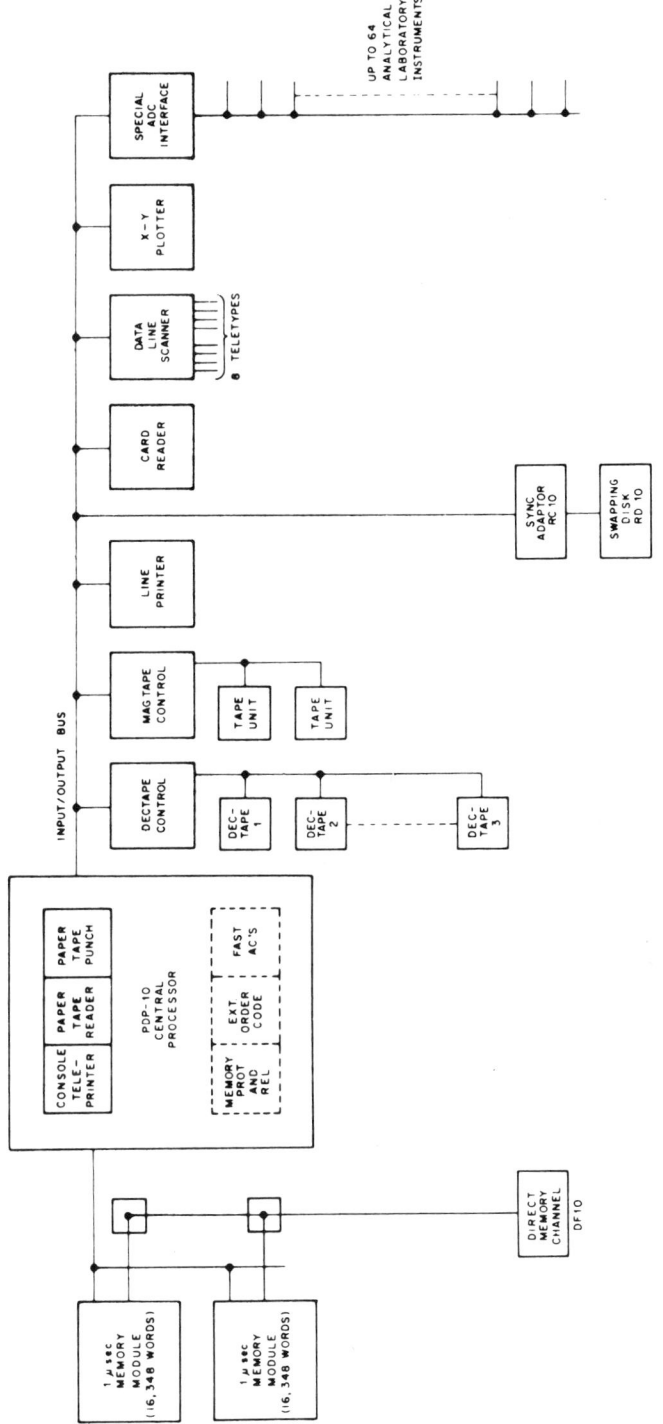

Fig. 1. PDP-10/50 Configuration at the Max-Planck Institute.

Another example of a time-shared laboratory system is that of the Max Planck Institut in Germany (Fig. 1) which employs a very large computer and monitor system designed for time-shared batch processing. The computer is the Digital Equipment Corporation PDP-10, which is a 36-bit machine implemented for this application with 32K of core memory (1-μsec cycle time) and the usual peripheral devices such as a line printer, a card reader, an incremental plotter, a paper tape reader/punch, magnetic tapes, and eight teletypes. It should be remembered that the monitor used for this application is also a time-shared batch-processing monitor that has been adapted for real-time applications. The computer is presently interfaced to a large number of gas chromatographs; a fast-scan, low-resolution mass spectrometer; and a pulsed NMR spectrometer.

The interface provides for slow-speed analog input lines. This definition of slow applies to analytical instruments that require variable data transfer rates of up to 40 points per second. The particular interface used on this system includes an ADC with a dynamic range of about 10^6 and a multiplexer capable of handling 32 analog channels. The maximum overall data rate for this device is 3.3 kHz with automatic range selection and 8 kHz in programmed range selection. This interface includes an internal clock and a preset counter which can be loaded by the software program, which is decremented every clock cycle, and which requests an interrupt to the processor if counted down to zero.

For fast analytical instruments (e.g., fast-scan mass spectrometers and pulsed NMR spectrometers) another interface can run at data rates of up to 20 kHz. To avoid trouble with the general time-sharing system, not more than one of the fast lines may be active at any given time. It should be remembered that such high-speed data are acquired usually for only very short periods of time, that is, several seconds. For instrument control, there are two other devices: a contact scanner to look at the status of 72 external relay contacts and a line driver to close or to open 72 external relay contacts. With these devices the software can execute start or stop requests; it can, for instance, switch a voltage at the spectrometer or stop an experiment.

As mentioned before, the monitor was developed mainly for use as a time-shared monitor for batch processing. It has been modified to allow the user to acquire both fast and slow data, to treat the data on-line as required, or to store the incoming data on a disk for processing at a later date. The latter feature is incorporated in the automation of a fast-scan mass spectrometer. Before initiating an experiment, the user may specify for this line whether he wants real-time smoothing and reduction of the data or intermediate transfer of the data onto a disk file for later processing.

As soon as a run has been initiated, the job associated with the line-number is brought into core to accept all of the data from the ADC. The real-time smoothing and data reduction, if requested, is done by a monitor routine which becomes active on interrupt level whenever one buffer of data is full. The data reduction job itself stays in core as long as the instrument run continues, that is, for a few seconds. Depending on the type of experiment, the data analysis may be accomplished by the data reduction job in core or by a separate analysis job to be brought into core at the end of the run.

For the IBM 1800 and PDP-10 systems, the monitors are about 10 K to 22K words in length depending upon the implementation requirements. For these systems, the programming for the data acquisition and control functions is often more difficult than it would be in a dedicated system because of the conflicts with the sophisticated time-sharing algorithms. However, the ability to have simultaneous foreground-background processing together with the ability to write the applications programs in a high-level language allows far easier programming of the analysis methods. Furthermore, these analysis programs can be more readily modified or exchanged than those coded in assembly language.

After examining two different time-shared laboratory systems with extensive foreground-background capabilities, we might reasonably examine a similar type of system that offers no background capability. Such a system has been implemented in the Analytical Chemistry Section at LRL. The system consists of a central processor, PDP-7 (18-bits), with 8K of 1.75-μsec core memory, a 15-bit ADC and multiplexer, a CRT, a 500K-word disk, and a write-only tape unit. Eight instruments have been interfaced to the system (Fig. 2). We are now in the process of interfacing several more instruments, 14 to 16 being the ultimate goal for simultaneous operation. All instruments operate asynchronously under a three-level monitor. Most of the instruments were designed or rebuilt for complete computer control, that is, the computer not only controls the instruments but also performs the required data acquisition functions and on-line computations.

The core memory of the PDP-7 cannot easily be expanded to more than 8K; therefore, to support 14 to 16 instruments simultaneously, it is necessary to write the applications programs in page format and store them on the disk. During real-time operations, the application programs will not be swapped in and out of core; instead, after a page has been executed, the next page of the program will be overlaid from disk into the same area.

This approach to time-sharing has several advantages and disadvantages. Among the advantages are the favorable time-response characteristics, the small core-memory requirements, and the rather simple monitor structure.

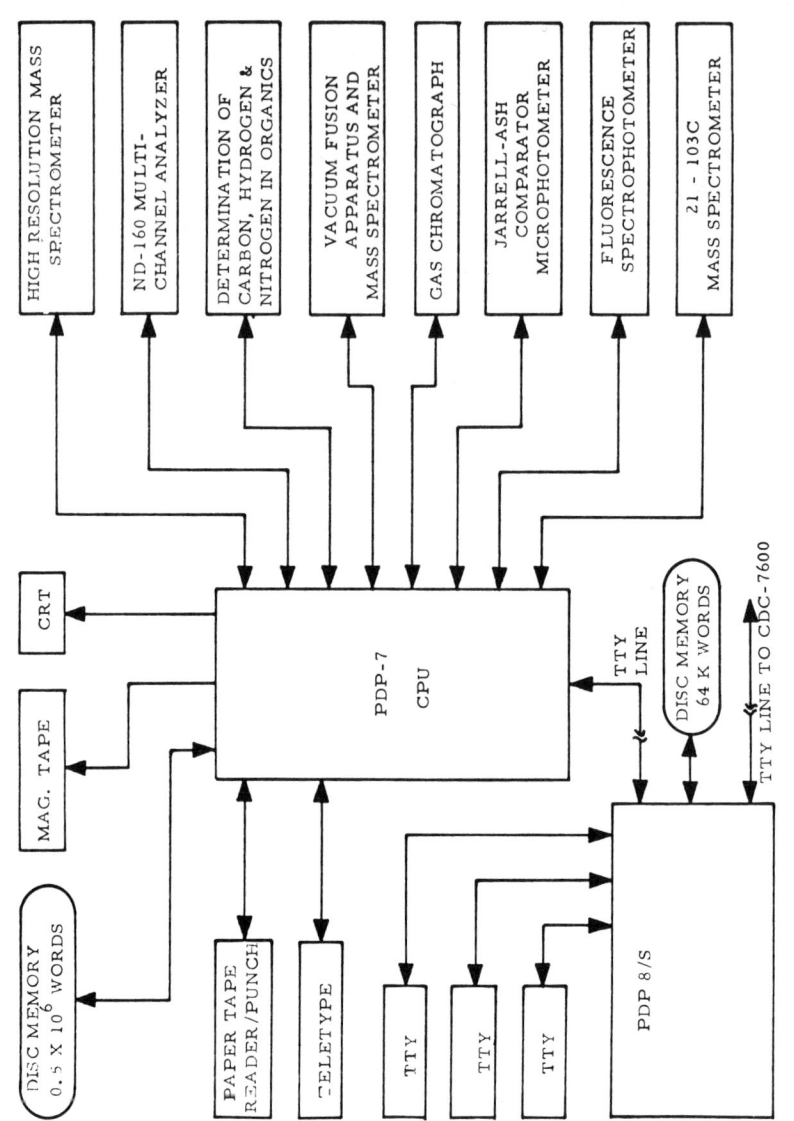

Fig. 2. PDP-7 time-shared system.

Of the several disadvantages of this approach, the worst is that all application programs must be written in assembly language. Since this computer automation had to be accomplished in addition to a normal work load, this disadvantage has notably retarded the implementation. Another disadvantage is the necessity for interrupting the real-time operations to assemble and debug new programs.

For easier and more convenient application programming for the PDP-7, it has been connected to a PDP-8/S. New programs are now being written much more easily. All control and data acquisition functions are being written in PDP-7 assembly language. However, all data that require mathematical manipulations in nonreal time will be handed off to the PDP-8/S. In the PDP-8/S, the mathematical programming is coded in FOCAL and executed in a four-user time-shared mode. This arrangement will greatly simplify the coding of applications programs.

Dedicated Systems

In the section on Software, we mentioned the software approach taken by Cooke in his use of a PDP-9 in the chemistry laboratory. His approach has been to use the computer in a dedicated mode. As much as possible, all applications programs have been written in high-level languages: FORTRAN when the programs are compiled and then executed, and FOCAL when the programs are executed interpretatively.

The computer application involves kinetic studies used for the characterization of enzyme reactions. The essential feature of the automation is that the computer can retrieve very useful information which is normally lost in the more conventional methods of recording. If the concentrations of the reactants are chosen to yield a reaction which is first-order with respect to the desired substrate, the observed transmittances (obtained from Hitachi-Perkin Elmer Model 139 UV-VIS spectrophotometer) yield a series of substrate concentrations, s. From consecutive values of s, corresponding values for the rate, r, can be determined and used to obtain the dependence of r upon s. This dependence is usually expressed by nonlinear plots such as $1/r$ versus $1/s$ or by the Michaelis-Menten constant K_m and the maximum velocity V_{max}.

The programs determine the derivative in order to obtain the rate. The presence of seemingly insignificant noise in such a system can destroy much of the information obtained in the derivative. It is therefore necessary to use some smoothing technique to minimize the effect of noise. For this application, a Chebyshev polynomial curve-fitting approach was chosen because essentially no advance information about either the data or the noise is needed. As an additional aid, the program is somewhat interactive, that is, the chemist can designate the sequential position of the first valid reading, thereby discarding points that appeared to be in error because of solution mixing or instrumental difficulties. The chemist can also terminate the calculations at will so as to avoid prolonged computation once acceptable results have been obtained.

An example of another approach to automation is that being implemented at Eli Lilly and Company in Indianapolis. An intensive effort in laboratory computerization was initiated to fill a need for on-line data acquisition and for calculations of results from a large variety of laboratory instruments. The company chose to approach the problem simultaneously from two directions. The first approach led to the lease of an IBM 1800 computer system with its gas chromatography software programs. The system was configured around a minimum 16K, 4-μsec CPU with one disk drive of 500K words. The system is presently interfaced with 25 gas chromatography instruments with up to 16 of them operating simultaneously.

The second approach involved the use of dedicated computers. The company purchased one Hewlett-Packard 2115A and eight additional 2114s. Since most of these systems are in areas that cannot be readily serviced by a larger central computer, each system is configured to handle the entire workload in each area for the present and near future. Some have greater expansion capabilities than others, but the systems can be easily reconfigured if unexpected expansion occurs in certain areas. Trouble-shooting is expedited by the ability to interchange components. The instruments served by these systems include multichannel autoanalyzers, a Technicon SMA 12/60 clinical analyzer, amino acid analyzers, CHN analyzers, scintillation counters, automatic titrators, and numerous specialized data acquisition applications. Projects now being developed or planned for the near future include gas chromatography, IR analysis, NMR spectrometry, mass spectrometry, automatic balances, automated microbioligical analyzers, biomedical sensors, and a variety of process-control applications.

Most of the systems are programmed in assembly language, since maximum core efficiency is needed to handle the multiple instrumentation programs in the 8K of core available. The company's success to date is attributable primarily to its approach to the training of its staff of scientists. Training has been provided primarily in two ways. A formal course in laboratory computerization was taught at the company by Dr. Sam Perone of Purdue University. This provided excellent background knowledge for a large group of candidate computer-oriented scientists. From this group, a number of individuals expressed a desire to have additional training, and a 6-month apprenticeship program was instituted. In this program individuals were encouraged to work full time with the small number of well-trained computer-oriented scientists available. A number of this newly trained group have been sufficiently well oriented to do most of their own computerization and also to serve as a nucleus for training others in their respective areas. Most of this group will have primary responsibility for one of the systems in their areas.

These dedicated systems have answered many of the company's needs. The obvious advantages are flexibility, involvement and training of a fairly large number of scientists in the basic elements of computer automation, and the absence of a need for sophisticated monitor systems. The main disadvantage has been the requirement that all applications programs be written in assembly language. In the future it will be possible to implement automation of this type via the use of powerful monitors into which are imbedded resident real-time interpreters similar to the system described under Software. As the automation increases, such monitors would allow the computers to be interconnected so as to separate easily from real-time functions such functions as file maintenance, laboratory management functions, and time-shared batch processing in one shared computer.

New Developments

The more advanced use of the small (12 to 18-bit) general purpose computer for both dedicated and time-shared functions appears to be via core-resident interpreters as described above. The use of interpreters operating within the framework of a general-purpose monitor should greatly enhance laboratory automation. These systems are just becoming operational; no laboratory automation has been fully implemented in this way (to our knowledge). The modification of the PDP-7 system as described above is an attempt to use some of the features of a core-resident interpreter within the framework of an existing system.

We have not yet implemented the full monitor/interpreter system, but we have used an interpreter with modified I/O handlers for automation via a dedicated computer. One application involves an interactive system for reading the photographic plates produced by spark-source mass spectrometry. The computer is a PDP-8/I with 8K core memory, 32K disk, a 12-bit ADC, and a 611 Tektronix oscilloscope. The plate reader is a Grant microphotometer comparator. The computer controls the comparator, performs the necessary data acquisition functions, displays the data on the oscilloscope, and allows the chemist to interact with the data to provide fast and effective data reduction.

A brief description of the analytical method will help to clarify the many

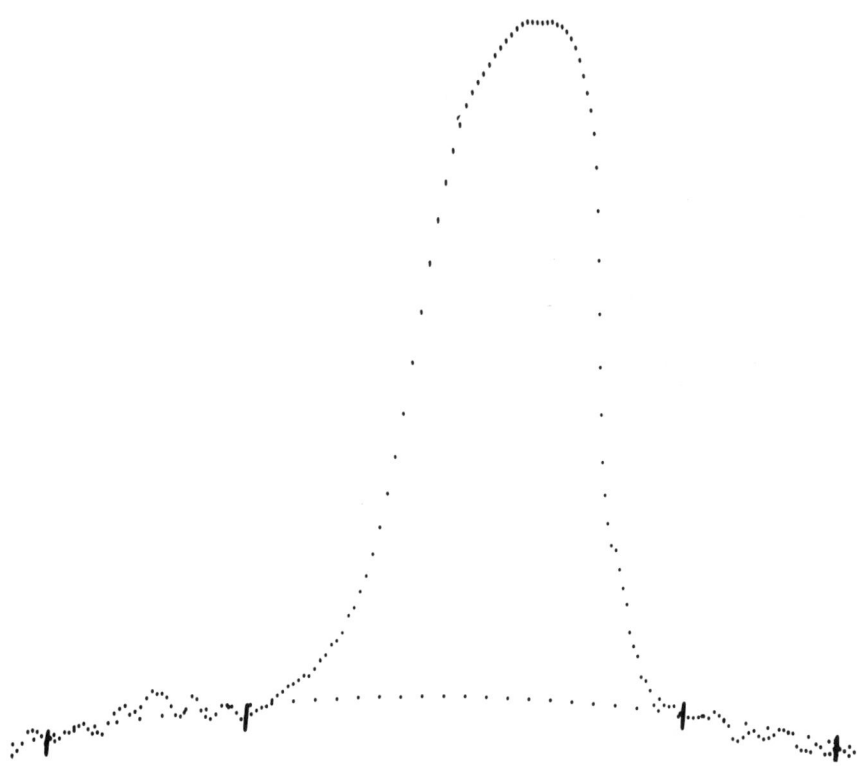

Fig. 3. Base line correction for mass spectra peak.

problems associated with automating this system. Take, as an example, the determination of 14 impurities in a rock specimen. The sample is ground to ~400 mesh or less, then mixed with powdered graphite and pressed into a pellet. The pellet is then carved into two electrodes and mounted in the source housing of the mass spectrometer. A high-voltage, pulsed radio-frequency applied across the electrodes produces a continuous spark that vaporizes and ionizes the electrode material. The ions are passed through a double-focusing mass analyzer, and the separate elemental species are brought into focus on a photographic plate. A dynamic range of 1 million can be obtained by varying the duration of the spark for several separate exposures on the same photographic plate. Quantitative data are obtained from comparison (with a microphotometer) of the optical densities of the mass lines with those of standard samples. The method requires calibration with a sample of similar matrix material and impurities.

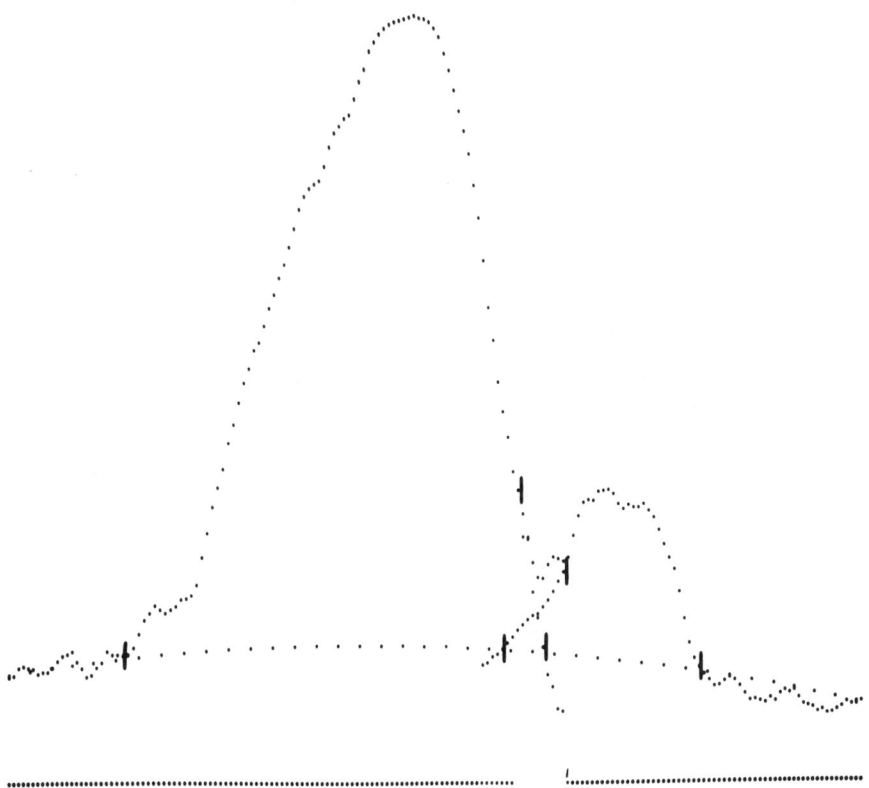

Fig. 4. Peak unfolding via interactive CRT program.

Manual reduction of the data on the photoplate can be time-consuming. The original computer automation of this unit was made with a PDP-8 by means of a noninteractive program [9], that is, the computer made all of the decisions by means of predefined algorithms. Although this method of data reduction saved labor, it was still very time-consuming. (Total time on-line to the computer ranged from 4½ to 8 hr). In addition, unusual samples often gave spurious or wrong results because their spectra fell beyond the boundary conditions of the programmed algorithms.

As one possible solution to this dilemma, we developed an interactive program whereby the chemist controls the plate reader via a "patchbox" and the computer. Mass spectral lines of interest are displayed on the CRT. By means of a cursor, the chemist selects representative portions of baseline on each side of the peak (Fig. 3). He next instructs the computer to project under the peak a linear or quadratic fit of the baseline projection

Table 2
Progress of Laboratory Automation

			Time-shared automation systems using hierarchical computer configurations ↓			
		Time-shared automation systems with foreground/background capabilities, applications programs written in FORTRAN ↓				
	Time-shared automation systems with no background capabilities, programmed in assembly language ↓					
				Dedicated computers using core-resident interpreters ↓		
Dedicated computers programmed in assembly language ↓						
1965	1966	1967	1968	1969	1970	1971

that appears to be correct. He then instructs the computer to integrate the area under the peak and print out the results in parts per million, as measured against the standard data on file in the computer. Similarly, peaks can be unfolded under computer control as shown in Fig. 4. For this case, the computer prints out the area of each peak defined by the chemists and the percent error between the total area under the composite peak and that under the sum of the two separate peaks. This approach not only decreased the total time for data reduction by factors of 10 to 20 but also improved the overall accuracy.

This is one of many applications in the chemistry laboratory for which interactive automation is the most effective solution. The applications programs were written in the interpretive language FOCAL.

DISCUSSION

We have surveyed the major methods of automation in use in the scientific laboratory, and for each case we have given several illustrations. Table 2 lists the different implementations that we have discussed. Historically, the first system to be implemented had one computer dedicated to a single experiment or two or more very similar experiments. These dedicated computers were programmed in assembly language, usually by the chemist in charge of the experiment. This method of automation is still preferred by many scientists and used quite extensively. We note, for example, the extensive automation at Eli Lilly Company via Hewlett-Packard 2114 computers operating in dedicated modes. However, as indicated in Table 2, automation via dedicated computer systems is beginning to assume a different character. In the more advanced systems, automation is now proceeding via a core-resident interpreter such as FOCAL or BASIC. To date, this implementation has generally been for one experiment at a time, as it was for the LRL plate-reader experiment.

Shortly after computers were introduced into the laboratory, time-shared real-time systems were being developed. These included the type of system implemented at LRL with a PDP-7. In this case, there were no background capabilities, and all of the programming was done in assembly language. Systems of this kind have several good characteristics: (1) fast response characteristics relative to I/O requirements, (2) relatively easy development of monitors as compared to TSX or MPX, and (3) relatively compact operating systems that require a minimum of core memory. On the negative side, this type of system has several disadvantages: (1) all the applications programs must be written in assembly language, (2) there are great possibilities for program interaction in the early debugging stages of any new system. At the same time that this type of system was being developed, some time-shared systems were being converted into real-time systems. Two of these were described above. Both have extensive foreground-background capabilities, and their applications programs can be written in FORTRAN. These are very important advantages; however, they too have some disadvantages: (1) the monitor can only be modified with great difficulty, (2) the monitors associated with these systems are very expensive in terms of core memory and in overhead timing characteristics, and (3) the systems generally have less desirable time-response characteristics unless they are provided with expensive demand-response interfaces.

If the monitor systems had been totally developed from the point of view of applications in a chemistry laboratory, several of these difficulties could have been greatly minimized. However, it should be noted that it is very difficult to incorporate into one computer both the characteristics required for a good real-time operating system in the laboratory and the characteristics required for an efficient batch-processing time-shared system. The two applications are mutually exclusive to some degree. In defense of these systems, it should be noted that *most* laboratory instruments demand only very slow data rates.

A new type of monitor system recently put into operation is a real-time monitor that allows one-user background capability and uses a core-resident interpreter. As we have shown, this monitor has many advantages: (1) most important, the application programs can be written in a high-level language, (2) an interpretive system is very solid and immune to user abuse, and (3) this monitor readily allows the construction of hierarchical computer structures.

This approach allows ready separation of the monitor requirements into two different computers, that is, the applications programs for data acquisition, control and many of the on-line calculations can be written in a high-level language on-line and, when required, readily changed and debugged, also on-line. When the user must have access to the capabilities of a batch-processing time-shared system, information can be handed off from one small computer to another small computer or to a larger central computer, whichever is required to perform the background tasks. For example, if the laboratory must have a large data base for such operations as mass spectrometry, IR spectrometry, or microwave spectrometry, such data bases can be handled easily by a small computer interlinked to several other small computers. The central computer merely performs the necessary sort-merge programs and handles the large data bases for interpretation and return of required information to the user's computer. If the laboratory

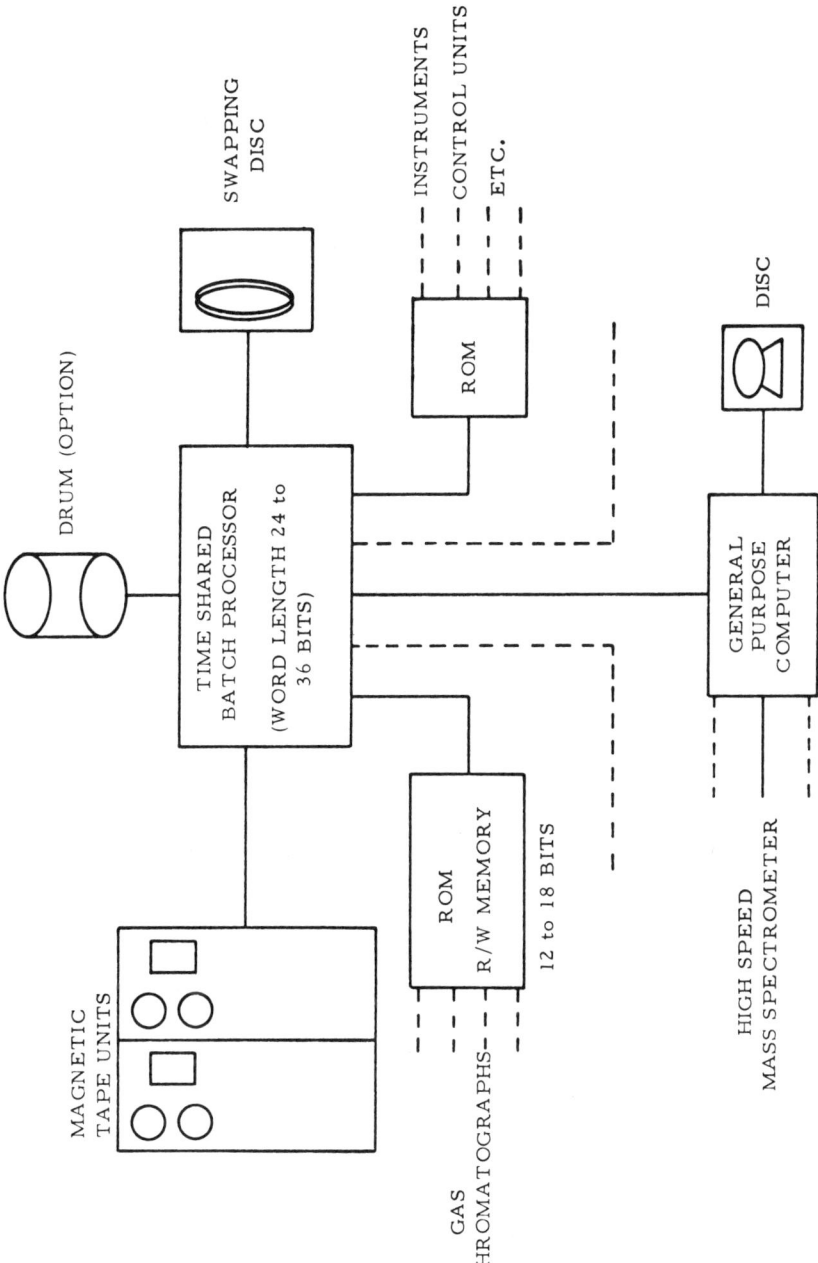

Fig. 5. Hierarchical time-shared system.

must have not only large data bases but also very heavy computational capabilities that require a larger computer such as the PDP-10, then the user's small computers can perform the data acquisition and control tasks. It can hand the required information to the large computer to be processed in a batch-processing mode, after which the output information can be returned to the user via teletypewriters or line printers in the user's area. Note that in this case the batch-processing is not a real-time requirement; in essence, it is performed off-line in a batch-processing mode, thereby utilizing the good operating characteristics of the monitors written for this type of processing. The tying together of the computers is a convenience; such a linkage greatly improves the time-response characteristics of the overall system as far as the user is concerned.

It is encouraging to note that computer automation in the scientific laboratory has progressed far in the past 5 years. Still, we are only beginning to learn how to use the computer properly and to implement it in such a way that it will give the best, most economical service to the user. We believe that in the future we shall be using more monitor systems that contain core-resident interpreters which will allow us to write our programs on-line and have them executed interpretatively. For most applications in the chemistry laboratory, such core-resident interpreters represent an ideal solution of the automation problem. Where they cannot be used, programs will still sometimes be written in assembly language but more often in FORTRAN or FORTRAN-like languages.

The systems of the future will require that more attention be given to the determination of such parameters as "system bandwidth" and "system response-time." These are not easily defined for time-shared systems, but when once defined for the specific system in question they provide the designer with useful information for choosing the proper hardware/software trade-offs. They also serve as an excellent means of evaluating the performance cost of a computer system.

It seems that for laboratory automation via the general purpose computer, future trends will be toward more extensive use of high-level languages as implemented by core-resident interpreters. In addition in the time-shared systems, we shall tend toward separation of certain functions, with the control, data acquisition, and minor calculations being performed by one computer and heavy calculations and file management processing being performed in a separate but interconnected computer. A hierarchical arrangement, similar to that shown in Fig. 5, will greatly reduce the software problems. The exact end points are still unsettled and rather vague. One factor that could greatly alter the above picture is the increased use of integrated circuitry to develop real-time high-level languages and monitor systems in hardware.

REFERENCES

[1] G. N. Salaita, *Semiautomated Method for Determining Kinetic Constants of Enzymes,* Thesis, Department of Chemistry, Univ. of Louisville, Louisville, Kentucky, 1969.
[2] C. E. Klopfenstein, Department of Chemistry, University of Oregon, Eugene, Oregon, private communications, 1970.
[3] B. Johnson, T. Kuga, and H. M. Gladney, *IBM J. Res. Develop.,* 13(1), 5 (1969).
[4] P. M. Grant, *IBM J. Res. Develop.,* 13(1), 5 (1969).
[5] D. M. Hannon, D. E. Horne and K. L. Foster, *IBM J. Res. Develop.,* 13(1), 79 (1969).
[6] H. M. Gladney, B. F. Dowden, and J. D. Swalen, *Anal. Chem.,* 41, (1970).
[7] J. Gayles, W. Honzik and D. Wilson, *IBM J. Res. Develop.,* 14 (1970).
[8] H. M. Gladney, *J. Comp. Phys.,* 2, 255 (1968).
[9] C. A. Bailey, R. D. Carver, R. A. Thomas, and R. J. Dupzyk, *A Computer-Controlled System for Automatically Scanning and Interpreting Photographic Spectra,* Lawrence Radiation Laboratory, Livermore, Preprint UCRL-70897 (1968).

Microprocessors in Chemical Instrumentation

F. Baumann / J. Hendrickson / D. Wallace

Varian Instrument Division, 2700 Mitchell Drive, Walnut Creek, California 94598, USA

Introduction

Microprocessors will proliferate over the next few years and will revolutionize analytical chemistry instrumentation. Instrument manufacturers will incorporate microprocessors in the basic organization of next-generation instruments. In addition, the research chemist and laboratory engineer will use these devices in experimental systems. This paper reviews microprocessor technology and applications to analytical chemical instrumentation.

The Technology

Era of LSI

The electronic industry has passed through three generations: vacuum tube, transistor and integrated circuit (IC). Over the years this technology has steadily advanced and more complex functions (registers, counters, memory arrays) are available on one IC. Today the electronic industry has evolved to a new generation – the era of large-scale integration (LSI). LSI circuits are small ceramic or plastic packages approximately an inch long, commonly called a chip. Each package contains a thin piece of silicon that has been impregnated with impurities to form electronic component patterns (resistors, transistors, diodes) necessary for a specific function. Tiny wires are attached to each chip for connection to other electronic circuits. A single LSI chip could conceivably contain more than 10,000 transistors.

The Microprocessor

Approximately three years ago MOS (metal oxide semiconductor) processes increased component densities making it possible to include enough functions to produce a central processor unit (CPU) of a simple 4 to 8 bit computer on a single or a few chips. This was the advent of the microprocessor.

Shown in Figure 1 are the characteristic elements of a microprocessor. The control section is the brain of the system and is used for decoding each instruction fetched from memory and in executing a sequence of hardware actions for each specific instruction. The ALU (arithmetic logic unit) is used for mathematical operations (addition, division, etc.), logical operations (and, exclusive or, etc.) and bit manipulation (rotate, test bit, etc.). The general registers are necessary for external communication to input and output (I/O) devices, for programmer use as intermediate storage, and for memory addressing. The data bus is important because it is the electrical path of information flow between memory, I/O devices and the microprocessor. External timing controls are also necessary for interfacing memory and other circuits to the microprocessor.

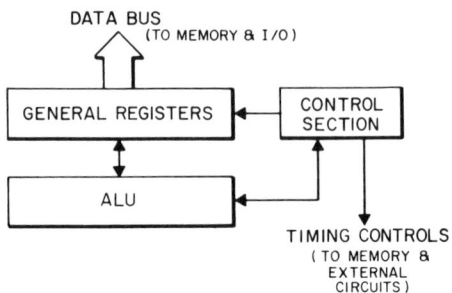

Fig. 1
- Basic elements of a microprocessor

Microcomputer

A microcomputer is obtained when memory and external circuitry for control of processor timing and data bus transfers are added to a microprocessor. (Figure 2.) Microcomputers are functionally similar to minicomputers but are distinguished from minicomputers by small size and from other LSI devices by programmable behavior. Memory is used to store both instructions and data. Data are usually stored in random access memory (RAM). RAM is read/write memory that has characteristics of random access and volatility. Volatile memory loses its contents when power is removed. Program instructions are usually stored in RAM, ROM or PROM. The proper choice of memory for program storage depends on the eventual application. ROM is read only, non-volatile memory having the advantage of low cost for high

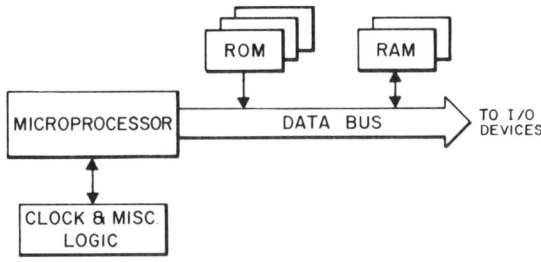

Fig. 2
- Basic elements of a microcomputer

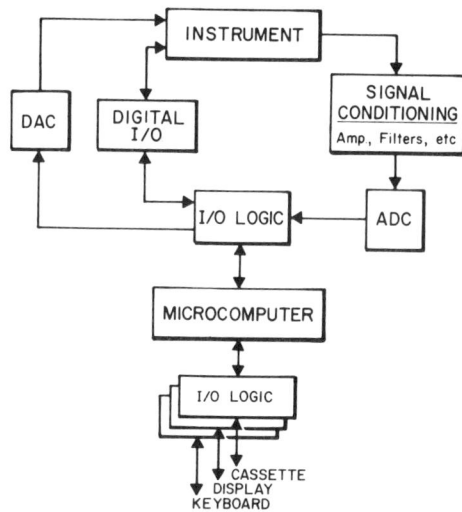

Fig. 3
- Basic elements of a microcomputer/instrument system

volume and the disadvantage of not allowing change. ROM prevents program loss when power is interrupted. PROM (programmable read only memory) is intermediate in cost and is reprogrammable.

The microcomputer promises a number of advantages in the chemical laboratory over its competition: hard-wired logic and the minicomputer. Because of the nature of LSI circuitry, microcomputers are small, low in cost, and consume low power. The miniaturization of the microcomputer allows their physical incorporation into smaller analytical instruments. Decreases in size should continue with further advances in LSI technology. The average price of currently available microprocessors is approximately $40 and is expected to decrease one-half by 1978 [1]. The average price of microcomputers is presently $325 and is also expected to decrease one-half by 1978 [2]. The reduction of wire interconnections improves reliability and further reduces design costs over conventional logic. In some chemical instruments, speed may be a limitation as instruction execution times are currently about 10 microseconds; by 1978 this should decrease to approximately 0.2 microseconds. Another significant advantage of this technology is the programmable nature which allows flexibility and facilitates adaption to specific applications in the laboratory.

Microcomputer System

When a complement of I/O devices is operated by the microcomputer, a system is formed. A general-purpose system for instrumentation applications (Figure 3) might include:

(1) modules such as ADC's, DAC's and their associated accessories (level converters, amplifiers, multiplexers, etc.)
(2) digital I/O timing and control modules such as flip flops, digital switches, relays and stepping motors
(3) operator communication and display devices such as a keyboard, CRT scope and a cassette tape

Standard interface chips of TTL are needed to interface the microprocessor with the I/O devices. The software program in the ROM or PROM determines the specific operation of the peripheral devices and analytical instrument which make up the system.

Design of Microcomputer Systems

The developmental requirements for microcomputer systems are similar to those for the minicomputer. In fact those experienced in developing minicomputer systems will be able to adapt rapidly to microcomputers. In general, though, microprocessor system development will be more difficult mainly because of the lack of support from the semiconductor vendors. Balancing this is the tendency to use microprocessors for simpler and small tasks. Pertinent work has been done recently by the ASTM Committee E-31 on computerized lab automation in recommending procedures for development of on-line systems [3, 4]. Key assertions include:

(1) A systematic and orderly approach to automation will reduce costs and increase the chances of success.
(2) The automation problem is highly interdisciplinary in nature and requires a thorough understanding of the chemical processes, instrumentation and operational procedures as well as the technologies of digital hardware, software, mechanical engineering and algorithm development. Many failures have been reported because a SYSTEM problem was mistakenly believed to be a computer problem.
(3) A guideline for system development has been established. This includes phases of definition, system specification, functional design, implementation design, implementation, evaluation and testing, and documentation.

The work of the ASTM E-31 is recommended reading before the construction of a microcomputer system is undertaken.

Hardware Considerations

A microcomputer system is reasonable to consider both for computing applications where minicomputers or programmable calculators are too costly or have other disadvantages, and for complex logical sequencing and control functions. In general, applications which require logical or arithmetic manipulation of data, more than several operations or tests, expandability or flexibility, or which would require more than 50–60 ICs are good prospects. The requirement for speed must be relatively low.

Once a decision to use microprocessors is made, the next step is the selection of the particular device. Microprocessors may be general purpose or optimized for either calculating or control. Calculator microprocessors generally are designed with emphasis on long data word size, BCD arithmetic and keyboard and display oriented I/O. Controller microprocessors emphasize speed in transferring data, interrupts, and bit and byte instruction sets. General-purpose microprocessors emphasize overall balance between these extremes and are similar to minicomputers. A more detailed examination of microprocessors requires checking the application requirements against the specific characteristics. The more important characteristics are shown in Table I. Because of large differences in the design of microprocessors, a straight comparison of execution times and number of instructions may be misleading. The best method of measuring and comparing microprocessors is with a benchmark program typical of the specific application. With this program, machine cycles and storage necessary for the application can be estimated for each microcomputer system.

Some microprocessors are microprogrammable. This should not be confused with programming the microcomputer. Instead, it means that the control section (Figure 1) is not fixed conventional logic but is implemented with programmable semiconductor memory. Since the control section determines the instructions set for the microprocessor, the user can tailor the instruction set to his needs.

Microcomputers are available in several stages of packaging: LSI components; printed circuit (PC) boards; and packaged systems. If the user purchases LSI chips, he must be prepared to design, debug and fabricate his own functional microcomputer and to test LSI components. A high degree of capability is required. With a complete, tested PC board, the user avoids these considerations but must still provide wiring to connectors, power supplies and packaging. With a complete packaged system the user receives all of the above plus a front panel and a Teletype interface and can concentrate on interfacing to peripheral devices.

Manufacturers of microprocessors are oriented toward support mainly for potential high-volume users. At the PC board and packaged system levels, support is more generally available from manufacturers and smaller microcomputer system companies. In general, however, these products are not offered with much peripheral equipment or applications software.

The choice of memories is important because they may be several times the cost of the microprocessor. LSI semiconductor memories provide simple interfacing, compatible storage capacities and low cost. RAM, ROM and PROM are available in standard packages from a number of semiconductor manufacturers. Chip storage capacity, access time, power requirements, interface and refresh requirements are important considerations.

Interfacing between the microcomputer and other system components commonly involves providing a required data rate considering both the hardware data transfer limitations and the software overhead time. Often simple I/O ports to the processor are provided. A data bus with prescribed word length, logic levels, timing and control offers lower cost and better flexibility where several system components are involved. General-purpose LSI and MSI components are now available which can simplify interfacing [5]. These include LSI first-in, first-out buffers (FIFO) to match data rates, and universal asynchronous receiver/transmitters (UART) for long-distance serial communication.

Table I. Microprocessor Characteristics

Word length	4-bit data word for control or BCD calculations, 8-bit for general purpose or character handling 12-bit for ADC resolution compatibility, and 16-bit. Longer word lengths increase efficiency and accuracy at a higher cost.
Instruction set	Sets of 40 to 100 instructions. Both the power and number of instructions affect ease of programming, program storage requirements, and flexibility of application. Calculating applications require stronger arithmetic capability; controller applications require decision and bit and byte instructions.
Memory addressing	Typical modes are direct (via an address pointing register or by other means), indirect, and immediate addressing (the instruction includes the data).
Execution time	Instruction execution times range from 2 to 60 microseconds; longer than for minicomputers due to limitations of MOS LSI. Times decrease in going from p-channel to n-channel to silicon-on-sapphire (SOS). Real-time control of fast instruments requires higher speed processors.
Interrupt capability	Either not provided or single level provided. With some, the program can determine the status of an I/O device by checking status lines. Interrupts are useful for real-time control applications.
Additional hardware	Microprocessors differ in the number of additional nonmemory integrated circuits required to implement a functional microcomputer. Typically, from 5 to 50 are required for timing, I/O control, buffering and interrupt control. A careful analysis of overall system requirements is necessary.
Software	Typically, an editor, assembler and a debug aid are provided. Program libraries and user groups will become available in time. Support critical as software may be the largest cost.
Availability	Microprocessors are commonly announced prior to actual availability. The status of a product not yet available must be carefully considered.

Software Considerations

The entire personality of a microprocessor-based system is determined by the software program stored in ROM, RAM or PROM and changes can be made by modifying memory rather than changing hardwired interconnections. This allows manufacturers and scientists to handle a multitude of functions and applications with common hardware and is an important advantage of microprocessor-based systems over hardwired systems. Because of greater standardization, other benefits occur such as reduced costs for inventory, training, documentation, installation and maintenance.

A word of caution – although the programmable nature of these devices are an advantage for most, it may present a problem for some. Programming differs greatly from operations encountered in conventional logic design. Writing the program is only a small part of the total problem. The program must be debugged, evaluated, tested and corrected. In general, software tools and support from microprocessor manufacturers are minimal. It is common for users to underestimate the complexity of software and overestimate the power of the hardware. Fortunately, because of previous experience with minicomputers, there will be many who will develop systems with convenience and ease.

Implementation steps for software (firmware) in a microcomputer system are shown in Figure 4. Before starting the programming task, definition, specification and design of the software system and the application algorithms should be done. Much algorithm development can be done on larger computers before implementing the programs in the microcomputer. To develop programs, the software programmer/engineer/scientist needs some or all of the following basic tools:

- Editor
- Assembler
- Compiler
- Simulator
- Memory Loader
- Memory Programmer
- Utility Programs
- Debugging Programs
- Prototype Hardware

Although programs for microcomputers can be written directly in machine code instructions (binary codes of 0's and 1's), this is a difficult and time-consuming task. Instead, techniques have been devised at varying levels of complexity for easier programmer use. The most machine-oriented programming technique is called assembly language. Assembly language programs are written with small groups of letters (mnemonics) and numbers – each line has a one-to-one correspondence to a machine instruction. Each instruction can be translated by a software package called an assembler to object code (machine instructions and addresses) which can then be directly used by the microcomputer.

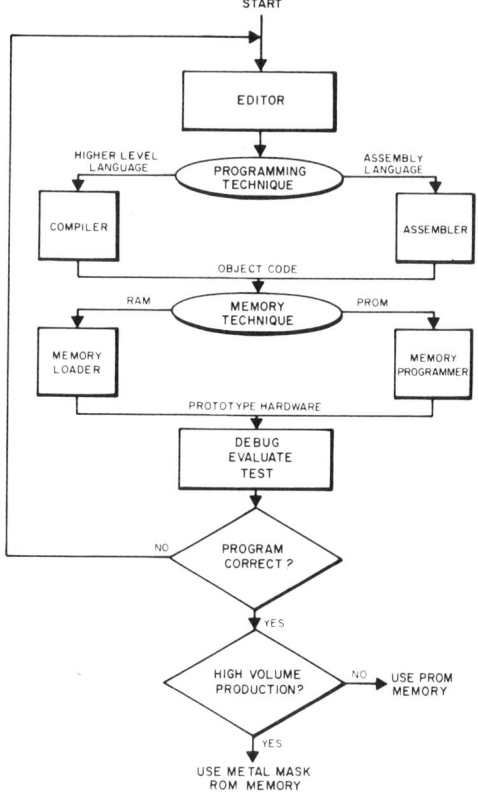

Fig. 4
- Software development sequence

So-called "hardware assemblers" are available on several PROM or ROM chips and can be executed by the microcomputer for which programs are being developed. More typical are "cross assemblers" which run on larger computers with higher speeds and more peripherals. Higher level languages have been developed to get the software translators to do more of the detailed work. These languages are typically algebraic in nature and each line of program is more powerful than assembly language code. Examples of higher level languages include FORTRAN COBOL and PL/I. Higher-level languages are translated into machine code by software programs called compilers. Compilers are almost always run on a computer other than the microcomputer.

Assembly programs offer the advantage of efficiency in terms of memory size and speed of program execution. Compiler-generated programs offer advantages in ease of use and maintenance and speed of development.

Editors are interactive software programs used to prepare programs and to facilitate changes. Editors may be available on minicomputers, CRT systems, batch systems, and time-sharing systems.

The output of the assembler or compiler is an object code usually on paper tape. In some cases it is possible to transfer the object code directly into semi-conductor memory using a transmission system. With the object code it is also possible to use a software simulator to observe the operation of the program as it is executed. However, most simulators must be run on another computer and therefore do not fully simulate a lab application which must run in real time and must interact with system interfaces and other equipment.

To load the object code into the prototype hardware system, a memory loader is used if the memory is RAM. This is a software program which reads successive computer words of the object code and copies them into successive locations in RAM. A different technique is necessary to load PROM. In this case a hardware memory programmer is used to encode bit patterns into the PROM. This can be a slow process; it is not unusual to require 30 minutes per chip. Once the PROM's are programmed, they can be inserted into designated sockets on the prototype system. Since programs of any reasonable complexity are never written without errors, prototype or breadboard hardware is necessary for debugging, evaluation and testing the program for the end product. The prototype hardware should include expanded memory facilities for storage of utility and debug programs, a TTY and associated interface to be used during the debug phase by the programmer, and a hardware control panel that displays registers and memory. Utility programs are general-purpose programs that run on the microprocessor and include commonly used arithmetic and I/O routines. Debug programs are necessary and include such capabilities as displaying and changing the program contents and data (if RAM is used), loading and starting programs, dumping memory contents and trapping through the program.

The debug, evaluation and test phase will determine the correctness of the program. After a number of iterations, the programming will be complete. Now a decision can be made to continue using PROM or RAM in the system or to send the object code to the semiconductor manufacturer for metal masking of the ROM's. RAM and PROM are favored for applications requiring reprogramming and/or small quantities.

An engineer or scientist undertaking a microprocessor-based project should be aware of the overall costs involved. It is not unusual for software costs to far exceed those of the hardware. Also the price of hardware will continue to decrease with time whereas the labor costs associated with software are rapidly increasing. Therefore the level of software support available may be a critical factor to the success of a microprocessor-based project. Many manufacturers have realized this and provide complete, packaged microcomputer systems to support customer development. These units include the microcomputer, PROM and/or RAM in various configurations, system monitor, assembler, editor, debugger, and a PROM programmer.

Applications

In general, automation in analytical chemistry is justified by two factors: favorable economics and data enhancement. Automation using microcomputer systems will be similarly justified except that the application will extend to:

1. Lower priced instruments and accessories which previously could not economically justify minicomputers.
2. Expensive and complex instruments in which automation can be optimized or improved through the use of microcomputer systems in combination with larger computers or in combination with other microcomputer systems.

As with other digital devices, microcomputer systems will be used for both control and computations. The fact that microprocessors are digital devices assures the general advantages of digital techniques and the highly developed state of digital technology. These are:

1. Highly accurate and precise calculations
2. Compatibility with other digital devices such as larger computers
3. Digital storage of data
4. Digital control of instruments and other devices
5. Digital data communication

At the present time, microcomputers will complement rather than replace larger computers. The most immediate applications will be the replacement of hardwired logic used in instrument or accessory design. The future will undoubtedly see microcomputers make inroads on minicomputers in analytical instrumentation applications. The remainder of this paper will review areas in which microprocessors are making the most immediate and largest impact.

Digital Accessories

The digital integrator is the most ubiquitous digital accessory in analytical chemistry. Previous to microprocessors, it had reached a high degree of development by implementation of conventional hardwired logic. Microprocessors are being used in these devices resulting in both improved performance and lower cost. These benefits achieved with digital integrators are a harbinger of things to come.

The general features of a microprocessor-based calculating digital integrator for chromatography are shown in Figure 5. Such systems are organized around a data bus system which handles the digital data in parallel fashion. The elements which make up such a system — ROM, keyboard, display, etc. — are interfaced through the data bus and become peripherals to the microcomputer system. The user interfaces the instrument through the display, printer and keyboard. The front panel of the

Microprocessors in Chemical Instrumentation 535

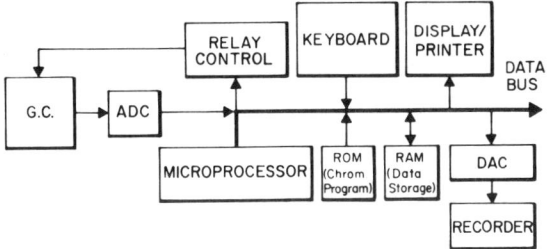

Fig. 5
- A stand-alone microcomputer system for chromatography data handling

instrument may still contain lights, dials, and switches but most of the data input is done digitally through a general-purpose keyboard.

The ROM contains the algorithms which determine the performance of the system. Typical ROM implementable algorithms for chromatography automation are listed in Table II. These algorithms control acquisition of the digital data and the reduction to meaningful analytical results. In addition, the dedicated, real-time microprocessor system can provide time events which control the internal functions of the instrument and external functions of valves, samplers, or other accessories. The display, reporting and communication of the data are also handled by the system. This includes transmission and receiving of data from computers and other microprocessor systems. Finally, the microprocessor system can continually check the data and its own operation and report any inconsistent or unreliable performance or data.

The first commercial application of LSI technology to digital integrators was the AutoLab System IV computing integrator for chromatography automation [6]. This device was transitional in that it used LSI memories but did not use a microprocessor. The CPU was a special-purpose minicomputer tailored to the application. This device automates the data handling of up to four chromatographs simultaneously and offers performance comparable to more expensive computer systems. All the algorithms for data acquisition, data reduction and time control are resident in ROM chips.

The AutoLab System I computing integrator is the first commercial microprocessor-based digital integrator [7]. It is a single-channel integrator which performs data acquisition, data reduction, and time control of chromatography automation. This instrument essentially extends sophisticated computer performance on a single-channel basis to low-priced chromatographs. The Hewlett-Packard 3380A is a similar single-channel integrator with data handling and time control capability for chromatography automation which, in addition, controls a printer/plotter for display of the analog signal. The CPU is a microprocessor of the calculator type.

These digital accessories are stand-alone devices which provide automatic data acquisition and reduction together with simple time control of events. More complete automation of low-cost analytical instruments is described in the following sections.

Table II. Algorithms for Chromatography Automation

Data Acquisition
    Digital smoothing
    Peak detection
    Integration
    Retention time
Data Reduction
    Spurious peak rejection
    Baseline calculation
    Area allocation:
        Perpendicular
        Tangent
        Formula
Relative or retention index calculation
Peak identification
Composition calculation
    Area percent
    Normalization
    Internal standard
    External standard
    Special
Calibration
Time-event Control
    Internal parameters:
        Solvent delay
        Sensitivity
        Area rejection
    External:
        Valves
        Autosamplers
        Accessories
Display/Reporting
    Instrumental parameters
    Data
    Edited reports
    Recalculation
    Printer/Plotter/Recorder Control
Communication Control
    TTY interfaces
    Other microprocessors
    Computer systems
System Checking
    Chromatograph operation
    Microcomputer operation
    Data validity

Intelligent Instruments

Total automation of low-cost instruments such as chromatographs, optical spectrometers, electro-analytical instruments, etc., is difficult to justify economically using minicomputers. Gas chromatography data handling has been extensively automated using minicomputers dedicated to multiple chromatographs [8]. However, two problems are encountered when trying to extend minicomputers to control multiple chromatographs as well: (1) the real-time problem of instrument control and (2) the dependence of multiple instruments on a single computer system. For these reasons the use of dedicated

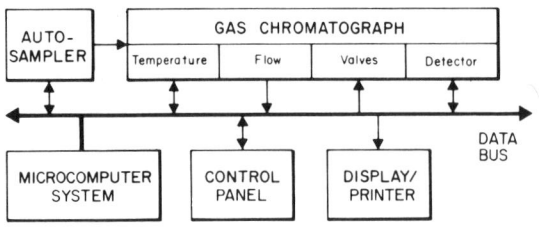

Fig. 6
• Automation of a gas chromatograph using a microcomputer system

microprocessors to control instrument operation on a one-for-one basis may be a more viable approach in some applications.

A generalized diagram of a microprocessor-based chromatograph is shown in Figure 6. The I/O data bus concept is used and the detectors, temperature controllers, etc. are peripherals to the microprocessor system. The microprocessor controls the entire analysis, sample injection, time events, data acquisition, data reduction, and reporting. In addition, the microprocessor can control the autosampler. Such control might include sample tray position, sample volume, and number of injections. Special information about the sample such as identification, analysis conditions, amount of standard, etc., might also be stored and used to calculate the report. Once the initial investment in a microprocessor is made, additional features are often relatively easy to implement.

The HP 5830A is a keyboard-controlled microprocessor-based gas chromatograph [9]. The gas chromatograph contains no dials, switches, or meters and G.C. parameters such as column oven, injector and detector temperatures, and detector parameters, are input through the keyboard and displayed through the printer. The flow rate is set manually using a flow controller and the flow rate is displayed digitally through the printer. The printer/plotter parameters − chart speed, attenuation, and zero control − are also set through the keyboard. The system uses set-point control of the gas chromatograph and printer/plotter parameters but temperature algorithms allow the processor to choose between three modes of temperature control: standard heating; room air; and cryogenic.

Closed-loop feedback control also will be an important use of microprocessors. As an example, an intelligent but optically conventional spectrometer with closed loop real-time feedback control might be configured as in Figure 7. The potential applications of a built-in microcomputer system are listed in Table III. The use of microprocessors to maintain the reference signal constant by controlling the slit width or photomultiplier dynode voltage is an example of the use of microprocessors in closed-loop feedback control.

Table III. Potential Applications of Microprocessors to Spectrometer Automation

Data Processing Functions
    Baseline correction
    Multiscan averaging
    Integrated spectra
    Derivative spectra
    Difference spectra
    Conversion between transmittance and absorbance
    Calibration
    Quantitative analysis
    Multi-component analysis
    Zero suppression
    Dark current compensation
    Stray light compensation

Instrument Control Functions
    Monochrometer rotation (prism or grating)
    Recorder synchronization and control
    Slit width to maintain constant reference signal
    Dynode voltage to maintain constant reference signal
    Automatic sample changer
    Lamp change
    Filter change

Vidicon detectors appear to have potential applications in multi-element [10] or fast-scan spectrometers [11]. The data handling problem is troublesome and expensive to solve with minicomputers. The coupling of microcomputer systems with vidicon detectors may provide both technical and economical solutions. The synergistic union of these two devices may produce a new spectrometer which is profoundly different in design, function and application from present-day spectrometers. Thus, the use of microprocessors will cause new concepts and approaches to be used in chemical instrumentation.

Distributed Systems

Instruments with high data rates or with complex data processing requirements are presently automated using dedicated minicomputers and/or large central time-shared systems. Examples of such instruments are mass spectrometers, X-ray spectrometers, NMR, etc. In some applications dedicated minicomputers are not justified and/or real-time demands place an I/O burden on central time-shared systems. One solution to these problems is to optimize the system by dividing the computational and control functions among several processors in a hierarchical distributed system. In this way, each processor has some portion of the total control and computational load. The use of microcomputers in distributed systems is shown in Figure 8. Microcomputer functions will include:

• real-time control of instruments
• local operator interface/intelligent terminal
• partial data processing, i.e., digital filtering, peak picking, etc.
• I/O handling of data: buffering and transmission

More or less data processing will be included with the microcomputer system depending on the situation. In general, the larger computer will handle storage of large

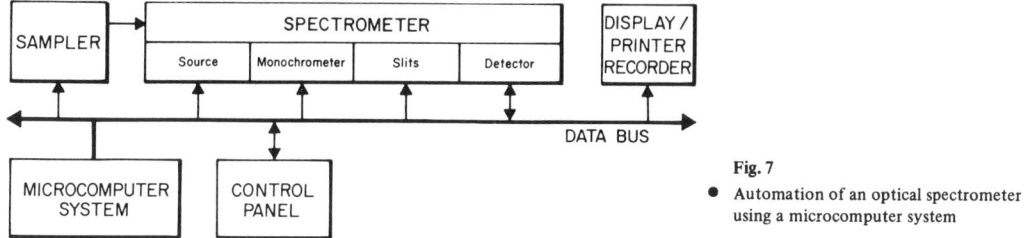

Fig. 7
● Automation of an optical spectrometer using a microcomputer system

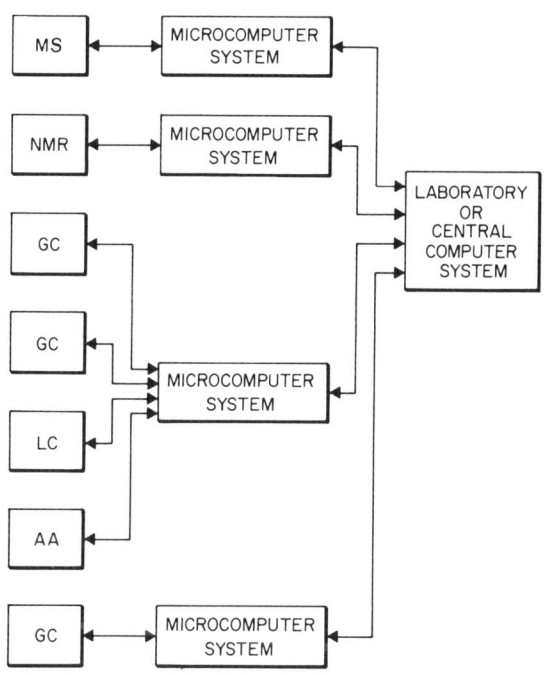

Fig. 8
● Inclusion of microcomputers in distributed systems

serial digital mode. This will eliminate problems of analog transmission and will reduce the cost of multiwire parallel digital data transmission. Serial transmission is easily expandable, is compatible with modems and can be used with standard phone lines. Standard interfacing hardware is becoming available and will ease the problems of interfacing.

Research Applications

In spite of the programming problem, research applications of microprocessors are being reported [12], particularly where minicomputers are under-utilized and hardwired logic is too inflexible. The low cost of microprocessors is attractive but the reader is cautioned that the ease of programming larger computers in high-level languages may compensate for this initial cost difference. The standard interfaces and components which are now available allow the scientist to build microcomputer systems in a modular fashion. A general-purpose configuration for research purposes is shown in Figure 3. A versatile configuration such as Figure 3 can be applied to a variety of applications by changing the stored program. The use of PROM or RAM for storage of programs is recommended in research applications where flexibility is important. The potential applications of microprocessors in research are vast and will involve many instrumental techniques.

amounts of data and complex computing tasks. Instruments with high data rates and/or complex control requirements such as NMR and MS will use dedicated micro- or minicomputer systems. Several instruments with lesser requirements may be multiplexed to a single microcomputer system. Instruments with built-in microcomputers will be capable of interfacing to larger computers or to other microprocessors.

Microcomputer systems used as intelligent terminals in distributed systems will be an important application. Besides providing the benefits mentioned above, the microprocessor will handle the transmission of data from "clusters" of instruments to the central computer in

Conclusions

Microprocessors will cause a profound change in the way that analytical instruments are designed and applied. Microprocessors will first replace hardwired logic but will eventually replace minicomputers in instrument control functions. The future will see microprocessors used in continually lower-priced instruments and even in portable instruments. Multi-microprocessors will be used to share workload, for modular design, to increase speed, and to functionally optimize systems for control, data processing and communications. More decentralization of distributed systems will occur as a result of microprocessors. Eventually, new principles and approaches to instrument design and application will be made practical by low-cost microprocessors. The large unit volume of microprocessors will create a demand for peripherals and a whole class of relatively low-cost "microperipherals" will become available to use in microcomputer systems.

Literature

[1] Quantum Science Corp. Industry Report on "Minicomputers and Microcomputers – The Squeeze is On". Copyright 1974, Samson Science Corp., Subsidiary of Quantum Science Corp., New York, p. 174.
[2] Ibid, p. 171.
[3] *Frazer, J. W.*, American Laboratory, Feb. 1973, p. 21.
[4] *Perone, S. P., Ernst, K.*, and *Frazer, J. W.*, American Laboratory, Feb. 1973, p. 39.
[5] *Dessy, R. E.*, and *Titus, J. T.*, Anal. Chem. **46**, 294A (1974).
[6] *Hettinger, J. D., Hubbard, J. R., Gill, J. M.*, and *Miller, L.A.*, J. Chromatog. Sci. **9**, 710 (1971).
[7] *Hettinger, J. D.*, and *Hubbard, J. R.*, Amer. Lab. **6** (2), 99 (1974).
[8] *Gill, J. M.*, J. Chromatog. Sci. **10**, 1 (1972).
[9] *Peterson, G. V.*, and *Poole, J. S.*, American Laboratory, May 1974, p. 70.
[10] *Mitchell, D. G., Jackson, K. W., Aldous, K. M.*, Anal. Chem. **45**, 1215A (1973).
[11] *Pardue, H. L., Milano, M. J.*, and *Cook, T.*, Presented: 167th ACS Natl. Mtg., Los Angeles, April 1-5, 1974.
[12] *Goedert, M., Wise, S. A.*, and *Juvet, Jr. R. S.*, Presented: Computers in Analytical Chemistry, Vienna, Sept. 24-27, 1974.

Bibliography on Microprocessors
Hardware

1. *Altman, L.*, "Single-chip Microprocessors Open Up a New World of Applications", Electronics, April 18, 1974, p. 81.
2. *Davis, S.*, "A Fresh View of Mini- and Microcomputers". Computer Design May 1974, p. 67.
3. *Lewis, D. R.*, and *Siena, W. R.*, "How to Build a Microprocessor", Electronic Design 19, Sept. 13, 1974, p. 60
4. *Murphy, J. A.*, "What's Available", Product Profile Microprocessors and Microcomputers, Modern Data, May 1974, p. 36.
5. *Rice, R.*, "Cost Perspectives Leading to Distributed Processing", I.E.E.E. International Convention and Exposition, March 1974, Session 10/1, I.E.E.E./Intercon, 3600 Wilshire Blvd., Los Angeles, Calif. 90010
6. *Shima, M.*, and *Faggin, F.*, "In Switch to n-MOS Microprocessor Gets a 2-μS Cycle Time", Electronics, April 18, 1974, p. 95.
7. "Understand the 8-bit μP: You'll See a Lot of It", EDN, Jan. 20, 1974, p. 48.
8. *Weisbecker, J.*, "A Simplified Microcomputer Architecture", Computer, March 1974, p. 41.
9. *Weissberger, A. J.*, "MOS/LSI Microprocessor Selection". Electronic Design 12, June 7, 1974, p. 100.
10. *Young, L., Bennett, T.*, and *Lovell, J.*, "N-channel MOS Technology Yields New Generation of Microprocessors", Electronics, April 18, 1974, p. 88.

Software

1. *Schultz, G. W., Holt, R. M.*, and *McFarland, J.*, "A Guide to Using LSI Microprocessors", Computer, June 1973, p. 13.
2. *Weiss, C. D.*, "Software for MOS/LSI Microprocessors", Electronic Design 7, April 1, 1974, p. 50.
3. *Weiss, C. D.*, "MOS/LSI Microcomputer Coding", Electronic Design, 8, April 12, 1974, p. 66.

Instruments/Applications

1. *Lee, R.*, "Microprocessor ICs Improve Instruments", Electronic Design 9, April 26, 1974, p. 150.
2. "Microprocessors and Their Applications", Three-part Microprocessor Course in EE/Systems Engineering Today, Part 1, Nov. 1973, p. 85; Part 2, Dec. 1973, p. 61; Part 3, Jan. 1974, p. 75.
3. "Microprocessor Runs Recorder", Electronics, May 16, 1974, p. 168.

4

Copyright © 1974 by Pergamon Press Ltd.

Reprinted from *Chromatographia* 7:539–546 (1974)

Application of an Inexpensive, General-Purpose Microcomputer in Analytical Chemistry

M. Goedert*/ S. A. Wise / R. S. Juvet, Jr.

Department of Chemistry, Arizona State University, Tempe, Arizona 85281, U.S.A.

Summary

Minicomputers are often underemployed and could be replaced by less powerful systems. The need for a product with characteristics intermediate between the minicomputer and hardwired systems resulted recently in the commercialization of inexpensive microprocessor units. A general purpose microcomputer has been designed around such a microprocessor unit, and the capabilities and limitations of this new device, when used to solve problems encountered in an analytical laboratory, are reported. Examples include: (1) on-line filtering of an analog signal at the output of a liquid chromatograph, (2) use as a buffer memory to rapidly store data from a GC/MS and then output it slowly on an inexpensive recorder, and (3) the analysis of noise from a new liquid chromatography detector using a conventional recorder as a storage oscilloscope. Applications for these inexpensive units are limitless, and we foresee their use in almost every field of instrumentation in future years.

Introduction

Commercially available minicomputer systems are generally not operated at full capacity when used on-line for process control or data handling. The Central Processor Unit (CPU) of a minicomputer is often used no more than 0.5 % of the time except during a period of data reduction. For the remainder of the time, the CPU is waiting for new data or performing Input or Output (I/O) operations with slow devices. Therefore except during the computational period of a problem, a minicomputer is underemployed and could be replaced by a less powerful system.

There is a genuine need for an inexpensive but versatile processor with less capability than commercially available minicomputers, and a product with characteristics intermediate between the minicomputer and hardwired systems has recently been placed on the market by several manufacturers of large scale integrated circuits (MOS). The purpose of this study is to determine the capability and limitations of these new devices when used to solve some problems encountered in an analytical chemistry research laboratory.

Reported in this paper are the main steps in the design of a general purpose microcomputer system around the CPU as well as several applications of this system—on-line digital filtering of an analog signal at the output of a liquid chromatography detector, use as a buffer memory between a mass spectrometer and an inexpensive recorder of low response time, thus eliminating the need for an expensive high-speed oscillograph for recording rapid-scan mass spectra, and the analysis of noise using a conventional recorder as a storage oscilloscope.

Miniature digital systems have been reported by others [1, 2], but like hardwired systems they can be used only for specialized problems. A more general system was designed by *Parker* and *Pardue* [3] from pocket calculator components. Their miniature digital computer included a small memory and an arithmetic logic unit useful for kinetic studies; however, the small instruction memory and the small number of mathematical operations available limited the versatility of the instrument for other purposes. The advantage of a microprocessor over previously reported systems is that the user can build at low cost a powerful and versatile system which fits his problem exactly and which can be easily modified by reprogramming.

System Description

In 1971 Intel Corporation (Santa Clara, California) was the only manufacturer to sell a microprocessor designed for use in desk calculators. In 1972 four different systems were available, and by the end of 1973 more than a dozen semiconductor manufacturers offered microprocessor units. There are two types of basic organization: *serial*, as in pocket calculators, and *parallel*, as in minicomputers. Serial microprocessors are less expensive but are slower. They are well suited for decimal arithmetic on small quantities of data and can work with long words. Parallel microprocessors are more expensive but are faster and more universally applicable than are the serial microprocessors. Words are 4, 8, 16, or even 32 bits in length. Four-bit microprocessors may be adapted to handle BCD numbers whereas 8-bit microprocessors are conveniently used for the transfer of ASCII characters or for data processing.

Most microprocessor manufacturers offer electronic memories called Read Only Memory (ROM) for program storage and Random Access Memory (RAM) for data storage. ROMs must be programmed at the factory. Another type of electrically reprogrammable Read Only Memory (PROM) is more suited for research laboratories since they can be programmed by the user by electrical charge transfer and erased by exposure to ultraviolet light [4].

---
* Present address: Laboratoire de Chimie Analytique Physique, Ecole Polytechnique, Paris, France.

Comparative studies of microprocessor units have been published [5-7]. Manufacturers include Intel, Fairchild, National Semiconductor, Rockwell, American Microsystems, Intersil, Signetics, Western Digital, Toshiba America, Zentec, RCA, Texas Instruments, and Motorola. These manufacturers offer, or will soon announce, libraries of programs such as assembler loaders and text editors. Some offer FORTRAN and PL/1 capabilities using larger computers for assembly.

The Intel Corporation has commercialized Microcomputer Systems (MCS) around two types of CPU — the 4004, a 4-bit processor and the 8008-1, a single-chip MOS 8-bit parallel CPU. The microcomputer described in this paper is designed using an Intel SIM 8-01 prototyping system, which utilizes the 8008-1.

The MCS8 uses the 8008-1 and at least 20 TTL chips to obtain an operating system. There are six general purpose registers and two address registers so that up to $16K \times 8$ bits of ROMs, PROMs, or RAMs can be addressed directly. There are 48 instructions, and the execution time is 20, 40, or 60 μsec depending on the instruction (12.5, 25, or 37.5 μsec with minor modifications). This microprocessor has an interrupt capability of one level to seven locations. Since this work was completed, Intel has commercialized another CPU, the 8080, with a 2 μsec cycle time, 74 basic instructions, multiple interrupt capability, and the possibility of addressing up to 65K bytes of memory.

A microprocessor is only a part of a microcomputer and consequently is not operational by itself, but must be interfaced with memory, clock, and other components. The complexity of the interfacing depends on the type of microprocessor and also on the versatility needed for the system. However, at least two steps can be defined: first, building a minimum operational system around the CPU and second, interfacing it with the process. The products offered commercially allow the user to either buy an operational general purpose microcomputer or to buy a prototyping card with a minimum operational system and interface it to the process.

By the end of 1972 Intel offered several prototyping cards: the SIM4 (4-bit CPU) and the SIM8 (8-bit CPU). These two printed cards include the CPU, RAMs, connectors for ROMs or PROMs, a clock generator, and several inputs and outputs. Recently these components have been grouped on several specialized cards such as Input, Output, RAM, PROM, and CPU cards. These printed cards form the parts of a family of general purpose microcomputers, the Intellec. Three microcomputers, the Intellec 4, Intellec 8 and Intellec 8/80 are presently available. These systems have been designed to help the user develop his microcomputer system including the programming of the PROMs.

When purchased complete, these microcomputers range in cost from $2500–$3900 and thus are almost as expensive as minicomputer systems. Microcomputers are generally slower than the minicomputers and are usually sold with less powerful software. Thus, these complete units probably still cannot compete with minicomputers but are very useful in developing both the hardware and the software for low cost specialized systems using microprocessors.

The total price of the components required for construction of the microcomputer described below is only $1388. This microcomputer contains $1K \times 8$-bit words of RAM memory and 512×8-bit words of PROM memory, power supplies, and the necessary 10-bit A/D and D/A converters for ready interfacing to chemical instrumentation. Those wishing more detailed wiring diagrams or copies of printed circuit boards developed may request them from the authors.

Design of the General Purpose Microcomputer System

The microcomputer built in our laboratory was designed for handling up to 10-bit data words. An 8-bit microprocessor was selected, and words of more than 8 bits are input on several machine cycles by multiplexing as many buffered inputs of 8 bits as necessary and storing their contents in memory. A 4-bit CPU could also have been chosen, but the complexity of the software when performing mathematical operations on data would have been increased. The general purpose microcomputer system was constructed around the commercially available prototyping card, SIM8-01, rather than attempting to build the entire system from the individually available CPU, RAM, and PROM components, thus saving considerable time and effort. Studies are now in progress to build specialized prototyping cards at even lower cost (ca. $400) which could be used for dedicated laboratory purposes since both the hardware and software can now be developed with the general purpose microcomputer described.

A simplified schematic of the SIM8-01 is given in Figure 1. This prototyping card includes the 8008-1 CPU chip, a clock generator, several buffered inputs and outputs, a memory, and the logic control. Information is exchanged between the CPU and the external components through an 8-bit bus during several cycles. Typically one 20 μsec memory cycle consists of five time states, and the execution of an instruction can take one, two, or sometimes three memory cycles. Two time states, T1 and T2, are necessary to address up to 16K words of memory. During T3, the instruction or data is present on the bus, and at T4 and T5 the instruction is executed. Multicycle memory instructions are performed in the same way, but only partial results or instructions are executed during intermediate memory cycles. All the internal or external operations such as input or output data and read or write in memory are selected according to the length of the instruction (1, 2, or 3 memory cycles) and the part of the cycle (T1 to T5). Four outputs and two inputs are included on the prototyping card. This number can easily be increased since the I/O device port, the bus, and the control lines are available on the card connector. As long as the ready line is low, the CPU is in the Wait state, and thus the ready line can be used to synchronize the CPU with any low-speed memory. Another application

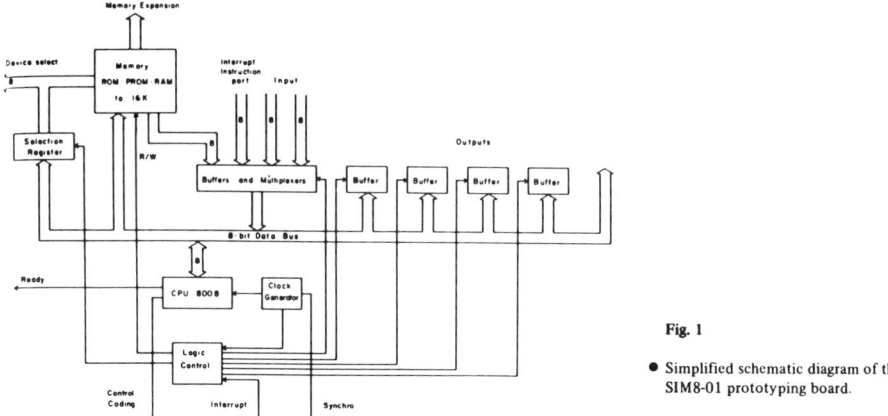

Fig. 1

● Simplified schematic diagram of the SIM8-01 prototyping board.

of the ready line is for stepping the computer cycle-by-cycle. This is particularly convenient in debugging a program or in input of new data into memory.

The interrupt instruction port is a 1-byte word that can be forced directly into the instruction register of the CPU whenever an interrupt occurs. When an interrupt occurs, the Program Counter (PC) is saved in the stack and the content of the interrupt instruction port is forced into the instruction register of the CPU. This interrupt instruction is supplied by the user and could be a jump to a subroutine for handling the interrupt or any other instruction. The PC content, which was saved in the stack register when the interrupt occurred, is transferred back into the PC at the end of the interrupt procedure, and the program continues until the next interrupt request occurs. The interrupt line has two different connections making it possible to recognize an interrupt when the processor is either in the Run or Stop state. The first possibility is the most efficient for interruption processing but is particularly difficult to handle with this microprocessor because there is no built-in subroutine to save the contents of all the registers, and consequently some values are lost. The interrupt is easier to handle when the CPU is in the Stop state since the content of the registers can be saved by the main program, but this is no longer an interrupt capability. In this case the interrupt line is used mainly to start the microcomputer or to continue the execution of the main program when an I/O operation has been performed. Then a no operation (NOP) can be forced in the interrupt instruction port and the program continues.

Microcomputer Description

Figure 2 shows a simplified schematic of the general purpose microcomputer built from the prototyping card previously described. The three basic parts are: the

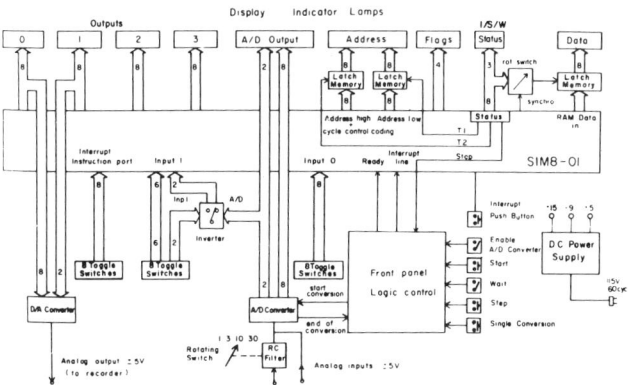

Fig. 2

● Simplified schematic diagram of the general-purpose microcomputer.

indicator lights, the input and output component parts, and the panel logic control.

Indicator lights display the content of the four outputs, the binary value of the A/D converter, 2 bits of cycle control coding and 14 bits of address, the status (Interrupt, Stop, or Wait), four flags (Carry, Parity, Zero, and Sign), and the binary value of the bus during any part of the memory cycle (T1 to T5) selected by rotating a switch on the panel control. Light Emitting Diodes (LED-MV55, Monsanto) were used as indicators because they are characterized by high reliability, long life, fast switching speed and TTL compatability. The power necessary for these indicators is not directly supplied by the prototyping card but by 7400N amplifiers or 7475N latch memories (Texas Instruments).

Input and output component parts consist of an Analog-to-Digital converter (Model ADC-D 10B, Datel Systems), a Digital-to-Analog Converter (Model DAC-49 10B, Datel Systems) and two 8-bit switch registers. The A/D converter has an output of 10 bits, a conversion speed of 50 μsec, and a specified accuracy of ± 0.05 %. By external strapping, the input is either unipolar (+ 10 V) or bipolar (± 5 V) and output coding can be straight binary, offset binary, or two's complement. The input voltage range of ± 5 V and two's complement output coding were employed in the instrument described. In this prototype one input is used for the lowest 8 bits and the 2 highest bits are shared through an inverter to the other input. Thus, this second input is shared between an eight switch register and the A/D converter (see Figure 2). When the A/D converter is not used, any data of 8 bits can be input from the toggle switches of Input 1. An active, low-pass band filter [8] rejects high frequency noise at the analog input to avoid certain problems in sampling [9]. The main use of this filter is to act as a time constant multiplier, and thus the filter can easily be miniaturized. The cut-off frequency of the filter can be adjusted to 1, 3, 10, or 30 cycles by means of a rotating switch on the front panel. If desired, the filter can be bypassed since there is a second direct input to the A/D converter.

The D/A converter has an input resolution of 10 bits with a settling time of 5 μsec. The input can be straight binary or two's complement and the output can be bipolar (± 5 V) or unipolar (± 10 V). Two's complement coding and bipolar voltage were employed in our system. The input of the converter is connected to two of the four 8-bit outputs of the prototyping card since 10-bit resolution is used, and the analog conversion is performed with two output instructions. The interrupt instruction switch register is a group of eight toggle switches used to force a byte into the CPU when the processor is interrupted.

The main purpose of the logic control is to interface the panel control switches and the A/D converter with the prototyping card. Figure 3 shows a timing diagram for data handling using the interrupt capability. This operation is only possible when the inverter sharing the Input 1 and the two highest bits of the A/D converter is on

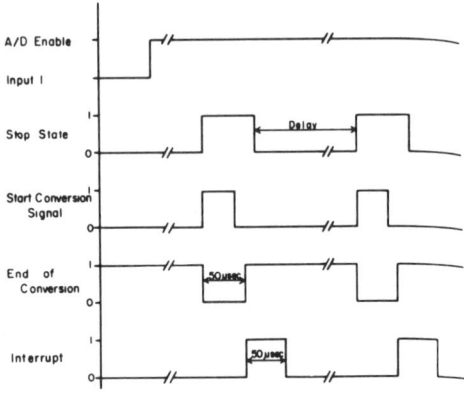

Fig. 3

● Timing diagrams for data handling using the interrupt capability.

"A/D Enable" and when the microprocessor enters the Stop state. Then the start conversion signal is high and the conversion begins. After the A/D conversion is performed (about 50 μsec later), an "end of conversion" is sent to the logic control which generates an interrupt signal, and the microprocessor leaves the Stop state. The Start switch, which is only effective when the A/D conversion is enabled by the inverter, also provides an interrupt.

The program may also be stepped one cycle at a time. When the Wait switch is "on", the ready line is low, and the microprocessor is in the Wait state. Each time the step switch is pushed, the ready line goes high, and the microprocessor executes one memory cycle.

Figure 4 shows a view of the completed microcomputer with all the indicators and control switches previously described. The prototyping card is inside a screened compartment in the bottom of the cabinet. The logic control cards, the A/D converter, and the D/A converter

Fig. 4

● The completed microcomputer showing indicators and control switches described in the text.

are contained in another screened box to avoid electrical interference. The three D. C. power supplies (± 15 V, Model PM555, Computer Products; −9 V, Model OEM-132-C and + 5 V, Model OEM-270-B, Viking Electronics) are fixed in the back of the cabinet, well separated from the logic cards. Natural cooling is sufficient and is provided by holes in the bottom and on the sides of the metal cabinet so that the unit may be left on without overheating the components. All connectors for the input and output of the analog signal and the interrupt line are located on the back of the cabinet providing ready access for interfacing to chemical instrumentation.

The total price estimation for the components required for the construction of this general purpose microcomputer is $1388. The prototyping card and the PROM memory cost about 74 % of the total price, the A/D and D/A converters are about 9 %, and the remaining 17 % is for power supplies, logic elements, LED displays, and other hardware items. Further cost reduction will require the design of a more simple prototyping card around the CPU ($180), RAM memory ($13.65 per 256 bits) and PROM memory ($67.50 per 256 × 8-bit words) and dedicated units with only sufficient components to perform the required task. Prices quoted should be considered as only approximate since they will vary with quantity purchased, transportation charges, etc.

The set of 48 instructions is well suited for binary arithmetic or for transfer of large amounts of characters or words of 8 bits, but there is a lack of basic instructions for accurate data processing. To read or write in memory or to execute a program already stored in memory, a minimum program called the bootstrap loader was written. When a Restart (RST) X is input from the interrupt instruction register, the address X is forced into the PC, and the corresponding subroutine is executed. With a RST 30_8 it is possible to fill an area reserved in memory by the initialization parameters with data entered from the switch register. A RST 40_8 allows the programmer to read the contents of any part of the PROMs or RAMs previously selected during the initialization subroutine. A RST 50_8 or RST 60_8 causes a program written in the PROMs or RAMs to be executed, respectively. A similar program is sold by Intel but is written for use with the commercially available MCS8 system and could not have been used directly with our configuration. PROMs may be programmed with an MP7-03 electrical programming terminal also available through Intel [10].

Selected Applications for the Microcomputer

1. Application for On-Line Filtering of Analog Chromatographic Signals

From a survey of signal to noise enhancement with instrumental techniques [11, 12], it would appear that D. C. signal filtering by analog filters is the simplest and the cheapest method of noise rejection. A major drawback of the analog filter is its lack of versatility, however. Response can be adjusted but only with a finite number of different values or in a relatively low range of frequencies; and as with all hardwired systems, its characteristics are not easily modified.

Digital filtering does not have these disadvantages but is generally more expensive. Curve fitting for chromatographic peaks gives accurate results [13]. Unfortunately, this method is used off-line, and as with other mathematical methods relating to simulation of gas chromatographic peak parameters [14, 15], requires a medium size computer. A simple method using convolute functions is the moving average method described by *Savitzky* and *Golay* [16] which has the advantage of on-line use without requiring a powerful computer.

A compromise between versatile but expensive digital filters and specialized but inexpensive analog filters is reported here using the microcomputer system to smooth an analog signal before recording on a standard graphic recorder. After digitalization, the signal is integrated for a finite number of samples according to the moving average method and, following D/A conversion, displayed on a graphic recorder.

The moving average method consists of calculating the mean of n successive data values. Then, by dropping the first of the n values and simultaneously adding the next successive value, a new average is computed, and the process is repeated. The mathematical description of the process is given by [16]:

$$Y^* = \frac{1}{n} \sum_{i=1}^{n} Y_{j-n+1} \qquad (1)$$

where j represents the running index of the ordinate data and n is the number of convoluting integers.

During initialization of the moving average program, the number n of data to be averaged and the sampling time t_s are input from the switch register. Then the RAMs, used for data storage, are cleared and the execution of the program begins. In the first step the content of all memories used for storage (stack register) are moved up one level to make room for new data. A new value is input from the A/D converter, all the data stored are added, and the result is divided by n. In order to simplify the software and save room in memory, n was restricted in our work to the values 1, 2, 4, or 8. Thus, division may be performed simply by rotating the binary sum 0, 1, 2, or 3 times to the right. The result is displayed on the graphic recorder after D/A conversion, and the program delays the jump to the first step for a time equal to the sampling period, t_s. The I/O operations and all calculations are performed in double precision. The complete program written in machine language occupies only 199 words of 8 bits, less than 1 PROM, and the data and parameters may be stored in the RAMs.

The effect of the moving average on peak-to-peak noise was studied using the output from the new, sensitive, liquid chromatography Spray Impact Detector (SID) [17]. Figure 5 shows amplified baseline noise in the output of

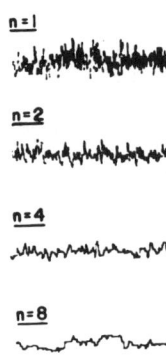

Fig. 5

• Effect of variation of n on the peak-to-peak noise level using the moving average method.

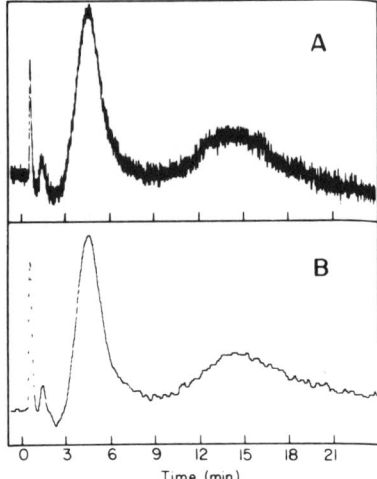

Fig. 6

• Reduction in noise by the moving average method in a liquid chromatogram of 18 ng octanoic acid, 300 ng nonanoic acid, and 200 ng decanoic acid using the Spray Impact Detector. A, direct output signal; B, microcomputer output, $n = 8$, sampling time, 2 sec. Column, Bondapak C_{18}/Corasil; mobile phase, distilled water at room temperature. Impurity peak follows sample injection.

the SID and the effect on the noise level of choosing different values for the convoluting integer, n. Noise rejection is a function of the square root of the number n of data averaged. With eight values, short-term noise is rejected by a factor approximately equal to $\sqrt{8}$, and only a low frequency signal remains. The difference between the decrease in noise calculated and that found experimentally in Figure 5 is less than 20 % in error. The step effect in Figure 5 for $n = 8$ is brought about by the limited sensitivity of the 10-bit A/D and D/A converters. A 12-bit A/D and D/A converter would improve the appearance of the outputted analog signal, and the total price of the system would be increased only modestly.

The chromatogram shown in Figure 6A is the output from the Spray Impact Detector in the separation of a mixture of nanogram quantities of the C_8, C_9, and C_{10} fatty acids using distilled water as mobile phase. The column used was two feet in length, packed with Bondapak C_{18}/Corasil (Waters Associates), and operated at room temperature. In order to test long distance transmission of the analog signal from the output of the detector electrometer (Keithley Model 610A, ± 10 V full scale) under the worst of conditions, the electrometer was connected to the computer A/D converter through 350 ft of coaxial cable (coaxial 8240 R. G. 58/U) strung near a row of fluorescent lights. No modification of signal could be detected in these remote experiments. The signal at the input of the computer was attenuated by a factor of two to fit the input specifications of the A/D converter. It is desirable that the density of measurement, that is the number of measurements within one standard deviation of the narrowest peak, be 5 or larger. This value is not as critical in these experiments as in previous accurate work [18], but measurement density should be large enough to eliminate a step effect on the time scale of the recording brought about by the A/D sample and hold conversion technique. The cut-off frequency of the analog filter before the A/D converter was adjusted to 1 cycle/sec which was a compromise between the smallest cut-off frequency of the filter, the sampling time, and the narrowest peak [9].

Figure 6B shows the result of averaging of eight data values. The sharp peak following sample injection is caused by impurities in the sample. Obviously, for this peak the density of measurement is not large enough. However, since this peak is not one of interest, the sampling frequency was not increased. If the signal to noise ratio of the three other peaks is considered, we note that it is improved by almost a factor of three. For example, the signal to noise ratio for the decanoic acid peak ($t_R = 14$ min) is equal to ca. 2.2 before averaging. With a moving average of 8 values, the ratio is increased to ca. 6. For analytical applications of chromatography we are interested in the effect any noise reduction technique may have on the area and on the position of the peak maximum. From Equation (1) the value of the integration after averaging is given by:

$$I^* = \frac{1}{n} \sum_{j=0}^{m} \sum_{i=1}^{n} Y_{j-n+1} \qquad (2)$$

Application of a Microcomputer in Analytical Chemistry 545

Without averaging, the integrated area is given by,

$$I = \sum_{j=0}^{m} Y_j \quad (3)$$

The difference between I^* and I depends mainly on the edge effect; that is, the value of the signal near the integration limits 0 and m. If this value is close enough to zero and if $m \gg n$, the result of the calculation is the same with and without averaging. Otherwise, the error depends on the smoothing parameters, the integration limits, and the particular values of the signal near the limits. In general, errors in integrated peak area can be made very small.

The position of the peak maximum after digitalization and averaging occurs when the change in integrated peak area is at a maximum. The signal is averaged on a bandwidth or time range Δt equal to:

$$(n-1) t_s = \Delta t \quad (4)$$

where t_s is the sampling time and n is the number of successive samples. For symmetrical peaks the apparent peak maximum is measured at the time,

$$(t_R)_{app} = (t_R)_{true} + \Delta t/2 \quad (5)$$

Therefore the absolute error is equal to half the value of the bandwidth of integration for peaks approximately symmetrical in shape. The relative error in the measurement of the maximum value of the peak is equal to,

$$\frac{(t_R)_{app} - (t_R)_{true}}{(t_R)_{true}} = \frac{\Delta t}{2(t_R)_{true}} \quad (6)$$

Thus the error introduced in retention time for each of the peaks in Figure 6B is 1 sec for $n = 2$, 3 sec for $n = 4$, and 7 sec for $n = 8$. Although the time range of integration tends to degrade the signal and to increase the apparent retention time if the value of n is made too large, in the worst case, $n = 8$, the error introduced is not worse than the ordinary graphical error. Moreover, any error introduced can easily be calculated and a correction made, if desired.

2. Application of the Microcomputer as a High-Speed Oscillograph in GC/MS

The microcomputer may be programmed to take in data at a rapid rate and then output these data at a rate slow enough to be followed by an ordinary graphic recorder, in effect functioning as an oscilloscope or oscillograph, but providing a considerably more accurate and permanent recording of the event. The sampling rate and the output rate were controlled by two delay loops in the program which could easily be modified by changing the initial parameters before execution of the program. Once the data are stored in RAM memory, they may be output any number of times at different speeds.

In the example given here only the eight least significant bits of the 10-bit data word from the A/D were stored in order to conserve space in the RAMs. This modification required the input range to be 0 to 2.5 V rather than ± 5 V, and the accuracy was thus limited to 1 part in 2^8 or 0.4 %.

Using this program, the microcomputer functions as a buffer memory and was used to record a portion of a mass spectrum of octadecanol. The mass spectrometer (Varian Atlas CH4-B) was operated at a rate of 3.3 sec/octave, and the sampling rate of the microcomputer was adjusted to approximately 180 data points/sec. Figure 7A compares the spectrum as recorded with a relatively expensive ($9,841 in 1967) oscillographic recorder (Honeywell 1508 Visicorder) in a period of 6 sec and the output of the inexpensive microcomputer, Figure 7E, as recorded on a simple recorder in 18 minutes. For publication purposes the microcomputer/recorder display has considerable advantage since the oscillographic recorder output is on constantly fading, light sensitive paper, making direct photographic reproduction extremely difficult. The quality of Figure 7A is obviously much poorer than 7B in spite of the greatest efforts in reproduction. Since our computer presently contains only 1 K RAM memory locations, the program employed allowed coverage of only *ca.* 100 mass units. An additional 4K RAM memory may be added for $750.

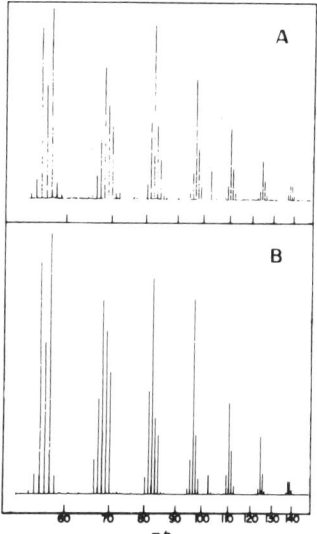

Fig. 7

● Portion of mass spectrum of octandecanol recorded using: A, oscillographic recorder with light sensitive paper, 6 sec scan; B, microcomputer/recorder output, sampling rate, 180 data points/sec.

57

3. Application of the Microcomputer as a Substitute for a Storage Oscilloscope in Instrumental Noise Evaluation

A system was designed to use the external interrupt capability of the microcomputer by triggering the microcomputer and a memory storage scope (Memo Scope Model 104, Hughes Products) simultaneously. This system was used to record the baseline noise of the new liquid chromatography Spray Impact Detector [17] in a detailed study of the contributing sources of instrumental noise. The data were stored at a rate of approximately 100 data points/sec and were output at a rate of 4 data points/sec to a simple recorder. A comparison of the output from the microcomputer/recorder combination and the storage scope is shown in Figure 8. This figure shows the amplified signal variation over a period of 10 sec. An input rate as large as *ca.* 4100 data points/sec can be used with this program in the study of higher frequency signals. Thus, the microcomputer could be employed in many applications where an oscilloscope is used to record a fast signal when a more permanent and accurate record is desired.

We foresee development of inexpensive microcomputer units for restricted and dedicated uses in the immediate future in almost every field of instrumentation.

Acknowledgment

The authors wish to thank *T. A. Kanneman* of the Arizona State University Technology Department for use of the MP7-03 programming board and Waters Associates for the gift of the Bondapak packing material used in the LC experiments. Financial support of the National Science Foundation under grants GP 25553 and GP 40830X is greatly appreciated.

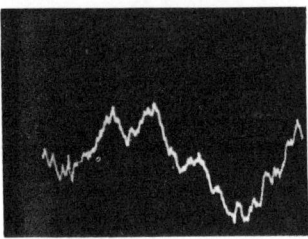

Fig. 8

● Comparison of noise evaluated simultaneously with a storage scope and with the microcomputer/recorder combination. Detailed study of noise from an LC Spray Impact Detector. 10 sec scan. Microcomputer input, 100 data points/sec; output, 4 data points/sec.

References

[1] *Crouch, S. R.*, Anal. Chem. **41**, 880 (1969).
[2] *Parker, R. A., Pardue, H. L.,* and *Willis, B. E.*, Anal. Chem. **42**, 56 (1970).
[3] *Parker, R. A.,* and *Pardue, H. L.*, Anal. Chem. **44**, 1622 (1972).
[4] *Frohman-Bentchkowsky, D.*, IEEE Journal of Solid-State Circuits, SC-6, No. 5, October 1971.
[5] *Wiener, H.*, Computer Decisions 5 (8), 8 (1973).
[6] *Gladstone, B.*, Electronics **46** (21), 91 (1973).
[7] *Holt, R. M.,* and *Lemas, M. R.*, Computer Design 13 (2), 65 (1974).
[8] *Bristow, Q.*, Electronics **45** (9), 102 (1972).
[9] *Goedert, M.,* and *Guiochon, G.*, Chromatographia **6**, 76 (1973).
[10] Intel Corp., MCS8 Microcomputer Set Users Manual, rev. 3, 1973.
[11] *Hieftje, G. M.*, Anal. Chem. **44** (6), 81A (1972).
[12] *Hieftje, G. M.*, Anal. Chem. **44** (7), 69A (1972).
[13] *Chesler, S. N.,* and *Cram, S. P.*, Anal. Chem. **45**, 1354 (1973).
[14] *Kelly, P. C.,* and *Harris, W. E.*, Anal. Chem. **43**, 1170 (1971).
[15] *Kelly, P. C.,* and *Harris, W. E.*, Anal. Chem. **43**, 1184 (1971).
[16] *Savitzky, A.,* and *Golay, M.*, Anal. Chem. **36**, 1627 (1964).
[17] *Mowery, R. A., Jr.,* and *Juvet, R. S., Jr.*, J. Chromatog. Sci., in press.
[18] *Goedert, M.,* and *Guiochon, G.*, Chromatographia **6**, 116 (1973).

Copyright © 1974 by the Royal Society of Chemistry

Reprinted from *Analyst* 99:838-852 (1974)

An Instrumentation-orientated Micro-computer: An Extremely Inexpensive Data Acquisition Computer Optimised for the Automated Laboratory*

By W. STEPHEN WOODWARD, THOMAS H. RIDGWAY†
AND CHARLES N. REILLEY

(Kenan Laboratories of Chemistry, University of North Carolina, Chapel Hill, North Carolina 27514, U.S.A.)

As the complexity and subtlety of phenomena regularly studied in the laboratory increase, a corresponding increase occurs in the sophistication of associated transducers and in the complexity of instrumentation required for monitoring and control of experiments. Commonly, this increasing requirement for sophistication in measurement has resulted in the extensive incorporation of electronics technology in scientific apparatus. This coupling of scientific technique to the electronics industry has naturally resulted in a number of combined effects. Among those of particular interest are the ways in which the economics of the industry have affected the fundamental ways in which instrumentation handles data. The enormous reductions in the marginal cost of complexity in electronic systems seen in the last two decades have been reflected in the intricacy of data manipulation routinely performed in measurement devices. This process has reached a culmination in approximately the last decade, with the direct incorporation into scientific instrumentation of the stored-program digital computer.

The introduction of the computer to scientific instrumentation has proved to be an area of unusual scope. Often, the advance of a particular field of measurement seems to depend on the rate at which it is combined with programmable digital data processing. The potential for enhancement of both the quality and clarity of presentation of acquired data via computer-aided experimentation results both from the flexibility of computer data manipulation and the ease with which control and processing algorithms (*i.e.*, programs) can be generated, tried and modified.

In the Chemistry Department of the University of North Carolina (UNC), we have accumulated over 4 years of experience with a general-purpose, multi-laboratory data acquisition and analysis system of considerable versatility and success. The project described in this paper capitalised on this experience, together with several very recent developments within the digital instrumentation industry, to produce and apply a type of integrated system for computer-based assistance with experimentation that possessed characteristics superior to those of previously existing techniques. Basically, the central purpose of the

---

* Due to unforeseen circumstances, this paper was not presented at the Centenary Celebrations.
† Present address: Department of Chemistry, Texas A & M University, College Station, Texas, U.S.A.

proposed work was to examine the technology loosely termed "computer-based laboratory automation" and, through the introduction of recent advances in the computer and electronics fields and of insights we at UNC had gained by our efforts at experiment control, to unify a portion of it. The aim of that unification was to modify the existing perspectives of laboratory automation in which the economic and technical logistics of computer technology often dominate the motivating objectives of instrumentation. This aim was to be accomplished through: (i) the completion of the development of a highly versatile set of experiment-control hardware, (ii) the support of this hardware via an equally versatile host computer and experiment-design software system, (iii) the application of the resulting technology to current experimental work in order to demonstrate and refine system capabilities and (iv) the use of the resulting system as a means for investigating basic issues in the computer assistance of experimentation.

The application of the digital computer in the chemical research laboratory has historically reflected the great diversity of the small computer industry. Virtually every computer vendor marketing systems that are suitable for real-time control and monitoring has representation at the instrument control level of laboratory automation. Obvious consequences of this variety in choice of hardware are duplication of design effort in the creation of instrumentation systems and difficulty in the transfer of technique beyond that which might reflect true differences in experimental method. A general lack of success in the inter-laboratory sharing of computer-based experimental technique as well as a large increase in the toteal costs o system development has frequently resulted from such hardware incompatibility.

Our aim is to provide for the research community a body of computer-based instrumentation technology that is highly versatile yet optimised to the requiremnts of thef laboratory. The objective is to offer an alternative to the continuing proliferation of types of laboratory automation systems.

Our current work represents the later stages of progress made during the past few years at UNC towards:

(i) The generation of an informal model of research-laboratory experiment - computer interactions with the intent of operationally defining the functional characteristics of programmable digital systems most relevant to effective instrumentation control and experimental data acquisition.

(ii) The design and development of a set of computer-orientated hardware components and of techniques optimised for the control and monitoring of research instrumentation according to parameters isolated in phase (i).

(iii) The configuration of a centralised computational resource for the support of experiment design and set-up and of data analysis and interpretation.

(iv) The implementing of a multi-access hierarchical laboratory automation system realising the characteristics identified in phase (i) with the systems and techniques developed during phases (ii) and (iii).

(v) The refinement and documentation of resulting methods and hardware into packages suitable for their transfer to other research facilities and disciplines.

At present, the characterisation of the research instrumentation system, as undertaken in phase (i), is well advanced. Over 4 years' accumulation of experience in computer-orientated control of experimental apparatus entailing data acquisition at rates ranging from less than 10 data per second (gas chromatography) to 5×10^6 data per second (television image capture) has provided considerable raw input for the modelling process. The image that has emerged indicates that instrumentation control systems demand considerable capability for versatile, rapid input - output. Computational demands placed upon an experiment-monitoring processor are generally associated with either pre-experiment set-up, or post-experiment data analysis.

Thus a model of an archetypal instrumentation-orientated computer system was derived that places great weight upon input - output flexibility and less upon rapid-response computation and decision making. This finding agreed well with the announcements in late 1971 and early 1972 (INTEL, ROCKWELL INTL, etc.) of monolithic central processors. We subsequently undertook the construction of a prototype laboratory instrumentation computer built around the INTEL MCS 8008 CPU component. This work culminated in the demonstration in March, 1973, of a basic working system. That system, together with the associated development of utility software and supporting systems, is described in this paper.

The UNC instrumentation-orientated micro-processor

Philosophy, capability, packaging and logistics—
One of the most important innovations to result from this work is the development and application of a specialised digital computer system intended to be used in the instrumental laboratory to perform management of real-time data collection. It was our intention to design a modular unit that would be sufficiently inexpensive and yet adaptable to various applications to possess the potential to become as ubiquitous as the research-quality oscilloscope or digital voltmeter. The system concept (Fig. 1) is characterised by the following features.

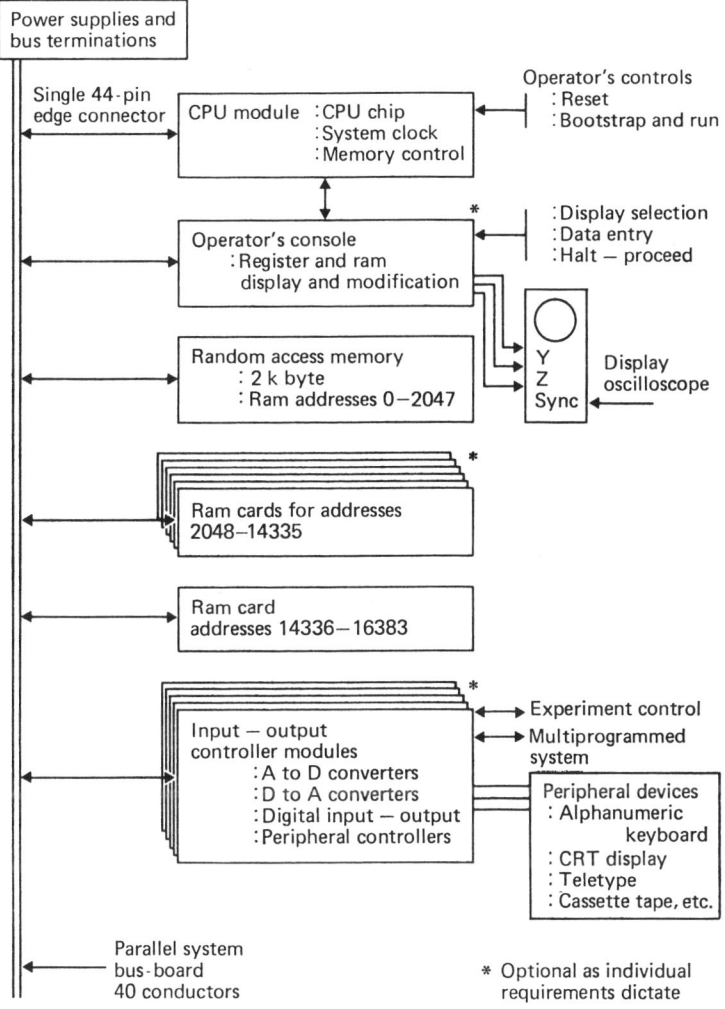

Fig. 1. Micro-computer system organisation

1. *Extremely low cost*—If the computer is to become a commonplace research tool, its cost to the experimenter must be compatible with existing general-purpose laboratory aids. A target figure of $1000 to 1500 for a complete, basic system capable of high-performance data acquisition and interactive analysis can be achieved through micro-computer technology.

2. *Modularity*—A viable computer-assisted data-acquisition research environment is often characterised by rapid changes in applications. Hence, a general-purpose data acquisition system is likely to be confronted by a succession of applications, each with differing requirements. A system concept suitable for this changing environment must either be capable of handling all application demands in some standardised configuration, or else must

lend itself with great ease to a "tailoring" process so that it may be "scaled" either up or down, as the occasion demands. In an effort to maximise the potential efficiency of capital investment, the system described here exploits the latter method to an advanced degree. Virtually all configuration options, including memory capacity, input - output peripherals and experiment interaction hardware, are packaged in an identical printed-circuit card format. The addition or removal of a system function consists solely of the insertion or deletion of one (or, in rare instances, more than one) printed circuit module into any one of an array of standardised edge-connector sockets soldered to a common "bus" board. This concept of organising system architecture around a standardised intercommunication bus is a technique that is coming into increasing use in the digital instrumentation industry. In the context of the UNC micro-processor, this organisation scheme offers a maximum in configuration versatility with very little in associated overheads. In addition, the standardisation of peripheral format reduces enormously the problems and cost of packaging and "pretty box" construction which often render impractical in-house construction and maintenance of digital equipment. This last factor is of particular importance because of the following system characteristic.

3. *Total in-house hardware responsibility*—The potential purchaser of a small-scale data-acquisition system may find himself in a dilemma when confronting the question of system maintenance. The manufacturers with the best bargains are seldom those with good service networks. Even the few vendors of small computers which have large service departments must charge considerable fees for maintenance contracts. Yet, if the purchaser contemplates in-house repair of computer-based instrumentation, difficulties arise. First, of course, is the fact that the digital computer, no matter how small, is a device of great complexity. Considerable specialised skill and familiarity with a particular model is necessary before even an experienced technician is comfortable working with it. Secondly, while the vendor's customer engineer can have at his disposal a number of sub-system components, which, by substitution for suspected parts, can greatly speed diagnosis and repair, capital expenditure considerations make this impractical for the average end user.

The solution to this problem, here proposed, is a rather drastic but nonetheless appealing one. If a suitable system construction format is chosen, one that minimises error-prone and time-consuming hand wiring, eliminates involved mechanical packaging and utilises only readily available mechanical and electronic components, it is feasible to construct the entire data acquisition system, including the computer, within house. The practicality of this approach for an average laboratory facility would have been difficult to defend as recently as a few years ago. The advent of medium and large-scale integrated circuit system components, however, makes possible the design of an entire data acquisition and experiment control system of low enough "package count" to make feasible in-house assembly by a support facility large enough only barely to maintain the usual type of computer system. This reduction in sub-system components also makes entirely feasible and economic the accumulation of enough substitutable parts to speed repair, as well as making possible a great increase in the number of systems owned and distributed within the facility.

4. *Optimisation of input - output structure for experiment control*—Careful examination of real-time applications in the laboratory indicates that a small computer dedicated to instrumentation will usually use up the available response time and data acquisition speed capability long before its computational capacity is exceeded. The computation associated with experiment control is often trivial or even non-existent. The computation that *is* needed is usually associated with post-experiment data interpretation and need be completed only in amounts of time commensurate with the patience of the researcher. The speed of actual data acquisition, however, is usually dictated by the physics of the system under study. Hence, if the computer system is too slow in that regard, the work simply cannot be performed.

For these reasons, the design of a minimal cost data-acquisition system must achieve a highly versatile input - output structure possessing speed and generality for real-time operations disproportionate to the computational power of the CPU. This property is achieved in the UNC micro-processor system through the inclusion of a direct memory access sequencing module. This input - output control device will manage the activity of system analogue ⇌ digital conversions and experiment control through the agency of a controlling program generated before initiation of the experiment by the local CPU and stored in the system memory. The relation of experiment interrogations and stimuli to the time stream is maintained with a high degree of ease and accuracy via reference to an internal time base of high stability. The over-all structure of the sequence controlling code is organised on lines compatible with envisaged experiment design and management languages.

The sequencing hardware conforms to the packaging guidelines of other peripherals. Hence, it is highly modular and may be excluded from terminal-system configurations not demanding such high-speed input - output capabilities.

5. *Compatibility with a hierarchical system*—Because the UNC micro-processors were configured from the outset as terminal components in a multi-laboratory system, features have been incorporated in their design that facilitate communication with a central host computer. "Bootstrapping" of the terminal computer into operation from the central system (*i.e.*, initial loading of controlling programs), transfer of data-acquisition control programs from the central library, data transfer and other facets of computer - computer communication are easily and efficiently performed. In addition to powerful performance specifications, an important aim in the design of the intercommunication hardware was simplicity of the connecting cabling. This detail, if improperly handled, can present great obstacles to the smooth expansion of a multi-laboratory installation. All transfer of information within the UNC system occurs along a single "multi-drop" "party-line" bus. In this way, connection of additional terminals is made maximally convenient. Wiring practices used on this bus were chosen so that vulnerability to signal interference and "ground-loop" noise were minimised.

6. *Availability of a wide variety of inexpensive, versatile peripherals*—A wide variety of input - output devices are needed to accommodate the diversity of experiment and instructional system requirements. The range of data type, volume rates, manipulation techniques and presentation methods encountered in the usual laboratory environment is reflected in a need for a rich repertoire of data acquisition, storage and operator-orientated input - output peripherals. In order to provide these types of resources, a wide choice of peripherals are made available for the UNC micro-computer system configuration.

SYSTEM MODULES AND COMPONENTS—

The list of system modules (Table I) indicates the extent of selection that is possible. The ready availability of sophisticated OEM peripheral components (Table II) and the simple input - output bus organisation of the UNC micro-computer system have combined

TABLE I

MICRO-COMPUTER SYSTEM MODULES, COMPONENT COST, ESTIMATED CONSTRUCTION TIME AND FUNCTION

Module	Cost/$	Time/man-days	Function
System housing (P.S.)	160	2	With power supply and motherboard
CPU card	200	1·5	Based upon Intel MCS 8008
Memory card	140	0·25	2048 byte, dynamic MOS, RAM 1-μs cycle
Communication interface card	25	0·6	To UNC Raytheon 706 computer
Teletype controller	28	0·4	Unmodified ASR-33
CRT graphic display controller	70	0·6	Hardware character generation and graphics
ADC card	180	0·6	12 bit, 20 μs, 8 channel
DAC card	160	0·5	12 bit, 1 μs, 2 channel
Sequencer card	40	0·75	Programmable high-speed experiment input - output (see text)
Magnetic tape controller	40	0·6	1–3 MDS drives (1·7–5·1 Mbyte; 4·8 kbyte s^{-1}
Alphanumeric keyboard controller	20	0·3	
TV controller	35	0·5	256-line raster display (colour capable)
Breadboard card	15	0·2	Universal peripheral prototyping card
Signature card	15	0·2	Identifies connected terminal
High-speed PT reader controller	20	0·4	120 byte s^{-1}
Large-capacity memory controller	35	0·6	Address management for up to 96 kbyte of data memory
Light-pen graphics interaction card	30	0·8	
Digital input - output card	40	0·4	16-bit digital input - output
Charge manipulation ADC card	160	0·6	Implements various integrating ADC methods (see text)
Programmer's console	50	1·0	Inspection/modification of CPU status and memory
Time-shared interface card	35	0·8	Full duplex communication (100 kbyte s^{-1})
Ensemble averager	25	0·3	Hardware multiple precision adder of ADC results

TABLE II
MICRO-COMPUTER PERIPHERALS

Device	Cost/$	Description
Teletype	925	ASR-33 TBE (with automatic paper-tape reader, punch)
Display scope	245	Telequipment S51B oscilloscope (8 × 10 cm flat screen)
Keyboard	98	Controls Research Co., 53-key, ASC11 encoded
Paper-tape reader	250	Addmaster 601 8-channel punched paper tape; 120 characters per second
Peripheral power supply	25	Provides +24-V d.c. power for paper-tape reader
Magnetic-tape cartridge drive	385	Mohawk Data Systems drive for 3M cartridge. 4-track, forward - reverse, read-after-write head; 25-ms start-time; 15-ms stop-time; 4800 byte s^{-1}; 1·7 Mbyte with 300-ft cartridge ($20)
Data memory	2800	Dataram core memory; 32 kbyte, 850 ns
Housing and power supply	200	−18 V at 6 A; +5 V at 12 A; for data memory

to make the design effort investment represented by this large repertoire surprisingly small. At the present time, considerable progress has been made towards the implementation of a prototype micro-computer-based instrumentation system of the format described and the support of this system on the UNC Raytheon 706 computer. Development has included the completion and successful demonstration of the following hardware:
1. Basic CPU with 4kbyte RAM.
2. Teletype controller.
3. Graphics display generator.
4. Alphanumeric keyboard controller.
5. 706 communications interface.
6. 12-bit, 20-kHz ADC module.
7. Dual 12-bit DAC module.

Some of the modules given in Table I are discussed further below.

Multi-purpose charge-manipulation ADC module—The multi-purpose charge-manipulation ADC module consists of a digitally controlled, dual-channel integrator-analogue comparator with associated counter, gate and 50-MHz frequency reference to form a complete voltage-to-time and time-to-number ADC. Strappable options on the module permit it to serve the following functions on each of two channels:
1. Dual-slope integrating ADC.
2. Dual-slope ratio-metric integrating ADC.
3. Direct logarithmic ADC.
4. Direct log-ratio ADC.

Functions 1 and 2 are of value in both signal-averaging applications (*e.g.*, free induction decay nuclear magnetic resonance spectroscopy) and in high-resolution applications (*e.g.*, wide dynamic range data such as gas chromatography). Functions 3 and 4 permit direct monitoring of photo-detector outputs. Conversion results are given by the equation $N = C \log (V_{in}/V_{ref})$ and, hence, are directly proportional to absorbance. This functionality is obtained without the usual log-ratio analogue circuitry and hence avoids the curious frequency response and limited accuracy often displayed by these devices. Conversion time is proportional to required resolution and is 0·64 μs for 5-bit, 20 μs for 10-bit and 20 ms for 20-bit dynamic range applications.

Television raster display generator—The television raster display generator calls raster description data and intensity control information from system RAM at a 0·67-Mbyte rate to generate a 256-line non-interlaced television frame to permit the use of minimally modified entertainment televisions as large-screen, low-cost displays. Signals generated include a luminance - synchronisation composite signal suitable for control of standard black-and-white televisions. Additionally, red - yellow, green - yellow and blue - yellow chroma signals are made available for colour control in suitably modified colour televisions so as to permit the use of the greater appeal and information capacity of multi-colour displays. This unit is of primary interest in classroom instructional applications. Approximately 8 kbyte of RAM are dedicated as a bit-map buffer for the 256 × 256 resolution frame. All character and graphics generation are performed by software preparation of the bit-map.

Large-capacity memory controller—The large-capacity memory controller provides interface and address management for the connection of large-capacity memory modules (up to 96 kbyte) to the micro-computer system bus for the high-speed acquisition of data blocks too large to be held in the standard configuration memory. Dataram 32-kbyte core memory stacks are used as the storage elements. The primary application envisaged is the high-speed ensemble averaging of pulsed data from free induction decay in nuclear magnetic resonance spectroscopy.

SOME MAJOR DESIGN CONSIDERATIONS OF THE UNC CPU MODULE—

Although supporting electronics for the MCS-8008 can be purchased ready made from a number of vendors (Intel, Control Logic, etc.), a considerably different implementation was chosen for the UNC micro-processor. Some of the more significant features which distinguish our design are given below.

1. *Compatibility with a system organisation based on a modular bus structure in which system modules communicate over a compact interconnection bus common to all*—This form of system organisation was deemed very important in view of the planned flexibility and ease in individual instrument configuration.

2. *Support of high-speed direct-memory access input - output*—Available implementations of MCS-8008 micro-computer systems depend upon program-managed data transfer for input - output operations. Because of the limited speed of the MCS-8008 in such functions (less than 5000 byte s^{-1}), program-managed input - output and, with it, MCS-8008 implementations available off-the-shelf, are inadequate for many high-speed experiment control applications. The UNC CPU card supports direct memory access input - output, which, once initiated by programmed commands, halts the CPU and permits external access to the system memory at rates limited only by memory cycle time (*i.e.*, up to 10^6 byte s^{-1}). Because MCS-8008 activity is inhibited during DMA input - output, CPU card-mounted addressing logic is made available for input - output control and need not be duplicated on the input - output device controller. This feature permits the DMA and programmed input - output interface on device controllers to be identical both functionally and physically. Thus, virtually any input - output device can be operated in either mode as circumstances may dictate.

3. *Hardware-implemented program bootstrap*—Because of the envisaged rapid turnover of application programming, all of a micro-system memory is composed of read - write storage, rather than partly read - write and partly read only. This organisation provides maximum flexibility in program implementation and change. However, the obvious requirement for a means of initial program entry in the event of memory loss is generated. The availability of the program-independent DMA input - output described previously, combined with a small amount of additional circuitry, permits the UNC micro-computer prototype to accept initial program input from a variety of devices, including 706 digital communications interface, teletype (both implemented), high-speed paper-tape reader and magnetic tape cartridge drive (planned). Specific "bootstrap device" selection is accomplished by a jumper wire connected from the CPU card to the appropriate device controller. The bootstrap or FILL mode is initiated automatically upon "power up," manually via a front-panel switch, or under program control. The FILL mode is terminated by end-of-record status within the bootstrap device controller.

4. *Support of high-speed, dynamic semiconductor memory*—Available MCS-8008 implementations typically utilise static semiconductor RAM memories (often the Intel 1101A) because of the simplicity of the associated circuitry. These compounds are not an optimum choice for large-capacity memory arrays such as those possible for the UNC instrument. The factors of cost per bit, power consumption, package count and cycle time favour the competitive dynamic memory technology. The UNC CPU card provides clock and refresh signals appropriate for the 1103-type dynamic memory.

5. *Derivation of all timing signals from a single internal 5-MHz crystal-controlled frequency reference or externally supplied standard*—Most available MCS-8008 implementations utilise multivibrator-derived CPU-clock signals, the limited stabilities of which are disadvantageous in high-precision experiment control. The UNC prototype derives all signals from an internal 0·01 per cent. accurate oscillator. For extremely demanding applications, the internal source can be disabled and an external reference (5 MHz) of arbitrary accuracy substituted.

INTEGRATION OF HIERARCHICAL LABORATORY AUTOMATION SYSTEMS INCORPORATING MICRO-PROCESSORS AND OTHER PROGRAMMABLE DEVICES

CENTRAL COMPUTATIONAL HOST—

A classic objection to the dedication of small computers to individual experimental application, apart from economic considerations associated with the duplication of the processors themselves, is the inability to consolidate such resources as mass storage, data bases and expensive high-speed peripherals. When the computational power owned by the laboratory organisation is diffused among a number of individual computers, it becomes impractical to provide industry-compatible magnetic tape drives, line printers, card readers and similar peripherals, each of which may cost more than each small CPU. This problem is only amplified by the availability of very low-cost instrumentation computers such as the UNC micro-processor. In contrast, by providing each remote, dedicated computer with access to a central computer, an economy of scale may be gained. Language processors of useful power, computational resources, program and data mass storage, and high-speed peripherals may thereby be made available and the associated costs amortised over the relatively numerous remote terminals. Some added benefits accrue, including greater ease in program and sub-routine sharing among the remote users.

Hence, a central, time-shared (multi-programmed) computing resource has been designed as a component of the UNC system (Fig. 2) to provide the following services.

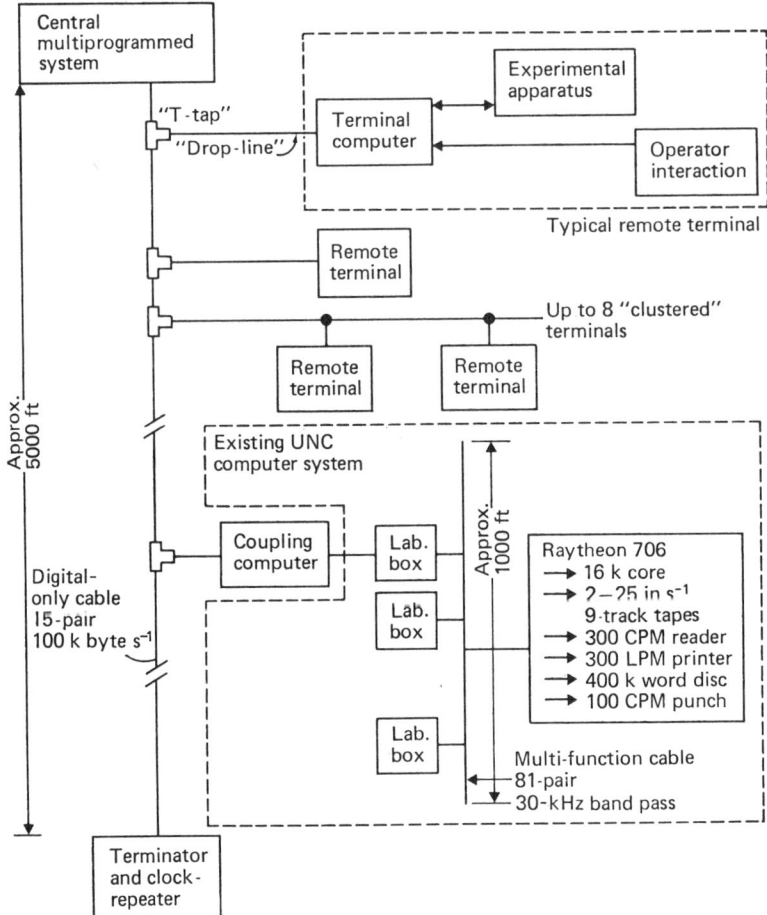

Fig. 2. Over-all proposed UNC data acquisition system

1. *A medium scale, 16-bit processor with 64 kwords of high-speed random access memory to serve as a common language and numerical analysis processor*—Specified CPU features include hardware multiply - divide, floating-point processing and multi-programming options to ensure adequate support of a many-user environment. The presence of powerful arithmetic hardware will aid the performance of those data analysis computations which are beyond the capabilities of the small remote CPUs. Multi-programming features, such as memory protection and hardware enforcement of system resource allocation (privileged restrictions on input - output operations, etc.), will reduce the danger of user - user interference.

2. *Large-capacity on-line mass memory*—An important feature of the UNC system is the ability of the remote user to develop, de-bug and access application software without leaving his remote terminal. The provision of this kind of service requires considerable rapidly accessed memory for the storage of program source statements and compiled code in a library-type format. In addition, a large store of this type can serve as a short-term file for user data undergoing collection or interpretation. Preliminary investigation indicated that a capacity of 5 to 10 Mbyte would provide adequate support of envisaged system usage. An economical implementation of this system resource can be accomplished through the installation of any one of the several variants on the IBM 2315 disc memory currently offered by independent vendors. These devices typically offer data-base media transfer rates of approximately 2 Mbaud and access times of less than 100 ms. Costs of 0·025 cent per bit capacity are typical.

3. *Line printer, card reader, industry-compatible nine-track magnetic tape*—Access to these peripherals often proves to be of great value, both to the local batch programmer and to the experimenter with need of hard-copy data output. While the prices of such devices are constantly falling, they represent substantial investment; hence, the central micro-processor support system does not include them, as they were already present on the UNC Raytheon 706 system. Access to them is provided via a link between the micro-processor host system and 706 system. Because the micro-host system is organised around a multi-programmed format, the fact that the Raytheon is not is no great handicap. Periods of access by the host to Raytheon peripherals occur as normal tasks in the Raytheon job stream. Delays in achieving access to Raytheon resources affect only the execution of the particular host - resident task initiating the request. Prior to the design of the micro-orientated system, considerable experience had been gained within the UNC Chemistry facility with such inter-computer sharing of peripheral resources (via existing data communication hardware) and considerable software had already been generated to enable the Raytheon 706 to provide such service to connected processors.

4. *Support of satellite mini-computers*—In addition to service as a central host to remotely located micro-computers, support is provided (in the form of access to mass storage, program library and high-speed peripheral resources) for various mini-computers located in the UNC Chemistry facility. Basic procedures and protocols for this support resulted from the design of resource-sharing techniques necessary for item 3 above. Techniques for minimisation of the amount of satellite - CPU-specific programming and special-purpose hardware required for each system added to the network, however, remained to be explored. The method investigated involves the use of micro-computer terminals serving as intelligent buffers between the high-speed, polled protocol of the shared communication bus, and a more easily handled parallel interface to the attached mini-computer. Use of micro-computers in this rôle greatly simplifies the hardware design and programming effort which would otherwise be required to bring individual mini-computers on-line. In addition, micro-computers serving in this rôle permit the use of low-cost micro-computer peripherals to be utilised by "symbiont" mini-computers.

Selection of a central host machine was an issue important to the aims of the central project. Essential support of individual micro-computer-based terminal systems can be adequately provided by any of a number of vendors of moderate-scale mini-computers. Considerable progress has already been made towards providing temporary support using the existing UNC Raytheon 706 system. A factor warranting consideration, however, was that the Raytheon 706 possesses an obsolescent configuration that failed to meet the specifications required for a fully effective central host.

It is largely for this reason that an independent central computer facility was chosen. The operating environment of the central system consists of a relatively straightforward multi-programming arrangement. Resident utilities include vendor-supplied system management (input - output handlers, operating monitors, resource allocation software), high-priority UNC-coded software for the control of intra-system communications and the link-editing, relocation and transmission of requested micro-processor load modules, re-entrant interactive computational packages for data analysis and interpretation and user-file management processors. Non-resident tasks consist of user-generated programs called into execution from remote access terminals (including the card reader - line printer facilities of the Raytheon 706).

In order to minimise the commitment of project personnel to system programming, care was used in selection of the central facility vendor so as to ensure the off-the-shelf availability of suitable advanced multi-access support software.

Intercommunication between the remote-terminal micro-computers, the existing Raytheon 706 system and the central time-share (multi-programming) system occur over a 100 kbyte s^{-1} common input - output bus. Allocation of the bus is done via a polling procedure maintained by the input - output hardware of the time-share CPU during periods of bus inactivity. Because of the high capacity of this bus, it should not constitute a limitation upon system utilisation in the foreseeable future. A key design feature of this inter-communication bus is toleration of considerable physical length and ease of extension of the area served when that is indicated. The cable length required for service of the existing chemistry facility is 1 mile. Connection of remote terminals to the intercom bus is made via active T-taps so as to avoid excessive loadings of the bus signal paths even in the presence of many scores of terminal connections. Selection of individual terminals is accomplished through the transmission of terminal-specific addresses during the polling procedure of the central system. Provision exists for up to 256 terminal addresses. Because of the large bus "band width," even 256 terminals can be polled ten times per second while using less than 8 per cent. of bus capacity for poll-response transactions.

Design procedures for micro-processor-based instrumentation systems—

Standardised design method—The design of computer-based instrumentation systems is a process of sufficient complexity to derive a considerable increase in efficiency from an orderly approach. The time and expense consumed in these activities became factors of great importance in view of the level of design activity encouraged by the availability of low-cost instrumentation-orientated micro-processors. In an effort to manage the magnitude of the resulting task, a standardised procedure was implemented in order to facilitate the definition, analysis and implementation of proposed systems.

The standardised procedure consists of eight phases. Of first concern is the generation of a concise statement of the proposed function of the instrumentation system under study. This "application description" includes the characteristics and capabilities of the planned instrument from an operational standpoint and serves as a definition of design aims.

Next comes an analysis of the experimental system to be controlled and monitored. In a large proportion of laboratory applications, most of the experiment-orientated portion of a proposed system will consist of an existing apparatus that is to be supplied with and interfaced to a suitable data system. In these instances, the analysis phase comprises a study of the input - output characteristics of the apparatus to be served. These requirements may usually be taken as given and thus serve to define instrumental parameters to the data-system level. In other instances, the over-all design task may originate much nearer to the phenomena level. In this latter circumstance, transducer and data-acquisition components of an instrumentation system will evolve together during the design process. Here the analysis phase of design must include a review of competing transducer technologies together with their impact upon data system requirements. In either instance, however, the analysis process terminates with an evaluation of input - output transactions that occur between experiment and computer during the experimental process. The resulting numbers specify system input - output requirements in both analogue and digital domains.

The third step in the design process is defined as a specification of the means to be provided for operator - system interaction. Included are an enumeration of operator-supplied experiment-control parameters and their methods of entry, and media and formats for the presentation of acquired data to the operator. Possible methods for parameter input range from switch and knob manipulation to keyboards and graphical pens. Output media include analogue and digital scalar displays (*e.g.*, meters), CRT and hard-copy plots and teleprinter copy. Optimal choices among these alternatives depend upon factors isolated in the previous two phases and through interview with the potential system users.

At this point it becomes possible to carry the design process to a fourth, synthetic, stage in which internal components of the data acquisition system are identified and configured.

    Investigator: *R. W. Murray and L. A. Simonson.*   Laboratory Location: *Kenan C347.*
    Application Title: *Micro-computer Control of Liquid Chromatograph.*
    Application Description: *Micro-computer will manipulate gradient elution and temperature programming and acquire chromatographic data to optimise electrosynthetic etc. mixture analysis.*
    Instrumental Inputs (data types: analogue, digital; acquisition rates, band pass; signal levels; impedances; required resolution): *Photodetector outputs: 10 to 100 samples per second, > 10 bits resolution. Chromatograph oven temperature monitor (high-linearity thermistor): 1 to 10 samples per second during oven temperature transitions.*
    Instrumental Outputs (same categories as inputs): *Analogue: strip-chart recorder drive. Digital: control word to solvent pump proportioning and rate controller (binary rate multiplier). Temperature control (integral-cycle proportional control of oven heaters).*
    Operator Interaction Modes (data displays and hard-copy presentations; manual control inputs): *Operator input: solvent gradient and temperature program, sensitivity correction for quantitative analysis. Output: strip-chart output of acquired chromatogram. TTY summary of peak areas and retention times.*
    Acquisition/Control Functions (total data volumes; required data reduction and/or signal processing algorithms, experiment control and optimisation functions; etc.): *Implementation of binary solvent gradients to be performed via computer control of binary rate multiplier hardware supplying input pulses to pump-driving stepping motors. Total flow-rate controllable by varying input frequency to rate multiplier. Use of computer in oven control circuitry will permit implementation of optimum response to column-oven temperature/time transfer function. Absorbance data acquisition will be performed by a direct-log conversion ADC.*
    Required Application-support Software:
      1. Suggested Controlling Program Functional Segmentation:
         A. *Operator specification of solvent gradient and temperature program (default option to be provided).*
         B. *Computer acquisition of chromatogram.*
         C. *Analysis program correction for species sensitivities and presentation of peak area* vs. *retention time data.*

      2. Operational Characteristics of Proposed System:
         A. *initialisation—largely default parameter oriented to permit rapid set-up for standard analysis.*
         B. *data acquisition—minimal interaction with TSS to permit efficient performance of long runs.*
         C. *data reduction and results display—numerical print-out with optional strip-chart plot of parts or whole of acquired chromatogram.*

    Hardware Requirements:
      1. Memory Capacity: Executive Programs: *4K*
         Data-handling Programs and Data Storage (maximum): *4K*
         TOTAL: *8K.*

      2. CPU Peripheral Devices and Controllers:
         C-M ADC Module.
         DAC Module.
         TTY Controller.
         2 Universal Modules (for pump control).

      3. Analogue Processing and Signal Conditioning Circuits:
         Pump controlling circuitry.

Fig. 3. Summary sheet for UNC Chemistry system configurator, showing results for one application

Data set volumes are computed, data reduction and enhancement algorithms developed and experiment excitation and control functions defined.

Having so modelled the structure of the experiment control and monitoring process, development of system software begins with the specification of a tentative basis for program segmentation together with identification of the function of and communication pathway between the resulting modules. This fifth phase provides both a basis for detailed program design and a start towards an estimate of the processor resources (*e.g.*, memory capacity) that will be needed.

Once the skeleton of system software is complete, a framework exists for design of detailed operational characteristics. This sixth design phase provides a view of the finished system as it will interact with both experiment, operator and host system in the processes of experiment set-up, execution and data interpretation.

At this level of design detail, an accurate estimate of system hardware requirements can be made. Three categories are identified: processor memory capacity, peripheral device complement and such special-purpose interfacing electronics as may be required.

Finally, actual system implementation begins. Always to be expected, of course, is the possible necessity of iteration of the earlier design phases in response to unexpected situations encountered before a working instrument emerges.

Design example—In Fig. 3, an example is given in which the results from the first seven phases of the design procedure are very briefly summarised. The format in which the design is presented was taken from a form generated to facilitate an orderly development and documentation of the instrumentation design.

The example cited is the computer-control of a high-pressure liquid chromatograph utilising two Waters Associates constant-flow solvent delivery pumps.

The wide adoption of digital computers in gas chromatography is now being extended to liquid chromatography for both data processing and experimental control. In this instance, the micro-computer will be used for control of the two pumps. Obviously, the micro-computer could generate an exceedingly large number of different gradients of binary solvents, including very complicated ones. Further, the computer approach would allow temperature and flow-rate gradients, *e.g.*, simultaneous binary solvent and temperature gradients would be possible. The micro-computer would also be used simultaneously to accumulate data from the photometer, to yield peak detection, peak position, peak intensity, peak symmetry, sensitivity compensation, statistical manipulations and, in some situations, direct concentration read-out for each component. Interestingly, the quoted price for a hard-wired gradient controller is comparable with the cost of the complete micro-computer data acquisition and pump control system.

Summation of initial applications—In Table III, twenty-four different applications resulting from expressed needs of various Faculty research groups in this Department, together with the hardware requirements for the twenty-eight micro-computer terminals to serve them, are summarised. These applications were each subjected to the standardised design method described above and represent the initial functions to be performed by our forthcoming multi-laboratory system. While it is certainly intended and hoped that these terminals will represent only the beginning of the growth and versatility of the system, they nevertheless constitute a very substantial beginning and, when fully supported and operational, will conclusively determine the value and power of the over-all plan.

Sixteen of the proposed applications represent direct application to chemical instrumentation. Two are education applications; three demonstrate the application of the micro-computer as system components serving to interface mini-computers to the central resource. Two more micro-computers are to serve exclusively as interactive computation terminals, and one is provided for system maintenance and development. We estimate that approximately 250 man-days will be required to construct the twenty-eight required micro-computers and associated modules cited in Table III.

The types of chemical instrumentation supported by the proposed terminals are extremely diverse. Data acquisition rates range from the 1-MHz rate required for high-sensitivity spectroelectrochemistry to the one sample per 10 s involved in potentiometric membrane studies. Data set volumes range from the tens of thousands of data samples required to characterise free induction decay transients in pulse nuclear magnetic resonance spectroscopy to the perhaps twenty numbers or so required for an amino-acid analysis. Experiment

TABLE III
INITIAL MICRO-COMPUTER APPLICATIONS

Number of modules and peripherals used*

Type of instrumentation serviced or function performed	Total cost/$	System housing	CPU card	Memory card	Communication card	Teletype controller	CRT graphic display controller	ADC card	DAC card	Sequencer card	Magnetic tape controller	Keyboard controller	TV controller	Breadboard card	High-speed "PT" card reader	Large-capacity memory controller	Light-pen graphics card	Digital input-output card	Charge manipulation ADC	Programmer's console	Ensemble averager	Teletype	Display scope	Keyboard	Fast paper-tape reader	Magnetic-tape drive	Data memory
Amino-acid analysis	2013	1	1	3	1	1	0	0	0	0	0	0	0	0	0	0	0	0	0	0	0	1	0	0	0	0	0
Classroom large-screen, interactive display	1500	1	1	6	1	0	0	0	0	0	0	1	1	0	0	0	1	0	0	0	0	1	1	1	0	0	0
Continuous flow kinetic apparatus	2723	1	1	4	1	1	1	1	1	1	0	0	0	0	0	0	0	0	0	0	0	0	1	0	0	0	0
Continuous wave nuclear magnetic resonance	2909	1	1	8	1	0	1	1	2	1	1	1	0	0	0	0	0	0	0	0	1	1	1	1	0	1	0
Data acquisition instructional laboratory	9478	4	4	16	4	1	4	3	3	1	4	4	1	0	1	0	1	0	0	2	1	4	4	4	0	5	0
Diagnostic and system development terminals and maintenance spares	5592	2	2	10	2	1	1	1	1	1	2	1	0	4	1	0	1	0	0	1	1	x	4	1	1	2	0
Electrochemical and kinetic studies on metal complexes	2177	1	1	4	1	1	1	1	1	1	1	1	0	0	0	0	0	0	0	0	0	1	1	1	0	1	0
Fourier transform pulse nuclear magnetic resonance	7338	1	1	6	1	1	1	1	2	1	1	1	0	4	0	0	0	0	2	1	1	x	1	1	1	2	2
High-sensitivity spectroelectrochemistry	1638	1	1	3	1	1	1	1	2	1	1	0	0	0	0	0	0	0	1	0	0	x	1	0	0	1	0
Interactive analysis and enhancement of spectral data	2123	1	1	4	1	1	1	0	1	0	1	1	0	0	0	1	1	1	0	0	1	x	1	1	0	0	0
Inter-system communication: R706 to time-share system	1500	1	1	2	1	0	0	0	0	0	1	0	0	0	0	0	0	0	0	0	0	0	1	0	0	0	0
Ion-cyclotron resonance mass spectroscopy	1607	1	1	3	1	1	1	1	3	0	0	1	0	2	0	0	1	0	0	0	0	x	x	0	0	0	0
Laser-light scattering from gases and low-temperature, wide-line nuclear magnetic resonance	1717	1	1	6	1	1	1	1	1	1	0	0	0	0	0	0	0	0	0	0	1	x	0	0	0	0	0
Liquid chromatography	1333	1	1	4	1	1	1	1	1	1	1	1	0	0	0	0	0	0	0	0	0	x	x	0	0	0	0
Low-resolution mass spectroscopy	1237	1	1	4	1	1	1	1	1	1	1	0	0	0	0	0	0	0	0	0	0	x	x	0	0	0	0
Magnetic susceptibility of condensed metal complexes at low temperature	1457	1	1	3	1	1	0	1	1	1	0	0	0	0	0	0	0	0	0	0	0	x	x	0	0	0	0
Photochemistry and photophysics	1952	1	1	5	1	1	1	1	1	0	0	0	0	0	0	0	0	0	0	0	0	x	1	0	0	0	0
Photo-ionisation studies	1213	1	1	3	1	1	1	1	1	0	1	1	0	0	0	0	0	0	0	0	0	x	1	0	0	0	0
Potentiometric membrane electrode response	1510	1	1	2	1	1	1	0	1	0	0	0	0	0	0	0	0	1	0	0	0	0	x	1	0	0	0
Programmable desk calculator	1160	1	1	3	1	0	1	0	1	0	0	1	0	0	0	0	1	0	0	0	0	0	1	1	0	0	0
Real-time front-end for small computer	1734	1	1	5	1	0	0	1	2	1	0	0	0	0	0	0	0	0	1	0	1	x	x	1	0	0	0
Spark source mass spectrometer photo-plate analysis	2228	1	1	4	1	1	0	0	1	0	1	0	0	0	0	0	0	0	1	0	0	1	0	0	0	0	0
Stopped-flow kinetic apparatus	2468	1	1	8	1	1	1	0	2	1	1	0	0	1	0	0	1	0	1	0	1	x	1	0	0	1	0
X-ray diffraction	1868	1	1	3	1	1	1	0	0	1	1	0	0	0	0	0	0	0	0	0	0	x	x	0	0	0	0
Total		28	28	119	28	18	22	14	28	7	14	9	3	10	2	2	5	3	10	3	6	5	15	9	2	17	2

* x means that this peripheral was already available and its cost was therefore not entered into the sum.

durations range from 4-hour chromatograms to 100-μs electrochemical transients. In short, these applications explore virtually the entire range of potential data acquisition environments.

SOFTWARE—

As in all computer systems, efficient generation of application programs is greatly aided by the existence of a diverse, powerful library of utility sub-routines. This is particularly true when, as in the case of the UNC micro-computer prototype system, these sub-routines can be referenced and automatically linked to an application program from a disc-based program library. Considerable effort has already been given to the task of generating for the Intel MCS-8008 a collection of such utilities and, so far, has resulted in the completion and de-bug of the software given in Table IV.

Still to be completed are sub-routines for higher mathematical functions (square root, trigonometric, logarithmic, exponential, etc.), preparation of experiment control packages, etc.

TABLE IV
MICRO-COMPUTER UTILITY SOFTWARE

Function	Software Name	Software Description
Host - satellite interaction	MICRODEF	An instruction definition package permitting the Raytheon 706 assembly processor to serve as a so-called "cross-assembler" for micro-computer programs
	MU.LINK	A "cross-system" link editor permitting micro-computer programs to be held in re-locatable form on the Raytheon 706 disc library. Micro-computer software catalogued in this way is handled by the same utilities used for 706-executable programs and make all the usual conveniences of re-locatable, linkable programs such as automatic sub-routine insertion and external referencing available to the micro-computer programmer
	MU.COLD, MU.QUES, MU.MASTER	A sequence of Raytheon 706 and MCS-8008 programs which handle the cold-start and initial program load of a micro-computer connected to the Raytheon 706 system through any laboratory interface box
	MU.CONVER	A communication handler for data exchange between 706 and micro-computer terminals connected by lab. boxes
Micro-computer input - output	MUOS	Micro-computer operating systems monitor
	MU.TTY, MU.SCPKY, MU.LBX	Micro-computer input - output driver routines for teletype, graphics display - keyboard, lab. box, respectively
	MU.FORM—	
	MU.DIGCO	String-oriented BCD input processor
	MU.BCD	BCD output generator
	MU.REALI	Real-number BCD input
	MU.REALG	Real-number F-format output
	MU.EXPGE	Real-number E-format output
	MU.INTGE	Integer output
	MU.INTIN	Integer input
	MU.IOCBS	Input - output control-block set-up
	MU.FIX24	24-bit fix
Multi-precision integer and floating-point mathematics	MU.IA, MU.IDM, MU.SHIFT	16-bit integer mathematical package including multiply - divide (currently 9, 3 and 5 functions, respectively)
	MU.FLA, MU.FLMD	32-bit floating-point mathematical package including multiply - divide (11 functions)
Program de-bug support	MUD, MUSAVE, MUREST	Programming support package permitting the inspection (dump), modification and re-execution of micro-computer programs under development

Some scepticism has been expressed to us regarding ease of programming of the MCS-8008 system. Having now had considerable experience with it, we can see no basis for such doubts. The MCS-8008 instruction format is, in many instances, because of the multiplicity of scratch registers and hardware sub-routine linkage processing, superior to some 12-bit mini-computers. A programmer, once familiar with the MCS-8008 system, has little difficulty in producing code as efficiently as is possible on larger computers. Inherent, however, is the inevitable abstruse nature of assembler code with its total machine orientation, rather than problem orientation. The limitation to assembly language when coding for the MCS-8008, as would be the case for any computer, results in restriction of programming activity to very experienced computer staff.

It has long been apparent that the removal of the requirement for familiarity with assembly-language computer programming for the design of high-performance computer-based instrumentation would undoubtedly enlarge the portion of the research community able to implement such systems. Many efforts have, in fact, been made towards the development of problem-oriented languages for data acquisition and process control (DEC, DATAK, INDAC, etc.), none of which has achieved widespread acceptance. Ideas pertaining to higher level experiment design languages have, however, been a concern of our group for some time and this project offers a rich opportunity to explore them in the context of hardware specifically designed for on-line control. The implementation of such a language as part of the initial system development would be, however, premature. The development of an experiment design language is a project which will be seriously considered after a mature productive system is achieved using more conventional coding techniques.

CHROMATOGRAPHY

Editors' Comments
on Papers 6 Through 10

6   GLENN and CRAM
    A Digital Logic System for the Evaluation of Instrumental Contributions to Chromatographic Band Broadening

7   SWINGLE and ROGERS
    Computer-Controlled Gas Chromatograph Capable of Real-Time Readout of High-Precision Data

8   DENTON et al.
    Computer Control of Data Acquisition, Reduction, and Display in Rapid Scanning Liquid Chromatography

9   DROMEY et al.
    Extraction of Mass Spectra Free of Background and Neighboring Component Contributions from Gas Chromatography/Mass Spectrometry Data

10 GATES et al.
    Automated Simultaneous Qualitative and Quantitative Analysis of Complex Organic Mixtures with a Gas Chromatography-Mass Spectrometry-Computer Sysem

Chromatography has benefited from computers as much as has any area of analytical chemistry. Control of temperature, timing, and data collection has improved the accuracy and precision of chromatographic data. The combination of the gas chromatograph and mass spectrometer would not be practical without a computer. The papers in this section represent significant contributions in the use of the computers in chromatography.

In Paper 6, Glenn and Cram report on the use of digital logic to evaluate instrumental contributions to band broadening. By careful selection and construction, including digital control of sampling and injection, a precision chromatographic system has been created that allows the precise determination of the statistical moments of chromatographic peaks. The use of such a system in examining band broadening is discussed and its performance is evaluated.

Swingle and Rogers (Paper 7) discuss the interfacing of a high-precision gas chromatograph to a minicomputer and the software necessary for real-time calculation of peak characteristics and control of the chromatograph. The authors are particularly interested in obtaining precise retention data. The most valuable part of the paper is the description of the system design. The hardware description is clear and instructive in giving reasons for various design considerations.

Papers 6 and 7 illustrate the gain that can be made in precision and control when a chromatograph is interfaced to a digital electronic system. In a related paper, Zwarg and Guiochon[1] have applied a microprocessor to the control of gas chromatographic ovens, using time-optimal control. A brief discussion of the usefulness of change of controller algorithm, depending on whether one is changing temperatures or maintaining the temperature constant, is included.

An area of active interest in analytical chemistry is the development of multichannel detectors for gas and liquid chromatography analogous to mass spectrometry. Denton et al. (Paper 8) describe one such system for liquid chromatography, using rapid scanning UV-visible spectrometry. High-pressure liquid chromatographic detectors, UV and otherwise, are continuing to evolve. This paper presents one solution to the problem of identification and resolution of components in a flowing liquid.

Another approach to a detector for liquid chromatography is that of Dessey et al.[2] They have detailed a minicomputer-automated linear photodiode array spectrometer system for high resolution liquid chromatography. Twenty spectra/sec can be recorded over the range 200–456 nm in a dual beam mode, using dual 256 element photodiode arrays as detectors.

Several papers on computerized enhancement of GC/MS data take different approaches to the problem of handling the large amount of data. Of course, the approach taken depends on the end use of the data and the needs of the user. The paper by Dromey et al. included here (Paper 9) treats the need for good spectra from GC/MS runs that may contain excessive background from sources such as column bleed or dirty ion sources as well as overlapping chromatographic peaks. The methods used in this paper take advantage of the enormous redundancy present in GC/MS data. In effect, one has a multichannel GC detector with the very helpful condition that as the mass spectral pattern of a molecule is fixed for the given operating conditions, the responses of the channels must remain proportional to each other for each component. Thus one can choose one ion as a model peak for each component that is free from interferences. This model peak can then be used to enhance the other ions in the spectrum, which may contain overlapping interferences. Figure 4 illustrates this idea.

The final paper in this section, Paper 10 by Gates et al., presents an approach to simultaneous qualitative and quantitative analysis of complex organic mixtures that uses computer interpretation of gas chromatography–mass spectrometry data. The algorithm used is centered on mass chromatograms rather than mass spectra. Selected designate ions are used to search a reference file in the region of the unknowns retention index. Confirming ions are used to improve compound identification. The quantitation of each component is also based on mass chromatograms, where the areas of selected model peaks are used in relation to the area of internal standards.

Stillwell and Stillwell[3] have described how GC/MS-computer methods can be used in the study of drug metabolism. They show how total ion current profiles can be enhanced by computer techniques based on location of maxima for individual mass chromatograms.

Clerc et al.[4] have published a paper on improving the efficiency of small gas chromatographic–mass spectrometric data systems with simple algorithms. In particular, they discuss automatic identification and show how selected mass chromatograms can be used to locate spectra of compounds with given structural constraints.

Nau and Biemann[5] have described the use of computer-assisted assignment of retention indices in gas chromatography–mass spectrometry. By coinjecting a set of normal alkane standards with the unknown mixture, a retention index for each component in the mixture can be assigned from the scale generated from the retention indices of the standards. The retention index values between those of two consecutive hydrocarbons are interpolated linearly. This is reasonably valid for temperature-programmed gas chromatography. They report that for finding the alkanes in the mixture data, searching only those regions where the standards were expected to elute, using a wide window of retention time was most universally applicable. Retention data is very useful information in automatic identification of organic compounds by gas chromatography–mass spectrometry, as Paper 10 illustrates.

REFERENCES

1. G. Zwarg and G. Guiochon, "Application of the Microprocessor and of Time Optimal Control Laws to the Control of Ovens," *Chromatographia*, **11**, 59 (1978).
2. R. E. Dessy, D. Reynolds, G. Nunn, C. A. Titus, and F. Moler, "New Minicomputer Automated Linear Photodiode Array Spectrometer System for High-Resolution Liquid Chromatography," *J. Chromatogr.* **126**, 347 (1976).
3. R. N. Stillwell and W. G. Stillwell, "Gas Chromatographic-Mass Spectrometric-

Computer Methods in the Study of Drug Metabolism," *J. Chromatogr.,* **126,** 547 (1976).
4. J. T. Clerc, M. Kutter, M. Reinhard, and R. Schwarzenbach, "Improving the Efficiency of Small Chromatographic-Mass Spectrometric Data Systems by Means of Simple Algorithms," *J. Chromatogr,* **123,** 271 (1976).
5. H. Nau and K. Biemann, "Computer-Assisted Assignment of Retention Indices in Gas Chromatography-Mass Spectrometry and its Application to Mixtures of Biological Origin," *Anal. Chem.,* **46,** 426 (1974).

6

Copyright © 1970 by Preston Technical Abstracts Company

Reprinted by permission from *J. Chromatogr. Sci.* **8**:46-56 (1970)

A Digital Logic System for the Evaluation of Instrumental Contributions to Chromatographic Band Broadening

by **Tom H. Glenn** and **Stuart P. Cram**, Deparment of Chemistry, University of Florida, Gainesville, Florida 32601.

Abstract

To evaluate the performance of chromatographic columns accurately, the peak area, peak width, and retention time must reflect only the behavior of the chromatographic column. A precision chromatographic system has been designed and developed to measure the first and second moments of a peak and the peak area, regardless of the shape of the effluent profile. The experimental moments are corrected for the influence of the sampling system, mixing volums, connecting tubing, and the detector and electronic time constants.

A digital logic system has been developed to accurately control the timing sequence to 1 part in 10^5 and the sampling and injection time to a standard deviation of 380 μsec. The GC-core storage interface includes a real time clock, an analog to digital converter, buffer storage, address advance, level converters, and sufficient analog filtering only to prevent "folding errors" in the analog to digital converter. The largest second moment contribution for this system is shown to be due to the time constant of the electrometer (5.3 msec.). The reproducibility of the system is limited by the stability and control of the pneumatic system.

Introduction

The development of the inherent capabilities of gas chromatography during the next decade will depend on precision measurements for the verification of theoretical concepts and the extension of the applicability of the technique. These developments will require improved data acquisition and reduction techniques and advances in the chemistry and dynamics of column technology.

The chromatographic system described here was designed to measure statistical moments of chromatographic peaks with high speed and precision and with instrumentation which could be applied to any chromatographic system, regardless of the elution profile, speed of separation, or peak width. This hardware system was designed for maximum versatility and remote and automatic operation. Such a logic system has the advantage of not requiring computer time or space, and yet it is stable, reliable, and may be readily modified. Thus the attention here is directed toward the front end hardware for increasing the dynamic capabilities of a digital chromatographic system. By measuring the instrumental or extracolumn contributions to chromatographic band broadening, it is possible to minimize these effects in the design of chromatographic instrumentation. In turn, fundamental contributions to band broadening then may be measured which reflect only the behavior of the chromatographic column.

The superior aspects of digital data acquisition over analog methods have been well documented outside the realm of chromatography (1-4). The ACS Symposium on Quantitative Gas Chromatography in 1967 reported several approaches to digital data acquisition and reduction for analytical GC (5-8). Baumann (9) and others (10-12) have investigated the role of data acquisition (including filtering, peak detection, and curve smoothing) and reduction on the quantitative interpretation of chromatographic data. Computer analysis of unresolved chromatograms by fitting Gaussian curves has been discussed by Westerberg (13) while a nine point least squares fit to the experimental curves has been shown to be applicable to skewed peaks (14).

1. Booman, G. L., Anal. Chem., **38**, 1141 (1966).
2. Flynn, G. J., Electronics, **37**, 58 (1964).
3. Brown, E. R., Smith, D. E., and DeFord, D. D., Anal. Chem., **38**, 1130 (1966).
4. Lauer, G., and Osteryoung, R. A., Anal. Chem., **38**, 1137 (1966).
5. Baumann, F., and Tao, F., J. Gas Chromatog., **5**, 621 (1967).
6. Karohl, J. G., J. Gas Chromatog., **5**, 627 (1967).
7. Shank, J. T., and Persinger, H. E., J. Gas Chromatog., **5**, 631 (1967).
8. McCullough, R. D., J. Gas Chromatog., **5**, 635 (1967).
9. Baumann, F., Herlicska, E., Brown, A. C., and Blesch, J., J. Chromatog. Sci., **7**, 680 (1969).
10. Hancock, H. A., and Lichtenstein, I., J. Chromatog. Sci., **7**, 290 (1969).
11. Crisler, R. O., Pittsburgh Conf. on Anal. Chem. and Appl. Spectr., Cleveland, Ohio, March 2-7, 1969.
12. Jones, K., McDougal, A. O., and Marshall, R. C., "Gas Chromatography 1966," A. B. Littlewood, ed., The Institute of Petroleum, London, 1967, p. 376.
13. Westerberg, A. W., Anal. Chem. **41**, 177 (1969).
14. Gladney, H. M., Dowden, B. F., and Swalen, J. D., Anal. Chem., **41**, 883 (1969).

A rapidly increasing number of workers have chosen to characterize chromatographic peaks by their statistical moments (15-19). The significance of these concepts has been reviewed (20,21). Kucera (17) has shown that for accurate measurement of the retention time, the center of gravity, as located by the first moment of the peak, should be used rather than the peak maximum. The second moment, or the variance of a Gaussian peak, depends on all the factors characterizing the transport of a given compound through the column as it is a function of radial and longitudinal diffusion, the equilibrium constants, the size and symmetry of the solid support, and the mass transfer coefficient. Because the second moment is a measure of the peak spreading, it may be used to calculate the optimum carrier gas velocity, diffusion coefficients in the mobile and stationary phases, partition coefficients, the rates of sorption and desorption, and other physical parameters. Giddings (22) has used the second moment to calculate plate height values and found that peak reproducibility requires a high precision system with a reproducible injection system.

Therefore, the accuracy and precision of measuring these moments is of fundamental importance. Oberholtzer and Rogers (16) have developed a high precision chromatographic system for measuring the retention time and peak area with a relative error of less than ±0.1 per cent. From this data they show that the HETP, variance (the statistical second moment), skew (obtained from the third moment and a measure of peak asymmetry), and excess (from the fourth moment) may be calculated with a relative error of ±0.2 per cent or less.

The importance of the extracolumn effects has not been previously emphasized experimentally. These effects become very important for accurate moment measurements because the second and higher moments weight heavily those contributions relatively far from the center of gravity of a peak. This becomes readily apparent when using digital techniques. This work is directed toward the correction of the measured second moments for the instrumental contributions in order to obtain a high precision system which will be capable of data acquisition and analysis for fast, high resolution or precision gas chromatography.

Sternberg (21) has developed equations for calculating the peak areas, center of gravity, and second moment for Gaussian, plug, exponential, reverse ramp, semi-parabolic, and isosceles triangle column input functions. The equations for peak broadening contributions from connecting tubing, mixing volumes, and the time constants for the detector and electronics are also considered. Calculation of the performance characteristics of a chromatographic system which will minimize the above extracolumn effects places stringent requirements on the system. Although these characteristics have not been previously realized in practice, they are a prerequisite to precision measurements for the study of column behavior.

Experimental

The chromatograph developed for this study was designed so that all components of the instrument, pneumatic and electronic, were of known specifications so that the band broadening contribution of each component could be calculated or measured. The system shown in Figure 1 used a capillary in place of a column so that all column effects were eliminated. The tubing was 0.023 in., i.d., and 13.5 in. long.

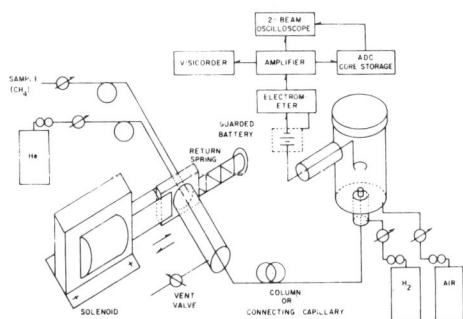

Figure 1. Chromatographic system for measuring the instrumental contributions to band broadening.

The two-stage He regulator (Model 8-350, The Matheson Co.) was operated at 60 psig in order to give a large pressure drop across the metering valve (Model B-2S, Nupro Co.), making the metering valve the predominant pneumatic resistance, which minimizes the effects of down stream pressure changes. Flow rates were measured with a soap-film bubble flow meter. All tubing connections except those in the sample carrying line were brazed rather than connected with tubing fittings, in order to eliminate dead-volumes and to accurately define the volumes and diameters of the instrumental system. The connecting capillary used special zero dead-volume o-ring sealed joints.

15. Grushka, E., Myers, M. N., Schettler, P. D., and Giddings, J. C., Anal. Chem., **41**, 889 (1969).
16. Oberholtzer, J. E., and Rogers, L. B., Anal. Chem., **41**, 1234 (1969).
17. Kucera, E., J. Chromatog., **19**, 237 (1965).
18. Vink, H., J. Chromatog., **20**, 305 (1965).
19. Yamazaki, H., J. Chromatog., **27**, 14 (1967).
20. Grubner, O., "Advances in Chromatography," Vol. 6, G. C. Giddings and R. A. Keller, eds., Marcel Dekker, New York, 1968, p. 173.
21. Sternberg, J. C., "Advances in Chromatography," Vol. 2, J. C. Giddings and R. A. Keller, eds., Marcel Dekker, New York, 1966, p. 205.
22. Ross, D. T., "Notes on Analog-Digital Conversion Techniques," A. K. Susskind, ed., The Technology Press, Cambridge, Mass., 1957, p. 2-1.

Sampling Valve

Several types of sampling valves have been evaluated previously under high precision chromatographic conditions (16). These valves are generally one of two major types: those employing a sliding seal between a series of ports, and those employing an elastomeric diaphragm to seal the ports. The first type generally offers the greatest accuracy, minimum leakage, and the most positive switching action. The second, on the other hand, offers faster operation and greater timing precision but lacks positive switching action. This contributes to the rise and fall times for the sample pulse and therefore to the peak spreading. Thus both types of values have distinct advantages, but neither offers all of the desirable properties of an ideal switching valve, and neither will yield good performance at high pressures. The sampling valve reported here is of unique design as the sample volume is electronically controlled and there are no sliding seals. There is essentially no dead-volume as the column (or connecting capillary) is built into the valve.

The valve, shown in Figures 1 and 2, consists of two 0.023 in., i.d., stainless steel capillary inlet tubes mounted in a block attached to the armature of an electrical solenoid. The centerline distance between the inlet tubes is 0.188 in. or about two column diameters. The column or connecting capillary is mounted in a brass bellows and is aligned with the He carrier gas inlet. The inlet tubes and the connecting capillary (or the column) are separated by a distance of r/2 so that the dynamic lines of flow of the vapor phase sample do not intersect the column axis except when the valve is switched for sampling. The sample venting rate is controlled by a needle valve (Model B-2S, Nupro Co.), which is mounted in the base of the bellows assembly.

The switching action of the valve takes place when the inlet tubes are moved horizontally so that the sample inlet tube is aligned with the column. The valve is operated by a solenoid (Guardian, Type 14) which is switched on and off under logic control. The sample size is determined by the amount of time the solenoid is actuated to align the sample inlet and the column and by the concentration and flow rate of the sample vapor. At the end of the sampling period the solenoid is cut off and a heavy coil spring returns the valve to its original position. The solenoid is switched by the driving circuit shown in Figure 3.

The coaxial alignment of the column with the carrier and sample inlets in the normal and injection positions is critical in order to achieve good reproducibility and to prevent sample bleed into the carrier gas line. However, the system is adjustable in all respects and was adjusted so that the background current from sample bleed into the flame detector was ca. 2×10^{-12} A. Two requirements must be met for high sampling precision and low background sample bleed onto the column:

1. The transit time for the carrier gas across the gap between the carrier inlet and the column inlet must be much smaller than the time required for diffusion of the sample vapor from the sample inlet to the axis of the carrier inlet-column assembly.

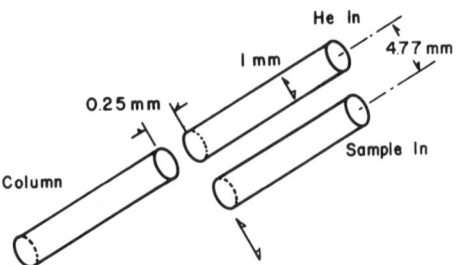

Figure 2. Exposed view of the sampling valve in the mounting block and bellows.

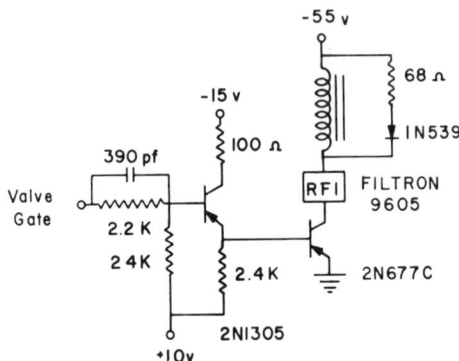

Figure 3. Solenoid driver circuit. Solenoid is Guardian type 14 (11 ohm coil).

2. There must be a sufficient excess of carrier gas flow to purge the interior of the bellows of the sample gas when the valve is in the normal position.

Thus the carrier gas flow into the valve should be 10 to 20 times the column carrier flow rate. Under these conditions the ratio of the flame detector response for a 0.77 μl sample of methane to the background current is greater than 3×10^4.

To evaluate the performance of the solenoid actuated valve, the position of the valve was measured as a function of time with a photocell. A flat, vane-type shutter was mounted on the solenoid armature. When the valve was actuated, the shutter uncovered a CdS photocell (response time of 1-2 msec) such that the area illuminated was proportional to the linear travel of the valve. The photocell output was amplified and displayed on an oscilloscope (Tektronix-549) or fed into the analog to digital converter for storage and later processing.

Detector and Amplifiers

The detector was designed to operate as either a standard flame ionization or flame temperature detector. The flame ionization mode was used in this work because of the sensitivity and speed of response. The detector jet was 0.023 in., i.d., stainless steel tubing, 3 mm long, and was grounded. The collector electrode was a 1 cm diameter ring of gold plated Kovar, which was mounted on a rod projecting from a side arm of the detector housing in order to keep the Teflon insulator cool. The polarizing potential for the collector electrode was supplied by a shielded 90 volt battery which was isolated from ground by a driven guard. The flame was operated with a hydrogen flow rate of 36 cc/min and an air flow rate of 650 cc/min.

Two electrometer amplifiers were available for signal measurement: 1) a Keithley Model 601 electrometer operated in the fast mode with the output taken from the unity gain output and the guard terminal, 2) an operational amplifier electrometer designed to provide very high speed over the range of currents most commonly encountered in GC. Both electrometer amplifier outputs provide a 0 to −10 v full scale output to the analog to digital converter as well as current outputs to a recording galvanometer (Model 1108 Visicorder, Honeywell).

Digital Control and Data Acquisition System

The data acquisition system has been designed to include three major sections: the real-time clock, the analog to digital converter (with its data buffer), and the data storage device. This system is shown in Figure 4. In order to make the most efficient use of the data storage space available, it is necessary to delay the beginning of data storage until the event of interest occurs. A synthetic chromatogram is shown in Figure 5 to illustrate the three distinct time periods to be considered: the readout delay time (R/O Delay), the data acquisition time (data time), and the equilibration time. The readout delay is the time between the injection and the beginning of the data time. The data time is the time during which data from the event of interest is stored, and the time following is the time required for the chromatographic system to reach equilibrium before another experiment is performed.

The real-time clock provides the timing, synchronization and control signals required by the experimental system. This includes the logic control facilities for the sampling valve, the preselected delay before data storage, and control signals for the Visicorder or oscilloscope. The analog to digital converter converts the data from an analog voltage between 0 to −10.23 v to an unsigned 10-bit binary number. These numbers are stored in octal-coded binary in the core memory of a Nuclear-Chicago Multichannel Analyzer (Model 34-27). Since the analyzer's internal organizations is 1-2-4-8 binary-coded decimal, the octal-coded binary numbers are stored in the first three bits of each memory location. While this does not utilize the maximum bit capacity of the core storage, it is the simplest method of avoiding the decimal carry from one decade to another. The result is storage of octal-coded binary data in a system organized for BCD with no additional hardware required for the conversion. The data logger was designed to operate as a single-cycle data break interface, a program interrupt interface, or a program controlled interface for the PDP 8/L laboratory computer (Digital Equipment Corporation) for on-line data reduction.

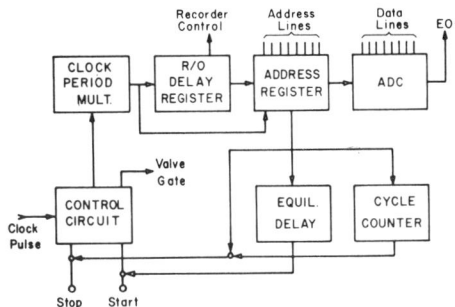
Figure 4. Block diagram of the data logging system.

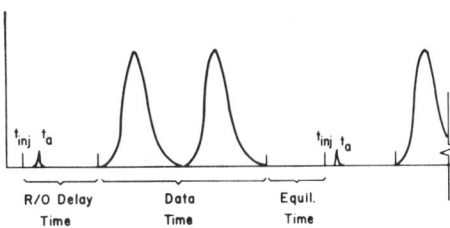
Figure 5. Logical time periods of the chromatogram.

The data logger was constructed from Digital Equipment Corp. R-Series logic modules, W-Series interface modules, and the A-Series A/D modules with the standard DEC harware. The logic levels were transmitted over twisted pair cables with one wire grounded in each pair, and pulses were transmitted over terminated 90 ohm coaxial cables. The symbols employed on the logic diagrams are those employed by DEC for the R-Series logic. A list of the module type numbers corresponding to the notations of the logic diagrams is given in Appendix I along with a list of the logic signal mnemonics, functions, and specifications.

The real-time clock includes three binary registers (the R/O delay register, the address register, and the cycle counter), a clock period multiplier, and the control circuit. The block diagram of the data logger of Figure 4 is expanded in the logic diagram of the control circuit in Figure 6.

The clock pulses are supplied by a pulse generator (Model PB-2, Berkley Nucleonics Corp.) which was calibrated with a Tektronix 184 Time Mark Generator. The calibration was made by transferring the desired pulse period using the delayed time base of a Tektronix 547 oscilloscope. The clock pulses are negative going from ground to −4 v and are converted to the standard DEC logic levels of ground and −3 v by Schmitt trigger ST1. After this conditioning the clock pulses pass to the stop-start synchronizer formed by flip-flops FF1 and FF2, and to the single-pulse synchronizer formed by FF3, FF4, and pulse amplifier PA3. The stop-start synchronizer provides manual control of the data logger through S1, S2, and their associated Schmitt triggers, ST2 and ST3, respectively. The synchronizer assures that manual control causes the data logger to begin operation only on integral clock periods. Similarly, the single-pulse synchronizer generates a single output pulse which initiates the valve gate signal (monostable MS1, IV1) and its complement. This valve operate pulse, VOP, is available at the front panel for triggering an oscilloscope time base or event marker. Flip-flop FF6 is set by the start pushbutton S2 and turns on the readout delay indicator (R/O delay). At the end of the R/O delay period, the address advance enable level change, AAE, resets FF6 and sets FF5. This turns off the R/O delay indicator, turns on the read indicator and enables the gate input of pulse amplifier PA2 which now passes 400 nsec delayed clock pulses, DCP, at the current clock period into the address register. Flip-flop FF5 also provides a front panel readout level signal, ROL, which may also be used to trigger an oscilloscope time base or it may be patched to a relay driver for controlling the chart drive or an event marker of the Visicorder. The complementary output of FF5 is buffered by bus driver BD2 and transistor Q1 to provide the parallel data transfer gate enabling signal, DTE, for the 400-word core storage of the multichannel analyzer. The delayed clock pulse train is buffered into the address scaler of the analyzer by bus driver BD1.

In order to provide a wide range of data time to the R/O delay, the clock period multiplier, Figure 7, is used to multiply the clock period by two, five, or ten. The clock period multiplier consists of a count of two scaler and a count of five scaler, which may be cascaded to obtain a count of ten. All control is accomplished by the internal gates of flip-flops FF7 through FF10. The clock period multiplier can be patched into the system so that either the R/O delay register or the address register, or both, may be operated at the increased clock period. The patching necessary to select one of these options is made directly on the wire-wrap pins of the module case to minimize degradation of the pulse rise time and to eliminate a cumbersome switching arrangement. This approach provides a wide range of R/O delay to data acquisition time ratios, which is an advantage in chromatographic work where the event of interest is a narrow peak which may occur a relatively long time after the initiation of the experiment.

The R/O delay register, Figure 8, is a 12-bit binary register (4096_{10}) which is incremented by the clock input pulses, CIP, (or the output of the clock period multiplier) up to a preset count level, beginning with the initiation of the experiment. It is composed of gates G1, G2, and flip-flops FF11 through FF22. An array of 12 single-pole, single-throw switches are used to connect the selected flip-flop outputs to the comparator gate inputs of G1. On satisfying the required count, the output of the comparator gate G1 goes to ground. This output provides the address advance enable signal to the analyzer and is inverted by G2 and used to disable the input gate of the register. The register is cleared at the end of a cycle by the power clear pulse, PCP.

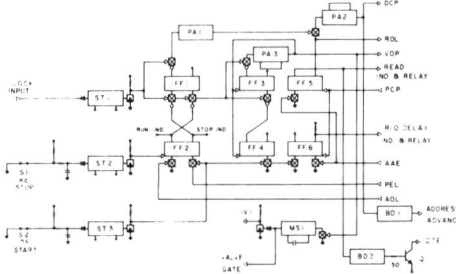

Figure 6. Logic diagram of the control circuit. Logic module type numbers and logic signal specifications are given in the Appendix.

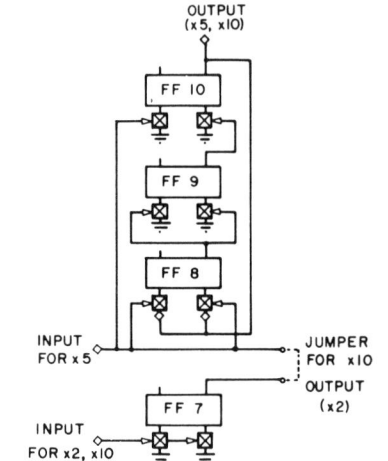

Figure 7. Logic diagram of the clock period multiplier.

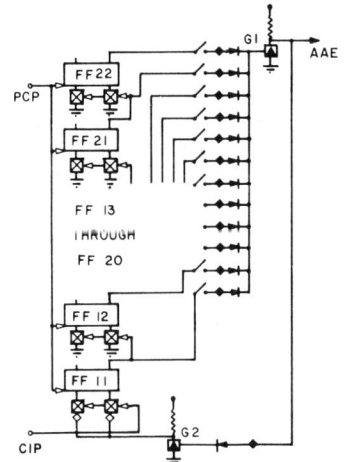

Figure 8. Logic diagram of the readout delay register.

The address register, Figure 9, is a 12-bit binary register identical to the R/O delay register, except that it includes a pulse amplifier, PA4, which generates the power clear pulse when the address overflow level, AOL, occurs. For normal operation with the 400-channel analyzer, the register is preset for a count of 401 for reasons which will become apparent in the discussion of the ADC and its buffer. The pulse amplifier PA4 has a manual input to one of its gates to allow manual resetting of the system after termination of an experiment. When the address register overflows, the PCP clears the address register and the R/O delay register, stops data storage, increments the cycle counter, and starts the equilibration delay. The equilibration delay is an FET monostable which may be set for delays up to about 10 minutes.

The cycle counter, Figure 10, is a four-bit binary counter which counts the number of times an experiment has been repeated up to a preset number (maximum of 16). The stop-start synchronizer is disabled by the program end level, PEL, when the preset count is reached and all of the registers are cleared. At this point the cycle counter must be manually reset before a new series of experiments can be recorded.

The analog to digital converter (ADC) used in this system is a 10-bit, successive approximation converter (Model C002, Digital Equipment Corp.) for inputs of 0 to -10.23 v. This converter has a nominal maximum conversion rate of 30kHz or 33 μsec per conversion. Since the converter is of the successive approximation type, the aperture time is also 33 μsec. To minimize the time required to convert and store a data point, a parallel transfer data buffer is provided as shown in Figure 11. After A/D conversion has been completed, the data word may be transferred into the data buffer in about 500 nsec by the end of conversion level EOC, supplied by the ADC. This allows the ADC to begin another conversion during the time required for the memory cycle to store the previously converted word. In practice it is more convenient to employ the delayed clock pulse, DCP, both to start the ADC and to strobe the data into the buffer. This yields a simpler logic timing system and eliminates possible jam-up between the ADC and the data buffer. It also results in a zero being stored in location zero of the memory storage. This provides a useful diagnostic since any other value stored in location zero indicates a spurious operation of the ADC, that the buffer was not cleared, or interference on the data line to the core memory.

Access to the core memory is in parallel form directly to the parallel data entry gates of the core buffer/arithmetic register. At the end of an experiment the data is printed and punched out on the Teletype (Model ASR 33) and displayed on an X-Y plotter (Model 850, Data Equipment Co.) or a scope. Since random access to the core storage is not required, the normal internal address scaler is incremented by a buffered version of the DCP. The only other control signal required for storing the data is the parallel data transfer gate enable, DTE, which is provided by BD2 and Q1 (saturated to -10 v @ 80 ma during the data acquisition time) in Figure 6. The data word is level shifted and buffered by the array of bus drivers shown on the data buffer diagram of Figure 11. The remote

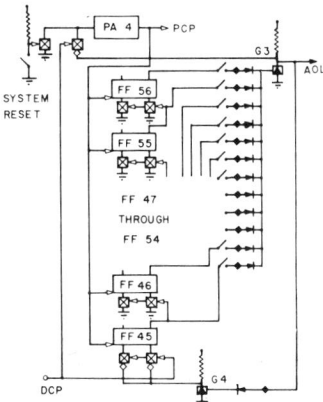

Figure 9. Logic diagram of the address register.

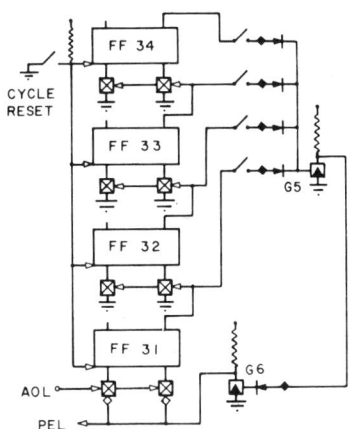

Figure 10. Logic diagram of the cycle counter.

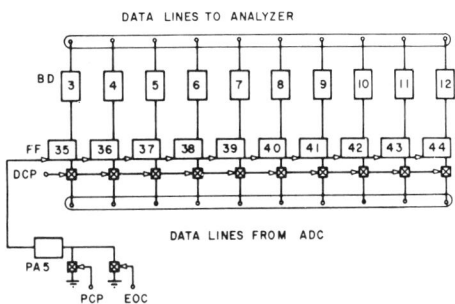

Figure 11. Logic diagram of the data buffer and bus drivers.

controls for the read and store modes provided on the control panel of the data logger, allow fully remote operation of the system from the data logger console at the experimental apparatus.

Results and Discussion

ADC Speed and Electrometer Time Constant

In considering the accuracy of digitized data from an ADC, it is particularly important to distinguish between the conversion rate, the maximum signal frequency, and the maximum frequency for a specified amplitude accuracy. The conversion rate is simply the maximum number of conversions which can be performed per second without regard to the signal. The maximum conversion rate for the converter employed in this work is 30 kHz and is determined by the delays and settling times required by the logic components which make up the converter. The maximum allowable signal frequency is the highest fundamental frequency which can be recovered from the sampled data. The Sampling Theorem (22) requires that at least two points per cycle be sampled for the highest significant frequency component of the signal in order to recover the signal without folding errors. For the Digital Equipment Corp. Model C002 ADC, this frequency is approximately 15 kHz. Finally, the most important specification for the type of work described here is the maximum frequency which may be sampled with a given amplitude accuracy. This specification depends on the resolution or precision desired and the aperture of the converter. For the C002 ADC and a resolution of 0.1 per cent, this frequency is approximately 8.5 Hz (23). Since it is difficult to find an interpretation of this specification in the frequency domain, it is much more convenient to consider this specification in terms of a rate limit. To obtain valid results at the maximum resolution of an ADC it is necessary that the signal not change by more than the value of one least significant bit during the aperture time of the converter. Hence the rate limit, L_r, of the converter may be calculated by

$$L_r = (E_{lsb}/T_a) \text{ v/sec} \qquad \text{Eq. 1}$$

where E_{lsb} is the value of the least significant bit, and T_a is the aperture time of the converter. For the 10 bit ADC C002 operating without a sample and hold amplifier, T_a is 33 μsec, and E_{lsb} is 19 mv, which gives a rate limit of ca. 3.3×10^2 v/sec, for a resolution of 0.1 per cent. With the addition of an A400 sample and hold amplifier the aperture time is reduced to 150 nsec. This yields an L_r of 6.7×10^4 v/sec, which is a very significant improvement in this specification. Since the behavior of an ADC in the region of its rate limit is nonlinear, calculation of the second moment contribution is very complex and will not be attempted here. The band broadening contribution may be considered to be negligible, however, because exceeding the rate limit will only serve to reduce the precision of the amplitude data.

The slope of the rising and falling edges of the detector output shown in Figure 12 is approximately 8×10^2 v/sec, which exceeds the rate limit calculated for

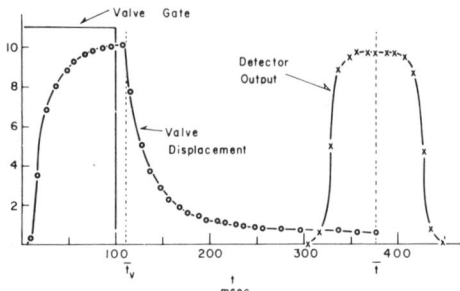

Figure 12. Displacement curve for the sampling valve and detector output for a peak injected by a 100 msec valve gate signal. For the displacement curve, 10 represents the full travel distance of the sampling valve, and 10×10^{-8}. A full scale for the detector output. Sample is methane, and the carrier flow rate is 1 cc/sec.

Table I. Summary of Instrumental Contributions to Band Broadening.

Contribution	2nd Moment sec²	Type of Contribution
Connecting Tubing	3.6×10^{-6}	Gaussian
Mixing Volume[a]	2.0×10^{-6}	Exponential
Electrometer and Cable	28.1×10^{-6}	Exponential
Detector	12×10^{-6}	Exponential

[a] Calculated for a flow rate of 1 cc/sec

the ADC. However, the experimental data still have significance since the above rate limit was calculated for a precision or resolution of 0.1 per cent. Since a successive approximations converter requires the same time for each step in the conversion, it is possible to estimate the precision of such a measurement by assuming the converter to be operating with a smaller number of bits but with the same amount of time required per bit. This has the effect of both reducing the aperture time and increasing the value of the least significant bit. In the case of the C002 converter operating with eight bits of significance the resulting rate limit is approximately 1.5×10^3 v/sec with a resolution of approximately 0.39 per cent. Hence the amplitude data on the rising and falling edges is valid to a precision of about 0.39 per cent.

In order to measure the time constant of the electrometer and its associated high impedance wiring, a 2×10^9 ohm resistor was connected in place of the detector. When the current through this resistor had reached its steady state value, the resistor was disconnected from the high impedance battery cable. The time constant of the electrometer was then obtained from the decay of the current measured at the output of the electrometer.

The electrometer time constant shown in Table I

---

23. "Digital Logic Handbook." Digital Equipment Corp., Maynard, Mass., 1968, p. 366.

was measured for the 10 x 10⁻⁸ A range. The sensitivity of the electrometer may be increased by employing a higher amplifier gain with the same measuring resistor with little increase in the time constant. An increase in sensitivity of 10 to 30 times is possible in this manner before noise in the electronics becomes troublesome. On the next most sensitive range, i.e., increasing the measuring resistor from 10^8 ohms to 10^9 ohms, the time constant is increased from 5.3 msec to 35 msec. It should be noted that these values of the electrometer time constant include the effects of the cables and the guarded battery.

Measurement of Detector Time Constant and Mixing Volumes

The method employed for the measurement of these constants was adapted from the method described by Kieselbach (24). The primary modification was the replacement of the graphical measurements on the peaks with the equivalent operations on the digitized data. The computation was done on a PDP 8/I computer employing the FOCAL programming system. The program was written to convert the data from the octal-coded binary form to decimal form and to punch a new data tape of the decimal data while proceeding with the calculation of the values to be plotted. The detector time constant was obtained from the plot of the half-widths of the chromatographic peak versus the reciprocal flow rate. These results are shown in Figure 13. The peak half-widths, t, were measured from the intersection of the tangents through the inflection points to a point on the tail at $1/e$ (0.37) times the height of the intersection. This half-width was expressed both in units of time and volume in order to obtain a measure of both the detector time constant and the mixing volume of the system. The mixing volume was obtained from the intercept of the plot of the half-width in volume versus the volumetric flow rate and is given in Figure 14. The interpretation of this data has been discussed by Kieselbach (24). The same program also calculated the area and the first and second moments of the digitized peaks from the finite sum form of the defining integral (in normalized form)

$$m_n = [\Delta t \; \Sigma \; t_i^n \; S(t_i)]/A \qquad \text{Eq. 2}$$

where m_n is the nth moment which locates the center of gravity t_0 for $n = 1$, and the second moment σ^2, Δt is the time interval between data points and t_i is taken as the time at which the peak amplitude of the detector output is $S(t_i)$. The program also introduces the proper scaling factors for the electrometer range settings and clock pulse period. The area A was concurrently calculated by Simpson's Rule and was introduced into the above equation when the sums had been completed.

In considering Kieselbach's method for determining the time constant of the detector, it is important to realize that the method yields a time constant which is the composite of the detector, cable and battery, and electrometer time constants. Hence other data is necessary to separate the effects of the individual components. Here it was most convenient to measure the electrometer-cable time constant separately and subtract (as its square) this value from the experimental value of Kieselbach's method. The second moment value of

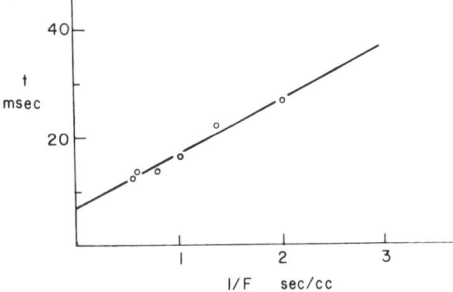

Figure 13. Measurement of the detector time constant. Valve gate width, 30 msec; sample, methane.

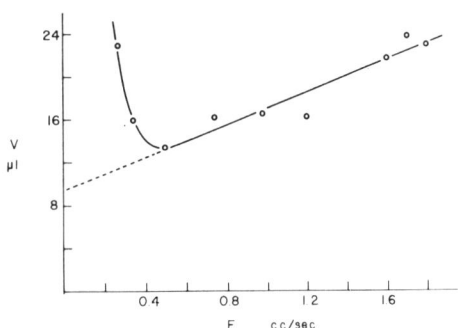

Figure 14. Measurement of mixing volume. Valve gate width, 30 msec; sample, methane.

12×10^{-6} sec² (time constant = 3.5 msec) is one of the smaller contributions and compares well with estimates of the mixing effect within the volume of the flame jet (approximately 1 μl) and the transport time (together approximately 2 msec) for the effluent from the end of the column to the collecting electrode. The greatest experimental problems in the use of this method are the control and measurement of the carrier gas flow rate. Most of the scatter of the data points and the error in the measurement seems to result from fluctuations in the flow rate and uncertainty in the measurement of the flow rate. Both the detector time constant measurement and the mixing volume measurement require that data be taken in the high flow velocity region. It is quite difficult to make this measurement of the flow rates with the required accuracy and without generating pressure drops which complicate the interpretation of the data.

One of the difficulties involved in digital computation of the time constant by this method is the time consuming and relatively inaccurate procedure re-

---

24. Kieselbach, R., Anal. Chem., 35, 1342 (1963).

quired to locate the true peak maximum by location of the inflection points and the intersection of the tangents through them. A procedure which could eliminate this difficulty has been suggested by Sternberg (21), who approximates the exponential time constant by an apparent time constant measured in the tail of the peak. More information will be required, however, to take advantage of the desirable properties of the technique, such as knowledge of the behavior of the appropriate profile functions in the presence of several independent exponential components. Another problem with the current method is the necessity of plotting the half-width data because of the limited size of the core in order to carry out extrapolations to obtain the final values. With alternative algorithms for locating the peak maximum and measuring the apparent time constant, it would be possible to use programmed procedures to eliminate diffusion limited points and to obtain a least-squares fit to the data. This would preserve the maximum accuracy of the data and would provide a faster and more reliable data analysis.

Table I summarizes the instrumental contributions to the band broadening for the system described here. In order to interpret the mixing volume data correctly, it is necessary to know the volume of carrier gas containing the sample at zero flow rate. Figure 14 gives a total apparent mixing volume of 9.5 μl. Part of this volume is due to the initial (diluted) sample volume and part is due to mixing in the gas stream. For a carrier gas flow rate of 0.27 cc/sec and a valve gate width of 30 msec, the sample initially occupies 8.1 μl, which leaves 1.4 μl of mixing volume. This value is to be compared with the approximately 1 μl volume of the flame jet and column end. The actual effect of this mixing volume will vary with the flow rate of the experiment although here the apparent mixing volume was corrected for the diluted sample volume at the lowest carrier flow rate measured. The broadening contribution due to transport down the capillary tube may be calculated by Golay's equation (25) for a nonsorbed species. Using a gas diffusion coefficient of 0.48 cm^2/sec for methane in helium and assuming no pressure drop for helium in the tube, Steinberg's approach (21) yields a value of 3.6×10^{-6} sec^2 for the connecting tubing at a carrier gas velocity of 1 cc/sec.

Evaluation of the Valve

The sampling valve was evaluated for reproducibility, vapor concentration profile, speed, and on-off ratio. The reproducibility of the valve was measured by automatically injecting eight samples of methane at five minute intervals. For this experiment the valve gate width was 100 msec, the carrier gas flow rate was 1 cc/sec with a split ratio of 10 to 1 (vent:column). The methane flow rate into the sampling valve was adjusted to yield a peak current of 8×10^{-8} A. The volume of methane was then estimated from the charge represented by the area of the peak and the detector response factor for methane (0.251 coulombs/mole) (26). This sample volume at STP was calculated to be 0.77 μl or 34×10^{-9} moles of methane.

Table II presents the area (expressed in μcoul), retention time and second moment of eight identical injections, which have been corrected for the second

Table II. Reproducibility Data for Sampling Valve.

Area[a] ($\times 10^9$ coulombs)	Retention Time (msec)	Second Moment (msec^2)
8.61$_4$	203.$_5$	780.$_2$
8.60$_1$	203.$_5$	781.$_3$
8.61$_2$	202.$_8$	771.$_2$
8.64$_0$	203.$_3$	780.$_4$
8.63$_2$	202.$_9$	770.$_7$
8.62$_4$	202.$_8$	776.$_5$
8.62$_3$	203.$_6$	784.$_1$
8.63$_3$	202.$_7$	773.$_4$
Average 8.62	203	777
Rel. Std. Dev. 0.15%	0.18%	0.64%

[a] Valve gate width, 100 msec; carrier flow rate, 1cc/sec; sample equivalent to 0.77 μl at STP.

moment contributions previously described. This data demonstrates the excellent reproducibility of the valve. A comparison of the values of the retention time and the second moment support the earlier indications of inadequate flow stability with the present system, i.e., peaks with retention times longer than the average are comparably broadened. The 5 minute interval between the injections was imposed by the Teletype output time of the analyzer and not the valve. With either more storage or a faster output device the valve could be operated much faster.

Figure 12 shows the time relationship of the valve gate signal, the mechanical displacement of the valve, and the vapor concentration profile resulting from the injection. Of particular importance is the delay of approximately 10 msec after the initiation of the valve gate signal before the valve begins to move and the position of the center of gravity, $\bar{t}_v$, of the valve displacement curve. The delay is due to the inertia of the moving parts and the time required to build up the magnetic field of the solenoid. Once motion of the solenoid begins, it accelerates until the return spring is highly compressed. It then decelerates until it reaches its limit position, where it remains until the magnetic field of the solenoids collapses. When the valve gate signal terminates, the drive current to the solenoid is cut off, but the energy contained in the magnetic field must be dissipated before the solenoid armature will be returned. The 68Ω resistor and the diode of Figure 3 serve this purpose. The most rapid discharge of the field would be obtained if the resistor were not present; however, the current which would flow through the diode would be sufficient to hold the solenoid pulled in for an additional 20 to 30 msec. The value of the re-

25. Golay, M. J. E., "Gas Chromatography 1958," D. H. Desty, ed., Academic Press, New York, 1958, p. 36.
26. Sternberg, J. C., Gallaway, W. S., and Jones, D. T. L., "Gas Chromatography 1961," N. Brenner, J. E. Callen, and M. D. Weiss, eds., Academic Press, New York, 1962, p. 231.

sistor was chosen to yield the most rapid discharge of the solenoid with the minimum additional hold-in and without exceeding the breakdown voltage of the transistor.

The relatively long return of the valve is due to the spring; however, increasing the spring tension led primarily to decreased rise-times, greater solenoid power requirements, and decreased reproducibility. As can be seen from the concentration profile, this slow return has relatively little effect on the sample profile because the sample line and the capillary tube first intersect at 0.88 on the ordinate, and thus the bulk of the sample enters the column during the time represented by the upper portion of the valve displacement curve. The valve is presently being modified to employ a solenoid for both actuation and return. In addition to the improved return characteristics, the solenoid return mechanism will reduce the load on the actuation solenoid so that the range of linear behavior of the valve with respect to the valve gate width (Figure 15) will be increased.

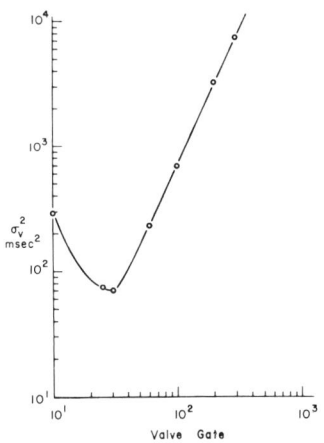

Figure 15. Measurement of the region of linear operation of the sampling valve. The σ_v^2 values plotted are corrected for the instrumental contributions.

The vapor concentration profile produced by the valve was examined as a function of gas velocity, as in the data of the previous section, and as a function of the valve gate width. No appreciable mixing volume could be ascribed to the valve as seen from the linearity of the data in Figure 14. Both the mechanical displacement and the detector output were measured for valve gate widths from 10 msec to 300 msec. The data was processed by an abbreviated version of the program previously described, such that only the area and first and second moments were calculated. For this experiment the flow rate was again 1 cc/sec with a split ratio of 10:1, and the methane flow rate was adjusted as before. Figure 15 shows the corrected sample size as a function of the valve gate pulse width. The valve is seen to give a linear response in sample size for pulses greater than 30 msec wide. The curve may be extended to gate widths of greater than 300 msec if larger sample sizes are desired and the solenoid can be kept from overheating. Below 30 msec the gate width is too short to fully activate the solenoid, and thus the mechanical effects limit the electronic control of the sample size.

The ratio of the methane concentration entering the column with the valve in the on and off positions was measured as the ratio of the respective detector currents. The current in the off position was measured as the difference between the current flowing with the methane supply turned on and off in order that the normal flame background current not bias the result. Since in this valve, unlike other designs, the sample source always has a physical path into the column, it is extremely important to know what the rejection of the sample is in the off position of the valve and to be able to maximize this factor. Figure 16 is a plot of the experimentally measured rejection ratio, V_{RR}, against the splitting ratio of the valve. V_{RR} is given by

$$V_{RR} = (I_{on}/I_{off}) \quad \text{Eq. 3}$$

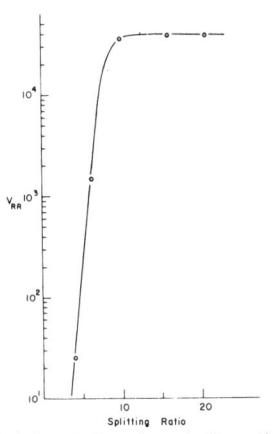

Figure 16. Variation of the valve rejection ratio with the splitting ratio. Measured with the valve static in both positions and with the column carrier flow rate constant at 1 cc/sec.

where I_{on} is the detector current with the valve in the on position, and I_{off} is the current due to the undesired bleed of sample into the column. The rejection ratio was measured as a function of the valve splitting ratio, maintaining the column flow rate constant at 1 cc/sec and the methane flow rate constant at the value previously set. This is effectively a plot of V_{RR} versus the carrier flow through the valve vent. This plot indicates that for a column flow rate of 1 cc/sec a split ratio of 10:1 or greater will yield a rejection ratio on the plateau. Higher column flow rates would allow this plateau to be reached at lower splitting ratios.

If the interior of the bellows (8 mm, i.d., x 63 mm) is regarded as a mixing chamber with a flow of 10 cc/sec purging it, a mixing chamber time constant of 315 msec results. The time required for methane to diffuse from the sample inlet to the axis of the column and carrier inlet (0.188 in.) is 270 msec. Hence the sharp break in the curve of Figure 16 is due to the mixing chamber time constant of the bellows becoming larger than the diffusion time between the sample inlet and the column axis. To obtain the maximum rejection ratio under any given set of conditions, it is necessary to maintain the mixing chamber time constant of the bellows smaller than the above-mentioned diffusion time. A reduction in the internal volume of the bellows would allow the maximum rejection ratio to be reached with a lower overall carrier gas flow rate. The small bleed of sample into the column remaining, when the valve is operated at the maximum rejection ratio, is probably due to entrainment of sample due to the close proximity of the two inlets and slight misalignment of the inlet tubes with the column.

With the availability of these digital systems for measurement and control, these experimental considerations become readily apparent in developing high precision chromatography and offer significant utility for the design and understanding of the dynamics of chromatographic systems.

Acknowledgment

The support of the United States Department of the Interior as authorized under the Water Resources Research Act of 1964, Public Law 88-379, and the National Aeronautics and Space Administration is gratefully acknowledged. ∎

Appendix

Logic Module List[a]

FF1-FF47	R202; Dual gated flip-flop
MS1	R302; Dual gated monostable
PA1-PA6	R602; Dual gated pulse amplifier
G1-G6	R111; Triple diode gate, with diode expander networks. R002 (includes IV1)
ST1-ST3	W501; Schmitt trigger
BD1-BD12	Bus driver, adapted from NAVWEPS Preferred Circuits, PSC-12. Modified to operate with DEC logic levels and power supplies.
Connectors[b]	W023, W028

[a]Specifications and an explanation of symbols is given in The Digital Logic Handbook (24).
[b]Connectors are not shown on the diagrams for clarity.

Logic Signal List

Mnemonic	Name	Assertion[a]
CIP	Clock input pulse	Gnd(p)
DCP	Delayed clock pulse	Gnd(p)
ROL	Readout level	−3 v.
VOP	Valve operate pulse	Gnd(p)
PCP	Power clear pulse	Gnd(p)
AAE	Address advance enable	Gnd
PEL	Program end level	−3
AOL	Address overflow level	Gnd
	Address advance	−10(p)
DTE	Data transfer enable	−10
	Read indicator drive	−3
	R/O indicator drive	−3
	Valve gate	−3
EOC	End of conversion	Gnd

[a]Pulses are designated by (p).

Manuscript received October 7, 1969.

Computer-Controlled Gas Chromatograph Capable of Real-Time Readout of High-Precision Data

R. S. Swingle and L. B. Rogers

Department of Chemistry, Purdue University, Lafayette, Ind. 47907

TWO HIGH-PRECISION GAS CHROMATOGRAPHS, capable of ±0.02% retention-time precision, have been reported in the literature (*1, 2*), and their utility has been well demonstrated (*3–7*). However, both of those instruments are limited by the fact that the high-precision results must be calculated using an off-line digital computer. Even the chromatograph described by Burke and Thurman (*2*), which uses a small dedicated computer for data acquisition, gives an on-line precision of only ±100 msec in the retention time. For high-precision work, those authors dump the raw data onto paper tape and perform further calculations on a larger computer. That requirement can often lead to long turn-around times and the nuisance of processing large amounts of paper tape.

In nonroutine applications and in cases where optimization of parameters is required, computer control of experimental variables is often desirable (*8–11*). In those cases, on-line capability for high-precision readout must be available since the ability of the computer to control the course of study depends upon how well it can evaluate the results from each experiment. Obviously, if the computer lacks the programming and/or the core-space to evaluate input with sufficient precision, uncertainties in the values for the parameters being optimized will be greater than necessary, and much of the overall capability of the system will be wasted.

To meet those objections, a high-precision chromatograph has been interfaced to a small digital computer. At the same time, software has been developed which permits precise peak-characteristics to be calculated in real-time and typed out within a few seconds after completion of a chromatogram. Hardware has been provided for computer control of temperature and flow, and for direct readout of column-inlet pressure and mass flow rate. The latter permits determinations of corrected flow rates at the column outlet, and, hence, calculation of precise retention volumes during the course of each chromatographic run. Because of these features, the instrument is easily used for rapid, precise determination of thermodynamic data, quantitation of peak shapes, and accurate qualitative identification of unknown peaks.

SYSTEM DESIGN

Gas Chromatograph. A block diagram of the gas chromatograph and its associated equipment is shown in Figure 1. The column oven, temperature controller, and temperature-measuring apparatus were essentially identical to those described by Oberholzer and Rogers (*1*). Helium carrier gas was purified by passing it through a quartz diffusion cell, Model SLM1 (Electron Technology Inc., Kearny, N. J.), fitted with a high-pressure outlet head. The 80 psig supplied from the tank regulator on that head was reduced to 50 psig by a Millaflow 2-stage pressure regulator (Millaflow Division, National Welding Co., Richmond, Calif.), and passed to a Brooks Model 8744 stainless steel flow controller (Brooks Instrument Division, Emerson Electric Co., Hatfield, Pa.). When operating the gas chromatograph under computer control in the automated mode, the flow controller was replaced by a Brooks Model 8504 ELF, NRS, stainless steel needle-valve with digital handle. A Hastings-Raydist Model LF-50 mass flow transducer (Hastings-Raydist, Inc., Hampton, Va.) and a Barocel Type 538-19 differential pressure sensor (Datametrics, Inc., Waltham, Mass.) were placed downstream from the flow controller. One side of the differential pressure sensor was vented to the atmosphere through two meters of 0.32-cm i.d. tubing. All the above equipment, with the exception of the diffusion cell, was housed in a thermostated box, held at 36.0 ± 0.1 °C.

Samples were injected into the carrier gas using a pneumatically operated Carle Model 2018 gas sampling valve (Carle Instruments Inc., Fullerton, Calif.) or a Seiscor Model VIII high-temperature liquid sampling valve (Seismograph Corporation, Tulsa, Okla.). The Carle valve was mounted directly on the front of the column oven while the Seiscor was placed in the oven itself. Each valve was connected to the column by approximately 15 cm of 0.058-cm i.d. capillary tubing.

A Varian 1800-series flame ionization detector (Varian Aerograph, Walnut Creek, Calif.) was modified for low dead-volume using a 7.2-cm length of 0.55-cm o.d. × 0.116-cm i.d. stainless steel tubing secured to the FID inlet with a Swagelok nut. About 0.9 cm of the 7.2-cm length of tubing was turned down slightly on a lathe to allow the hydrogen to sweep over the end of the tubing and carry the sample to the flame. A 10-cm length of 0.058-cm i.d. capillary tubing was silver-soldered to the 0.55-cm o.d. tubing and extended directly into the oven of the chromatograph. This modified FID was insulated with glass wool and heated with nichrome wire. The polarizing voltage for the detector was supplied from a solid-state Computer Products Model PM538 170-V power supply (Computer Products Component Division, Fort Lauderdale, Fla.).

---

(1) J. E. Oberholtzer and L. B. Rogers, ANAL. CHEM., **41**, 1234 (1969).
(2) M. F. Burke and R. G. Thurman, *J. Chromatogr. Sci.*, **8**, 39 (1970).
(3) J. E. Oberholtzer and L. B. Rogers, ANAL. CHEM., **41**, 1590 (1969).
(4) A. K. Moreland and L. B. Rogers, *Separ. Sci.*, **6**, 1 (1971).
(5) L. J. Lorenz, R. A. Culp, and L. B. Rogers, ANAL. CHEM., **42**, 979 (1970).
(6) R. A. Culp, C. H. Lochmüller, A. K. Moreland, R. S. Swingle, and L. B. Rogers, *J. Chromatogr. Sci.*, **9**, 6 (1971).
(7) D. Macnaughtan and L. B. Rogers, ANAL. CHEM., **43**, 822 (1971).
(8) S. P. Perone, D. O. Jones and W. F. Gutnecht, ANAL. CHEM., **41**, 1154 (1969).
(9) G. E. James and H. L. Pardue, *ibid.*, p 1618.
(10) L. Ramaley and G. S. Wilson, *ibid.*, **42**, 606 (1970).
(11) R. G. Thurman, K. A. Mueller, and M. F. Burke, *J. Chromatogr. Sci.*, **9**, 77 (1971).

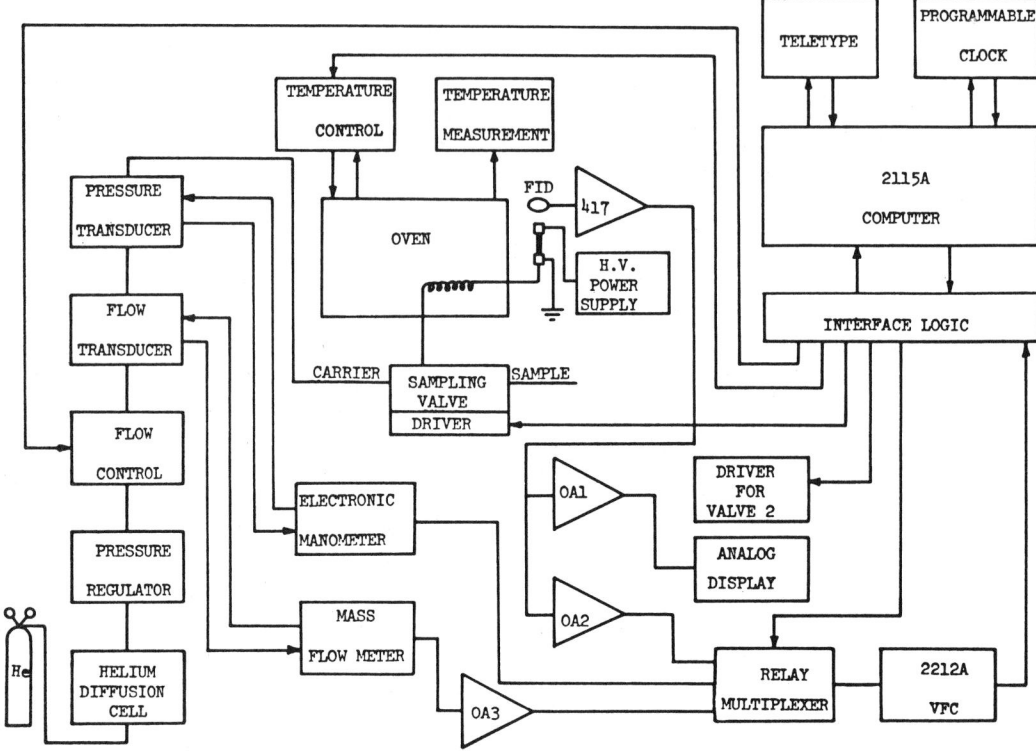

Figure 1. Block diagram of the instrument

To help damp high-frequency vibration, the remote head of a high-speed Keithley 417 picoammeter (Keithley Instruments, Cleveland, Ohio) was mounted together with the FID on a special rack that rested on rubber cups. Picoammeter input capacitance was minimized by connecting the FID directly to the remote head with a 4.2-cm section of low-noise cable. Thus, full-scale response times of about 100 msec at 10^{-11} ampere and above were possible.

Interface and Computer Hardware. A Hewlett-Packard Model 2115A computer, (Hewlett-Packard, Palo Alto, Calif.) equipped with an 8192-word core memory, an extended arithmetic unit, a teleprinter, and a high-speed paper-tape reader was used for data acquisition, processing, and control operations. Program assembly and/or compilation was done using a Hewlett-Packard 2116A computer with 8192-word core memory, teleprinter, high-speed paper-tape reader, high-speed paper-tape punch, and a Model 2020 magnetic tape unit.

Analog-to-digital conversion was accomplished in part with a Hewlett-Packard Model 2212A voltage-to-frequency converter, which provided an output frequency of 100 KHz for full-scale inputs of 1.0, 0.1, or 0.01 volts. The precise timing used in data acquisition was achieved using a DEC Model B405 10-MHz crystal clock (Digital Equipment Corporation Maynard, Mass.). Software-programmable decades from 100 KHz to 0.01 Hz as well as options for dividing an interval by 2 or 5 were available (12). All interface logic was constructed using DEC R-, W-, and A-series logic cards.

Two Model HDUM-125-16-10 bi-directional stepping motors with Model 14DUM-A-K81 stepping motor drivers (United Shoe Machinery, Harmonic Drive Division, Beverly,

Mass.) were used for computer control of oven (column) temperature and flow rate. The stepping-motor resolution was 2000 steps per revolution, while the stepping rate was variable up to 3000 steps per second. Computer control of column temperature was achieved through attachment of one stepping-motor shaft to 10 K- and 200-ohm Model 7222, double-shafted Helipot potentiometers (Helipot Division, Beckman Instrument Inc., Fullerton, Calif.) mounted in tandem. The 200-ohm potentiometer was wired to the programmer input of a Melabs temperature controller, permitting computer control over a range of approximately 100 °C above the Melabs set-point. Computer control of column flow rate was achieved by having the second motor turn, in tandem, a 10-Kohm potentiometer as above, and the nonrising stem of the Brooks needle valve mentioned previously. In both temperature and flow arrangements, the 10-K potentiometers were used with mercury cells to determine relative motor position. Both assemblies were mounted on flat aluminum tooling plate with shaft alignments machined to 0.0025-cm tolerance. Zero backlash flexible couplings (Pic Engineering, New York, N. Y.) were used for all shaft connections.

A schematic diagram of the signal-conditioning operational amplifiers is shown in Figure 2. To reduce noise pickup, the analog signals from the chromatograph were amplified to a nominal 10-V full scale. The type 1014A Datametrics Electronic Manometer, used in conjunction with the Barocel differential-pressure sensor, supplied 10-V full scale with no modification. An attenuator switch on the manometer provided full-scale output for ranges from 100 to 0.1 psid.

The Keithley picoammeter supplied 3-V full scale for its current ranges. The output was amplified fourfold using an Analog Devices Model 232J chopper-stabilized operational amplifier (Analog Devices Inc., Cambridge, Mass.).

---

(12) S. P. Perone and J. F. Eagleston, *J. Chem. Educ.*, in press.

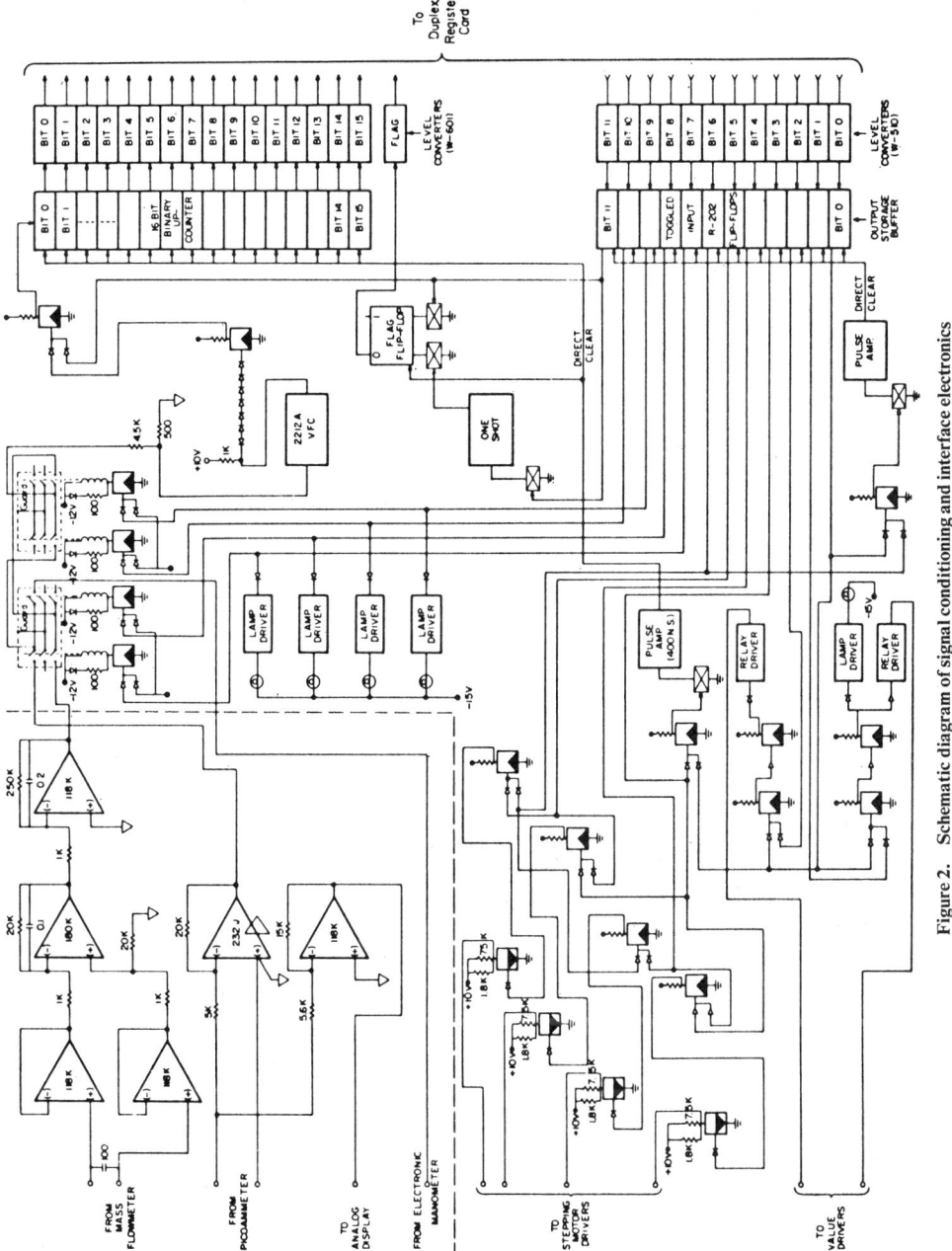

Figure 2. Schematic diagram of signal conditioning and interface electronics

The 12-V full scale took advantage of the 150% linear over-range capability of the voltage-to-frequency converter. A separate operational amplifier was used to drive the analog display, usually a recorder, and provided signal isolation from any back emf generated by the display.

The Hastings mass flowmeter used with the flow transducer provided only 2 mV for a 50 ml/min mass flow rate. Furthermore, although the flowmeter output was pressure independent, it did have superimposed on it a 100-mV peak-to-peak 10-KHz ac signal from the bridge excitation source. Thus, while that signal was adequate for operation with a mV recorder, operation with a digital data-acquisition system required additional circuitry. The circuit shown in Figure 2, used Analog Devices components and effectively filtered out the carrier wave and amplified the 2-mV dc signal to 10 V. It should be noted that while that circuit was relatively slow, the flow transducer itself had about a 15-sec full-scale-response time for a step-change of 0-30 ml/min of helium. Furthermore, most gas chromatographic systems require at least 30 seconds to equilibrate after a change in flow rate. Only in fast pressure programming studies would a slow response lead to difficulties, and in those cases perhaps the flowmeter designed by Thurman and Burke (13) would be more suitable. Unfortunately, those authors gave no response times and said only that the flowmeter response was "fast."

All resistors in the above circuits were 1%, 5 ppm/°C wire-wound precision resistors (Precision Resistor Co., Hillside, N. J.). The analog circuitry was housed in a 36.0 ± 0.1 °C thermostated box.

Figure 2 also shows the interface logic. The logic symbols are those used in Reference *14*. The analog signals were multiplexed through DEC A-111 guarded-relay multiplexer switches to a 10:1 voltage divider located at the 1-V input of the voltage-to-frequency converter (VFC). The pulse train from the VFC passed through a modified DEC W-510 level converter to achieve 0 and −3-V logic levels. A NAND gate, when enabled, passed the pulse train to a 16-bit binary counting register. Upon execution of the appropriate computer commands, the contents of this register were transferred through the W-601 level converters into the duplex register card and into the accumulator of the computer.

Table I summarizes the output control functions and their corresponding octal codes. Computer control of the stepping motors, valves, multiplexer, and interface itself was accomplished using output bits 0–11 of the duplex register card. The output of each of those bits passed through a modified W-510 level converter whose output, in turn, toggled a J-K flip-flop. The flip-flop served as temporary output-storage buffers, so that different output operations could be accomplished independently during the same time period. For example, if the multiplexer was being held in a given state, the binary register could be cleared without clearing the multiplexer or without including the multiplexer control code in the register-clearing output code. As will be seen later, this buffer greatly facilitated the control programming.

Figure 2 shows the output functions for all 12 bits. Bits 0–2 were used to actuate valve-drive circuitry discussed previously (*15*). Bits 0 and 6 were gated together and combined with a pulse amplifier for direct clearing of all output bits. Bits 1 and 2 performed the same clearing operation on the binary counting register, but in that case, the pulse amplifier was jumpered to provide a 400-msec pulse and prevent "carries" in the counting register. Remote actuation of the stepping-motor drivers required a negative-going pulse of at least 10-V magnitude, 1-μsec fall-time, and 20-μsec duration. The gating illustrated satisfied those requirements and, for example, when bits 3 and 5 were set, stepping motor number one took a step in the counter-clockwise direction. Bits 7–10 controlled the multiplexer. Bit 11 controlled the binary-register gate and the flag bit. When the gate was disabled, the one-shot provided a 1-μsec delay to allow complete settling of the counting register before setting the flag and strobing the contents of the counting register into the computer.

Software. To facilitate the development of original programs for this gas chromatograph, an assembly-language relocatable subroutine library was compiled. Subroutines in the library are first, second, and third order, and for the most part, perform specific functions related to the hardware described above. For example, a simple "CALL REGCL" instruction, results in a direct clear operation being performed on the binary counting register. A list of the subroutine names along with a short description of the purpose of each is shown in Table II. The routines grouped together at the bottom of the table are used in the programs described below so as to permit data acquisition under interrupt control,

Table I. Octal Codes for Computer Control of Experiment

Function	Code (octal notation)
Value 1 (sampling valve)	3
Value 2	5
Stepping Motor 1, Forward	30
Stepping Motor 1, Reverse	50
Stepping Motor 2, Forward	120
Stepping Motor 2, Reverse	140
Clear Binary Counting Register and Flag	110
Clear Output Bits	101
Multiplexer	
Flame Detector	2200
Flow Rate	1200
Column Head Pressure	2400
Oven Temperature	1400
Gate control for VFC pulse train	4000

Table II. Subroutine Names and Functions

Name	Functions
CLEAR	Clear output bits
FLO	Read mass flowrate
PRESS	Read inlet pressure
RATE	Get data acquisition rate from TTY
REGCL	Clear 16-bit binary up-counter
1STPF	Step motor one forward
1STPR	Step motor one reverse
2STPF	Step motor two forward
2STPR	Step motor two reverse
STOP 3	Disable programmable clock
VALVE	Activate sampling valve
STRT	Establish interrupt linkage on data channel
GO	Initialize pointers, and take first 200 data points
CKFLG	Rotate data into working buffer
SERVE	Service interrupt on data channel

and pseudo "real-time" data processing. A system of common buffers is used for temporary storage of integer data points. Those data are then floated and transferred to a working buffer in blocks of fifty, replacing data that have been processed.

A general-purpose program, GOUT1, was written in Hewlett-Packard Fortran for use in operating the chromatograph in a semi-automated mode. Data are taken under interrupt control and processed in, essentially, real-time. A Savitsky and Golay (*16*) eleven-point quadratic derivative smooth is used in conjunction with consistency tests to detect peaks. Peak area is determined by summing the data points from the beginning to the end of the peak. A linear least-squares fit to about 100 data points both before and after each peak is applied for area correction. Less than 100 points are used if the program detects an absolute value of the base-line derivative that exceeds a specified slope.

The retention time of the peak mean, t, is defined by the first moment of the peak (*17*, *18*). The peak variance, V, is found by calculating the second moment about the mean and is further related to the number of theoretical plates by the equation $N = (t)^2/V$. Program GOUT1 makes no provision for overlapped peaks other than to drop a perpendicular to the base line if the second peak is detected below a threshold of 1.2 times the base line.

---

(13) R. G. Thurman and M. F. Burke, *J. Chromatogr. Sci.*, **9**, 181 (1971).
(14) "Logic Handbook," Digital Equipment Corp., Maynard, Mass., 1968.
(15) J. E. Oberholtzer, ANAL. CHEM., **39**, 1627 (1964).
(16) A. Savitzky and M. J. E. Golay, ANAL. CHEM., **36**, 1627 (1964).
(17) O. Grubner, "Advances in Chromatography," J. C. Giddings and R. A. Keller, Ed., Marcel Dekker, New York, Vol. 6, 1968, pp 173–209.
(18) E. Grushka, M. N. Myers, P. D. Schettler, and J. C. Giddings, ANAL. CHEM., **41**, 889 (1969).

PAUSE								
SCL1	SCL2	TAMB	TOVEN	PBAR	TRIG	PKMIN	LENGTH	PITCH
3	8	18	126.32	747.7	1	1000	100.0	20

DATA RATE =
10

	AREA	AREA(UC)	MEAN R.T.	MEAN R.V.	−LOG AREA
1	43802.83	.219011403E − 01	20.6869	2.9956	.18033712E + 01
2	31429.27	.15714623E − 01	24.8156	3.5934	.18033712E + 01
	8.6883	.8204			

	VARIANCE	PLATES	PLATES/SEC	HETP
1	.4995	1272.8916	50.4809	.0786
2	.7977	1146.9033	37.9169	.0872

PAUSE

Figure 3. Teletype input/output for program GOUT1

The operation of GOUT1 is best described by looking at the input and output shown in Figure 3. The initial pause permits the experimenter to enter any "passive" parameters such as name, date, run-identification number, etc. using the teletype "local" mode. Upon pushing "Run," the experimenter is prompted to enter the following active parameters. SCL1 and SCL2 are the picoammeter attenuation and range settings, respectively, used for calibration of peak area in microcoulombs. TAMB is the ambient temperature, and TOVEN the oven (column) temperature, PBAR is the barometric pressure, to the nearest 0.1 mm, and is assumed to be the column-outlet pressure. TRIG is the derivative trigger value used for peak detection. PKMIN is the minimum area a peak must have before GOUT1 will recognize it, LENGTH is the column length in cm, and PITCH is the number of points at the beginning of a run which the experimenter does not wish to have processed. This last parameter is necessary because sample injection using an automated sampling valve sometimes gives an uneven base line at the beginning of a chromatogram which can be interpreted by GOUT1 as a peak. Finally, the experimenter is prompted for the data-acquisition rate, in points per second. After all manual input has been completed, the program samples the mass flow rate and the column-inlet pressure, and then halts so as to permit any last-minute adjustments, should they be necessary.

Upon pushing "Run," the sample is injected and data acquisition and processing begin. After the experimenter is satisfied that the chromatogram has been completed, he sets bit 1 of the switch register up. GOUT1 terminates data acquisition, finishes processing the resident data, and prints on the teletype for each peak detected the peak number, area in counts, area in microcoulombs, corrected retention time, corrected retention volume, and negative logarithm of the area. The outlet flow rate in ml/min corrected for temperature and pressure, and the James and Martin gas-compressibility coefficient, j, are then printed, followed by the peak number, variance, number of plates, plates-per-second, and height equivalent to a theoretical plate for each peak in the chromatogram. The program then cycles back to its beginning and again pauses. Should a second run be desired, the experimenter may elect to enter a new set of active parameters, or may set bit 15 up and bypass the input stage of the program. In that case, GOUT1 assumes the active parameters to be the same as before, and it proceeds to ask for a new data-rate.

A second Fortran program, OPCON, operates the gas chromatograph in its most highly automated mode. As such, OPCON incorporates essentially all of GOUT1 plus the programming necessary for computer control of the column temperature and flow rate. A simplified flow chart is shown in Figure 4. Because OPCON is being used in resolution studies, it is presently set up only for two-component mixtures. It could, however, be changed easily to accomodate any resonable number of components.

In using OPCON, the operator sets up the initial experimental conditions at the lowest flow rate and column temperature to be run. OPCON prompts the experimenter for various control parameters in addition to the input for GOUT1 described above. These include the number of temperatures to be run, not to exceed seven; the number of steps between adjacent temperatures, the number of flow rates to be run at each temperature not to exceed seven; the number of steps between adjacent flow rates; the number of replicates to be run at each set of experimental conditions; the maximum run time per chromatogram; the delay, in minutes, between changing the flow and initiating the next run; the delay, in minutes, between changing the temperature and initiating the next run; the total maximum number of steps allowed for increasing the temperature, not to exceed 19,500; the total maximum number of steps allowed for increasing the flow, not to exceed 19,500; and the slope of the calibration curve for number of steps vs. temperature. The operator is then prompted for the data-acquisition rate, and, after it has been entered, OPCON begins execution.

The flow rate is sampled by the computer at two-second intervals until the flow rate calculated from two successive readings differs less than 0.5%. The column-inlet pressure is then read in a similar manner until the j factor varies less than 0.05% for two successive readings. If the flow is equilibrated, both parameters are fixed in less than 10 seconds. The sample is injected and data acquisition begins. The computer continues to acquire and process data until either two peaks have been eluted or the maximum run-time has been exceeded, following which the data for each peak are dumped on the teletype. OPCON then decides whether all replicates at all experimental conditions have been run as specified in the control input. If they have not, the program makes the decision as to what changes, if any, in the experimental conditions are appropriate, makes those changes, waits for system equilibration, and returns to read the flow rate, etc. Changes in flow rate are made in a pseudo-random order. Likewise, after all desired flow rates have been run at a given temperature, the temperatures are varied in a pseudo-random fashion.

The OPCON output for each chromatogram contains the same information about each peak as described previously for GOUT1. In addition to the outlet flow rate and j factor, OPCON writes the oven temperature in °K, calculated from a calibration curve that will be described below, the resolution between the two peaks calculated from

$$R_s = \frac{(t_2 - t_1)}{2(\sigma_2 + \sigma_1)}$$

and a pressure correction factor, f_1 (*19*) calculated from

$$f_1 = \frac{9}{8} \frac{(P^4 - 1)(P^2 - 1)}{(P^3 - 1)^2}$$

where P is the inlet-to-outlet pressure ratio.

---

(19) J. C. Giddings, S. L. Seager, L. R. Stuki, and G. H. Stewart, ANAL. CHEM., **32**, 867 (1960).

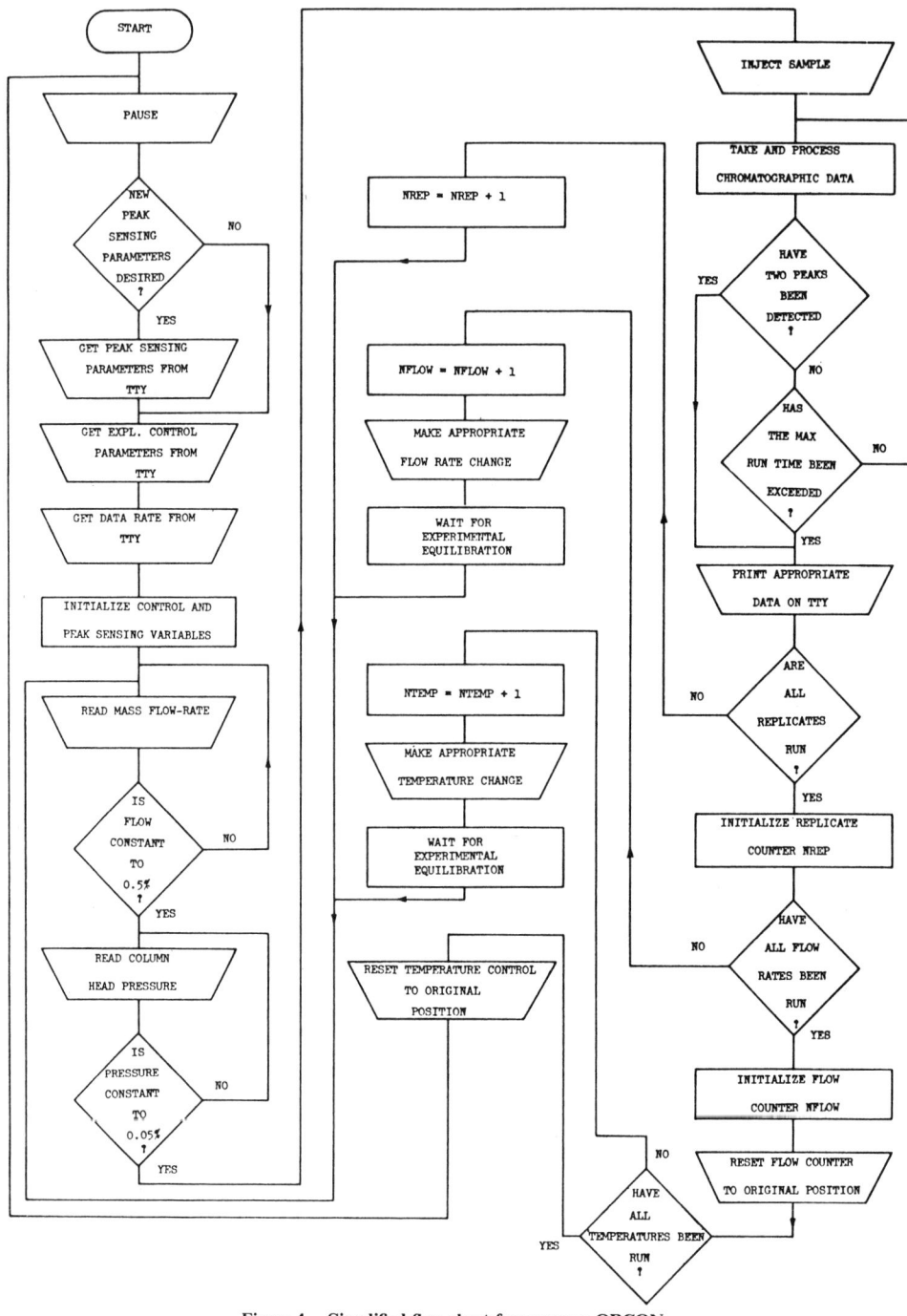

Figure 4. Simplified flow chart for program OPCON

A simple assembly-language program, TANDR, was written for use in calibration of flow rate, inlet pressure, and stepping-motor settings. The program acquires a number of data points at a given rate from one multiplexer channel, and then dumps the individual points and their average on the teletype. Both the data-acquisition rate and the number of points are specified by the experimenter. It should be noted that the data-acquisition rate must be the same as that used under normal operating conditions to ensure identical integration times for all data.

Calibration. The digitized output of the mass flowmeter was calibrated against a soap-film bubblemeter. These

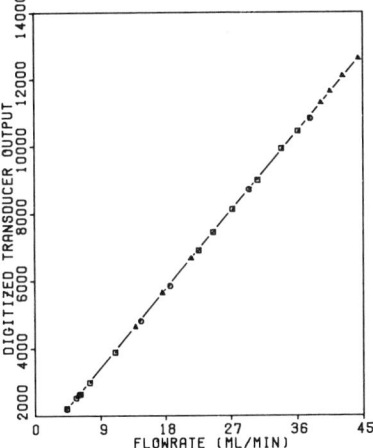

Figure 5. Output of mass flowmeter vs. flow rate from soap-film bubblemeter

    □ Day one
    △ Day two
    ○ Day three

Figure 6. Output of electronic manometer vs. differential pressure from Heise guage

    + Day one
    △ Day two

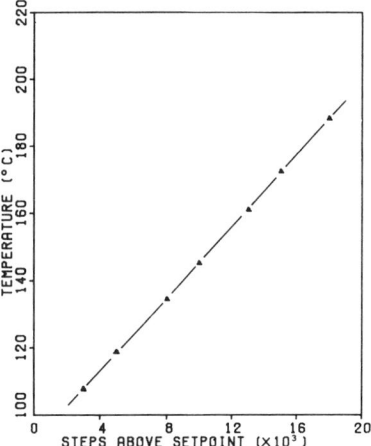

Figure 7. Temperature of platinum resistance thermometer vs. stepping motor position

results are shown in Figure 5. The different symbols indicate data taken on different days. Each point represents the average of 600 readings taken at 10 points/sec. The short-term fluctuations over an approximate 1- to 2-min period were found to be less than 0.1%. First- through fourth-order polynomial equations were least-squares fitted to the data and, as expected from the nearly linear relationship, a cubic equation having very small coefficients for the squared and cubed terms was found to be satisfactory. The standard deviation of the points from the fitted line was 0.095 ml/min, over a range of 3 to 48 ml/min, or about 0.2% of full scale. This was the best that could reasonably be expected from the system because the Brooks flow controller was rated at 0.3% stability. Furthermore, 0.2% of the 2-mV output of the mass flowmeter corresponded to measurements of voltage differences in the μV range. Because the accuracy of the measurement depended upon proper zeroing of the operational amplifier circuit, and because an accuracy of 0.1 ml/min was desired (and attained), temperature thermostating of all of the apparatus for controlling and measuring the flow was necessary. The flowmeter output was observed to be independent of pressure over a 1 to 50 psid range.

The 30-psid range of the Datametrics electronic manometer was calibrated against a Heise Model C, 75 psig, bourdon-tube gauge (Heise Bourdon Tube Co., Inc., Newtown, Conn.). The data are shown in Figure 6 with different symbols, again, denoting data taken on different days. Short-term variations in the manometer output were less than 0.01%. A linear least-squares fit to the data gave a standard deviation of 0.004% over the range from 3 to 30 psid. This was not surprising considering the rated specification of ±0.01% accuracy. In this case, the pressure manometer output itself was more reliable than the technique used for calibration. As indicated earlier, pressure ranges other than 30 psid were available, but inlet pressures above 30 psid have not been encountered in our study.

Whenever the computer made a change in column temperature, the new temperature was calculated from the calibration curve shown in Figure 7 in which the temperature in °C is plotted against the number of steps the stepping motor has turned a 200-ohm potentiometer wired to the temperature controller. The linear equation that was least-squares fitted to those data gave a standard deviation of ±0.25 °C. The slope of the line was independent of the initial setpoint of the temperature controller, as expected, since a platinum sensing element was used. The programming range of 100 °C provided by the 200-ohm potentiometer could be increased to 300 °C by substituting a 500-ohm model, which would also change the slope of the calibration curve.

Chemicals. The chromatographic column was a 100-cm × 0.32-cm i.d. stainless steel tube packed with 5% OV-1 on 80/100-mesh Chromosorb G.

A sample stream consisting of a binary mixture of n-pentane and n-hexane (Matheson Coleman & Bell, Chromotoquality Grade) was prepared in the following manner. Each of two glass gas-saturation assemblies was half-filled with one of the hydrocarbons, and placed in separate water baths. Each water bath was thermostated at a different temperature to ±0.1 °C with a Sargent Model S-W temperature con troller (E. H. Sargent & Co., Chicago, Ill.). Two Brooks Model 8743 flow controllers provided steady streams of high-purity nitrogen (PrePurified Grade, Air Reduction Co.) through each saturator. The streams were then mixed and passed to the sampling valve. By changing the temperatures of the water baths and the flow rates of the nitrogen streams, binary samples of any relative concentration could be produced. Further dilution of the mixture with pure nitrogen permitted changes to be made in the absolute levels of the concentrations. To prevent sample condensation, all connecting tubing between the sampling valve and the water baths operated above room temperature, was heated to about 60 °C using heating tape and was insulated with glass wool and aluminum foil. Similarly, a sample stream of methane (Matheson Co., C.P. Grade) in high-purity nitrogen was prepared in order to determine the void volume of the column.

RESULTS

The peak-sensing portions of programs GOUT1 and OPCON were tested by interrupting the program operation at an appropriate point and entering, through the switch register, synthetic data for a base line and for a normal distribution curve at every 0.1 standard deviation unit. The program was then restarted and the computer output for the synthetic peak compared to hand-calculated values. Agreement within 0.01% was found for area, mean retention time, and peak variance, thereby indicating that the programs were functioning correctly.

Using the pentane–hexane sample stream, five replicate chromatograms were run at several different sets of experimental conditions. Table III shows the precision found for retention time, variance, and HETP for one set. Note that the standard deviations of ±0.02% for retention time and ±0.04% for peak variance were comparable to those obtained using an off-line computer (1). Presumably, higher peak moments could also have been comparable had they been calculated. It should be mentioned that standard deviations for peak characteristics depend a great deal upon the peak shape. Badly tailed peaks could not be characterized with the precision reported above because of difficulties in detecting reproducibly the end of the peak.

Table II also gives data for the column-outlet flow rate, corrected for temperature and pressure drop, and the gas compressibility coefficient. Although the ±0.35% uncertainty in flow rate was expected from the calibration procedures and was comparable to that taken with a soap-film bubblemeter, the present system was considerably faster and more convenient. Programs GOUT1 and OPCON printed only four significant figures of the gas compressibility constant, and, since the measurement was better than that, precision was reported simply as <0.01%. One further measure of the system reliability can be seen from the retention-volume data for methane which are shown at the bottom of Table III. Note that the data were obtained at two different column temperatures, inlet pressures, and flow rates and that retention volumes agreed within ±0.4%.

Because of a long time-constant for equilibration of the sampling system (~30 min), no attempt was made to achieve completely reproducible sample sizes in these runs. Consequently the precision of peak area in Table III is not representative of the gas chromatograph, but indicates the nonequilibrium in the sampling system. From the results found with the normal curve, it was felt that with a base line constant to about ±2 counts, reproducibility of peak area would always depend upon sampling and not the chromatograph or the peak-detection software.

Table III. Precision of Chromographic Data

Parameter	Units	Value	% Rel.
Flow rate	ml/min	14.42 ± 0.05	±0.35
j factor	...	0.6761 ± 0.001	<±0.01
Area			
(n-Pentane)	μcoul	0.02853 ± 0.00017	±0.61
(n-Hexane)	μcoul	0.03137 ± 0.00051	±1.63
Retention time			
(n-Pentane)	sec	9.000 ± 0.003	±0.03
(n-Hexane)	sec	15.946 ± 0.003	±0.02
Variance			
(n-Pentane)	sec^2	0.2265 ± 0.0001	±0.04
(n-Hexane)	sec^2	0.4002 ± 0.0004	±0.09
HETP			
(n-Pentane)	mm	0.642 ± 0.001	±0.16
(n-Hexane)	mm	0.786 ± 0.001	±0.13
Retention volume			
(Methane)	ml	2.227 ± 0.009	±0.40

DISCUSSION

To date, our primary objective has been to demonstrate that high-precision chromatographic data can be obtained using a small, on-line, digital computer. However, significant extension of the existing software is severely limited by the availability of only 8 K of core memory. A special compilation procedure, combined with deletion of almost all Holarith output, was necessary to put OPCON into 8K. At present, there are less than 10 uncommitted core locations. By rewriting some of the programs, a small gain in "free" core would be possible. However, a major increase in core locations is required to include desirable features such as direct computer readouts of column temperature to ±0.01 °C and column-outlet (or barometric) pressure, computer-controlled feedback for each of the major variables, and software capability for modification of experimental design. There is little question that all of the above goals could be accomplished in 8K of core by writing all, or nearly all, of the software in assembly language, but the use of a higher-level language was considered to be highly desirable for one-of-a-kind research applications because the programs could be written and debugged in shorter times. For that reason and because the programs could be adapted more easily to other computers, GOUT1 and OPCON were written in Fortran.

Fortunately, more memory will soon be available on a different computer system. As a result, the two hardware extensions mentioned above will be incorporated in addition to an interface modification. That last change will involve the gate between the binary counting register and the VFC, which is controlled directly from the computer. The timing for enabling and disabling this gate was critical in order to obtain meaningful analog-to-digital conversion. This meant that the program had to avoid the chance that the computer might have just started executing a time-consuming uninterruptable instruction. For example, a floating-point divide requires 560 μsec, during which up to fifty "extra" counts could be added to the counting register. For that reason, the programmable clock was set up to provide ten flags for each data point. The interrupt service routine counted the flags and, after the ninth, disabled the interrupt and waited for the tenth. When it came up, the gate was immediately disabled, thereby preventing any extra pulses from being counted. However, as a result, one tenth of the computer time was lost through waiting. Even so, data rates up to 25 points per second could be handled in real-time. Since the new computer system will operate in a time-shared mode among several instruments, the clock will have to be wired directly to the gate in order to free the "lost" computer time.

Similarly, the availability of more memory will permit desirable software modifications to be included. For example, GOUT1 uses an eleven-point first-derivative quadratic smooth for peak detection. That choice will not always be the best, and it would be beneficial to provide a program option whereby the computer could choose one from a number of such functions, the choice being based upon data-acquisition rate, base-line noise, peak variance, etc. Unsophisticated program deconvolution of overlapping peaks, and the calculation of the third and fourth central moments of each peak would also be welcome additions.

One of the most productive uses of this high-precision gas chromatograph will be in automatic empirical determinations of optimum chromatographic conditions for particular separations. A first step in this direction has been taken by Thurman, Mueller, and Burke (11). Their approach involved, however, only single-component systems, and a computer programmed to increase the flowrate between successive chromatograms until a minimum was found in the van Deempter plot. Furthermore, no absolute determination of the column-outlet flow rate was possible. To arrive at a better solution to the optimization problem and, at the same time, to realize the full capabilities of our instrument, a program is being written which will include a statistical design for locating, in minimum time, the optimum in a multidimensional plot of the parameters to be optimized against resolution per unit time.

At present, optimization studies are being performed, but the decision about how to proceed after each set of runs is made by the operator who then has to enter by teletype new instructions for the next series of runs. Even so, there is a tremendous saving, both in turn-around time between data acquisition and processing, and in operator time.

ACKNOWLEDGMENT

The authors are indebted to Dr. S. P. Perone for the use of the two computers, provided through the National Science Foundation, and to Dr. J. W. Amy for his many helpful discussions.

RECEIVED for review December 14, 1970. Accepted February 24, 1971. Supported in part by the U. S. Atomic Energy Commission through Contract AT(11-1)-1222. One of the authors (RSS) is grateful for a Fellowship from Phillips Petroleum Company.

8

Copyright © 1977 by the American Chemical Society
Reprinted from *J. Chem. Inf. Comput. Sci.* 17:238-242 (1977)

Computer Control of Data Acquisition, Reduction, and Display in Rapid Scanning Liquid Chromatography[†]

MARK S. DENTON, BARBARA A. JEZL, W. R. HEINEMAN, and HARRY B. MARK, JR.*

Department of Chemistry, University of Cincinnati, Cincinnati, Ohio 45221

Received June 16, 1977

> An outline of the on-line computer system and supporting software used in a preliminary study on utilizing a rapid scanning spectrometer as a liquid chromatography (RSS/LC) detector is given. Refinements of this system in data acquisition, reduction, and display that were made out of convenience or necessity are then described in more detail. The present system allows computer control of the RSS, variable CATing, automatic absorbance calibration, rapid plotting of absorbance (A) vs. wavelength vs. time (3D chromatograms) and A vs. time (conventional chromatograms), and, finally, an automated integration routine for quantitative measurements.

INTRODUCTION

In a preliminary paper, the feasibility and advantages of an oscillating mirror rapid scanning ultraviolet–visible spectrometer as a detector for liquid chromatography were demonstrated.[1] It was felt that the optimum spectrometric detector for liquid chromatography (LC) would monitor the entire UV–visible region throughout a chromatogram, thereby facilitating qualitative and quantitative measurements. The coupling of an oscillating mirror rapid scanning spectrometer (OMRSS) with a high-performance liquid chromatograph has proven quite successful in moving toward meeting this objective. The emphasis of this paper was on the actual experimental techniques used, the RSS/LC system, and the resulting chromatograms from a standpoint of hardware capability. This paper describes the computer system and supporting software required for practical application of this spectrometer as a routine detector.

The marriage of rapid scanning spectrometry and liquid chromatography led by necessity to computer interaction for two reasons: first of all, for on-line experimental control and storage of the massive amounts of data generated, and, secondly, for post-run data analysis and display. A description of the minimum capabilities needed for computer interaction in RSS/LC will be made, followed by an outline of what was added or refined out of necessity and convenience.

PRELIMINARY STUDY

Experimental. A Raytheon 704 minicomputer with 16K memory and 16 bit word length was used for all RSS/LC studies. The peripheral devices include the following: two direct memory access (DMA) Raytheon magnetic tape drive units, a cartridge disc unit for all driver routines and LC programs, a high-speed paper tape reader, a teletype or Tektronix T-4002 graphic computer terminal, a Tektronix 601 memory oscilloscope, an 8 channel multiplexed ADC and 4 channel DAC, and X–Y recorder display units for plotting

[†] Symposium on Computer Assisted Chemical Research Design—Joint United States Japan Cooperative Science Program, Washington, D.C., Aug 1976.

three-dimensional (3D) chromatograms.

A second generation Harrick rapid scan spectrometer (RSSB) was used for all of the preliminary studies. This instrument and the chromatographic system used were described previously.[1]

Data Acquisition. In the original system, the Raytheon 704 triggered a scan of the RSSB galvanometer mirror (GM) each second (while not the upper limit of repetition rate, this is more than sufficient for most LC applications). The computer merely caused a transistor to short the trigger input of the Harrick RSS signal processing module to ground using a +10-V pulse. Control of GM during the scan itself, however, was done by the Harrick processing module. The absorbance signal from this same module was then amplified 5× by standard circuitry and inputed to the ADC by the Raytheon. Initially a separate baseline spectrum was taken prior to each chromatogram. This greatly increases the number of runs necessary, and, to avoid this, the first 20–100 scans of a chromatogram (solvent background) were averaged and used as a baseline. This will obviously not work if the retention time of any compound is very small. The software routines were, therefore, altered so that an averaged baseline spectrum is taken only once for a particular solvent system and is then automatically subtracted from each spectrum. More will be said along these lines in later sections.

Data Reduction. The original software for data manipulation and display has been described previously.[2,3] The software for data reduction was limited to routines for averaging baseline spectra and subsequent subtraction before plotting spectra on a X–Y recorder.

Data Display. Examples of the first 3D liquid chromatograms can be found in Figures 2–5 of ref 1. Each of the 3D plots represent 3–5 h of work with the initial software. The procedure involved the use of two mag-tapes and several teletype commands. One mag-tape held the stored data while the averaged baseline was written on the second tape. This baseline was then subtracted from the data and stored in core ready for plotting. Usually 1000–1500 spectra were taken each run, and this procedure had to be repeated for each spectrum that was to be plotted.

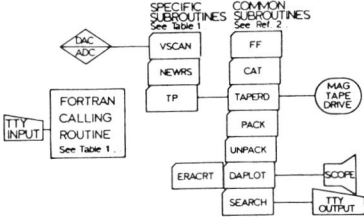

Figure 1. RSS/LC software block diagram.

The single wavelength plots shown in Figures 3–5 of ref 1 were an even more time consuming and difficult procedure. At the time, these were plotted out point-by-point by hand. There was also a considerable degree of uncertainty in these plots since peaks of lower retention times often obscured later peaks. For RSS/LC to become a routine analytical method rather than a novelty, it was obvious that the new software proposed in the preliminary paper had to become a reality.

REFINEMENTS OF SYSTEM

This section will be limited for the most part to software changes as the hardware modifications of the electronic and optical systems have also been quite extensive.[4] There has been two basic hardware changes worthy of mention at this point. First, the RSS is completely compatible with any commercial or component high-performance liquid chromatograph. Second, since we are undertaking types of chromatography other than ion exchange, we now employ an 8-μL flow cell rather than the original 87-μL cell. The larger cell was completely adequate for the initial ion exchange separations of nucleotide derivatives[1] since the detector volume was less than 10% of the LC peaks of interest in all cases.[6] A smaller volume cell has become necessary for the high-performance, reverse-phase separations we are doing on dyes, vitamins, and polyaromatic hydrocarbons. Table I lists and describes each of the new routines and subroutines in the RSS/LC software. How these interact with the overall system and each other is illustrated in Figure 1.

Data Acquisition. The changes made in the software designed for collection of data (VCAT and VSCAN of Table I) were made, first of all, out of necessity as a changeover was made to a specially modified first generation RSS (RSSA)[4] and, secondly, out of a desire to have the capability of a variable CATing routine. With the RSSA, the Fortran calling routine (FCR), VCAT, calls upon VSCAN which, not only signals the galvanometer to start a scan, but also steps it through a preset number of positions (number of data points, NPT) set by a teletype command. VSCAN accomplishes this task by simultaneously outputting a variable sawtooth (−10 to +10 V) wave form on the DAC and acquiring one channel of spectral data on the ADC and storing it on magnetic tape. This procedure is repeated until the designated number of scans has been made. Each scan cycle is stored as a data buffer and the total number of cycles requested as a data file. Unique to VCAT is the ability to signal average, for example, 10 spectra at 10 spectra/s and store this as one CATed spectrum, rather than the usual storage of one unaveraged spectrum/s. Any wavelength region from 200 to 930 nm and any wavelength range up to 730 nm can be achieved by setting a variable parameter (variable sawtooth). By setting yet another variable parameter (32 767 to 22 μs, time between points, TBP) along with the number of points, one can easily select the scan rate. The repetition rate is set by a combination of all of the above; NPT, TBP, the number of spectra CATed, and the number of spectra stored on mag-tape. For 1000 points of data, the maximum rep-rate is 12.8 spectra/s and the max-

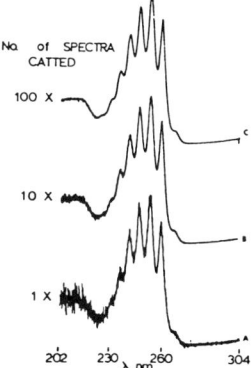

Figure 2. Benzene in hexane: effect of CATing. (A) 1 spectrum taken, VCAT not used; (B) 10 spectra, signal averaged; (C) 100 spectra, signal averaged.

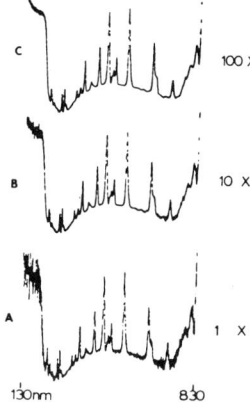

Figure 3. Ho$_2$O$_3$ in perchloric acid: wavelength standard. (A) 1 spectrum taken, VCAT not used; (B) 10 spectra, signal averaged; (C) 100 spectra, signal averaged.

imum scan rate is 45.5 spectra/s. At this time, these maxima are totally limited by the rate in which the data can be CATed and dumped onto mag-tape, not by the spectrometer itself.

The design of the new Fortran calling routine, VCAT, had several implications for our system. First, it takes more core space since it is a more complex routine and it uses double precision numbers. Second, the original triangular wave form was to be replaced by a sawtooth wave. Previously on the rising edge of the triangular wave, up to 500 data points could be taken, and another 500 points taken on the falling edge. The two resulting 500 point spectra are mirror images, but are not superimposable owing to hysteresis effects. Half of every spectra was unusable, and as a result, the resolution greatly suffered. The sawtooth has demonstrated its worth by thus increasing the resolution, and by simplifying the software used to retrieve data from a mag-tape. Finally, VCAT will enhance the signal-to-noise ratio and even make otherwise unusable data qualitatively and quantitatively meaningful. Especially note the shorter wavelengths of spectra in ref 1 and Figure 2A of this paper for examples of spectra done without this routine. Figure 2 generally demonstrates its utility on spectra of benzene in hexane. Figure 3 is a similar example which also illustrates the wide wavelength range obtainable thus far in RSS/LC only by this system.

Data Reduction and Display. Preceding each set of routine daily chromatographic runs, one first records spectra of air

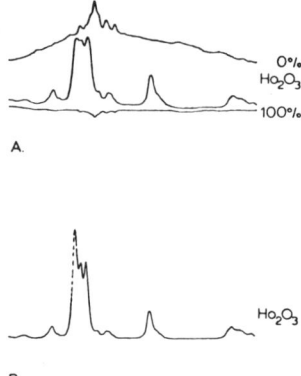

Figure 4. (A) Intensity spectrum of Ho$_2$O$_3$ filter (note distortion of spectrum at wavelengths where the Xe arc has strong emission peaks). (B) Absorbance spectrum illustrating automatic absorbance calibration routine (1 V = 1 au)

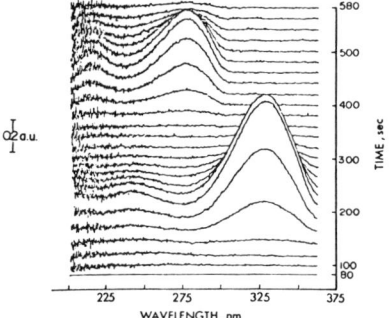

Figure 5. 3D chromatogram for separation of 4-thiouridine (λ_{max} 245 and 331 nm) and cytidine·0.5H$_2$SO$_4$ (λ_{max} 211 and 280 nm): Not signal averaged; Aminex A-4 cation-exchange resin, 10-cm column, 7-mm o.d. × 2.6 mm i.d.; solvent 1 M NH$_4$Cl/0.1 M HCl at 1.0 mL min^{-1}; repetition rate 1 spectrum/s; 940 data points/spectrum; baseline, 80 signal-averaged spectra.

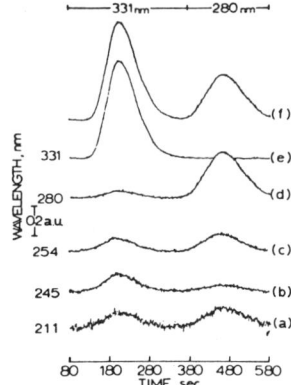

Figure 6. Single wavelength (A vs. t) chromatograms from the 3D plot in Figure 5; 500 data points/spectrum: (a) 211 nm, (b) 245 nm, (c) 254 nm, (d) 280 nm, (e) 331 nm, (f) combination 331 and 280 nm.

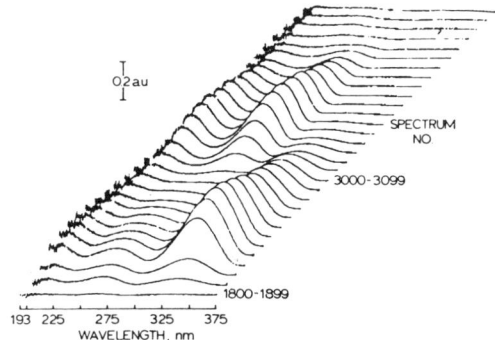

Figure 7. 3D chromatogram for separation of 4-thiouridine (λ_{max} 245 and 331 nm) and cytidine·0.5H$_2$SO$_4$ (λ_{max} 211 and 280 nm): Aminex A-4 cation-exchange resin; 4-cm column, 7 mm o.d. × 2.6 mm i.d.; solvent 1 M NH$_4$Cl/0.1 M HCl at 1.0 ml min^{-1}; repetition rate 10 spectra/s; 1000 data points/spectrum. A total of 100 spectra were signal averaged and baseline subtracted before plotting each spectrum; i.e., each spectrum represents 10 s of the chromatogram; baseline: 1200 signal-averaged spectra.

vs. air (or solvent vs. solvent) as 100% intensity (I) or the baseline. Then, blocking the sample PMT, 0% or the Xe arc background emission is recorded. A wavelength standard (Hg emission lines or holmium oxide solution) is then recorded, and finally, the data, which must necessarily lie between 0 and 100% I, is stored. This procedure allows later conversion from intensity to absorbance spectra, thereby eliminating the necessity for an absorbance standard (e.g., K$_2$CrO$_4$). For the details of this routine see ref 5. The routine, NUHS, with its subroutine, NEWRS, uses this method to rapidly skim through intensity tapes and temporarily convert the spectra to absorbance vs. wavelength, baseline corrected plots on a storage scope; see Figure 4. Once the operator is certain that the run is of value, the intensity spectra can be converted and permanently stored on a separate mag-tape by the routine TABS and its subroutine TP.

The two Fortran-calling routines that allow rapid plotting of the 3D chromatograms are SKIM (absorbance vs. wavelength) and MPIC (absorbance vs. time). The utility of SKIM can be seen in Figure 5, a 3D chromatogram of two nucleotide derivatives. As opposed to the extremely time-consuming original plotting software, this plot represents an investment of only 20–30 min. Figure 6 demonstrates the power of MPIC to plot any number or combination of single wavelength spectra (conventional chromatograms). Spectrum f shows how the plotting of a combination of optimum single wavelengths at

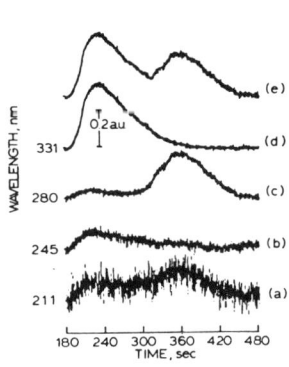

Figure 8. Single wavelength chromatograms of the raw 3D data.

the λ_{max} of each component enhances sensitivity in qualitative and quantitative measurements. As there are spectral differences between the two components, a degradation of chromatographic resolution (see Figure 7) could be tolerated

Table I. Routines in RSS/LC Software

INTENSITY TAPE:	SAMPLE(I_S)-REFERENCE(I_R)	
	FORTRAN(F)	ASSEMBLER(A)
ROUTINE		FUNCTION
R2DAM (F)	FORTRAN CALLING ROUTINE (F.C.R.) FOR PLOTTING A SINGLE λ FROM ABSORBANCE TAPES. ABS. VS. TIME.	
R2DAS (F)	F.C.R. FOR SKIMMING THROUGH AND PLOTTING ABS. TAPES. A VS. λ.	
NUHS (F)	F.C.R. FOR SKIMMING THROUGH INTENSITY TAPES (CHANGES TEMPORARILY TO ABS.). A VS. λ. FOR FAST VIEWING OF DATA JUST ACQUIRED TO DETERMINE VALUE OF RUN.	
NEWRS (F)	SUBROUTINE OF NUHS FOR CONVERTING I → A.	
MPIC (F)	F.C.R. FOR SKIMMING THROUGH AND PLOTTING SINGLE WAVELENGTHS FROM ABS. TAPES. A VS. T. CAN BE USED WITH I TAPES WHICH HAVE BEEN CORRECTED FOR BASELINE AND CONVERTED TO A TAPES. MULTI-RUNS WITH SINGLE BASELINE ENTRY POSSIBLE.	
R2DAI (F)	F.C.R. INTEGRATION ROUTINE FOR ABS. TAPES.	
TABS (F)	F.C.R. FOR CONVERTING I TAPES TAKEN DIRECTLY FROM THE RSS TO ABSORBANCE (BASELINE CORRECTED) TAPES.	
TP (A)	SUBROUTINE OF TABS FOR STORING CONVERTED I → A DATA ON MAG TAPES.	
VCAT (F)	DATA ACQUISITION: COMPUTER AVERAGED TRANSIENTS ROUTINE (F.C.R.) FOR INTENSITY DATA.	
VSCAN (A)	SUBROUTINE OF VCAT FOR OUTPUTTING A VARIABLE SAWTOOTH WAVE FORM ON THE DAC AND COLLECTING DATA ON THE ADC.	
SKIM (F)	R2DAS WITH BOTH BASELINE OPTIONS. (1) ADD BASELINE TO CORE. (2) 0.0 FOR BASELINE (BASELINE ALREADY SUBTRACTED FOR I → A CORRECTED SPECTRA).	
INTG (F)	R2DAI WITH BOTH BASELINE OPTIONS. ABS. TAPES OR CORRECTED I → A TAPES. MULTI-RUNS WITH SINGLE BASELINE ENTRY.	

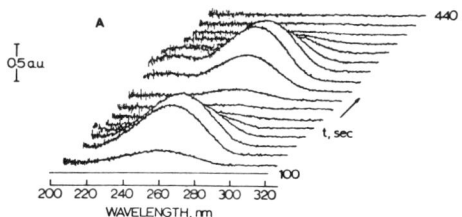

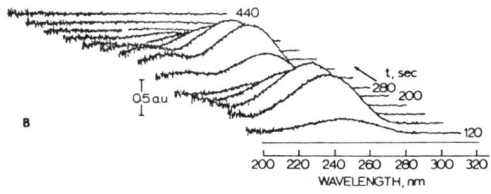

Figure 10. 3D chromatogram for separation of uracil (λ_{max} 260 nm) and xanthine (λ_{max} 231 and 260 nm): no signal averaged; Aminex A-4 cation-exchange resin; 10-cm column, 7 mm o.d. × 2.6 mm i.d.; solvent 1 M NH_4Cl/0.1 M HCl at 1.0 ml min^{-1}; repetition rate 1 spectrum/s; 940 data points/spectrum; baseline; 20 signal-averaged spectra. (A) Right 45° angle view; (B) left 27° angle view.

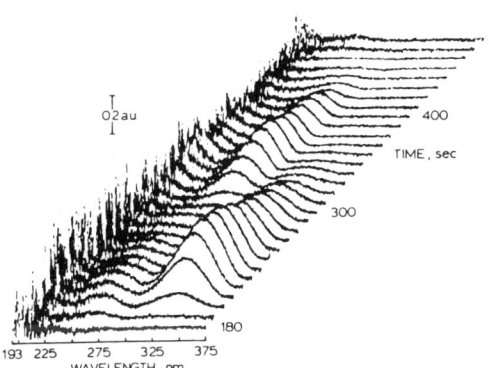

Figure 9. Same conditions as Figure 7 except spectra are plotted every 10 s without signal averaging.

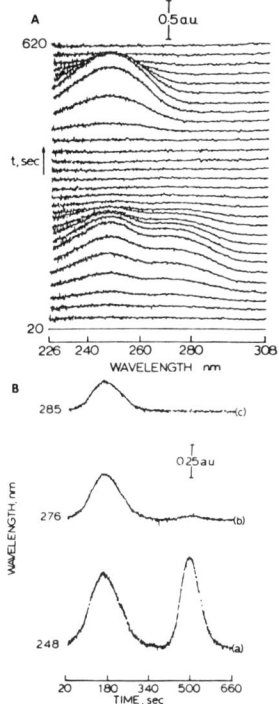

since useful information could still be obtained from single wavelength plots (Figure 8). The tailing found here is a result of overloading the 4-cm column. Figure 9 is the same chromatogram run under particularly noisy conditions and plotted without signal averaging. It should be obvious from this example that the signal-to-noise ratio (S/N) can be greatly enhanced by using this technique. Even the high-frequency noise at low wavelengths in Figure 7 could be eliminated by filtering the ... RSS signal prior to the ADC. Figures 10 and 11 demonstrate the fact that, even with little or no chromatographic resolution, a 3D chromatogram gives, in one run, all the information needed to determine which single wavelengths are useful for qualitative identification of the components even where there is only a single spectral difference. Figures 10A and 10B further demonstrate the possibility of plotting the chromatogram at any slope, which facilitates the viewing of peaks obscured by more major peaks at higher wavelengths. Finally, Figure 11B illustrates how one can

Figure 11. (A) 3D chromatogram for separation of guanine (λ_{max} 248 and 276 nm) and hypoxanthine (λ_{max} 248 nm): not signal averaged; Aminex A-4 cation-exchange resin; 10-cm column, 7 mm o.d. × 2.6 mm i.d.; solvent 1 M NH_4Cl/0.1 M HCl at 1.0 ml min^{-1}; repetition rate: 1 spectrum/s; 940 data points/spectrum; baseline, 20 signal-averaged spectra. (B) Single wavelength chromatograms taken from the 3D plot; ~640 data points/spectrum: (a) 248 nm, (b) 276 nm, and (c) 285 nm (hypoxanthine eliminated).

rapidly select the optimum wavelength to monitor a single component without the interference of another. This could have significant implications in many fields of analysis.

All of the software described above has been for qualitative determinations. INTG is a simple routine for doing quantitative measurements within these 3D chromatograms. The operator selects the wavelength and the time span of interest and the routine drops perpendicular lines to the baseline and integrates the peak.

CONCLUSIONS

The use of a minicomputer to control data acquisition, reduction, and display in RSS/LC has come a long way from being a difficult and time-consuming task to a relatively simple and routine one. Although additional software ideas exist to further simplify the technique, a working system is at hand, and some of the unlimited applications available can now be studied in detail.[4] Investigations presently underway include separations of vitamins, dyes, and polyaromatic hydrocarbons.

ACKNOWLEDGMENT

The authors gratefully acknowledge the financial support provided by National Science Foundation Grants GP-35979, CHE76-04321 (H.B.M.), CHE74-02641 (W.R.H), and a U.C. Graduate Council Scholarship and Twitchell Fellowship (M.S.D.).

REFERENCES AND NOTES

(1) M. S. Denton, T. P. DeAngelis, A. M. Yacynych, W. R. Heineman, and T. W. Gilbert, "Oscillating Mirror Rapid Scanning Ultraviolet-Visible Spectrometer as a Detector for Liquid Chromatography", *Anal. Chem.*, **48**, 20–4 (1976).

(2) A. M. Yacynych, Ph.D. Thesis, University of Cincinnati, Cincinnati, Ohio, 1975.

(3) H. B. Mark, Jr., R. M. Wilson, T. L. Miller, T. V. Atkinson, H. Wood, and A. M. Yacynych, "The On-Line Computer in New Problems in Spectroscopy: Applications to Rapid Scanning Spectroelectrochemical Experiments and Time Resolved Phosphorescence Studies" in "Information Chemistry; Computer Assisted Chemical Research Design", S. Fujiwara and H. B. Mark, Jr., Ed., University of Tokyo Press, Tokyo, 1975, pp 3–28.

(4) M. S. Denton and T. W. Gilbert, "Rapid Scanning Liquid Chromatography Detector: Instrumentation, Software and Applications", manuscript in preparation.

(5) A. M. Yacynych and H. B. Mark, Jr., "Automatic Absorbance Calibration Routine for a Computerized UV-Visible Rapid Scanning Spectrometer", *Chem. Instrum.*, in press.

(6) M. S. Denton and T. W. Gilbert, *J. Chromatogr.*, manuscript in preparation.

9

Copyright © 1976 by the American Chemical Society
Reprinted from *Anal. Chem.* 48:1368-1375 (1976)

Extraction of Mass Spectra Free of Background and Neighboring Component Contributions from Gas Chromatography/Mass Spectrometry Data

R. G. Dromey,[1] Mark J. Stefik, Thomas C. Rindfleisch,* and Alan M. Duffield[2]

Departments of Computer Science, Genetics, and Chemistry, Stanford University, Stanford, Calif. 94305

An effective, minicomputer-based method is described for systematically extracting resolved mass spectra of mixture components from GC/MS data. Using tabular peak models derived directly from the raw data, the spectra have column bleed background removed and are corrected for interference from neighboring elutants and peak saturation. Individual components are detected in the data by means of a pair of histograms which statistically characterize the positions of mass fragmentogram peak modes. These data-adaptive corrections avoid costly iterative numerical procedures and allow obtaining representative mass spectra from GC/MS data of complex mixtures on a routine basis. Using this approach, components that elute within less than two spectral scan times of each other can be detected and their mass spectra well resolved.

With the increasing application of gas chromatography/mass spectrometry (GC/MS) systems to mixture component identification in biomedical research (1, 2) and other areas (3), it has become important to be able to systematically isolate and identify minor components in the complex mixtures being analyzed. Because of instrumentation limitations, the mass

---

[1] Present address, Research School of Chemistry, Australian National University, Canberra, A.C.T., Australia.
[2] Present address, School of Physiology and Pharmacology, University of New South Wales, 2033, Australia.

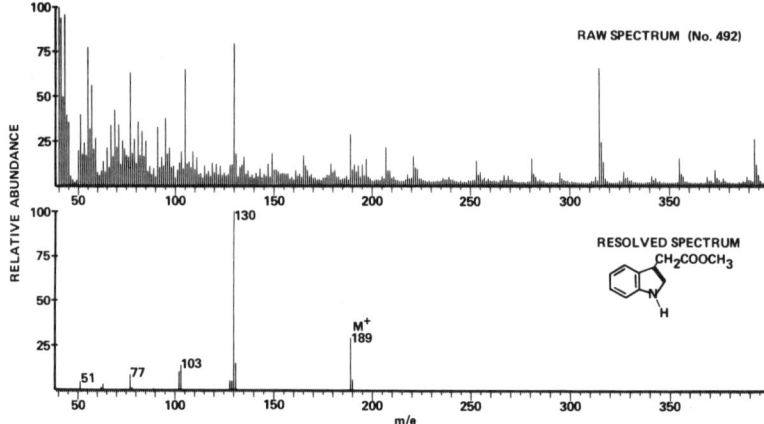

Figure 1. (a) Spectrum of indole acetic acid 3-methyl ester from a GC/MS analysis of human urine before processing. (b) Resolved spectrum of indole acetic acid 3-methyl ester

spectra obtained from a GC/MS analysis of a complex mixture are often markedly different from the spectra of the corresponding pure compounds. Differences may be caused by contributions from unresolved neighboring components during partial separation and also from GC septum and column bleed. These extraneous contributions may severely distort the relative intensities of ions in the mass spectrum of a particular component as well as contribute peaks that are not characteristic of the component being examined. Characterization and removal of these spurious ion contributions is especially important in the analysis of minor constituents where the mass spectra of interest may be substantially masked or distorted.

Our objective has been to implement a solution to these problems which is general and can systematically and reliably resolve GC/MS data with a minimum of human intervention. At the same time we have constrained the design so that the programs can run on a laboratory minicomputer. The first of these objectives has necessitated the use of a relatively complex mathematical treatment of the GC peak profile analyses as compared to that previously reported by Biller and Biemann (4). Both the present approach and that of Ref. 4 are based on analyses of mass fragmentogram profiles (4–6), a method which has been in use in various laboratories including our own, for a number of years. The method described here, however, differs substantially in the extraction of information from the profiles and thereby avoids several serious limitations inherent in the system described previously (4). By using tabular models of the elutant peak shapes together with a polynomial approximation to the GC background, and by deploying the elutant location and multiplicity information gained in analyzing individual fragmentogram profiles to assist in analyzing the others, we can achieve significant advantages in the quality of the reduced data. These include better final GC resolution, the proper assignment of ions to resolved elutant spectra (whether or not they are shared between neighboring components), more accurate spectral amplitudes free from background contributions, and the recovery of usable information from distorted data as in saturated peaks. We feel these improvements are important to a system which can reliably extract component spectra of sufficiently high quality from GC/MS runs to enable more definitive library matching, easier human interpretation of unknowns, and even the addition of extracted spectra to a library as authentic spectra. In our experience, these are essential assets for a GC/MS data system which is to be routinely applied in medical research and amply justify the complexity of the analysis.

EXPERIMENTAL

The GC/MS computer system used in this investigation consists of a Finnigan 1015 Quadrupole mass spectrometer interfaced to a PDP-11/20 minicomputer system for data acquisition. In one frequent mode of operation a complete mass scan (from mass 40 to 450) is completed each 3.7 s and 600 consecutive mass spectra are collected during a typical GC/MS analysis. Our initial experience of comparing the experimental mass spectra from a complex GC/MS analysis with a library of known mass spectra produced very poor results because of contamination of the experimental data by spectra of column bleed and of neighboring, unresolved components. Tolerable matches were only achieved when a component was present in large quantity in the GC/MS analysis. In order to overcome these problems, we have developed a computer program capable of systematically extracting from the raw GC/MS data, spectra representative of the pure elutant compounds.

The raw mass spectrum (Figure 1a) of indole acetic acid 3-methyl ester obtained from a GC/MS analysis of the acidic fraction (after methylation) of human urine typifies this situation. This component elutes at or near spectrum number 492 in the total ion plot (TIC) shown in Figure 2. Closer examination of Figure 2 shows that this component is submerged both in mass spectral contributions from neighboring components and background from GC column bleed. For comparison, a library spectrum (7) of indole acetic acid 3-methyl ester is shown in Figure 3. Figure 1b shows how, after processing the raw GC/MS data by the method described below, we can retrieve a high quality mass spectrum of indole acetic acid 3-methyl ester free from the environmental perturbations present in Figure 1a.

Thus, in the systematic analysis of GC/MS data the problem is first to detect where in the GC trace each component shows its maximum ion intensity and then to extract from these regions representative spectra for each of the detected components. The extracted mass spectra should be as free as possible from intensity distortions relative to their library counterparts, and from the presence of extraneous ions (e.g., peaks from either neighboring components or gas chromatographic column bleed).

DESCRIPTION OF METHOD

To obtain a reliable solution to these problems, it is necessary to analyze a number of spectra on either side of the ion current maximum for each elutant. A basic assumption of our approach is that the mass spectra of two neighboring unresolved elutants can be distinguished; that is, there exist some masses for which ions occur in the mass spectrum of one component but not in the other and vice versa. A schematic representation for two closely spaced elutants is given in Figure 4. By locating the "resolved" or singlet fragmentogram peaks at such masses (detected on the basis of profile mor-

1370 R. G. Dromey et al.

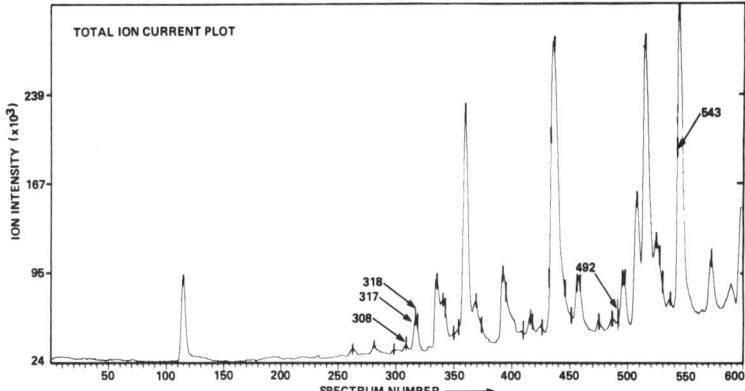

Figure 2. Total ion current plot for a GC/MS analysis of a urine sample. Components were found at vertical bar marks on TIC

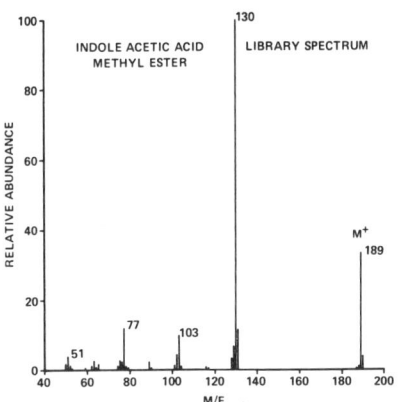

Figure 3. Mass spectrum of indole acetic acid 3-methyl ester taken from a library of biological compounds

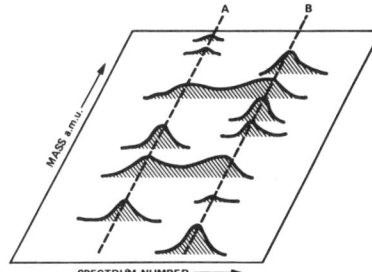

Figure 4. A schematic representation of the set of partial mass fragmentograms for two closely spaced elutants. Components A and B have some masses in common

phology), one can infer directly the positions of the elutants present and derive tabular models of the individual peak shapes. These models can be used subsequently to separate the unresolved fragmentogram complexes. The use of tabular peak models derived from the data itself accurately accommodates the a priori unknown peak profiles of particular elutants without solving for multiparameter, nonlinear model functions. Since the data are sampled often enough to satisfy the sampling theorem (8), these tabular models contain the necessary information to reconstruct a continuous peak envelope and can therefore be used as if they were continuous analytical models. For the typical peak shapes encountered, the collection of 5–10 mass spectra per singlet elutant peak represents a sampling frequency greater than twice the Fourier bandwidth of the peak. In addition, the mass by mass analysis of the fragmentogram peak complexes facilitates the mass dependent subtraction of background. (The large variation in background levels for different masses is a function of both the type of GC column used and the mixture being analyzed.)

By addressing the problem in this way, we have been able to produce accurate intensity information for the processed mass spectra and simultaneously distinguish with greater confidence which masses contribute to particular elutant spectra. We have been able to distinguish reliably elutants coming off within one and a half to two spectral scan times of each other. The succeeding sections discuss in more detail the procedures used to detect and resolve the mass spectra of unique elutants.

Detection of Elutants in GC/MS Data. Elutant detection involves finding the location of each mixture component in the GC/MS data, even if it does not have a corresponding peak maximum in the overall total ion current trace. Ideally for a given elutant, the fragmentograms for all its ion masses will show maxima at the same time and, in practice, this holds for well-resolved materials. However, for partially resolved mixtures, the complicating factors of peak overlap and background contributions can cause fragmentogram maxima for neighboring components to show significant variation in their positions on the time axis. Reliable position information for each elutant is best derived from the fragmentogram profiles containing singlet peaks for that elutant, that is, from fragmentograms at those spectral masses unique to the elutant relative to its neighbors.

The approach used for elutant detection is to compute two histograms of candidate singlet peak positions and to select as elutant locations significant histogram maxima. The first histogram measures the number of singlet mass fragmentogram profiles which reach maxima in each time interval. The second histogram measures the total singlet ion intensity above background at these maxima. These two types of histogram contribute complementary information for judging elutant locations. At a given elution time, the histograms include fragmentogram peak maxima from all masses over seven spectra. The position of each maximum is determined by a parabolic least squares interpolation about the top five points in the sampled peak data. If the intensities of the five points contributing to the maximum are Y_{-2}, Y_{-1}, Y_0, Y_1, and Y_2, then the expression for the time coordinate of the maximum is

$$t = \frac{7(2Y_{-2} + Y_{-1} - Y_1 - 2Y_2)}{10(2Y_{-2} - Y_{-1} - 2Y_0 - Y_1 + 2Y_2)}$$

The time coordinates of maxima are estimated to one third of the time to collect each spectrum in order to separate very close neighbors. Because we measure peak locations to one third of a spectral scan time, appropriate shifts are also included to account for the fact that higher masses are measured later in each spectral scan than lower masses. To build these histograms, the program examines the profiles of each mass fragmentogram in the data. Only peaks with intensities above a prescribed threshold are added into their appropriate time positions in the histograms. Peaks that are obvious multiplets (multiple extrema) are not incorporated into the histograms but are marked for later resolution. After all of the histogram information is collected for a given region, components are defined to be detected at locations where both the intensity and peak count histograms show maxima that are above a threshold. This statistical approach, looking for "clusters" of fragmentogram peaks in the histograms, does not depend upon a correct decision for each peak but rather on a preponderance of good decisions looking over all of the data. It will fail to resolve elutants very close together which do not have enough distinguishing mass spectral components as described above. In general, however, using this approach we are able to detect and resolve spectra reliably that elute with a separation in time as small as one and a half to two spectral scan times. (Two scan times correspond to 25% of a typical GC peak width at the scan rate we use.) Elutants this close often do not show multiple extrema in the fragmentogram profiles of masses common to both spectra and could not be separated properly except for this type of procedure. If two elutants are separated by less than 1.5 scan times, resolution becomes less certain depending on their relative concentrations and mass spectral distinctness.

Estimation of Spectral Intensities and Background for Well Resolved Elutants. Once the locations of elutants in the GC effluent have been determined, we proceed to compute a resolved spectrum for each material. To illustrate the principles involved in spectral amplitude and background estimation, we consider the simple case of an elutant that is well separated in time from its nearest neighbors. This analysis will be extended to the more complicated case of multiplet resolution in a later section. By "well-separated" we imply only that there are no maxima in the elutant detection histograms for three or four spectra on either side of the elutant under consideration. In such a situation, each of the mass fragmentogram profiles in the vicinity of the elutant will consist of a background on which is superimposed a peak with amplitude representative of the elutant spectral component at that mass. The background (contributed by both GC column bleed and possible tailing from nearby, high-concentration elutants) is distinguished from the elutant peak by the fact that it varies much more slowly with time. Reasonable estimates can be made by assuming that, for any particular mass fragmentogram, the background amplitude varies at most linearly with elution time in the vicinity of a given elutant. This approach to background determination, using the actual fragmentogram characteristics around each elutant, automatically tracks changes in the bleed levels observed during a run. It should be noted that our model is a first-order approximation subject to some error. A more accurate approximation would involve representing the background variations over a larger span of spectral scans than we are able to manage with the current program organization and computer memory limitations. We feel that the linear estimate is justified, however, in that it produces results within the error limits from other data uncertainties.

To complete the estimation process, we use a model peak to determine the contribution of each mass fragmentogram to the elutant spectrum. Much work has been done on the analytic approximation of gas chromatographic peak shapes (9, 10). Our experience has been that relatively simple models do not adequately approximate the range of shapes encountered and more complex models require large amounts of computing to determine model parameters. Noting that a separate model must be developed for each elutant and with a view toward obtaining the peak shape and definition necessary for multiplet resolution within reasonable computing resources, we have approached the problem by using tabular peak models taken from the data itself. Such models, defined at discrete sample points, can be evaluated at any required intermediate point by interpolation (since the sampling theorem is satisfied) and automatically reflect any peak asymmetries which may be present. For a given elutant, the model will be independent of mass, assuming that relative molecular fragmentation probabilities do not change with elutant pressure within the mass spectrometer. A number of criteria should be satisfied by the tabulated model peaks. They should be singlet peaks superimposed on as small a background as possible and they should be relatively intense in order to ensure a good signal-to-noise ratio and good definition of peak skirts.

Candidate singlet peaks may be distinguished from doublet or background peaks by the feature that they are relatively sharp. One way to measure peak "sharpness" is to use a logarithmic rate function defined as follows:

$$\text{rate} = \sum_{t=1}^{3} \left[\frac{(Y_{t-1} - Y_t)}{Y_t} + \frac{(Y_{-(t-1)} - Y_{-t})}{Y_{-t}} \right]$$

where the Y_t are evaluated at equal scan widths at each side of the mode of the peak. It can be seen that this rate will be large for peaks which are sharp and smaller for peaks which are broad. The rate as defined is also independent of amplitude for peaks of identical shape. A peak with a computed rate below a threshold appropriate to the experimental conditions is considered to be either an artifact of the gas chromatograph (background peak) or a multiplet and is not included in the detection histograms.

During the process of computing the detection histograms, a list is kept of the unimodal fragmentogram peaks having the highest rate factors in the region under analysis. When a component is detected in a given region, a model peak is then immediately in hand that can be used in the peak height estimation and background removal process. The local minima just on either side of the model peak are used as estimates of the local background (a straight line through the greatest of these minima is removed before the model peak is used for analysis). The selection of the peak with the highest rate factor as our model peak has worked well in producing models which are singlets and suffer least from interference by background and neighboring fragmentogram peaks.

Given the fragmentogram peak model for this case of a well-separated elutant, we can now correct the individual mass fragmentograms for background and estimate true mass spectral intensities for the elutant. For the fragmentograms exhibiting peak maxima "near" the location of this elutant (see below for detailed selection criteria), each peak in the set is quadratically interpolated to align it on a common time origin (this removes the time shift between collection of low and high mass data). This is done by fitting a parabola through successive groups of three points near the peak mode and interpolating to give four equally spaced points about the mode, separated by one spectral scan time. With the peaks in this standard form, they are ready for the least squares analysis below. Assuming a linear background model over the region of 5 to 10 scan intervals under consideration, the local background B_t at time t is approximated by

$$B_t \sim c + dt$$

where c is the background offset and d is its slope. The interpolated elutant peak model is normalized to unit area and has amplitudes P_t at times t. Then for a given mass fragmentogram, the amplitude of the actual fragmentogram profile Y_t at time t can be approximated by

$$Y_t \sim pP_t + (c + dt)$$

where p measures the elutant amplitude above background. Note that this model assumes a superposition principle based on the earlier assumption of constant relative fragmentation probabilities and a linear encoding of ion current information. If ion current data are obtained from nonlinear electronic systems or read from film, the peak model itself would be amplitude dependent and this linear analysis could not be applied until appropriate amplitude linearization corrections were made. From the above model, we can derive a least squares estimate for the elutant amplitude p and the background parameters c and d by minimizing the error function

$$E = \sum (Y_t - pP_t - c - dt)^2$$

according to the conditions

$$\frac{\partial E}{\partial p} = \frac{\partial E}{\partial c} = \frac{\partial E}{\partial d} = 0$$

The summation in the error function is over all available points in the peak profile as well as the neighboring background points within the window of scans contained in the computer's memory. These conditions yield three linear equations in the three parameters which can be solved by standard techniques (11). From the solution of these equations for the value of p, we get the spectral intensity for the mass under consideration. This analysis is applied to all mass fragmentograms with maxima near the elutant location to obtain the complete, intensity-corrected spectrum. It is worth noting that this method, using a tabular model peak derived from the data and elutant locations obtained from the detection histogram analysis, reduces the calculation for each mass spectrum intensity to the solution of a set of linear equations. Specifically, this avoids iterative methods for determining the parameters of a theoretical peak model and for determining elutant time positions.

Fragmentograms are selected for this analysis on the basis of several criteria. Given the nominal elutant position from the detection histogram analysis, a fragmentogram is excluded (mass spectrum assigned zero intensity) if it has no local peak maximum or if its maximum is displaced from the reference elutant position by more than two thirds of a spectral scan time on either side. Each fragmentogram peak meeting this test must also have an acceptably high rate factor, to be included in the analysis. For peaks of masses greater than 200 amu, we require a rate factor greater than 25% of the rate for the model peak. This restriction is useful for eliminating contributions caused by peaking in column bleed components. In carefully examining GC/MS data sets, we can observe that masses characteristic of the spectra of column bleed components show maxima in their mass fragmentograms just prior (one to two spectra) to the elution of an actual component. In essence, the component appears to "push" the bleed out ahead of itself. Because these peaks are formed by a different process than normal elutant mass fragmentogram peaks, they usually have a much broader shape. Consequently their rate factors will be significantly reduced and they can be eliminated by the rate threshold criterion. The combination of the fragmentogram peak location criterion together with the minimum rate criterion effectively discriminates against extraneous contributions to the intensity-corrected spectra without removing authentic mass peaks.

Extraction of Poorly Separated Elutant Spectra. Many instances arise in the analysis of GC/MS data where two or more elutants are poorly resolved by the gas chromatograph. The resulting mass spectra in such a region exhibit ion intensity distortions which reflect the interactions (overlap) between adjacent elutants in addition to the ion contributions of background. The extension of the above procedures to the general case is not difficult. Through the histogram detection and model procedures, one can extract normalized peak models $P, Q, R, \ldots$ for the various elutants present. Then with the assumption of a linear background, the elutant contributions to each fragmentogram profile Y can be estimated by minimizing the error function

$$E = \sum (Y_t - pP_t - qQ_t - rR_t - \ldots - c - dt)^2$$

with respect to the elutant amplitudes $p, q, r, \ldots$, and the background coefficients. Sets of linear equations result for each mass to extract the resolved spectra. In practice, we have not implemented this full procedure beyond the doublet case. Through the following approximations, reasonable results are achievable within available minicomputer resources. Using the histogram method described earlier, neighboring elutants are handled with a "look ahead" procedure. That is, information about an elutant that has just been detected is stored and the detection algorithm is applied to the data in the immediate neighborhood by extending the range over which the detection histograms are calculated. If by including this extended region an additional elutant is detected, we record the position of its mode, select a model peak for this second elutant using the rate criterion, and initiate a doublet resolver algorithm. At present, the extended histograms project four spectral scan widths beyond the position where the first elutant of the multiplet was detected (limited by computer memory). The same criteria are applied as in the singlet case to decide which fragmentogram peaks belong to the pair of detected components. The model used to process the composite fragmentogram peaks (many of which may be singlets belonging to either elutant) assumes that there are two overlapping peaks superimposed on a linear background. The doublet model represents an oversimplification of some situations as, for example, in the case where 3 components elute within a very brief interval. By applying it, however, to successive pairs of elutant peaks (taking first-order account of peak tail contributions from any earlier elutant), it provides acceptable accuracy and peak resolution effectiveness. As indicated above, a fit of the two peak models P and Q with a linear background to the fragmentogram profile Y may be described by the approximation

$$Y_t \sim pP_t + qQ_t + c + dt$$

Minimizing an error function analogous to the earlier singlet case results in 4 linear equations in the peak amplitudes p and q and the background parameters c and d. This set of equations again can be solved by standard methods.

In cases where peaks are actually singlet peaks, the solution should yield zero for the amplitude of the missing component. In practice, for such cases the amplitude of the second component is a very small positive or negative value which is representative of how well the model fits the data. Amplitude results for masses that belong to the second component of the doublet are stored temporarily until this component is moved into the processing window at which time they are incorporated into the analysis of the newly detected component.

Reconstruction of Saturated Peaks in Elutant Spectra. From a practical viewpoint, a fairly common occurrence in

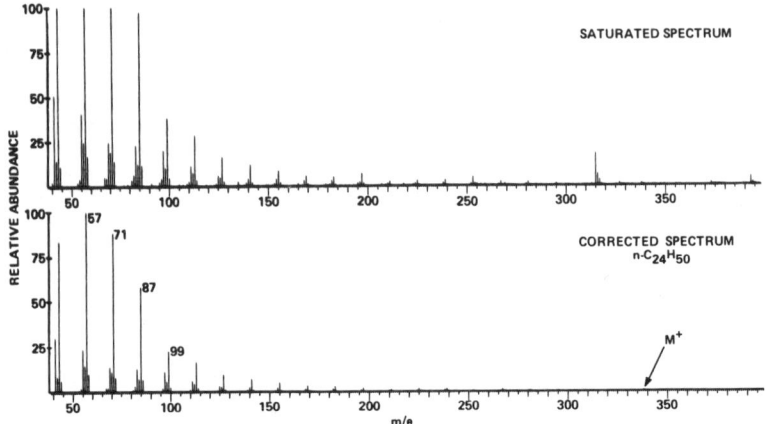

Figure 5. (a) Saturated unprocessed mass spectrum for tetracosane. (b) Spectrum of tetracosane after processing and correction for saturation

GC/MS data collection systems is the problem of mass peak saturation. Saturated peaks occur when the concentration of a component in the ion source is such that for one or more ion masses the detection system analog-to-digital converter becomes overloaded. Saturated peaks are easily detected because of their characteristic flat tops which have an amplitude determined by the overload limit of the detection system (e.g., the saturation value is 4095 for a twelve-bit analog-to-digital converter).

To obtain accurate amplitudes for component spectra that include saturated mass peaks, we must reconstruct these peaks to estimate their true amplitudes. A convenient way to do this in the singlet case is to use the least squares model that we derived in the preceding sections. To actually apply it for reconstruction of saturated mass fragmentogram peaks, we need to make a small modification to the equations. Instead of summing over all the points in the peak, we sum over only those points that are not saturated in the fragmentogram. As an estimate of the peak mode, we use the mode of the intensity histogram for the component being analyzed. An example of reconstruction of a mass spectrum with saturated ion intensities is given in Figure 5. Figure 5a shows a saturated spectrum of tetracosane (spectrum number 545 in Figure 2) and Figure 5b is the corresponding reconstructed spectrum. It is clear that the reconstructed spectrum will give a far better match with a library spectrum than the saturated spectrum which is badly saturated at masses 43, 57, and 71.

Before leaving the discussion of saturation, we should point out that we have not in practice extended the procedure for saturation correction of singlet peaks to the doublet case as we believe that it would be inadequate for reliable intensity estimates. If too many points are overloaded, there will be insufficient data to accurately estimate the amplitude of each multiplet component. Despite such correction algorithms, there is no substitute for the collection of good quality raw data at the start.

RESULTS AND DISCUSSION

The program based on the algorithm outlined in the preceding sections has been tested on a wide variety of biological samples. It fits comfortably into a DEC PDP 11/45 computer (with 28K words of memory) and takes approximately 8 min to analyze a raw GC/MS data set of 600 mass spectra (scanned from masses 40 to 450). Much of this time is spent in reading the raw data from the disk and other input–output operations. Copies of the program, which is written in FORTRAN, are available from the authors. Currently, this program forms part of an automated analysis system for the GC/MS analysis of urine and blood samples. The program reduces the raw GC/MS data set of approximately 600 spectra to a set of about 60 resolved elutant spectra which are then matched against a library of mass spectra of biological compounds. This whole process takes about 20 min and produces an analysis of the sample, with known compounds in the mixture identified and the remaining unknown set marked for further study by chemists or other DENDRAL programs (12–14).

In evaluating performance of the program, a major issue is how well it is able to detect elutants in the data. The vertical bars on the TIC (Figure 2) indicate all the places where the program detected and isolated a component from the raw GC/MS data. The program's power of detection is illustrated for example by the elutant detected near spectrum number 492 in the total ion current plot shown in Figure 2. Although there is no evidence of a maximum in the TIC in the region near 492, the program was able to detect and isolate a good quality spectrum of indole acetic acid methyl ester (Figure 1). In the raw data, this spectrum is clearly submerged in background and overlapping contributions. A comparison of the resolved spectrum (Figure 1b) with a library spectrum (Figure 3) shows that the basic spectral intensity profiles are very similar even including the very low intensity ion of mass 89. Some very small ions (of intensity less than 5% relative abundance) are absent from the resolved spectrum because they have been lost in the background noise. It is worth noting that there are no peaks present in the resolved spectrum that are not in the library spectrum, that is, the extraneous mass spectral peaks in the raw data including peaks at masses 105, 253, and 315 are not included in the resolved spectrum. The relative intensities of the mass spectral peaks at masses 51, 62, 65, and 77 have been changed significantly from their levels in the raw data. This illustrates the importance of correcting the intensities for background. The mass spectral peaks at masses 51, 52, 63, 78, and 129 appear to maximize near spectrum 496 in the raw data instead of spectrum 492 because of the overlapping contributions of a poorly resolved elutant. Similar examples of the power of this technique exist in other parts of the GC profile in Figure 2.

The detectability of unresolved elutants is clearly a function of their amplitude relative to neighboring components and background. One way to characterize this is to measure the ratio of the total ion intensity (sum of the mass spectrum amplitudes) in the resolved spectrum compared to that in the unprocessed spectrum including background and overlap effects. The mass spectrum of the processed component at

1374 R. G. Dromey et al.

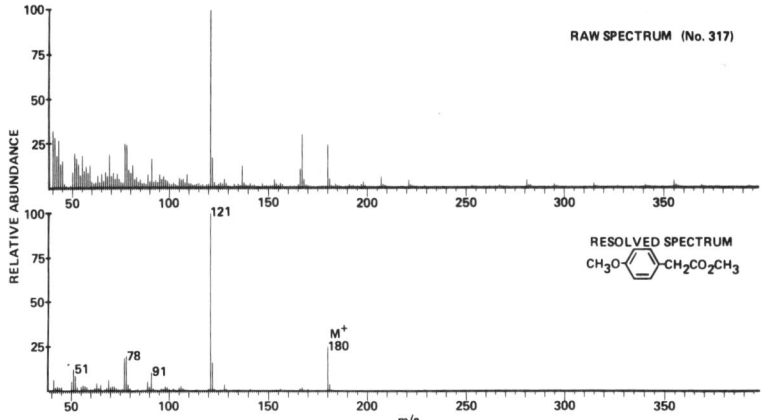

Figure 6. (a) Mass spectrum of 4-methoxyphenylacetic acid methyl ester before processing. (b) Resolved mass spectrum of 4-methoxyphenylacetic acid methyl ester

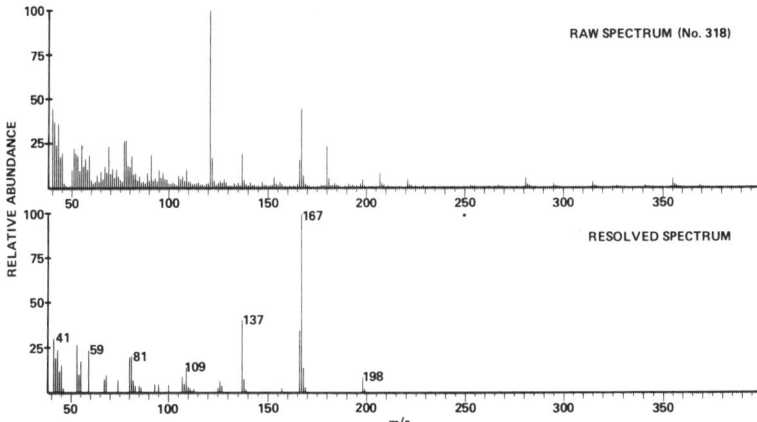

Figure 7. (a) Spectrum of an unknown aromatic ester before processing. (b) Resolved spectrum of unknown ester in Figure 7a

spectrum number 492 comprises only 4% of the total raw ion current. It can be expected that there will be problems detecting components with an ion current ratio that falls much below a level of 4%. Also if two compounds elute within less than 1.5 to 2 spectral scan times of one another, there is an increasing chance that the program will make the wrong decision as to whether there is one or actually two elutants present. Such errors are dependent on the ion current ratio between adjacent elutants, the similarity of their mass spectra, and the stability with which peak positions can be determined.

As an example of doublet resolution, consider the region near spectrum numbers 317 and 318 in Figure 2. The program detects that there are two elutants present and Figures 6 and 7 illustrate the raw and resolved spectra at these locations. The spectrum in Figure 6b is a good representation for 4-methoxyphenylacetic acid methyl ester. The other component is an unknown aromatic ester.

We have evaluated the efficiency of background removal for singlet elutants by examining their mass fragmentograms. After calculating the least squares peak and background levels, we concluded that the computed results are consistent (within 5–10%) with human estimates. They tended to be less accurate for very weak peaks whose shapes were more sensitive to noise distortions.

For the multiplet case, where the peak profiles can be considerably more complex, there is a stronger possibility that the model will not produce accurate amplitude information. In such cases, as when there are three rather than two elutants present, there is a danger that background contributions will be incorrectly estimated particularly with the limited number of scans that can be held in our minicomputer memory at one time. We feel however that use of a more complex model for triplets is not likely to be able to guarantee much greater precision. A sequential application of the doublet model has produced acceptable results in our experience with the program. Problems most frequently occur when a small amount of an elutant occurs just prior to, or just after, an elutant of high concentration. The intensities of peaks in the small elutant that are common to the large elutant tend to be less accurately calculated than singlet peaks and sometimes may even be discarded as negligible if their intensity relative to the large peak falls much below 10%. This may be especially important for the molecular ions of compounds with the same molecular weight which would be expected to elute near each other.

Comparison with library mass spectra has indicated that correction of intensities for saturated singlet peaks is satisfactory. However, as expected, the accuracy of the calculation decreases as peaks become more heavily saturated. In our case, we are working with model peaks that extend over nine points (i.e., nine scan widths). If more than four of a peak's nine points are saturated, we can expect that its estimated intensity will have only limited accuracy because there is insufficient information left to accurately characterize its shape.

Conditions arise in the raw GC/MS data for which it is not possible to extract resolved mass spectra unambiguously. One case is when the elutant-to-background ratio falls significantly below 5%. In these cases, the very weak intensity ions, including isotope ions, usually do not appear in the resolved mass spectra. The other difficulties arise when it is not possible to detect the presence of multiple elutants because they occur within less than one mass spectrum scan time of each other. In this case, the processed spectrum represents the mixture of the two elutants.

In general, we have found that the present system works very well and is capable of detecting and isolating high quality representative mass spectra in GC/MS experiments involving complex biological mixtures.

ACKNOWLEDGMENT

We thank D. Smith, W. Pereira, and W. Yeager who have contributed in a major way to the continued refinement of the computer programs implementing these algorithms and to the critique of results from their operational use in our laboratory. We also acknowledge the work of B. E. Blaisdell of Juniata College, Huntingdon, Pa., on an exploratory alternative approach to the present problem.

LITERATURE CITED

(1) R. A. Hites and K. Biemann, *Anal. Chem.*, **42**, 855 (1970).
(2) C. C. Sweeley, N. D. Young, J. F. Holland, and S. C. Gates, *J. Chromatogr.*, **99**, 507 (1974).
(3) W. H. McFadden, "Techniques of Combined Gas Chromatography/Mass Spectrometry: Applications in Organic Analysis", Wiley Interscience, London, 1973.
(4) J. E. Biller and K. Biemann, *Anal. Lett.*, **7**, 515 (1974).
(5) R. E. Summons, W. E. Pereira, W. E. Reynolds, T. C. Rindfleisch, and A. M. Duffield, *Anal. Chem.*, **46**, 582 (1974).
(6) R. N. Stillwell, 22nd Annual ASMS Conference—Mass Spectrometry, Philadelphia, Pa., 1974, p 454.
(7) "Mass Spectra of Compounds of Biological Interest", *U.S. At. Energy Comm. Rep.*, No. TID-26553, S. P. Markey, W. G. Urban, and S. P. Levine, Ed.
(8) See for example: R. S. Ledley, "Digital Computer and Control Engineering", McGraw-Hill Book Co., New York, 1960, p 742.
(9) E. Grushka, M. N. Myers, and J. C. Giddings, *Anal. Chem.*, **42**, 21 (1970).
(10) C. D. Scott, D. C. Chilcote, and W. W. Pitt, *Clin. Chem.* (Winston-Salem, N.C.), **16**, 637 (1970).
(11) See for example: S. D. Conte and C. de Boor, "Elementary Numerical Analysis: An Algorithmic Approach", McGraw-Hill Book Co., New York, 1972, p 241 and following.
(12) A. Buchs, A. B. Delfino, A. M. Duffield, C. Djerassi, B. G. Buchanan, E. A. Feigenbaum, and J. Lederberg, *Helv. Chim. Acta*, **53**, 1394 (1970).
(13) R. E. Carhart, D. H. Smith, H. Brown, and C. Djerassi, *J. Am. Chem. Soc.*, **97**, 5755 (1975).
(14) R. G. Dromey, B. G. Buchanan, D. H. Smith, J. Lederberg, and C. Djerassi, *J. Org. Chem.*, **40**, 770 (1975).

RECEIVED for review September 15, 1975. Accepted April 30, 1976. This work was supported by grants (Nos. RR-612 and GM-20832) from the National Institutes of Health and (No. NGR-05-020-632) from the National Aeronautics and Space Administration.

10

Copyright © 1978 by the American Chemical Society

Reprinted from *Anal. Chem.* **50**:433-441 (1978)

Automated Simultaneous Qualitative and Quantitative Analysis of Complex Organic Mixtures with a Gas Chromatography–Mass Spectrometry–Computer System

S. C. Gates,[1] M. J. Smisko,[2] C. L. Ashendel,[3] N. D. Young,[4] J. F. Holland, and C. C. Sweeley*

Department of Biochemistry, Michigan State University, East Lansing, Michigan 48824

A system has been developed which uses retention indices in performing an off-line reverse library search of selected mass chromatograms from the repetitive scanning GC-MS analysis of complex mixtures. More than 100 components in a typical mixture of organic acids from urine are automatically identified and quantitated at a rate of one compound each 6 s. Typical of the analytical results obtained in this study are an observed precision of retention index determination of 0.2%, a lower limit of detection of 10 ng injected, a GC-MS precision of 8% upon duplicate determinations of the same sample, and a 1000-fold linear range of quantitative analysis.

Analysis of multicomponent organic mixtures is of interest

[1] Present address, Department of Chemistry, University of Michigan, Ann Arbor, Mich. 48104.
[2] Present address, Aeroquip Corp. 300 S.E. Ave. Jackson, Mich. 49203.
[3] Present address, McArdle Laboratory, University of Wisconsin, Madison, Wis. 53715.
[4] Present address, Upjohn Company, Kalamazoo, Mich. 49001.

in a number of specialities within the field of analytical chemistry. Classically, analytical procedures have been designed for the analysis of one or a very small number of compounds in these mixtures, but a more recent trend has been to develop a means of measuring the complete "profile", or analyte pattern. The utility of such systems is particularly apparent to forensic, atmospheric, and clinical chemists, all of whom occasionally analyze multicomponent mixtures when it is not known in advance what compounds will be of most importance.

A common feature of most such profiles is the relative lack of qualitative variation from sample to sample, despite the considerable complexity of the samples. Thus, for example, the clinical chemist analyzing urine samples encounters virtually the same set of compounds in each sample, with relatively minor variations. The two features of interest, then, are quantitative differences in one profile compared to a profile from another source or the same source at a different time, and the appearance of one or more highly unusual constituents in a particular sample. The task of analyzing the profile is therefore greatly simplified because a massive library search of comparison features (spectra or other physical properties) need not be undertaken to identify every compound in each such sample. A "local" library of features that are particularly pertinent to the type of sample being analyzed can thus be maintained and referenced conveniently on a conventional laboratory minicomputer system.

A typical approach in the determination of profiles has been a chromatographic separation of the components, followed by spectroscopic analysis of the separated sample. When 100 or more components are present in a given sample or fraction, the method of choice has most frequently been gas chromatographic separation followed by mass spectrometric identification of individual GC peaks (1–4). Some use has also been made of liquid chromatographic separations to obtain profiles (5, 6), but such efforts have been hampered by the lack of a detection system capable of uniquely distinguishing a wide variety of closely eluting substances in a liquid medium.

Two problems associated with profile analysis by gas chromatography–mass spectrometry (GC-MS), however, have been the lack of a means of identifying substances that are incompletely resolved, and an inability to quantitate the large number of minor components typically present in biological mixtures. Traditional library search techniques, even on small data bases, are poorly suited to the identification of individual components when a mass spectrum may contain contributions from as many as five or more compounds. Furthermore, selected ion monitoring, usually the method of choice for quantitative analysis by GC-MS, cannot be used for analyses of more than a dozen or so compounds, at least with presently available instruments.

Hence, most published GC and GC-MS profiles of biological fluids have not identified or quantitated more than the major components in the mixture (1–4, 7, 8). An approach taken by some laboratories to circumvent this problem has been to group compounds together and quantitate clusters of unresolved compounds by GC peak area (7, 8). A more satisfactory solution has been the use of capillary GC columns (9–11); these have been especially effective when used with suitable data processing techniques to allow compound identification (12). However, even capillary columns do not fully resolve all components, nor do they provide completely unambiguous identification of substances without the use of a mass spectrometer.

Hence, the goal of this laboratory has been to develop a low resolution GC–low resolution MS–computer system which would provide automated quantitative and qualitative analysis of 100 or more components in a complex organic mixture. The system that was developed to meet this goal is described in this paper.

EXPERIMENTAL

Urine Separation. Organic acids are separated from human urine by a modified version (13) of the procedure of Thompson and Markey (7). In brief, this procedure consists of precipitation of polybasic inorganic salts from 1 to 2 mL urine with barium hydroxide, oxime formation with hydroxylamine hydrochloride (J. T. Baker Chemical Co.), and separation of the neutralized urine on a column of DEAE-Sephadex (Pharmacia) in the acetate form. After an aqueous wash of the column (50 mL), acids are eluted with 40 mL of 1.5 M pyridinium acetate and the eluate is lyophilized to dryness. The residues are treated with 250 μL of bis(trimethylsilyl)trifluoroacetamide (BSTFA), containing 1% trimethylchlorosilane (TMCS) (Pierce), in dry, redistilled pyridine (4:1, v/v). Trimethylsilylation is carried out at 80 °C for 1 h, after which the samples are stored in sealed, silanized glass capillaries at 4 °C.

Gas Chromatography. Gas chromatography is performed on a Varian 2100 GC equipped with dual flame ionization detectors and Varian A-25 recorders. Aliquots (2 μL) of the trimethylsilylation reaction mixture are chromatographed on 12 ft by 2 mm i.d. glass columns containing 5% OV-17 on 80/100 mesh Supelcoport (Supelco). Conditions of analysis include injector and detector temperatures of 300 °C, amplifier gain of 10^{-10} A/V, attenuation at 2, and recorders at 1-mV full scale. The column oven is temperature programmed from 60 to 290 °C at 4°/min, with no initial isothermal period. Gases used are helium as carrier at 40 mL/min, hydrogen at 30 mL/min and air at approximately 300 mL/min.

Mass Spectrometry. Mass spectral data are obtained on an LKB-9000 gas chromatograph–mass spectrometer (LKB Produktur) with a Digital Equipment PDP 8/e-based data system (14). The gas chromatograph of the LKB contains a 10 ft by 2 mm i.d. coiled glass column packed with 5% OV-17 on 80/100 mesh Supelcoport. During normal operation, the column is temperature programmed from 50 to 260 °C at 4°/min; the sample (2–8 μL) of the silylated organic acids fraction (described above) is injected when the column reaches 60 °C and data collection begins approximately 7 min later. Other conditions are: ion source temperature, 290 °C; GC injection port, 150 °C; gain 8 on the multiplier; scans at constant 4-s intervals at scan speed 8 over the range m/z 49 to m/z 550; accelerating voltage, 3.5 kV; trap current, 65 μA; filament current, approximately 4A; and ionizing voltage, 70 eV. Calibration of nominal mass against perfluorokerosene masses and checking of system noise levels are performed once each day. A test sample of urinary organic acids is also injected at the beginning of each day to check the performance of the entire GC–MS–computer system.

When the system has met the test specifications, the GC column of the LKB is pre-treated by two injections of the BSTFA–TMCS silylating mixture and the column cooled to room temperature. An aliquot (0.5 μL) of a mixture of 8 straight-chain hydrocarbons (with 10, 11, 12, 14, 16, 18, 20, and 24 carbon atoms) in hexane is withdrawn into a 10-μL syringe, followed by a 0.5-μL air "spacer" and 2 to 8 μL of the derivatized urine sample. The sample capillary is discarded immediately, even if sample remains. The sample is analyzed on the LKB under the above conditions, after which the run is validated by brief manual inspection of a few mass chromatograms, and the data are then transferred to a PDP 11/40 computer (Digital Equipment) for subsequent processing. During the transfer, which takes 6 to 10 min, the data are converted to the standard mass spectral data format used on the PDP 11/40 in this laboratory (15). The PDP 11/40 system consists of a 16-bit, 56 000-word core memory minicomputer with two 1.2-million word removable disks, a 7-track magnetic tape drive, DECwriter, Tektronix 4010 scope display unit, and a Tektronix 4610 hard copy unit. All programs on the PDP 11/40 are designed to be used with the Digital Equipment time-sharing system, RSX-11D (Version 6B). Programs are written in assembly language, Fortran IV, or a mixture of both. MSSMET (described later) occupies approximately 8100 words of core memory in the PDP 11/40, exclusive of a 12 000-word library of general purpose system and Fortran subroutines.

ANALYSIS OF GC–MS DATA

Once the GC–MS data are collected and transferred, they can be processed at any later time by the mass spectral metabolite program (MSSMET), copies and details of which are available from the authors. This program is used to convert GC–MS data to analyte identities and relative or absolute concentrations. In brief, it does this by using a reverse library search of individual ion profiles (mass chromatograms (16)) in the region of the expected gas chromatographic retention index (17) of each compound of interest. If a small set of 2 to 8 pre-selected differentiating ions are all found to apex at the same location and in the proper ratio, as specified by the library entry for that compound, the substance is considered positively identified. Each compound is quantitated by calculating the ratio of an ion peak area and peak height of the compound relative to the ion peak area and peak height of a quantitative internal standard. This search of m/z intensities is performed automatically by the computer for each library entry and the results are printed or stored for further statistical analysis.

Program operation is flow-charted in Figure 1; details are published elsewhere (18). In general, MSSMET is designed to

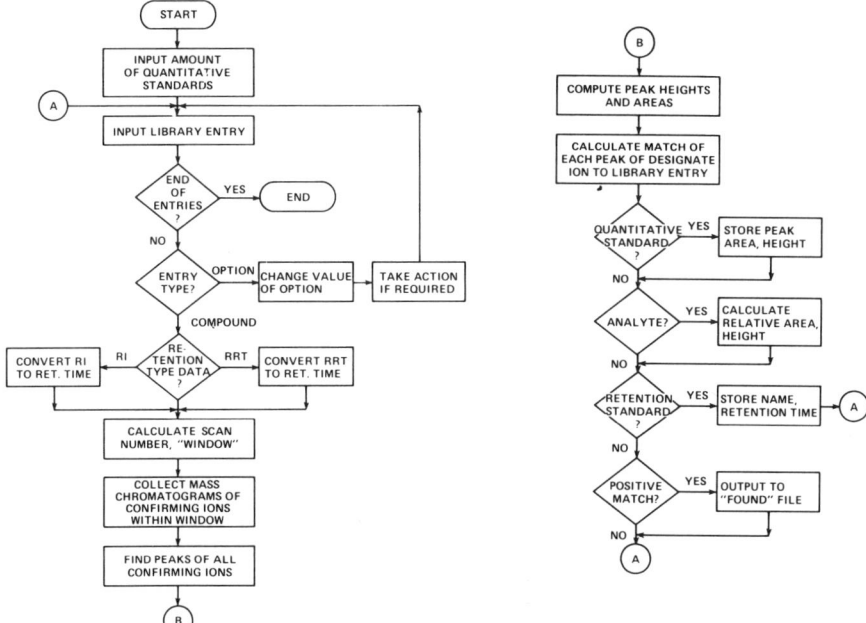

Figure 1. MSSMET flowchart

be completely automatic; hence, all but the initial set of commands are entered from a library file which resides permanently on the disk and which is read sequentially by the computer. All entries in the library are of one of two forms: either a command to change the value of a particular program variable ("option") or a set of information about a particular compound. Each of these types of entries may also be entered manually from the teletype during the program execution, if desired.

A typical analysis by MSSMET begins with the operator specifying the names of the two disk files containing the library entries and the mass spectral data. In addition, the operator is prompted to provide information about the amount of internal standard added to the original sample, the volume of sample (urine) extracted, and the amount of any other normalizing factor to be used in quantitation (e.g., creatinine concentration). Once these data are entered, the first entry in the library file is read. Usually, the first several entries provide the initial numerical values of the 34 program options. The information on the first compound in the library is then read in the library file. This four-line entry includes the compound's identifying number and name, if known; its nominal (expected) retention time; the m/z of the "designate" ion and a quantitation factor, and the set of "confirming" ions and their expected relative intensities. The designate ion is that ion which is expected, based on previous experience, to be most differentiating of the compound at its anticipated elution position. The ratio of the area of the designate ion of the compound to the area of the designate ion of the internal standard is converted to actual concentration using the quantitation factor and the following formula:

$$C = k \frac{A_x}{A_{is}} \frac{W}{V} \qquad (1)$$

where C is concentration in mg/mL, A is the area of the designate ion, W is the weight (mg) of the internal standard added to the sample, and V is the volume (mL) of sample extracted.

The quantitation factor, k, is determined experimentally utilizing a pure reference compound. The accuracy of concentration (C) obtained from Equation 1 depends on the reproducibility of the designate ion intensity, relative to total ionization. The confirming ion set includes those ions that must peak at the same time as the designate ion, each paired with its intensity relative to the other confirming ions. By definition, the designate ion must be included in the set of confirming ions.

Once a library entry for a compound is read into core memory, the computer calculates an expected retention time for the substance. Library retention data may be expressed as retention time in minutes and seconds, relative retention time, or retention index; regardless of the form, it is converted to the scan number in the GC-MS data. A time "window" is then calculated within which the substance is expected to elute. The width of the window is based on the value of a library option; it is typically 120 to 200 s wide in our system, and is centered at the expected retention time. Mass chromatograms are collected of all of the confirming ions within the window, and the computer finds any peaks of the individual ions.

After the mass chromatogram peaks are located, an area and height are calculated for each. The ratios of the areas (or heights) of all of the confirming ions are calculated and compared to the library ratios using a slightly modified version of a formula by Grotch (19):

$$MC = \left\{ 1 - \left[\frac{\sum_{j=1}^{N} |I_j^F - I_j^D|}{\sum_{j=1}^{N} (I_j^F + I_j^D)} \right] \right\} 100 \qquad (2)$$

where MC = match coefficient, N = number of confirming ions, I_j^F = intensity of jth ion in library entry, I_j^D = intensity of jth ion in data.

The "match coefficient" calculated by this method is used

as one measure of the match between the library entry and the spectra from the GC-MS analysis. The other measure is the deviation of the retention index (or other measure of retention behavior) observed experimentally from that predicted by the library entry. The match coefficient and the retention index deviation must both be within certain limits, set by library options, for a positive identification ("+" match category). Marginal matches between library and sample data are noted ("?" match category), as are negative matches ("−" match category), and a file containing only compounds considered to be positive matches is established; it is this "found" file which is generally retained for manual or statistical evaluation. Both types of files contain the name of the compound; the match coefficient and the match category (+, ?, or −); peak area and height; uncorrected relative concentration; retention time; retention index; deviation in time and retention index units of the compound from its nominal value; and the scan numbers where the peak was found. Match coefficients and relative concentrations are computed from both peak heights and peak areas. Any or all of this information may be suppressed using the appropriate value of the "print" option.

Peak and Baseline Determination. Two critical parts of the program deal with the detection and quantitation of peaks in the individual mass chromatograms. Peak detection consists of a series of decisions based upon the critical parameter values provided by the library. The program consecutively detects three regions of each peak. Initially, the intensity at each succeeding scan within the window is examined until a pre-set number of increasing points, or a critical slope value, is surpassed. Then, in the second region, the program examines points until a maximum has been detected and the peak intensity falls below a predetermined threshold, which may be either a fixed intensity value or, more usually, a fraction of the height of the peak. In the last region, each point is examined to see if the intensity has begun increasing again; if it has, or if the end of the window is encountered, the peak is considered ended.

When all peaks of the confirming ions within the window have been located, a second pass through the data is used to select up to 20 points for a baseline. All data points within the window are examined as potential candidates for the baseline. They are discarded from consideration if they meet any of the following criteria: occurrence before the beginning of the first peak; occurrence after the end of the last peak; occurrence within the boundaries of a peak, except for the first and last points of the peak; occurrence at a point common to two unresolved peaks, unless that point is lower than the previous baseline point; occurrence at the first point of the window, unless lower than the next baseline point; or occurrence at the last point of the window, unless lower than the previous baseline point.

The baseline points selected by this process are used to determine an unweighted least-squares fit to an nth order curve, where n is set by the library to a value between 1 and 5, inclusive. A complete baseline is then interpolated from these data and the baseline subtracted areas of all peaks are determined. Unresolved peak areas are divided at the minimum between the two peaks. No further attempt is made to deconvolute the peaks.

Retention Index Determination. Although gas chromatographic retention behavior may be expressed in terms of elapsed time, relative retention time, or retention index, the last measure has been most commonly used with MSSMET. Two similar methods for computing retention indices have been tried. The first relies exclusively on the retention data of the straight-chain hydrocarbons co-injected with the sample. In this method, one of the hydrocarbons is located using an estimated retention time and a very wide search window. All other hydrocarbons are located using estimated retention times relative to the first hydrocarbon located. Retention indices of components of the mixture are then calculated by linear interpolation between the appropriate pair of flanking hydrocarbons.

The second approach is to locate a hydrocarbon standard by estimated retention time, then locate two neighboring sample components (not hydrocarbons) by relative retention time. These two are then used as retention index standards to locate, by linear extrapolation, several other sample components, which are, in turn, added to the list of retention index standards. Retention indices of other compounds are then calculated by linear interpolation between flanking retention index standards, exactly as when using hydrocarbon standards.

Selection of Library. The library is built using the standard text editor on the PDP 11/40. Each library entry is based on studies of the mass spectra of reference compounds, or mass spectra from the type of sample to be analyzed. Retention indices are determined empirically and periodically updated as more samples are analyzed.

Two methods have been used to select designate and confirming ion sets for use in the library. One of these is strictly intuitive, based on knowledge of the general types of spectra involved. The second, and more recent, approach is to use an algorithm (MSSDSG) which compares the complete library spectrum to spectra taken from a sample of the type to be analyzed. The key feature of this comparison is that the library spectrum is compared to spectra taken from the region of the sample where it would be expected to occur, based on the retention index of the library spectrum. Thus, for example, a mass spectrum of the trimethylsilyl derivative of lactic acid might be compared to 16 mass spectra taken during GC separation of a urine extract in the region around the nominal retention index of the TMSi derivative of lactic acid (1101 on 5% OV-17). Based on this comparison, ratios are computed which compare each ion of the normalized library spectrum to the corresponding ion in the normalized sum of the sample spectra:

$$R_m = \frac{L_m{}^q}{S_m} \qquad (3)$$

where R_m is the ratio for ion mass m, L_m is the normalized intensity of ion of mass m in the library spectrum, S_m is the normalized intensity of the ion mass m in the summed sample spectra, and q is a factor (typically 1.05) used to weight intense ions more heavily.

The ratios are calculated for as many as 10 different samples and the ratios computed for each sample are then ranked by another program (MSSCHS) and summed. The highest set of ranked ratios are selected and the ion with the highest ratio is chosen to be the designate ion. Up to 7 other ions are selected to complete the confirming ion set. Only one ion from each isotope cluster is chosen.

Once library entries have been selected by either method, they are tested against several samples. Using the "debug" option to examine the actual ratios of the confirming ions, the library entry is modified until it adequately functions in finding the library compound in several biological samples. The library determined by this method for organic acids in human urine is available from the authors and is included as an appendix in Ref. 18.

RESULTS AND CONCLUSIONS

Baseline and Peak Area Determination. The performance of MSSMET was evaluated in part by comparing the results obtained by this method with data calculated from manually determined baselines and areas. As illustrated in

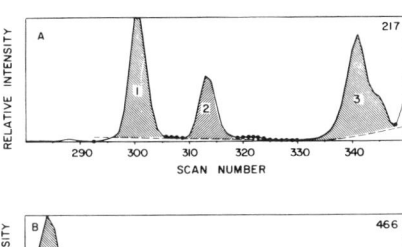
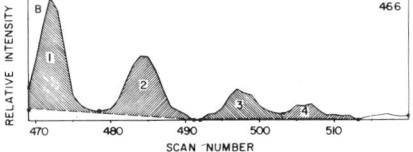
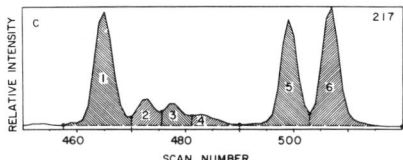

Figure 2. Peak and baseline determination by MSSMET. Each black circle represents a baseline point which would be selected by MSSMET. The dotted line represents a second-order least squares fit of the baseline points

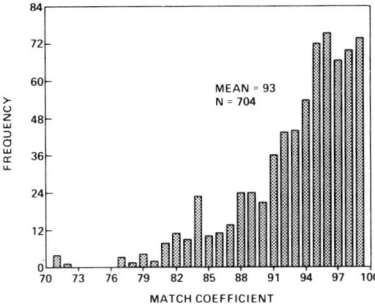

Figure 3. Distribution of individual match coefficients. Peak height and peak area match coefficients were calculated for compounds found in 4 urinary organic acid samples by MSSMET

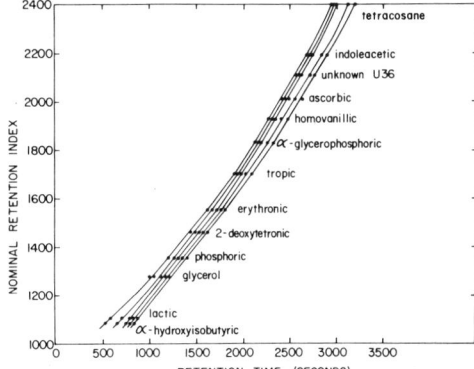

Figure 4. Variability of retention times of retention index standards. Retention times were found by MSSMET for each of the retention index standards in 5 samples of urinary organic acids analyzed on the LKB-9000 over a 1-year period

Figure 2, the MSSMET algorithms are capable of resolving complex peaks, detecting changing or nonlinear baselines, and integrating even small peak areas. Unlike other peak resolving algorithms (20), there is no limit to the number of components which can be resolved in a cluster of peaks, as long as the confirming ions of each compound are at least partially resolved from one another. The determination of baseline can use up to a fifth-order least-squares fit, but a second-order fit is generally most satisfactory for a window width of 30 scans (120 s). The typical mass chromatogram peak of a designate ion detected by MSSMET is 15 scans, although peaks from 4 to 28 scans are routinely detected within the 30-scan window.

Match Coefficient. Previous studies by us with pure compounds have indicated that the match coefficient is sensitive and reliable down to the limit of detection of the designate ion (21). Since the limit of detection varies with the substance and the intensity of the ion chosen, as well as with the sensitivity of the GC-MS at the time of analysis, no definite detection limit can be established. However, for most samples, about 10 ng must be injected before match coefficients above 80 are observed. When program options are set so that either the peak height or the peak area match coefficient must exceed 80 to permit a positive match, the 157 compounds found in a series of urine samples had a mean match coefficient of approximately 93, as shown in Figure 3. Experience with the system has shown that retention behavior and peak areas for compounds with mean match coefficients below 86 are unreliable; these low match coefficients are observed with substances that are present in very small amounts, and with substances for which an optimal library entry has not yet been obtained.

Retention Index. As noted by Nau and Biemann (22) and independently by our work with metabolic profiling (21, 23–25), retention indices are an extremely precise measure of compound identity, despite the fact that GC-MS data points are recorded for each mass only once every scan (4 s in our system). Pure compounds are found with the most precision; the mean standard deviation for multiple determinations of several pure compounds was 2.20 ($n = 156$). Compounds in a highly complex biological sample such as urine are found somewhat less precisely (mean standard deviation was 2.79 for 652 determinations). If substances in the sample are used instead of hydrocarbons as retention index standards, precision improves considerably (mean standard deviation was 2.38 for 784 determinations). Retention times and relative retention times are less precise than retention indices. As shown in Figure 4, retention times, even when they are reasonably reproducible over short time periods, can vary markedly over longer periods of time (1 year in Figure 4), with some variations as well in the shape of the retention time vs. nominal retention index curves.

Quantitative Analysis. MSSMET has been tested on a variety of pure samples to evaluate quantitative precision and accuracy (24, 25). Subsequent studies on urine samples have confirmed that the repetitive scanning technique provides linear results over approximately a 1000-fold range. In complex mixtures such as the urinary organic acid fraction, the technique therefore yields reliable quantitative data even for substances representing 0.01% or less of the dry weight of the total mixture. The reproducibility of relative peak area determinations with the organic acids fraction from urine samples is illustrated in Figure 5. MSSMET was used to quantitate the relative peak areas of 106 urinary organic acids found in two injections of the same sample. Individual data are plotted in order of increasing retention index. The sample marked with triangles was analyzed on the GC-MS one week after the sample marked with circles; data from both samples are plotted as the percent each is of the mean value of the

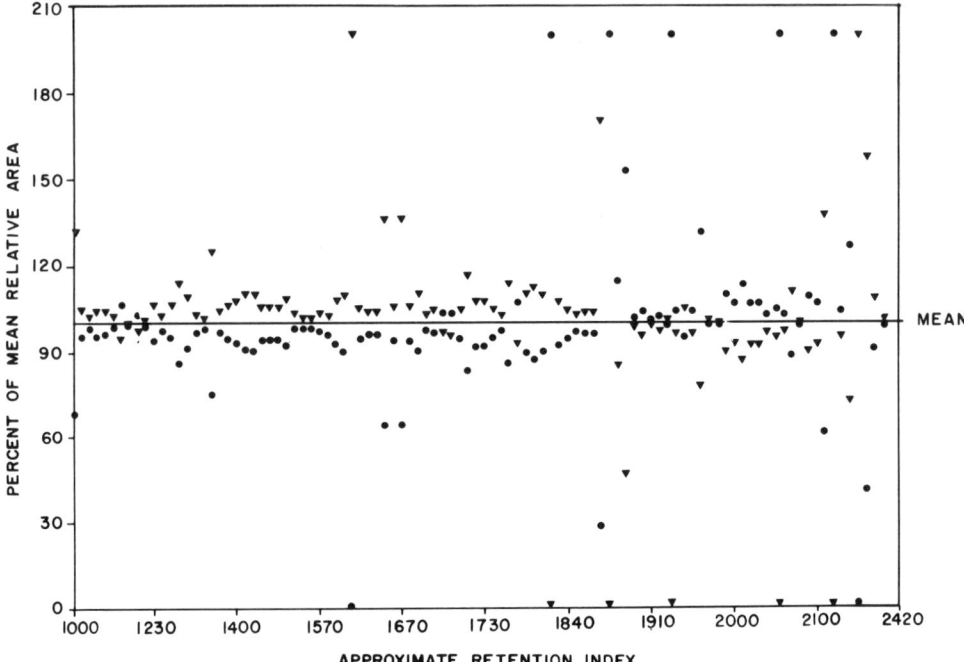

Figure 5. Reproducibility of repetitive scanning GC-MS on urine samples

two samples. Of the 14 compounds with a relative standard deviation greater than 35%, one is an artifact peak, 2 are substances just above their limits of detection, 3 have a retention index more than 6 retention index units from the library value (and 2 of these also have one of their match coefficients below 80), and 4 show evidence that the designate ion peak is poorly resolved. The problems with the remaining four substances are not explainable on the basis of the data contained in the MSSMET outputs. Analysis of the data suggests that use of multiple quantitative standards would help improve quantitative precision.

Analysis of Urine Samples. Routine MSSMET analysis of the acidic fraction of urine samples has indicated that approximately 100 ± 30 components are reliably detected and quantitated by MSSMET with a version of the library that does not contain some 40 unidentified substances for which library data are presently being obtained. As shown in Figure 6, many of the substances are incompletely resolved by the GC. The analysis is therefore dependent on the use of differentiating ion mass chromatograms. It should be noted that many of the substances in Figure 6 are compounds for which pure standards are not yet available; these substances are treated no differently by MSSMET. Provisional identities in some cases are based on spectra in the literature (26, 27).

DISCUSSION

MSSMET is a reliable method for both quantitative and qualitative analysis of complex mixtures by GC-MS techniques. Once a suitable library has been assembled and tested, MSSMET operates rapidly and requires little or no decision-making by the analyst. Typically, with the 157-compound library, less than 6 s is required for the computer to identify and quantitate each compound. Excluding the time required to set up files and locate the retention index and quantitation standards, the time required is approximately 4 s per compound. Most of this time is required to transfer data from the disk to core memory.

Occasionally, one or more of the retention time standards is not found properly because it deviates too far from its expected retention time. In these cases, it is a simple matter for the operator to change the expected retention time and have the computer reprocess the library entry for the standard. Generally, once the retention standard is properly located, no further difficulties are encountered. After the quantitative internal standard is located, no further operator intervention is required. About 4 min per sample is required at the teletype for overall supervision of the process.

Reverse Library Search Using Retention Indices. The use of retention indices and a reverse library search with a small set of differentiating ions has resulted in a much higher degree of precision of identification than most conventional library searches that are based solely on forward searches of mass spectral data. Retention indices often provide more precise information about compound identity than do mass spectral fragmentation patterns, especially when dealing with large numbers of relatively similar mixtures. In our system, retention index precision is such that the compound can be assumed to occur within a region of 4 to 5 scans (about 9 RI units) on either side of its nominal retention index; hence, the use of the retention index narrows the search for a compound to a region representing less than 2% of the total GC-MS run. Even including the somewhat larger "window" needed to integrate the peak area adequately, less than 4% of the spectra are examined to find a particular compound.

Abramson has pointed out (28) that reverse search procedures, even without using retention indices, avoid the problem of supression of important ions from one compound by the presence of ions from a large, overlapping compound. McLafferty has also advocated the use of the reverse search method (29) and has provided a means for selecting appropriate ions to use in such searches (30). Nau and Biemann originally developed a method for assigning retention indices to GC-MS data (22), which they used to help interpret results from GC-MS analyses (31). Reimendal and Sjovall (32) used

Analysis of Complex Organic Mixtures • 439

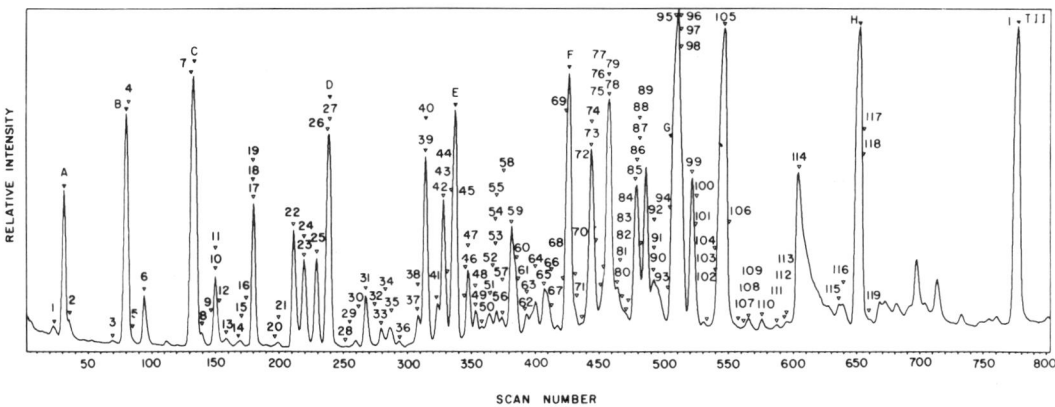

Figure 6. Urinary acids identified by MSSMET in urine sample from a "healthy" adult subject. Substances labeled A through I are straight-chain hydrocarbons with 10, 11, 12, 14, 16, 18, 20, 24, and 28 carbon atoms per molecule, respectively. The other substances are: (1) U-1, (2) U-50 (3) α-hydroxyisobutyric, (4) lactic, (5) U-2, (6) glycolic, (7) β-hydroxybutyric, (8) U-4 (pyruvic oxime), (9) U-79, (10) U-5 (cresol), (11) oxalic, (12) U-6, (13) U-51, (14) U-7, (15) glycerol, (16) levulinic, (17) malonic, (18) methylmalonic, (19) U-RA 183, (20) U-OXB1, (21) U-9 (2-methylglyceric), (22) phosphoric, (23) U-10 (deoxyerythronic), (24) benzoic, (25) U-11 (deoxythreonic), (26) succinic, (27) fumaric, (28) phenylacetic, (29) nicotinic, (30) U-54 (deoxytetronic), (31) U-14 (deoxytetronic), (32) U-56 (deoxythreonic), (33) U-57 (threonolactone), (34) glutaric, (35) 3,3-dimethylglutaric, (36) citramalic, (37) malic, (38) U-58 (3-methyl glutaconic-peak 1), (39) U-16 (erythronic), (40) U-80 (3-methylglutaconic-peak 2), (41) U-59 (threonolactone), (42) U-17 (threonic), (43) mandelic, (44) adipic, (45) 3-methyladipic, (46) o-hydroxybenzoic, (47) U-60, (48) α-hydroxyglutaric, (49) U-61, (50) β-hydroxy-β-methylglutaric, (51) U-21, (52) U-82, (53) m-hydroxybenzoic, (54) pyroglutamic, (55) U-83 (hydroxymethylfuroic), (56) U-22, (57) o-hydroxyphenylacetic, (58) U-84, (59) tropic (internal standard), (60) arabonolactone, (61) α-ketoglutaric oxime, (62) p-hydroxybenzoic, (63) m-hydroxyphenylacetic, (64) U-24, (65) p-hydroxyphenylacetic, (66) ribonolactone, (67) arabonic, (68) suberic, (69) β-glycerophosphoric, (70) U-64, (71) U-87, (72) U-65, (73) α-glycerophosphoric, (74) U-26, (75) cis-aconitic, (76) U-66, (77) U-67, (78) U-68, (79) citric, (80) azelaic, (81) terephthalic, (82) vanillic, (83) U-89, (84) U-29, (85) homovanillic, (86) galactono-1,4-lactone, (87) p-hydroxyphenylbydracrylic, (88) veratric, (89) U-30, (90) o-coumaric, (91) hexuronic, (92) gluconic, (93) p-hydroxyphenyllactic, (94) U-72, (95) vanilmandelic, (96) ascorbic, (97) U-91, (98) hexuronic, (99) hexuronic, (100) hydrocaffeic, (101) U-74, (102) U-2071, (103) palmitic, (104) U-75, (105) hippuric, (106) caffeic-peak 1, (107) U-76 (hydroxydecanedioic), (108) U-77, (109) U-37, (110) indoleacetic, (111) U-NE8, (112) caffeic-peak 2, (113) urocanic, (114) uric, (115) U-41, (116) m-hydroxyhippuric, (117) U-42, (118) 3,4,5-trimethoxycinnamic, (119) 5-hydroxyindoleacetic

mass chromatograms in a semi-automated reverse search procedure which later included some quantitative analysis (*33*). However, it has been the combination of retention indices and reverse library search, developed in this laboratory (*21, 23–25*) and later used elsewhere (*20, 34*), which has proved to be the most precise means of identifying multiple components in complex mixtures.

Retention indices also allow a considerable extension of the type of technique used by McLafferty (*30*) to select appropriate ions and weighting factors for computer decisions about whether a spectrum from a particular sample represents a sufficient match to a given library spectrum. As shown in Figure 7, the intensities of ions at a given area of the GC-MS run may vary markedly from the overall distribution of ions in the sample analyzed. This variation frequently can be observed even on a scan-to-scan basis, so that the quality of the reverse library search is improved considerably by knowing, in advance, the approximate set of "interfering" substances which may be present at the same set of scans as the compound of interest. Thus, selection procedures such as MSSDSG and MSSCHS can be used with retention indices to identify the region where the compound is expected to elute in any sample; the reference spectrum can then be compared to the spectra of exactly the milieu in which it must elute, if it exists in the sample.

Baseline and Peak Area Determination. Originally, the peak area within a mass chromatogram was determined by integrating the designate ion intensity from the starting point of a peak to its end, with the baseline determined by linear interpolation between the two points. Comparison to manually chosen integration limits and baseline values indicated that this approach was inaccurate in many cases. Hence, a series of increasingly sophisticated algorithms was tested, each one being modified as anomalous results were detected.

The principal limitation to accurate peak area determination appears to be the narrow width of the window. Most of the special provisions within the baseline determination algorithm are a consequence of the difficulty in processing peaks which either begin prior to the start of the window or finish after the end of the window. Frequently, a baseline value must be extrapolated for such peaks, as shown in Figure 2. Fortunately, if the window is properly centered, the peak of interest occurs at the center of the window, and integration is much more accurate.

It is important to note that the proper choice of designate and confirming ions plays a critical role in peak area determination. If these ions are chosen so that they are the most well-resolved ions for that compound, the integration will be correspondingly accurate. The need for well-resolved mass chromatograms peaks is not unique to MSSMET; the Stanford group's "cleanup" procedure (*20*) is equally dependent upon the detection of such peaks. It may be possible to devise methods that are capable of functioning without the need for well-resolved ion sets; one such method has been developed in this laboratory and will be published elsewhere (*35*).

In our experience, it is almost always possible to find at least 2 ions of reasonable intensity that are well resolved (unique for a particular compound). Cases such as the one illustrated in Figure 8, where a number of compounds elute at almost the same time, are extremely common; qualitative and quantitative analysis are thus both dependent on the correct selection of designate ions. Practically the only exceptions are those in which two isomeric substances elute very near each other. In these cases, confirming ions of one substance may be partially obscured by the ions of the other. Depending upon the degree of overlap of the two substances, MSSMET either groups the two substances together (if overlap is severe) or computes separate areas for each.

Quantitation. In addition to providing rapid qualitative

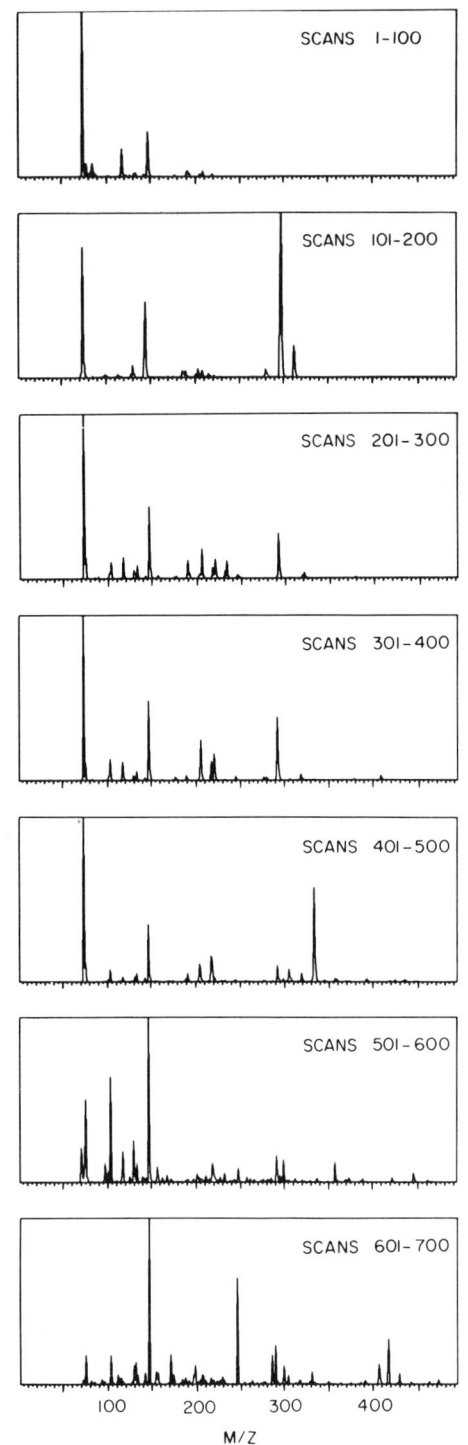

Figure 7. Dependence of ion distribution upon the region of the GC-MS analysis of trimethylsilyl organic acids of human urine. Ion intensities have been summed for selected scans within the scan range marked on each plot

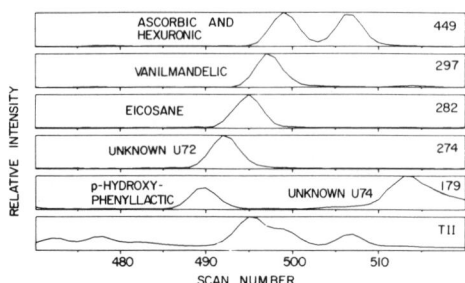

Figure 8. Identification of closely-eluting substances by MSSMET

analyses, MSSMET routinely calculates the areas of designate ions, relative to an internal standard designate ion, of each of the 100 or more substances present in a typical sample of organic acids in urine. Since k factors have not yet been determined for most of the library compounds, a value of 1.00 is usually used for k in Equation 1. Hence, the values reported are uncorrected for percents of ionization or recovery during the extraction procedure. The uncorrected concentrations suffice for most profiling purposes, but inter-laboratory or inter-method comparisons of data would require absolute values.

An alternate method has been described by Smith et al. (36) for qualitative and quantitative analysis of complex mixtures. It utilizes techniques similar to MSSMET, in that a reverse library search is made with specific regions of mass chromatograms, and identifications are based on both gas chromatographic retention indices and mass spectral patterns. However, this technique differs from MSSMET in that all of the ions belonging to a single compound are used for the quantitation, after spectra are "cleaned up" of contributions from background and other interfering peaks. Quantitation is not corrected for incomplete recoveries in the pre-purification process or for differences in total ion intensities among compounds, since it is relative to an external standard.

Direct comparison of the results obtained by Smith et al. (36) with data obtained by MSSMET is not possible, since different sample extraction techniques were used, and because accuracy, precision, sensitivity, and linear range of the method of Smith et al. have not yet been reported (36). However, their system can resolve doublets but not multiple overlapped mixtures of substances, whereas MSSMET has successfully and routinely handled peak envelopes containing contributions from 5 or more substances (Figure 8). Since multiply overlapped peaks can be quite common in some complex mixtures (Figure 6), this can be a significant problem. To our knowledge, no other automated system is currently capable of resolving all of the components shown in Figure 6 by GC-MS. In addition, methods that base quantitation upon an estimate of total ion intensity (called "areal total ion current" in Ref. 36) are very dependent upon accurate resolution of the most intense ions; however, in the case of trimethylsilyl derivatives, the most intense ions are frequently those produced by the derivatizing group, and hence mass chromatograms of these ions are usually the least well-resolved by the GC. Dependence upon approximate methods for resolving such ions may result in significant inaccuracies during quantitation. In contrast, MSSMET uses designate ions which are selected specifically because they are well-resolved and unique; hence, quantitation is based almost exclusively on ions for which there is minimal inaccuracy. If several ions for a given compound are well-resolved, precision (not accuracy) may be improved by using all such ions during quantitation; however, the more complex the mixture, the smaller the number of well-resolved ions will be.

In general, the results obtained with MSSMET over a period of more than two years lead us to believe that it has a very good potential for metabolic profiling studies. While our experience has been limited to analyses of the organic acids and steroid fractions of human urine, it appears that MSSMET can easily be adapted to a variety of other fractions and sample types. We have begun to use MSSMET "found" files as the basis for statistical analysis of urine samples (13) and to develop other methods for increasing the size of the MSSMET library (35, 37). Preliminary results on these projects suggest that at least 140, and perhaps more, organic acids can be monitored routinely in urine, and that the challenge for users of MSSMET and related systems will be the interpretation of the wealth of new data provided by these techniques.

ACKNOWLEDGMENT

The authors gratefully acknowledge the technical assistance of J. Harten and N. Dendramis, and D. Byrne for typing the manuscript.

LITERATURE CITED

(1) E. C. Horning and M. G. Horning, J. Chromatogr. Sci., 9, 129 (1971).
(2) F. Hutterer, J. Roboz, L. Sarkozi, A. Ruhig, and R. Bacchin, Clin. Chem. (Winston-Salem, N.C.), 17, 789 (1971).
(3) E. Jellum, O. Stokke, and L. Eldjarn, Scand. J. Clin. Lab. Invest., 27, 273 (1971).
(4) J. C. Crawhall, O. Mamer, S. Tjoa, and J. C. Claveau, Clin. Chim. Acta, 34, 47 (1971).
(5) D. S. Young, Am. J. Clin. Pathol., 53, 803 (1970).
(6) W. W. Pitt, Jr., C. D. Scott, W. F. Johnson, and G. Jones, Jr., Clin. Chem. (Winston-Salem, N.C.), 16, 637 (1970).
(7) J. A. Thompson and S. P. Markey, Anal. Chem., 47, 1313 (1975).
(8) A. M. Lawson, R. A. Chalmers, and R. W. E. Watts, Clin. Chem. (Winston-Salem, N.C), 22, 1283 (1976).
(9) B. J. Kimble, R. E. Cox, R. V. McPherron, R. W. Olsen, E. Roitman, F. C. Walls, and A. L. Burlingame, J. Chromatogr. Sci., 12, 647 (1974).
(10) W. Bertsch, R. A. Chang, and A. Zlatkis, J. Chromatogr. Sci., 12, 175 (1974).
(11) J. A. Luyten and G. A. F. M. Rutten, J. Chromatogr., 91, 393 (1974).
(12) A. B. Robinson and L. Pauling, Clin. Chem., (Winston-Salem, N.C.), 20, 961 (1974).
(13) S. C. Gates, N. Dendramis, and C. C. Sweeley, unpublished results.
(14) N. D. Young, J. F. Holland, and C. C. Sweeley, unpublished results.
(15) C. Ashendel, N. Young, J. F. Holland, and C. C. Sweeley, unpublished results.
(16) R. A. Hites and K. Biemann, Anal. Chem., 42, 855 (1970).
(17) E. Kovats, Helv. Chim. Acta, 41, 1915 (1958).
(18) S. C. Gates, "Automated Metablic Profiling of Organic Acids in Human Urine by Gas Chromatography-Mass Spectrometry", Ph.D. Dissertation, Michigan State University, East Lansing, Mich., 1977.
(19) S. L. Grotch, Anal. Chem., 45, 2 (1973).
(20) R. G. Dromey, M. J. Stefik, T. C. Rindfleisch, and A. M. Duffield, Anal. Chem., 48, 1368 (1976).
(21) S. C. Gates, N. D. Young, J. F. Holland, and C. C. Sweeley, "Advances in Mass Spectrometry in Biochemistry and Medicine, Vol. II", A. Frigerio, Ed., Spectrum Publications, New York, N.Y., 1976, p 171.
(22) H. Nau and K. Biemann, Anal. Lett., 6, 1071 (1973).
(23) C. C. Sweeley, N. D. Young, and S. C. Gates, J. Chromatogr., 99, 507 (1974).
(24) S. C. Gates, N. D. Young, J. F. Holland, and C. C. Sweeley, "Advances in Mass Spectrometry in Biochemistry and Medicine, Vol. I", A. Frigerio and N. Castagnoli, Ed., Spectrum Publications, New York, N.Y., 1976, p 483.
(25) C. C. Sweeley, S. C. Gates, R. H. Thompson, J. Harten, N. Dendramis, and J. F. Holland, "Quantitative Mass Spectrometry in Life Sciences", Elsevier, Amsterdam, 1977, p 29.
(26) S. P. Markey, W. G. Urban, and S. P. Levine, "Mass Spectra of Compounds of Biological Interest", National Technical Information Service, U.S. Department of Commerce, Springfield, Va., 1974.
(27) G. Lancaster, P. Lamm, C. R. Scriver, S. S. Tjoa, and O. A. Mamer, Clin. Chim. Acta, 48, 279 (1973).
(28) F. P. Abramson, Anal. Chem., 47, 45 (1975).
(29) F. W. McLafferty, R. H. Hertel, and R. D. Villivock, Org. Mass Spectrom., 9, 690 (1974).
(30) G. M. Peysna, F. W. McLafferty, R. Venkataraghavan, and H. E. Dayringer, Anal. Chem., 47, 1161 (1975).
(31) H. Nau, H.-J. Förster, J. A. Kelley, and K. Biemann, Biomed. Mass Spectrom., 2, 326 (1975).
(32) R. Reimendal and J. B. Sjövall, Anal. Chem., 45, 1083 (1973).
(33) M. Axelson, T. Cronholm, T. Curstedt, R. Reimendal, and J. Sjövall, Chromatographia, 7, 502 (1974).
(34) B. E. Blaisdell, Anal. Chem., 49, 180 (1977).
(35) B. E. Blaisdell and C. C. Sweeley, unpublished results.
(36) D. R. Smith, M. Achenbach, W. J. Yeager, P. J. Anderson, W. L. Fitch, and T. C. Rindfleisch, Anal. Chem., 49, 1623 (1977).
(37) S. C. Gates and C. C. Sweeley, unpublished results.

RECEIVED for review August 8, 1977. Accepted December 6, 1977. This investigation was supported by Grant RR-00480 from the Biotechnology Resources Branch of the National Institutes of Health. All figures are taken from the Ph.D. Dissertation of S.C.G. (Ref 18).

SPECTROSCOPY

Editors' Comments on Papers 11 Through 15

11 HOLLAND, TEETS, and TIMNICK
   A Unique Computer Centered Instrument for Simultaneous Absorbance and Fluorescence Measurements

12 CUSHLEY, ANDERSON, and LIPSKY
   Computer Controlled Fourier Transform Nuclear Magnetic Resonance System for Carbon-13 and Phosphorus-31 Spectrometry

13 PERRY, BRYANT, and MALMSTADT
   Microprocessor-Controlled, Scanning Dye Laser for Spectrometric Analytical Systems

14 VENKATARAGHAVAN, KLIMOWSKY, and McLAFFERTY
   On-Line Computers in Research. High-Resolution Mass Spectrometry

15 SWEELEY et al.
   On-Line Digital Computer System for High-Speed Single Focusing Mass Spectrometry

Spectrometers have long been targets of computer automation. The papers presented in this section illustrate the wide variety of design and control used in interfacing computers and spectrometers. Articles are presented from NMR, mass spectrometry, UV-visible, and fluorescence spectroscopy.

Paper 11, by Holland, Teets, and Timnick, treats computer control of a combined spectrophotometer-spectrofluorimeter. Fluorescence techniques have a great inherent sensitivity but have lacked reproducibility and accuracy. In this paper computer correction and control of the measurements provide a means of improving the reproducibility and the accuracy.

Cushley, Anderson, and Lipsky (Paper 12) report on the computer control of a carbon-13 and phosphorus-31 NMR spectrometer. Interface electronics and software are discussed and improvements in sensitivity and resolution noted when the instrument is run in the Fourier transform mode.

Recent applications of computers to NMR spectroscopy include those by Goedde, Ader, and Neff. Goedde et al.[1] have described a computer assisted method for measuring NMR relaxation times. Stable pulses from any of the pulse generators of a pulsed NMR spectrometer are conditioned by a small circuit that allows reproducible selection of trigger pulse length and trigger delay time. An algorithm is outlined for the calculation of any of the characteristic nuclear spin relaxation times.

Ader et al.[2] have used an Intel SBC 80/10 microprocessor to control a pulsed NMR spectrometer for measurement of relaxation times. The program to direct the sequence of pulses and receive operator commands is programmed in 3 1K PROMs (programmable read-only memory). Data collected is stored in random access memory prior to transfer to a PDP-10 for further processing.

Neff et al.[3] have described a program for fully automatic correction of phase and amplitude distortions resulting from signal-conditioning circuits and from anomalies in quadrature phase detection in NMR spectrometry.

Perry et al. (Paper 13) demonstrate how a microprocessor can use its strongest characteristic—control—to direct the procedures needed to switch dyes to scan different spectral regions in a scanning dye laser spectrometer. The interface and control circuits are presented in reasonable detail.

Papers 14 and 15 delineate the use of on-line computers in mass spectrometry. Venkataraghavan, Klimoski, and McLafferty (Page 14) review the application of on-line computers to high resolution mass spectrometry, while Sweeley et al. (Paper 15) set forth in a detailed manner a system designed to produce bar graphs and/or tables of mass versus relative intensity of selected compounds eluted from a gas chromatograph into a single focusing mass spectrometer. These papers form a complementary pair on the use of computers in mass spectrometry. The former is more general, while the latter presents detailed descriptions and considerations involved in computerized data acquisition and reduction from a gas chromatograph–mass spectrometer.

Fausett and Weber[4] have attacked the problem of qualitative and quantitative analysis of mixtures using low resolution mass spectra. Three methods are evaluated, all of which require a set of reference spectra of pure compounds. The three methods evaluated are minimization of (1) the sum of the squares of the residuals (least squares), (2) the sum of absolute values of the residuals, and (3) the largest absolute value among the residuals (Chebyshev approximation). The authors conclude that choice of method depends on the needs of the user.

New multichannel detectors continue to appear. Horlick and Codding[5] have reported on a simultaneous multi-element and multiline atomic absorption analyser using a computer-coupled photodiode array spectrometer. A region of 130 Å is observed as a 256-point array, which is digitized and stored on a PDP 8/e in about 5 msec. As pointed out in the article, one of the primary goals is the implementation of simultaneous multi-element analyses.

Felkel and Pardue[6] have designed and evaluated a random access vidicon-Echelle spectrometer that is applied to multi-element determinations in atomic absorption spectrometry. Thirty elements with six wavelengths per element can be handled. Measurements are corrected for dark current and stray light by subtracting from the peak signal the average value of the background signal on either side of the peak.

A computer controlled vidicon spectrometer has been developed and characterized by Rieman and Enke.[7] Under computer control, the readout beam can be deflected to any channel at random or made to scan sequentially the 230 nm spectral region. A PDP 8/I minicomputer with 12K of memory is used.

In other applications of computers to spectrometers, Eaton and Stuart[8] have described a microcomputer assisted, single-beam photocoustic spectrometer system for the study of solids. A schematic of the data acquisition circuit is provided.

Chrisman et al.[9] have described the design and construction of a computer controlled high sensitivity digital Raman difference spectrometer for studies of solutions of biological molecules. The interface to a Data General 16K minicomputer is described in detail, as is a typical operating session.

Crepeau et al.[10] have described a novel use of an on-line computer system for UV laser scanning and fluorescence monitoring of analytical ultracentrifugation. Sedimentation is monitored by fluorescence, and the data is fitted to equations to determine sedimentation coefficients. Preliminary studies by the authors indicate that the technique will improve the sensitivity of measurements and the discrimination between sedimenting species and will thereby enhance the range of experimental possibilities in ultracentrifugation.

REFERENCES

1. A.O. Goedde, M. F. Froix, and D. J. Williams, "A Computer Assisted Method for Measuring NMR Relaxation Times," *Chem. Instrum.*, **7**, 179 (1976).
2. R. E. Ader, A. R. Lepley, and D. C. Songco, "Utilization of a Microprocessor in a Pulsed NMR Spectrometer," *J. Magn. Reson.*, **29**, 105 (1978).
3. B. L. Neff, J. L. Ackerman, and J. S. Waugh, "Fully Automatic Software Correction of Fourier Transform NMR Spectra," *J. Magn. Reson.*, **25**, 335 (1977).
4. D. W. Fausett and J. H. Weber, "Mass Spectral Pattern Recognition Via Techniques of Mathematical Programming," *Anal. Chem.*, **50**, 722 (1978).

5. G. Horlick and E. G. Codding, "Simultaneous Multielement and Multiline Automic Absorption Analysis Using a Computer-coupled Photodiode Array Spectrometer," *Appl. Spectrosc.*, **29**, 167 (1975).
6. H. L. Felkel, Jr. And H. L. Pardue, "Design and Evaluation of a Random Access Vidicon-Echelle Spectrometer and Application to Multielement Determinations by Atomic Absorption Spectrometry," *Anal. Chem.*, **49**, 1112 (1977).
7. T. A. Rieman and C. G. Enke, "Development and Characterization of a Computer Controlled Vidicon Spectrometer," *Anal. Chem.*, **48**, 619 (1976).
8. H. E. Eaton and J. D. Stuart, "Microcomputer Assisted, Single Beam, Photoacoustic Spectrometer System for the Study of Solids," *Anal. Chem.*, **50**, 587 (1978).
9. R. W. Chrisman, J. C. English, and R. Stuart Tobias, "A High Sensitivity Digital Raman Difference Spectrometer for Studies on Solutions of Biological Molecules with On-line Computer Control of Data Acquisition and Reduction," *Appl. Spectrosc.*, **30**, 168 (1976).
10. R. M. Crepeau, R. M. Conrad and S. J. Edelstein, "UV Laser Scanning and Flourescence Monitoring of Analytical Ultracentrifugation with an On-line Computer System," *Biophys. Chem.* **5**, 27 (1976).

A Unique Computer Centered Instrument for Simultaneous Absorbance and Fluorescence Measurements

John F. Holland, Richard E. Teets, and Andrew Timnick
Department of Chemistry and Department of Biochemistry, Michigan State University, East Lansing, Mich. 48823

APPLICATIONS OF FLUORESCENCE to chemical analysis have been widespread, but an adequate understanding of the manifested observations has not been completely attained to this date. An accurate model for fluorescence is of particular importance to the chemical analyst who is faced with the problems of lack of uniform standardization, relatively poor accuracy, and in many cases extreme dilutions.

In fluorescence, measured quantities do not reduce to a simple ratio as in spectrophotometry and there are basic geometric and wavelength dependent factors which are unique for each fluorescence instrument. In spite of the fundamental complexities of fluorescence analysis, it is felt that greater accuracy and better correlation between measurements made with different instruments can be attained with more effective control of the measurement variables. The factors involved in fluorimetric measurement can be divided into two categories, instrumental and photophysical. The instrumental factors pertain to the equipment employed and the conditions of observation, while the photophysical aspects concern the absorption, re-emission, and competitive losses of energy within the molecule, solvent cage, or solution occurring during the period of observation.

The major instrumental variables for 90 degree observation of steady state fluorescence from dilute solutions are variations of the source intensity, spectral distribution of the source radiation, efficiency of the excitation and emission dispersion units, geometry of the cell system, sensitivity of the detector, and gain and linearity of the readout system.

The photophysical variables for any fluorescence measurement include all forms of absorption by the sample, quantum efficiency of the fluorophore, quenching effects, refractive index of the solution, light scattering by the solution, and any anisotropic effects.

Much work has been done in recent years to correct for the effects of the instrumental variables which has resulted in the development of double monochromator spectrofluorimeters with which corrected or true fluorescence spectra can be recorded. These systems involve the following general considerations: conversion of the intensity of the excitation beam incident upon the sample into units of energy or of quanta, conversion of fluorescence into units of energy or of quanta and comparison of these values to yield a quotient which, in effect, is independent of many instrumental variables of the measuring system. Sensitive thermopiles or bolometers are generally used as the detectors in systems designed for energy correction (*1–5*), while quantum counter solutions or screens are usually employed in quantum corrected instrument systems (*6–8*). Corrected excitation spectra are readily attained by these systems to a fair degree of success since the emission monochromator is not changing and thus the emission detector sees only light of a fixed wavelength.

Emission scans are considerably more difficult to correct. Because of the low levels of fluorescence radiation, linear energy measuring devices and linear quantum measuring devices are too insensitive as detectors and therefore extremely sensitive photodetectors are generally required. As the emission monochromator changes, the photodetector sees light of varying wavelength to which it has varying sensitivity. The efficiency of the emission monochromator and the observational geometry of the cell also introduce variables which may need correction. In cases where corrected emission spectra are to be attained, tedious point by point calculations are made for the wavelength dependence of the photodetector and the other variables in the emission system. These calculations are based on data obtained with the assistance of various combinations of calibrated sources, calibrated monochromators, reflecting and scattering standards, and detectors of known response (*9, 10*). In addition to the various instrumental techniques employed to obtain these wavelength dependent correction factors, a series of standard fluorophores has been advocated to use their known fluorescence properties to calibrate emission systems (*11, 12*).

In recent years, digital computers have been utilized to perform tedious calculations necessary to produce corrected spectra. However, computer applications to date have been primarily concerned with "off-line" data processing (*13–15*)

---

(1) American Instrument Co., Silver Spring, Md., Bull. 2392 D (1967).
(2) G. K. Turner, *Science*, **146** (364), 183 (1964).
(3) H. K. Howerton, "Fluorescence," Marcel Dekker, New York, N.Y., 1967, Chapter 5.
(4) W. Slavin, R. W. Mooney, and D. T. Palumbo, *J. Opt. Soc. Amer.*, **51**, 93 (1961).
(5) P. R. Lipsett, *ibid.*, **49**, 673 (1959).
(6) *Instrum. News*, **21** (2), 12 (1970).
(7) C. A. Parker and W. T. Rees, *Analyst (London)*, **85**, 587 (1960).
(8) B. Witholt and L. Brand, *Rev. Sci. Instrum.*, **39**, 1271 (1968).
(9) W. H. Melhuish, *Opt. Soc. Amer.*, **52**, 1256 (1962).
(10) J. Lee and H. H. Saliger, *Photochem. Photobiol.*, **4**, 1015 (1965).
(11) R. Argaver and C. E. White, ANAL. CHEM., **36**, 2141 (1964)
(12) C. A. Parker, *ibid*, **34**, 502 (1962).
(13) H. V. Drushel, A. L. Sommers, and R. C. Cox, *ibid.*, **35**, 2166 (1963).
(14) P. Byron and J. B. Hudson, *Talanta*, **15**, 714 (1968).
(15) R. M. Dagnall, S. S. Pratt, R. Smith, and T. S. West, *Analyst (London)*, **93**, 638 (1968).

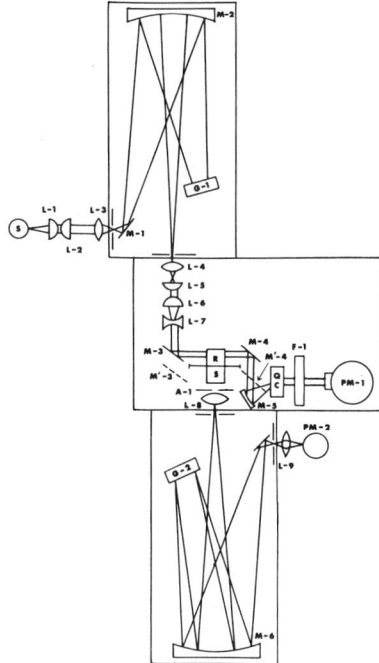

Figure 1. Optical system for simultaneous measurement of absorption and fluorescence

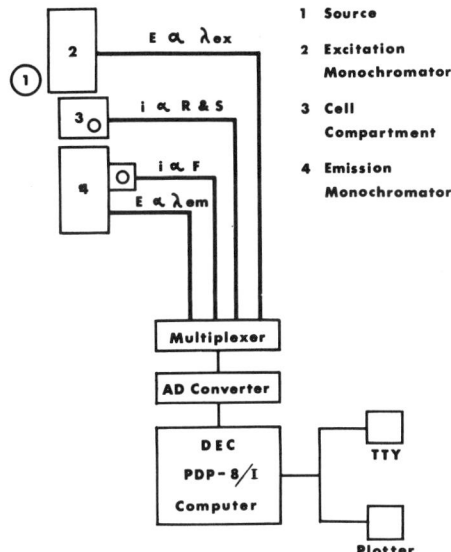

Figure 2. Block diagram of total system illustrating source and nature of analog signals supplied to the computer

rather than with using the computer "on-line" to enhance the accuracy of measurement and to expand the scope of the correction processes.

In making all of these corrections, the coveted goal is to obtain from an excitation scan, a curve that is identical in form to the absorption curve for the same fluorophore (*16*) and from an emission scan a curve whose integrated area will be linearly related to the total quantum efficiency (*17*) of the fluorophore.

At this present state, most of the instrumental variables have been automatically controlled or corrected but little attention has been given to the automatic correction for photophysical variables. Mathematical corrections have been made for refractive index influences (*18*), and the effects of scattered light, when significant, have been attenuated by use of filters or measured and subtracted manually (*19*). Corrections for absorption, re-absorption, and other inner filter effects have been applied to collected data, but often the effects of the photophysical variables have been assumed to be either constant, reproducible, or negligible. The accuracy and reproducibility of fluorescence measurements still are significantly less than what would be expected from a method with such great inherent sensitivity. Obviously further corrective measures must be directed to the photophysical variables.

---

(16) R. Rusakowica and A. C. Testa, *J. Phys. Chem.*, **72**, 793 (1968).
(17) C. A. Parker and W. T. Rees, *Analyst* (London), **87**, 83 (1962).
(18) J. J. Hermans and S. Levinson, *J. Opt. Soc. Amer.*, **41**, 490 (1951).
(19) J. M. Price, M. Kaikara, and H. K. Howerton, *Appl. Opt.*, **1**, 521 (1962).

Of the photophysical variables, the one that causes the first concern is the absorption by the sample system. This process attenuates the excitation beam with the result that the detector observation geometry changes as the concentration of the absorbing species increases. Unlike the fixed refractive index correction or the single wavelength scatter correction, this variable can introduce serious and unknown wavelength dependent errors into fluorescence measurements, especially in excitation spectra where the absorption is changing during the scan. The resulting excitation spectrum, to be free from this absorption effect, must be corrected at each wavelength. This requires knowledge of the absorption as well as the fluorescence at each of these data points. This indicates clearly the great desirability of simultaneous absorption and fluorescence measurements with one instrument system. By using this simultaneous measurement technique, high confidence can be placed in the interrelationship between absorption and emission, since both measurements are made with the same solution and with the same optical system. This feature is highly desirable for accurate quantum efficiency studies.

Because of the complexity of the instrumentation and the multiplicity of the corrections that must be made, it was a rather directed decision to integrate the spectrofluorimeter with a computer into one system that could record the fluorescence and absorption measurements and calculate and apply the various corrections. The following report described the structural details and performance of such an instrument system.

INSTRUMENTAL

Optical System. Figure 1 illustrates the optical system designed and constructed specifically for simultaneous absorption and fluorescence measurements. The source S is an Osram XBO 150-watt Xenon arc powered by a Sola Electric 7.5-ampere 21-volt power supply. Radiation from the source is focused on the entrance slit and directed through the excitation monochromator. The exit lens combination collimates the beam and directs it onto a front surface mirror M-3 with an incident angle of 45°. The reflected beam passes through the cell compartment and impinges on M-4, again at an angle of 45° from which it is directed onto the focusing mirror M-5 which focuses the beam onto the quantum counter QC. The fluorescence from the quantum counter solution is monitored by the photomultiplier PM-1. The cutoff filter F-1 assures that only the desired emission line of the fluorescence is observed and excludes stray light, especially that which may originate from the original beam. The solution in the quantum counter, sensitive in the range 250 to 600 nm (20), consists of 8 grams of Rhodamine B per liter of ethylene glycol.

The beam as shown in Figure 1 passes through the reference cell R. The mirrors, M-3 and M-4 are attached to the top of a vibrating bridge assembly from a Beckman DB spectrophotometer. The top of this bridge moves back and forth in a direction parallel to the optical axis of the excitation monochromator exit beam. The solid line drawing indicates one extreme position of mirror movement while the dashed lines indicate the other extreme. Note that when the mirrors are in the position M'-3 and M'-4, the beam passes through the sample cell S.

Light fluoresced by the sample is gathered by lens L-8, and directed into the emission monochromator. The functions of the internal components in this monochromator are exactly the same as in the excitation case, except that the direction of light travel is reversed. The selected wavelength is brought to a focus at the exit slit and directed to the detector PM-2 by lens L-9. Both monochromators are B and L 33-86-45 high intensity grating monochromators.

Each monochromator is driven during scans by a selsyn motor, ARMA Type 1F. This motor functions as a slave to a transmitting selsyn, Bendix 3H9, that is driven by a shaded pole induction motor, Dayton 3M098, through a variable gear and pulley train. This method of activating the wavelength drive system of the monochromator was chosen for two reasons. First, one of the characteristics of the selsyn or "synchro" motor is that when not activated, it has negligible friction thus permitting free manual adjustment of the monochromator. This would not be possible if a synchronous motor were attached directly to the drive chain. The second reason for use of selsyns lies in the capability of such a system to drive both monochromators at exactly the same rate which permits simultaneous scanning. In this mode of operation, the system can be used to measure scattered light.

The monochromator wavelength drive is linked to the selsyn motor through a combination spur gear and pulley train. The wavelength encoding device, a Keltron Corporation infinite resolution linear potentiometer Model LP625769, is attached directly to the grating rotating arm and is thus unaffected by any gear play in the linkages of the motor drive train.

The detectors used in this instrument are RCA 1P28 photomultiplier tubes. Each derives its dynode voltage from a Kepco model ABC 0–1.5 kV power supply. The photomultiplier signals are amplified and critically filtered to accommodate the bridge vibration rate.

Figure 2 identifies the components of the complete system and indicates the nature of the analog signals involved. Four analog signals, two photomultiplier outputs and two wavelength encoder outputs, are connected to a multiplexer–A to D converter combination, Digital Equipment Corporation model AF01. Under program control, the computer can switch the multiplexer to admit any one of these voltages to the A to D converter where it is digitalized and from which it can be read into the computer, a Digital Equipment Cor-

poration PDP-8/I. The central processor contains 8K of core, memory multiplexer, and the extended arithmetic unit. Inputs to the computer other than the I/O buss are a KSR-35 teletype and a high speed paper tape reader. The outputs from this unit include the teletype, a high speed paper tape punch, and an incremental plotter, Houston Model 3660. The signal specifications for the A to D input are 0 to +10 volts and the conversion time is 40 μsec. Expanded memory provisions include two DECTAPE tape drive units and two 32K magnetic disks.

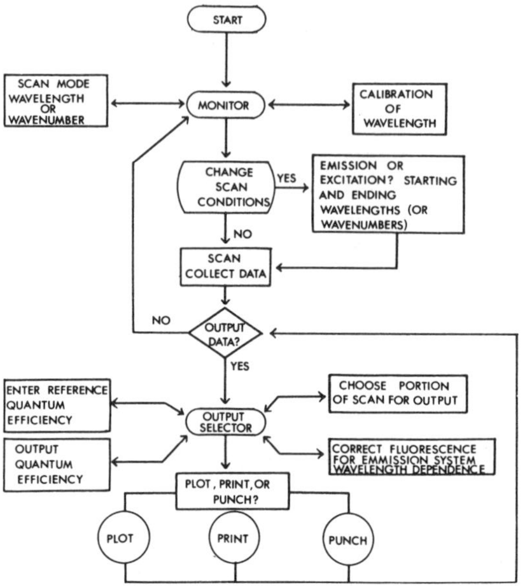

Since a quantum counter is used to monitor R and S, the magnitude of the respective signals will be related to the number of quanta remaining in each beam. Status flags enable the computer to distinguish between the reference and the sample beam.

In the functioning of the flag system, the driving power to the vibrating bridge activates a single pole double throw relay, the position of which is synchronized to the position of the bridge. Each relay closure triggers a logic circuit which enables the computer to identify the position of the vibrating bridge. Since the emission detector sees light only when the excitation beam is irradiating the sample, the flag for the sample beam also serves for the fluorescence measurement period.

The magnitude of the signal received from PM-2 is dependent upon the number of photons fluoresced by the solution, the cell geometry, the efficiency of the monochromator, and the wavelength sensitivity of the detector. Since R and S are obtained in terms of quanta, it is desirable to express fluorescence in terms of quanta. Conversion of the PM-2 signal into terms of quanta is accomplished by the computer under program control as will be described below.

General Program. Figure 3 illustrates the flow diagram for the computer program developed to collect data, apply corrections, and produce the desired outputs. All of the programming for this instrument was done on the PDP-8/I and written in its machine program assembly language, PAL. The monitor is the central or control part of the program. For convenience it has been shown in Figure 3 as being in two parts, one for pre-scan and the other for post-scan options.

Figure 3. Flow chart of the computer program

---

(20) J. Yguerabilde, *Rev. Sci. Instrum.*, **39**, 7 (1968).

Table I. Output Formats

$R \propto$ quanta in reference beam
$S \propto$ quanta in sample beam
$F \propto$ quanta fluoresced

Quantity	Code		Operations
Raw sample	SS	=	$S \left[\dfrac{k_n}{k_d}\right]$
Raw reference	RR	=	$R \left[\dfrac{k_n}{k_d}\right]$
Raw fluorescence	FF	=	$F \times \left[\dfrac{k_n}{k_d}\right]$
Transmittance	TR	=	$\dfrac{S}{R} \times \left[\dfrac{k_n}{k_d}\right]$
Absorbance	AB	=	$\log \dfrac{R}{S} \left[\dfrac{k_n}{k_d}\right]$
Uncorrected fluorescence	UF	=	$F \times P_{esc} \times \left[\dfrac{k_n}{k_d}\right]$
Corrected fluorescence	CF	=	$F \times P_{esc} \times \dfrac{1}{R} \times \left[\dfrac{k_n}{k_d}\right]$
Partial quantum efficiency	PQ	=	$F \times P_{esc} \times \dfrac{1}{R-S} \times \left[\dfrac{k_n}{k_d}\right]$
Quantum efficiency (print only)	QE	=	$\int_{\lambda_1}^{\lambda_2} F \times P_{esc}\, d\lambda \times K_s \dfrac{1}{R-S}$
Background	BA	=	$[R - S + 200] \times \left[\dfrac{k_n}{k_d}\right]$

The CALIBRATE, a pre-scan routine, enables the computer to set up the relationship between the output of the voltage encoder on each monochromator and the wavelength of that monochromator. The computer reads the voltage from each encoder, calculates the constants for the linear equation, $\lambda = mV + b$, and stores these constants for subsequent use in calculating wavelength, λ from voltage, V.

The subroutine WAVELENGTH automatically sets the program pointers for wavelength dependent data collection and presentation while the subroutine WAVENUMBER sets the program pointers for wavenumber data collection and output.

The routine CHANGE accommodates the dialogue between the computer and the operator during which the emission or excitation scan, starting wavelength or wavenumber, and ending wavelength or wavenumber are selected. During scans, data are collected at wavelength intervals of 0.25 nm. If another collection interval is desired, a special routine can accommodate this change.

The independent variable of the data collection algorithm is wavelength or wavenumber. After the SCAN routine is initiated, the computer monitors the selected monochromator until the desired starting wavelength is reached. A set of data points is then taken in the following sequence. The computer looks for the reference flag, waits until it senses it, and then digitalizes the voltage from PM-1 four times during the flag interval with a delay of 0.5 millisecond between each point. The average of these four points is stored and the computer then looks for the sample flag. When it is sensed, the above process is repeated to obtain the data point for the sample beam. The computer then switches the multiplexer channel to PM-2 and immediately repeats the above process and stores a point for fluorescence. This sequence is repeated ten times during a data collection interval. These values are averaged and a single precision data word is stored for each of the three R, S, and F. The location in core of the data points determines the wavelength for that particular set of points. The computer then switches back to the channel of the monochromator which is scanning and awaits the wavelength at which the next point is to be taken. The above process is then repeated which results in another set of data points. These points are stored in the next successive position in each of the data files, the wavelength is incremented, and the computer awaits the next data point. This process is repeated until the computer senses that the end of the selected wavelength range has been reached at which time it exits from the scan routine.

A QUANTUM CALIBRATE routine used only after emission scans calculates the system constant for the specific optical and electronic conditions employed while scanning a known fluorophore which has been selected for the reference compound in the comparison method used to evaluate quantum efficiency.

The OUTPUT RANGE routine enables the operator to select any portion of a scan to be outputted. With this option, undesired portions of collected data may be deleted.

The NEW routine sets up a background subtraction file from the results of a prior scan made with identical solutions in the reference and sample cells. This offset file is used to correct for absorption differences between the sample and reference cells. A routine called AVERAGE permits the averaging of several background scans in the creation of the NEW offset file described above. The SUBSTRACT routine subtracts the offset file from the sample files accumulated during later scans.

A special routine corrects for the wavelength dependence of the emission system before proceeding to the reduction and output routines. The computer sets up a new file for the fluorescence by applying at each wavelength a sensitivity coefficient, P_{esc}, retrieved from a correction table. The resulting corrected fluorescence value will be directly proportional to the number of quanta fluoresced by the sample.

Output Formats. Table I defines the various output formats for this program. The left column indicates the quantity desired. The terms in boxes are typed into the computer by the operator. The remaining terms are retrieved by the computer from the data files. It must be emphasized that the experimentally collected data files remain unaltered during all of these routines. Thus trial calculations can be repeated until meaningful presentations are obtained, and the decision for long term storage can be made after the significance of the data has been evaluated.

In all routines, the data points are presented in single precision arrays against wavelength or wavenumber. When using the printer or the paper tape punch, outputs of the data files of S, R, and F are recorded as collected while other quantities from these numbers must be scaled so that the resulting values may be stored with optimum significance in single precision.

Since all calculations are performed in floating point, scaling constants need not be entered, but when the incremental plotter is to be used, a scaling numerator and denominator are entered from the keyboard to provide maximum flexibility with the ordinate axis such that measurable, meaningful plots may be obtained. These scaling factors, k_n and k_d, are shown in Table I as the last term in the data processing sequence.

For the raw sample, raw reference, and raw fluorescence outputs, the computer plots, prints, or punches the data as collected. These formats provide the outputs for the most efficient long term storage.

For the transmittance output, the computer calculates the values of S/R for the scanned range. This ratio is the standard output for double beam spectrophotometers to which the results of this instrument are directly comparable.

To output absorbance, the computer calculates the log R/S.

For the uncorrected fluorescence, the computer multiplies the measured value by emission system correction factor, P_{esc}. Since the only scan variable used in this output is the fluorescence, F, the results will be uncorrected fluorescence in terms of quanta and should correlate with results of other systems using similar sources and detectors.

In the corrected fluorescence format, CF, the data are multiplied by the correction factor, P_{esc}, as above. This product is divided by the number of quanta, R, in the excitation beam. This operation corrects or normalizes the observed fluorescence for the intensity of the excitation radiation which in effect compensates for variations in source intensity.

The unique quantity this instrument system is capable of outputting after an excitation scan is the partial quantum efficiency, PQ. The computer calculates the quotient $F/R - S$ and outputs this value as a function of the excitation wavelength. It is apparent that at any given point during an excitation scan, PM-1 will give a value of R and S in terms of quanta. The difference, $R - S$, will give the number of quanta absorbed by the sample. The output from PM-2 after computer correction will represent the number of quanta fluoresced within the bandpass of the emission monochromator. This represents only a small part of the total emission resulting from the absorption. To obtain the total quantum efficiency, it is necessary to scan through the entire emission spectrum with the excitation wavelength fixed and obtain the area under the emission curve in terms of quanta. Note however, that if the emission spectrum is invariant with excitation wavelength, this part or partial quantum efficiency obtained will be linearly related to the total quantum efficiency (7).

The quantum efficiency routine, QE, is based on the comparison method as first proposed by Bowen (21) for obtaining quantum efficiencies. In operation a known fluorophore is scanned throughout its entire emission spectrum with the excitation monochromator set to an appropriate absorption wavelength and the total emission is integrated by the computer. The known value of its quantum efficiency is then introduced from the keyboard and the computer solves for the system constant, K_s in the expression shown in Table I. The emission system correction table factor P_{esc}, accounts for the wavelength dependent variables in the measurement and this constant, K_s, should account for all the wavelength independent variables. The unknown fluorophore is then scanned throughout its entire emission spectrum, its total emission is integrated, and its quantum efficiency is calculated by the computer. As indicated previously, the term $R - S$ carries the significance of the number of quanta absorbed by the fluorophore. Since the integral under the fluorescence curve produces a number related to the number of quanta fluoresced, this division will satisfy the classical definition of quantum efficiency.

The systems constant, K_s while independent of wavelength, is definitely dependent on instrumental variables such as slit settings, photomultiplier voltages, and amplifier gain control positions. For accurate quantum efficiencies, these must be invariant during the scans of the comparison reference and the solutions to be measured. Since this instrument has been constructed to measure intensities in terms of quanta and since grating monochromators are used, wavelength dependent collection is designated in all scans from which values for PQ or QE are to be processed.

An additional format called background, BA, outputs the values of R minus S as a function of scan wavelength. This routine is used primarily as a diagnostic program to evaluate the relative intensities of the sample beam and the reference beam with the reference solution in both cells.

Auxiliary Routines. A program, EMISSION SYSTEM CORRECTION, has been developed that employs light scattering data obtained with substances of known scattering properties. With this information, the computer creates a wavelength dependent sensitivity correction table for the emission system by which the output from PM-2 can be expressed in terms of quanta. In addition to the efficiency of the detector, this method also corrects for monochromator efficiency and the observation cell geometry.

Presently, information for the computation of the correction table is obtained by the method of reflection from a substance of known light reflection properties (22, 23). Using a freshly prepared plate of magnesium oxide in the sample cell and with the reference cell empty, the desired wavelength range is scanned with both the excitation monochromator and the emission monochromator in synchrony. During the scan, the computer compiles two data files as a function of wavelength. In one file is stored the output of PM-1 (reference, R) and in the second the output from PM-2 (fluorescence, F). The output of PM-1 is directly proportional to the number of quanta in the excitation beam. Since the reflector intercepts a beam of the same intensity and wavelength as that which passes through the reference side, PM-2 should detect a fraction of the quanta in the incident beam. This fraction will be proportional to the reflection efficiency of the magnesium oxide and the observational geometry of the emission system. In this technique, reflection from the standard is assumed to be independent of wavelength through the range employed. The computer calculates the multiplier constant that will produce a linear relationship between R and F. These calculated multipliers are arranged into a table and inserted into core to be used as the wavelength correction coefficients for future fluorescence scans.

An ASSAY program was developed in which the fluorescence from a standard series of known concentrations of a specific fluorophore are measured one at a time under fixed instrumental conditions. After each measurement, the concentration is entered on the keyboard. At the completion of this series, the computer calculates the best least squares straight line through the more dilute points. The line defines an idealized linear relationship between fluorescence and concentration. The computer then calculates the factors necessary to place the measured fluorescence for each standard on this best fit line. Once these factors have been determined, they are related to the absorption, $R - S$, and are stored in computer memory for subsequent use in fluorimetric assays. This program enables the fluorophore to be assayed more accurately in higher concentrations than otherwise would be practical, and permits assays for fluorophores in the presence of chromophores, often eliminating the necessity for dilutions or other procedures to remove the effects of interfering chromophores.

A subroutine, ABSORPTION CORRECTION, performs analogously to ASSAY, but linearizes the interrelationship between F and $R - S$. Since this interrelationship is theoretically linear, this subroutine when applied, should correct the fluorescence for absorption effects.

A modification of the DEC TOS program that operates the tape drive unit on the computer has been developed. The entire program for the fluorimeter, including the emission system correction table is stored on the tape and can be read directly into the core when called.

RESULTS AND DISCUSSION

Performance as a Spectrophotometer. A series of experiments were conducted to evaluate the performance of this instrument as a double beam spectrophotometer. The linear reciprocal dispersion of the monochromators used

---

(21) E. J. Bowen, *Trans. Faraday Soc.*, **50**, 97 (1954).

(22) G. Weber and F. W. J. Teale, *Trans. Faraday Soc.*, **53**, 646 (1957).

(23) S. Vavilov, *Z. Phys.*, **23**, 266 (1924).

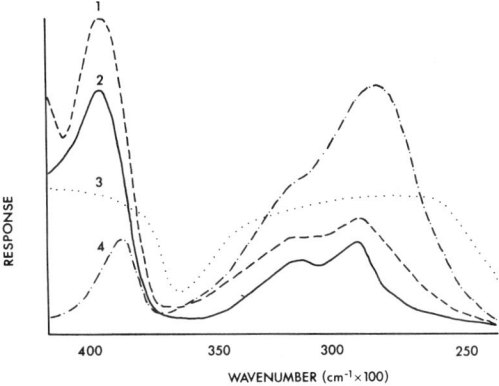

Figure 4. Multiple output from a single fluorescence excitation scan for $10^{-5} M$ quinine bisulfate in $0.1 N$ sulfuric acid

1. Corrected fluorescence
2. Absorbance
3. Partial quantum efficiency
4. Uncorrected fluorescence

in this system is 3.3 nm per mm. For high resolution work, band passes of less than 1 nm are readily attainable. With such a narrow band pass, the resolution compares favorably with that obtained on commercial double beam scanning spectrophotometers.

Photometric accuracy tests on this system were performed with the aid of colored glass and neutral density filters. Table II indicates the data obtained from a comparison of per cent transmittance scans on a colored glass from this instrument and the Hitachi EPS-3T. Agreement in these results lies well within the specified accuracy of the commercial instrument. This and photometric accuracy tests performed with neutral density filters indicated that the system has an accuracy of ±0.002 absorbance unit over the absorbance range of 0–2.0.

Because of the high quality of the monochromators and the high levels of radiation usually measured, stray light effects are negligible. However, for excitation in the range between 575 nm and 600 nm, a cutoff filter must be placed in the excitation beam for accurate photometric results.

Performance as a Spectrofluorimeter. Several experiments have been designed and conducted to evaluate the various instrumental parameters of the fluorescence measurements. It is of great importance that the intensity response of PM-1 be linear with respect to that of PM-2. Repeated outputs of part of the uncorrected and the corrected fluorescence scans for a known fluorophore in which the excitation monochromator slit width was varied to produce relative intensity variations of up to 64-fold were compared. The normalized or corrected fluorescence, F/R varied less than 1% while the uncorrected varied directly with the excitation intensity. This test demonstrates the linearity of the two detecting systems and dramatically illustrates the analytical significance of corrected fluorescence.

Corrected fluorescence excitation spectra of several compounds compared favorably to other published corrected spectra.

In general, the emission spectra obtained from this instrument correlate well with the published results from other instruments; however, there is a better agreement in the excitation comparisons than in the emission comparisons. This

Table II. Comparison of % Transmittance Values of a Standard Filter at Selected Wavelengths

	% T	
Wavelength, µm	Commercial instrument	Spectrofluorimeter, spectrophotometer mode
374	48.5	48.4
380	46.5	46.0
403	60.0	60.0
474	41.8	42.0
560	48.0	48.0

is to be expected since the excitation intensity levels are sufficient so that thermopiles or quantum counters can be used to compensate effectively for wavelength dependence in the excitation system. Unfortunately, the relatively low sensitivity of these linear devices precludes their use in emissions systems.

The favorable results of these experiments have given the instrument a confidence level upon which other experiments have been conducted that employ the versatility of the computer for data processing and outputting. Figure 4 shows the shapes of the various output formats of the system from a single excitation scan on quinine sulfate in $0.1 N$ sulfuric acid. Since all of these outputs are calculated from the same set of data points taken simultaneously during the scan, they are plotted on a common energy axis. The detection of small shifts in energy is readily accomplished and the correlation of fluorescence to absorbance is greatly facilitated by these multiple outputs.

Partial Quantum Efficiency. In accordance with the definition of the partial quantum efficiency, PQ, as given in the program section, we should expect from theory that this quantity will be flat across an absorption band and independent of the concentration of the fluorophore provided no quenching effects occur upon dilution and no impurities with significant absorption of fluorescence are present. Near ideal behavior has been observed in several cases and is illustrated by the dashed line traversing the absorption band, indicated by the solid line, in Figure 5A. Other typical responses are similar to those indicated by curves B through F in which the dashed line represents the theoretical and the solid line the experimental PQ. Each of these cases shall be considered in the light of current understanding attained from the analysis of numerous experimental results.

Curve B and those similar to it have been obtained when the concentration artifact was significant. The theoretical magnitude of PQ is not realized because of the nonlinearity of the fluorescence response produced by the concentration artifact. This effect attenuates PQ the most where the absorption is the greatest, producing a definite depression at the wavelength of maximum absorption. This effect is readily diagnosed by dilution, the result of which will cause the curve to more closely approximate the theoretical.

Curves similar to that of C are attributed to the presence of a second fluorophore, whose individual theoretical PQ is also shown, and which fluoresces with a greater quantum efficiency than the primary fluorophore. In the case shown, the secondary fluorophore has a narrower absorption band than the primary. Note that the second transition does not reach its full value of PQ since the total absorption always includes the absorption of the broader band. The PQ will rise as the absorption of the second transition becomes a larger part of the total absorption and falls when it becomes a lesser part. If

Table III. Precision of Quantum Efficiency Determinations for $10^{-5}M$ Quinine Bisulfate

Time, hr	Uncorrected	Background corrected
0	0.546	0.546
1	0.552	0.546
2	0.591	0.549
3	0.620	0.547
0	0.546	0.546
1	0.590	0.547
2	0.640	0.544
3	0.625	0.548

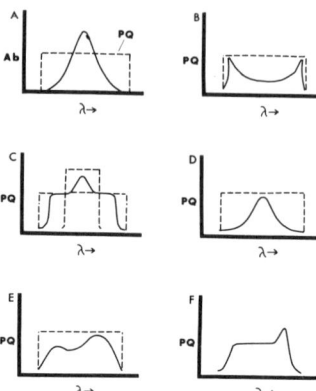

Figure 5. Types of observed partial quantum efficiency curves

the interfering fluorophore has a smaller quantum efficiency, the resulting PQ curve will be depressed where the present maximum exists.

A large number of curves appear much like D. In this case, the PQ curve will resemble the shape of the absorption spectrum of the fluorophore; however, it will be truncated to varying degrees. This response occurs whenever the sample contains a chromophore that absorbs in the excitation region and does not fluoresce within the wavelength range of the emission window. As the absorption of the nonfluorescing species increases, the denominator in the PQ expression becomes larger and the quotient is reduced. The extent of the truncation is obviously dependent on the relative magnitudes of the two absorptions. The less the chromophore absorbs, the closer PQ approaches its theoretical value. The greater the second absorption, the more PQ resembles the absorption spectrum of the fluorophore.

The type of curve shown in E illustrates the effect of a chromophore with a narrow absorption band on the PQ of a fluorophore with a broader absorption band shown in D. The presence of this combination yields a curve that appears to approximate theory in certain areas but shows a loss of PQ in the region of the absorption by the chromophore. This loss need not be symmetrical but will usually give a fair indication of the shape of the absorption band of the chromophore.

By far the most persistent and vexing responses are those similar to F. This type of output has been observed in many cases and its origin is very difficult to assess. Since it deviates upward on the low energy side of the absorption band, it occurs at the most critical point in the measurements of excitation scans. At this point, during the scan, in many samples the absorption is approaching zero. This causes the denominator, $R - S$, in the partial quantum calculation, to approach zero. The computer places a lower limit of one on this difference, but in the region where it is approaching one, large variations in PQ are produced by the computer round-off processes and random noise. This greatly reduces the reliability of PQ measurements in the low energy region of an absorption band. Likewise, any unbalance between the two beams will produce the greatest error in this region where S approaches R. A novel BACKGROUND subtraction routine has been developed to correct for this imbalance to within the 12-bit precision of the computer system. Another complication arises at this point in the scan with compounds whose emission band overlaps with the high wavelength end of its excitation band. This introduces the possibility of scattered light which also would result in the production of curves such as F. A system similar to C, indicative of a second transition, or one similar to E, indicative of a second absorption, can also produce this type of curve. And last, the concentration arti-

fact can also cause this type of response. Needless to mention, great care must be taken when this type of variation occurs either to remove it by more careful control of experimental parameters or to account for it with chemical explanations.

If a generalization can be made at this point, it does appear that a single fluorophore, free from interference or contamination, and at a proper dilution will produce a flat partial quantum efficiency curve as predicted by theory. However, minor exceptions have already been noted and the tenure of this assumption will depend upon the results of research that will continue for some time and that hopefully will lead to accurate delineation of the source of the observed nonlinearities.

The conclusion that the partial quantum efficiency is independent of concentration over its optimum range has been supported and the acceptable range of concentration has been found to be limited on the upper end by the absorption artifact and limited in the lower end by the accuracy of $R - S$ as S approaches R. In practice, this has set a lower limit of approximately 0.01 A. This independence of concentration is in accord with accepted theory and should hold for any pure fluorophore that does not undergo conformational, bond, or solvation changes within this range of dilution.

Any variable that will affect the intensity of the fluorescence radiation or the magnitude of the absorption processes will alter PQ. Therefore, variations in this partial quantum efficiency during an excitation scan can enhance further insight into the chemical nature of the sample solution and the competing excitation and relaxation processes. PQ also has been used to indicate the presence of conformational and bond changes in various chemical systems (24, 25).

Total Quantum Efficiency. The procedure employed in the evaluation of the total quantum efficiency is neither new nor unique. What represents an innovation is that the absorption and the fluorescence measurements are made on the same sample solution at the same time and within the confines of one instrument system. This arrangement produces the most meaningful pertinence of the absorption measurement to the efficiency ratio. However, this output format has been the most demanding upon the instrument system. Since the absorption is being measured simultaneously with the fluores-

(24) L. Bieber, F. Schroeder, and J. Holland, *J. Antibiotics*, **24**, 846 (1971).
(25) F. Schroeder, J. F. Holland, and L. L. Bieber, *Biochemistry*, **11**, 3105 (1972).

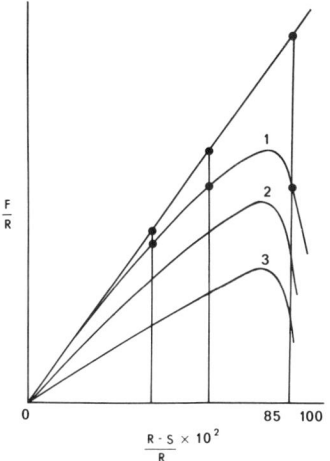

Figure 6. Relationship between fluorescence intensity and absorption by three individual fluorophores and correction for the absorption artifact

1. Quinine bisulfate in 0.1N sulfuric acid
2. Coumarin in ethanol
3. Bovine serum albumin in water

cence on dilute solutions, the absorption is per se a relatively small value, often only equal in magnitude to the imbalance between the two beams, thereby necessitating extremely effective beam balance corrections. Any variation in this averaged $R - S$ value will carry directly to the quotient. For this reason, the comparison method is employed and the samples are run immediately after the calibration compound.

Table III illustrates the results of a precision study clearly showing the value of the BACKGROUND subtract routine for removing variations due to beam unbalance and source fluctuations. The precision of this system compares extremely favorably with that of other systems where variations between successive determinations are commonly between 5 to 12% with precision down to 2 to 5% possible only by averaging several runs. The results in Table III are a typical set of single runs. Repeated runs taken in succession consistently vary less than ±0.002 in quantum efficiency. The overall precision obtained from this approach to quantum efficiency determinations indicates a great potential for the simultaneous measurement technique. The total quantum efficiency may be obtained for a compound in a few minutes compared to other methods, which may take several minutes, hours, or even days.

Table IV lists a comparison of total quantum efficiencies obtained on this instrument with a range of values found in the recent literature. In each compound studied, our values are within the range of the published values and the values shown have been selected because of their general acceptance as being fairly accurate. This by no means constitutes a stringent evaluation of the accuracy of the system but could at least define a range in which acceptable values must lie.

Absorption Correction. Factors which produce a decrease in the observed fluorescence by means other than through competitive deactivation are called the inner filter effects. These may be divided into three categories, absorption of the emitted fluorescence by the molecules of the fluorophore in the solution, absorption of an amount of the exciting radia-

Table IV. Quantum Efficiency Values Obtained by the Comparison Method

	Φ	Range of reported values
Quinine bisulfate $1 \times 10^5 M$ in $1N$ H_2SO_4	0.546	0.46 (*16*)–0.58 (*26*)
Anthracene $1 \times 10^{-5}M$ in ethanol	0.274	0.22 (*13*)–0.28 (*7*)
Rhodamine B $1 \times 10^{-6}M$ in ethanol	0.711	0.69 (*7*)–0.97 (*22*)

tion sufficient to produce a viewing artifact in the emission measurement system, and concentration dependent reactions, such as dimerization, that could affect either the absorptivity or the quantum efficiency. In general, this effect is manifested by a nonlinearity between the amount of radiation absorbed and the amount fluoresced as the concentration of the fluorophore is increased.

In order to present a reasonable approach to a study of this effect, the contribution of the reabsorption and chemical reaction were precluded by carefully selecting compounds whose absorption and emission bands did not overlap and compounds that were also thought not to undergo changes in species throughout the range of concentrations employed. Use of this type of compound permitted an investigation into the effect of absorption only on the measured values of fluorescence. Under these conditions, in the ideal case if all of the fluoresced radiation could be seen by the detector, the relationship between the observed fluorescence and the absorption by the fluorophore would be linear. In practice, the detector is not so efficient since it views only a fraction of the total emission. Any changes in the geometry of the emission detection system will destroy the linearity between fluorescence and absorption. In this case the nonlinearity can be called the absorption artifact and has been the object of study in recent years (*27, 28*).

Results of a study using the computer program ASSAY on several fluorophores are summarized in Figure 6. Thes data represent a study of the relation of $R - S$, the number of quanta absorbed, and F, the number of quanta fluoresced, for three different fluorophores. Note that all three curves are similar. This observation indicates that the absorption artifact is independent of the wavelength of the excitation radiation and independent of the nature of the absorbing species. This would support a general conclusion that this effect is caused exclusively by the attenuation of the excitation beam by absorption within the cell. Assuming this specific cause and effect, it now becomes possible to make corrections for this particular phenomenon.

The computer was used to correlate and normalize the absorption and fluorescence relationships for various concentrations of a fluorophore and produce one curve, the geometry of which is similar to all the others. The top curve of Figure 6 illustrates this curve and the least squares line fit determined from the small $R - S$ values where the artifact is negligible. A correction curve was then calculated to produce linearity between $R - S$ and F. This curve supplies the factors that can be used to correct measured values of fluorescence for the absorption occurring within the cell during the period of observation. The specific factor used will in each case depend

---

(26) S. W. Eastman, *Photochem. Photobiol.*, **6**, 55 (1967).
(27) W. E. Ohnesorge, *Anal. Chim. Acta.*, **31**, 484 (1964).
(28) R. A. Passwater and J. W. Hewitt, *Fluorescence News*, **4** (4), 9 (1969).

Table V. Comparison of Relative Errors on Determination of Quinine Bisulfate in 0.1N H$_2$SO$_4$ between Normal Corrected and Absorption Corrected Fluorescence

Concn taken, moles/liter	Concn found,[a] normal	Relative error, %	Concn found[a] absorption corrected	Relative error, %
1.0×10^{-7}	$1.000_5 \times 10^{-7}$	-0.05	$1.000_5 \times 10^{-7}$	-0.05
0.5×10^{-5}	0.386×10^{-5}	-22.8	0.494×10^{-5}	-1.2
1.0×10^{-5}	0.760×10^{-5}	-24.0	0.993×10^{-5}	-0.7
0.5×10^{-4}	0.160×10^{-4}	-68.0	0.522×10^{-4}	$+4.4$
0.25×10^{-3}	0.040×10^{-3}	-84.0	0.261×10^{-3}	$+4.41$
0.5×10^{-3}	0.0281×10^{-3}	-94.4	0.535×10^{-3}	$+7.0$

[a] Determined by program assay.

on the magnitude of this absorption, $R - S$, a quantity obtained simultaneously with the fluorescence as described above.

This ABSORPTION CORRECTION routine has been introduced into the fluorescence ASSAY program where it enhances the analytical capability of accurate determinations of wide ranges of concentrations. Table V lists a comparison of fluorescence assays with and without the ABSORPTION CORRECTION. Note that even in the most extreme case, where the absorption is greater than 95%, there is a 10-fold reduction in the relative error.

It must be emphasized an absorption artifact is caused by any absorbing species which results in the attenuation of the excitation beam. Since most samples on which fluorescence measurements are made are not free from additional chromophores, neglect of this artifact may be one of the major causes of the relatively poor accuracy of such fluorescence determinations. The term absorption corrected fluorescence is proposed to identify the technique that automatically compensates for this artifact. To correct fluorescence from a system with both chromophore and fluorophore absorption, a complex correction scheme has been developed which compensates for the combined absorption effects (29).

(29) J. F. Holland, R. E. Teets, G. K. Sindmack, and A. Timnick, ANAL. CHEM., **45**, in press.

SUMMARY

As a general conclusion, the application of a small dedicated computer for in-line, real time data collection and reduction appears to be exceedingly successful in the evaluation and control of many of the variables which affect fluorescence measurement. The ability of the computer to present varied and convenient output formats is a great aid to the chemist and the method of simultaneous measurement of absorption and fluorescence, although limited in concentration range, represents a powerful new approach to quantum efficiency studies. The automatic correction for the effects on the excitation beam and on the observation geometry by absorption leads to a greater analytical capability, especially in those areas where samples are often contaminated by other absorbing species, or where the fluorophore itself may be in a relatively high concentration. These combined capabilities have indicated that fluorescence measurements may indeed be made with a greater accuracy than previously feasible.

RECEIVED for review February 7, 1972. Accepted August 14, 1972. Presented as paper 115 at the Pittsburgh Conference on Analytical Chemistry and Applied Spectroscopy, Cleveland, Ohio, March 1971.

Computer Controlled Fourier Transform Nuclear Magnetic Resonance System for Carbon-13 and Phosphorus-31 Spectrometry

R. J. Cushley, D. R. Anderson, and S. R. Lipsky
Section of Physical Sciences, Yale University School of Medicine, New Haven, Conn. 06510

A high resolution NMR spectrometer has been modified for pulse-Fourier spectrometry (FFT-NMR). Data acquisition and data handling are accomplished by means of an IBM 1800 computer with 24K of 4 μsec core storage and numerous peripheral devices. The NMR free induction decay signal (up to 8192 data points) can be digitized at rates up to 20 KHz. Spectra of ^{13}C, ^{31}P, and ^{1}H nuclei have been determined and time savings of 100-fold or sensitivity enhancement of 10- to 20-fold have been realized.

THE COMPLEX Fourier relationship:

$$f(j\omega) = \int_{-\infty}^{\infty} f(t)e^{-j\omega t} \times dt \quad (1)$$

and

$$f(t) = \int_{-\infty}^{\infty} f(j\omega)e^{j\omega t} \times d\omega \quad (2)$$

shows that the frequency response function and the time response function form a Fourier transform pair. Application of the Fourier transform technique to proton NMR spectrometry was demonstrated in 1966 by Ernst and Anderson (1). These authors outlined the theory of a spin system subjected to a periodic sequence of pulses. In pulse NMR spectrometry, the time response function—the free induction decay (FID) signal—can be recorded in times on the order of T_2* (transverse relaxation time plus field inhomogeneity effects), thus making possible recording of spectra with a time saving of 100-fold, or, by use of a time-averaging computer, sensitivity enhancement on the order of 20-fold compared to continuous wave (cw) measurements.

We wish to describe a versatile computer arrangement for fast Fourier transform NMR spectrometry (FFT-NMR). The experimental system consists of a high resolution spectrometer modified for pulse-Fourier and directly interfaced to an IBM 1800 computer with 24K of 4 μsec core storage and numerous peripheral devices for data handling and data presentation.

The importance of sampling with a computer of large memory capacity is determined by the bandwidth and resolution requirements of the spectrum being measured. The sampling theorem (2) states simply that, the observed spectral bandwidth is limited to one half the digitizing frequency. Thus, for nuclei exhibiting large chemical shifts (*e.g.*, ^{13}C and ^{31}P) digitizing rates of 10 KHz or greater are needed. Since spectral resolution varies inversely with total observation

---

(1) R. R. Ernst and W. A. Anderson, *Rev. Sci. Instrum.*, **37**, 93 (1966).

(2) H. S. Black, "Modulation Theory," D. Van Nostrand, Princeton, N. J., 1953, Chapter 4.

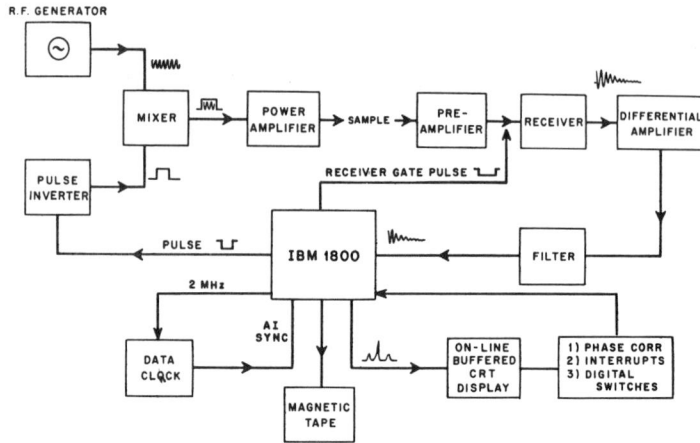

Figure 1. Block diagram of IBM 1800-based spectrometer system for Fourier transform spectrometry

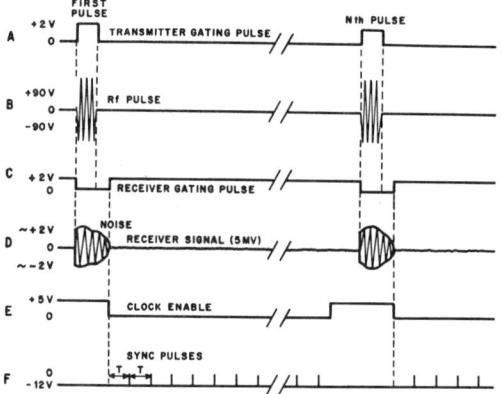

Figure 2. Experimental timing chart

time, $\Delta = 1/T_{obs}$, a memory array of 8192 points digitized at 10 KHz yields minimum line widths of 1.2 Hz.

The 1800 is capable of sampling up to 8192 FID double precision or 16,384 single precision data points at rates up to 20 KHz.

Using this system, we have thus far recorded spectra of ^1H, ^{13}C, and ^{31}P nuclei and substantial time saving or enhanced sensitivity has been realized.

INSTRUMENTATION

The basic configuration of the Fourier transform spectrometer is depicted in Figure 1. It consists of a high resolution Bruker HFX-3 nuclear induction spectrometer operating at 21.5 kilogauss, equipped with a B-SV2 power amplifier for proton noise decoupling and a Bruker 20-watt power amplifier for the excitation pulse. An internal field-frequency lock is provided at 84.7 MHz (^{19}F) or 90.0 MHz (^1H). The rf excitation pulse is created at the mixers (HP 10514) which are gated by logic levels from an IBM Electronic Contact Operate (ECO) register. The transmitter gating pulse and the resulting rf pulse are shown in parts A and B respectively, Figure 2.

The FID signal is detected using the spectrometer low noise, tuned preamplifier (gain $\approx$ 1000) and receiver; however, the final field demodulation stage is omitted. Another mixer is used at the output of the preamplifier to gate out the large noise burst created by the rf excitation pulse in the FID signal, thus preventing saturation of succeeding amplifier stages. The receiver gating pulse is also generated by the 1800 ECO (Figure 2C and 2D). After further amplification and phase detection, the signal is conditioned for the ±5-volt input level of the 1800 ADC Multiplexer with an HP-8875A Differential Amplifier followed by a single section RC bandpass filter with selectable cutoff frequencies for use at several sampling rates. Transmission noise between spectrometer and computer (approximately 25 feet) was negligible when the RC filter was placed immediately before the ADC unit.

The IBM 1800 data acquisition system was designed primarily for use in applications involving relatively slow processes running asynchronously with the computer. Because of this design concept, the ADC makes use of external sampling pulses (sync pulses) to synchronize the computer with an external process. These sync pulses were initially provided by a Tektronix Time-Mark Generator–Type 184. However, since the Time-Mark Generator is free-running and asynchronous with the computer, and the computer-generated exciting pulse with its resulting FID signal is synchronous with the computer, phase incoherence exists between the rf pulse and the onset of sampling (i.e., the time between the rf pulse and the first sync pulse is random). Random phase prevents coaddition of the FID, thus resulting in decreased signal to noise ratios. In the absence of programming to circumvent this problem, the first sync pulse occurs at any time less than the digitizing period (e.g., 100 µsec at 10 KHz) after the rf exciting pulse. This jitter has been minimized to ±4 µsec by holding the computer in a dynamic wait mode until a Time-Mark Generator initiated interrupt continues program flow. It is probably impossible to reduce this jitter further by programming means because of hardware variations in interrupt servicing response. These variations depend upon the status of the 1800 CPU instruction execution at the time the interrupt occurs. To eliminate this jitter completely, a computer controlled data clock was constructed utilizing the basic 2 MHz oscillator of the 1800. A schematic diagram of the data clock is given in Figure 3.

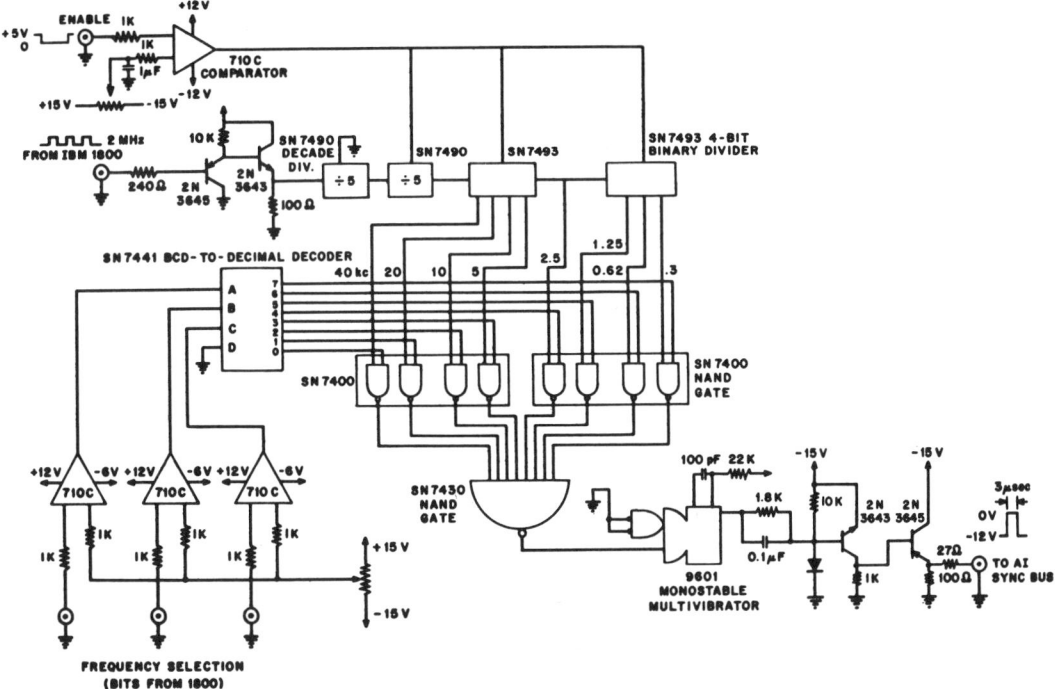

Figure 3. IBM 1800 compatible data clock for digitization of FID's (all voltages and most grounds are deleted for simplicity)

The data clock is enabled by an ECO voltage level change which is generated coincident with the trailing edge of the receiver gating pulse (see Figure 2E). The enable pulse is sensed by means of a Fairchild μA710c comparator which converts the input pulse to TTL compatible levels. The input circuit for the 2-MHz 1800 frequency is a transistor network designed as a buffer amplifier to isolate the data clock from the 1800 logic (input impedance > 1 Megohm). The 2-MHz frequency is first counted down by a factor of 25 using two TI SN7490 decade dividers. The resultant 80-KHz frequency is counted down by a series of 8 flip-flops (two TI SN7493 4-bit binary counters) to yield digitizing rates from 40 KHz to 312 Hz.

The desired frequency is determined by comparator sensing of 3 ECO rate selection bits, set by the controlling program. This is then translated using a TI SN7441 BCD-to-Decimal decoder to NAND the particular frequency desired. The output circuit consists of a retriggerable one-shot (Fairchild TTμL-9601 Monostable Multivibrator) to shape the sync pulse to the proper width for the IBM 1800 ADC ($\sim$3 μsec). There follows a buffered level-translator to give the sync pulse the necessary amplitude for the ADC (-12 to 0 V) and an emitter-follower with a current limiting resistor at the sync pulse output jack (output impedence >1 Megohm).

The results to be derived from adding the data clock are evident from Figure 4 which shows oscilloscope traces of the sync pulses. The oscilloscope was externally triggered in each case on the leading edge of the uninverted transmitter gating pulse. Figure 4A shows the relationship between the sync pulses created by the data clock and the uninverted gating pulse. Figure 4B shows the sync pulse jitter present when using the Time-Mark Generator. The exposure time for

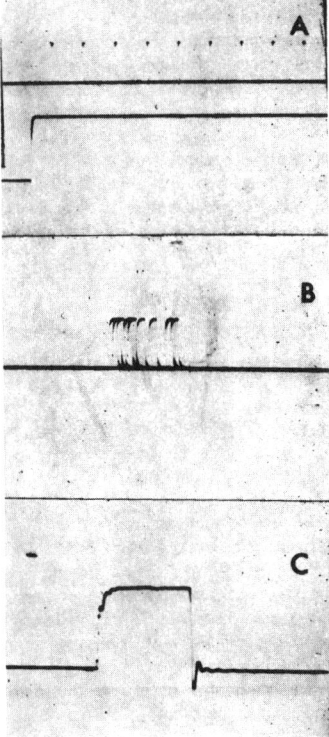

Figure 4. Oscilloscope traces showing sync pulses

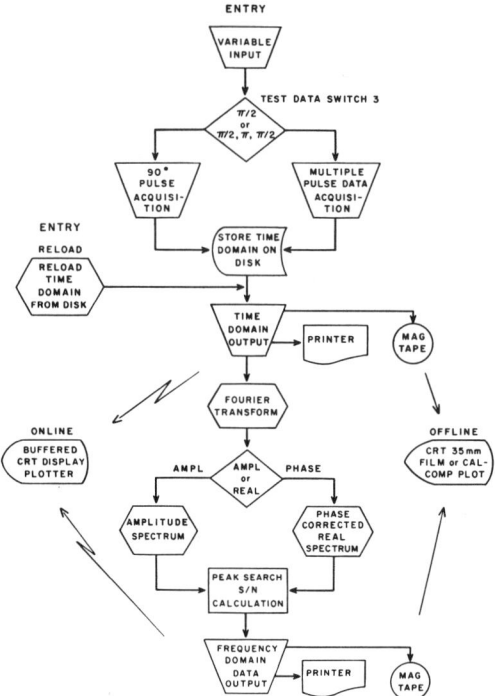

Figure 5. Flow diagram of Fourier transform programs

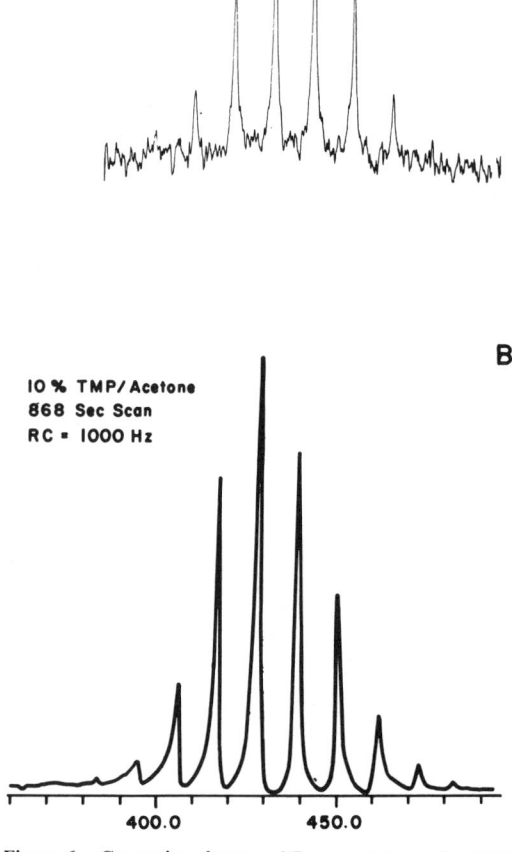

Figure 6. Comparison between ^{31}P cw spectrum of a 10% trimethylphosphite solution (A) and the correspondng Fourier transform spectrum (B)

Figures 4A and 4B was 0.5 sec. Finally, in Figure 4C, a 1-sec exposure of an expanded data clock generated sync pulse is shown. The pulse width is 3 microseconds and, although exposure time is twice that for Figure 4B, no jitter is apparent.

Located at the spectrometer console is a 4K Fabri-Tek 1064 hardwired computer fitted with a Tektronix RM504 oscilloscope and a HP Moseley FAM X-Y recorder. The Fabri-Tek has been digitally interfaced with the 1800 and serves as an on-line buffered CRT-plotter.

SOFTWARE

A flowchart of the Fourier transform programs is depicted in Figure 5. These programs have been coded in Fortran IV and Assembler. The data acquisition routines (NIN1O and NIN2O) must be coded in Assembler to digitize and time average data up to 20 KHz, while routines such as that used in interactive phase correction are necessarily in Assembler, so that some degree of real-time data processing may be maintained. In this case, the Assembler coded routine takes 1 sec to phase correct 4096 frequency domain data points, while the Fortran version takes over 1 min. There has also been a heavy emphasis on integer mode programming wherever loss of precision will not occur or is insignificant.

All routines are stored on disc in core image format. The data acquisition and Fourier routines are divided into six core loads with each block overlaying the previous block. The computer is remotely controlled through the 1816 typewriter, interrupts, and analog inputs at the NMR console.

Data acquisition is initiated by interrupting the 1800, which calls for the initial core load, INPUT. INPUT prompts *via* the typewriter for all variables necessary for data acquisition and transformation and stores them in Fortran Common. INPUT is then overlayed with the pulsing and digitizing core load, NIN1O. Depending upon the particular pulsing program selected, data may be digitized and summed at rates up to 10 KHz (8K two-word integers, or 16,384 one-word integers; 11 bit resolution) or at 20 KHz (4K two-word integers; 11 bit resolution). The rf pulsing is started at the NMR-CRT console by an operator initiated interrupt. The creation of the transmitter gating pulse by the 1800 ECO is a novel feature of the system since most other units use the computer output as a trigger only. The method consists of an assembly language instruction generating the leading edge of the pulse, a precisely timed iteration creating the pulse duration, and an instruction generating the trailing edge. Pulse duration can be varied from 20 microseconds to 16 milliseconds in 500 nanosecond increments. The program can be made to generate periodic pulses, or any multiple pulse

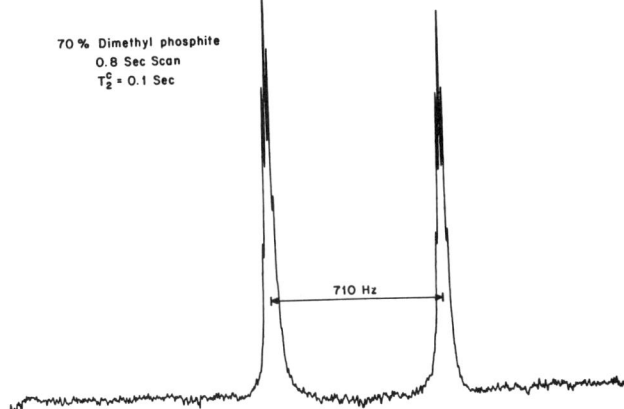

Figure 7. Phosphorus-31 Fourier transform spectrum of dimethylphosphite

combination such as DEFT (3), SEFT (4), and ZRSE (5). The receiver gating pulse is also generated at this time; the leading edge synchronous with the leading edge of the transmitter gating pulse, and the trailing edge after a period equal to the width of the transmitter gating pulse plus some variable delay equal to the preamplifier recovery time (20 μsec to 2 msec). At this point the data clock and the ADC are enabled. As each sync signal occurs, the ADC samples the FID and cycle steals the data via data channel into one of two 128-word chained buffers in core. As each buffer is filled with incoming data, its contents are added to the running summation in the FID array. After collecting the desired number of samples along the FID curve, the ADC and clock are disabled. The program may be delayed at this time up to 60 seconds in a carefully timed loop to allow relaxation of nuclei. After the specified number of FID's are averaged, the final step in the data acquisition program is to subtract any base-line offset and normalize to one-word integers if data have been collected in double precision mode.

An error condition during data acquisition causes the computer to print a message and take appropriate action: digitizer overload—erase memory and reinitialize for new pulse sequence; spectrometer flux stabilizer off—save FID on disc and exit; core overflow—exit to Fourier Transform via the time domain output coreload, ECHO. The rationale for accepting data in which core overflow has occurred is that in all probability only a few data points near the beginning of the FID array will overflow during the scan in which it is detected, thus causing minimal spectral distortion.

After completion of data acquisition, the ECHO coreload is entered. This program first stores the FID array on disc, and then dumps the array to cards, printer, magnetic tape, or the on-line Fabritek depending on which options have been specified. If convolution is desired (i.e., multiplication of the time domain by e^{-t/T_2^c}), it is performed in this coreload.

The data are next transformed to the frequency domain spectrum in the FFOUR coreload using the Cooley-Tukey radix two FFT algorithm (6). Because this routine is partially written in Fortran, the time required to transform 8192 real integer data points is 4 min. The resulting frequency spectrum may be presented as either the phase corrected real or amplitude (power) spectrum. This is accomplished in the PHASE coreload. In the real spectrum case, phase correction is performed with the 1800 and the operator at the NMR-CRT console in an interactive mode. A series of six potentiometers at the CRT console provide appropriate angles for phase correction. Pots 1-5 provide angles for frequency dependent phase correction, with pot 1 representing ϕ at 0 frequency, pot 2, ϕ at $F/4$; pot 3, ϕ at $F/2$, etc. Intermediate phase values are obtained by interpolation. Potentiometer 6 provides a frequency independent phase angle. Although all frequency correction could be done with the five frequency dependent pots, these settings need to be changed much less often than the frequency independent phase angle; thus a separate pot is desirable for operating convenience. Finally, the frequency domain output coreload, OUTPUT, is entered. After calculating RMS base-line noise, line positions, and S/N ratios of each resonance line, the various output options may be selected.

RESULTS

As stated earlier, frequency domain spectra are presented in two ways. The amplitude spectrum is given by

$$Y(\omega) = \sqrt{Y_r^2(\omega) + Y_i^2(\omega)} \qquad (3)$$

where $Y_r(\omega)$ and $Y_i(\omega)$ are the real and imaginary coefficients of the Fourier transform, respectively. $Y(\omega)$ is the amplitude of the power spectral density of the time domain data. This method of data presentation possesses the advantage of being independent of the phase. The drawback in using this method is a broadening of resonance signals.

The real spectrum is defined as

$$Y(\omega) = Y_r(\omega) \cos \phi + jY_i(\omega) \sin \phi, \qquad (4)$$

where ϕ is the appropriate phase angle needed to yield an absorption mode spectrum (1). The real spectrum gives resonance lines which more nearly approach the Lorentzian line shape. After careful evaluation of the iterative phase correction techniques sometimes employed (1, 7), these were rejected in favor of an operator-IBM-1800 iteractive method. In this method, frequency dependent and frequency indepen-

---

(3) E. D. Becker, J. A. Ferretti, and T. C. Farrar, J. Amer. Chem. Soc., **91**, 7784 (1969).
(4) J. S. Waugh, J. Mol. Spectrosc., **35**, 298 (1970).
(5) A. Allerhand and D. W. Cochran, J. Amer. Chem. Soc., **92**, 4482 (1970).
(6) J. W. Cooley and J. W. Tukey, Math. Comput., **19**, 297 (1965).

(7) R. R. Ernst, J. Magn. Resonance, **1**, 7 (1969).

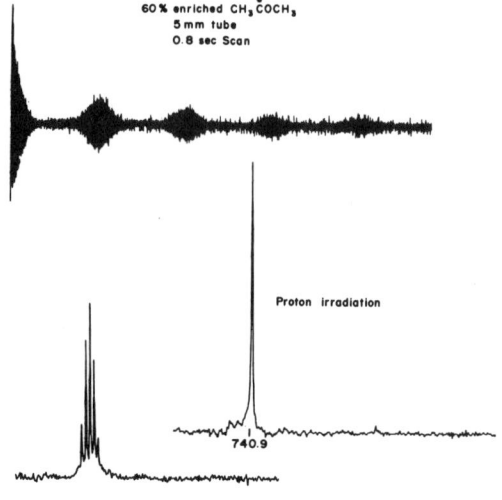

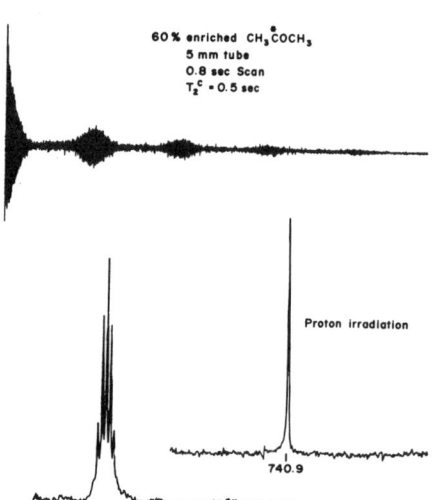

Figure 8. FID and Fourier transform of ^{13}C-enriched acetone showing effects of proton noise irradiation and convolution (lower traces)

dent phase angles are supplied as a series of six potentiometer controlled voltages at the CRT console, with the phase angles being changed, a new phase correction performed, and the resulting spectrum displayed. This process is continued until a satisfactory absorption mode spectrum is obtained.

Some of the advantages accrued from the use of the interactive method over iterative methods are that the interactive technique allows an absorption mode spectrum to be obtained even at low S/N ratios while the ability of iterative techniques to converge on an absorption mode spectrum is dependent on S/N ratio and that both frequency independent and frequency dependent phase correction may be achieved much faster by the interactive method.

For presentation of spectra we have chosen examples of ^{31}P and ^{13}C nuclei since these nuclei possess large chemical shifts and are, respectively, 6×10^{-2} and 1.8×10^{-4} times less sensitive than ^{1}H nuclei in natural abundance for equal numbers of nuclei at the same field strength. All spectra presented will be phase corrected real spectra.

Figure 6 depicts the sensitivity enhancement realized with ^{31}P nuclei. Figure 6A shows a normal continuous wave (cw) spectrum of 10% trimethylphosphite in acetone. The scan time was 1000 sec and a filter time constant of 0.5 sec was used.

The multiplet contains only 6 of the 10 lines associated with the phosphorus resonance. Figure 6B contains the FFT-NMR spectrum of the same sample using identical spectrometer conditions. The 10 lines of the ^{31}P multiplet are clearly visible. The conditions of the FFT-NMR experiment were: total time requirement less than that of the cw experiment (868 sec vs. 1000 sec) while covering the same frequency range; use of a bandpass RC filter ($f_c = 1000$ Hz); and no computer manipulation (i.e. convolution or apodization) on the FID signal was performed.

From a comparison of spectra in Figures 6A and 6B a conservative 10-fold enhancement of sensitivity is claimed for ^{31}P nuclei. It should also be pointed out that, although trimethylphosphite contains a number of lines, it is not an ideal candidate for the FFT-NMR method. Studies using the pulse-Fourier technique indicate a $T_1 \sim 3$ times T_2 for trimethylphosphite.

The alternative advantage of using FFT-NMR over conventional methods is the large time-savings realized. In Figure 7 is displayed a ^{31}P spectrum of dimethylphosphite (70% dimethylphosphite:30% acetone). The large coupling observed is the directly bonded P–H coupling while the smaller splittings are due to the three-bond P–O–C–H coupling. The spectrum depicted in Figure 7 was recorded in 0.8 sec. Samples of ^{31}P compounds in greater than 50% concentration (13-mm sample tubes) are routinely run in times less than 1 sec.

We have studied several biological phosphates and find their salts to be particularly well suited to study by the pulse-Fourier method. The best conditions for enhanced sensitivity by the Fourier method arise when $T_1 = T_2$. This allows extremely rapid pulse periods, hence, time-averaged data will be accumulated rapidly. We have found, for instance, that for the disodium salt of polyadenyllic acid, $T_1 \leq T_2 = 500$ milliseconds and have determined the ^{31}P spectrum for a 3×10^{-5} molar solution in approximately 14 hours. We believe much lower concentrations, concentrations that are "biologically significant," can be achieved in our laboratory.

Some ^{13}C results are presented in Figure 8. The sample consists of acetone enriched to 60% in the carbonyl position and contained in a 5-mm sample tube. The 5-mm tube is mounted co-axially in a 13-mm tube containing hexafluorobenzene for field-frequency locking. The spectrum was recorded in 0.8 sec. The upper portion of Figure 8 is comprised of the ^{13}C FID resulting from a 90° pulse (pulse width = 250 μsec). The resulting phase corrected real spectrum appears below left while a transformed spectrum due to the ^{13}C resonance decoupled from the methyl protons appears at lower right. In the lower portion of Figure 8, the effects of convolution of the time domain data are given. The time domain data are multiplied by the function e^{-t/T_2^c}, where $T_2^c = 0.5$ sec, and the resultant decrease of the noise present in the signal trace is apparent. The corresponding transformed spectra are included for comparison.

Finally, Figure 9 shows a FFT spectrum of a proton decoupled ^{13}C natural abundance spectrum of diethylphthalate. The spectrum is "routine" in that no special efforts

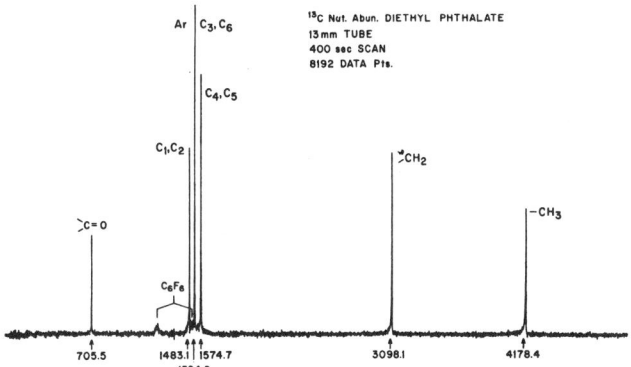

Figure 9. Natural abundance ^{13}C Fourier transform spectrum of diethylphthalate

were taken to optimize conditions prior to recording the spectrum. The solution was contained in a 13-mm tube with approximately 25% C_6F_6 added for field-frequency locking. The digitizing rate was 10 KHz with an rf pulse width of 50 μsec. Pulse widths of 50 μsec were generally used for ^{13}C nuclei since these yielded an optimum flip angle. Satellites present with some peaks are spinning sidebands. The average signal to noise ratio for the peaks is greater than 100 to 1.

The cutoff characteristics of a single section low-pass RC filter with f_c = Nyquist frequency allows down-conversion ("aliasing") of high frequency noise. All spectra except that of natural abundance diethylphthalate (Figure 9) were determined using an RC filter. The filter used in Figure 9 was a 4-pole Tchebyscheff active filter. The active filter was constructed from two experimental components (Analog Devices No. 9171, No. 9172) which allowed six f_c's to be selected under computer control in a manner identical to the rate selection in the data clock. Generally, a filter bandwith corresponding to one-half the current digitizing rate is used. The Tchebyscheff transfer function achieves an extremely sharp roll-off per number of poles. The filter phase shift is greater than 180° throughout the frequency range and becomes distinctly nonlinear near f_c. The present method of phase correction handles such characteristics extremely well. Substitution of the active filter for the RC filter has increased S/N by a factor of two.

Although we have software for forming multiple pulse sequences, a power amplifier which will yield π/2 pulses of 10 μsec or less is needed, and extensive modification of the Bruker high resolution probe is necessary to withstand the high voltages developed. Multiple pulse experiments have therefore been curtailed. However, rapid accumulation of periodic pulses is shown to give significant improvements which warrant extensive use of the pulse-Fourier system. We are currently modifying a Collins 30L-1 linear amplifier (500 watt) to decrease the pulse width necessary for a 90° pulse.

ACKNOWLEDGMENT

We thank Mr. L. Berman for his assistance in the design of some of the electronic circuitry. We should also like to express our appreciation to Dr. Dieter Ziessow for his aid in certain aspects of the computer programming.

RECEIVED for review January 25, 1971. Accepted June 7, 1971. Presented at the 160th National Meeting of the American Chemical Society, Chicago, Ill., September 13–18, 1970. This work was supported by a National Institutes of Health Grant, RR 00356, Biotechnology Resources Branch.

Microprocessor-Controlled, Scanning Dye Laser for Spectrometric Analytical Systems

James A. Perry,[1] Melton F. Bryant,[2] and Howard V. Malmstadt*

Department of Chemistry, School of Chemical Sciences, University of Illinois, Urbana, Illinois 61801

A dye laser has been developed which automatically tunes to any desired wavelength from 360 to 650 nm upon command from a keyboard. A microprocessor controls the wavelength selection by adjusting the angle of a diffraction grating and moving one of several dyes into the laser cavity. A motorized, multiple dye cell carriage provides rapid change of up to 16 previously aligned dye cells. An intracavity telescope provides spectral linewidths as narrow as 0.06 nm. The 360–650 nm region may be scanned at speeds up to 20 Å/s with an added 5 s required whenever a change to the next dye solution is needed. The microprocessor system can either collect analytical data directly, or be used in a slave mode and incorporated into a larger instrumental system run by a minicomputer. Microprocessor-obtained atomic fluorescence data and minicomputer-obtained molecular fluorescence data demonstrate the analytical utility of the system.

The potential of the tunable dye laser has been acclaimed in many recent review articles (1-3), and several commercial systems are presently available. However, most systems can only be tuned over a spectral region covered by one dye, typically 20 to 40 nm, and then require a cumbersome manual change of dye solutions. As a result, the vast majority of dye laser applications reported in the literature have been limited to experiments that can be performed in a wavelength region as narrow as one laser dye. Furthermore, the very desirable feature of scanning the dye laser over large spectral regions is rarely used because of the inconvenience of changing dye solutions. This problem and the complexity of wavelength calibration and adjustment have left the dye laser a powerful but cumbersome spectral source.

Changing dye solutions in the middle of a scan is a difficult operation because of the stringent alignment requirements of a laser cavity. Large wavelength scans over a region such as 360–650 nm require as many as 13 different dye solutions for good efficiency throughout the scan. Thus, after each dye solution is originally aligned, the controlling circuitry must be capable of repeatedly placing it back in exactly the same position. Accurate knowledge of the laser wavelength as determined by the grating angle is also a necessity, and many experimental situations call for the ability to quickly change the excitation wavelength from one precisely known value to another. These instrumental complexities combined with the scientist's need for flexibility in designing experiments make the scanning dye laser a prime target for automation and computerization.

The microprocessor is an excellent device for controlling an automated instrument such as a dye laser. The tedious tasks of adjusting the grating angle and changing dye solutions can be easily programmed into these new, inexpensive, but very powerful integrated circuits. With a small keyboard and display added to provide a communication link, calibration and adjustment of the laser wavelength become a matter of fingertip control. In similar fashion, the microprocessor may be used as a slave device to a minicomputer or time-sharing terminal. Such a combination can greatly speed an analysis by allowing the more powerful computational devices to spend their time calculating, printing, and storing data while the time-consuming wavelength adjustment is performed by the microprocessor.

In this work, the construction of a multiple dye cell, microprocessor-controlled dye laser is discussed. The nitrogen laser that is used as a pumping source is also described. A few experiments which take advantage of the instrument's wide tuning range are then presented.

INSTRUMENT DESIGN

The basic components which make up the scanning dye laser are shown pictorially in Figure 1. A complete description of each element, its function, and interrelationship to the overall design is given. Helpful information on dye laser design (4-6) and on nitrogen lasers (7-9) is available in the literature.

Nitrogen Laser. The nitrogen laser is based on the design of Metcalf and Schenck (7), but with a number of modifications. These include removal of the bandsaw blade from the laser channel which was found to be unnecessary, and various component and circuit modifications for higher voltage operation as indicated in Figure 2. All components of the

---

[1] Present address, Central Research and Development Department, E. I. DuPont de Nemours and Co., Experimental Station, Wilmington, Del. 19898.
[2] Present address, Department of Chemistry, University of Georgia, Athens, Ga. 30602.

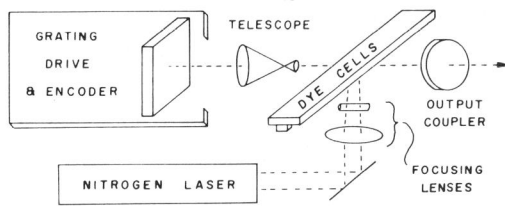

Figure 1. Scanning dye laser

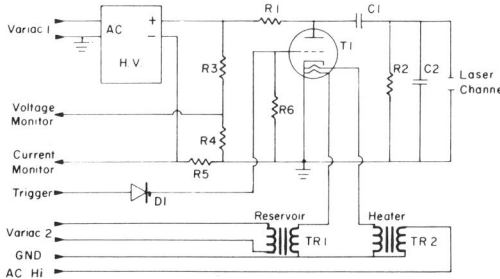

Figure 2. Circuit diagram of nitrogen laser

10 Hz, 20 kV, and 50-Torr dynamic pressure. The laser output energy for these conditions is 1.35 mJ/pulse which corresponds to 135 kW for a 10-ns pulse.

In order to shield against the large quantity of radio frequency interference (RFI) which accompanies the high voltage discharge, the entire nitrogen laser and associated circuitry has been placed in an 18-gauge steel box (110 × 20 × 20 cm). The steel dampens the RFI appreciably, but any metal connections placed within a few inches of the box will still pick up the noise. Therefore, the trigger signal must be delivered through a 50-cm fiber optic cable in order to isolate the digital electronics from the RFI, as illustrated in Figure 3. Adequate RFI shielding is an extremely important consideration if the laser is to be used near digital circuitry.

The nitrogen laser is physically located approximately 70 cm below the optical bench which holds the dye laser, and the output beam is folded up through a hole in the instrument table. Once above the table, the 337-nm laser beam passes through a $2^{1}/_{2}$-inch diameter quartz spherical lens with focal length of 10 inches, then through a hole in the aluminum optical bench, and finally is focused onto the dye cell with a 50-mm focal length quartz cylindrical lens. This optical configuration condenses the nitrogen laser beam which is originally 5 mm × 37 mm into a sliver of radiation just long enough to cover the 1-cm dye cell. Concentrating the UV pumping source into a very narrow line has proven to be important for obtaining good efficiency and a narrow bandwidth from the dye laser.

Dye Laser. A photograph of the dye laser is shown in Figure 4. The unit is 30 cm in length and is composed of five basic elements exclusive of the nitrogen laser used as a

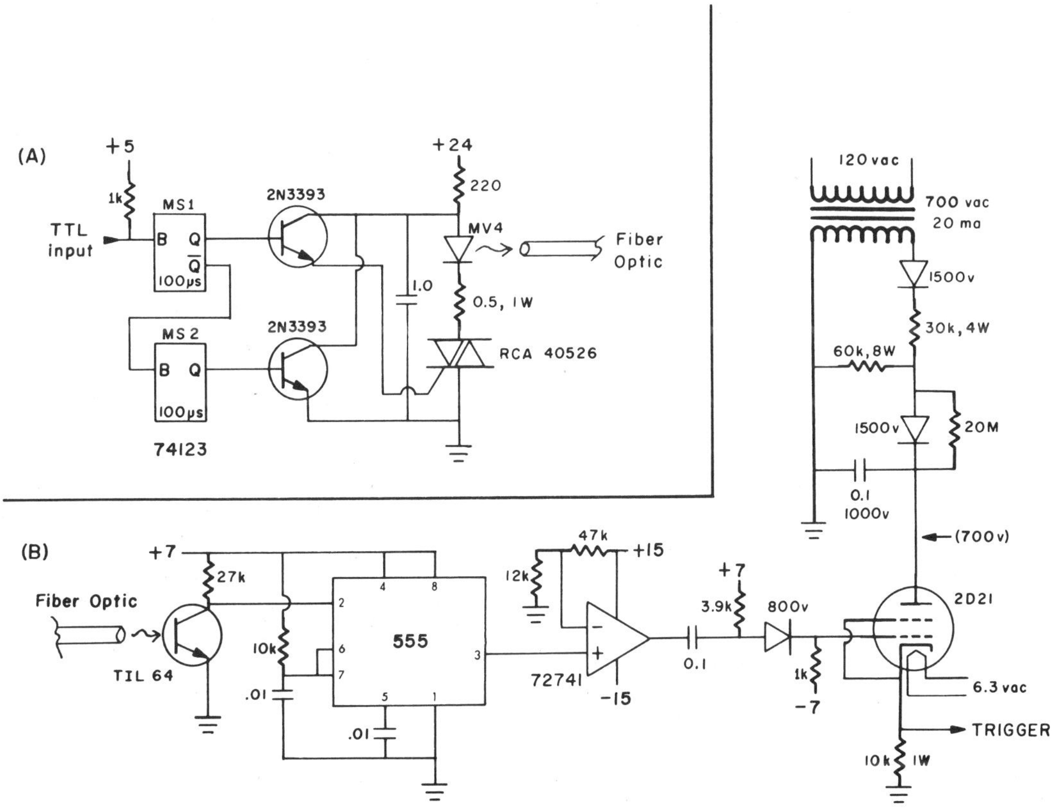

nitrogen laser and their manufacturers are listed in Table I. The circuit provides for operation of the laser at repetition rates up to 25 Hz with an applied voltage of 25 kV. Provisions have been made for monitoring both current and voltage. For dye laser pumping, the nitrogen laser is typically operated at

Figure 3. Nitrogen laser trigger circuit at (a) microprocessor, and (b) nitrogen laser

Table I. Nitrogen Laser Components

Component	Manufacturer
H. V. Power Pack Model 25C, 25kV	Hipotronics, Inc., Brewster, N.Y.
R1 = 200 k ⎱ Wire wound power resistor R2 = 10 k ⎰	Ohmite, Skokie, Ill.
R3 = 125 MΩ, 25 kV (57 × 2.2 MΩ, 1W ⎫ R4 = 50 k, 1/2 W ⎪ R5 = 500, 1 W ⎬ Carbon resistor R6 = 150, 1.0 W ⎭	
C1 = 0.02 μF, 50 kV EB-203-MX-1	Condenser Products, Brooksville, Fla.
C2 = 30 × 500 pF, 30 kV	Centralab Electronics Division, Milwaukee, Wis.
D1 = 30 kV PIV, 70GF-S300	Sarkes-Tarzian, Bloomington, Ind.
T1 = 5949A Glass Hydrogen Thyratron	ITT Electron Tube Division, Easton, Pa.
TR1 = 6.3 V ac @ 10 A	
TR2 = 6.3 V ac @ 20 A	

Figure 4. Photograph of scanning dye laser. (A) output coupler mount, (B) dye cell carriage, (C) telescope, and (D) grating

pumping source. These include a diffraction grating, a beam expanding telescope, a quartz dye cell, an organic dye solution, and an output coupler. Each of these elements has important properties which affect the dye laser output.

Grating. A 1200 lines/mm reflection grating (50 mm × 50 mm) blazed for 500 nm is the tuning element of the dye laser. The efficiency of the grating is an important consideration. Grating efficiency reaches a maximum at the blaze wavelength and typically drops to one half of the maximum value at 2/3λ and 3/2λ where λ is the blaze wavelength ("Diffraction Grating Handbook", Bausch and Lomb, Inc., Rochester, N.Y.). The grating used in the dye laser has approximate efficiencies of 40% at 360 nm, 90% at 500 nm, and 80% at 650 nm. For maximum efficiency from the dye laser at a specific wavelength, one would obviously want to use a grating blazed as close as possible to the desired wavelength. However, if a system is to be capable of scanning a large wavelength region, some compromises must be made between the grating blaze wavelength and the other components of the dye laser. Gratings are typically blazed for either 400 or 500 nm in the visible and the choice depends on where the best efficiency is desired.

Beam Expanding Telescope. This element serves to spread the narrow beam of dye fluorescence over a larger area of the grating. Since the resolution obtainable from a grating is proportional to the width of the grating covered by radiation, expanding the beam will narrow the bandwidth of the tuned radiation. Theoretically, a 10× telescope should narrow the laser bandwidth by a factor of 10, providing that the grating does not become overfilled. The beam expander used in this work is a commercial unit (Oriel Corp., Stamford, Conn.) based on the Galilean telescope design. An output lens with 46-mm aperture can be combined with a choice of input lenses to give expansion powers of 10×, 20×, or 40×. In conjunction with the 1200 lines/mm grating used in 1st order, the telescope yields bandwidths of 0.25, 0.12, and 0.06 nm, respectively. All lenses are anti-reflection coated which helps to prevent broadband lasing from occurring between the telescope input lens and the output coupler. The telescope also serves as a focusing element in the laser cavity directing a larger portion of the fluorescence radiation back into the narrow, excited region of the dye solution for amplification.

Dye Cells. The most important feature in the dye cell design was an accurately reproduced geometry such that all cells could be aligned in the laser cavity without having to adjust the telescope or output mirror. Compactness was also an important consideration in designing a multicell system. In addition, the use of a dye cell with windows perpendicular to the region of dye excitation was found to be unacceptable since broadband lasing will occur between the cell walls. The answer to these requirements was a special quartz cell made especially for dye lasers which uses a three-degree angle as shown in Figure 5A. The cells, which are very similar to those available commercially (Molectron Corp., Sunnyvale, Calif.), were sliced into 6-mm sections with a common glass grinding wheel. Teflon sides were etched (Hi-D No. 40 Fluorocarbon Etchant, Cadillac Plastic, Chicago, Ill.) (and then glued) (Epoweld #13003, Hardman Inc., Belleville, N.J.) to the sections to yield cells with volumes of approximately 0.5 mL. Teflon tubing was also etched and glued into holes in the Teflon side pieces so that the dye solution could be easily changed or flowed to provide maximum output at high repetition rates. The dye cells also have an anti-reflective coating which helps to reduce cavity losses.

Dye Solutions. Many laser grade dyes are now available commercially from a number of sources (Molectron Corp., Sunnyvale, Calif.; Eastman Organic Chemicals, Rochester, N.Y.; Exciton Chemical Co., Dayton, Ohio; New England Nuclear, Watertown, Mass.). The specific dyes used to cover the spectral region from 360 to 650 nm are listed in Table II. The tuning range and maximum efficiency is different for each dye. The obtained output energy, however, depends not only on the dye but also on the various other components of the dye laser and on the energy in the pumping source. When the dye solution is stationary, repetition rate is an important factor. The output from three of the dyes (Table II) will fade rapidly to a steady-state level at a repetition rate of 10 Hz. All dyes lose pulse energy at repetition rates above 15 Hz and should be flowed through the dye cell if maximum output energy is required. However, flowing of dye solutions has not been necessary for any investigations to date.

Output Coupler. It is desirable to choose an output coupler that lets the maximum amount of laser radiation out of the cavity while providing enough feedback for tuning the dyes pump energy. Thus, the optimum reflectivity for the output coupler may change greatly from one wavelength to the next. However, changing output couplers while scanning from one wavelength to the next would add to the system's complexity. The high gain of the dye solutions coupled with an intense pumping source allows the use of an uncoated quartz substrate as the output coupler. The nominal 4% reflectivity of a single quartz-air interface used in conjunction with the other components described in this work has proved to be adequate for tuning dyes in the 360–650 nm range.

Multiple Dye Cell Carriage. An aluminum dye cell carriage capable of holding up to 16 dye cells is shown schematically in Figure 5, B and C. Each cell is individually adjustable in both the x–z and x–y planes to allow proper alignment in the laser cavity. Independent alignment of each cell is important in order to compensate for minor differences in the quartz cells and to correct for variations in refractive index of the various dye solutions, particularly those using different solvents. The carriage is suspended from a support structure and is moved when a dc motor acts upon a $1/4$-inch diameter threaded rod. The position of the carriage is monitored by a relative encoder with resolution of one quarter turn corresponding to 0.3 mm ($1/80$ inch). This has proved to be very reliable for aligning the dye cells which are 6 mm ($1/4$ inch) wide and spaced 12.5 mm ($1/2$ inch) apart.

The control circuitry for the dye cell carriage is shown in Figure 6. The encoder circuit uses the 555 integrated circuits as wide-range Schmitt triggers for noise immunity. Care must still be taken, however, to avoid opening or closing the direction-determining relay when power is applied through transistor Q_1. Any such attempt to instantaneously change the direction of the motor produces a large surge of RFI and generates false encoder pulses. The proper timing needed to meet this requirement is quite simple with the programmable microprocessor.

Thirteen dye cells are mounted on the carriage and rolled into the laser cavity as needed to tune the region from 360 to 650 nm. Only 5 s are required to change between dye solutions which are adjacent to each other on the carriage. As an added option, one of the vacant dye cell positions has been fitted with a folding mirror so that the 337 nitrogen laser beam may be directed down the same optical axis.

Grating Drive and Control. In order to provide precision adjustment of the grating, the sine bar-leadscrew mechanism of a commercial monochromator was used (700 Series Monochromator, GCA/McPherson Instrument, Acton, Mass.). Only a few modifications were required which included removing the front cover and turning the grating around 180° on the sine-bar mount so that the grating faces out the front of the monochromator. A slight realignment was then needed to make the grooves of the grating orthogonal to the sine-bar movement. Adjustment in the length of the sine bar was also needed to convert the unit from a Czerny–Turner to a Littrow configuration. The appropriate grating equation is:

$$m\lambda = 2a(\sin \theta)(\cos \phi)$$

where $\phi = 17°$ for GCA Czerny-Turner and $\phi = 0°$ for Littrow. Thus, the sine-bar length must be increased by the factor cos (0°)/cos (17°). The GCA monochromator comes equipped with an encoder and TTL control circuit that allows for slewing through the wavelength region in either direction and stepping the grating at 0.01 nm per step.

Microprocessor Control. An Intel 8080 microprocessor is used to control the dye laser system as illustrated in Figure 7. All of the microprocessor components enclosed in the dotted lines are part of the Intel System Design Kit (Intel Corp., Santa Clara, Calif.). The printed circuit board has room for 48 lines of parallel input/output and up to 4k of PROM (programmable read-only memory). With the exception of

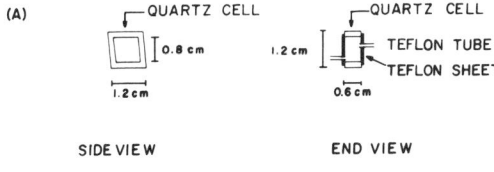

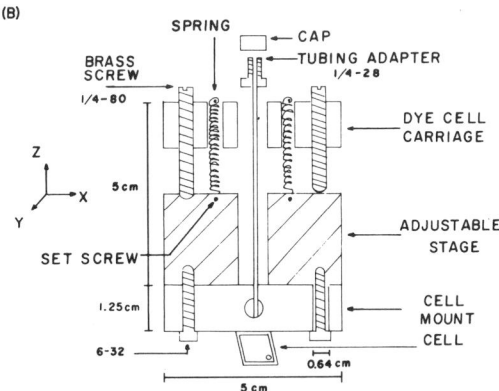

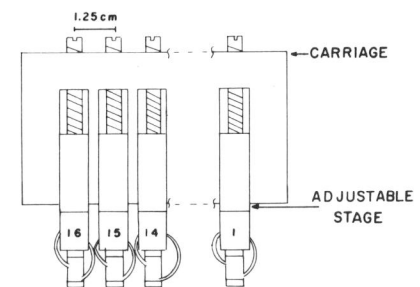

Figure 5. (A) Quartz dye cell, (B) cross-section of dye cell carriage, (C) front view of dye cell carriage

Table II. Laser Dyes Used to Cover 360–650 nm

Cell No.	Tuning range, nm	Programmed wavelengths	Dye
1	337	000–360	Nitrogen Laser
2	360–399	360–379	PBD[a]
3	373–399	379–394	BBQ[a]
4	391–411	394–405	PBBO[a]
5	396–416	405–414	DPS*[a]
6	411–430	414–429	BIS-MSB[a]
7	420–457	429–447	C-120[b]
8	440–478	447–473	C-1[b]
9	453–495	473–487	C-102[b]
10	475–547	487–531	C-500[c]
11	515–583	531–570	C-153*[b]
12	568–605	570–599	R6G[a]
13	594–643	599–636	RB[b]
14	628–651	636–652	RB + CVP*[b]

[a] Molectron Corp., Sunnyvale, Calif. [b] Eastman Organic Chemicals, Rochester, N.Y. [c] Exciton Chemical Co., Dayton, Ohio. "*" indicates dyes which faded markedly to a steady-state level when stationary solutions were pumped at 10 Hz.

over their full lasing range. The necessary reflectivity depends on the dye solution, the efficiency of the grating, and on the

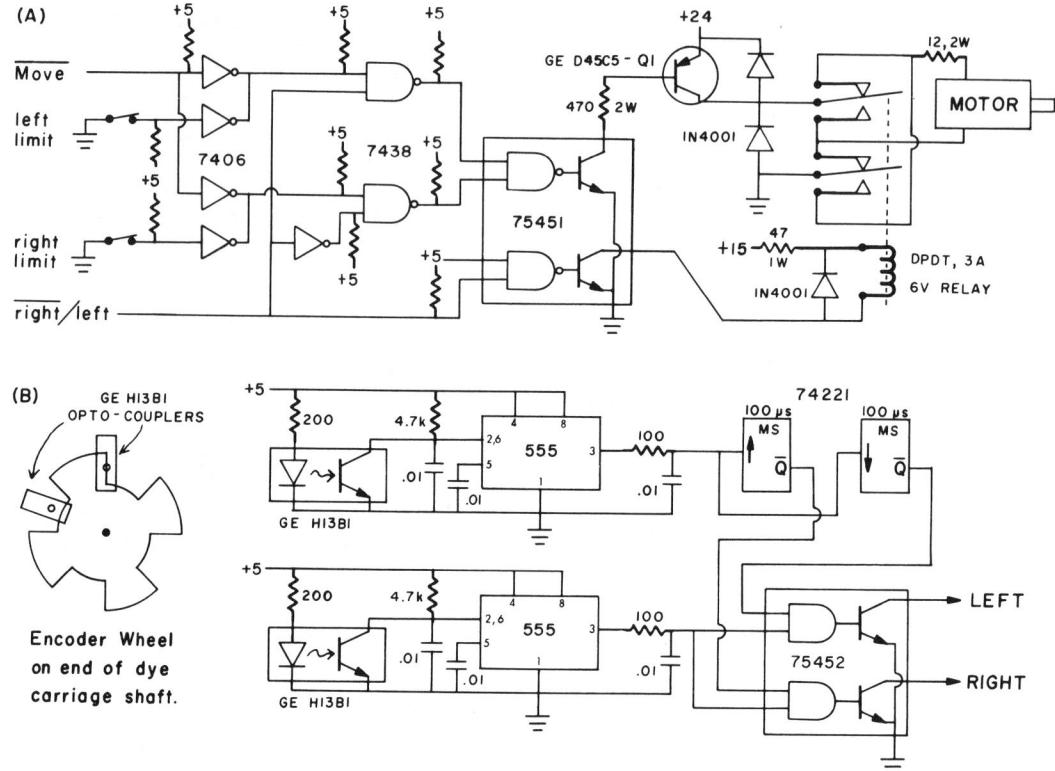

Figure 6. Dye cell carriage control circuit. (A) motor control, (B) encoder circuitry

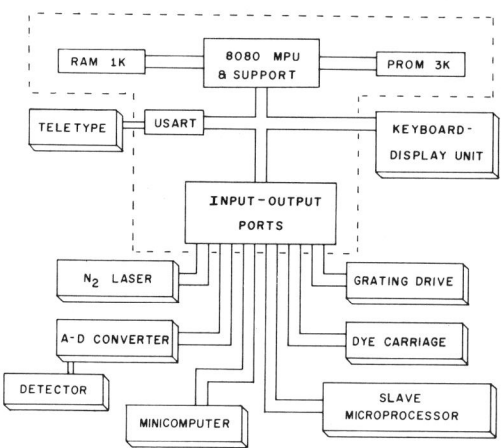

Figure 7. Microprocessor controller

the teletype, all external devices are linked to the microprocessor through the parallel I/O ports. The keyboard-display unit consists of a small 20-key keyboard and nine 7-segment displays, and is used to provide a communication link between operator and processor. Thus, the instrument does not require a relatively expensive teletype unless data-taking routines are being run and hard copy of the data is needed. A copy of the keyboard-display unit circuit is given elsewhere (10). The interface circuit between the microprocessor and all other components is shown in Figure 8 and consists primarily of buffering circuits needed to drive the connecting cables.

The microprocessor system is responsible for selecting the best dye solution and locating the correct grating angle for any given wavelength. This is accomplished by sending the appropriate signals to the control motors and "observing" their motion by monitoring the encoder signals with the microprocessor interrupt system (Figure 8). Calibration of the grating and dye cells is accomplished by running the mechanisms to their lower limit and applying a predetermined value stored in PROM.

For small scale analysis that requires minimal data manipulation, the microprocessor can also fire the laser and acquire the data through an 8-bit analog-to-digital converter. Results can be shown on the keyboard-display unit or, if a teletype is available, hard copy of wavelength and data can be provided. Analysis of complex samples often requires the more sophisticated data handling capabilities of a minicomputer. Under such circumstances, data acquisition is handled by the minicomputer while the microprocessor is dedicated to controlling the dye laser wavelength. This combination can greatly speed sequential analysis. After gathering data for one wavelength, the minicomputer processes the raw data and prints the results while the microprocessor is busy locating the next wavelength. The processor–computer communication link is established in a manner that gives the minicomputer priority over the microprocessor keyboard only during the few microseconds required to transfer an instruction. Thus, keyboard control of the dye laser remains an available option at all times.

The programmable feature of a microprocessor makes the dye laser an extremely versatile research tool. General

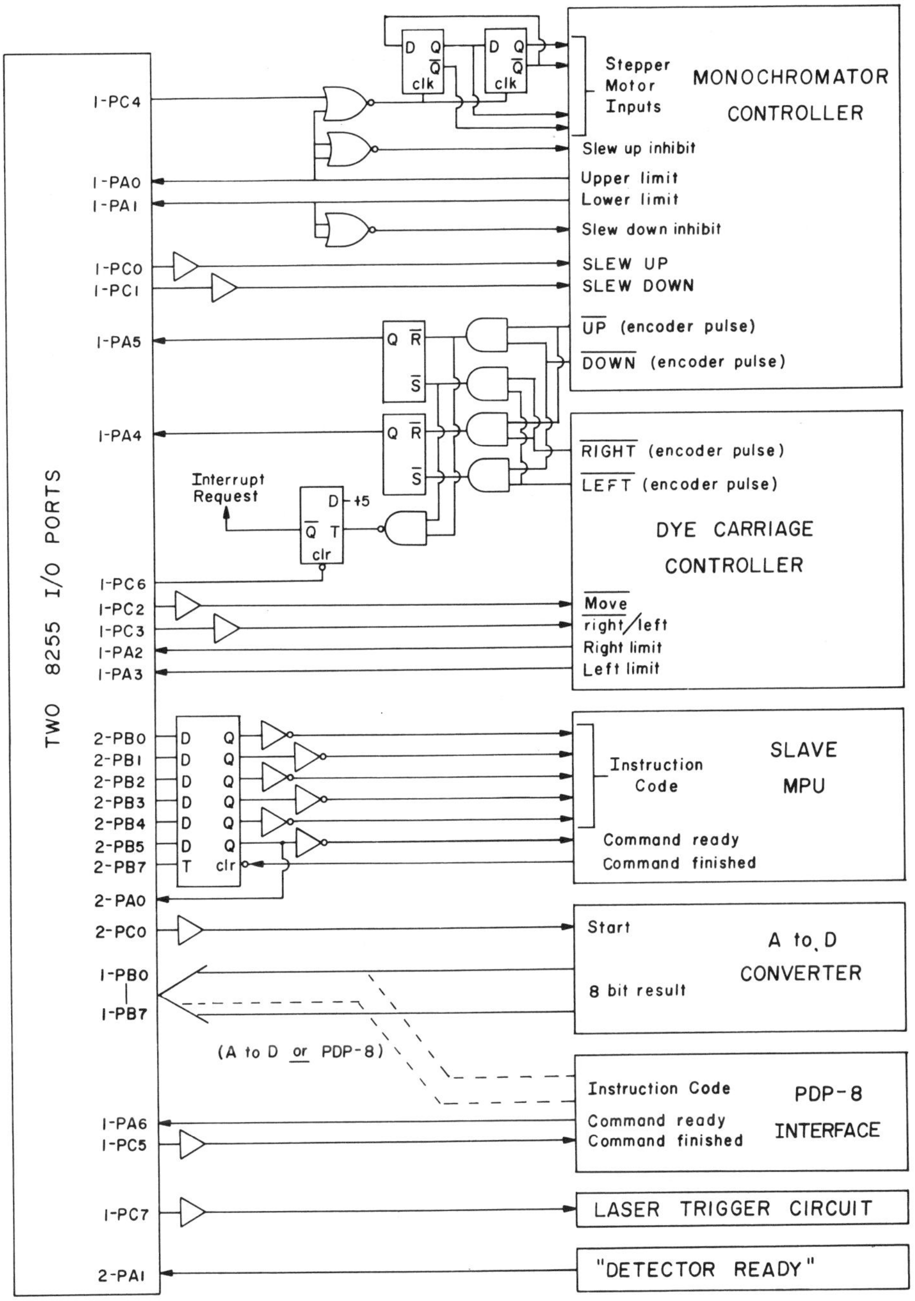

Figure 8. Microprocessor interface

programs are stored in PROM's and are easily called up from the small keyboard the moment power is supplied. These programs include: (1) setting the dye laser output to within 0.01 nm of any specified wavelength; (2) scanning between any two extremes at a given rate; (3) peak finding; (4) acquiring, averaging, and printing data; and (5) automatic calibration of the grating angle and dye cell position. New programs are generally written in assembly language on a PDP-8 minicomputer which produces binary code on paper tape. This allows the program to be loaded into the microprocessor RAM (random access memory) through the teletype paper tape reader. A short monitor routine, stored in PROM, then permits the user to examine, modify, and execute the test program by using the microprocessor's small keyboard and display. Once a program is "debugged", it may be written into a PROM to provide a permanent copy which will not be lost when power is turned off.

Detection of Equipment Problems. In addition to

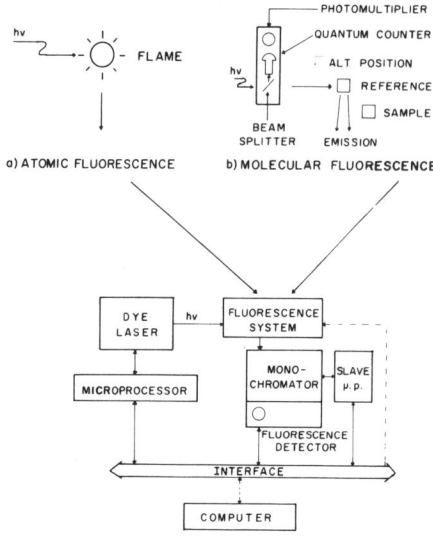

Figure 9. Experimental configuration for collecting (a) atomic fluorescence with microprocessor and (b) molecular fluorescence and background under minicomputer control

control, the microprocessor is also capable of detecting equipment malfunctions. Programming this type of intelligence into the microprocessor can save the operator from wasting both time and precious samples on a malfunctioning instrument. Costly repair bills and time consuming searches for the cause of a problem may also be avoided. One example would be the detection of a mechanical problem in the dye cell carriage or monochromator mechanism. Since the encoding scheme for the dye cell and grating movements provide a closed-loop of communication, the microprocessor expects to receive interrupting encoder pulses within a short period of time after activating a motor. Absence of encoder pulses might indicate that a mechanical freeze-up has occurred and that a control motor is in danger of being burned out. The closed-loop system makes it possible for the control software to detect such situations, and avoid damaging expensive equipment. If encoder pulses are not received within 1 s after a "motor on" signal has been sent, the microprocessor will deactivate the motor, ring the teletype bell 12 times to alert the operator, and flash a number on the keyboard displays pinpointing the source of the malfunction. In a similar fashion, the microprocessor will indicate an unplugged control cable, and signal the operator if either the grating or dye cell drive has reached the limit of its travel. Data-taking routines also monitor the detector response to be sure that the laser actually fires when it is instructed to do so.

EXPERIMENTAL

The instrumental arrangement used for collecting the fluorescence data is shown in Figure 9, and was used for both atomic and molecular information. The atomic data were gathered entirely by the microprocessor while a minicomputer was used to control the more complex molecular system. A second microprocessor is used to control the wavelength of the emission monochromator. The "slave" microprocessor is sent instructions from either the master microprocessor or from the minicomputer, and operates in a manner identical to the master microprocessor when receiving instructions and when adjusting the emission monochromator grating.

The device used to capture the short burst of fluorescence radiation is a transient integrating detector (TID) which was designed at the University of Illinois (11). The detector is capable of integrating the current from a 1P28 photomultiplier tube for periods of 2–100 ns. An integration period of 50 ns is generally used to catch the entire burst of fluorescence while excluding virtually all background and dark current.

Multielement Atomic Fluorescence. A relatively bright air-acetylene flame from a premixed burner was used to gather the atomic fluorescence data. The gas mixture was not optimized for any particular element. A composite solution containing 10 ppm Ca, In, Mn, Sr, and 1 ppm Na was prepared from stock solutions and aspirated into the burner at approximately 1 mL/min. Entrance and exit slits on the emission monochromator were opened completely to 2000 μm in order to provide maximum throughput. No other light gathering optics were used. The fluorescence data were then generated by scanning the dye laser across each of the atomic lines. The starting point of each elemental scan was manually loaded into the microprocessor through the small keyboard and followed by an execution key. The master microprocessor then adjusted both the laser and the emission monochromator to that wavelength, and fired the laser ten times at 10 Hz while reading the detector response through an 8-bit A-to-D converter after each pulse. After printing the wavelength and average of the 10 points, the laser and emission monochromator were both stepped forward by 0.02 nm and the data-taking process was repeated until the atomic line or lines were covered. The next starting point was then entered and the routine continued until all elements were analyzed.

Molecular Fluorescence. The molecular fluorescence spectrum of fluorescein (disodium fluorescein, Matheson Coleman and Bell, in 0.05 M disodium phosphate) was obtained by the use of a double-beam recording spectrofluorometer. The reference and sample cells were placed in a cell holder that was moved on a sliding stage by a computer controlled air cylinder (Model TF Tiny Tim, J. N. Fauver Co., Schaumburg, Ill.) as shown in Figure 9b. The computer automatically positioned the reference or sample cell (1-cm fluorometric cells) in the excitation path and recorded the 90° fluorescence intensity. A dual-beam detector system shown in Figure 9 was used to measure the fluorescence and monitor the change in excitation intensity as the dye laser was scanned. The reference detector consists of a rhodamine B quantum counter–photomultiplier combination similar to that reported by Mielenz et al. (12). Ratioing the fluorescence intensity to the reference detector signal provides a quanta-corrected spectra when scanning the dye laser. A background-corrected spectrum was obtained during this scan by measuring the sample and blank fluorescence at each wavelength and recording the difference. This system is described in more detail elsewhere (13).

Dye Laser Energy. Measurements of the dye laser output energy from 360 to 650 nm were made with a thermopile detector (Model 36-0001 Laser Power Meter, Scientech, Inc., Boulder, Colo.) and microvoltmeter (Model 115 Null Detector Microvoltmeter, Keithley Instr., Cleveland, Ohio). The thermopile has a virtually flat spectral response over this region and measures average power with a sensitivity of 100 mW/V. The data were taken under microprocessor control with the meter output of 0–1 volt read through an A-to-D converter. Measurements were taken by firing the laser for 40 s at 10 Hz, performing 100 A-to-D conversions of the meter voltage, and storing the results in memory. After a 40-s "off" period, another set of conversions was taken and subtracted from the first in order to correct for long-term drift in the thermopile. Wavelength and intensity were recorded on a teletype paper tape punch for later input to a minicomputer plotting routine. Energy measurements were taken at 1.0-nm intervals from 360 to 430 nm, and at 2.0-nm intervals from 430 to 650 nm.

RESULTS

Laser Characteristics. The pulse energy scan from the dye laser is shown in Figure 10. Maximum output reaches 68 μJ/pulse which corresponds to a peak power of 6800 Watts assuming an 10-ns dye laser pulse. At a 10-Hz repetition rate, the average power is 0.68 mW. For high resolution studies and atomic excitation, the power/unit bandwidth becomes an important consideration. Based on a 0.10-nm halfwidth, the dye laser output may be represented as 68 kW/nm peak

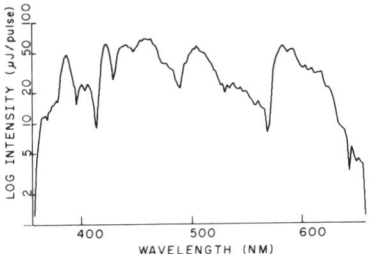

Figure 10. Dye laser pulse energy from 360 to 650 nm

or 6.8 mW/nm average power. Low points in the energy spectrum correspond to change-over points between dyes. Three of the dyes indicated in Table II fade rapidly to a steady-state intensity level. Improved performance can be obtained by attaching a pump and reservoir to the dye cell tubing and flowing the dye solution. This has not been done, however, and the data shown in Figure 10 were taken with all solutions stationary in the 0.5-mL dye cells.

Pulse-to-pulse reproducibility of the laser is another important characteristic. Measurements made with a 2-inch photovoltaic cell monitored by an oscilloscope indicate a variation of approximately 3%. This fluctuation is the result of similar variation in the nitrogen laser pumping source, and must be taken into consideration if precision results are to be obtained.

The dye laser linewidth can be determined by examining the width of the peaks produced by scanning an elemental line. Since the atomic absorption bandwidth is much smaller than the laser linewidth, the width of these peaks gives an accurate reflection of the width of the laser line. This procedure allowed the linewidth to be determined to within 0.01 nm, which is slightly better than the resolution obtainable with standard monochromators. When a dye solution is fresh, the typical linewidth is 0.10 to 0.12 nm. However, a scan of 10 ppm Sr taken after the coumarin 1 dye had been used continually for 4 h at 10 Hz indicated a linewidth of 0.06 nm. This narrowing is easily explained. As the dye fades slightly over time, the gain in the fringes of the optical bandwidth determined by the telescope and grating drops below the lasing threshold, and only the narrower spectral region with sufficient gain continues to lase.

Atomic Fluorescence. The data recorded from a multielement scan are shown in Figure 11. The narrow linewidth of the laser is apparent in the scans of the manganese triplet and the sodium doublet. Complete scans of the elemental lines would not be necessary in routine analysis. Time could be saved by taking only peak-finding and background measurements requiring 15 to 20 s. Considering the additional time needed to slew the grating from one element to the next and to change dye cells, the five elements shown here could be analyzed in 2.5 min requiring 2 to 3 mL of solution. The ability to make background and scatter corrections with the dye laser is an important advantage not readily available with hollow cathode atomic fluorescence systems which emit only at the elemental line. Such information is easily obtained by tuning the laser to a point just off the elemental absorption line.

Molecular Fluorescence. The excitation spectrum of fluorescein is shown in Figure 12A. Production of these data requires scanning the laser through three dye solutions as indicated in Figure 12B, and illustrates the importance of automating the laser for recording excitation spectra. The ability of this system (13) to compensate for variations in source intensity and thus provide corrected spectra is also illustrated in Figure 12. Note that the maximum corrected

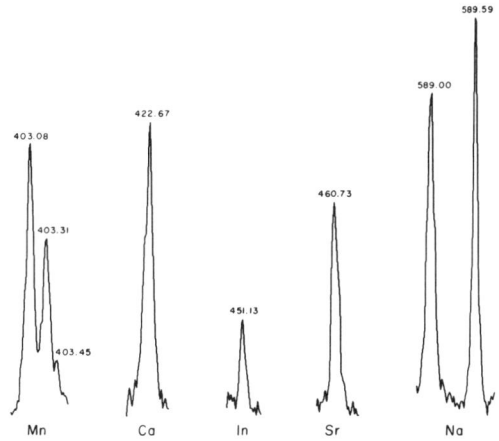

Figure 11. Multielement scan from composite solution of 10 ppm Mn, Ca, In, Sr, and 1 ppm Na

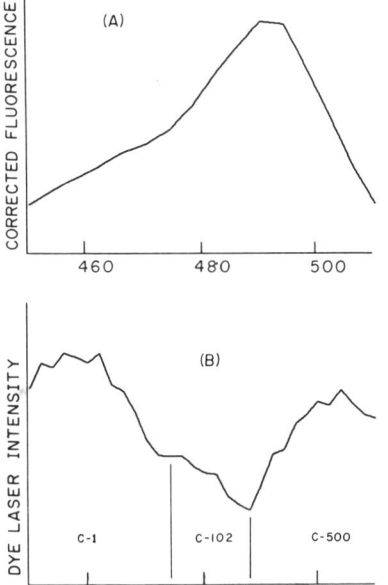

Figure 12. (A) Fluorescein excitation spectrum corrected for dye laser intensity. (B) Dye laser intensity through the region of fluorescein spectrum shown in (a)

fluorescence intensity is measured at the minimum laser intensity.

LITERATURE CITED

(1) J. R. Alkins, *Anal. Chem.*, **47**, 753A (1975).
(2) J. I. Steinfeld, *Crit. Rev. Anal. Chem.*, **5**, 225 (1975).
(3) J. P. Webb, *Anal. Chem.*, **44** (6), 30A (1972).
(4) J. E. Lawler, W. A. Fitzsimmons, and L. W. Anderson, *Appl. Opt.*, **15**, 1083 (1976).
(5) T. W. Hansch, *Appl. Opt.*, **11**, 895 (1972).
(6) G. Capelle and D. Phillips, *Appl. Opt.*, **9**, 2742 (1970).
(7) P. Metcalf and H. Schenck, *Appl. Opt.*, **12**, 183 (1973).
(8) B. W. Woodward, V. J. Ehlers, and W. C. Lineberger, *Rev. Sci. Instrum.*, **44**, 882 (1973).
(9) M. Geller, D. E. Altman, and T. A. DeTemple, *Appl. Opt.*, **7**, 2237 (1968).
(10) James A. Perry, Ph.D. Thesis, University of Illinois, Urbana, Ill., 1977.

(11) Paul C. Dryden, Ph.D. Thesis, University of Illinois, Urbana, Ill., 1975.
(12) E. D. Cehelnik and K. D. Mielenz, *Appl. Opt.*, **15**, 2259 (1976).
(13) Melton F. Bryant, Ph.D. Thesis, University of Illinois, Urbana, Ill., 1977.

RECEIVED for review April 4, 1977. Accepted June 29, 1977. One of the authors (J.A.P.) expresses his appreciation for a Uniroyal Fellowship for part of this work, and the authors are grateful for partial support of the work by the NIH under Grant HEW PHS GM 21984, and by the NSF under Grant NSF MPS 74-12248.

On-Line Computers in Research. High-Resolution Mass Spectrometry

Rengachari Venkataraghavan, Richard J. Klimowski, and Fred W. McLafferty

Experimentation in chemistry has been revolutionized in the last two decades through the direct application of sophisticated instrumentation, such as spectrometers, chromatographs, and electronic devices, for measuring, recording, and controlling (Figure 1). A new general development which promises to cause an equivalent impact is the incorporation of the high-speed digital computer *into the experimental system* (Figure 2). Such "on-line" operation involves the direct interfacing (coupling) of the computer to the instruments or other devices sensing the experimental outputs (results and conditions) to acquire and process the data, to return control signals to the experiment, and to correlate, display, and record the results in the form which is most meaningful to the experimenter. All of this is accomplished in "real time" (during or shortly after completion of the experiment). The data system not only relieves the experimenter of the tedium of observing, calculating, and recording the results, but it can yield impressive improvements in accuracy, speed, and other aspects of performance, often permitting experiments which otherwise would be impossible.[1]

Complete computer-interface systems designed to handle a variety of instruments simultaneously cost $200,000 or more, an expense that has seriously limited application in most chemical research laboratories. The recent development of the small but fast computer has made possible systems which are tailored for specific applications, such as gas chromatography, nuclear magnetic resonance spectroscopy, mass spectrometry, and X-ray crystallography, for $40,000–100,000. However adequate systems can be constructed for considerably less, as the computers alone cost only $8,000 to $25,000. Indeed, we feel that for many experiments the chemist should design, assemble, and/or modify his own hardware (equipment) and software (program instructions); the incentives are similar to those which have induced chemists to become familiar with the operation and maintenance of specialized instruments and the interpretation of various types of spectra.

A number of reports have appeared recently illustrating the awakening interest in this field. Perone and coworkers[2] employ a small digital computer on-line to a stationary electrode polarograph. The computer continuously monitors the experiment, performs real-time calculations on the output, and supplies feedback signals to set the interrupt potential and delay time between voltage sweeps. Pardue and his students[3] utilize a small digital computer for on-line processing of reaction rate data for quantitative analysis; preliminary reaction rate calculations are used to optimize data acquisition rates and other operating parameters. Several concurrently operating gas chromatographs can be connected on-line to a digital computer for data reduction to yield improved accuracy and resolution.[4] To illustrate the nature and merits of such systems we will describe some applications of computers to high-resolution mass spectrometry.[5-8]

Information Available from Modern Instrumental Techniques. A basic reason for the usefulness of spectroscopic and chromatographic techniques is that they yield two dimensions of information: the ordinate shows a characteristic response (for example, absorptivity or thermal conductivity) at each of a number of positions on the abscissa (for example, wavelength or retention time). The amount of information about a sample in its spectrum or chromatogram thus depends on the dynamic range over which the ordinate response can be measured and the number of positions (resolution increments) that are distinguishable on the abscissa. By this definition of information content high-resolution mass spectrometry far surpasses other molecular techniques such as infrared spectroscopy, nuclear magnetic resonance spectroscopy, and gas chromatography. Although sample requirements are <1 μg, a molecule of molecular weight 1000 will typically exhibit more than 100 peaks in its mass spectrum. These peaks represent the masses of the molecular and fragment ions formed from the sample by electron bombardment. The mass spectrum is a *line* spectrum; because the mass of an ion is defined exactly by its elemental composition, the essential limit of the narrowness of a peak is the resolving power of the instrument. By measuring the mass of an ion peak with an accuracy of 1 mmu (millimass unit) there are approximately 10^6 mass locations at which each of the peaks of a mass spectrum can be located, and the abundance of each of these can be measured over a dynamic range of $>10^4$; this compares to $<10^2$ resolution increments and $<10^3$ dynamic range for most other methods.

(1) Applications of computers in chemical experimentation have been discussed recently by J. W. Frazer, *Anal. Chem.*, **40** (8), 26A (1968); G. Lauer and R. A. Osteryoung, *ibid.*, **40** (10), 30A (1968).
(2) S. P. Perone, D. O. Jones, and W. F. Gutknecht, *ibid.*, **41**, 1154 (1969).
(3) G. E. Jones and H. L. Pardue, *ibid.*, **41**, 1618 (1969).
(4) A. W. Westerberg, *ibid.*, **41**, 1595, 1770 (1969).
(5) J. H. Beynon, "Mass Spectrometry and its Application to Organic Chemistry," Elsevier Publishing Co., Amsterdam, 1960.
(6) K. Biemann, *Advan. Mass Spectrom.*, **4**, 139 (1968).
(7) F. W. McLafferty, *Science*, **151**, 641 (1966).
(8) Computer-mass spectrometry systems are available from Perkin-Elmer, Inc., Norwalk, Conn.; Picker-Nuclear Corp., White Plains, N. Y.; and Varian Associates, Palo Alto, Calif.

Figure 1. Schematic representation of the use of instrumental techniques to sense the conditions and results of the experiments. The researcher observes these, computes more meaningful results where necessary, and uses the results to modify the experiment.

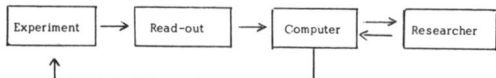

Figure 2. The on-line computer can assume the routine functions of data acquisition, data reduction, and control of the experiment. The research man makes high-level decisions and communicates these to the computer for execution.

Table I
Common Ions of Nominal Mass 43

Composition	Typical structure	Exact mass[a]
CHNO	–CONH–	43.0058
C_2H_3O	CH_3CO-	43.0184
CH_3N_2	$CH_3N=N-$	43.0296
C_2H_5N	$-CH_2NHCH_2-$	43.0421
C_3H_7	$(CH_3)_2CH-$	43.0547
C_2H_5D		43.0532
$^{12}C_2{}^{13}CH_6$		43.0502

[a] Sum of the exact nuclidic masses of the atoms based on ^{12}C = 12.00000.

One of the most serious handicaps in utilizing this tremendous amount of information is the great effort necessary to measure the spectral data and from this to calculate the exact masses and abundances. The elemental composition of each ion peak can be assigned uniquely if the mass can be determined with sufficient accuracy (Table I). Accuracy of mass measurement is paramount for larger molecules, as the number of possible elemental combinations which will yield a particular nominal mass (± 0.5 mass unit) goes up exponentially with molecular weight. For example, chemically logical (including isotopic) combinations of C, H, N, and O yield 12 common ions of nominal mass 43, but 500 ions of nominal mass 430.

Exact mass measurement in most instruments is done by "peak matching." Images of the unknown peak and a selected reference peak are displayed alternately on an oscilloscope by a rapid scan over a small range of the ion accelerating voltage; the mass difference of these peaks is determined by the exact differential voltage necessary to make the peaks coincide. Although accuracies of 1–10 ppm are attainable by this method, it requires 1–5 min/peak. One seldom determines more than a few peaks per spectrum by this method.

Another serious problem is operation of a complex mass spectrometer to obtain a maximum amount of useful information with a minimum amount of sample. The operator must control many instrumental parameters simultaneously, such as rate of sample volatilization, ion focusing, and slit settings to maximize resolution or sensitivity. Special measurements, such as for metastable ions utilizing a defocused electrostatic analyzer,[9] require additional adjustments, as will be discussed later. The operator must control these parameters from experience or quick calculations based on information from instrument meters and the spectral recorder; such feedback obviously has severe limitations in speed and accuracy.

Photoplate-Recorded Mass Spectra

In a mass spectrometer with Mattauch–Herzog geometry all of the ions over a wide mass range (for example, m/e 25–700) can be focused simultaneously on a photographic plate. In this way the complete spectrum can be recorded in minutes. However, manual measurement of the distances between the centers of the ion lines of a complex spectrum requires several hours with an accurate comparator. To reduce the time, tedium, and human errors associated with these measurements an automatic comparator–microdensitometer was designed and constructed.[10] In its operation a light beam defined by a narrow vertical slit is focused on the photoplate to produce an image with the same orientation as the ion lines. The photoplate spectrum is moved through the light beam, and the transmission of light is measured by a photomultiplier at 0.25-μm intervals along the 330-mm length of the mass axis. These analog transmisson (T) values are transformed with an analog-to-digital (A/D) converter to digital values containing 3 significant decimal figures. To conserve storage space, only $(1 - T)$ values above a preset threshold are saved, along with a value from a shaft position encoder on the drive screw to define the displacement of the photoplate. Only 12 min is required to measure and threshold the 10^6 data points and store the significant information.

Improvement in Mass Measuring Accuracy. Not only does this system shorten dramatically the time of data acquisition, but the results are much more detailed. These data provide information concerning the distribution of ions in each individual line; experimentally this corresponds closely to a Gaussian function (Figure 3). To find the center of an ion line, the computer data-reduction program smooths the corresponding transmission data, fits the results by a least-squares procedure to a Gaussian function, and calculates the center for this function. The observed precision of this dynamic measurement method is ± 0.25 μm, which is far superior to the precision of >1 μm found for the static mechanical or optical analog methods used in manual operation of the comparator. Optimization of this method has reduced average mass measuring errors from ± 2 or 3 to ± 0.5 mmu.[11]

Improvement in Resolving Power. The ability to resolve ion peaks of nearly the same mass is one of the most important specifications for a high-resolution mass spectrometer. If two or more ion lines overlap on the

---

(9) M. Barber and R. M. Elliott, ASTM E-14 Conference on Mass Spectrometry, Montreal, 1964, p 150; T. W. Shannon, T. E. Mead, C. G. Warner, and F. W. McLafferty, *Anal. Chem.*, **39**, 1748 (1967).

(10) R. Venkataraghavan, F. W. McLafferty, and J. W. Amy, *ibid.*, **39**, 178 (1967).
(11) R. D. Board, Ph.D. Thesis, Purdue University, 1969.

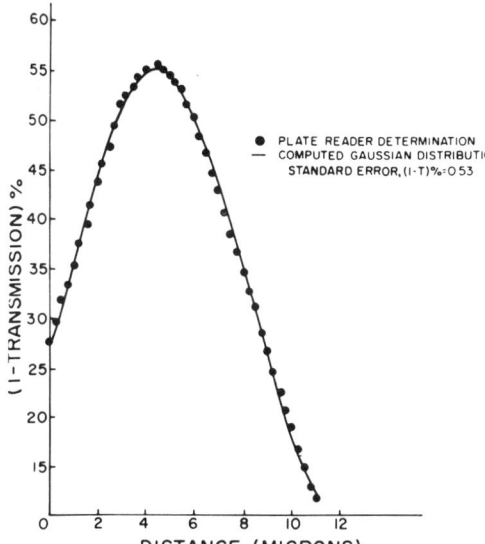

Figure 3. Experimental points from the automatic comparator-microdensitometer for a single ion line in a photoplate mass spectrum. The line is a best-fit Gaussian distribution generated by the computer.

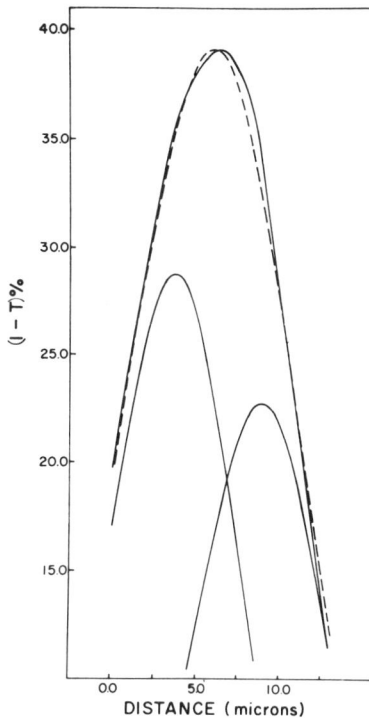

Figure 4. Deconvolution of data from overlapping ion lines ($C_{10}{}^{13}CH_9$ and $C_{11}H_{10}$, $\Delta m/e = 0.0045$ amu). Upper curves: solid line, experimental data from comparator reading of photoplate mass spectrum; dashed line, best-fit Gaussian distribution generated by the computer. Lower curves: component ion distribution curves determined by the computer, whose measured mass difference is within 0.0004 amu of the true value. The deconvolution routine was automatically activated because the peak width at half-height exceeded specifications.

photoplate, the resulting transmission profile will represent the sum of their individual Gaussian distributions. If the computer program cannot fit the transmission data for a particular ion line to a Gaussian distribution within predetermined limits, or if it finds the width at half-height too large (see Figure 4), the program deconvolutes the peak into the component Gaussian functions by an iterative routine. Deconvolution is imperative for the comparator system, as the data are convoluted by the slit width of ca. 3 μm used as a compromise of resolution, sensitivity, and edge diffraction errors. Despite this, the ability of the overall photoplate system to distinguish overlapping peaks is substantially improved over the performance of the mass spectrometer in the conventional magnetic scanning mode of operation employing a ~1-μm ion exit slit. Of even more importance is the resulting improvement in mass-measuring accuracy. A common source of error is the overlap of ^{13}C isotope peaks, as the mass difference between ^{13}C and ^{12}CH is only 4.5 mmu. A small contribution of this kind shifts the apparent centroid of the peak; even when the contribution is sufficient to cause a resolved doublet, the positions of these maxima are displaced from the true peak positions.[10,11]

Data Reduction and Display. After the basic data have been acquired by the computer, it can perform a variety of more conventional reduction operations. The computer program determines the exact positions and areas for each ion line in the spectrum, locates the lines due to perfluorokerosene which is added as a mass reference, interpolates between these using a higher order polynomial to assign exact masses to the positions of the peaks from the sample, and computes the possible elemental compositions corresponding to each mass. The data can be displayed in a variety of forms for interpretation, such as a topographical element map (Figure 5).[12] The computer can also check the data against a file of reference spectra for possible identification, or it can attempt to interpret the spectrum.[13-15]

On-Line Photoplate Data Acquisition System. Biemann and his coworkers[6] have coupled a photoplate comparator on-line to a medium-sized computer (IBM 1800). Transmission values at 0.5-μm intervals are digitized and stored in core memory. When a memory section of 1000 locations is filled, the computer starts to process these data, and the incoming data are stored in a new section. Data processing during the scan in-

(12) R. Venkataraghavan and F. W. McLafferty, *Anal. Chem.*, **39**, 278 (1967).
(13) R. Venkataraghavan, F. W. McLafferty, and G. E. Van Lear, *Org. Mass Spectrom.*, **2**, 1 (1969); A. M. Duffield, A. V. Robertson, C. Djerassi, B. G. Buchanan, G. L. Sutherland, E. A. Feigenbaum, and J. Lederberg, *J. Amer. Chem. Soc.*, **91**, 2977 (1969); and references cited therein.
(14) M. Senn, R. Venkataraghavan, and F. W. McLafferty, *ibid.*, **88**, 5593 (1966).
(15) K. Biemann, C. Cone, B. R. Webster, and G. P. Arsenault, *ibid.*, **88**, 5598 (1966).

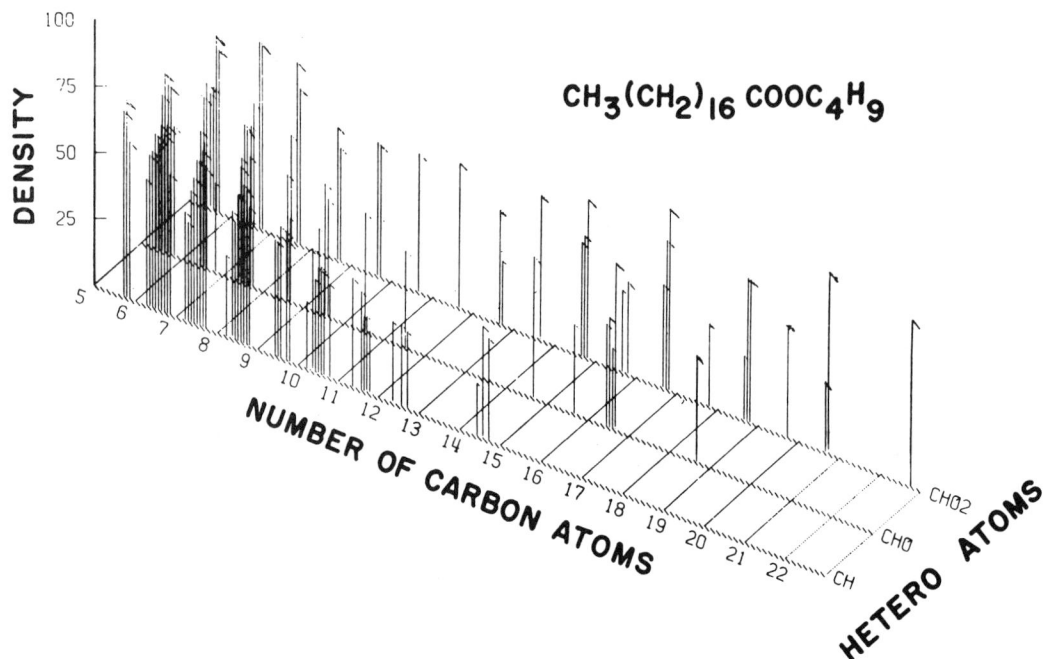

Figure 5. Topographical element map of butyl stearate. Note that the highest mass peak in the CHO row is C_{18}, indicating that four carbon atoms must be lost before an oxygen atom is lost.

cludes the calculations of background, peak centers, and simple deconvolution. As soon as the scan is completed the computer uses the stored peak data to determine the elemental compositions of all of the ions in the spectrum. Thus the reading and complete processing of the photoplate spectrum require an average elapsed time of only 20 min.

However, the delay caused by photoplate development does not make it feasible to utilize the data while the sample is still in the mass spectrometer. To achieve this, systems have been developed in which the output from the electron multiplier during magnetic scanning is processed on-line by a digital computer.

Real-Time Computing Systems

A further major reason for incorporation of a computer into the experimental system is to be able to utilize the results to optimize the experiment while it is in progress. For this either the operator can make adjustments based on the results (man in the loop), or the computer itself can be programmed to send appropriate control information to the apparatus (closed loop).

The design factors for an on-line computer system for high-resolution mass spectrometry are similar in many ways to those encountered in systems for a wide variety of chemical experiments, despite the fact that some requirements, such as data rate and data storage, may vary widely. It is imperative that the experimenter himself becomes closely involved in the design; in most successful systems the computer "looks at" the process and resulting data in the same way as the experimenter would. Although it may be necessary to call on experts for many technical problems of software and hardware, the success of the overall system will depend directly on how well the goals and other problems of the experiment are met.

Time-Shared vs. Dedicated Computers. It is usually possible to justify a larger computer and a greater number of specialized personnel if the same computer can serve several projects. This can be done by running each experiment at a separate time, with manual connection and disconnection of the computer for each run; different interface hardware as well as program software may be necessary for each. For greater convenience and flexibility a computer can service a number of experiments on a nearly simultaneous basis by "time sharing"; however the time-sharing operation itself requires substantial memory capacity and sophisticated software. Problems arise also if the computer and the experiment are separated; a sufficiently capable terminal for communication with the computer (input–output devices) must be located at the experiment, and transfer of high frequency analog data over distances >25 m is not advisable due to possible signal distortion. Transfer of digital signals often requires expensive special equipment. Therefore in designing an on-line system the small dedicated computer should be given serious consideration.

Interface Design. The interface makes possible the transmission of data from the experiment to the computer, and *vice versa*. To accomplish the former, the interface must accept one or more analog signals from the experiment and make these suitable for computer processing. The interface may include special hardware devices which carry out simple high-rate operations to reduce the load on or increase the efficiency of the central processing unit, thus reducing the size requirements of the computer. The major components generally found in the interface are the signal-conditioning equipment, analog multiplexer, A/D converter, and time-base generator. The signal-conditioning equipment converts the analog output of the instrument into a form that is acceptable (in voltage level and range, amperage, etc.) to the A/D converter or that is better for computer processing (filtered to improve the signal-to-noise ratio, for example). The multiplexer can receive analog signals from a variety of sources for sequential introduction to the A/D.

The major criteria for the selection of an A/D converter are the data rate and the dynamic range of the information obtainable from the experiment. For example, with most gas chromatographs a sampling rate of 10 Hz should be sufficient to describe gc peaks adequately, even for studies of peak skewing or deconvolution, although a dynamic range of 10^6 may be necessary. On the other hand, temperature-jump experiments on a microsecond time scale require sampling rates exceeding the rates of the fastest A/D converter available commercially. In the design of a general purpose interface suitable for various instruments it is advantageous to incorporate an A/D converter with sufficient flexibility to handle data of different rates and magnitudes.

Information on the abscissal position (indicative of frequency, mass, retention time, etc.) must also be supplied to the computer with each ordinate measurement. This can come directly from the instrument being monitored; for the photoplate reader described above this was done by triggering the A/D converter at constant distance intervals sensed by a shaft encoder on the drive screw (4000 pulses/revolution). The instrument abscissa can also be driven at a reproducible rate with the triggering supplied by a time-base pulse generator. Such generators are of various precision and frequency. Crystal clocks are available with a precision of $1/10^8$. RC (resistor–capacitor) clocks are less precise, but much less expensive. The computer itself can be used to generate the time base if the demand for its attention from the experiment is not excessive.

Feedback Control. Communication from the computer back to the experiment must also go through the interface; a wide variety of experimental functions, such as heating, sample or reagent flow, pressure, slit width, position, or emergency shutdown can be placed under feedback control. Most of the special devices to perform these functions can be operated by an electrical analog signal. This is supplied by simple on–off relays or a digital-to-analog converter (D/A) in the interface, using a multiplexer if there are a number of control devices in the experiment.

Specifications of the Computer. The selection of the central processing unit (CPU) and its associated peripheral devices for a real-time data acquisition and reduction system depends on the experiment itself. The CPU of modern low-cost digital computers has a fast processor and a high-speed (0.7–2.0 μsec) random-access core; their size and speed basically determine the rate of data acquisition and the complexity of the computations that can be performed. Increasing the size of the CPU is relatively expensive, so that alternatives should be sought. Although the CPU is ideal for data storage, magnetic tape can store large volumes of data at high speeds more economically. Paper tape is much slower, but requires an even smaller investment. Tape data can be processed later by the CPU, but this processing must mainly be in a sequential fashion. If it is necessary, for example, for feedback control to compare data taken at noncontiguous times, the computer will have to carry out a time-consuming search of the tape. An excellent alternative bulk storage device is the magnetic disk (or drum). Although it is more expensive than tape for the storage of large volumes of data, all of the data stored on disk can be accessed randomly in a relatively short time (ca. 20 msec). Further, by using a disk to store part of the computer program for transfer ("program swapping") at appropriate times to the CPU, the processing capacity of the CPU can often be greatly increased. In essence the software can treat the disk as slow core for programming and storage purposes. Thus the optimum balance of investment in CPU, disk, and tape in the computer system depends on the rate and volume of data and how they are to be processed, stored, and retrieved.

The most common data output device is the teletype. Although relatively inexpensive, it is slow (10 characters/sec) compared to high-speed printers (300–1200 *lines*/min). For many experiments the cathode ray tube provides an elegant output device, but sophisticated software and hardware are often required to drive it.

A High-Resolution Mass Spectrometry System Incorporating a Small On-Line Computer

A number of types of systems coupling high-resolution mass spectrometers to computers have been developed.[8,16] A pioneering system is that of Burlingame and coworkers[17] which uses the powerful XDS-930 and XDS-Sigma-7 computers. The computer compresses the digitized raw data by deletion of all intensities below a preset threshold, displays the resulting data on a cathode ray tube to allow operator interaction with the system, and stores the accepted data for later computer processing. The system developed in our laboratory, described below, is similar in principle, but employs a much smaller computer which necessitates more complex hardware interfacing.[18]

(16) Reviewed by A. L. Burlingame, *Advan. Mass Spectrom.*, **4**, 15 (1968); W. J. McMurray, S. R. Lipsky, and B. N. Green, *ibid.*, **4**, 77 (1968); H. C. Bowen, E. Clayton, D. J. Shields, and H. M. Stanier, *ibid.*, **4**, 257 (1968).

(17) A. L. Burlingame, D. H. Smith, and R. V. Olsen, *Anal. Chem.*, **40**, 13 (1968); A. L. Burlingame, D. H. Smith, F. Walls, and R. V. Olsen, Proceedings of the 17th Annual Conference on Mass Spectrometry, Dallas, Texas, May 1969, p 28.

This system produces elemental composition information accurately and rapidly and controls the mass spectrometer for acquisition of metastable ion data. This also illustrates a system approach which should be very attractive for many other chemical experimentation problems.

Hardware Characteristics. A block diagram of the system is shown in Figure 6. The interface has an A/D converter which can convert the analog signal to a digital word of 12 bits (range of 1 to 4096) 20,000 times/sec (20 kHz), matching the 12-bit word length of the computer. The signal-conditioning equipment includes a logarithmic amplifier so that the larger dynamic range of the mass spectrometer (ca. 1/20,000) will not be reduced by the A/D, and a filter to reduce the noise level of the signal. A 20-kHz crystal clock (precision of $1/10^8$) triggers the A/D and drives a counter with a 24-bit buffer which registers the elapsed scan time. A divider switch on the clock makes lower conversion rates possible.

The interface also includes hardware logic for detection and thresholding. In the usual high-resolution mass spectrum the peaks are so far apart and narrow that significant data constitute only a few per cent of the total of the approximately 10^6 measurements emerging from the A/D converter. By allowing the transmission of only those peaks exceeding a preset threshold value, the data storage and processing requirements in the computer are reduced correspondingly. Although the CPU could discard the base-line information, to do this it would have to determine whether each datum is above or below noise level, a decision requiring 20 μsec once every 50 μsec (at a 20-kHz digitization rate). Thus the CPU can do other calculations until the interface signals that significant data have been gathered from the mass spectrometer. Further CPU operation economies are made possible by the interface because the computer does not have to trigger the A/D or update the clock register every 50 μsec, or record an abscissal (time) value with each ordinate (ion abundance) value. The latter values are taken at equivalent intervals, so that the time is only recorded once in a sequence of abundance values. The interface also incorporates a 10-bit digital-to-analog converter which enables the computer to send control signals back to the mass spectrometer.

The computer used[19] has 4096 12-bit words of core memory, the automatic multiply–divide option, teletype, and a 32,000 word random-access disk. The incorporation of the disk makes it possible, despite the small core memory, to store the relatively large amount of data from the mass scan and to process it in a continuous fashion. The disk also stores most of the

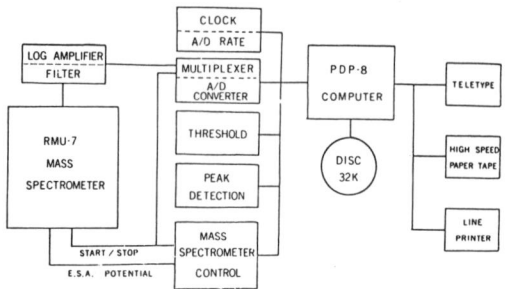

Figure 6. On-line data-system for high-resolution mass spectrometry utilizing a small (4096 core words) computer.

software, so that sections are brought into core memory as needed during the operation. The execution of different phases of the program thus is automatic and continuous. A high-speed paper tape reader-punch is convenient for development and maintenance work, but is not necessary for routine operation.

System Operation. When the start-of-scan button on the mass spectrometer is pushed, this is sensed by the computer and the clock is started. The clock in turn triggers the A/D converter continuously at the preselected rate. The threshold logic compares the digitized value of the ion current to the preset threshold requirement. When the value goes above this level, the computer is interrupted so that the succeeding values can be transmitted to it and stored. The peak detection logic of the interface checks each value and terminates collection when four consecutive sub-threshold values are transmitted. The logic then generates an interrupt to read the clock value, which has been accumulating in a clock register of 24-bit capacity (14 min at 20 kHz). The value read thus corresponds to the elapsed scan time at which the last datum point was collected. Figure 7 shows typical values generated by a large peak.

The data are stored sequentially in buffer tables in the core memory. When each table is filled it is transferred as a data block to the disk as soon as there is sufficient time (ca. 35 msec) between peaks. This overlapping operation of data storage continues until the scan is stopped by pushing the stop button on the mass spectrometer. This signal is sensed by the computer to initiate the next data reduction step.

Software Characteristics. Effective utilization of hardware in an experimental system depends to a large degree on the organization of the software. High-level languages such as Fortran, Focal, Basic, etc., which have been developed for off-line applications are less suitable for on-line systems because the object codes generated by their compilers are not very efficient. When the processing time or number of core locations is critical, it is better to write the programs in a lower level language such as the Assembler language. Application programs for this system are written in Macro-8 along modular lines to use the full potential of the disk.

---

(18) R. Venkataraghavan, J. W. Amy, R. D. Board, R. D. Brown, R. J. Klimowski, and F. W. McLafferty, Proceedings of the 16th Annual Conference on Mass Spectrometry and Allied Topics, Pittsburgh, Pa., May 1968, p 114; R. J. Klimowski, Ph.D. Thesis, Cornell University, 1969; R. J. Klimowski, R. Venkataraghavan, F. W. McLafferty, and E. B. Delaney, submitted for publication.

(19) PDP-8 computer, Digital Equipment Corporation, Maynard, Mass.

```
0147  0202  0226  0239  0259  0258  0235  0222
0258  0278  0310  0370  0447  0562  0671  0742
0774  0803  0831  0890  0947  0983  1066  1183
1306  1394  1459  1514  1571  1622  1655  1690
1730  1730  1698  1694  1722  1774  1802  1819
1831  1855  1887  1890  1862  1807  1738  1675
1626  1590  1555  1526  1498  1471  1446  1407
1367  1331  1295  1242  1170  1102  1035  0979
0946  0918  0898  0870  0818  0742  0658  0578
0523  0478  0410  0338  0258  0190  0142  0098
0062  0026  0002
00059476
```

Figure 7. Typical output values from the on-line system describing the profile of a major peak, followed by a decimal value of the elapsed time.

REL.AB.	MASS	CALC.	ERR.	C12/C13	H	N	O
1.1	193.0711	193.0725	-1.41	8/0	9	4	2
	193.0773	NO HIT					
.0	193.0816	193.0805	+1.09	8/1	10	3	2
100.0	194.0811	194.0803	+.74	8/0	10	4	2
5.4	195.0829	195.0836	+.75	7/1	10	4	2

CAFFEINE $C_8H_{10}N_4O_2$ mw = 194

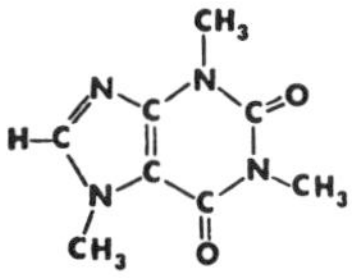

Figure 8. Partial listing of computer-generated elemental composition in vicinity of molecular ion for caffeine. NO -HIT signifies that no composition is found within the specified error tolerance.

The data reduction steps of the software are divided into five phases: (I) data acquisition; (II) peak center calculation; (III) reference peak identification; (IV) exact mass calculation; and (V) elemental composition calculation.

Before the scan is initiated, the operator answers questions posed by the computer concerning the sample, reference compound, scan speed and range, threshold, and instrument conditions. From this information the computer sets the desired operating parameters in the interface. Phase I acquires the data and stores the resultant compressed spectrum on the disk system as described above.

When the scan-stop button is pushed, the computer causes the phase II program to be "swapped" into core from the disk, i.e., this program is transferred to the core locations of the phase I program. Phase II recalls the stored spectral data from the disk, rescales it from the logarithmic to a linear function, and determines the center of each ion peak and the area under its profile; these calculations usually require less than 60 sec for a typical spectrum. The last operation of the phase II program is to swap the phase III program into core. This phase establishes a time–mass relationship (dispersion curve) by automatically identifying the reference peaks using an extrapolation technique. In phase IV the time values of the sample peaks are converted to mass values. The exact mass of each peak is calculated using four reference lines with a Lagrange polynomial equation; this requires about 5 sec for the whole spectrum. In phase V all possible elemental compositions within the prescribed error tolerance and element limits are assigned to the individual ion peaks in the spectrum using an iterative routine employing valence rules.[10,20] An example of phase V output is shown in Figure 8 for the molecular ion region of caffeine.

Results

Resolution and accuracy are primary criteria of overall system performance. In Figure 9 the mass separation between the C_5H_5N and $C_5{}^{13}CH_6$ peaks is 1 part in 10,000, indicating that resolution comparable to that achieved in conventional operation of this mass spectrometer is possible with good mass measuring accuracy. The latter is heavily dependent on mass spectrometer performance, but in general accuracy limits of ±3 mmu are readily achievable in single-scan measurements at 10,000 resolution. We find that the major part of this error is due to random fluctuations in the signal, as shown by the error convergence in Figure 10 achieved on multiple scans. At 10,000 resolution the averaging of ten scans improves the accuracy of mass measurement to ±1 mmu. Similar results have been reported by Burlingame, et al.,[17] using a different mass spectrometer and on-line data acquisition system. Thus the computer of this system can also act as a powerful signal-averaging device.

Time Requirements. One of the most dramatic advantages of the on-line system as compared to peak-matching or photoplate recording is the rapid availability of the reduced data. Output of results begins in 1–2 min after completion of the scan. Because of the slow teletype output, the amount of data determines the total time required, usually 10–20 min. We have recently installed a moderately priced ($10,000) high-speed printer. For a compound such as caffeine (Figure 8) this reduces the time required after completion of the scan to *approximately 2 min.*

Closed-Loop Feedback Control. The mass spectrometer–computer system described above can also be used to measure metastable ions by the Barber–Elliott–Major technique.[9,21] Lowering the energy of the electrostatic analyzer to m_2/m_1 of its normal value allows passage of ions of mass m_2 formed in the first field-free drift region by the reaction $m_1 \rightarrow m_2$. This also eliminates the normal energy ions, so that metastable ions can be detected over a dynamic range of 10^4–10^5.[21] By controlling the electrostatic analyzer potential, the computer can thus measure selected metastable transitions during the magnetic scan. Alternatively, all possible values of m_2/m_1 can be scanned repeatedly during a slow magnetic scan to record all of the meta-

(20) D. M. Desiderio and K. Biemann, Proceedings of the 12th Annual Conference on Mass Spectrometry, Montreal, June 1964, p 433.

(21) F. W. McLafferty, J. Okamoto, H. Tsuyama, Y. Nakajima, T. Noda, and H. W. Major, *Org. Mass Spectrom.*, **2**, 751 (1969).

REL.AB.	MASS	CALC.	ERR.	C12/C13	H	N	O
6.5	78.0345	78.0343	+ .14	5/0	4	1	0
5.2	78.0456	78.0469	−1.24	6/0	6	0	0
100.0	79.0425	79.0421	+ .33	5/0	5	1	0
.6	79.0503	79.0502	+ .14	5/1	6	0	0
1.7	80.0466	80.0454	+1.14	4/1	5	1	0

Figure 9. Partial computer listing of elemental compositions for a mixture of benzene and pyridine. The mass separation at m/e 79 $(M/\Delta M)$ is 1/10,000, scan rate 60 sec/decade. The relative abundance for m/e 80 is low because no correction has been made for the peak area below the preset threshold value.

stable ions.[22] This system saves many hours per spectrum over the extremely laborious manual method.

Other possible instrument parameters for which feedback control would be valuable are the filament current, bombarding electron energy, accelerating voltage, source and exit slit widths, ion source temperature, and magnet current. However, sophisticated hardware and a complex network of software are required for such a fully automated data-acquisition system.

Laboratory Data System

The on-line system described above provides speed, accuracy, resolution, and limited feedback functions at a relatively reasonable price; very similar data-handling systems could be of substantial utility for a wide variety of instruments and experimental techniques. However, a more powerful computer is required for many desirable on-line operations with mass spectrometer systems, such as the feedback controls and deconvolution of overlapping peaks described above. Much faster processing of data is advantageous when the mass spectrometer is directly coupled to a gas chromatograph, and direct identification of components by comparison against standard reference spectra stored in the computer can be envisaged. Computer interpretation of mass spectra[13–15] also has much higher core requirements. We have recently proposed[23] the direct determination of amino acid sequences in the mixtures of peptides resulting from the degradation of proteins and other polypeptides. Analysis of more complex mixtures appears feasible with repeated high-resolution scans and defocused metastable determinations during sample vaporization; however the necessary on line reduction of the large volume of data produced would require a much more powerful computer system.

As noted earlier, justification of a large computer usually necessitates that it serve more experiments. Such a multiple-use system, especially if it utilizes "time sharing," generally requires much more complex hardware and software, so that justification is questionable for tasks that can be done by dedicated small computers. However, additional applications of potential similar to those enumerated above for mass

(22) This computer system for metastables was developed by J. E. Coutant in this laboratory.
(23) G. E. Van Lear and F. W. McLafferty, Ann. Rev. Biochem., 38, 289 (1969); F. W. McLafferty, R. Venkataraghavan, and P. Irving, Biochem. Biophys. Res. Commun., in press.

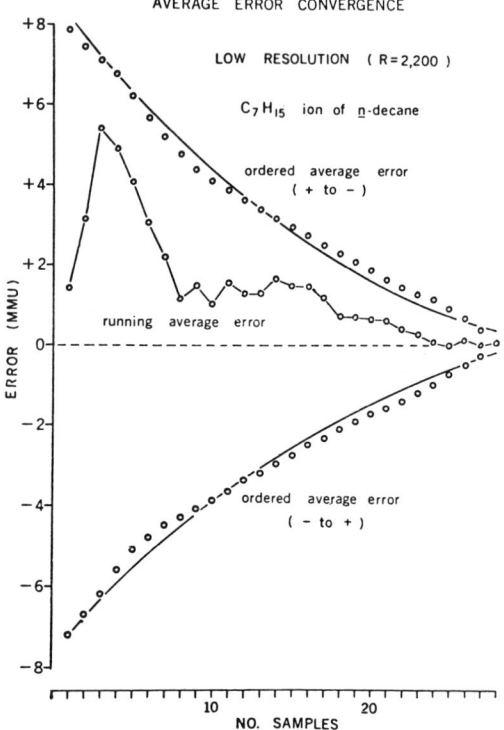

Figure 10. Improvement in mass-measuring accuracy resulting from multiscan averaging utilizing the on-line computer.

spectrometry can be cited for a wide variety of chemical research problems. We feel that the logical next step is to incorporate a number of such small dedicated computers handling a variety of such experimental devices into a larger on-line laboratory data system organized around a more flexible and powerful computer. The capability of the large computer for handling large volumes of data and more complex calculations (such as deconvolutions and Fourier transformations) would be available both on-line and off-line to any part of the system where necessary, but the small computers would handle primary data processing and experiment control. Worthy of special note is the potential of graphics-display devices such as the plotter and cathode ray tube (CRT) which can greatly facilitate man-in-the-loop type interaction of the investigator with the results of his experiment or of his theoretical calculations. For example, a larger computer can display molecular structures or potential energy surfaces on the CRT with rotation or modification of the image under direct control of the investigator for study of three-dimensional molecular interactions, interpretation of diffraction data, or location of potential minima.

We are indebted to D. H. Bessel, F. L. Hiltz, R. E. Hughes, R. A. Plane, D. A. Usher, and other members of the Cornell Chemistry Department staff who helped plan the laboratory data system that is described, and to the U. S. Public Health Service (National Institutes of Health Grant GM-16609) for generous financial support.

15

Copyright © 1970 by the American Chemical Society
Reprinted from *Anal. Chem.* 42:1505-1516 (1970)

On-Line Digital Computer System for High-Speed Single Focusing Mass Spectrometry

Charles C. Sweeley, Bruce D. Ray, William I. Wood, and John F. Holland
Department of Biochemistry, Michigan State University, East Lansing, Mich. 48823

Micah I. Krichevsky
Environmental Mechanisms Section, National Institute of Dental Research, National Institutes of Health, Bethesda, Md. 20014

INSTRUMENTS with gas chromatographic columns coupled directly to single focusing or high resolution mass spectrometers are available commercially. A notable feature of these instruments is the massive information that can be obtained in a relatively short time (*1*). Manual data reduction is not only time consuming and laborious but leads to a relatively high error rate. Recently, high speed mass marking systems have become available (*2*); although they improve the process of peak identification, they have not shortened the time considerably in converting data in the form of an oscillographic recording into useful information such as tables of nominal mass and intensity or bar graph plots. Several digital data handling systems have already been described for high resolution (*3–8*) and single focusing mass spectrometry (*9–11*).

This paper describes an on-line digital computer system for data acquisition and reduction of mass spectral information from a single focusing mass spectrometer coupled with a gas chromatograph inlet. Major considerations in the development of the system were ability to track scans of 1 second duration from m/e 0 to 500; to process the scans in real-time; to analyze the data without the inclusion of an internal standard in the sample; to do bookkeeping, file manipulation, data reduction, and output locally without interaction with a computer center; and to enable output of data at times most convenient to the laboratory schedule by storage of finished data on a magnetic tape.

(1) R. A. Hites and K. Biemann, ANAL. CHEM., **40**, 1217 (1968).
(2) P-A. Jansson, S. Melkersson, R. Ryhage, and S. Wikstrom, *Arkiv. Kemi*, **31**, 565 (1969).
(3) R. Venkataraghavan, F. W. McLafferty, and J. W. Amy, ANAL. CHEM., **39**, 178 (1967).
(4) D. D. Tunnicliff and P. A. Wadsworth, *ibid.*, **40**, 1826 (1968).
(5) W. J. McMurray, B. N. Greene, and S. R. Lipsky, *ibid.*, **38**, 1195 (1966).
(6) A. L. Burlingame, D. H. Smith, and R. W. Olsen, *ibid.*, **40**, 13 (1968).
(7) R. J. Klimowski, R. Venkataraghavan, and F. W. McLafferty, unpublished data, 1970.
(8) R. Venkataraghavan, R. J. Klimowski, and F. W. McLafferty, *Accounts Chem. Res.*, **3**, 158 (1970).
(9) R. A. Hites and K. Biemann, ANAL. CHEM., **39**, 965 (1967).
(10) *Ibid.*, **42**, 855 (1970).
(11) B. Hedfjall, P-A. Jansson, Y. Marde, R. Ryhage, and S. Wikstrom, *J. Sci. Instrum.*, **2**, 1031 (1969).

METHODS AND MATERIALS

Overall Configuration. The system is designed to produce, under operator control, finished bar graphs and/or tables of mass *vs.* relative intensity for mass spectra of selected compounds eluted from a gas chromatograph directly into a single focusing mass spectrometer. The data are acquired and preliminary processing is done in real-time so that the core memory of the computer is continually ready to receive data from subsequent magnetic scans of the mass spectrometer. The system, outlined in Figure 1, consists of a LKB 9000 Gas Chromatograph-Mass Spectrometer with a mass marker and a heated membrane inlet (LKB Instruments, Inc., Rockville, Md.). The three-pen oscillograph records data received from a 14-stage electron multiplier which provides ion intensity information. A Hall Effect Probe yields effective magnetic field strength (*i.e.*, a signal related to m/e). Signals from the electron multiplier and the Hall Effect Probe are fed through 5 channels of a computer-controlled multiplexor to a 12-bit, 28.6-KHz analog to digital (A/D) converter. (In this system, the actual A/D conversion rate used is 40 μsec/conversion or 25 KHz when scan data are being acquired.)

The digital information is processed by a PDP8/I (Digital Equipment Corp., Maynard, Mass.). The synchrony of operation of the mass spectrometer and data acquisition with the computer is ensured by initiating and terminating both operations from the "scan start/stop control" of the mass spectrometer.

The data acquired from a scan are read onto the disk, which will accommodate a maximum of 62 scans of relatively few peaks each, or down to 31 scans with up to 1000 peaks/scan. When a particular scan is to be used for calibration, the data are read back into core, the calibration is performed, and calibration coefficients are written onto the system tape. Alternatively, if the data are for a regular mass spectrum, they are converted into relative intensities and nominal masses and written onto a scratch tape for storage prior to output. The output devices are a KSR35 Teletypewriter (Teletype Corporation, Skokie, Ill.) and an incremental plotter (Model 6650-006, Houston Omnigraphic, Bellaire, Texas). The whole process is under program control. Options are chosen by an interactive dialogue with the operator through the KSR35.

Sequence of Operation. A flow chart of the sequence is shown in Figure 2. Before a mass spectrometer run is initiated, control parameters (run name, intensity channel, noise threshold, etc.) are entered from the KSR35. The scan and data acquisition are initiated from the scan start control. Output from the selected intensity channel is digi-

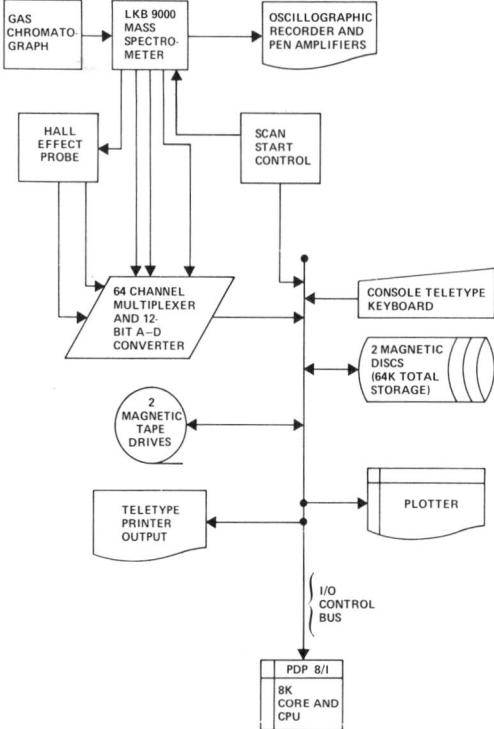

Figure 1. Major components and information flow for on-line mass spectrometry system. Interfaces and unused components are not shown. The direction of data flow and command signals are shown by the arrows

tized and followed until the summit of a peak is found. At this point the computer switches the multiplexor to one of the Hall Effect channels, takes the median of five successive readings, stores it along with the intensity of the peak, and returns to searching for the next peak.

At the completion of the scan, the summit intensities and their corresponding Hall Effect values are stored on the fixed-head disk to await further processing. The system is either reinitialized for another scan or the data conversion process is begun.

To determine the masses and relative intensities of peaks in an unknown mass spectrum, the Hall Effect values are converted to exact masses by interpolation, using cubic coefficients obtained in the calibration procedure. The peak intensities are normalized and the exact masses are then converted to nominal masses before output. A table of these results is listed on the KSR35 and/or plotted as a bar graph on the incremental recorder.

If the scan is that of a reference standard such as a hydrocarbon or perfluorokerosene (PFK), the observed peak values of ion intensity and their corresponding Hall Effect probe values are listed from the disk onto the KSR35. Up to 100 peaks are selected by the operator and their exact masses typed into the computer to the nearest 0.1 amu. A table of cubic coefficients, relating exact mass to Hall Effect voltages, is generated by the technique of "spline fitting" (12). The table of coefficients is stored on the system tape.

---

(12) R. H. Pennington, "Introductory Computer Methods and Numerical Analysis," The Macmillan Company, New York, N. Y., 1965, pp 404–411.

Details of the equipment configuration and the computer programs will be described below.

Equipment Modifications and Interfaces. The initiation and termination of a scan is communicated to the PDP8/I through the "Scan Start/Stop Interface" diagrammed in Figure 3. Thus, the status of the mass spectrometer is sensed by the computer under program control without the use of the program interrupt. Minus 12 volts appears on the scan light when a scan is under way. This level is used to drive a Schmitt trigger, the output of which is inverted and put through a NAND gate to the skip buss. Consequently, the skip buss is grounded whenever the mass spectrometer is scanning at the time of a skip instruction. This flag is used to start the data collection, and is checked after each pair of data points. Failure to skip causes an exit from the collection routine. The normal operation of the mass spectrometer for an automatic scan is followed, with no special techniques needed to accommodate the data collection.

The ion intensity information is brought into the analog multiplexor through a modified preamplifier and amplifier (Figure 4). Replacement of the preamplifier and amplifier section of the LKB 9000 gave an appreciably quieter signal. The analog signals that carry the intensity information, at gain levels of $1\times$, $10\times$, and $100\times$, were not taken directly from the pen amplifiers, but rather from a second set of solid state amplifiers that parallels the pen amplifiers. This method serves to isolate the output to the multiplexor from the generator action of the galvanometer coils after responding to high intensity peaks. Otherwise, this artifact would produce shadow peaks which the computer could identify and list. The parallel amplifiers eliminated another problem. If the pen amplifiers were not isolated from the multiplexor in some manner, turning them off induced a large voltage transient in the analog lines which destroyed the input field-effect transistors of the multiplexor. As shown, some filtering of the signal was included to maximize signal stability without appreciable loss of peak resolution.

The magnetic field strength transducer, *i.e.*, the Hall Effect probe, is interfaced as shown in Figure 5. This technique gives sufficient resolution of the signal for accurate determination of higher mass numbers (See Discussion).

Scan Algorithm. The operator types in the run name of the series of scans, the intensity channel to be used, and the threshold value for a valid peak intensity. The scan routine (Figure 6) is entered when the "scan-start" button is pressed. A search is made for a valid base line by digitizing the ion intensity in blocks of twelve A/D conversions. This string is searched until six successive numbers are found within ± 1.0 of each other. These six conversions are averaged. This average is added to the operator-set noise threshold and the sum is subtracted from all subsequent conversions. Upon finding the base-line level, the intensity is continually monitored for the occurrence of a peak by comparing each converted value (about the threshold) with the last until the negative slope of the peak is detected. The value of the highest point is retained as the summit. The analog multiplexor is immediately switched to the first Hall Effect probe channel and one A/D conversion performed. If its value is 4095 (in decimal notation, or 7777 in octal), then the channel pointer is set so that the multiplexor uses the second Hall Effect probe channel for the remainder of the scan. In either case, the Hall Effect output is digitized four more times, and the median value is found by a sorting routine interleaved with the conversions.

If the scan is not finished (ascertained by checking the status of the "scan start/stop flag"), the computer returns to searching for the next peak. Alternatively, if the scan is finished, the peak summits and their corresponding Hall Effect numbers are stored on the disk. The program is automatically reinitialized to accept another scan. At this time, the op-

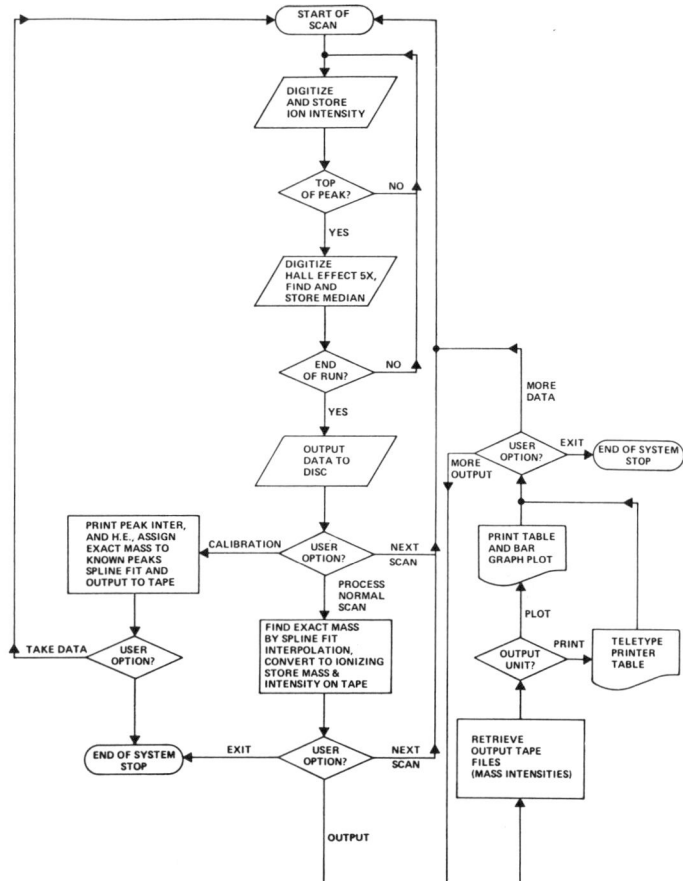

Figure 2. Flow chart of the overall operations accomplished by the mass spectrometer–computer system

erator can (a) start another scan using the same file name and parameters (the scan number is automatically incremented); (b) change either file name or parameters or both; or (c) return to the monitor.

Calibration Algorithm. The desired final output from the data processing is the nominal mass and relative intensity of each valid peak detected in a scan. To determine nominal mass, it is necessary to relate the Hall Effect probe voltage to exact mass. This is the first step of the data processing.

Mass spectral data are collected according to the above scheme with a reference compound such as PFK. The threshold noise level is usually set quite high so that only the major peaks are stored. As an option, it is possible to use two threshold levels, with the change from one to the other at the transition point in the Hall Effect probe interface (described below), so that smaller peaks in the upper mass range are recorded. The peak intensities and Hall Effect numbers are printed on the KSR35, and the operator compares this list with an oscillographic record of the scan. Mass numbers are assigned to as many as 100 Hall Effect values, selected by reasoned comparison of intensities recorded on disk and those visually observed on the oscillographic recording. The Hall Effect numbers are segmented into discrete ranges for convenience of dialogue with the

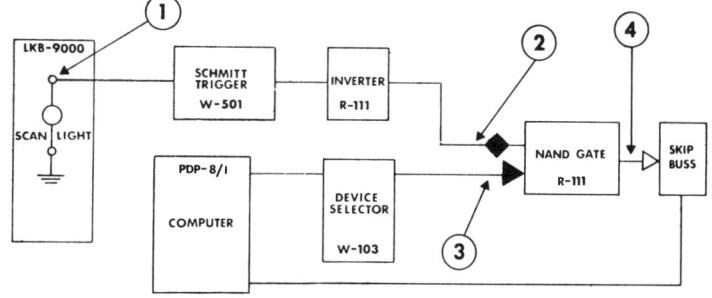

Figure 3. Scan start/stop interface. All component numbers are Digital Equipment Corp. parts. The circled numbers refer to the following notes

1. −12 volts when scan on; 0 volts when scan off
2. −3-volt level during duration of scan
3. −3-volt pulse under program control
4. Ground whenever AND logic fulfilled

1508 C. C. Sweeley et al.

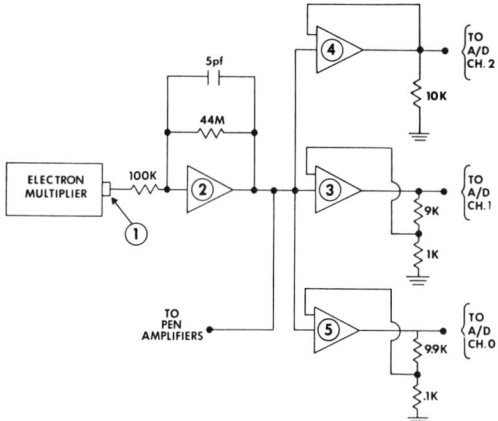

Figure 4. Interface between the electron multiplier and the A/D converter. The circled numbers refer to the following notes

1. Electron multiplier anode
2. Philbrick/Nexus-1408 amplifier (Philbrick/Nexus Co., Dedham, Mass.)
3,4,5. Zeltex Co. (Concord, Calif.) Zel-1 amplifiers (only channel 1 is currently implemented. The others have been tested)

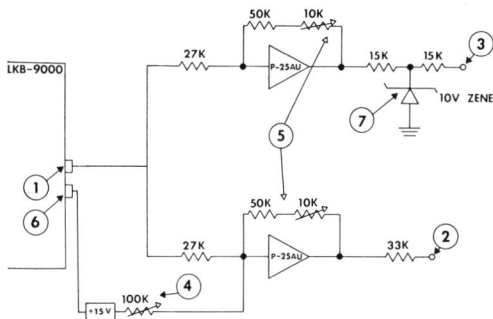

Figure 5. Expanded resolution interface between the Hall Effect probe amplifier and the A/D converter. Both inverting amplifiers have gains of 2. The second amplifier is offset by 10 volts, thus effectively increasing the resolution twofold. The circled numbers refer to the following notes

1. Hall Effect output on back of mass spectrometer
2. Output -10 volts to 0 volts during 0 to about 250 m/e, 0 volts to 10 volts from about 250 m/e to maximum
3. Output 0 volts to 10 volts from m/e 0 to about 250; saturated above m/e 250
4. Adjusts 10-volt offset on output 2
5. Adjusts $2 \times$ gain of both Philbrick/Nexus amplifiers
6. $+15$ volts output on back of mass spectrometer
7. Limits output to 10.5 volts to prevent leakage of field effect transistor to A/D converter input

Table I. Copy from the Initial Section of a Dialog between Computer and Operator

The information on the right would actually appear below that on the left

SPLINE FIT
ENTER RANGE, NO..1665,3.   1705
                                  MASS = ?43.0
ENTER RANGE, NO..1710,1.   1857
ENTER RANGE, NO..2250,4.   MASS = ?51.0
ENTER RANGE, NO..2700,2.   2163
                                  MASS = ?S
ENTER RANGE, NO..3050,2.   2165
                                  MASS = ?69.0
ENTER RANGE, NO..3450,3.   2167
ENTER RANGE, NO..3800,3.   MASS = ?S
                                  2517
ENTER RANGE, NO..4095,3.   MASS = ?93.0
ENTER RANGE, NO..300,3.   2612
                                  MASS = ?100.0
ENTER RANGE, NO..650,3.   2853
ENTER RANGE, NO..900,3.   MASS = ?119.0
ENTER RANGE, NO..1200,4.   2995
                                  MASS = ?131.0
ENTER RANGE, NO..1600,6.   3130
ENTER RANGE, NO..1900,4.   MASS = ?143.0
ENTER RANGE, NO..2100,4.   3334
                                  MASS = ?162.0
ENTER RANGE, NO..2300,6.   3408
                                  MASS = ?169.0
520
MASS = ?S   3528
522   MASS = ?181.0
MASS = ?4.0
1374   3643
MASS = ?28.0   MASS = ?193.0

it is not one of those selected. An example is shown in Table I.

The array of Hall Effect voltages and the corresponding exact masses is used to establish an interpolation table and a table of cubic coefficients for spline fitting essentially by the method described by Pennington (12). This is a curve-fitting technique that derives the coefficients of a set of cubic equations that conform to a specific function. These coefficients are used to interpolate the desired function. The interpolation table is in two parts. Part one is derived from the Hall Effect values taken from the amplifier corresponding to the lower range of m/e; the section for the higher range is derived from the digitized values obtained from the offset Hall Effect amplifier. The transition point (ca. m/e 240) is handled by spline fitting from one range to the other across the transition point.

Data Reduction. The reduction of the original scan data into a form suitable for final output (Figures 7 and 8) is accomplished by file manipulation and calculations controlled by the operator, as follows. The Hall Effect voltages from scans not treated as calibration scans are called, one at a time, from the disk (Figure 7). The spline fit coefficients obtained in the calibration procedure, outlined above, are read into core from the systems tape. The Hall Effect voltages are converted to exact masses by Pennington's interpolation method.

The observed exact masses are next converted to nominal masses. The logic of this conversion process takes advantage of the fact that even the largest mass defects (due to hydrogen) result in exact masses that differ by less than 0.1% from the nominal integer mass (or halfway between integer values for doubly charged ions). It also allows adjustment in any given scan for drift in the system (11). For each range of $n \times 100$ to $(n + 1) \times 100$ mass units ($n = 0, 1, 2 \ldots 10$), the observed deviation of the exact

computer. Normally, 3 to 7 selected peaks are incorporated into each range. However, the total number of peaks (denoted N) within the range must include the selected peaks and all the others that have equal or greater magnitude.

Within each range, N peaks are found in order of decreasing intensity. The Hall Effect voltage for each peak is presented in turn (and in order of increasing Hall Effect voltage) on the KSR35. After each number is typed out, the operator either types in the assigned exact mass or skips the peak if

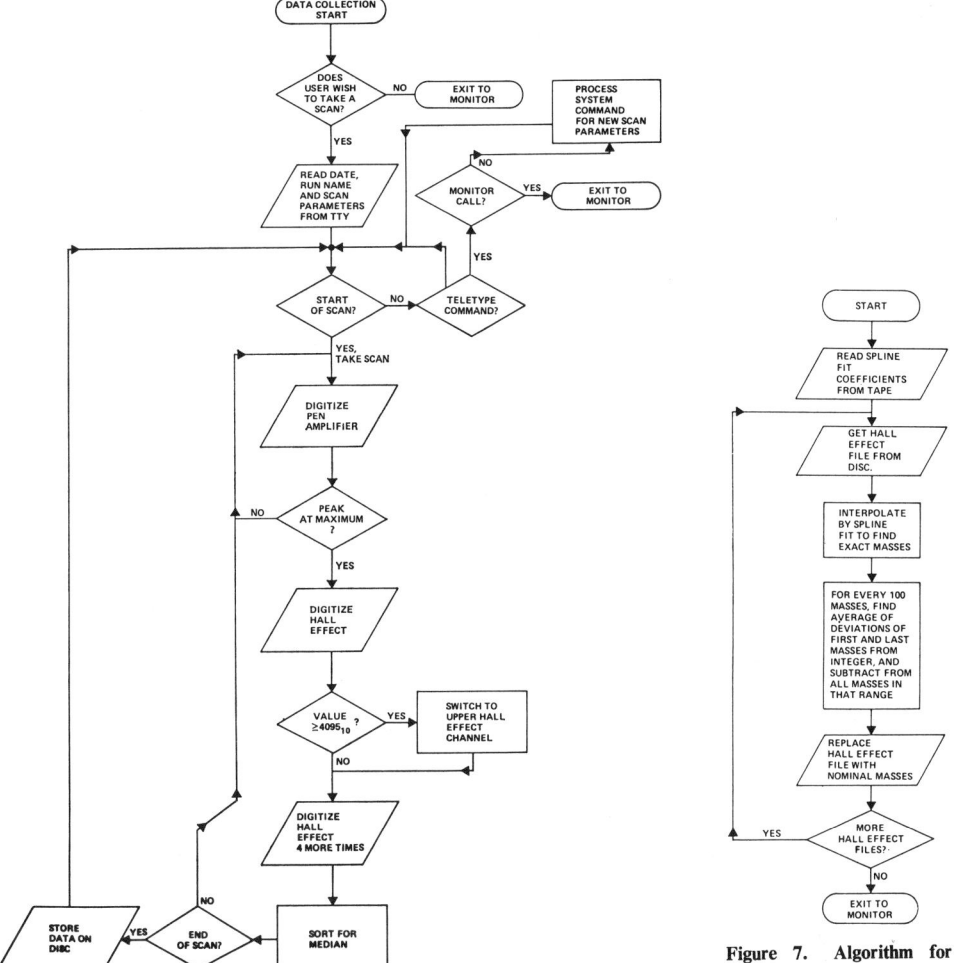

Figure 6. Flow chart of scan data acquisition algorithm

Figure 7. Algorithm for conversion of exact mass to nominal mass

mass from the nearest integer mass (after correction for accumulated error, as described below), of the first and last peak within the range is determined. If the corrected deviation of either peak is greater than 0.25 mass unit, the penultimate peak is tested. The testing is continued towards the center of the range until both peaks satisfy the criterion. The corrected deviations of the two selected boundary peaks are averaged, the average deviation is combined with the accumulated corrected average deviation of all previous ranges, and the sum is subtracted from the exact mass found for each peak in the range. The resultant value is stored as the nearest integer. This section of the system processes the data for each scan in a minimum of 2 seconds, 4 seconds for a scan having 100 peaks, and a working maximum of 8 seconds for scans having 300 to 400 peaks.

After adjustment to nominal masses, the operator has the option (Figure 8) of having the scan processed alone or averaging it with others (presumably, but not necessarily, replicates). The first step in averaging is to arrange the intensities in the scans being averaged by nominal masses. Next, the peaks are compared for occurrence at equal masses. The intensities are averaged at equal mass points. Lone peaks are divided by the number of scans being averaged. Because of the core limitations, reflected in the software, the averaging option can be exercised only once in any single named file of scans.

As a further option, peaks can be subtracted which are due to the background found from scans obtained when the gas chromatograph output (as monitored by the total ion current) showed a base-line value. The method used is similar to that for averaging scans except that the background peaks are subtracted instead of averaged.

The above methods were evaluated at various fast scan speeds available in the LKB 9000 in order to test the limits of the entire system. The scan speed control on the mass spectrometer is marked in integers from 1 through 11. In these studies, most of the experiments were performed at scan speeds 6 through 11. The time required to scan from m/e 0 to 500 for these scan speeds is approximately (i.e., triplicate determinations with a stop watch): 6 = 8.0 sec; 7 = 4.2 sec; 8 = 2.3 sec; 9 = 1.3 sec; 10 = 0.8 sec.

RESULTS

The first step in developing the mass spectrometry system was to establish the relationship between the Hall Effect volt-

Figure 8. Algorithm for averaging scans and preparation of output files

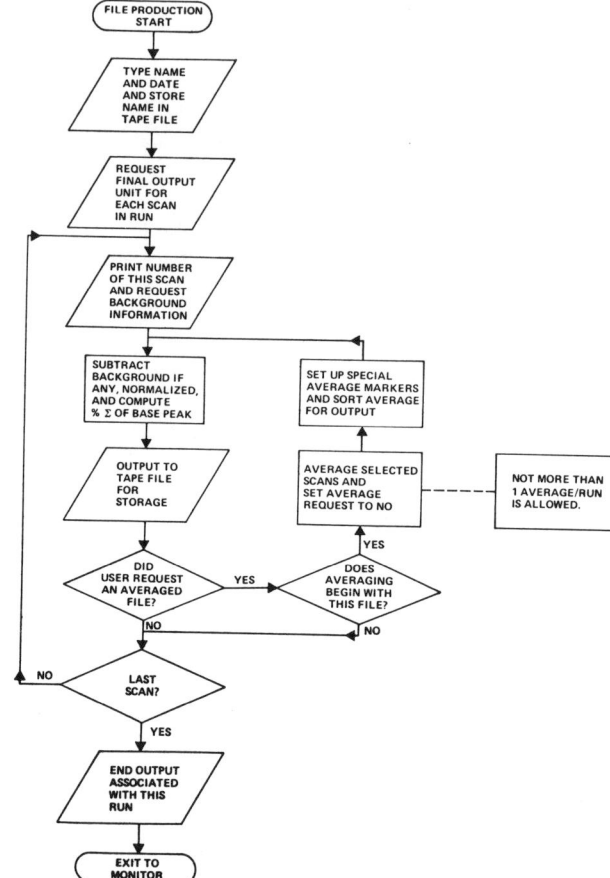

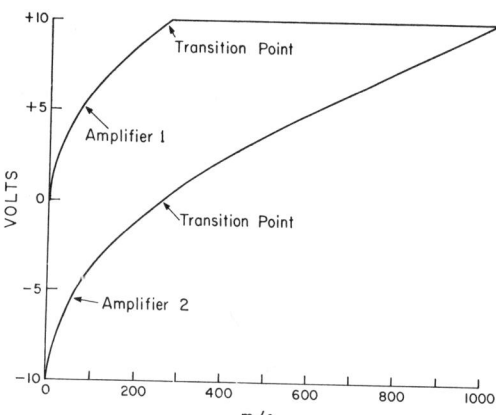

Figure 9. Relationship between Hall Effect voltages and observed m/e. The circuit diagram for the interface is shown in Figure 5

ages from the two amplifiers and the peaks obtained during a calibration scan of PFK. The overall range in the A/D converter is 0 to 10 V which results in an input range of 0 to 4095 (decimal). Two channels were used for all Hall Effect data, giving an effective 0 to 20 V range corresponding to 0 to 8191 (decimal). The average resolution of the system is slightly more than 22 (decimal)/amu between m/e 31 and 39. Near the transition point (m/e 240), the resolution drops to about 9 (decimal)/amu and in the mass range between m/e 681 and 693, it is still 4.5 (decimal)/amu. This relationship is further illustrated in Figure 9.

The next step was to find a method for assigning exact mass numbers to observed Hall Effect voltages. Initially, various methods of polynomial curve fitting were tried but none gave the required precision. After evaluation by the method of Wampler (13), it was concluded that the word size of the PDP8/I (i.e., 12 bits) was insufficient. Therefore, it was necessary to choose a method that could make use of an effective resolution of at least 13 binary bits. The internal representation of 13 bits in a 12-bit word size is accomplished as follows. Let N be the number of bits per computer word. Then, restricting ourselves to one word per data element,

1. For random data with variable sign, the data may be represented to the precision of $N - 1$
2. For random data with constant, known sign, the data may be represented to the precision of N
3. For monotonically increasing or decreasing data of one sign only, the data may be represented to the precision $N + 1$, provided that one register of storage is used to save the count at the point at which the most significant bit, represented as

---

(13) R. H. Wampler, *J. Res. Nat. Bur. Stand., Sect. B*, **73**, 59 (1969).

Table II. Intermediate Results of Hall Effect Conversions

Numbers (decimal) shown are from a
section of data from Scan 1, Figure 10

ORIGINAL HALL EFFECT	CALCULATED EXACT MASS	CALCULATED NOMINAL MASS
2611	+0.9621067E+02	96
2625	+0.9724540E+02	97
2651	+0.9917865E+02	99
2717	+0.1041284E+03	104
2733	+0.1053374E+03	105
2744	+0.1061709E+03	106
2771	+0.1082259E+03	108
2783	+0.1091436E+03	109
2796	+0.1101410E+03	110

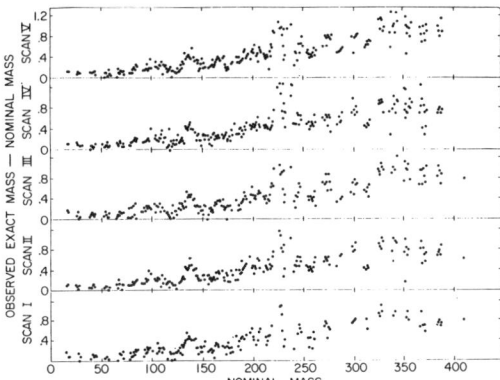

Figure 10. The divergence of observed exact mass (in amu) from nominal mass

Five replicate scans of cholesterol were performed at scan speed 7, slit settings in mm, 0.07/0.2 for exit slit/collector slit

bit -1, changes status from 0 to 1. Note that the most significant bit is not part of the word in storage

4. For monotonically increasing or decreasing data with a change in sign, the data may be represented to the precision N, provided that one register is used to save the count at the point where the sign changes.

Extension of this logic would allow even greater precision because of the monotonically increasing nature of the data being represented in this case.

Spline fit interpolation (12) was the method chosen to achieve the requisite resolution. Its use is illustrated in Table I. The computer types "ENTER RANGE, NO..". The operator responds with the Hall Effect voltage and a selected number of peaks within this range (usually only the largest ones). This continues until the last of the scan is covered. The computer then types out each of the selected Hall Effect voltages and "MASS=?". The operator responds with "S" for skip or with the exact mass of the peak if it is known and has been selected for use in calibration.

Since the algorithm for finding peaks and their maximum intensities trades speed and simplicity for sensitivity to noise in the original signal, it became necessary to evaluate that noise. At the preamplifier output, it was found to be between 50–120 mV (peak to peak, with a 10-volt full scale), and to have 60-Hz modulation, extreme microphonics and intermittent oscillations. In addition, the shadow peaks described under "Methods and Materials" were found following very intense peaks. By replacing the preamplifier (Figure 4), the residual noise at this point was decreased to <1 mV peak to peak. The final output to the A/D converter with the original amplifier was 12–20 mV (peak to peak) with intermittent oscillations. The residual after modification was <4 mV peak to peak. Most of this was 60-Hz hum. This level of noise, added to the noise specifications of the A/D converter, corresponds to errors from electronic sources of somewhat less than 0.1% full scale in determining intensities. Even if the summit of the peak is found exactly, however, the reproducibility of mass spectral analyses at high scan speeds is seldom better than ±5%.

To evaluate the spline fitting procedure, a special utility routine was written which prints a table of data of the form shown in Table II. The difference between calculated exact mass and nominal mass vs. the nominal mass itself is shown in Figure 10 for five successive scans of cholesterol. The variation in the exact mass from scan to scan seldom exceeded 0.3 amu. While the system showed good precision from scan to scan, the exact mass minus nominal mass relationship was not as smooth or as small as would be expected solely on the basis of mass defect. The most likely explanation is a combination of machine drift (calibration had been performed two days previously) and anomalies in the magnet itself and/or its current supply. It should be noted that the localized bumps in the data lie between calibration points and thus cannot be compensated for by any curve fitting technique. For example, the bump centered around m/e 138–139 is bounded by calibration points which were chosen at m/e 131 and 143. Had a calibration point been selected at m/e 138, the bump would not have been observed. The anomaly in the range of m/e 220–230 is probably due to the approaching transition point between the two amplifiers. This is easily overcome by adjusting the transition point in the amplifiers. (Again, the hump in the data was bounded by calibration points, i.e., at m/e 219 and 231.)

The minimum amount of data required for the conversion of exact mass to nominal mass was determined by performing seven replicate scans of cholestane at 70 eV. For the first five scans the multiplier high voltage supply setting was ×6 (the multiplier high voltage supply has 11 integer settings; each integer increase in setting represents an approximate doubling in sensitivity). The noise threshold was set at 35 mV initially and raised by 25 mV for each successive scan. It remained at 135 mV for the last two scans, while the multiplier was set at ×5 and ×4, respectively. The number of peaks detected for the seven scans were: 82, 60, 50, 44, 38, 25, and 7. No errors in assigning nominal masses were detected.

Another special utility routine was used to determine the shape of individual peaks at various masses and scan speeds as follows. When a scan is initiated, A/D conversions are performed on a selected channel at the maximal rate of the converter (35 microseconds/conversion). Each 3 successive values are averaged and the average stored in a table. When the table is filled (1000 points, maximum ca. 6000), the values are plotted as equal increments. After the graph is completed, the routine is reinitialized awaiting the next mass scan. These experiments were done because it was noted in the course of these studies that relative peak intensities in the lower mass range (m/e up to 80) were diminished at faster scan speeds (>7). Furthermore, small peaks were not being detected at the faster scan speeds, especially when they occurred immediately after a large peak (i.e., the isotopic form with ^{13}C, ^{17}O, etc.).

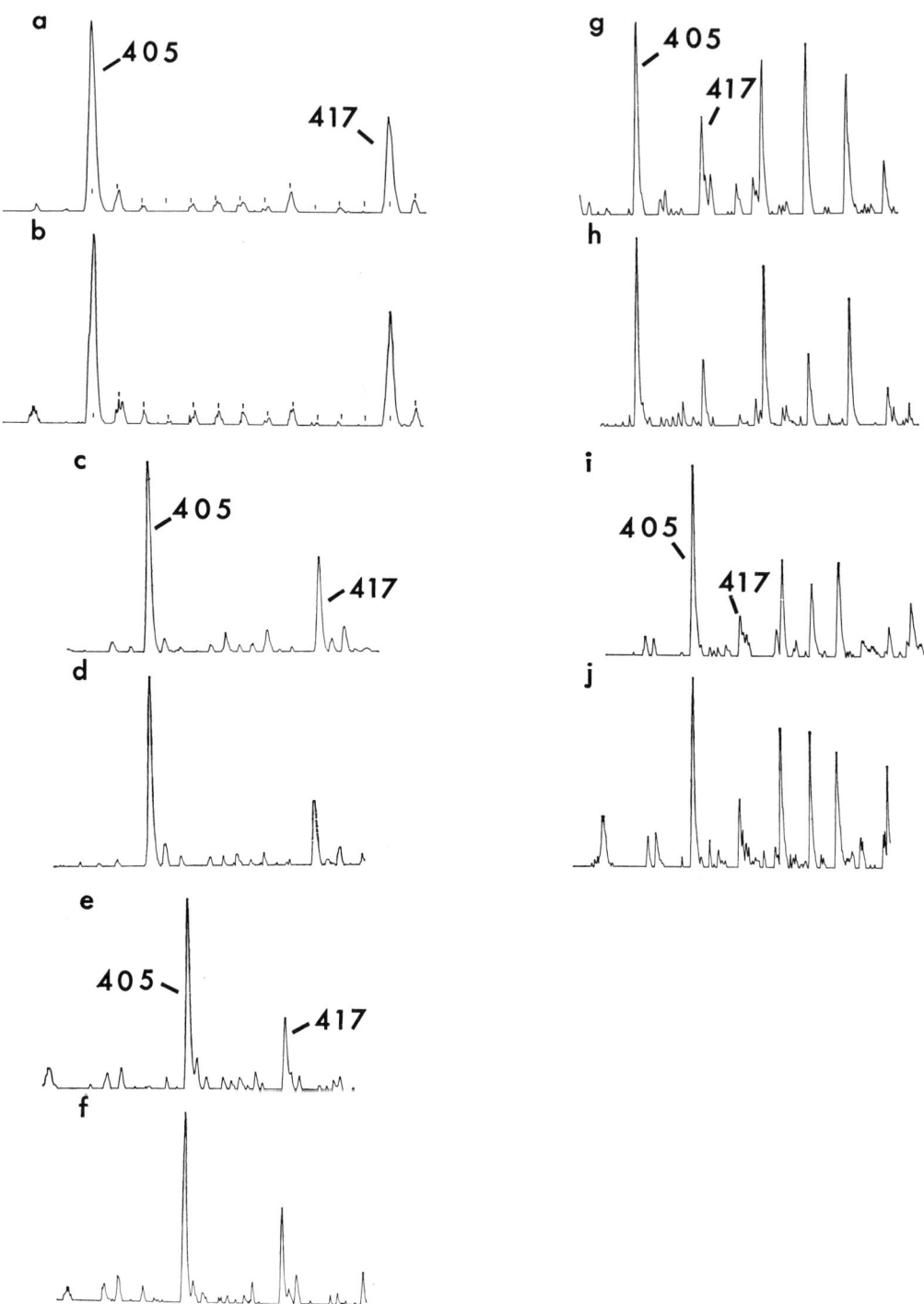

Figure 11. Peak shapes and resolution at scan speeds 6 through 10

Successive scans of PFK were performed with and without the 5-pf filter capacitor (Figure 4) in place. All scans were done with slits set at 0.03/0.10. *a.* Scan speed 6 with capacitor. *b.* Scan speed 6 without capacitor. (Vertical lines discribed in text.) *c.* Scan speed 7 with capacitor. *d.* Scan speed 7 without capacitor. *e.* Scan speed 8 with capacitor. *f.* Scan speed 8 without capacitor. *g.* Scan speed 9 with capacitor. *h.* Scan speed 9 without capacitor. *i.* Scan speed 10 with capacitor. *j.* Scan speed 10 without capacitor. Peaks at m/e 405 and 417 are shown for each scan speed

Figure 11 illustrates the between peak separation and the peak shapes obtained at scan speeds 6 through 10, with and without the filtering capacitor in place. At scan speed 6, the peaks at m/e 405, 406, and 407 are quite distinct. Furthermore, the electron multiplier voltage returns to base line between peaks. However, there is more discernible ripple in the tracing when the capacitor is removed, as would be expected. With the very simple summit finding routine used in this study, the summit position would be assigned either at the location of the first or the second high spot on the bimodal m/e 411 peak with the capacitor (Figure 11a). Neither assignment would be exactly correct. The error is about 0.1 amu. While this magnitude error is not serious at m/e 400, it may well be serious at higher masses because of the compression of the data with increasing mass. Bimodal peaks can result in two peaks being determined for one amu if the Hall Effect voltage increases by 1 or more (binary) between the two summits. If it does not increase, the higher of the two summits is recorded as the peak. At scan speed 6, the peak shapes are relatively symmetrical and unaffected by the presence of the filtering capacitor.

The short verticle lines on both recordings for scan speed 6 (Figure 11a,b), denote equal spacing for each amu (i.e., the integer values for each amu when the summits for 405 and 417 amu are taken to have error free nominal or integer values). The deviation for each of the summit locations from the integer values was measured for the tracing of Figure 11a to be: +0.08 at 406 amu; 0 at 407; +0.11 at 409; +0.07 at 410; +0.11 at 411; +0.12 at 412; +0.06 at 413; and 0 at 415.

The data for scan speed 7 (Figure 11c,d), shows that the electron multiplier voltage still has time to return to base line between peaks. In contrast, the valley between peaks did not descend to base line at scan speed 8 (Figure 11e,f). (Note the doublet at m/e 409 in Figure 11f.) Scan speed 9 (Figure 11g,h) is fast enough to merge small peaks into preceding large peaks. With the filter capacitor in place (Figure 11g), the peak at m/e 406 was essentially unresolved from that at m/e 405. This was less so with no filtering. Scan speed 10 (Fig 11i,j) showed little resolution of m/e 405, 406, and 407.

As is well known, narrowing the slits and changing other focusing parameters affects peak resolution (albeit at the sacrifice of sensitivity) so that if fast scans are required, adjusting the mass spectrometer will help. Thus, by using the utility program in this way, it is possible to optimize the machine parameters quickly and sensitively. However, it would appear that scan speed 9 is marginal for the detection of small peaks with any degree of reliability (see Discussion).

The truncation problem (i.e., the lowering of peak intensities at low mass numbers with increasing scan speeds) is ascribable to the narrow shape of the peak observed at low masses. Since the program does not find the area under the peak, any effect resulting in determining a lower summit would produce lower intensities. With an extremely sharp peak, the summit of the peak may occur between successive A/D conversions. At the highest possible scan speeds, this will happen most of the time. The peak shapes do not change appreciably over the range of scan speeds 6 through 9 regardless of the presence or absence of the filter capacitor (Figure 12). However, the distance (i.e., time) between m/e 15 and 18, and hence the time that it takes a peak to sweep by, is shortened considerably with increasing scan speed.

A comparison was made of the plot obtained with the on-line computer system with that made from intensities measured manually on an oscillographic recording of the same scan. There were no detectable differences in the two plots (Figure

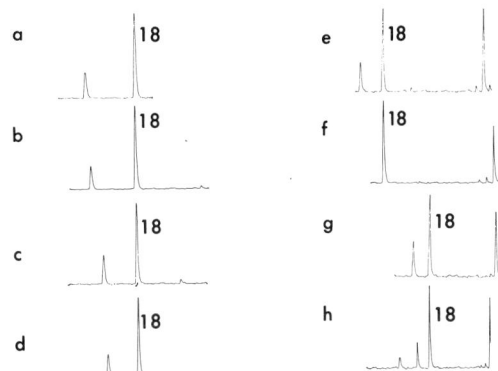

Figure 12. Peak shapes at low m/e at scan speeds 6 through 9

Successive scans of PFK were performed with and without the 5-pf filter capacitor (Figure 4) in place. The slit settings were 0.08/0.24. The peak at m/e 18 is labeled in all cases. *a.* Scan speed 6 with capacitor. *b.* Scan speed 6 without capacitor. *c.* Scan speed 7 with capacitor. *d.* Scan speed 7 without capacitor. *e.* Scan speed 8 with capacitor. *f.* Scan speed 8 without capacitor. *g.* Scan speed 9 with capacitor. *h.* Scan speed 9 without capacitor. In all traces, m/e increases from left to right.

13, *A* and *C*), indicating that all of the mass assignments by the computer were correct and that the intensities of large and small peaks had been accurately found. The results of repetitive scans of cholestane at scan speeds ranging from 6 to 9 showed no particular difference due to either noise or varying scan speed (Figure 13, *B–E*). The relative intensities were nearly identical over the entire mass range for scan speeds 6, 7, and 8 but at scan speed 9, systematically lower intensities were observed in the low-mass range up to about m/e 100 and the isotopic peaks in the higher mass range were abnormally high, presumably because of the incomplete resolution described above.

DISCUSSION

Like many instruments which were not designed specifically to interface with a high speed digital computer, the standard output of the LKB 9000 was found to be excessively noisy. The preamplifier and amplifier were designed to drive galvanometer pens and do so adequately. However, since the data sampling rate of the A/D converter is so high (35 μsec/conversion), there is no equivalent of the integrating effect of the galvanometer pen inertia. Therefore, the preamplifier–amplifier section of the LKB 9000 was replaced with one having greater stability and less noise.

Errors due to noise are usually minimized by either analog methods (i.e., filtering) or digital techniques (i.e., data smoothing). In the system under discussion, analog filtering was chosen for the ion intensity information because of the design parameter that the system be quickly responsive in real-time. This is to accommodate as fast a scan speed as possible.

Most of the noise observed in the LKB 9000 seemed to be of higher frequency than the mass peaks themselves. Thus the filtering requirement was reduced to choosing a filter network with a time constant sufficient to diminish the noise but not enough to introduce appreciable capacitive skewing of the mass peaks themselves. With the circuitry used in this study, it appeared that machine parameters (slit width, magnet focusing, etc.) in the mass spectrometer itself were primarily responsible for the peak skewing. The presence or absence of

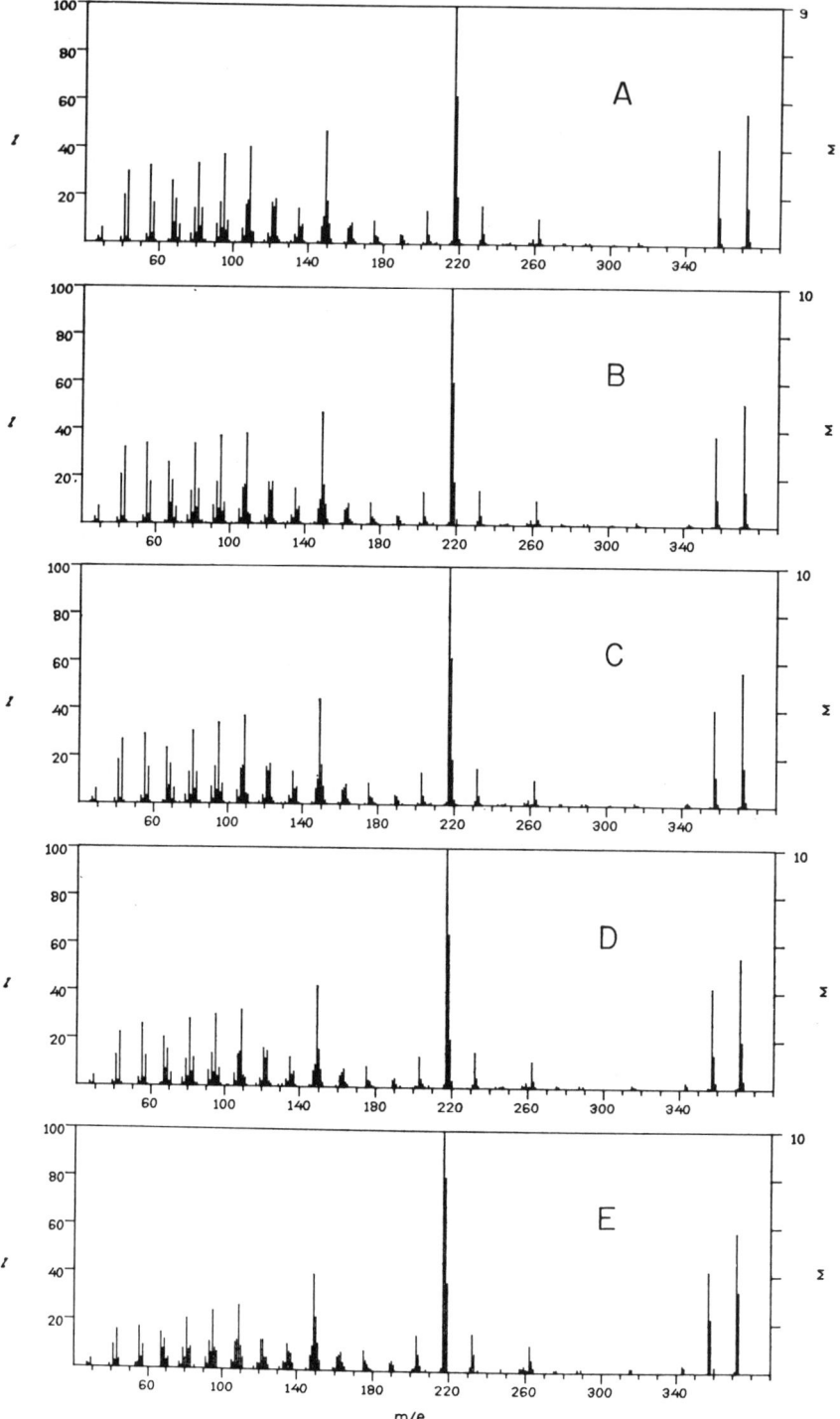

Figure 13. Comparison of oscillographic output and computer-generated graphs

Successive scans of cholestane were performed at increasing scan speeds. The oscillographic recording at scan speed 7 was normalized and plotted (A). Panels B,C,D,E are for scan speeds 6,7,8, and 9, respectively, as reduced by the computer system. Thus, panels A and C are derived from the same scan

the filter capacitor affected only the smoothness of the line, which is the desired result.

The computer was interfaced directly with the Hall Effect probe instead of the digital logic of the mass marker for several reasons. First of all, we wanted the calibration parameters to be performed by the computer rather than by manual adjustment of the network of potentiometers in the mass marker system. Second, calibration should be inherently much faster with this technique. Finally, it is generally good practice to have as few components in a data acquisition sequence as is possible.

The consideration of noise in the Hall Effect probe was treated by digital means in the computer itself. This resulted in no loss of response speed since the five digitizations and interleaved median sort for the Hall Effect voltage requires only 150 microseconds, which is less than the time required for any but the smallest peaks to return to base line.

If one considers the resolution required to locate a mass peak to the nearest 0.1–0.2 mass unit over the total range of 1000 mass units, it is evident that a single span of 12 binary bits (4096 decimal) is not sufficiently precise. By splitting the output of the Hall Effect probe between two amplifiers of similar gain, an effective resolution of 8192 parts is achieved. The stability and noise characteristics of the Hall Effect probe are such that a third range could be introduced (i.e., a total of 12,288 parts) should it be required at the higher mass numbers. By adjusting the gains of the amplifiers, it is possible to change the resolution for a given segment of the scan without affecting the others. By changing the value of the offset, the transition point can be changed as well. Thus, three amplifiers, used in this way, could result in a flexible, precise interface.

The use of the median of five successive readings of the Hall Effect probe was found to be more reliable than the mean. The time required to perform the five digitizations and the sort is less than the descending side of a mass peak at scan speeds through 10. Thus, the assignment of a mass unit to a Hall Effect voltage at one scan speed is valid within the limits required to determine nominal masses at the other scan speeds.

In order to accomplish the calibration of the Hall Effect voltages in terms of mass, the technique of spline interpolation was chosen after polynomial curve fitting was found to have inadequate resolving power. Further, spline fitting is able to follow the apparent anomalies and discontinuities in the relationship, such as those due to using multiple interface amplifiers and/or magnet asymmetries.

The system has a remarkably versatile range of scan speeds which can be used. We have not tested lower ranges but there is no reason to believe there will be difficulties as the scan time is longer since resolution will be improved. Intensity errors and occasional losses of small peaks due to lack of resolution are the observed problems at scan speeds of 9 or higher. The effective limit is therefore about 2.3 seconds for a scan from m/e 0 to 500, but even when the time for this mass range is decreased to 0.8 second, these were the only errors and the nominal masses were correctly determined. Thus, when accurate intensities and small following peaks are not of major significance, the system can be used satisfactorily at the highest available scan speed. The truncation of ions in the low mass range would presumably not occur if this computer system were used with exponential scanning of the mass spectrometer.

Experience has shown that the calibration procedure described in "Methods and Materials" is somewhat cumbersome. The long term stability of Hall Effect voltages is within the capability of reassignment by the computer in the calibration procedure, unless the mass spectrometer magnet or the Hall Effect probe is repositioned. Thus, we are investigating possible algorithms for doing the calibration essentially under computer control.

The observed mass that is determined by the application of spline interpolation is not the integer mass (or "nominal" mass). The observed value deviates from the integer because of a combination of the inherent mass defect (due to the packing fractions of the various atoms comprising the ion), the drift in the system as a whole, and the error of a single determination. The arithmetic value of the mass defect varies with the species of ion. Its magnitude is small, even in the worst case of a hydrocarbon wherein the mass defect of any ion never exceeds approximately 0.1 % of the nominal mass of that ion. This contribution to the observed mass is consistent enough to be treated as part of the overall drift of the system. If the system is not calibrated with reasonable frequency, drift errors can easily accumulate to 1 or 2 mass units by m/e 500. The variation in successive determinations appears to be less than 0.3 mass unit for any given peak, and it may be neglected in the correction of observed mass to nominal mass except as a decision constraint in the detection of doubly-charged ions. If properly designed, the correction algorithm should be highly reliable.

Hedfjall et al. (11) have described an algorithm which monitors the deviation from nominal mass based on the time of occurrence of a mass peak after a mass marker pulse. The test for deviation is performed each time a mass peak more intense than a preset level is detected. The found deviation (if it is less than 0.25 mass unit greater than the last detected shift) is subtracted from all subsequent peaks until a new calibration peak is found and the new deviation value is determined.

The above type of correction algorithm was not adopted for two reasons. Complicated interface hardware and programming are required to detect and acquire the mass marker pulses. Furthermore, in scans without many mass peaks, the difference in observed masses between valid mass peaks might be greater than 0.25 mass unit, in which case the algorithm would fail.

An analysis of the relationship between observed and nominal masses led us to consider that a procedure based on the average rate of drift of segments of the scan would be simpler and more accurate. The first attempt is that described in "Methods and Materials." It was found to be generally satisfactory. However, it was sensitive to errors caused by determining the drift rate with peaks which were abnormally high or low in divergence from nominal mass. Furthermore, no attempt was made to compensate for local anomalies in the data (Figure 10). At this writing, a new algorithm is being tested which eliminates these problems. It uses the median of the divergence of a set of successive mass peaks to obtain the value used in calculating the average rate of drift. The detection of doubly charged ions of odd integer mass, which occur at "half mass" values of m/e in a magnetic scan, may possibly be accomplished by comparing the divergence from nominal mass, after correction for drift, of each peak to those four immediately surrounding it (also drift corrected). If it deviates by more than 0.35 mass unit from the median of the surrounding peaks, it is considered a candidate for half mass status. If there is a peak at twice this value, it is listed as a half mass. Otherwise, the value of the observed mass is listed separately as an anomalous peak to be interpreted as either due to noise or a contaminant.

Another modification planned for the system is the interfacing of the total ion current to the A/D converter. This will allow each scan to be associated with the total ion current observable at the time the scan was obtained. Further, it will allow the intensities observed during a scan to be adjusted for the change in total ion current taking place during the course of the scan itself (*14*). Another benefit of this interface will be the ability of the computer, under program control, to select the appropriate intensity channel and threshold value.

---

(14) C. H. Sederholm, IBM Scientific Center, Palo Alto, Calif., private communication, 1970.

ACKNOWLEDGMENT

The authors are grateful to Jack Harten for technical assistance.

RECEIVED for review March 15, 1970. Accepted August 5, 1970. This investigation was supported in part by research grants from NIAMD-NIH (AM 12434), DRRF-NIH (RR 00480) and (FR-05656) of the PHS as well as National Science Foundation (GB 7856 and GU 2293). It is published with the approval of the Director of the Michigan Agricultural Experimental Station as Journal Article 5060.

ELECTROCHEMISTRY

Editors' Comments on Papers 16, 17, and 18

16 CREASON, LLOYD, and SMITH
Evaluation of a Computerized Sampling Technique for Digital Data Acquisition of High-Speed Transient Waveforms: Application to Cyclic Voltammetry

17 SCHWALL et al.
High Speed Synchronous Data Generation and Sampler System: Application to On-Line Fast Fourier Transform Faradaic Admittance Measurements

18 HANAFEY et al.
Analysis of Electrochemical Mechanisms by Finite Difference Simulation and Simplex Fitting of Double Potential Step Current, Charge, and Absorbance Responses

The first paper in this section, Paper 16 by Creason, Lloyd, and Smith, evaluates a sampling technique for acquiring transient waveforms. The method is developed with respect to cyclic voltammetry, but is applicable to any repeatable, transient signal. Basically the technique consists of shifting the origin of the sample and repeating the data collection to yield a nondegenerate set of data points from which the original signal can be reconstructed. A three orders of magnitude increase in effective bandpass is obtained for the method. By starting the sampling process at a slightly different point for each scan, sufficient data are acquired to reconstruct the transient waveform. Of particular interest are the description of the circuits used to implement the method and the discussion of signal conditioning.

Schwall et al. (Paper 17) apply an improved version of the above high-speed synchronous data generation and sampling system to on-line fast Fourier transform faradaic admittance measurements. The interest in high-speed sampling is that high data rates are needed to provide the total cell admittance from which the faradaic components of this admittance can be determined by frequency domain analysis. The authors describe, in effect, a generalized transfer function measurement system with a large bandpass. The high data acquisition rates are achieved by using sixteen 1024 x 1 bit shift registers in parallel for each of two memories, which can be used to store sampled waveforms until they are transferred to disk storage.

The concepts applied in this paper have been discussed more fully in an instrumentation article by Smith in *Analytical Chemistry* on data processing in electrochemistry, in which he describes the acquisition of electrochemical response spectra by on-line fast Fourier transform.[1] Smith points out that there are several characteristics of electrochemical measurements that make the fast Fourier transform a natural approach to data analysis. Among these characteristics is that the frequency domain of electrochemistry is much lower than optical spectroscopy, making possible direct digital recording of the time domain waveforms. Another convenient feature of electrochemical responses is that they do not contain the complexity of detail frequently found in spectroscopy. He discusses appropriate strategies for acquisition of data and surveys typical instrument requirements for FFT electroanalytical measurements.

Note that another paper describing an application of the methods presented in Paper 17 appears in the same issue of *Analytical Chemistry*. Schwall, Bond, and Smith[2] have delineated how on-line fast Fourier transform faradaic admittance measurements may be used for real-time deconvolution of heterogeneous charge transfer kinetic effects to improve analytical accuracy in electrochemical assays. The authors examine various experimental systems and demonstrate how the reversible response can be extracted from observed quasireversible faradaic admittance spectra.

In Paper 18 Hanafey et al. present a method for analysis of electrochemical mechanisms by finite difference simulation and simplex fitting of double potential step current, charge, and absorbance responses. This lengthy paper generates a large set of accurate response data for various electrochemical mechanisms, which can be used as a reference file for determining the mechanisms of electrochemical reactions. The discussion of the approach used is detailed and thorough, as is the analysis of the results.

Other work applying computers to electrochemistry includes a series of four papers[3-6] by Angerstein-Kozlowska, Klinger, and Conway that describe a computer simulation of the kinetic behavior of surface reactions driven by a linear potential sweep. They treat 1-electron reactions with a single adsorbed species, sequential reactions of adsorbed species, monolayer formation by a nucleation and growth mechanism, and kinetic behavior of a nucleation and growth controlled surface process.

Another computer simulation is that by Barradas et al.,[7] who have described a computer simulation of the voltammogram corresponding to the two-dimensional progressive nucleation and growth of a passivating film on an electrode surface. Using equations for the surface coverage and current as a function of applied potential and time, they simulate the linear potential sweep as the sum of a large number of potential increments, while considering the surface coverage as remaining constant for each increment. The effect of ohmic overpotential and the effect of variation of initial surface coverage on the peak potential and current are examined.

Simulations of the measurement process serve a useful purpose in helping direct experimental investigations. Electroanalytical chemistry studies like this are of importance in the study of solid electrodes under conditions where surface passivation is a problem.

In other applications of computers to electrochemistry, Zipper et al.[8] have reported on computer controlled monitoring and data reduction for multiple ion-selective electrodes in a flowing system. Up to five electrodes are monitored simultaneously. Bond and Grabaric[9] have reported on differential pulse polarography and voltammetry with a microprocessor controlled polarograph and a pressurized mercury electrode. The advantage of being able to store and later subtract a blank from the sample is illustrated.

REFERENCES

1. D. E. Smith, "Data Processing in Electrochemistry," *Anal. Chem.*, **48**, 221A (1976).
2. R. J. Schwall, A. M. Bond, and D. E. Smith, "On-Line Fast Fourier Transform Faradaic Admittance Measurements: Real-Time Deconvolution of Heterogeneous Charge Transfer Kinetic Effects for Thermodynamic and Analytical Measurements," *Anal. Chem.*, **49**, 1805 (1977).
3. M. Angerstein-Kozlowska, J. Klinger, and B. E. Conway, "Computer Simulation of the Kinetic Behavior of Surface Reactions Driven by a Linear Potential Sweep. Part I. Model 1-Electron Reaction with a Single Adsorbed Species," *J. Electroanal. Chem.*, **75**, 45 (1977).
4. M. Angerstein-Kozlowska, J. Klinger, and B. E. Conway, "Computer Simulation of the Kinetic Behavior of Surface Reactions Driven by a Linear Potential Sweep, Part II. Sequential Reactions of Adsorbed Species," *J. Electroanal. Chem.*, **75**, 61 (1977).
5. M. Angerstein-Kozlowska, B. E. Conway, and J. Klinger, "Computer Simulation of the Kinetic Behavior of Surface Reactions Driven by a Linear Potential Sweep. Part III. Monolayer Formation by a Nucleation and Growth Mechanism," *J. Electroanal. Chem.*, **87**, 301 (1978).
6. M. Angerstein-Kozlowska, B. E. Conway, and J. Klinger, "Computer Simulation of the Kinetic Behavior of Surface Reactions Driven by a Linear Potential Sweep, Part IV. Kinetic Behavior of a Nucleation and Growth Controlled Surface Process Under Potentiostatic Conditions and Comparison with Conclusions for Potentiodynamic Conditions," *J. Electroanal. Chem.*, **87**, 321 (1978).
7. R. G. Barradas, F. C. Benson, and S. Fletcher, "A Computer Simulation of the

Voltammogram Corresponding to the Two-Dimensional Progressive Nucleation and Growth of a Passivating Film" *Electrochim. Acta,* **22,** 1197 (1977).
8. J. J. Zipper, B. Fleet, and S. P. Perone, Computer-Controlled Monitoring and Data Reduction for Multiple Ion-Selective Electrodes in a Flowing System," *Anal. Chem.,* **46,** 2111 (1974).
9. A. M. Bond and B. S. Grabaric, "Differential Pulse Polarography and Voltammetry with a Microprocessor-Controlled Polarograph and a Pressurized Mercury Electrode," *Anal. Chim. Acta.,* **88,** 227 (1977).

Copyright © 1972 by the American Chemical Society
Reprinted from *Anal. Chem.* **44**:1159-1166 (1972)

Evaluation of a Computerized Sampling Technique for Digital Data Acquisition of High-Speed Transient Waveforms: Application to Cyclic Voltammetry

Sam C. Creason, Robert J. Loyd, and Donald E. Smith[1]

Department of Chemistry, Northwestern University, Evanston, Ill. 60201

THE UTILIZATION OF DIGITAL DATA acquisition systems based on on-line laboratory digital computers ("minicomputers") is becoming increasingly widespread in electrochemical experimentation. The rather impressive advantages afforded by this measurement concept have been discussed widely and numerous computer-enabled experimental innovations have been reported (*1-11*). For the majority of electrochemical measurements, the capabilities of even relatively modest minicomputer systems normally exceed experimental requirements with regard to speed, accuracy, etc. However, in electrochemical relaxation measurements involving a transient response, a ubiquitous difficulty is encountered whenever the transient becomes sufficiently short-lived. The problem arises because of the finite digital data acquisition period (the time required for analog-to-digital conversion and storage of a data point). Unless the duration of the transient is much longer than the data acquisition period, sufficient data points cannot be obtained during one transient to adequately define the waveform. Further, if the transient is "X-Y" in nature, as in cyclic voltammetry, the problem is compounded in that two channels of data must be sampled at precisely the same point in time. In this report we present an evaluation of a technique whereby the effective sampling rate of a computerized digital data acquisition system can be increased manyfold if the transient waveform can be repeatedly reproduced. The acquisition of single-cycle cyclic voltammograms (SCCV) is employed for evaluation purposes.

The data acquisition technique in question is based on a stroboscopic principle similar to that used in the sampling oscilloscope (*12-14*). Specifically, the transient is generated and repeatedly sampled as rapidly as possible for the duration of the transient, then generated again and sampled as before except that the sampling process is started at a slightly different time in the life of the transient. As the process is repeated, each repetition yields a non-degenerate set of data points and, eventually, sufficient data are acquired to construct the transient waveform with adequate precision. The provision to acquire multiple data points per single transient cycle (when possible) is the only basic difference between the data sampling concept just described and the sampling oscilloscope's data acquisition mode which is limited to one point per cycle. However, this difference is significant because by removing the restriction of one point per cycle, the data acquisition system always is permitted to reconstruct the waveform at the maximum possible rate, whereas a single point per cycle scheme will "waste time" whenever the transient is significantly longer than the data acquisition period. This is accomplished by the multiple-point per cycle scheme while retaining the same bandpass advantages characterizing the single-point schemes, because the same factors determine the ultimate bandpass achievable in each case (see below). Of course, as the transient cycle time approaches the data acquisition period, the approach discussed here automatically reduces to the sampling oscilloscope's one point per cycle mode.

It should be evident from the discussion to follow that the basic measurement concept examined, as well as the specific means for its implementation, is generally applicable to any repeatable transient signal of either electrochemical or non-electrochemical origin. Cyclic voltammetric measurements are used to demonstrate the measurement principle primarily because the cyclic voltammogram represents one of the more challenging electrochemical transients. It requires simultaneous acquisition of current and potential signals and the voltammogram structure often is more complex than encountered in other common transient measurements.

---

[1] To whom correspondence should be addressed.

(1) G. Lauer, R. Abel, and F. C. Anson, ANAL. CHEM., **39**, 765 (1967).
(2) G. Lauer and R. A. Osteryoung, *ibid.*, **40**(10), 30A (1968).
(3) S. P. Perone, J. E. Harrar, F. B. Stephens, and R. F. Anderson, *ibid.*, p 899.
(4) S. P. Perone, D. O. Jones, and W. F. Gutknecht, *ibid.*, **41**, 1154 (1969).
(5) F. B. Stephens, F. Jakob, L. P. Rigdon, and J. E. Harrar, *ibid.*, **42**, 764 (1970).
(6) W. F. Gutknecht and S. P. Perone, *ibid.*, p 906.
(7) D. O. Jones and S. P. Perone, *ibid.*, p 1151.
(8) L. B. Sybrandt and S. P. Perone, *ibid.*, **43**, 382 (1971).
(9) H. E. Keller and R. A. Osteryoung, *ibid.*, p 342.
(10) J. Lawrence and D. M. Mohilner, *J. Electrochem. Soc.*, **118**, 259 (1971).
(11) "Computers in Chemistry and Instrumentation," J. S. Mattson, H. D. MacDonald, Jr., and H. B. Mark, Jr., Ed., Vol. 2, M. Dekker, New York, N.Y., all articles, in press.

(12) W. M. Grove, *Hewlett-Packard J.*, **15**(8), 5 (1964).
(13) W. E. Bushor, *Electronics*, **32**, 69 (July 31, 1959).
(14) J. J. Amodei, *ibid.*, **33**, 96 (June 24, 1960).

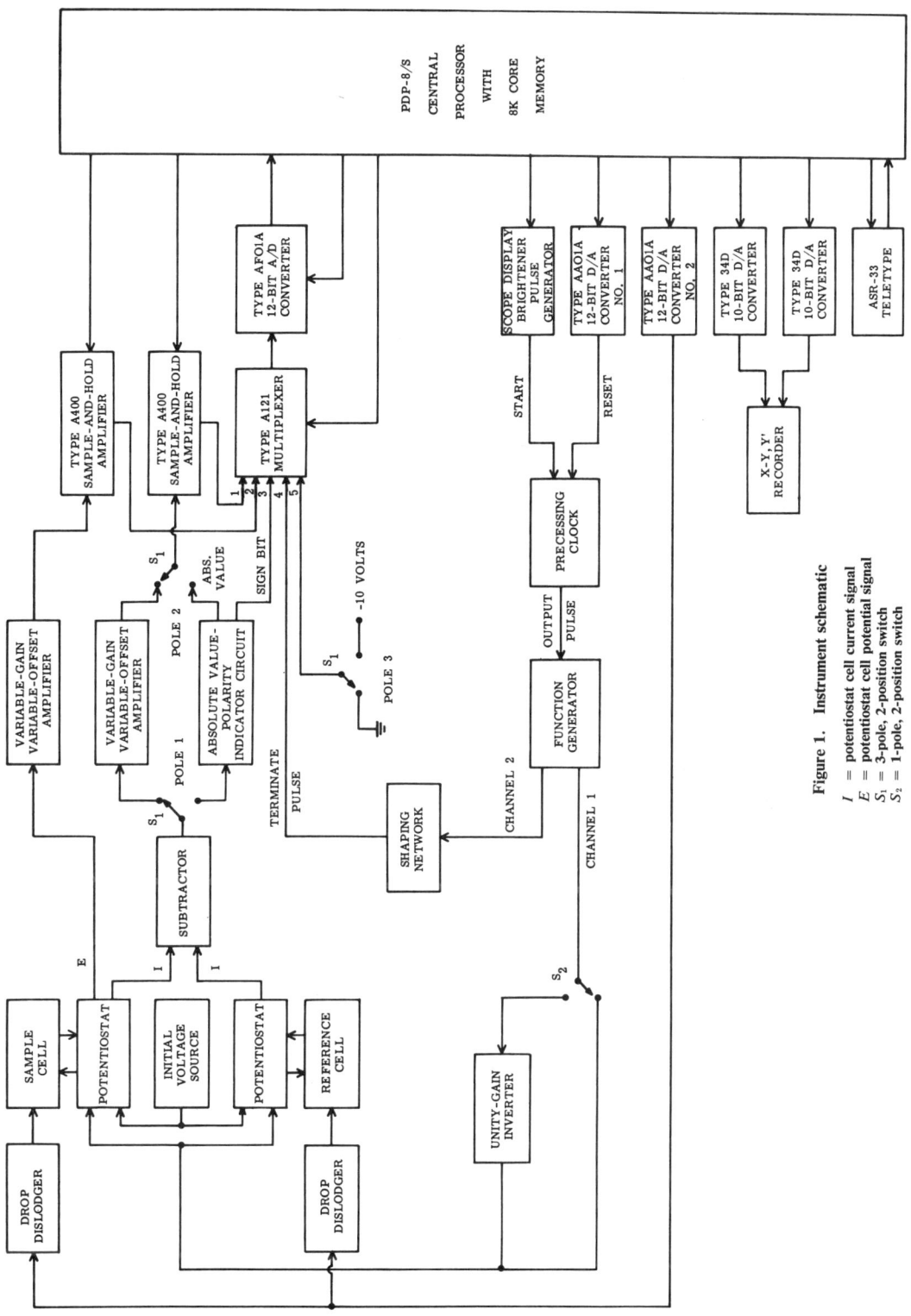

Figure 1. Instrument schematic

I = potentiostat cell current signal
E = potentiostat cell potential signal
S_1 = 3-pole, 2-position switch
S_2 = 1-pole, 2-position switch

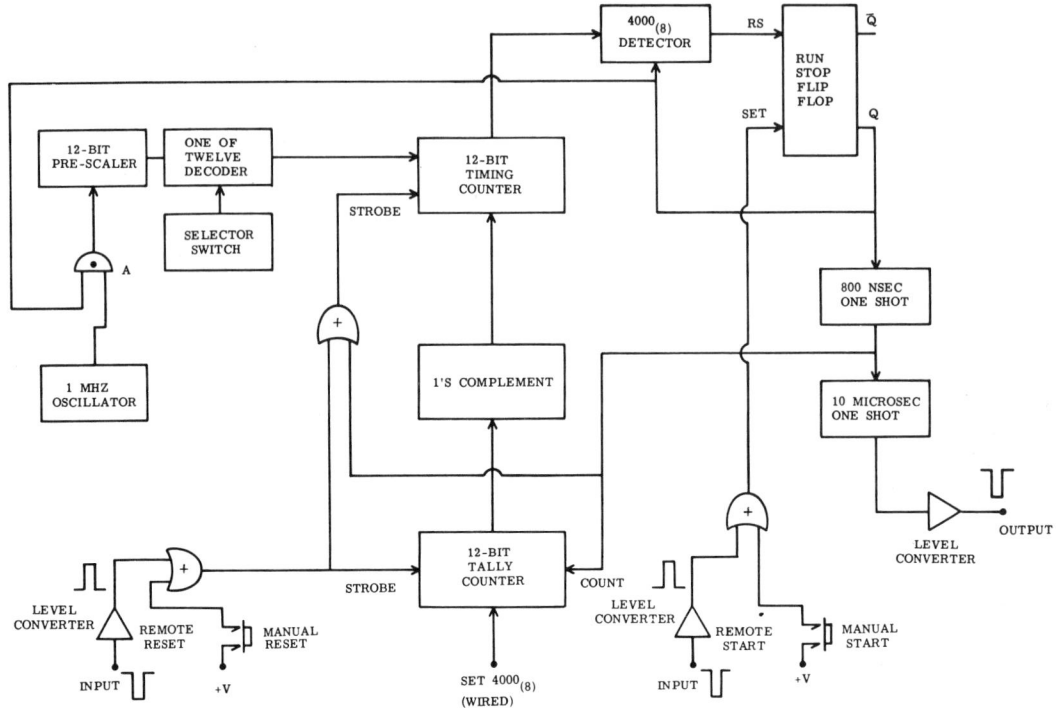

Figure 2. Precessing clock schematic

The most convenient basis for repeating an electrochemical transient experiment is found in the dropping mercury electrode (DME) at which one can readily generate a reproducible transient once per drop life with the aid of appropriate timing circuitry (2, 3, 15, 16). The present work exploited this characteristic of the DME.

EXPERIMENTAL

Instrument Components. A block diagram of the instrument system is shown in Figure 1. The minicomputer subsystem was manufactured by Digital Equipment Corporation (DEC). It features a Model PDP-8/S central processor with an 8K, 12-bit word core memory and an 8-microsecond cycle time. Its peripherals include an ASR-33 Teletype with a paper tape reader and punch, a DEC Type AF01A 12-bit analog-to-digital (A/D) converter, a DEC Type A121 20-channel multiplexer, two DEC Type A400 sample-and-hold (S/H) amplifiers activated by a DEC Type AC01 sample-and-hold control, two DEC Type AA01A 12-bit digital-to-analog (D/A) converters and a DEC Type 34D scope display control. The scope display control components were not utilized for their originally intended purpose in this work, as indicated in Figure 1. The 10-bit X and Y scope display D/A converters were used to drive an Electro Instruments Model 480 X-Y,Y' recorder equipped with Model 468 (X-axis) and 420 (Y-axis) plug-in modules. The display brightener pulse was used to start the external precessing clock.

When evaluating the success of the data sampling technique under consideration, it is important to recognize that the aforementioned minicomputer subsystem represents one of the slowest available from the viewpoint of both data acquisition and arithmetic manipulation. The computer cycle time of 8 microseconds is much longer than more typical values for minicomputers of around 1 microsecond. Although A/D conversion alone requires only 35 microseconds, the necessary accompanying operations of activating S/H amplifiers, data storage, memory address incrementation, flag interrogation, etc. requires many computer cycles so that the maximum data acquisition rate is about one X-Y point every 600 to 800 microseconds. Because of these characteristics, it is fair to state that unusually stringent demands are placed on the data sampling technique under consideration with the minicomputer employed in this work.

The key element controlling the timing of the experimental sequence is the *precessing clock* (PC). Its basic purpose is to control the time in the life of the SCCV at which the data sampling process starts. The interval between a start pulse to, and an output pulse from the PC depends on the number of start pulses which have been applied subsequent to a reset pulse. When the first start pulse is applied, a selectable interval of between one and 2048 microseconds (the precession time) elapses before an output pulse appears. When the second pulse is applied, twice the interval elapses before an output pulse appears, etc. The method whereby this is accomplished may be understood by consulting Figure 2, which provides a block diagram of the PC. The device is basically a crystal-controlled clock and a counter which issues a pulse when a particular count is reached. The precessing action is obtained by starting the count at successively lower numbers. At the start of an experiment a reset pulse is applied which causes a count of $4000_{(8)}$ to be strobed into the 12-bit tally counter. The one's complement is taken, and the result is strobed into the 12-bit timing counter. Hence, before the first start pulse is received, the PC is in a stable state with the timing counter set to $3777_{(8)}$. When the first

---

(15) D. E. Smith, in "Computers in Chemistry and Instrumentation," Vol. 2, J. S. Mattson, H. D. MacDonald, Jr., and H. B. Mark, Jr., Ed., M. Dekker, New York, N.Y., in press.

(16) E. R. Brown, T. G. McCord, D. E. Smith, and D. D. DeFord, ANAL. CHEM., **38**, 1119 (1966).

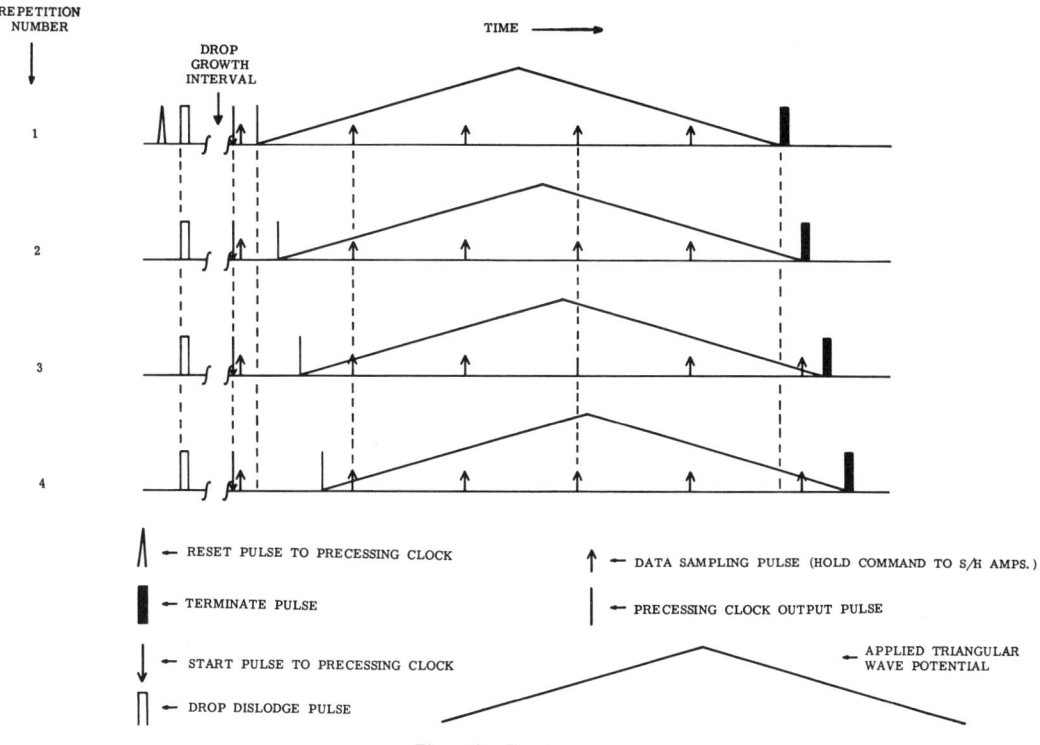

Figure 3. Event sequence

start pulse is received, the stop-run flip-flop is triggered to the run state, enabling the clock via gate A. The timing counter counts one clock pulse and reaches $4000_{(8)}$ which activates the $4000_{(8)}$ detector, in turn triggering the stop-run flip-flop to the stop state, which disables the clock. In this instance, the direction of the state-change of the flip-flop is such that the 800-nanosecond and ten-microsecond one-shots are triggered. Consequently, an output pulse is generated via the level conversion circuit, the tally counter is incremented to $4001_{(8)}$ and the one's complement, $3776_{(8)}$, is strobed into the timing counter. The PC is then in a stable state, awaiting the second start pulse. When the latter is applied, a similar process occurs except that the timing counter starts at $3776_{(8)}$, thereby counting two clock pulses before triggering the $4000_{(8)}$ detector. In this way, the interval between start and output pulses is increased by the duration of one clock period for each start pulse applied, provided a reset pulse is not received. The choice of precession times of one to 2048 microseconds is obtained from the 1-MHz crystal oscillator with the aid of a 12-bit pre-scaler which divides the clock frequency by the power of two selected by means of a front panel switch which controls the one-of-twelve decoder. A detailed circuit diagram of the PC, which is constructed from commercially-available integrated circuits, may be obtained from the authors on request.

It should be noted that the action of the PC can be implemented by software alone—*i.e.*, the minicomputer can be programmed to duplicate the timing sequence so that an external precessing clock would be unnecessary. However, for the case of a PDP-8/S minicomputer, the minimum precession time available through this scheme would be about 28 microseconds which is inadequate in many instances. Nevertheless, this concept is a viable alternative to the use of an external hard-wired PC, particularly when a faster computer is available.

The dual potentiostat, which features positive feedback compensation of *iR* drop and subtractive elimination of double-layer charging current, the initial voltage source, the cell, and the electrodes have been described previously *(16-18)*. The drop dislodgers were constructed from a Metrohm Model E261 unit with a driving circuit which is described elsewhere *(19)*.

Signal conditioning of the voltage and current signals from the potentiostat is accomplished with the aid of a variable-gain, variable-offset amplifier (VGVOA) and either a VGVOA or an absolute value-polarity indicator circuit (AVPIC), respectively. The devices are standard operational amplifier circuits which have been given elsewhere *(15-19)*. They are designed to meet the requirements of the Type AFO1A A/D converters which demand inputs in the range 0 to -10 volts. In addition to amplifying the potentiostat signals to take advantage of the full-scale signal capability of the A/D converters, the signal conditioning circuits also convert the normally bipolar potentiostat signals to the unipolar (negative) voltages required by the A/D converters. The VGVOA simply biases the potentiostat signal to the point where all VGVOA output signal excursions are confined to negative values. The AVPIC provides a suitably amplified negative output, regardless of input polarity. It also provides a second output (polarity indicator) from a trigger circuit whose output state indicates the actual sign of the input signal. The AVPIC provides twice the signal resolution of the VGVOA because, for a given gain factor,

(17) E. R. Brown, D. E. Smith, and G. L. Booman, ANAL. CHEM., **40**, 1411 (1968).
(18) E. R. Brown, H. L. Hung, T. G. McCord, D. E. Smith, and G. L. Booman, *ibid.*, p 1424.
(19) S. C. Creason, Ph.D. Thesis, Northwestern University, Evanston, Ill., 1973.

input signal excursions can be twice as large. However, the AVPIC causes a 50% slower data acquisition rate, relative to the VGVOA because three A/D conversions are required per data point, rather than two. Consequently, the choice of current signal conditioning mode depends on whether speed or precision is the more crucial requirement in a particular experiment.

A Hewlett-Packard Model 3300A function generator with a Model 3302A trigger/phase lock plug-in accessory provided the single-cycle triangular wave (Channel 1) which drives the potentiostat. The terminate pulse, which signals the completion of each repetition of a SCCV is derived from a square-wave taken from the second channel (Channel 2) of the function generator. The pulse is formed by a passive circuit (19) which differentiates the square wave, widens, and half-wave rectifies the resulting pulse train. To maintain proper phasing of the square-wave channel of the function generator, and thereby prevent termination of data acquisition at the midpoint of the voltammogram, the 3302A plug-in phase control must be set to the −90° position. In this state the triangular wave is initially positive-going. Consequently, a Burr-Brown Model 1527 operational amplifier, configured as a unity-gain inverter, is inserted in the output line of the function generator when an initially negative-going triangular wave is required (Switch S_2).

An Electro Instruments Model 480 X-YY' recorder with Model 468 and 470 plug-ins was utilized for on-line computerized plotting of the cyclic voltammograms. A Hewlett-Packard Model 141B oscilloscope with Model 1400A and 1402A plug-ins aided in system testing and real-time analog monitoring of the SCCV. Direct current voltages were monitored with the aid of a Hewlett-Packard Model 5243L electronic counter with a Model 5265A digital voltmeter accessory.

Data Acquisition Sequence. The overall data acquisition process may be understood by consulting Figures 1 and 3. The former indicates the signal paths and directions (arrows) between the various instrument components, while the timing diagram of Figure 3 outlines the event sequence. The first data acquisition cycle (repetition 1) commences when the computer issues a *reset* pulse to the PC(AA01A D/A converter 1) and a *drop dislodge* pulse (AA01A D/A converter 2) which initiates formation of a fresh drop. The computer then waits a specified time by means of a programmed countdown cycle while the fresh mercury drop grows (drop growth interval). When the desired time has elapsed, the computer issues a *start* pulse (from the scope display brightener pulse source) to the PC. Activated by the *start* pulse the PC, after the pre-selected delay, emits an *output* pulse which starts the function generator whose triangular wave output (Channel 1) is applied to the potentiostat to produce the SCCV. After completion of the triangular wave cycle the *terminate* pulse is generated via signal generator channel 2. Meanwhile, after generating the *start* pulse, the computer begins taking data from the potentiostat current and voltage channels as rapidly as possible until the *terminate* pulse is sensed on multiplexer input channel 4 by the A/D converter. At this point the computer ceases taking data points, issues a *drop dislodge* pulse and a new repetition ensues. Each new repetition proceeds in the same manner as its predecessor, except for the increased delay between the *start* pulse to the PC and the *output* pulse from the PC.

Measurement of a single data point involves three basic phases: (a) First, a *hold* command is issued to the S/H amplifiers by the computer (data sampling); (b) then A/D conversion is effected sequentially on the two S/H amplifier outputs; (c) finally, the computer checks multiplexer channel 4 for a *terminate* pulse before repeating steps (a)–(c). S/H amplifiers are used to ensure that both the current and voltage signals are measured at effectively the same point in time.

One should note from Figure 3 and the above description that data points are acquired at the same time, relative to the computers *start* pulse (and also the *drop dislodge* pulse), in

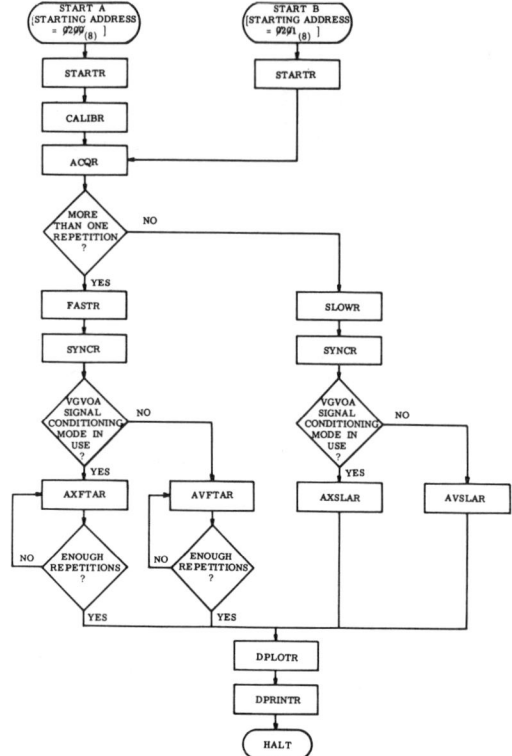

Figure 4. Program subroutine flow diagram

each successive repetition. It is the point in time where the triangular wave is applied which varies relative to the *start* pulse on successive repetitions enabling the acquisition of nondegenerate data. This approach proved to be the most effective of several alternatives, but it does present an apparent flaw in the fact that the triangular wave is applied at a slightly different time in the life of the mercury drop with each new repetition. Consequently, one can argue that the mercury drop area is not identical during each repetition. However, the attendant error is negligible because whenever more than one repetition is truly necessary to generate sufficient data points, the relevant time scale is quite short. Specifically the point of application of the triangular wave will not differ by more than 600 to 800 microseconds (the single point data acquisition period) between the first and last repetitions. One convenience of the timing sequence employed is that the drift of the instrument system is automatically monitored if the chosen precession time happens to be greater than that required for the computer to execute a *hold* command (about 30 microseconds). Under these conditions the initial data point is acquired before the triangular wave potential sweep ensues, as indicated in Figure 3. Thus, this particular data point, representing the initial state of the electronic system, is obtained with each successive repetition. Clearly, if drift in the system is negligible, the initial data point should exhibit the same coordinates on each repetition, and vice versa.

Program Description. A flow diagram of the program, which is written in PAL-III assembly language (20), is presented in Figure 4. The major functions of each of the sub-

---

(20) Digital Equipment Corporation, "Introduction to Programming," Digital Equipment Corp., Maynard, Mass., 1969.

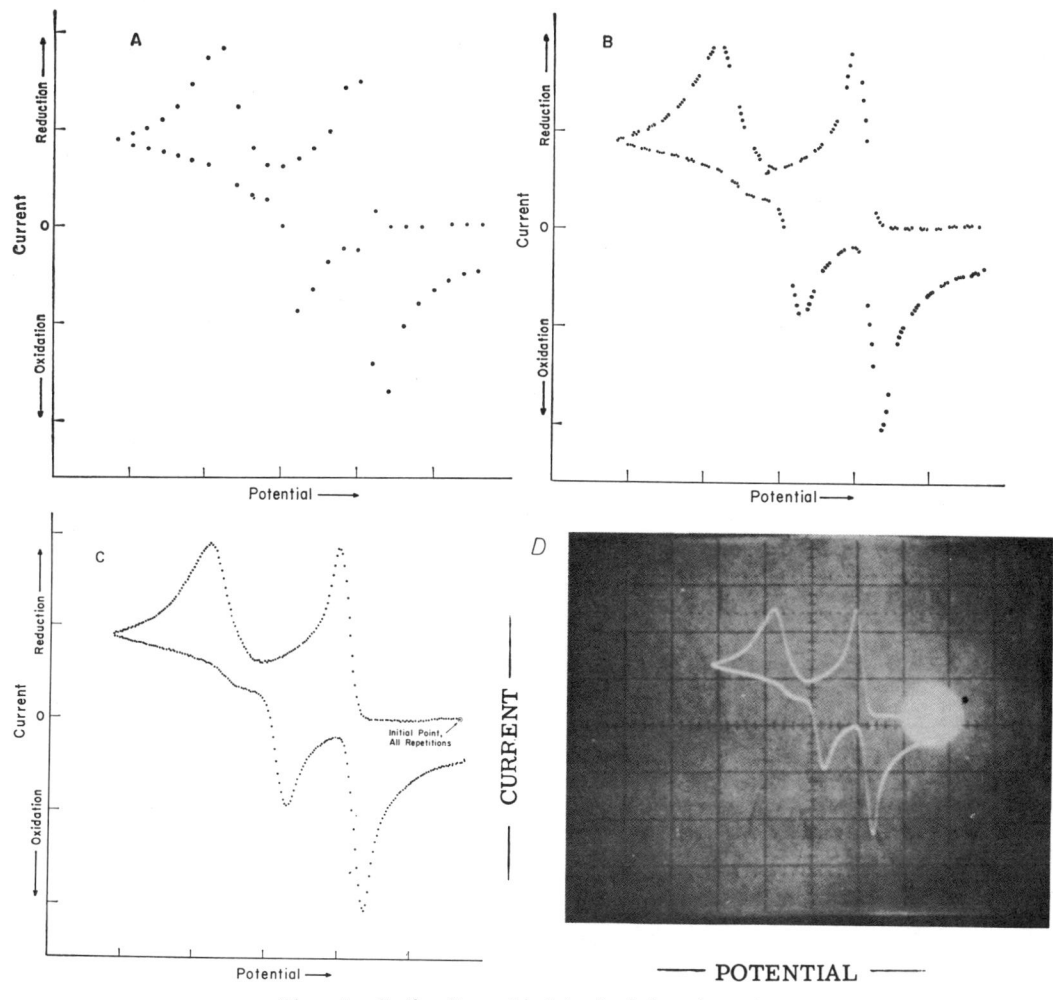

Figure 5. Cyclic voltammetric data of cadmium–zinc system

System: $1.0 \times 10^{-3} M$ Cd^{2+} + $1.0 \times 10^{-3} M$ Zn^{2+} in $0.5M$ Na_2SO_4
Applied: Triangular wave potential of 20 Hz
Readout:
    A. Plot of digital data points after one repetition
    B. Plot of digital data points after four repetitions
    C. Plot of digital data points after eight repetitions
    D. Photograph of oscilloscope trace of cyclic voltammogram (abscissa and ordinate uncalibrated)

routines are summarized in Table I. The program listing and the source and binary tapes (punched paper tapes) may be obtained from the authors on request.

The program may be entered at one of two starting points, STARTA or STARTB, depending on whether or not one wishes to carry out the calibration routine (CALIBR). Once loaded, program entry is effected by appropriate manipulation of the computer console controls (20).

Subroutine STARTR simply reads the voltage state of pole 3 of switch S_1 (multiplexer channel 5) which informs the computer regarding the signal conditioning mode employed in the current signal path. This information is stored for utilization in subsequent subroutines.

CALIBR is the subroutine by which the current and potential channels are calibrated. After application of a series of specified voltages to a channel, a linear least-squares calibration curve is computed for that channel and stored for utilization in the data output subroutine, DPRINTR. Preliminary work indicates that the calibration step can be completely automated using the D/A converters as sources of the calibration voltages. However, the D/A converters available for this work provide only negative outputs, whereas positive calibration voltages are frequently required. Hence, a pair of low-drift, level-shift (biasing) amplifiers is required. In the interest of economy, a manual calibration routine was used in the work described here. However, since channel gains are changed frequently, auto-calibration would provide a worthwhile time savings whenever the system is used on a routine basis.

The data acquisition portion of the program is directed primarily at implementation of the multi-repetition data sampling technique which is the subject of this report. How-

Table I. Description of Subroutine Functions

Subroutine	Function(s)
STARTR	Determines current channel signal conditioning mode in use. Stores this information.
CALIBR	Allows application of calibration voltages to signal channels. Computes least squares calibration curves and stores calibration curve parameters.
ACQR	Operator inputs, via Teletype, triangular wave frequency, desired number of data points, and time in the drop life at which SCCV is to be generated. Computer calculates required number of repetitions.
FASTR	Computer recommends the optimum precession time, and recomputes the number of repetitions required based on actual precession time chosen by operator (input via Teletype).
SLOWR	Computer calculates the required delay time between acquisition of individual data points when desired number of points can be obtained during one repetition.
SYNCR	Simulates the data acquisition routine except that no data points are taken. Allows a final check of system before data acquisition. Exit from this routine is effected by toggling the switch register.
AXFTAR	Data acquisition routine for use when the VGVOA is in the current signal channel and more than one repetition is required.
AVFTAR	Same as AXFTAR except that AVPIC is in the current signal channel.
AXSLAR, AVSLAR	Same as AXFTAR and AVFTAR, respectively, except that enough data points can be obtained during one repetition.
DPLOTR	Plots the data.
DPRNTR	Prints out the data via Teletype.

ever, the program also provides for those experiments in which the desired number of data points can be obtained during a single repetition. Subroutine ACQR obtains from the operator via Teletype a variety of operating parameters, including the information required to calculate the necessary number of repetitions. Following ACQR, the program branches according to whether one or more repetitions are required. If only a single repetition is necessary, subroutine SLOWR introduces the correct additional time delay into a countdown routine which is inserted between successive *hold* commands to avoid acquiring more than the requested number of data points—*i.e.*, data are not acquired as rapidly as possible in this instance. In the case of primary interest where more than one repetition is required subroutine FASTR is entered in which the computer recommends via Teletype the optimum precession time to provide an evenly-spaced set of data points. Then, because only a set of discrete precession times is actually available, none of which may match the optimum value precisely, the operator inputs (Teletype) the actual precession time selected and, based on this information, the required number of repetitions is recomputed.

To double-check most aspects of the system operation, a subroutine SYNCR repeatedly activates the data acquisition sequence, except that data points are not stored. This phase of the program is primarily designed to permit scrutiny of the SCCV on the oscilloscope, to ensure proper operation of the cell, and to allow final adjustment of the initial voltage, triangular wave amplitude, mercury column heights in the two cells, etc.

Depending on the current channel signal conditioning mode employed and whether more than one repetition is required, one of four acquisition routines, AXFTAR, AVFTAR, AXSLAR, and AVSLAR is chosen by the computer (see Table I). The data acquisition sequence discussed above gives an adequate description of the operations involved in each of these four data acquisition routines, since they differ

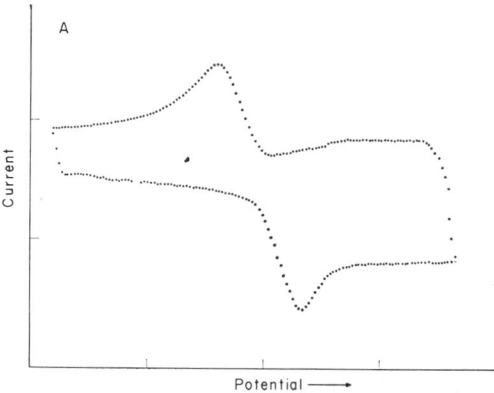

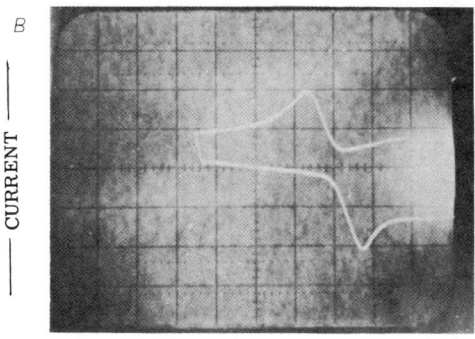

—— POTENTIAL ——

Figure 6. Cyclic voltammetric data of cadmium system

System: $3.0 \times 10^{-3} M$ Cd^{2+} in $0.5 M$ Na_2SO_4
Applied: triangular wave potential of 500 Hz
Readout:
A. Plot of digital data points after 90 repetitions
B. Photograph of oscilloscope trace of cyclic voltammogram

only in the number of repetitions and number of data channels interrogated per data point.

After acquisition, calibration factors generated in CALIBR are applied to the raw data and the results are output in the form of a plot and a table by subroutines DPLOTR and DPRINTR.

Electrode Processes. Measurements involved the aqueous redox systems $Fe(C_2O_4)_3{}^{3-}/Fe(C_2O_4)_3{}^{4-}$ in $0.5M$ $K_2C_2O_4$, $Cd^{2+}/Cd(Hg)$ in $0.5M$ Na_2SO_4, and $Zn^{2+}/Zn(Hg)$ in $0.5M$ Na_2SO_4. Solutions were prepared using reagent sources and precautions given previously (*16, 18*). Three different solution compositions were prepared: (a) $1.0mM$ Fe^{3+} in $0.5M$ $K_2C_2O_4$ (the "iron system"); (b) $3.0mM$ Cd^{2+} in $0.5M$ Na_2SO_4 (the "cadmium system"); (c) $1.0mM$ $Cd^{2+} + 1.0$ mM Zn^{2+} in $0.5M$ Na_2SO_4 (the "cadmium–zinc" system). Measurements on the iron system, which is characterized by reversible (diffusion-controlled) behavior under the conditions employed, were performed at $24.2 \pm 0.1\,°C$. Measurements on the cadmium and cadmium–zinc systems were made at ambient room temperature, 23–27 °C, and at scan rates that produce quasi-reversible behavior.

RESULTS AND DISCUSSION

The digital acquisition of a cadmium–zinc system SCCV taken with a triangular wave frequency of 20 Hz using 8 repetitions is shown in Figure 5. The oscilloscope trace of the SCCV is

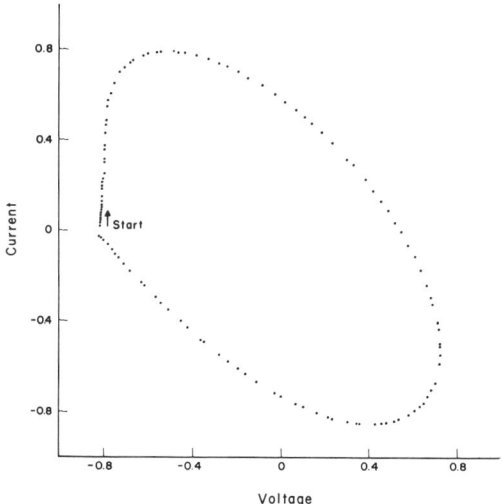

Figure 7. Lissajous pattern of phase-shifted single cycle sine waves

Applied: Sine wave potentials of 10 kHz
Readout: Plot of digital data points after 117 repetitions (abscissa and ordinate uncalibrated)

included for comparison (Figure 5D). Data from only the first repetition are shown in Figure 5A, which is the most detailed SCCV that could be acquired by our PDP-8/S system under the indicated conditions, if a non-repetitive acquisition approach were used. The gross structure of the SCCV becomes apparent after four repetitions, as shown in Figure 5B. However, for the SCCV in question, at least eight repetitions are required for a satisfactory quantitative characterization (Figure 5C). Figure 5 is a typical illustration of the manner whereby a high-speed SCCV is "developed" by the multi-repetition data sampling technique and it clearly demonstrates the technique's effectiveness in increasing the effective bandpass of the data acquisition system.

The ability of the system to acquire an SCCV at very high frequencies is demonstrated in Figure 6. A digital SCCV of the cadmium system taken at 500 Hz and an oscilloscope trace of the SCCV are shown. At this frequency the computer can acquire only 2 or 3 data points per repetition, so that 90 repetitions were required to produce the 250 data points shown. Because the potentiostat employed cannot produce a completely undistorted SCCV at this frequency, due to bandwidth limitations, no attempt was made to compensate for double-layer charging current or solution iR drop when generating the results shown in Figure 6.

The quantitative fidelity of the data sampling system is demonstrated in Figures 5 and 6 by the fact that, within our ability to compare, the digital SCCV's (Figures 5C, 6A) match the corresponding analog traces (Figures 5D, 6B) precisely. Because an oscilloscope pattern is not a particularly high resolution standard on which to evaluate accuracy, this characteristic also was assessed with satisfactory results using

the iron system as a model for diffusion-controlled behavior. The iron system yielded digital SCCV parameters, such as anodic and cathodic peak separation and ratio, which were in excellent agreement with theoretical predictions for a diffusion-controlled SCCV.

Phase shift distortion of input signals by the DEC A400 S/H amplifiers becomes significant at frequencies above 1 kHz and determines the bandpass limit of the data sampling technique with available hardware. However, we have performed tests which show that the digital portion of the system functions well up to *at least* 10 kHz. A simple illustration is given in Figure 7 which shows a Lissajous pattern generated by directly injecting a single-cycle "sine wave" into the voltage channel, while it was applied to the current channel via a 0.03-mF capacitor to produce some phase shift. The deviation from a perfect ellipse at the start of the cycle results from the fact that the signal applied to the current channel is not purely sinusoidal at first because of the transient response in the RC phase shifting network. At the frequency employed in Figure 7, only one data point is acquired per repetition. Nevertheless, even under these extreme conditions, the digital networks are clearly adequate, indicating that with the use of faster S/H amplifiers (available commercially), a bandpass of *at least* 10 kHz would characterize the data acquisition routine under investigation. Since the data acquisition bandpass of our system using the standard single repetition scheme is considerably less than 20 Hz (cf. Figure 5A), a bandpass gain exceeding three orders of magnitude can be attributed to the data sampling technique evaluated here. The actual bandpass improvement one can envision far exceeds the value we are able to demonstrate with the circuitry available for this work. In principle, the sampling technique bandpass is determined by the time-resolution associated with the precessing clock and S/H amplifiers, and not by the on-line computer. State-of-the-art hardware can provide these functions with time resolution in the nanosecond range so that actual bandpass magnitudes in the megacycle range should be achievable by simply substituting faster versions of the PC and S/H amplifiers.

Given a computerized data acquisition system, the stroboscopic data sampling technique described above can be implemented at little cost (cost of PC) or none at all (when software is used to obtain the PC action). Considering this and the substantial data acquisition bandpass improvement demonstrated, the approach must be viewed as economically highly advantageous relative to obvious alternatives (*e.g.*, acquiring ultra high-speed A/D converters and direct memory access ports to core memory). Of course, like many "data enhancing schemes" the method examined here demands that the experiment be conveniently repeatable.

ACKNOWLEDGMENT

The authors are indebted to Donald E. Glover for writing the plotting subroutine, DPLOTR.

RECEIVED for review September 30, 1971. Accepted February 15, 1972. S. C. Creason was a NSF Trainee (1971-72) and a NDEA Fellow (1970-71). This work was supported by National Science Foundation Grant GP-16281.

High Speed Synchronous Data Generation and Sampler System: Application to On-Line Fast Fourier Transform Faradaic Admittance Measurements

R. J. Schwall, A. M. Bond,*[1] R. J. Loyd, J. G. Larsen, and D. E. Smith*

Department of Chemistry, Northwestern University, Evanston, Illinois 60201

A *synchronous data ge*neration and *s*ampler (SYDAGES) system constructed to assist broadband FFT electrochemical relaxation measurements is described and evaluated. SYDAGES consists of a digital-to-analog converter, two analog-to-digital converters, three 1024-word shift register memories, and control circuitry. It functions as a programmable signal generator and two signal averaging data acquisition channels which are synchronized up to data rates of 500 kHz. SYDAGES is run directly by a minicomputer without manual control. Its performance characteristics are demonstrated here using dummy cell and electrochemical cell admittance data. In the latter instance, good quality cell admittance data are obtained to 125 kHz, enabling dynamic nonfaradaic measurement and compensation to reveal the faradaic admittance up to 40 kHz.

Previous work in this laboratory (1-5) has demonstrated the benefits of on-line multiple frequency Fast Fourier Transform (FFT) faradaic admittance measurements using a particular applied pseudo-random waveform. The benefits include the ability to acquire kinetic and thermodynamic parameters with unprecedented speed and precision and to characterize reactions with very high rates. The particular waveform suggested has the following important properties.

(a) It is periodic.
(b) It contains a limited number of frequency components of approximately equal amplitude which are selected specifically to cover the frequency range of interest without being overly redundant.
(c) All frequency components are *odd* harmonics of the lowest.

[1] On leave from Department of Inorganic Chemistry, University of Melbourne, Parkville, Victoria 3052, Australia.

(d) The phases of the various frequency components are randomized as a function of frequency and measurement pass.
(e) The signal is generated by a digital-to-analog converter (DAC) from a data array of 2^n points (n is an integer).
(f) The applied waveform generation is synchronized to the analog-to-digital converter (ADC) operations which acquire the reference-working electrode potential and the cell current signals.

In reported work the signal generation and sampling was controlled directly by a minicomputer, which could repeat a complete conversion cycle (two sampling and ADC steps, one DAC operation, and data storage) on a period no shorter than 100 μs. The effects of solution ohmic resistance (R_s) and double-layer capacitance (C_{dl}) were removed by analog techniques restricted to liquid-metal electrodes under strictly controlled conditions, including accurate potentiostat operation. One overall result of this strategy was an effective measurement bandpass limitation of a few kiloHertz.

A far more general method of acquiring the faradaic admittance is the frequency domain analysis of the total cell admittance, as recommended originally by Sluyters (6), but modified to include correlation of measured input and response waveforms as invoked by deLevie (7) and Pilla (8). This approach has the advantage of providing, in the presence or absence of a faradaic component, rapid, dynamic measurement of R_s and C_{dl} which will detect changes in these variables with time. This can be very important in many situations, such as when monitoring reaction kinetics as a mercury electrode grows (9), or with solid electrodes, especially outside the clean laboratory environment. Of course, the acquisition of the nonfaradaic component magnitude enables convenient vectorial subtraction of these contributions from the total cell admittance to reveal the faradaic response. However, for many chemical systems this total cell admittance analysis requires that the data be acquired at significantly higher frequencies than a 100-μs conversion cycle will allow. Another stimulus toward performing higher-frequency measurements is the possibility of monitoring still faster rate processes. Yet another is the prospect of acquiring information at high frequencies which complement data at lower frequencies, thus providing confirmation checks on proposed mechanisms (10).

In response to these stimuli we undertook the construction of a device capable of generating arbitrary signals from one 1024-word memory through a DAC, while synchronously sampling two signals via two ADCs and storing these points in two additional 1024-word memories with provision for time-domain signal averaging. No such device exists commercially, to our knowledge, nor can a reasonable compromise be realized using combinations of commercial data recorders for less than four times the cost (including construction time assessments) of the unit described here. Our approach provides a generalized transfer function measurement system with quite high frequency capabilities.

INSTRUMENTATION

A block diagram of the *synchronous data generation and sampling* (SYDAGES) system is shown in Figure 1. The device has only four external connections: one output waveform generated through the DAC; two input signals sampled through the buffer amplifiers, sample-and-hold circuits and ADCs; and a 26-wire data bus to the computer. SYDAGES has no manual controls. By executing a single input/output instruction, the computer can effect any one of the following functions:

(A) Read 1 word from memory A;
(B) Read 1 word from memory B;
(C) Load 1 word into memory G;
(D) Read DELAY and PASS counter word;
(E) Read STATUS word;
(F) Load PERIOD into clock;
(G) Load DELAY and PASS comparator registers;
(H) START data sequence;
(I) ERASE entire A and B memories;
(J) Return all memories to HOME position;
(K) CLEAR interrupt;
(L) DISABLE interrupt;
(M) ENABLE interrupt.

Commands A, B and C each result in all three memories being shifted one position. The memories are basically shift registers, are accessed only sequentially, and always shift together. The numbers put in the DELAY and PASS comparators (1 to 255) are used to control the number of waveform cycles to be used when START is executed. One "cycle" is a memory cycle of 1024 words (one waveform period) with the period between individual words being defined by the word in the CLOCK register. During DELAY cycles the output signal is generated through the DAC, but the A and B memories are shifted unchanged. This is to allow startup transients to become negligible, validating the steady-state approximation (11) and improving the accuracy of the FFT algorithm (12). During PASS cycles, output signal generation continues, but converted data from the signal inputs are added into the A and B memories, in the fashion of a digital signal averager.

Commands D and E allow the computer to monitor the progress of SYDAGES through its automatic operations, which are initiated by commands H, I, and J. The done flag is a bit in the STATUS word (read by command E) which goes on when an automatic sequence is completed. If command M has been given, every appearance of the done flag causes an interrupt in the computer. This feature allows the computer to execute other programs while SYDAGES is executing an automatic sequence.

A rather typical command sequence program used in running SYDAGES follows (upper case letters designate commands in the foregoing list):

(1) K and L;
(2) J, to send memories to HOME position;
(3) E repeatedly until DONE flag appears, then K;
(4) I to erase, M and adjust computer's interrupt software to go to step 6 when interrupted;
(5) Proceed with program;
(6) Execute C 1024 times, transferring output waveform from computer to SYDAGES G memory;
(7) F to put shift period into CLOCK register;
(8) G to fill DELAY and PASS comparators;
(9) H, K, M to start, adjust interrupt software to point to step 11;
(10) Proceed with program;
(11) L and K to free computer's interrupt hardware;
(12) Execute A 1024 times to read all of A memory into computer memory;
(13) Execute B 1024 times to read all of B memory into computer;
(14) Go to step 4.

"Proceed with program" may include such tasks as generating new time domain waveforms, and transferring sampled waveforms to bulk storage (e.g., disc).

The STAR BOX mentioned in Figure 1 is a general input-output interface adaptor for the Raytheon computers. It was designed and built locally by Drake and Schwall (13).

MASTER CONTROL is the part of SYDAGES that interprets the computer's commands, controls the interface gates, and monitors the status of the major control lines. It also generates the single CONVERT pulse each time a command A, B, or C is executed.

The INTERFACE operates under the command of the signals on the *interface control* cable to gate data between the computer and the various elements of SYDAGES only when the appropriate computer command is executed. The CLOCK is a 10-MHz crystal oscillator with programmable frequency division controlled by the clock register. Available clock periods are of the form $A \times 10^B$ seconds, where A is 1, 2, or 5 and B is an integer from −6 to 0. A and B are entered into the clock register in coded form. Under control of RATE signal from MASTER CONTROL, the CLOCK will run at a 500-kHz rate independent of the clock register. This is used for commands I and J. The CONVERT signal is exerted once for each pulse during any automatic operation and once on any A, B or C command. The ADCs then immediately begin conversion. While they are converting, they hold EOC (end of conversion) signals false, thus locking the sample-and-hold modules in the hold state. When the conversion is complete, both EOC signals become true and the SYNC module shifts the memories. The memories are always shifted together and the SHIFT COUNTER keeps count of the shifts and exerts the HOME signal once in 1024 shifts.

The memories are built from 1024 × 1-bit CMOS shift registers and 4-bit adders. A and B memories are 16-bits wide. On each shift they will add the input data word to the word currently at the output port, or they will force zeros into the current word, depending on the state of the ERASE signal. The G-memory is 10-bits wide and either will simply circulate data past the output port, or will force in data from the input port, depending on the state of the LOAD signal.

The buffers are unity gain subtractor circuits built from DATEL Model 100 operational amplifiers. The sample-and-hold units are DATEL Model SHM-2. The ADCs are DATEL Model ADC-G10B4C 10-bit types with extra gating circuitry to force zeros onto the output if the GATE signal is not true. The DAC is a 10-bit DATEL Model DAC-V10B3D. All other circuitry is home-built from the TTL integrated circuits. All circuitry is mounted on twenty-one

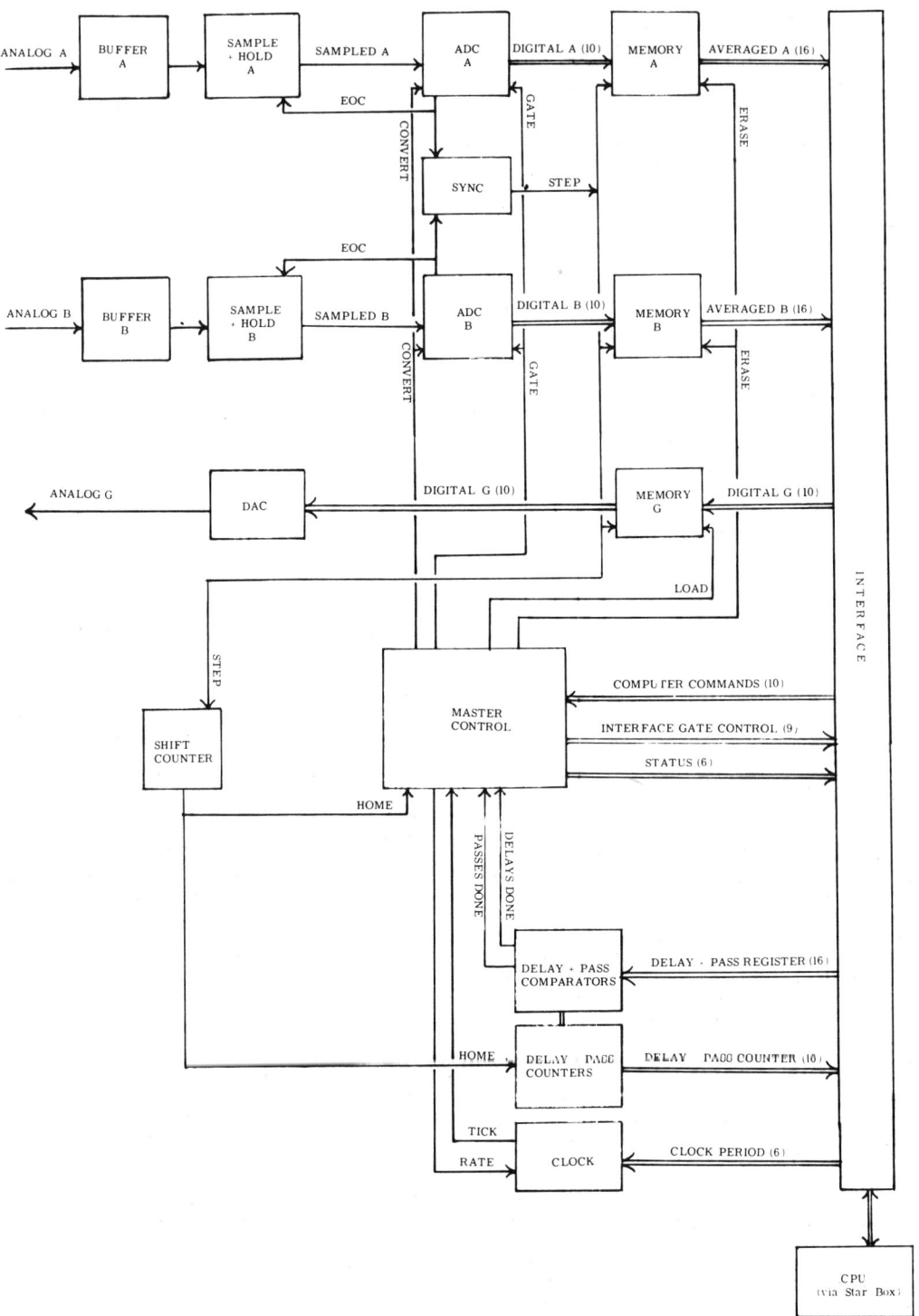

Figure 1. Block diagram of the SYDAGES device. Numbers in parentheses above double-lines are numbers of bits in multibit signals

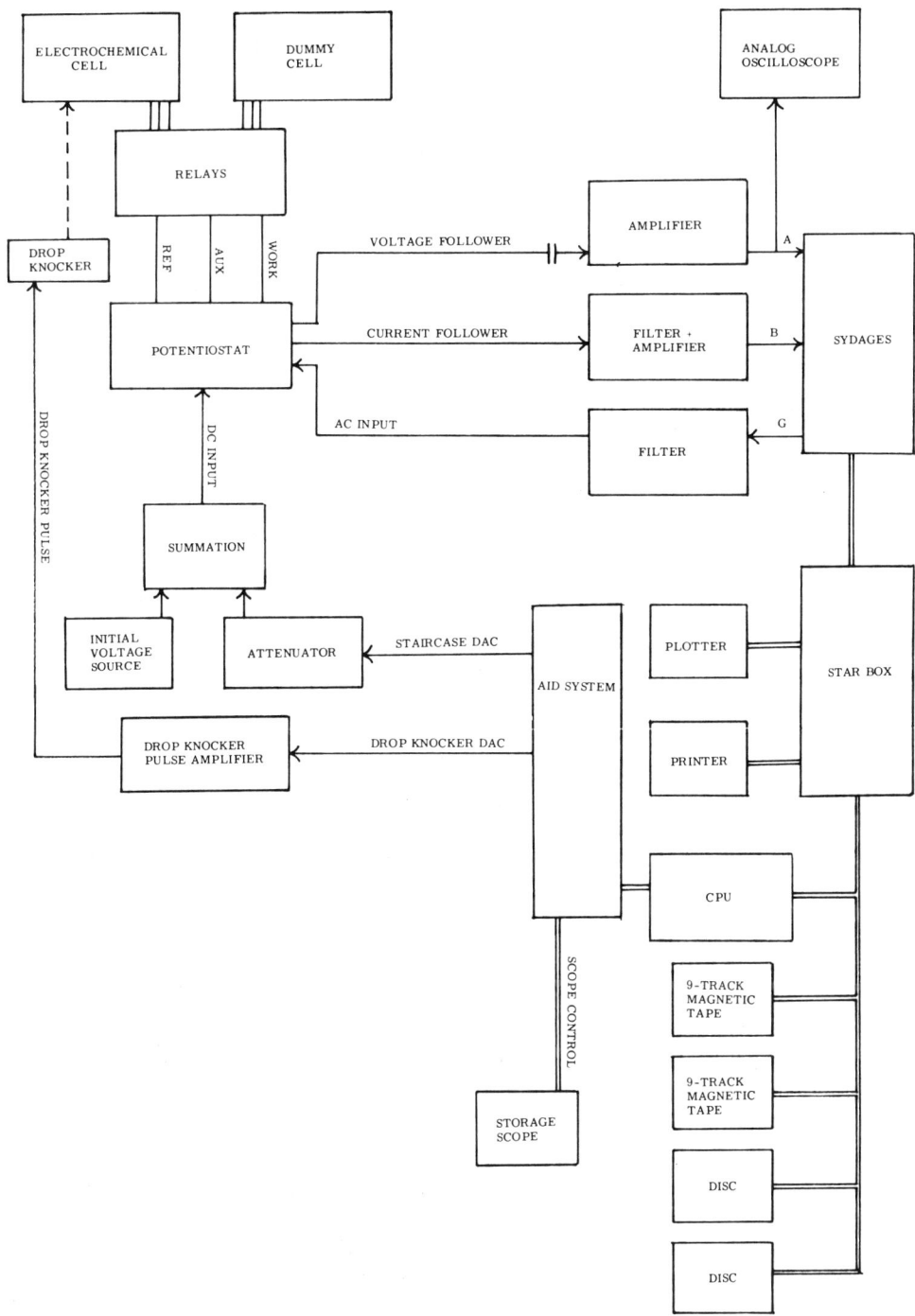

Figure 2. Block diagram of the overall experimental installation

11 × 11.5 cm Fiberglas circuit boards which plug into a backplane structure. Detailed circuit diagrams are available elsewhere (14).

The overall experimental setup is shown in Figure 2. The potentiostat, cells, relays, dc voltage sources, and drop knocker have been described elsewhere (3, 15). The computer is a Raytheon RDS-500 with a 32K core memory, two 1.2-Megword discs, two 9-track tape drives, and an AID system which includes two DACs and a storage display scope. The interfaces between the Star Box and the Centronix 500 printer and

Complot DP-1 plotter were provided by Drake (13). The analog oscilloscope is a Hewlett-Packard Model 141B. The filters are 2-pole Butterworth low-pass filters with a cutoff frequency usually at about $(4T_s)^{-1}$ Hz, where T_s is the SYDAGES sampling period. They are built from standard operational amplifier circuits (16). The "Amplifiers" are operational amplifier inverting circuits with a gain of about 500. The capacitor in the voltage follower signal path is 5 μF and is used to block the large dc potential component.

The system described above is capable of data acquisition rates up to 500 KHz (input *and* response). At this maximum rate, the 1024-word memory cycle is completed in 2.048 ms, which represents the minimum waveform acquisition time. Unfortunately, the reciprocal of the minimum waveform acquisition time does not even closely approximate the effective repetition rate for spectrum acquisition, because of delays in data storage operations, mostly associated with the disc file acquisition times. Consequently, one finds that the effective admittance spectrum repetition rate is about 10 s^{-1}, which is still quite satisfactory for many studies involving the monitoring of temporal variations of the faradaic admittance.

COMPUTER PROGRAMS

Only a brief survey of software development is given here. Full details are available elsewhere (14). There are three main programs named ACPOL1, ACPOL2, and ACPOL3.

ACPOL1 requests experimental parameters from the operator and acquires raw time domain data. The parameters requested include:

(1) which harmonics of the base frequency are to be used;
(2) the number of DELAYS and time-domain PASSES for SYDAGES to use, and the CLOCK period;
(3) the initial, final, and increment values for the dc potential scan;
(4) whether the experiment to be performed is ac polarography at the dropping mercury electrode (DME) or ac cyclic voltammetry at a stationary electrode (17);
(5) the number of replicate measurements (drops) to take at each dc potential.

After the parameters are entered, the program proceeds to its data collection sequence automatically. Here it generates a new waveform for each mercury drop by modifying the applied waveform phase array (except in the ac cyclic mode) and proceeds through an experimental loop as described above under "Instrumentation". The time domain data is stored on disc and the program effectively interleaves disc, SYDAGES, and program execution. The FFT processing of data is not performed at this point to retain compatibility with the ac cyclic experiment (17) which acquires new waveform arrays so often (up to 10 s^{-1}) that there is insufficient time for mathematical operations. It is important for calibration purposes to take a data set with an appropriate dummy cell before or after taking real chemical data sets, without changing any instrument settings (see below).

ACPOL2 reads the time domain data from disc, applies the FFT to it, and divides the current signal spectrum by the voltage spectrum to get a dimensionless transfer function (an "apparent" admittance). If the data are obtained with a dummy cell (calibration run), the theoretical admittance of the dummy cell is divided by the observed transfer function to obtain the deconvolution function. If the data belongs to a chemical cell, it is multiplied by this "deconvolution function" to acquire the actual cell admittance. Thus, both deconvolution functions and cell admittances are in units of mho (ohms^{-1}). This deconvolution process should, in principle, correct the initial transfer function for contributions to its spectrum arising from any of the electronic components,

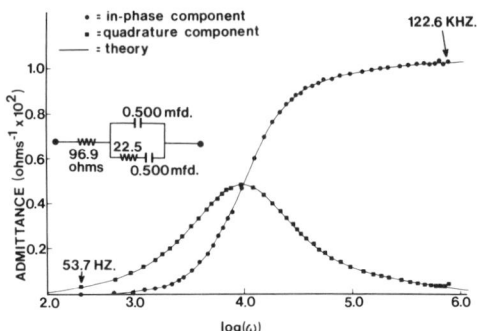

Figure 3. Theory–experiment comparison of transfer function, measurement using precision dummy cell

including the potentiostats and analog circuitry in SYDAGES, by calibrating the system for admittance at all frequencies. The raw transfer function, deconvolution function and/or the final admittance may be stored on magnetic tape. ACPOL2 also averages the transfer functions if replicate measurements were taken. The latter operation includes an option to discard replicates which are found to be more than two standard deviations from the average. Such catastrophic events are most often caused by the occasional failure of the drop knocker to dislodge a mercury drop. The averaging referred to here is "frequency domain" averaging which is not to be confused with the "time domain" averaging which automatically occurs in the SYDAGES data acquisition cycle. In other words, the SYDAGES-based system described in Figure 2 is capable of two signal averaging modes; frequency domain and time domain.

Program ACPOL3 performs in a highly interactive manner many mathematical and display operations on the frequency-domain data read from tape. It performs the high-frequency fits to get R_s and C_{dl} after the method of deLevie (7) and then subtracts the contributions of these components from the total cell admittance to reveal the faradaic response. ACPOL3 can perform various data conversions such as computing the phase angle cotangent and/or total admittance. It can plot or print data in a variety of formats, some of which appear in this paper. Other operations, such as FFT digital filtering (18), also are provided.

RESULTS AND DISCUSSION

A. Dummy Cell Results. After initial debugging, the SYDAGES-based measurement system was subjected to performance evaluation using precision dummy cells. Typical results are given in Figures 3 and 4, and Table I. Figure 3 shows a transfer function theory–experiment comparison for the dummy cell shown in the figure. The result invoked the deconvolution procedure described above. The data depicted extends to 122.6 kHz, which differs negligibly from what we consider the measurement system bandwidth upper limit (maximum data rate/4). The data set in Figure 3 requires two measurement operations, each encompassing a portion of the observed frequency band. Figure 4 provides a graphical illustration of the effects of the deconvolution operation on the Fourier spectrum obtained with the dummy cell described in the figure. Figure 4A shows the raw transfer function. Figure 4B shows the deconvolution function computed with this network and the Figure 4A data set. This graph would be a horizontal line, in-phase only, if the filters and amplifiers in the voltage and current measuring circuits all had a flat ("ideal") frequency response. The deviation from the horizontal response is a measure of the various rolloffs in these circuits. Figure 4C is the admittance computed by multiplying

Table I. Comparison of Theoretical and Measured Admittances of RC Network[a]

Angular frequency, krad/s	Theoretical admittance, mmho		Measured admittance,[b] mmho		Measured admittance,[c] mmho	
	In-phase	Quadrature	In-phase	Quadrature	In-phase	Quadrature
0.6136	0.2008	0.1770	0.2053	0.1785	0.2029	0.1767
1.841	0.2256	0.5298	0.2269	0.5343	0.2246	0.5269
3.068	0.2746	0.8787	0.2753	0.8770	0.2699	0.8753
4.295	0.3473	1.221	0.3471	1.226	0.3426	1.216
5.522	0.4426	1.555	0.4433	1.552	0.4375	1.541
7.977	0.6956	2.189	0.6905	2.187	0.6895	2.172
11.66	1.205	3.030	1.205	3.025	1.194	3.006
17.79	2.270	4.084	2.266	4.077	2.256	4.057
26.38	3.856	4.863	3.864	4.863	3.842	4.831
38.66	5.750	5.037	5.763	5.032	5.739	4.988
57.06	7.545	4.515	7.572	4.513	7.538	4.459
81.61	8.742	3.672	8.761	3.647	8.713	3.579
118.4	9.504	2.756	9.556	2.735	9.471	2.669
155.2	9.828	2.176	9.854	2.139	9.815	2.070

[a] Network circuit as shown in Figure 4A. [b] Using dummy cell with R_s = 124.1 Ω, C_{dl} = 0.5 μF, R_f = 4960 Ω for deconvolution. [c] Using 219-Ω resistor for deconvolution.

the data in Figure 4A by a deconvolution function computed from a nominally different dummy cell network (R_s = 124.1 ohm, C_{dl} = 0.500 μF, R_f = 4960 ohm). Table I shows a comparison of the theoretical and measured admittances for this dummy cell.

The third data set (right-hand) in Table I shows how the admittance is disturbed if a simple pure resistive dummy cell is used to acquire the deconvolution function. One notes that the theory–experiment agreement is more than acceptable for many purposes, but that the accuracy of the quadrature components is slightly, but noticeably degraded at higher frequencies, relative to the results in Figure 4C. This is an example of a general result we have found: that measurements at high harmonics of the base frequency are much improved if the high frequency behavior of the dummy cell used to evaluate the deconvolution function mimics that of the unknown cell. It is the relative frequency that is important; for, if the same absolute frequency is measured with a smaller SYDAGES period (higher base frequency), the effect is much smaller. We believe the phenomenon results from the fact that the waveform-generating DAC puts out a signal consisting of small steps, rather than smooth variations. This signal quantization produces "false" high-frequency components. If f_s (=1/T_s) is the sampling frequency and f_D is the frequency which was to be output, the false frequencies are

$$f_x = nf_s - f_D$$
$$f_x = nf_s + f_D$$

These frequencies all are above the maximum which can be correctly sampled, according to the sampling theorem, and they all will be indistinguishable from f_D when sampled (19)! The low-pass filters in Figure 2 are designed to reduce these high-frequency components, but cannot remove them completely, since no real low-pass filter has an infinitely sharp cutoff. In particular, if f_D is near $f_s/2$, the maximum frequency which can be generated, then one f_x also will be near $f_s/2$, which will be very hard for a low-pass filter to remove. As a result, the transfer function at a high f_D really is a weighted average of the true transfer functions at f_D and the f_x's. However, if the dummy cell used for deconvolution is a good match to the high-frequency behavior of the unknown, this "leakage" effect will distort its transfer function in the same manner as the unknown's and the deconvolution process will tend to cancel the effect. With experience we have found that by keeping all f_D's below $f_s/4$ and using low-pass filters with a cutoff of about $f_s/4$, we can measure admittances within ±1% if the dummy and unknown cell R_s's are matched within

Table II. Comparison of Solution Resistances and Double-Layer Capacitances Measured in 1 M $ZnSO_4$-0.18 M H_2SO_4, with and without 1 mM $CdSO_4$ Added[a]

Potential, V vs. Ag/AgCl	Without Cd^{2+}		With Cd^{2+}, 10^{-3} M	
	R_s, Ω	C_{dl}, μF	R_s, Ω	C_{dl}, μF
-0.4590	89.75	0.3277	89.83	0.3283
-0.4700	89.77	0.3199	89.93	0.3244
-0.4820	89.90	0.3117	90.11	0.3230
-0.4940	89.87	0.3111	90.16	0.3170
-0.5060	89.93	0.2977	89.86	0.3105
-0.5180	89.98	0.2928	89.58	0.2974
-0.5300	89.93	0.2879	89.50	0.2888
-0.5420	90.08	0.2839	89.92	0.2865
-0.5540	89.91	0.2754	89.82	0.2779
-0.5650	89.99	0.2762	89.96	0.2725
-0.5770	90.00	0.2680	89.97	0.2664
-0.5890	89.99	0.2643	89.93	0.2608

[a] Conditions as in Figure 7.

about 50% and the C_{dl}'s within a factor of two. This normally is easy to do by making a quick preliminary run at a very few dc potentials to obtain an initial estimate of the chemical cell parameters.

Casual evaluation of the agreement in the two measured data sets in Table I may make the attention paid the differences and small systematic error seem quixotic. However, reducing the error from 5 to 1% in the total admittance quadrature term, for example, has a substantial influence on the accuracy with which the nonfaradaic components can be evaluated which, in turn, markedly influences the accuracy of nonfaradaic compensation and the effective faradaic component measurement bandwidth. For the dummy cell used in Figure 4A and Table I, the high-frequency fit programs yield a computed R_s = 96.7 ohms and a C_{dl} = 0.303 μF, in excellent agreement with the actual values.

B. Electrochemical Cell Results. The fidelity of the deLevie algorithm approach to obtaining the faradaic and nonfaradaic component magnitudes when applied to SYDAGES-acquired data was carefully evaluated. Results obtained are most satisfactory as Tables II and III illustrate. Table II compares the R_s and C_{dl} values at dc potentials near the peak of the Cd^{2+} faradaic admittance wave, in the presence and absence of Cd^{2+}. The nonfaradaic magnitudes computed from the two solutions are essentially identical within experimental error. Table III compares the Cd^{2+} faradaic admittance at one frequency in the same media, obtained by two methods. With both methods the R_s effects first were removed by the deLevie calculation. Following this, Method

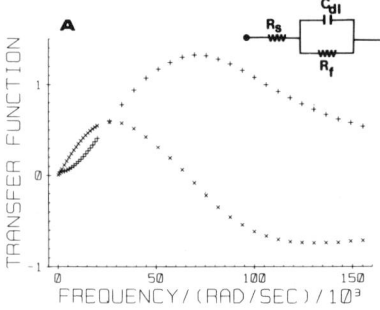

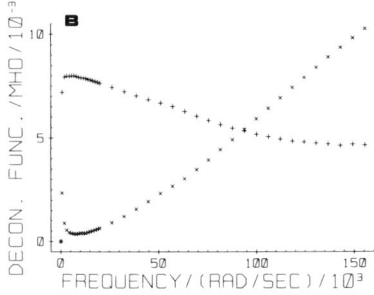

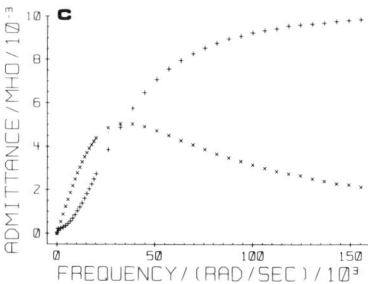

Figure 4. Illustration of deconvolution function effect on raw transfer function. (A) Dimensionless transfer function of the RC network shown in figure where R_s = 96.9 Ω, R_f = 4960 Ω and C_{dl} = 0.300 µF. SYDAGES step time = 10 µs, 10 DELAY periods, 64 time-domain passes, and 10 complete replicates (frequency domain averages). + = in-phase component, x = quadrature component. (B) Deconvolution function computed from above data. (C) Corrected admittance of this RC network after multiplying by a deconvolution function measured using R_s = 124.1 Ω, R_f = 4960 Ω, and C_{dl} = 0.500 µF

Table III. Comparison of Cd^{2+} Faradaic Component at 488 Hz after Subtraction of Non-Faradaic Effects by Two Different Methods[a]

Potential, V vs. Ag/AgCl	Method 1 admittance mmho		Method 2 admittance, mmho	
	In-phase	Quadrature	In-phase	Quadrature
−0.4590	0.1340	0.1296	0.1407	0.2167
−0.4700	0.3175	0.2994	0.3184	0.3662
−0.4820	0.7633	0.6876	0.7700	0.7363
−0.4940	1.709	1.437	1.713	1.482
−0.5060	3.286	2.461	3.297	2.501
−0.5180	4.866	3.171	4.875	3.214
−0.5300	4.785	2.800	4.790	2.848
−0.5420	3.266	1.836	3.269	1.882
−0.5540	1.765	0.9948	1.767	1.047
−0.5650	0.8542	0.4826	0.8580	0.5354
−0.5770	0.3601	0.2113	0.3665	0.2778
−0.5890	0.1526	0.08828	0.1545	0.1559

[a] Conditions as in Figure 7.

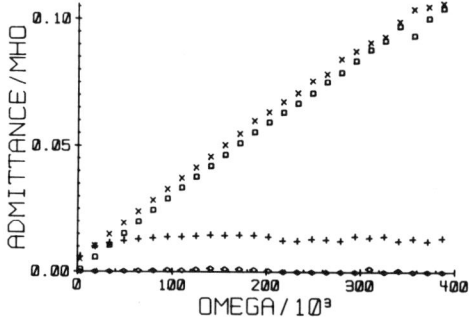

Figure 5. Cell admittance spectrum after computations to remove effects of uncompensated solution resistance. System: DME–aqueous 1.0 M ZnSO₄, 0.18 M H₂SO₄ at 25 °C, with and without 1.0 × 10⁻³ M Cd^{2+}. Applied: Pseudo-random, odd harmonic ac waveform (3) with 1.5 mV per frequency component, 26 components; superimposed on dc voltage = 0.523 V vs. Ag/AgCl (satd NaCl) (Cd^{2+} peak potential at low frequencies). Measured: ◊ = in-phase component, no Cd^{2+}. □ = quadrature component, no Cd^{2+}. + = in-phase with Cd^{2+}. x = quadrature component, with Cd^{2+}. All run at 3.0 s in life of DME, 10 replicates

1 computed the Cd^{2+} faradaic admittance by subtracting the double-layer admittance obtained with supporting electrolyte alone. Method 2 used the deLevie high-frequency evaluation of C_{dl} in presence of the faradaic component, as the basis for double-layer admittance compensation. The results of the two approaches are in quite satisfactory agreement, especially in the crucial dc potential region near the peak of the faradaic wave.

Figure 5 shows admittance spectra at the Cd^{2+} wave peak potential for the two solutions addressed in Table II, after removal of R_s effects. This figure allows one to visualize the magnitude of the double-layer admittance compensation problem as a function of frequency which one encounters in extracting the faradaic data. It should be noted that the quadrature component is linear with frequency and the in-phase component is negligible for the data without Cd^{2+}, as

predicted for a pure capacitor. This is further verification of accurate ohmic resistance compensation in a real electrochemical environment to quite high frequencies (62 kHz in Figure 5). No experimentally significant inductive or other frequency dispersion effects have been noted to date in this or any other studies performed with this experimental apparatus.

Figure 6A shows a plot of phase angle cotangent for the Cd^{2+} faradaic admittance. A linear response with $\omega^{1/2}$ is obtained to far higher frequencies (20.5 kHz) than has ever been achieved in this laboratory. The slope of this line yields k_s = (8.5 ± 0.5) × 10⁻² cm s⁻¹, using an α-value of 0.35, computed from the difference between $[E_{dc}]_{peak\ cot\phi}$ and $E_{1/2}^r$. It is important to recognize that measurement of cot φ requires accurate appraisal of both the in-phase and quadrature faradaic admittance components, thus demanding precise double-layer subtraction. The situation is less difficult in evaluating the in-phase faradaic component magnitude, which is orthogonal to the double-layer admittance. For many systems, the same will be true for the total cell admittance, which is primarily in-phase at the higher frequencies. Thus, for the Cd^{2+} system depicted in Figure 6A, good total cell admittance data are realized to even higher frequencies (41.4

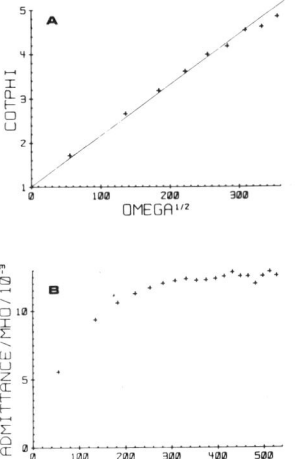

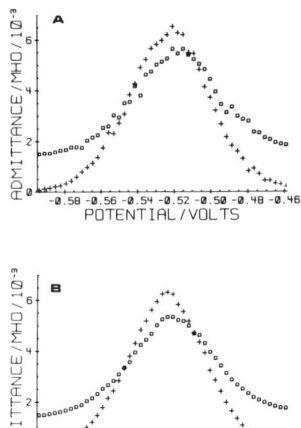

Figure 6. Peak Faradaic admittance spectrum of $Cd^{2+}/Cd(Hg)$ system after nonfaradaic component compensation. Conditions as in Figure 5 with Cd^{2+} present. (A) Phase angle cotangent spectrum. (B) Total admittance spectrum

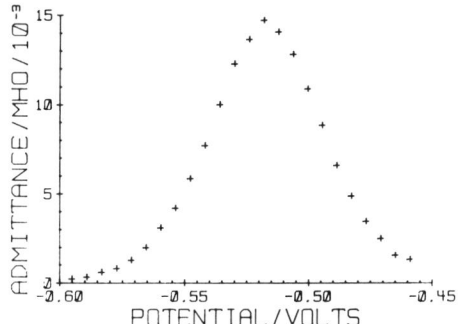

Figure 7. Total admittance polarogram of $Cd^{2+}/Cd(Hg)$ at 22.4 kHz. Conditions as in Figure 5 with Cd^{2+} present using staircase dc potential scan

Figure 8. Comparison of phase-selective Faradaic admittance polarograms of $Cd^{2+}/Cd(Hg)$ using odd-harmonic and all-harmonic pseudo-random waveforms. System: 1.0×10^{-3} M Cd^{2+} at DME-aqueous 1.0 M $ZnSO_4$ (unacidified) at 25 °C. Applied: Pseudo-random ac waveform with 1.5 mV per component and (A) all harmonics (60 components), (B) odd harmonics (30 components); superimposed on staircase dc voltage scan. Measured: In-phase (+) and quadrature (□) admittance at 880 Hz after compensation for ohmic resistance (i.e., includes double-layer admittance) at 3.0 s in life of DME, 10 replicates

kHz), as illustrated in Figure 6B. Data for the Cd^{2+} system at frequencies beyond those shown in Figure 6 degrade with frequency. The cot φ results show evidence of systematic error, due to small imperfections in double-layer admittance subtraction. The total admittance data show simply a larger stochastic component, which can be eliminated by more extensive signal averaging. We estimate that the in-phase faradaic admittance can be accurately measured to at least 60 kHz.

Figure 7 shows a typical high frequency admittance polarogram obtained with the SYDAGES-based measurement system. As can be seen, random noise and systematic background signal levels are quite adequately suppressed.

A primary reason for in-house construction of the SYDAGES system was the desire to apply an unusual, commercially unavailable test signal waveform which operated synchronously with dual channel data acquisition. Otherwise a conventional broadband signal source (pulse or white noise generator) and dual channel data acquisition system could have been obtained commercially. The advantages of phase-randomization of the waveform in question (to minimize faradaic nonlinearity contributions) together with nonstochastic harmonic amplitudes are adequately documented (2, 3, 5). However, published work has not as impressively demonstrated the desirability of using only odd-harmonics, rather than all possible frequencies. This has occurred because published comparisons of waveform efficiency (3) were confined to peak potential spectra where, for the systems in question, the major faradaic nonlinearity contributions (second harmonic) are negligible. Of course, this is not true at all potentials along a faradaic admittance wave. Consequently, much more striking evidence in favor of limiting the test signal to odd harmonics is revealed by comparing single-frequency admittance polarograms obtained with "odd harmonic" and "all harmonic" waveforms, as shown in Figure 8. This advantage of the odd-harmonic test signal is crucial, not simply for reducing ensemble averaging times (3), but for permitting experiments where satisfactory data after one measurement pass are required, such as with FFT ac cyclic voltammetry and other situations involving temporal variations of the faradaic admittance.

CONCLUSIONS

The SYDAGES system has been found to provide a marked advance in the ability to accurately measure the faradaic admittance. We have extended our measurement bandwidth for *faradaic data* by more than an order of magnitude, without loss of accuracy, and have achieved automatic, dynamic measurement of and compensation for nonfaradaic effects. In addition to providing for nonfaradaic compensation, the nonfaradaic information can be valuable on a day-to-day basis in verifying the conditions of the electrodes and solution. The extended measurement bandwidth value cannot be overestimated in terms of its ability to check and verify reaction mechanisms in a very short measurement time. The SYDAGES system described here has been in routine use in this laboratory for six months, as documented in several other reports (10, 17, 20) which encompass successful applications in aqueous and nonaqueous media, with liquid and solid metallic electrodes and with inorganic and organic reactants.

It should be recognized that, whereas the described use of SYDAGES is limited to the measurement protocol for which it was designed, the device's applicability is of much broader scope. Essentially any input–response measurement scheme, electrochemical or nonelectrochemical, can be serviced by the device. Thus, electrochemical applications to pulse or square-wave polarography, cyclic voltammetry, etc. are readily invoked.

Finally, the 500-kHz data rate upper limit is a result primarily of our conservative attitude in approaching higher frequency measurement. Having encountered none of the anticipated special problems, such as solid electrode frequency dispersion, insidious electronic component nonlinearities, and the like, it now appears that a faster SYDAGES would have been worth considering. With proper substitution of solid state components, the basic SYDAGES should adapt to data rate increases of about an order-of-magnitude. This possibility is being considered.

LITERATURE CITED

(1) S. C. Creason and D. E. Smith, *J. Electroanal. Chem.*, **36**, A1 (1972).
(2) S. C. Creason and D. E. Smith, *J. Electroanal. Chem.*, **40**, A1 (1972).
(3) S. C. Creason, J. W. Hayes, and D. E. Smith, *J. Electroanal. Chem.*, **47**, 9 (1973).
(4) S. C. Creason and D. E. Smith, *Anal. Chem.*, **45**, 2401 (1973).
(5) D. E. Smith, *Anal. Chem.*, **48**, 221A (1976).
(6) M. Sluyters-Rehbach and J. H. Sluyters, in "Electroanalytical Chemistry", Vol. 4, A. J. Bard, Ed., M. Dekker, New York, N.Y., 1970, Chap. 1.
(7) R. deLevie, J. W. Thomas, and K. M. Abbey, *J. Electroanal. Chem.*, **62**, 111 (1975).
(8) A. A. Pilla, in "Computers in Chemistry and Instrumentation", Vol. 2, J. S. Mattson, H. D. MacDonald, Jr., and H. B. Mark, Jr., Ed., M. Dekker, New York, N.Y., 1972, pp 139–181.
(9) K. Matsuda and R. Tamamushi, *J. Electroanal. Chem.*, **75**, 193 (1977).
(10) R. J. Schwall, A. M. Bond, and D. E. Smith, *J. Electroanal. Chem.*, in press.
(11) D. E. Smith, in "Electroanalytical Chemistry", Vol. 1, A. J. Bard, Ed., M. Dekker, New York, N.Y., 1966, Chap. 1.
(12) E. O. Brigham, "The Fast Fourier Transform", Prentice-Hall, Englewood Cliffs, N.J., 1974.
(13) K. F. Drake, Doctoral Dissertation, Northwestern University, Evanston, Ill., 1978.
(14) R. J. Schwall, Doctoral Dissertation, Northwestern University, Evanston, Ill., 1977.
(15) S. C. Creason, Doctoral Dissertation, Northwestern University, Evanston, Ill., 1973.
(16) "Applications Manual for Computing Amplifiers", George A. Philbrick Researches, Inc., Boston, Mass., 1966.
(17) A. M. Bond, R. J. Schwall, and D. E. Smith, *J. Electroanal. Chem.*, in press.
(18) J. W. Hayes, D. E. Glover, D. E. Smith, and M. W. Overton, *Anal. Chem.*, **45**, 277 (1973).
(19) R. Saucedo and E. E. Schiring, "Introduction to Continuous and Digital Control Systems", Macmillan, New York, N.Y., 1968, Chap. 4.
(20) R. J. Schwall, A. M. Bond, and D. E. Smith, *Anal. Chem.*, following paper in this issue.

RECEIVED for review March 11, 1977. Accepted July 1, 1977. The authors are indebted to the National Science Foundation (Grant No. MPS74-14597) and the Australian Research Grants Committee for support of this work.

18

Copyright © 1978 by the American Chemical Society

Reprinted from *Anal. Chem.* 50:116-137 (1978)

Analysis of Electrochemical Mechanisms by Finite Difference Simulation and Simplex Fitting of Double Potential Step Current, Charge, and Absorbance Responses

M. K. Hanafey, R. L. Scott,[1] T. H. Ridgway,[2] and C. N. Reilley*

Kenan Laboratories of Chemistry, University of North Carolina, Chapel Hill, North Carolina 27514

Double potential step chronocoulometry, chronoamperometry, and chronoabsorptometry are often useful techniques in the determination of mechanistic pathways and associated chemical rate and equilibrium constants. The analysis may be done by performing a series of double potential step experiments in which the switching time is varied over a suitable range and comparing the results to the theoretical response ratios for various mechanisms in an attempt to find a satisfactory match. A prerequisite is a large set of accurate theoretical responses for a wide range of mechanisms in a convenient form for comparison with experimental results by some systematic fitting procedure. One problem is that the number of possible mechanisms is large and theoretical responses are difficult to generate. Analytical solutions have been obtained only for the simpler cases, thus the bulk of the working curve generation must be done numerically. Although the explicit finite difference simulation (EFDS) technique has successfully been applied to a number of mechanisms under a variety of boundary conditions, predicted double potential step current, charge, and absorbance response ratios have been reported for relatively few mechanisms. More importantly, reported responses usually are in the form of a small graph and therefore are practically useless for quantitative curve matching. Often a suitably broad range of rate constants was not considered, and the gaps between the calculated points are too wide for accurate interpolation.

Table I defines some electrochemical mechanisms which have appeared in recent publications. The list is not exhaustive and contains only the more common pathways. Many references to mechanism are made throughout the rest of the paper either by letter or name (left hand column), or data set number (right hand column) which also indicates the mechanisms covered in this work. The first and second parameters referred to throughout the paper are defined in the parameters column. Table II summarizes the availability of theoretical responses for this limited set of pathways. Although Table II is undoubtedly incomplete, it must be emphasized that the prime deficiency is the unsuitability of reported responses for quantitative use.

We hope to remedy the situation by providing a set of working curves with the following characteristics: (a) broad range of mechanisms, (b) broad range of kinetic parameters, (c) sufficiently high point density for adequate characterization, and (d) sufficient accuracy, as well as a suitable procedure for extracting useful mechanistic information from experimental results using the working curve data set.

SYSTEM RESTRICTIONS

In order to reduce the amount of work to a reasonable level, a number of restrictions on the complexity of the system had to be made. Severe restrictions immediately follow when one considers computation time requirements. Even using a relatively fast minicomputer (Modcomp II, floating point hardware) around 3.5 min was required to simulate a single double step experiment. There are two approaches to the fitting problem. Response ratios can be computed by EFDS upon demand by the fitting routine. In the two-parameter case, at least 15 trials are usually needed for adequate convergence, or around 0.75 h per mechanism fitted. Because the idea was to find the best fit from a number of mechanisms, this would be intolerably slow. The alternative is to compute the response ratios for a set of mechanisms and store them in a high speed mass storage device. A reasonable point density for a two-parameter mechanism is around 250, corresponding to about 14.5 h of computer time. The addition of one more parameter puts the problem in the unreasonable domain of at least 150 h for each three-parameter mechanism. With only two-parameter flexibility, other restrictions follow immediately. All diffusion coefficients were assumed equal (D_i/D_j eliminated), no adsorption was considered (Γ_i eliminated), and potentials were stepped so as to reduce the surface concentration of electroactive species to zero (k_s, α, $E-E°$ eliminated). Planar, semi-infinite diffusion was the only mode of mass transport. Table I lists the mechanisms that were included along with the parameters needed to uniquely define a response ratio. These parameters are established by the following linear transformations of the differential equation system describing the mechanism

$\theta = k't$ (dimensionless time)

$y = \sqrt{\dfrac{k'}{D_A}}\,x$ (dimensionless distance)

$\sigma_i = [i]/[A]_b$ (dimensionless concentration).

where $k' = k$ (for 1st order cases) or $k[A]_b$ (for 2nd order cases). During the first potential step, the surface concentration of the left hand member of each electrochemical couple was reduced to zero, while during the second potential step of equal duration the surface concentration of the right hand members of each couple was reduced to zero. Although in certain situations "novel" potential stepping can be useful (68-70), these cases were not simulated.

---

[1] Present address, Instrument Applications Laboratory, Dow Chemical U.S.A., Midland, Mich. 48640.
[2] Present address, Department of Chemistry, University of Cincinnati, Cincinnati, Ohio 45221.

Table I. One- and Two-Parameter Electrochemical Mechanisms

Mechanism identifier	Reaction scheme[a]	Parameters	Mechanism data set no.
A. Stable radical	$A^b = B$		
B. EC, first order	$A^b = B$; $B \underset{k_b}{\overset{k_f}{\rightleftharpoons}} C$	$\log(k_f\tau), k_b/k_f$	1
C. EC, radical–radical dimer	$A^b = B$; $2B \underset{k_b}{\overset{k_f}{\rightleftharpoons}} C$	$\log(k_f[A]_B\tau), k_b/k_f[A]_B$	2
D. EC, parent–radical dimer	$A^b = B$; $A^b + B \underset{k_b}{\overset{k_f}{\rightleftharpoons}} C$	$\log(k_f[A]_B\tau), k_b/k_f[A]_B$	3
E. CE, prekinetics first order	$A^b \underset{k_b}{\overset{k_f}{\rightleftharpoons}} B^b$; $B^b = C$	$\log(k_f\tau), k_b/k_f$	4
F. EC, second-order disproportionation	$A^b = B$; $2B \underset{k_b}{\overset{k_f}{\rightleftharpoons}} A^b + C$	$\log(k_f[A]_B\tau), k_b/k_f$	5
G. ECE, first order	$A^b \overset{n}{=} B$; $B \underset{k_b}{\overset{k_f}{\rightleftharpoons}} C$; $C \overset{n}{=} D$	$\log(k_f\tau), k_b/k_f$	6
H. ECE, nuance	$A^b \overset{n}{=} B$; $B \overset{k}{\rightarrow} C$; $C \overset{n}{=} D$; $A^b + D \overset{K_{eq}}{\rightleftharpoons} B + C$	$\log(k\tau), -pK_{eq}$	7
I. ECE, radical–radical dimer	$A^b \overset{n}{=} B$; $2B \underset{k_b}{\overset{k_f}{\rightleftharpoons}} C$; $C \overset{n}{=} D$	$\log(k_f[A]_B\tau), k_b/k_f[A]_B$	8
J. ECE, parent–radical dimer	$A^b \overset{n}{=} B$; $A^b + B \underset{k_b}{\overset{k_f}{\rightleftharpoons}} C$; $C \overset{n}{=} D$	$\log(k_f[A]_B\tau), k_b/k_f[A]_B$	9
K. ECEC, radical–radical dimer	$A^b \overset{n}{=} B$; $2B \overset{k_1}{\rightarrow} C$; $C \overset{n}{=} D$; $D \overset{k_2}{\rightarrow} E$	$\log(k_1[A]_B\tau), k_2/k_1[A]_B$	10
L. ECEC, parent–radical dimer	$A^b \overset{n}{=} B$; $A^b + B \overset{k_1}{\rightarrow} C$; $C \overset{n}{=} D$; $D \overset{k_2}{\rightarrow} E$	$\log(k_1[A]_B\tau), k_2/k_1[A]_B$	11
M. ECEC, first order	$A^b \overset{n}{=} B$; $B \overset{k_1}{\rightarrow} C$; $C \overset{n}{=} D$; $D \overset{k_2}{\rightarrow} E$	$\log(k_1\tau), k_2/k_1$	12
N. EC, first-order regeneration	$A^b = B$; $B \underset{k_b}{\overset{k_f}{\rightleftharpoons}} A^b + C$	$\log(k_f\tau), k_b[A]_B/k_f$	13
O. EC, second-order regeneration	$A^b = B$; $B + C^b \overset{k}{\rightarrow} A^b + D$	$\log(k[A]_B\tau), [C]_B/[A]_B$	14
P. EC, second order	$A^b = B$; $B + C^b \overset{k}{\rightarrow} D$	$\log(k[A]_B\tau), [C]_B/[A]_B$	15
Q. ECE, second order	$A^b \overset{n}{=} B$; $B + C^b \overset{k}{\rightarrow} D$; $D \overset{n}{=} E$	$\log(k[A]_B\tau), [C]_B/[A]_B$	16
R. ECC, second-order disproportionation	$A^b = B$; $2B \overset{k_1}{\rightarrow} A^b + C$; $C \overset{k_2}{\rightarrow} D$	$\log(k_1[A]_B\tau), k_2/k_1[A]_B$	17
S. ECC, heterogeneous disproportionation	$A^b = B$; $B \overset{k_1}{\rightarrow} C$; $B + C \overset{k_2}{\rightarrow} A^b + D$	$\log(k_1\tau), k_2[A]_B/k_1$	18
T. ECC, homogeneous disproportionation	$A^b = B$; $B \overset{k_1}{\rightarrow} C$; $2C \overset{k_2}{\rightarrow} A^b + C$	$\log(k_1\tau), k_2[A]_B/k_1$	19
U. ECC, parent–radical parent dimer	$A^b = B$; $A^b + B \overset{k_1}{\rightarrow} C + D$; $A^b + C \overset{k_2}{\rightarrow} E$	$\log(k_1[A]_B\tau), k_2/k_1$	20
V. ECC, first- and second-order regenerative	$A^b = B$; $B \overset{k_1}{\rightarrow} A^b + C$; $B + C \overset{k_2}{\rightarrow} A^b + D$	$\log(k_1\tau), k_2[A]_B/k_1$	21
W. ECE, first-order regenerative	$A^b \overset{n}{=} B$; $B \underset{k_b}{\overset{k_f}{\rightleftharpoons}} A^b + C$; $C \overset{n}{=} D$	$\log(k_f\tau), k_b[A]_B/k_f$	22
X. One-half regeneration	$A^b = B$; $B \overset{k}{\rightarrow} \frac{1}{2} A + C$	$\log(k\tau)$	b

[a] A superscript "b" indicates a species present in the bulk. A "$=$" ("$\overset{n}{=}$") indicates electrode electron transfer involving n (undefined) electrons. [b] The one-half regeneration mechanism (X) is a limiting form of the heterogeneous (S) and homogeneous (T) disproportionation mechanisms. As k_2/k_1 in S and T approaches infinity, the chemical kinetics of these mechanisms becomes equivalent to the kinetics in X when $2k_1$ of S and k_1 of T equals k of X.

Table II. Electrochemical Mechanisms, Excitations, and Responses Reported in the Literature[a]

Mechanism	Ramp current	Cyclic current	Single Potential Step			Double Potential Step		
			Current	Charge	Absorbance	Current	Charge	Absorbance
A. Stable radical	20, A26, A49, 61	1, 20, A26, A49, 60	18, 40, 58	3, A33	46, 59	2, 40	3, A33, A54	46
B. EC, first order	20, 24, A29, 30, 34, 35	20, 24, A29, 30, 34, 35	2, 3, 4, 40, 55	3, 4	4, 7, 59	2, 3, 4, 5, 40, 62, *	3, 4, 5, 6, A54, 62, *	4, *
C. EC, radical-radical dimer	30, 31, 39, 50	30, 31, 39, 50	40			5, 40, *	5, *	*
D. EC, parent-radical dimer		63				5, *	5, *	*
E. CE, pre-kinetics	20, A28, 42	20, A28, 42	14, 18	14	59	*	*	*
F. Second-order disproportionation	37, 38, 41, 53	37, 38, 41, 53	1, 8, 19, 22, 43, 56	4, 43	59	45, 56, 64, *	4, *	*
G. ECE, first order	1, 21, 23, 30, 37, 51, 65	21, 30, 37, 51	1, 4, 9, 10, 56, 58	4	4, 11	4, 5, 6, 56, *	4, 5, 6, A12, *	4, *
H. ECE, nuance	32, 51, 52, 53, 57	32, 51, 52, 53, 57	27, 43, 56, 57	43, 57		56, *	*	*
I. ECE, radical-radical dimer	48	48	43, 44	43		*	*	*
J. ECE, parent-radical dimer	48	48	43	43		5, *	5, *	*
K. ECEC, radical-radical dimer						*	*	*
L. ECEC, parent-radical dimer						*	*	*
M. ECEC, first order			4	4	4	4, 6, *	4, 6, *	4, *
N. EC, first-order regeneration	20, 25	20, 25	17, 18	3	17, 46, 59	*	3, *	46, *
O. EC, second-order regeneration				47		*	*	47, *
P. EC, second order	30	30				*	*	*
Q. ECE, second order	30	30, 66				*	*	*
R. ECC, second-order disproportionation	37, 38, 41, 53	37, 38, 41, 53	1, 8, 19, 22, 43, 56, 59	4, 43	36, 59	45, 56, 64, *	4, *	*
S. ECC, heterogeneous disproportionation			4			*	4, *	*
T. ECC, homogeneous disproportionation			4			*	4, *	*
U. ECC, parent-radical-parent dimer						*	*	*
V. ECC, first- and second-order regenerative			67			*	*	*
W. ECC, first-second-order regenerative			67			*	*	*
X. One-half regeneration			13, 14, 15, 16	4, 13, 14, 15	13	13	4, 13	13

[a] An "A" preceding a reference indicates that absorption was considered. An "*" indicates responses reported here.

Although not extensively applied, the more complicated finite difference procedures based on the Laasonen implicit scheme (71), or variable space dimensions (nonlinear transformation of the space variable) followed by an explicit or Crank-Nicholson implicit finite difference method (72) can save significant computer time and extend the kinetic range. The implicit schemes described (71, 72) do not realize the full potential of the implicit method because they treat the chemical kinetics explicitly. The implicit treatment of kinetics increases the complexity of the method because a coupled set of nonlinear simultaneous equations results. Preliminary work with the equations indicates they can be solved iteratively. Nonlinear transformation of the time variable should be beneficial for the same reason that justifies the space dimension transformation, but no results have been published.

FINITE DIFFERENCE SIMULATION OF WORKING CURVES

Working Curve Generator. A working curve generating program was written (FORTRAN IV) based on the explicit finite difference technique described by Feldberg (1). The input consists of which kinetic parameters to vary and what values

Table III. Second Parameter Values (see Table I) for Which Response Ratios Were Simulated

Mech. No.	Working Curve No.											
	1	2	3	4	5	6	7	8	9	10	11	12
1	0.0	0.01	0.05	0.1	0.2	0.4	0.8	1.6	3.2	6.4	10.0	25.0
2, 3	0.0	0.05	0.1	0.2	0.4	0.8	1.6	3.2	6.4	10.0	50.0	100.0
4	0.01	0.05	0.1	0.2	0.4	0.6	0.8	1.6	3.2	6.4	10.0	25.0
5, 6	0.0	0.01	0.05	0.1	0.2	0.4	0.8	1.6	3.2	6.4	10.0	25.0
7	−6.0	−4.0	−3.0	−1.0	0.0	0.30103	0.60206	1.0	2.0	3.0	4.0	5.0
8, 9	0.0	0.01	0.05	0.1	0.2	0.4	0.8	1.6	3.2	6.4	10.0	25.0
10–12	0.001	0.01	0.05	0.1	0.2	0.4	0.8	1.6	3.2	6.4	10.0	25.0
13	0.0	0.01	0.05	0.1	0.2	0.4	0.8	1.6	3.2	6.4	10.0	25.0
14–21	0.001	0.01	0.05	0.1	0.2	0.4	0.8	1.6	3.2	6.4	10.0	25.0
22	0.0	0.01	0.05	0.1	0.2	0.4	0.8	1.6	3.2	6.4	10.0	25.0

they should take on, a specification of the details of the mechanism (first- and second-order reactions can be handled either as a finite rate reaction or as an equilibrium), and a number of simulation parameters. The results are stored on a direct access moving head cartridge disk file. The following information is stored for each mechanism at 252 points in the two-parameter space along with identifying and kinetic parameter information.

$$Q_R = [Q(\tau) - Q(2\tau)]/Q(\tau) \quad \text{(i)}$$

$$I_R = -i(2\tau)/i(\tau) \quad \text{(ii)}$$

$$A_{R,j} = [A_j(\tau) - A_j(2\tau)]/A_j(\tau) \quad \text{(iii)}$$

$$A_j(\tau) \quad \text{(iv)}$$

where $Q \equiv$ charge, $i \equiv$ current, $j = 1, 2 \ldots \#$ species except those in the bulk, and $A_j \equiv$ transmission absorbance, jth species. The relative absorbance (assuming all species have the same molar absorptivity) of each species at time τ, $A_j(\tau)$, is needed to calculate the absorbance ratio in cases when the measured absorbance is the result of absorbance by two or more species

$$A_{R,\text{composite}} = \frac{\sum_i^{\text{all species}} (\epsilon_i/\epsilon_j) A_i(\tau) A_{R,i}}{\sum_i^{\text{all species}} (\epsilon_i/\epsilon_j) A_i(\tau)}$$

where $A_{R,\text{composite}}$ is the measured absorbance ratio, ϵ_i/ϵ_j is the molar absorptivity of the ith species relative to the jth species, $A_{R,i}$ is the absorbance ratio if only the ith species absorbs, and $A_i(\tau)$ the relative absorbance at time τ. In all cases the first parameter in Table I was varied from −2.0 to 2.0 in steps of 0.2 (21 values). Table III lists the 12 values of the second parameter for which results were simulated. Storage requirements for one two-parameter mechanism is 24K bytes.

The problem of standard response curve storage received a considerable amount of attention in an effort to optimize the compactness and ease and speed of retrieval of data.

The first approach was to take the 21 response ratio points for each standard response curve and subject them to a polynomial curve fitting routine so that instead of storing all 21 data points it would only be necessary to store a few polynomial coefficients. Some of the standard response curves could be handled very nicely with this technique but others gave poor fits even with polynomials as high as 10th order. Some of these ill-fitting data sets could be handled by pretreating the standard response curves with various exponential functions before subjecting them to the polynomial routine. In other cases a satisfactory fit could be obtained only by splitting the range of log ($k\tau$) values into two parts, treating each part of the curve separately. The end result was a data base which had not been significantly compressed, and with data in a form not convenient to use. In addition, accuracy is inevitably lost in the fitting procedure and this is partic-

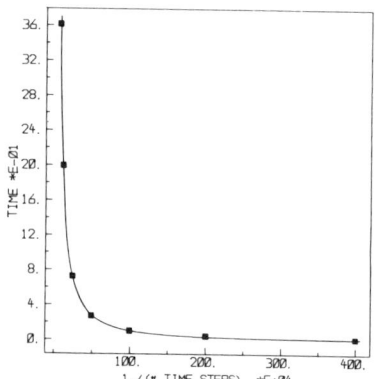

Figure 1. Execution time (in seconds) vs. the time step size for the first-order regeneration mechanism. This should be contrasted with the rate of convergence as shown in Figure 3

ularly undesirable when one considers the expense of generating the numbers. The decision was to store unmodified all of the simulated results.

Another decision was whether to use sequential or direct access storage. The former has the advantage of requiring the least space and being both magnetic tape and disk compatible, but the direct access mode allows much faster random accessing of the data set members. It was found that the moving head disk search time for a mechanism near the middle of the file was longer than the time needed to fit experimental results to the retrieved data when the sequential storage mode was used. It was decided to organize the data set for direct access.

Finite Difference Simulations. The details of the explicit finite difference technique have been treated in detail elsewhere (1, 4, 73). The discussion here deals only with a few points which have not been adequately discussed, as well as defining precisely some of the variable simulation parameters.

Preliminary work involving a fourth-order convergent formula for the second derivative clearly demonstrated the profound effect roundoff has on the accuracy of finite difference approximations (because subtraction of approximately equal numbers is involved). All of our computations are as a consequence done in 38-bit precision (six-byte floating point) using the usual second-order convergent second derivative finite difference approximation.

One of the more important simulation variables is the number of time steps. Two considerations limit the number of time steps. The roundoff problem is discussed later, but in practice the real limitation is the simulation time. Convergence occurs roughly linearly with the reciprocal of the number of time steps (that is, with Δt, which is a dimensionless time obtained by dividing the time step by the duration of the potential step), but as Figure 1 shows, execution time increases drastically as Δt is decreased. Based on the results of an error analysis described later, it was decided to vary the number of time steps between 400 and 1500 (on each potential step), the latter number being used when rate constants were at their largest. Large rate constants require more work because the relative amount of reaction in a time step must be small and because rapid kinetics leads to very sharp concentration profiles which leads to greater inaccuracy in the diffusion calculation. To extend the kinetic range much beyond our limits would demand the technique proposed by Ruzic and Feldberg (74) on the heterogeneous equivalent for handling compact reaction layers.

Another important parameter is

$$\lambda = D \Delta t / \Delta x^2$$

where D is a diffusion constant and Δt and Δx are the time and space discretization elements. From the numerical analysis of the convergence of the pure diffusion case, it can be shown (75) that 1/6 is a good choice for λ in that this particular value increases the order of convergence with respect to Δx by two. The optimum value of λ can be derived based on a different criterion. At short enough time, solution chemical kinetics has had an arbitrarily small effect. The first simulated current thus should be virtually independent of the mechanism, and clearly this first point does not correspond to the time origin. It seems reasonable to choose the first point to correspond to $\Delta t/2$ (convenient for integration) and force the first simulated current to equal that calculated from the Cottrell equation, thereby fixing the remaining free variable Δx. The resulting values for λ are

$1/2\pi = 1/6.28$ 1st-order flux equation   (i)

$9/32\pi = 1/11.2$ 2nd-order flux equation   (ii)

The flux equations are discussed later.

A variable of relatively minor significance is how far into the bulk of the solution to carry the simulation (that is, how many x points spaced by Δx to use). With the explicit finite difference method, the profiles grow at the rate of one x point for each time step. However, when the number of time steps is large, this approach results in excessive unnecessary computation. For example, at the 1000th time step, theoretically 1000 space points should be involved, but after around 100 space points, convergence to bulk values is sufficiently complete for the computation to be terminated without altering the results. We used a generous algorithm for determining how many space points to use and this factor should have no consequence in the accuracy of the simulations.

Accuracy of the Simulations. The most satisfactory check on the accuracy of a numerical solution is to compare the results to the analytic solution. However, two problems arise. Only for the simpler cases have analytic solutions been obtained, making a test of the complex cases by this method impossible, and in any event the analytic solution itself must be computed using numerical methods. If the analytic equations involve nothing more than the frequently used functions (for example, square root, exponential, and error function in the first-order catalytic case), they are easily calculated to more precision than is needed. However, if the solution involves infinite series of confluent hypergeometric functions (EC case), then considerable effort is required to ensure that the numerically computed "analytic" results are accurate.

If it is assumed that the explicit finite difference method is convergent and stable (this evidently has not been proved for the system resulting from an arbitrary electrochemical mechanism), then it might be concluded that very accurate reference results could be obtained in a few cases simply by using a very fine spatial and temporal grid mesh and investing

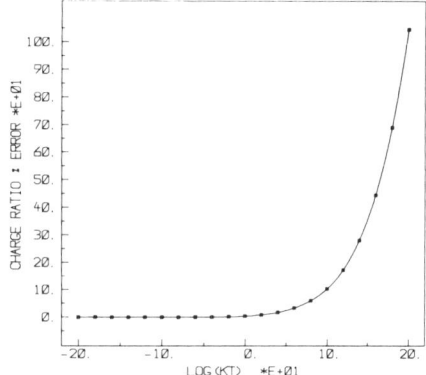

Figure 2. Percent error in the charge ratios of the first order regeneration mechanism as a function of log $(k\tau)$

a lot of computer time. The problem with this approach is that the finite difference equations are a good example of the errors introduced into most digital computations because of finite word length. Although mathematically

$$\left(\frac{\partial^2 C}{\partial x^2}\right)_a = \lim_{\Delta x \to 0} \left[\frac{C(a - \Delta x) - 2C(a) + C(a + \Delta x)}{\Delta x^2}\right]$$

in numerical computing Δx can only be reduced so far before the reduction in convergence error is overwhelmed by the increase in roundoff error, and in an involved calculation it is not obvious where the point of minimum total error lies.

A better approach is the convergence plot. In this analysis the same calculation is performed using a number of different grid meshes and the results are plotted as a function of the time step size. As predicted, these plots are roughly linear, so extrapolation to the infinitely fine grid mesh result provides an accurate estimate of the true value. Three extrapolation techniques were compared. Least susceptible to gross error is plotting and visual extrapolation. Because the base points are virtually noise-free, exact matching of a cubic polynomial was tried and found to be quite successful in most cases, although obvious errors occur in some cases if the distance of extrapolation becomes comparable to the range of points used. Most reliable appears to be linear regression extrapolation of data for $\Delta t = 1/800, 1/1000, 1/1200, 1/1500$.

Because the analytic solution of the 1st-order catalytic regeneration mechanism can readily be computed to ten digit precision, this mechanism was chosen to check the validity of the extrapolation technique. The results appear in Table IV (Mechanism 13). The relative error in the extrapolated finite difference result is dependent on the size of log $(k\tau)$, increasing in the charge ratio case in a roughly exponential way from a negligible 0.7 ppm at −2.0 by about 4 orders of magnitude at +2.0. The rather abrupt loss of accuracy in the explicit simulation as the homogeneous kinetics becomes faster is shown in Figure 2. The regenerative nature of this mechanism makes it particularly sensitive to systematic errors in the kinetic step of the calculation. Fortunately the accuracy of the extrapolated result is readily estimated from the slope of the ratio vs. Δt function.

Table IV also summarizes a number of other comparisons used to establish the reliability of our simulations. Unfortunately the reference numbers are not ideal. The analytic solution of the EC first-order case is difficult to evaluate numerically, and comparison with the results of a different numerical method or the extrapolated results of the same numerical method clearly leaves some doubt. Overall it appears the accuracy of the numerical method will not be the

Analysis of Electrochemical Mechanisms • 121

Table IV. A Comparison of Analytic, Extrapolated Simulated, and Simulated Response Ratios for a Number of Mechanisms

Mech. No.	k_2	$\log(k_1\tau)$	A. Charge Ratio	A. % Difference	B. Current Ratio	B. % Difference	C. Absorbance of B Ratio	C. % Difference	D. Absorbance of C Ratio	D. % Difference	Source
13	0.	-2.0	0.581102	Reference	0.285927	Reference	0.590474	Reference			Analytical[a]
			0.581102	0.735×10^{-4}	0.285925	0.544×10^{-3}	0.590476	0.303×10^{-3}			Extrapolated finite difference[b]
			0.581139	0.635×10^{-2}	0.286381	0.159	0.590239	0.398×10^{-1}			Finite difference (400 time steps[c])
		0.0	0.305095	Reference	0.397655×10^{-1}	Reference	0.867332	Reference			Analytical[a]
			0.305104	0.300×10^{-2}	0.397647×10^{-1}	0.175×10^{-2}	0.867305	0.319×10^{-2}			Extrapolated finite difference[b]
			0.304997	0.319×10^{-1}	0.398495×10^{-1}	0.211	0.867360	0.000000			Finite difference (600 time steps[c])
		1.4	0.195169×10^{-1}	Reference	0.261171×10^{-13}	Reference	1.000000	Reference			Analytical[a]
			0.189713×10^{-1}	2.79	0.267666×10^{-13}	2.48	1.000000	0.000000			Finite difference (739 time steps[c])
		2.0	0.497512×10^{-2}	Reference	0.000000	Reference	1.000000	Reference			Analytical[a]
			0.499782×10^{-2}	0.456	0.000000	0./0.	1.000000	0.000000			Extrapolated finite difference[b]
			0.445575×10^{-2}	10.4	0.000000	0./0.	1.000000	0.000000			Finite difference (799 time steps[c])
1	0.	-2.0	0.580803	Reference	0.287936	Reference	0.590193	Reference	-0.822104	Reference	Analytical[d]
			0.5808	0.0000	0.1841	0.562	0.5901	0.0000	-0.8224	0.360×10^{-1}	Finite difference
			0.580846	0.740×10^{-2}	0.288389	0.157	0.589956	0.402×10^{-1}	-0.822319	0.262×10^{-1}	Extrapolated finite difference[b]
		0.0	0.274644	Reference	0.552379×10^{-1}	Reference	0.855379	Reference	-0.406484	Reference	Analytical[d]
			0.2746	0.0000			0.8593	0.0000	-0.4064	0.307×10^{-1}	Finite difference
			0.274410	0.852×10^{-1}	0.553367×10^{-1}	0.179	0.859432	0.617×10^{-2}	-0.406349	0.332×10^{-1}	Extrapolated finite difference[b]
2	0.	-0.3	0.406119	Reference	0.183072	Reference	0.636037	Reference	-0.230488	Reference	Numerical[f]
			0.405986	0.332×10^{-1}	0.183111	0.210×10^{-1}	0.636031	0.423×10^{-3}	-0.229284	0.522	Finite difference (800 time steps[c])
		0.0	0.321039	Reference	0.135738	Reference	0.661934	Reference	-0.183285	Reference	Numerical[f]
			0.320867	0.536×10^{-1}	0.135761	0.169×10^{-1}	0.661861	0.110×10^{-1}	-0.182242	0.569	Finite difference (800 time steps[c])
3	0.	-2.0	0.583531	Reference	0.290308	Reference	0.589332	Reference	-0.151656×10^{1}	Reference	Extrapolated finite difference[e]
			0.583562	0.521×10^{-2}	0.290762	0.161	0.589099	0.402×10^{-1}	-0.151293×10^{1}	-0.239	Finite difference (400 time steps[c])
		0.0	0.428348	Reference	0.138172	Reference	0.803713	Reference	-0.913578	Reference	Extrapolated finite difference[e]
			0.427968	0.892×10^{-1}	0.138296	0.903×10^{-1}	0.803774	0.763×10^{-2}	-0.910975	-0.285	Finite difference (500 time steps[c])
		2.0	0.297484×10^{-1}	Reference	0.104773×10^{-18}	Reference	1.00000	Reference	-0.277297×10^{-1}	Reference	Extrapolated finite difference[e]
			0.291604×10^{-1}	1.98	0.104979×10^{-18}	0.204	1.00000	0.	-0.274777×10^{-1}	-0.909	Finite difference (700 time steps[c])

[a] Reference 3 for A. Differentiation of Reference 3 for B. Reference 46 for C. [b] 800, 1000, 1200, 1500 time steps[c]. [c] Time step numbers refer to each potential step; the total number of times steps for a double step experiment is twice the number given. [d] Reference 54 for A. Reference 4, p 302 for C and D. [e] 200, 400, 800, 1200 time steps[c]. [f] Reference 40.

limiting factor in the usefulness of fitting experimental results to the data set since experimental numbers under optimum conditions are expected to be on the order of 1% inaccurate.

Besides providing reliable reference numbers, the extrapolation procedure can yield other useful information. Although still classified as an explicit finite difference method, a number of relatively minor differences exist between various versions. For example, the usual term that allows for the following kinetic perturbation

$$A + B \underset{k_b}{\overset{k_f}{\rightleftharpoons}} C \tag{A}$$

is

$$\Delta[C] = \{k_f[A][B] - k_b[C]\}\Delta t \tag{1}$$

However, it it could be assumed that diffusional and kinetic mass transport were uncoupled, then Equation 1 can be viewed as a truncation of

$$\Delta[C] = (k_f[A][B] - k_b[C])\left\{\frac{1 - e^{-k_f\{[A] + [B] + (k_b/k_f)\}\Delta t}}{k_f\left\{[A] + [B] + \frac{k_b}{k_f}\right\}}\right\} =$$

$$(k_f[A][B] - k_b[C])\left\{1 - \frac{k_f([A] + [B]) + k_b}{2}\Delta t + \ldots + \ldots\right\}\Delta t \tag{2}$$

Equation 2 is an approximation of the integrated rate equation for reaction A, valid so long as

$$|\Delta[C]| \ll [A] + [B] + \frac{k_b}{k_f} \text{ or } [\Delta C]^2 \ll$$

$$[A][B] - \frac{k_b}{k_f}[C]$$

The utility of Equation 2 depends on the relative error contributions of (i) the finite difference second derivative approximation, (ii) the coupling of the diffusion and solution kinetics, and (iii) numerous other sources of truncation and roundoff error, to the approximation (under the stated assumption) leading from Equation 2 to Equation 1, this of course being dependent upon the size of the rate and diffusion constants.

The usefulness of the integrated rate equation approach in the catalytic regeneration case is shown in Figure 3. Because the computation time at a given Δt is almost independent of the method, the smaller the absolute value of the slope, the more efficient the method. The integrated rate equation approach was used by Booman and Pence (22) although the system of differential equations being solved numerically was different from those considered here because nonlinear transformations had been applied. Flanagan and Marcoux (58) briefly mention the advantage of the integrated rate law and suggest that when integration is complicated a numerical technique such as Runge–Kutta be applied. This does not appear justified because integration of the kinetic side of the differential equation already involves the approximation of diffusion and kinetics being uncoupled. To be reasonable, even the integrated kinetic perturbation must be small, and it is more efficient to apply this restriction which results in a simple approximate integrated rate equation for reversible second-order kinetics.

The term in braces in Equation 2 provides a convenient parameter for monitoring the validity of the simulation of the

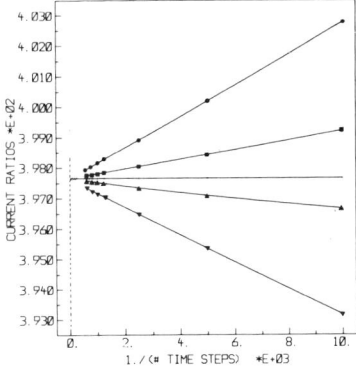

Figure 3. Convergence plot for the first-order regeneration mechanism, $\log(k\tau) = 0$.

	Second order flux	Integrated rate equation
▼	yes	no
▲	yes	yes
■	no	yes
●	no	no

The unmarked line is the theoretical result

kinetics; the more it differs from unity, the greater the error in the kinetic approximation. At the fast kinetic extreme corner of the region covered in this study, this term in some cases became negative and this positive feedback situation leads to completely invalid results; as a consequence this term is monitored and arbitrarily restricted to being greater than 0.5. Another method to improve the accuracy of the unavoidably crude estimates when large rate constants are involved is to ensure that the kinetic perturbation never leads to a negative concentration. This involves much more computation at an inner loop level than simply testing for negative concentrations at the end of the kinetic calculation and is thus quite costly in terms of computer time. However, it is more accurate in that the latter method does not correct the concentration of the species which were increased by the reaction which led to the negative concentration.

Figure 3 also shows the difference between first- and second-order convergent flux formula. Although the difference is not striking, an interesting point should be made regarding the treatment of the interfacial region. The usual approach (73, 74, 76) is to set up the spatial grid mesh as follows

Electrode |←$\rho\Delta x$→|←Δx→| ... Solution
$\quad\quad\quad\quad\quad C_0 \quad\quad C_1 \quad\quad C_2$

where $\rho = 1/2$, and to approximate the flux with

$$Z = D\frac{C_1 - C_0}{\Delta x/2} \tag{3}$$

and the second derivative at element 1 with

$$\left(\frac{\partial^2 C}{\partial x^2}\right)_1 \approx \frac{\frac{C_2 - C_1}{\Delta x} - \frac{Z}{D}}{\Delta x} \tag{4A}$$

$$= \frac{C_2 - 3C_1 + 2C_0}{\Delta x^2} \tag{4B}$$

Although approximation 4B intuitively seems convergent, a rigorous derivation using the Taylor series approach reveals

$$\left(\frac{\partial^2 C}{\partial x^2}\right)_1 = \frac{C_2 - 3C_1 + 2C_0}{\Delta x^2} + \frac{1}{4}\left(\frac{\partial^2 C}{\partial x^2}\right)_0 + 0(\Delta x) \quad (5)$$

where $0(\Delta x)$ are terms of first and higher order in Δx. That is, Equation 4B is divergent if the second partial distance derivative is nonzero at the interface, which certainly is the case when solution kinetics is involved.

Using Taylor series it can be shown that the correct formula in the case of unequally spaced points is

$$\left(\frac{\partial^2 C}{\partial x^2}\right)_1 = \frac{2[\rho C_2 - (1+\rho)C_1 + C_0]}{(\rho + \rho^2)\Delta x^2} + 0(\Delta x) \quad (6)$$

where

$$0(\Delta x) = \frac{2}{(\rho + \rho^2)}\left[\left(\frac{\partial^3 C}{\partial x^3}\right)_1 (\rho - \rho^3)\frac{\Delta x}{3!} + \left(\frac{\partial^4 C}{\partial x^4}\right)_1 (\rho + \rho^4)\frac{\Delta x^2}{4!}\right]$$

which in the specific case of $\rho = 1/2$ becomes

$$\left(\frac{\partial^2 C}{\partial x^2}\right)_1 = \frac{4}{3}\left[\frac{C_2 - 3C_1 + 2C_0}{\Delta x^2}\right] - 0(\Delta x) \quad (7)$$

Equation 6 shows why it is desirable to have $\rho = 1$; the order of convergence is increased by one in this case. The validity of Equations 5 and 7 was confirmed with calculations on a number of transcendental functions.

Equation 4A is convergent so long as Z is approximated with a higher than first-order convergent approximation. Because current and charge are directly affected by the flux approximation, it appears reasonable to improve Equation 3. The second order analog of Equation 3 is (from Taylor series expansions)

$$Z = \frac{9C_1 - C_2 - 8C_0}{3\Delta x} \quad (8)$$

and Equation 4A becomes the convergent Equation 7 when Equation 8 is substituted.

Equilibrium. There are two basic approaches to equilibrium restrictions. The method detailed by Feldberg (1) for the ECE nuance case involves equilibrating the system after doing the diffusion perturbation and then calculating the kinetic perturbation by algebraically combining the equilibrium restriction with the rate equation. Because we wanted a working curve generator capable of handling a variety of mechanisms specified as input data, we decided to use the alternative approach of calculating the diffusion perturbation, equilibrating, calculating the slow kinetics perturbation, and finally equilibrating again. This is significantly slower than the former approach, but is fairly easily incorporated into a general program, so long as equilibria coupled through a common species are excluded. It was shown that as $\Delta t \rightarrow 0$, the two approaches are identical, but the details of the proof will not be given.

COMPUTER ANALYSIS OF EXPERIMENTAL DATA

The intention in undertaking this project was to develop a large data base of double potential step response ratios which could be used to characterize experimental data according to mechanism (more realistically probable mechanisms) and to help determine the associated kinetic parameters. Although the computer programs in one sense obviate the need for graphic working curves, the latter are nonetheless useful and informative in that they allow the investigator to perceive differences and similarities between the mechanisms in manner too complex to program at the computer level. Figures 4–9 provide a fairly comprehensive summary of double step results in a very compact form ideal for qualitative comparison but virtually useless for quantitative work. The supplementary material contains the same information in tabular form. The calculation of these results required over 350 hours of CPU time (ModComp II with floating point hardware). It is clear that an automated computer analysis of experimental data is highly desirable when one considers that information is being extracted from around 50 000 numbers. In order for the simulated responses to be useful it is necessary to have a system which will compare experimental data to the simulated responses, choose values for the various parameters to optimize the fit of the experimental points to the simulated curve, and provide the investigator with a measure of the degree of fit. Traditionally this has been done by manually plotting data points on copies of published curves and varying the rate constant in the ordinate of the plot to achieve the best fit, a tedious and time-consuming task which can be automated.

Because at most two parameters are involved, and because of the ease with which constraints can be specified, the simplex optimization procedure described by Deming and Morgan (77) was used as a base on which to construct a computer program which reduces the analysis procedure to punching the experimental results on cards and giving the computer a few simple commands. Either the entire mechanism set or a few selected members can be fitted to, and the results displayed in a number of convenient formats. The remainder of this section briefly discusses the more important points regarding the analysis.

Interpolation. One characteristic of the simplex procedure is that it requires the system response at points which usually do not correspond to the base points, necessitating a bivariate interpolating procedure. Because there is very little noise associated with the base points, a combinational smoothing-interpolating procedure is not indicated. The problem falls into the category of surface interpolation, the theory of which is still being actively researched. Figure 10 illustrates the situation. The dots are the base points and the response at (X,Y) has been demanded by the simplex. A surface needs to be fitted to the base points in order to approximate the function at (X,Y). The simplest surface is a plane through the surrounding three points, but since the gaps between base points are large, such a low order approximation is expected to be unsatisfactorily inaccurate. If the 16 points surrounding (X,Y) are used, considerable computation is needed to solve the fairly large system and round off can result in significant error. The more usual approach is to successively apply univariate interpolation methods.

One of the more simple approaches is to fit cubic polynomials to the four points nearest to the open rectangles on the $X = X_1$, X_2, X_3, and X_4 lines (Figure 10). These polynomials are then used to approximate the function along the Y line at the four X values (that is, at the open rectangles). These cubic in Y interpolated results are then fitted to a cubic in X polynomial which can then be used to approximate the function at (X,Y). At the edges of the region of base points, a less than optimal distribution of base points must be used, the interpolated point being bracketed on one side by three points and on the other side by one point. Cubic interpolation is entirely satisfactory with a sufficiently high base point density, this being defined loosely as meaning that the number of points in any subregion is more than an order of magnitude

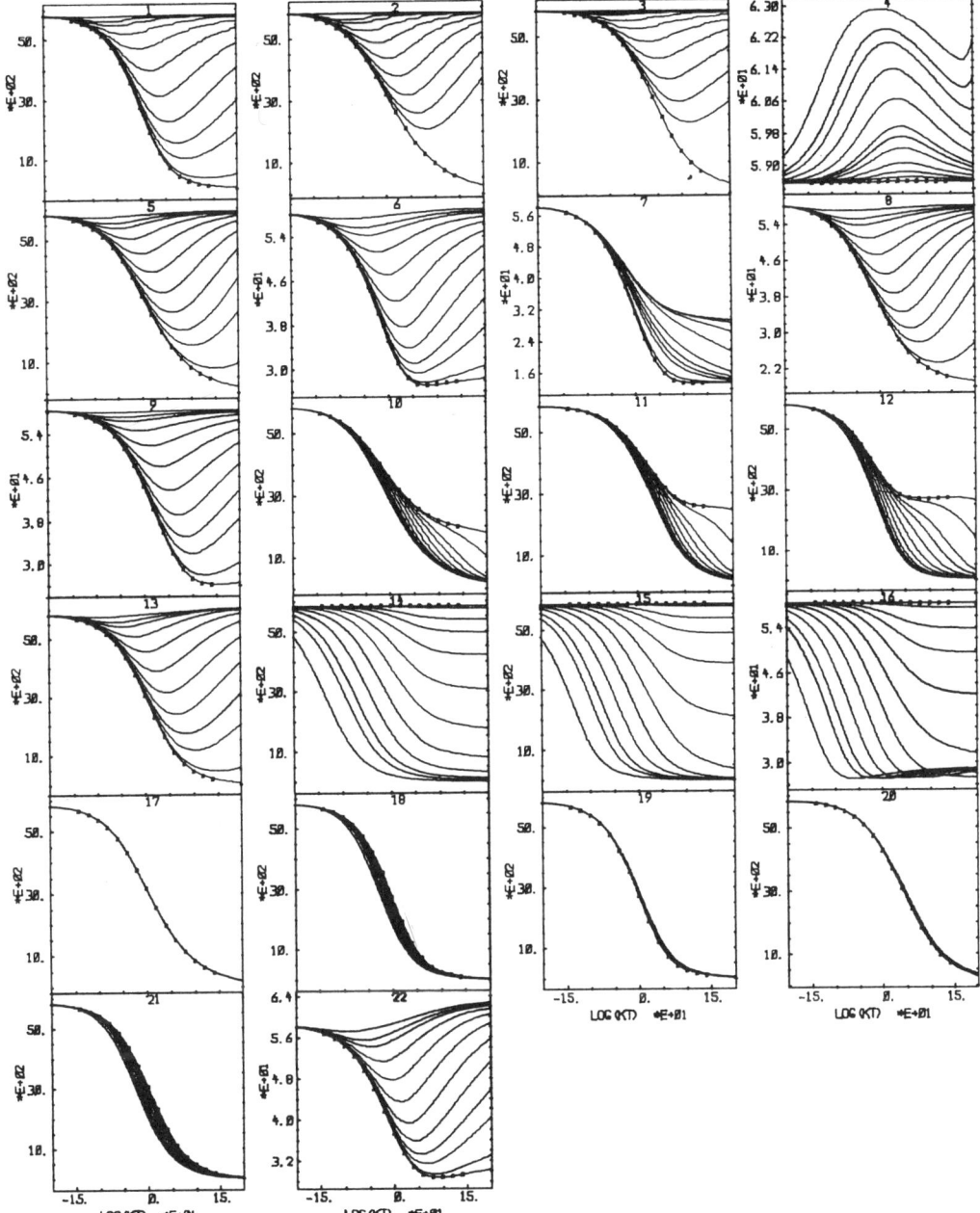

Figure 4. Charge ratios. The individual curves correspond to different values (given in Table III) of the second parameter (defined in Table I). The closed circles mark the first curve of the set

larger than the number of inflection points expected in the fitted curve for that region, and that there are no "abrupt" changes in the expected second derivative (78). If this condition is not satisfied, the cubic interpolated result shows "obviously erroneous" (based on visual analysis of a plot of the points and the cubic fit) excursions between points. In addition, piecewise cubic fitting to more than four points ensures only that the function is continuous at the nodes, which in some cases results in obvious discontinuities in the first and second derivatives at the nodes.

In cases where cubic interpolation is unsatisfactory because sparse data result in a fit which intuitively is not sufficiently "smooth", the cubic spline approach is indicated. Spline functions constitute a relatively new subject in analysis, the concept being introduced in 1946. In 30 years spline interpolation has developed into a large, well developed field, the details of which can be found in a number of books (79, 80). The most important features of cubic spline interpolation are apparent from its definition. Given a set of mesh points (not necessarily equally spaced) and the associated function values,

203

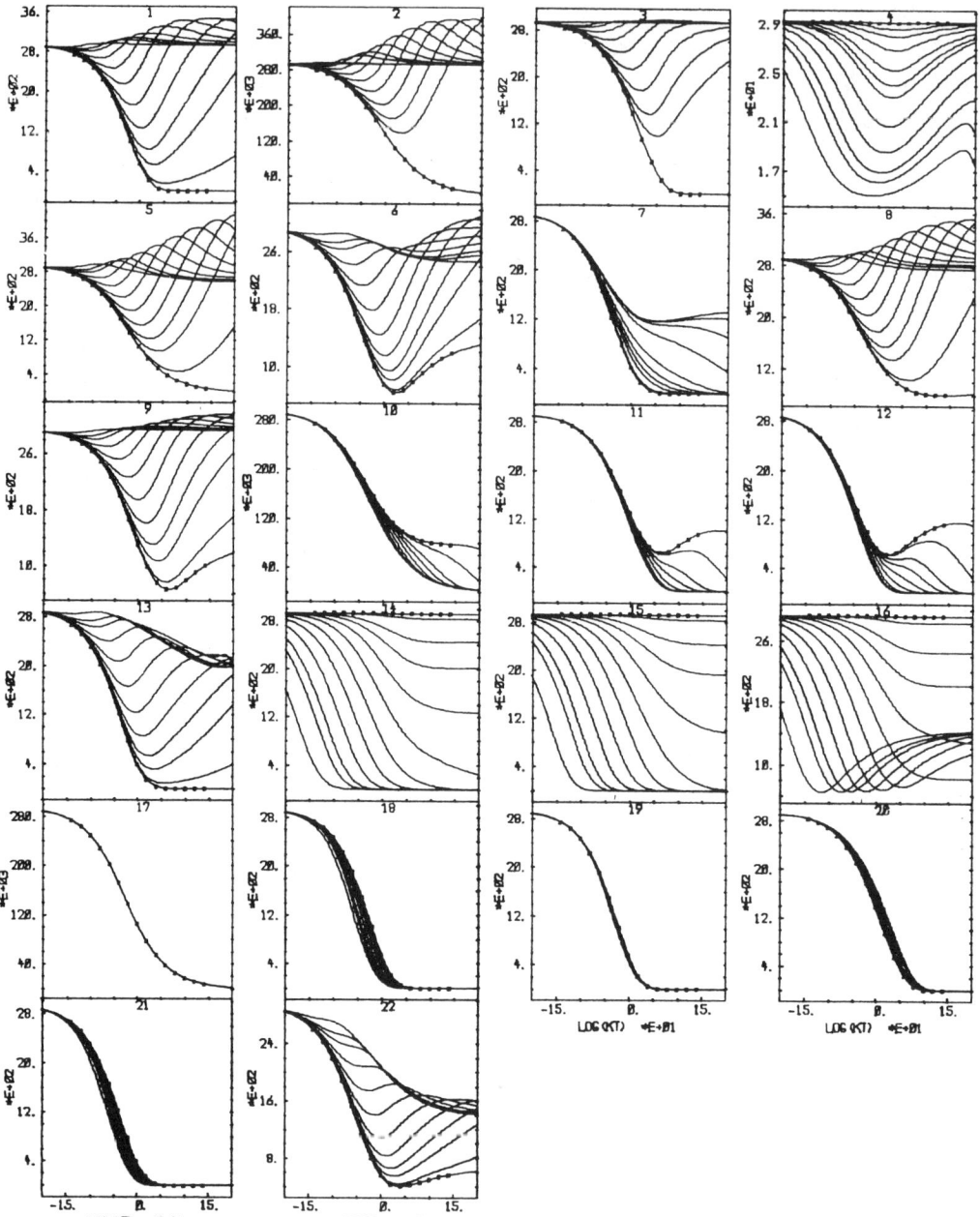

Figure 5. Current ratios. The individual curves correspond to different values (given in Table III) of the second parameter (defined in Table I). The closed circles mark the first curve of the set

the cubic spline is the function which is continuous together with its first and second derivatives on the entire interval (in particular, at the nodes), coincides with a cubic polynomial in each subinterval (i.e. between adjacent points), and passes through each base point. It turns out that this definition leaves two degrees of freedom, allowing the curvature at the end points to be independently specified, a frequent choice being zero (in this case it is called a natural cubic spline). This particular spline has the property that the integral of the square of its second derivative is the minimum amongst the set of all functions which pass through the base points. In this sense it is the smoothest possible fit to the base points. Although the spline is a global rather than local curve (as is the case with cubic interpolation), that is, altering a single base point affects the spline throughout the interval, the effect diminishes rapidly as the distance from altered point increases, leading to a well conditioned tridiagonal matrix easily solved even when hundreds of points are involved.

For simplicity the discussion of splines has been restricted to single dimension interpolation, but the concept is readily

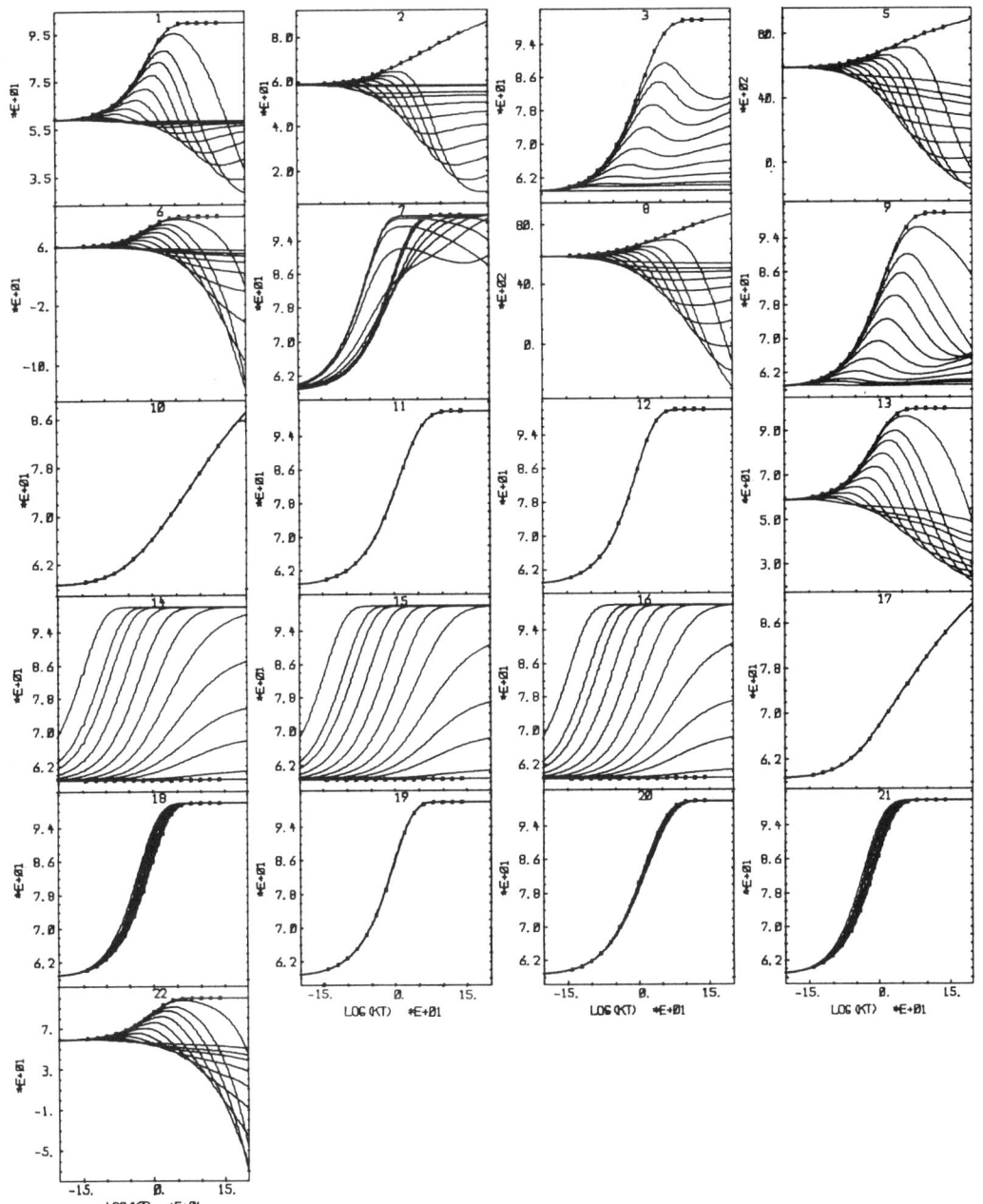

Figure 6. Absorbance ratio of B. The individual curves correspond to different values (given in Table III) of the second parameter (defined in Table I). The closed circles mark the first curve of the set

extended to many dimensions. Bicubic spline interpolation is more complex than piecewise cubic interpolation and as a consequence consumes more memory and time.

The accuracy of both interpolation schemes was tested using the EC, radical–radical dimer case because this mechanism has a region $(0. < \log (k_f[A]_b \tau) < 2., 0 < k_b/k_f[A]_b < 0.05)$ in the charge response ratios where interpolation is expected to be least accurate, as well as regions where the spacing between the base point responses is more typical (see Figures 4–9). The correct values at a number of points between the base points were obtained by digital simulation using the same simulation parameters as were used in generating the base points. Some of the results are summarized in Figure 11. The natural bicubic spline is more accurate only in the region where the function is changing slowly (and where both methods are more accurate than the base points justify), and significantly less accurate in the most difficult interpolation region. Considering the information in Figure 11, as well as simplicity and speed,

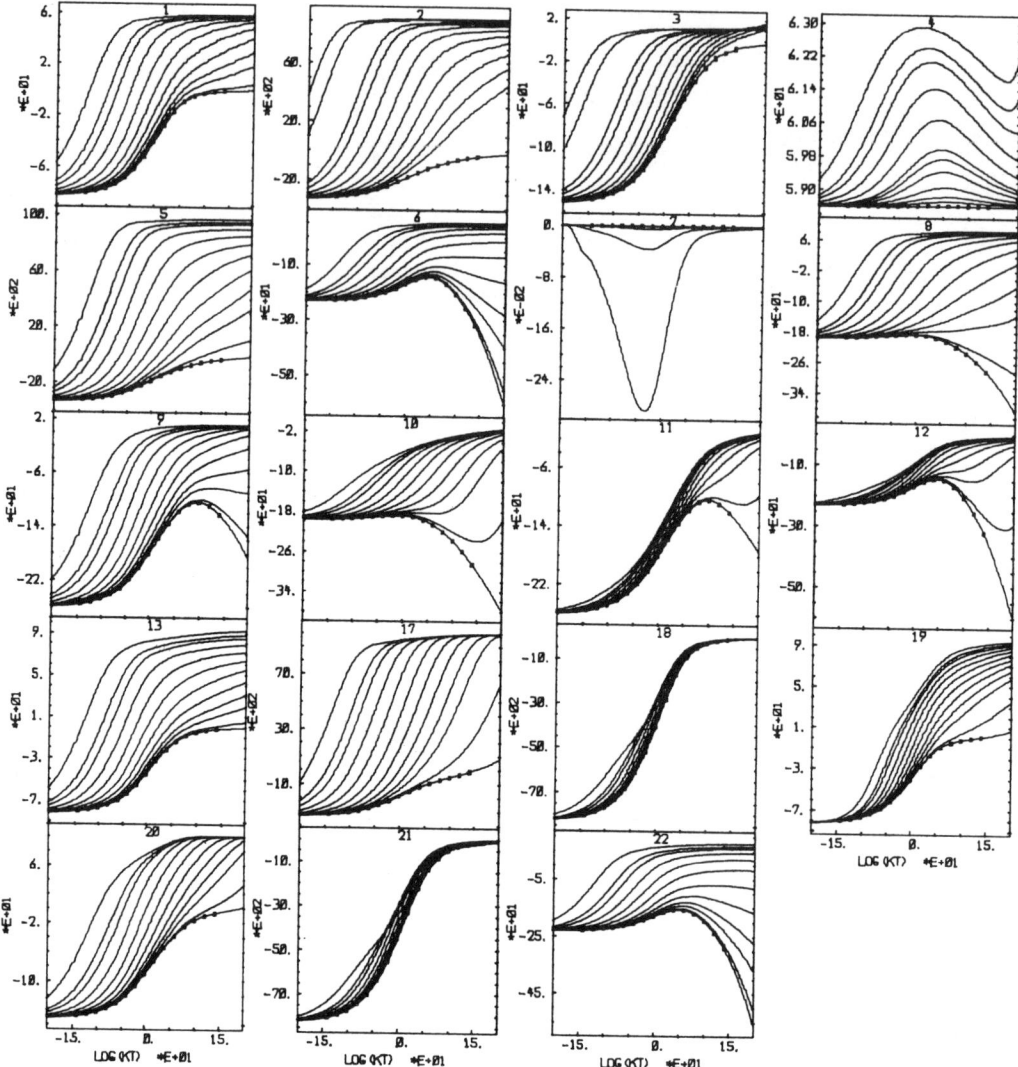

Figure 7. Absorbance ratio of C. The individual curves correspond to different values (given in Table III) of the second parameter (defined in Table I). The closed circles mark the first curve of the set

simple cubic interpolation was chosen for the mechanism fitting package.

Examination of the working curves in Figures 4–9 shows that the change in response ratio between successive working curves is roughly a constant at each value of the first parameter. In contrast the second parameter typically doubles for successive working curves. (Although the working curve presentation in Figures 4–9 is convenient in some respects, it can cause confusion when interpolation is being considered. The working curves are contour plots in which the axes are the first independent variable and the function value and the contours correspond to a fixed value of the second independent variable.) This observation implies that a nonlinear transformation exists which would tend to linearize the base points making interpolation more accurate. Although a systematic search for the optimal transformation was not undertaken, an intuitive guess at the appropriate function turned out to be fairly successful in reducing the interpolation error in those regions that needed an improvement the most. This transformation is applied to the second parameter and is defined as

$$Z = \begin{cases} \sqrt{Y} & Y \geq 0 \\ -\sqrt{-Y} & Y < 0 \end{cases} \quad (9)$$

where Y is the second parameter and Z is its transformation. The improvement in interpolation accuracy resulting is shown in Figures 11 and 12. In order to compress Figure 11, the results for $k_b/k_f = 0.015$ with the square root transformation are plotted with the sign reversed. The transformation overcorrects, actually decreasing the accuracy around log $(k[A]_b \tau) = 1$, but improving the accuracy at log $(k[A]_b \tau) = 2$ by a factor of about 2 (Open squares vs. closed diamonds). In the more typical region for $k_b/k_f[A]_b = 0.15$, the improvement is more impressive. The transformed result is close to the almost negligible error for the interpolation at $k_b/k_f[A]_b = 0.4$. This procedure worked well with all mechanisms except

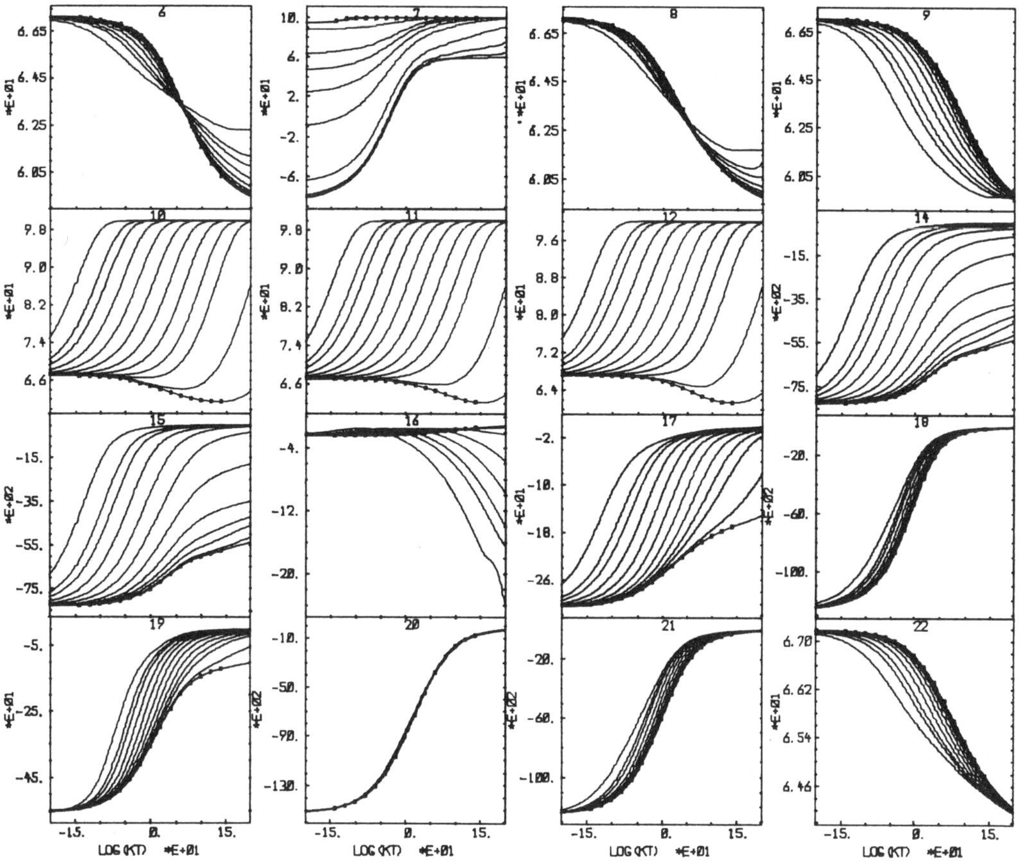

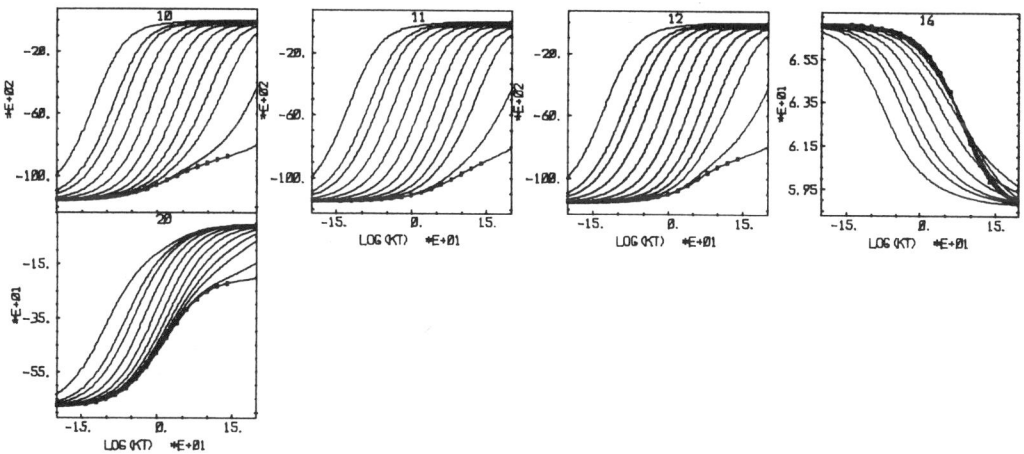

Figure 8. Absorbance ratio of D. The individual curves correspond to different values (given in Table III) of the second parameter (defined in Table I). The closed circles mark the first curve of the set

Figure 9. Absorbance ratio of E. The individual curves correspond to different values (given in Table III) of the second parameter (defined in Table I). The closed circles mark the first curve of the set

the ECE nuance. When interpolation with respect to $\sqrt{K_{eq}}$ is attempted for large K_{eq}, obviously erroneous excursions of the polynomial occur between the final two base points. This problem was handled by working with $-pK_{eq}$; in this case the log transformation is effective in linearizing the base points and otherwise overwhelming interpolation errors are reduced to within acceptable limits.

Local Minima. The surface (which is the average square deviation between the working curve and the experimental points as a function of two kinetic parameters) over which the

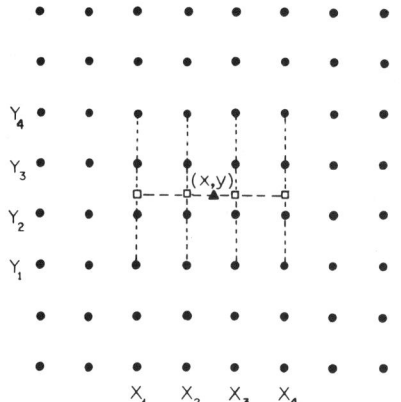

Figure 10. Successive cubic interpolation of the function value at (x,y) from the base points surrounding the point. The base points are not necessarily equally spaced. It does not make any difference if the first interpolation is along y_i lines. ●, base points; ▲, point to be interpolated; □, results for interpolation along the x_i lines

simplex moves is determined by both the working curves and the experimental data, complicating characterization. Experimentally it was found that the typical set of working curves along with a typical set of experimental results often showed at least two minima. In these cases the outcome of a simplex fit is dependent on the position of the starting simplex, and an automated system must guard against converging on other than the absolute minimum. Moreover, a poor initial guess significantly increases the number of iterations required for convergence and even with a relatively fast computer the convergence time is not negligible. Another problem is that the kinetic parameters can vary over a few orders of magnitude, making it essential that the program examine the range of switching times in order to estimate a reasonable starting position. After considerable trial and error development using CRT plotting to monitor the movement of the simplex, it became clear that considerable computer time is needed in order to make the finding of the absolute minimum acceptably probable. The problem turned out to be sufficiently complex that the solution had to be made flexible so that the user could decide how much time to spend looking for the absolute minimum. The following scheme seems to be a satisfactory solution of the fitting problem.

The first step is to survey the surface by sampling the response at a user specified density. A grid is set up as shown in Figure 13 and the degree of fit at each grid point (dots in Figure 13, equally spaced N_x by N_y matrix) using the grid point parameters obtained. The initial guesses are made consistent with the length of the switching times by defining a log average time, τ_{av}

$$\tau_{av} = 10^{\frac{1}{N}\sum_i^N \log(\tau_i)}$$

where N is the number of data points and τ_i is the switching time. The N_x first parameter values do not include the extremes of the data set range because from the definition of τ_{av} the experimental data fall on both sides of the selected points and extrapolation of the data set is not permitted because it is unreliable. Reliability is significantly improved when the extremes of the second parameter are included in the sampling. The important features of the grid are that it is scaled to the input and that it spans the range of the data set. The finer the mesh, the more probable it is that one of the grid points will lie close to the absolute minimum, but the more computer time is involved in the survey. It turns out that in many cases the absolute minimum lies at the bottom of a fairly narrow well and even fairly fine meshes (e.g., 10 × 10) often are not sufficiently close to the absolute minimum for the fit at the grid point nearest the minimum to show the smallest deviation. When the simplex is simply started from the best point of the survey, it is sometimes nearer a local minimum and the absolute minimum escapes detection. It was found not to be practical (based on computation time) to simply use a sufficiently fine mesh to bring the probability of finding the absolute minimum to an acceptable level. Starting the simplex from a user specified number of the better grid points is more efficient. Figure 13 shows a 4 by 7 mesh and the three possible kinds of starting simplex. The size of the starting simplex is consistent with how close to the minimum it is likely to be, which is dependent upon the grid spacing. In many cases starting from the better four points leads to the same place but cases show up where the first three starts lead to local minima while the fourth best grid point of a 3 by 5 mesh leads to the optimum fit. The 3 by 5 mesh with four starts appears to be reliable without taking excessive computer time; typically 7.5 min is required to fit to a 22 member data set, around 20 s per mechanism.

Just as interpolation accuracy can be improved by suitable transformation, so can the performance of the simplex be optimized by transforming the surface into a form suitable for simplex stepping. Fortunately the appropriate transforming function has properties similar to the one used to improve interpolation and a single transformation suffices. The optimum simplex surface is one in which the step size is reasonable at all points on the surface. Without the square root transformation, a simplex of reasonable size at one end of the k_2/k_1 axis (e.g., a $\Delta(k_2/k_1)$ of 0.01 at the low end spans two working curves) is completely inappropriate at the other end $((k_2/k_1)_1 = 11.0, (k_2/k_1)_2 = 11.01$ and the distance between the working curves in that region is 15.0). It is better to vary the transformed second parameter. With the first parameter, the first thought was to let the simplex vary the rate constant to optimize the fit but, for the reasons given above, better results are obtained if the simplex chooses trial values of log $(k\tau_{av})$.

The simplex continues to move according to the rules of the modified simplex (77) until the vertices have coalesced to within a specified tolerance to a point.

The simplex algorithm is particularly adaptive to constraints. Because extrapolation beyond the range of the data set is unreliable, it is not permitted. If the simplex demands the response at a point that would require extrapolation with respect to the second kinetic parameter, the interpolation–standard deviation subroutines simply return a very poor degree of fit which always results in the next simplex move being a contraction through the centroid and therefore a move out of the forbidden area. The extrapolation question with the first kinetic parameter is complicated by the fact that a given rate constant often requires that only some points be extrapolated while those at the other end of the switching time range are within the data set range. In this case, if the fraction of points which would require extrapolation exceeds a limit set by the user, a poor fit is assigned; otherwise the standard deviation is calculated using only those points within the data set range.

Tests of the Simplex–Interpolation Fitting. A number of tests were used to evaluate the accuracy of the simplex–interpolation fitting scheme. The first test demonstrated the ability of the simplex to find the minimum in the absence of interpolation error. This was done by inputting the central 13 points of one working curve from each mechanism as experimental data and then fitting to that mechanism and all others in the data set. In general the mechanism from

130 • M. K. Hanafey et al.

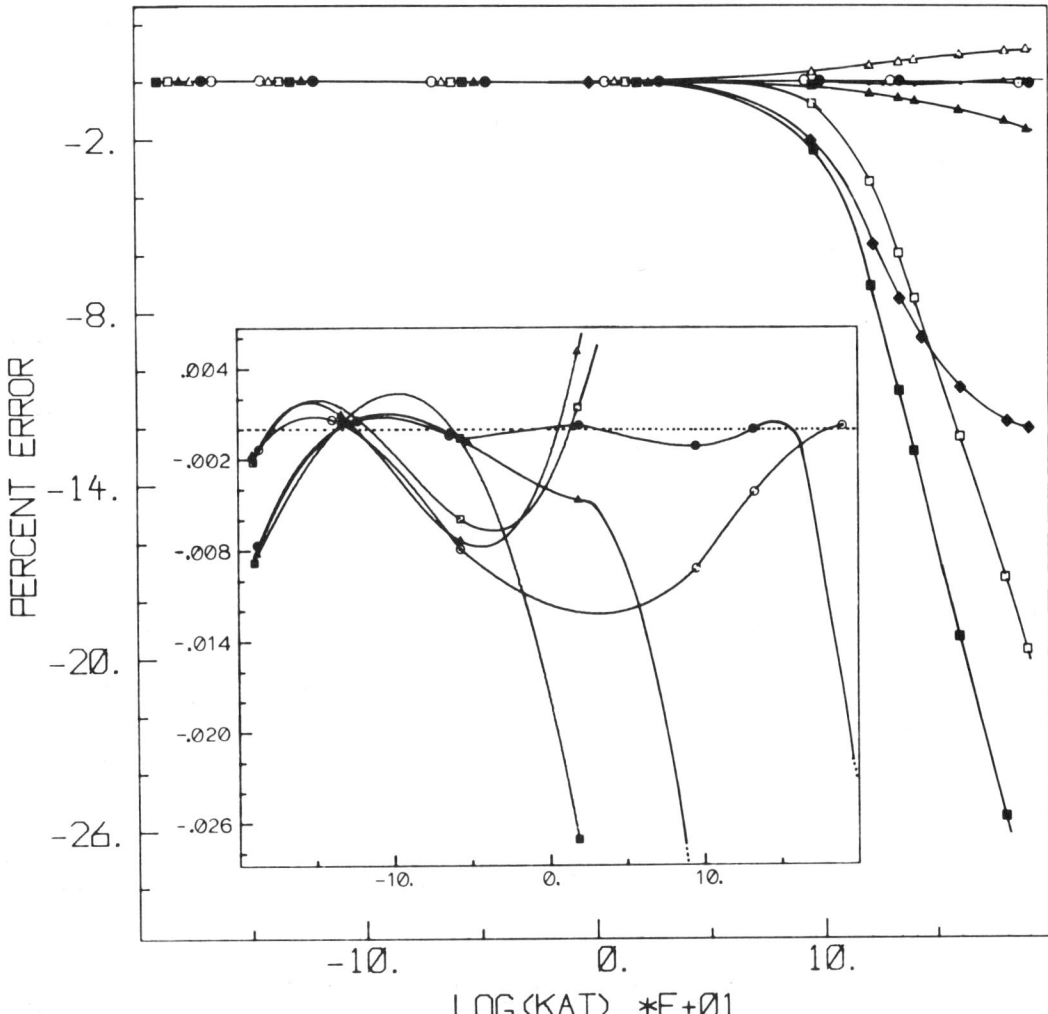

Figure 11. Interpolation error in charge ratios as a function of log $(k[A]_b\tau)$ for the EC radical–radical dimer mechanism for three values of $k_b/k_f[A]_b$ and a number of interpolation methods. ■, cubic spline, $k_b/k_f[A]_b = 0.015$; □, simple cubic, $k_b/k_f[A]_b = 0.015$; ♦, simple cubic with square root transform of $k_b/k_f[A]_b$, $k_b/k_f[A]_b = 0.015$ (the negative of the actual result is plotted, the transformation overcompensates) ▲, cubic spline, $k_b/k_f[A]_b = 0.15$; △, simple cubic, $k_b/k_f[A]_b = 0.15$; ·, simple cubic with square root transformation of $k_b/k_f[A]_b$, $k_b/k_f[A]_b = 0.15$ (this line is difficult to see because it is close to the ○ and ● line about which it oscillates; ●, cubic spline, $k_b/k_f[A]_b = 0.4$ (no interpolation of $k_b/k_f[A]_b$); ○, simple cubic, $k_b/k_f[A]_b = 0.4$ (no interpolation of $k_b/k_f[A]_b$). The data were simulated from -1.9 to $+1.9$ log $(k[A]\tau)$ for a total of 21 points, these points thus lie at random places between the working curve log $(k[A]_b\tau)$ values. The insert is same information but with a reduced dynamic range to reveal the nature of the small errors in regions more suited to interpolation. The square root transformed results differ little from the untransformed results and are not shown

which the data came gave the best fit by more than three orders of magnitude and the kinetic parameters were in error by typically less than a few parts in ten thousand.

In order to evaluate the error in kinetic parameters as a result of interpolation error but nevertheless an idealized test in that the data were noise free, response ratios were simulated at points not coincident with those in the data set. The EC radical–radical dimer case was chosen because its working curves seem to span the range of behavior in the data set. The results are in Table V. It is clear that considerable error in kinetic parameters results from errors in interpolation, and for this mechanism, most of the error results from the spacing of the second parameter $(k_b/k_f[A]_b)$. Figure 12 shows more clearly the nature of the interpolation error. The reason for the large errors for interpolation in the interval $0. < k_b/k_f[A]_b < 0.05$ is clearly that the response ratio varies rapidly in this interval. The implication is that the double step methods (except for absorbance of C) are very sensitive to whether the radical coupling is irreversible or only nearly irreversible, although the usefulness of this distinction is limited.

The final test of the fitting scheme demonstrates the effect of random noise on the outcome of the analysis. The interior 13 points of one charge working curve from each mechanism were randomly altered by relative noise levels of 1, 2, and 5%. The resulting set of points can be considered an idealized set of experimental data suitable for providing two kinds of information—how much error can be expected in kinetic parameters as a consequence of random noise (interpolation effects and the accuracy of the base points are eliminated in the idealized experimental data), and secondly an incomplete

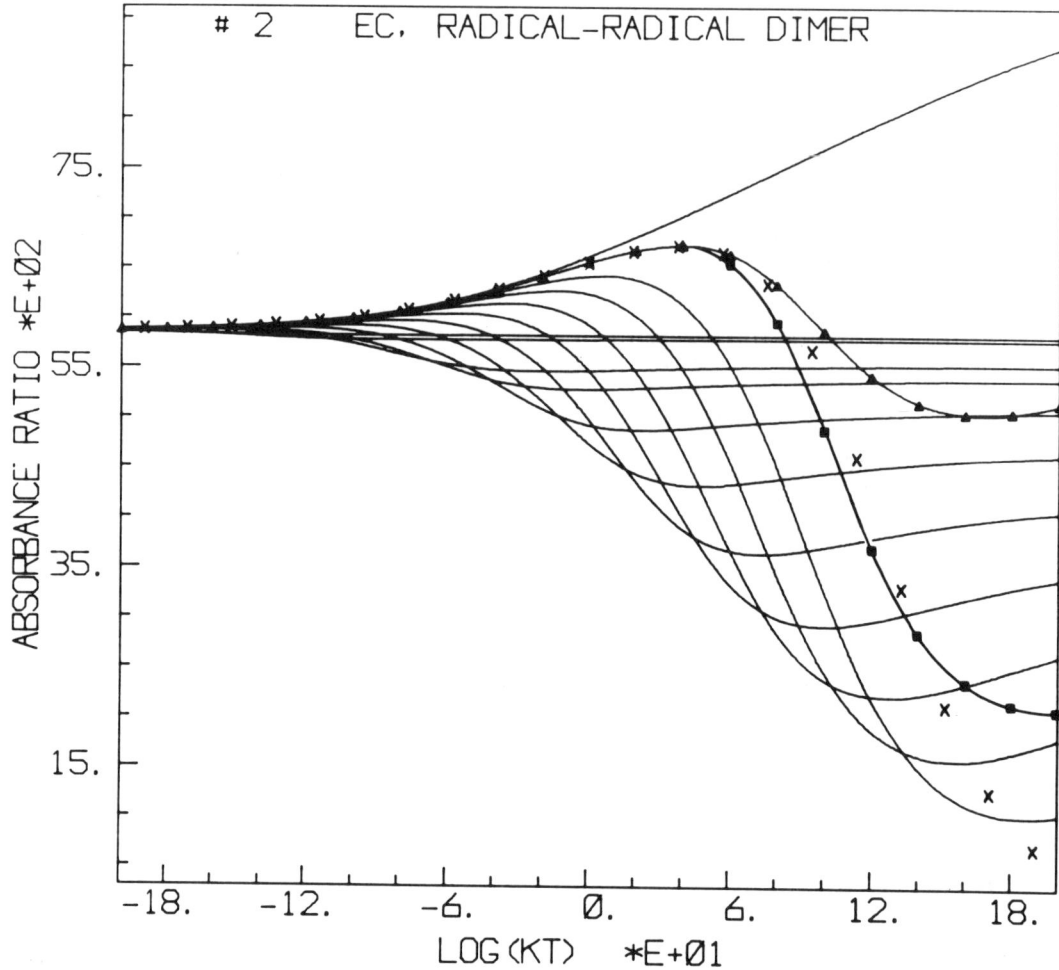

Figure 12. Fitting results with and without transformations. The user of the package optionally can plot the results of a fit. In this case data (plotted as X) were simulated in the region between the upper (in the region of log $(k[A]_b\tau) = 0$.) two curves of the EC radical–radical dimer mechanism (the working curves for #2 are the solid lines without plotting symbols). The simulated data was fitted to that mechanism without the square root transformation (▲) and with square root transformation of $k_b/k_f[A]_b$ (■). In the absence of interpolation error the fitted (interpolated) curve would pass through the X's. The advantage of the transformation is obvious

answer to the question of how distinguishable the various mechanisms are by double step information. The information is obtained by fitting each set of noisy data to all mechanisms in the data set (actually a subset of the mechanisms in Table I because of additions after this study) which amounts to over a thousand simplex optimizations. The information regarding mechanism differentiability is incomplete for a number of reasons, and the results must be viewed with these limitations in mind. The analysis is restricted to charge ratios. Only a single working curve from each mechanism is represented, and obviously a complete study would evaluate the differentiability as a function of the kinetic parameters as well as noise level. The results are derived from one distribution of noise at each level, and since only 13 points are involved, a more rigorous study would involve the average result for a number of noise distributions. The effect of different noise distributions was investigated with one mechanism and it appears that averaging the results of many noise distributions would not significantly affect the results which are summarized in Table VI. The letters in Table VI correspond to mechanisms as defined in Table I. Following the letter for the mechanism from which the data fitted came (first column) is the percent error in the kinetic parameters in parentheses. Under the column labeled "best" is the mechanism that fitted the data with the smallest mean square deviation; under the column labeled "within ×2" are all mechanisms that fitted the data with a mean square deviation no larger than twice the deviation of the best fit, and so on.

Figure 14 indicates the different information content of the various response ratios. The charge, current, and absorbance ratios from the ECE nuance mechanism (#7) were perturbed with the indicated noise level and then fitted to the entire data set. The abscissa in Figure 14 is the logarithm of the ratio of the average square deviation of each mechanism to the average square deviation of the best fit. Mechanisms are indicated by data set number as defined in Table I. This presentation spreads out the mechanisms which fit reasonably well while compressing those that fit poorly. The results for charge and current ratio data at ~1% noise (A and B) and ~2% noise (J and K) show some interesting features. In this

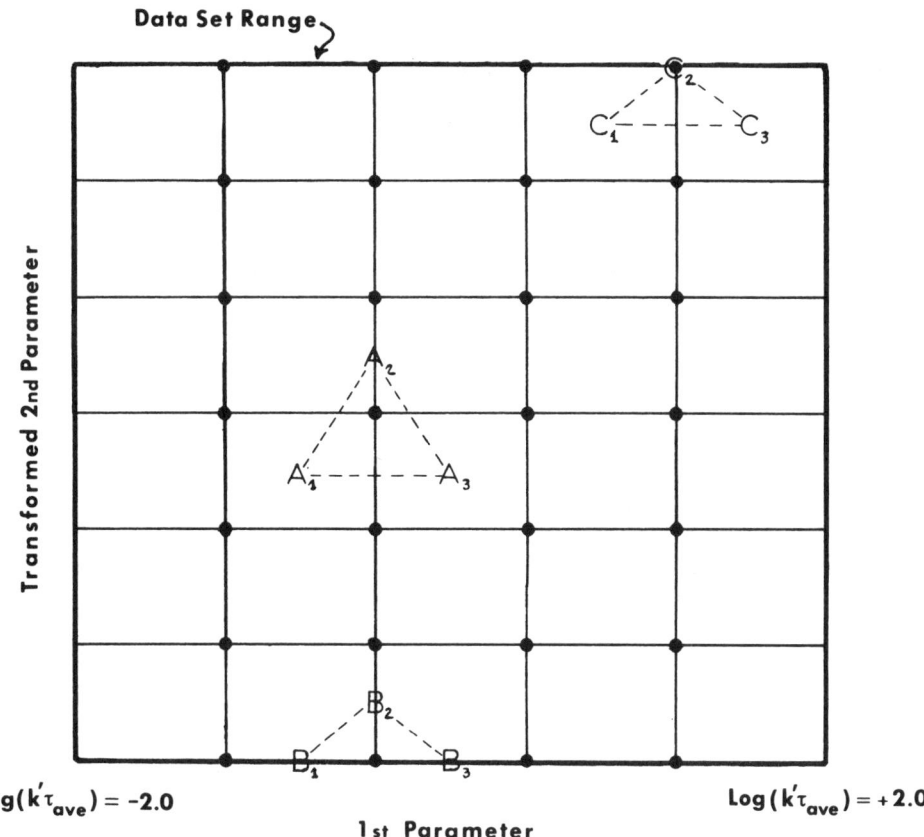

Figure 13. Example of a 4 by 7 survey mesh with the simplex starting from three points chosen to indicate the three possible starting sizes. The border defines the region covered in the data set. The mesh is arranged to span this region, the points being equally spaced in the transformed second parameter. The response is obtained at each dot and the deviations are ordered. The starting simplex spans an area roughly equal to the discretization area

case, charge and current information are in a sense complementary. Mechanisms 12 and 20 are significantly more distinct from the behavior of 7 when charge rather than current ratios are employed. The reverse is true for mechanisms 2 and 5. Interestingly, mechanism 14 fits the noisy data better than the mechanism from which the data were synthesized for both charge and current. Since 14 is a second-order case in which two components are present in the bulk, it can be distinguished by varying the bulk concentration ratios. The comparison of the absorbance data is complicated by the fact that the absorbance of any species in the ECE nuance scheme must be tested for confusion with all other species of all the remaining mechanisms. It is not feasible to test cases where more than one species contributes to the absorbance because of the vast number of cases that would have to be considered to make the study meaningfully complete. The absorbance ratio of B vs. B (C and L) does a reasonable job of discriminating mechansim 7; 14, 15, and 16 are all cases where the second parameter is a concentration ratio and thus only appear like 7 if the concentrations are made to have a specific ratio. Mechanism 20 is close enough at the 2% noise level to cause concern. Rows D, E and M, N show how well the absorbance of species B in the ECE nuance mechanism can be accounted for assuming species C and D of any other mechanism were the absorbing species. In Figure 14, cases where all mechanisms are a 1000 times worse than 7 are not shown (for example B vs. E).

An Example of the Use of the Simplex-Interpolation Fitting Scheme. In order to illustrate the steps involved in an analysis of experimental data using the simplex-interpolator, the program was applied to some double potential step charge ratios reported by Van Duyne (6) for the relatively well characterized first-order EC benzidine rearrangement of azobenzene reduced to hydrazobenzene in strong acid. The following twelve directives were entered into the computer; their meanings and the results obtained are described below.

1. DATA <READ = 'CR> LIST PAGE
2. DATA FROM REF. 6, P 212 COLUMN B. REPORTED RESULT IS EC IRREVERSIBLE, K, = 4.27 ± 0.5
3. TAU. .0098, .01, .01, .025, .05, .1, .15, .25, .275, .275, .3, .3, .35, .4, .4
4. RATIO .559, .551, .56, .531, .461, .427, .366, .288, .234, .276, .251, .227, .204, .177, .15
5. CHARGE <MESHY = 6, STARTS = 3>
6. DISPLAY LIST PAGE
7. CHARGE 1, 0.
8. PLOT FIT <MISS = 13> (Figure 15)
9. CHARGE <ABSRES = .00001, RELRES = .01> EXCEPT 4, 14, 15, 16, 7
10. DISPLAY BAR <CUTOFF = 0.5>
11. CHARGE 16 START <K1 = 1., K2 = 1.5, K3 = 1.7, R1 = 4., R2 = 4.3, R3 = 4.>

211

Table V. Percent Errors in the Kinetic Parameters of the EC Radical–Radical Dimer Mechanism (C) as a Consequence of Interpolation Error[a]

Actual		Charge				Current				Absorbance of B				Absorbance of C			
		k_f		k_b/k_f		k_f		k_b/k_f		k_f		k_b/k_f		k_f		k_b/k_f	
k_f	k_b/k_f	a	b	a	b	a	b	a	b	a	b	a	b	a	b	a	b
1.0	0.015	1.75	−1.48	24.3	−20.7	−4.19	−4.00	44.3	−26.5	99.5	−43.7	62.6	136.0	−11.5	4.49	39.3	−19.9
1.0	0.15	−0.644	0.0152	−1.65	0.143	−1.63	−0.0877	0.520	0.814	−0.728	−0.0221	4.22	1.60	2.75	−0.439	−3.47	0.168
1.0	0.40	−0.0156	−0.0091	−0.0398	0.0127	−0.032	0.00105	0.0391	0.0264	−0.00341	0.0221	−0.0379	−0.0034	0.0647	0.00296	−0.0389	0.00423

[a] This table should be viewed in conjunction with Figures 4–7 and Table III. Column a is interpolation with respect to k_b/k_f; column b is interpolation with respect to $\sqrt{k_b/k_f}$.

Table VI. Summary of the Ability of Charge Response Ratios to Differentiate among Mechanisms and Reveal Kinetic Parameters in the Presence of Three Noise Levels

Mech	Best (% error in parameters)	1% Noise				Best (% error in parameters)	2% Noise				Best (% error in parameters)	5% Noise		
		Within ×2	Within ×4	Within ×10	Within ×100		Within ×2	Within ×4	Within ×10	Within ×100		Within ×2	Within ×4	Within ×10
B	B (0.7, −0.07)		G		MKOLRD	B (−0.13, 2.2)	G			QPOH	B (−5.3, −4.4)	PRKOMLD	BH	OPQH
C[a]	C (0.41)			P	BCNHTX	C (−0.86)	P	MKOLR	D	BHNTXS	C (0.41)	OD (1.7)	NHCTXS	NTXH
D[a]	D (−0.016)	P	KMLO	R		D (−1.3)	PKOML	RB	NCHT	XSHJGQ	P	KMLBR		
E	E (−7.5, −7.9)				BG	E (−37, 163)			B	GJLDHMNIK..	E (150, −90.)	B	GJLD	HMNIKXTR
G	G (1.0, −0.25)	B	H (1.2, 7.2)		MKB	G (−0.096, 4.2)	B			OPQ	G (−6.7, −7.7)	H (9.0, 12.) P	BKM	OPQH
H	O	H	P			O	H (2.7, 10.) P		B	KMCILRDJ..	O			CLRID
I[a]	I (0.41)	H	PO		JBQ	I (−1.7)	PHO	J		BQGMKLC	H	POI (0.41)	KM	BJHQMGK
J[a]	J (0.41)	OPQ		G	IHBKM	J (−2.5)	OQPGH	IB		KMHLCRDNT.. PIB	H	OGQJ (2.1)		LCR
K	K (0.97, −5.0)	L		MR	PCDOB	K (−0.80, −2.0)	RM	PCD	O	LBNHTXS	K (−3.2, 22.)	LRPMDOC	B	NHTXS
L	L (0.46, −1.8)		M	DPKO	RNBCT	L (−2.1, 3.5)	MDPO	KRNB	CTX	SHJGQ	P	MLODKNBRT (−10., 0.006)	XCHS	
M	M (0.55, −1.0)			LD	PKRO	M (−1.0, 0.20)	L	DPK	ORNC	BTXSH	M (−1.6, 6.3)	LDPOKR	NCBTX	S
N[a]	N (0.41)	P	LM	OT	XBSDKR	N (−0.86)	PLMO	TXB	SD	KRC	P	N (−0.016)	MLTOXB	SDKR
O	O (0.40, −0.24)		P	C	RKLMDB	O (−0.78, −0.46)	P	KCR	LMDB	HNTHXS	O	PKRC	LDBMH	NTX
P	P (0.66, −2.1)		DO	K	LMRNBCT	P (−0.64, −1.3)	DO	KLM	RNB	CTXSH	P (−2.1, 12.)	ODLKMN	RBT	XCS
Q	Q (0.77, −0.55)	PO	H		GJB	Q (−0.98, −4.2)	POH		G	J	Q (−5.4, 11.)	POHG	JB	I
R	R (0.25, 50.)		PK	OM	CDLBN	R (−0.98, −11.)		PKO	MDC	LBNTHXS	P	KR (0.14)	LB	NTHXS
S	S (0.86, −6.3)	M	B		TPXNL	S (−1.5, 14.)	M		BTP	XNLODKR	S (−4.8, 61.)	ODMC MBTPX	N	LO
T	T (0.49, 19.)	P	XBM	S	NLO	T (−0.47, 12.)	N		LO	DKRC	T (−1.1, 50.)	PBMXSN	LO	DK
X	X (0.41)	PTM	B	SNL	OD	X (−0.86)	SN		LO	DKRC	M	PX (−0.44) TBSN	LO	DK
	Hits = 18 Misses = 1					Hits = 17 Misses = 2					Hits = 11 Misses = 8			

[a] These mechanisms were treated as being irreversible ($k_b/k_f = 0$).

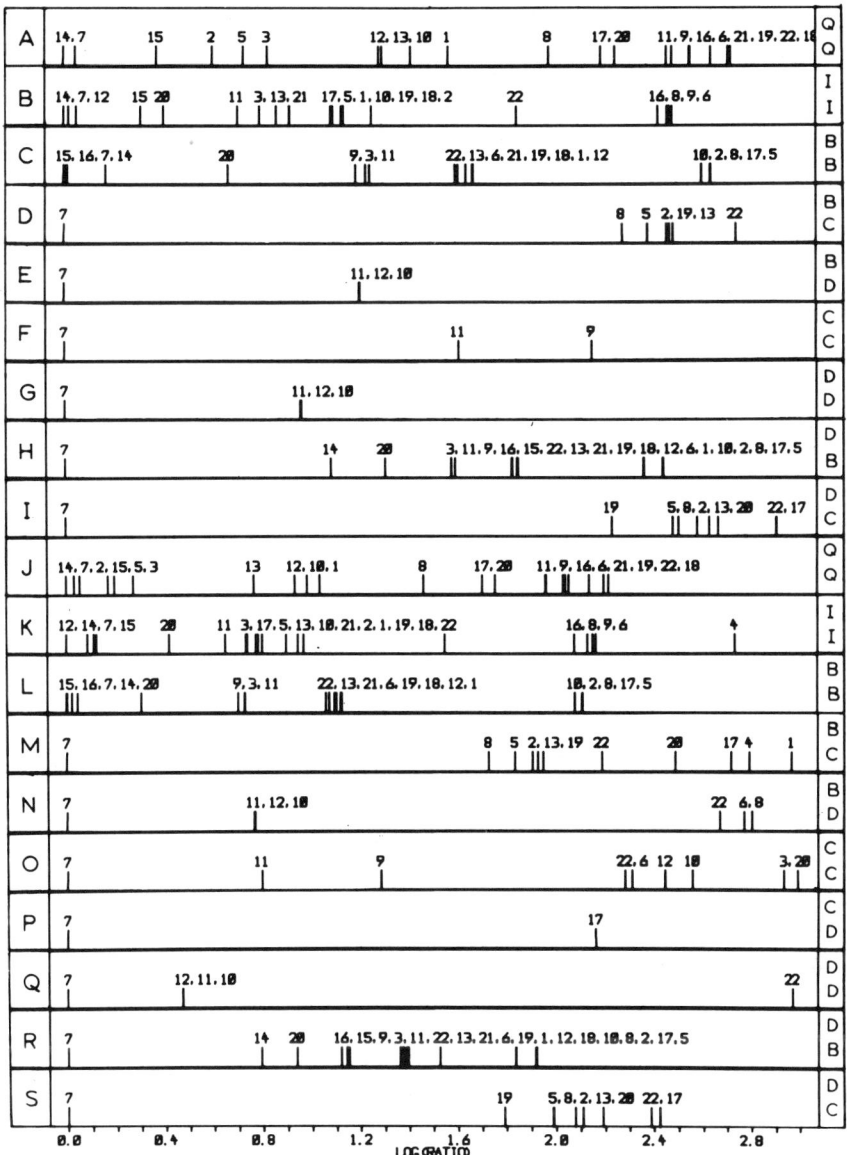

Figure 14. Influence of noise level and type of monitored response on mechanism differentiability. The 13 interior points from mechanism #7 were perturbed with the indicated noise level and then fitted to all mechanisms in the data set using a 3 by 6 survey mesh and starting the simplex from the better four points. The absorbance of each species in #7 were fitted to absorbance responses of all species of all other mechanisms. Absorbance B vs. absorbance C should be interpreted as absorbance B vs. absorbance C of mechanisms other than #7 and absorbance of B of #7. These results for a mechanism are not shown if the fit with #7 was three orders of magnitude better. The letters on the left are row markers used below; the letters on the right define in shorthand the information given below ($^Q_Q \equiv$ charge vs. charge, etc.). The numbers above the bars are data set numbers as defined in Table I. The ratio plotted on the abscissa is defined as (mean square deviation)/(minimum mean square deviation). Percent errors in k and K_{eq} for #7, and the mean square deviation of the best fit are given in parentheses: A. Charges vs. charge, 1% noise (1.2, 7.3, 7.5 × 10^{-6}). B. Current vs. current, 1% noise (0.076, 9.1, 8.8 × 10^{-6}). C. Absorbance B vs. absorbance B, 1% noise (0.67, 3.3, 7.3 × 10^{-6}). D. Absorbance B vs. absorbance C, 1% noise (0.67, 3.3, 7.3 × 10^{-6}). E. Absorbance B vs. absorbance D, 1% noise (0.67, 3.3, 7.3 × 10^{-6}). F. Absorbance C vs. absorbance C, 1% noise (0.19, 0.0020, 5.5 × 10^{-6}). G. Absorbance D vs. absorbance D, 1% noise (0.59, 0.18, 8.3 × 10^{-6}). H. Absorbance D vs. absorbance B, 1% noise (0.59, 0.18, 8.3 × 10^{-6}). I. Absorbance D vs. absorbance C, 1% noise (0.59, 0.18, 8.3 × 10^{-6}). J. Charge vs. charge, 2% noise (2.7, 10., 2.5 × 10^{-5}). K. Current vs. current, 2% noise (4.7, 37., 1.9 × 10^{-5}). L. Absorbance B vs. absorbance B, 2% noise (-2.1, 1.5, 2.6 × 10^{-5}). M. Absorbance B vs. absorbance C, 2% noise (-2.1, 1.5, 2.6 × 10^{-5}). N. Absorbance B vs. absorbance D, 2% noise (-2.1, 1.5, 2.6 × 10^{-5}). O. Absorbance C vs. absorbance C, 2% noise (0.82, 0.0051, 3.8 × 10^{-5}). P. Absorbance C vs. absorbance D, 2% noise (0.82, 0.0051, 3.8 × 10^{-5}). Q. Absorbance D vs. absorbance D, 2% noise (-3.5, -0.061, 2.7 × 10^{-5}). R. Absorbance D vs. absorbance B, 2% noise (-3.5, -0.061, 2.7 × 10^{-5}). S. Absorbance D vs. absorbance C, 2% noise (-3.5, -0.061, 2.7 × 10^{-5})

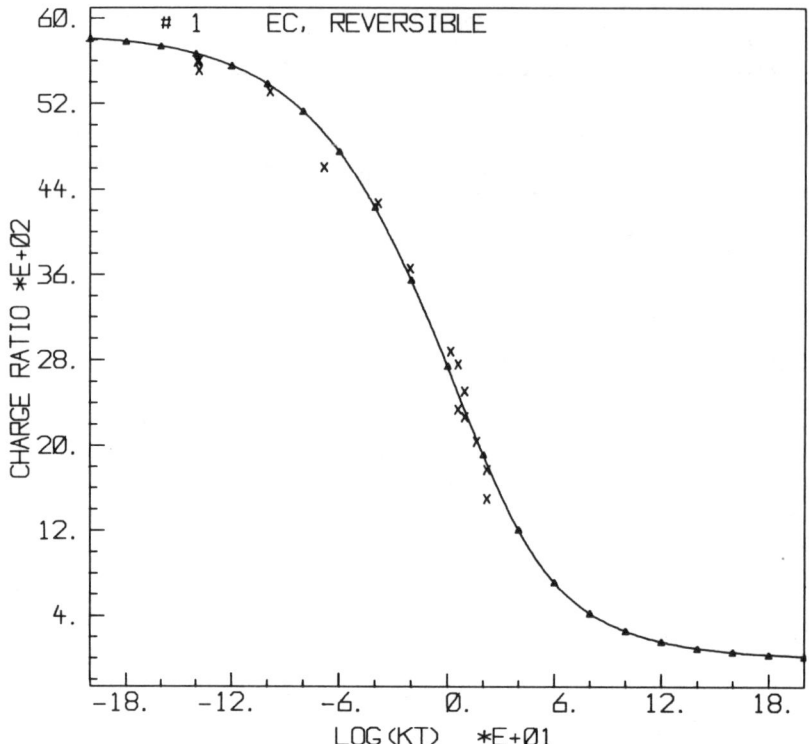

Figure 15. Graphic display of the degree of fit. ▲, interpolated best fitting working curve from the EC irreversible mechanism (a constrained optimization from the EC reversible set). ×, experimental charge ratios

12. CHARGE 16 START <R1 = 3., R2 = 3.3, R3 = 3.>

1. The experimental data is input from the card reader and echo checked on a new page.

2, 3, 4. The input is identifying information, switching time, and response ratio (free format).

5. The experimental data just read in is fitted to the charge ratio working curves of every member of the mechanism set, changing the Y mesh and number of starts parameters as indicated. All other parameters remain set at their defaults. MESHY and STARTS will remain as set until reentered or reset.

6. An ordered (increasing average square deviation) list of the results of the fitting is produced; the first four rows of the table are shown below

MECH # 18.	HETEROGENEOUS DISPROPORTIONATION
KF1 = 4.15634	$K_2[A]/K_1$ = 0.100643E-02, ERROR = 0.288374E-03
MECH # 1.	EC, REVERSIBLE
KF1 = 4.16191	KB/KF = 0.775327E-05, ERROR = 0.289584E-03
MECH # 12.	ECEC, IRREVERSIBLE
KF1 = 4.43815	K_2/K_1 = 1.38711, ERROR = 0.290225E-03
MECH # 15.	EC, SECOND ORDER
KF1 = 1.02275	[C]/[A] = 4.34328, ERROR = 0.302743E-03

The best fit is the heterogeneous disporportionation mechanism with a small value for $k_2[A]/k_1$, which reduces in the limit as $k_2[A]/k_1$ goes to zero to the EC irreversible case. The next best fit is the EC reversible mechanism with a small value of k_b/k_f, consistent with the chemical reaction being considered irreversible. The rate constants in both cases (4.16 s^{-1}) is close to that obtained by Van Duyne (4.27 s^{-1}). The EC, second order result is also reasonable. Evidently the concentration ratio is large enough that the response is nearly pseudo first order, as inferred from Figure 4 (mechanism 15) where it is seen that as the concentration ratio parameter is increased beyond about 1.6, the only effect is a horizontal shift of the working curve.

The double step information does not eliminate the ECEC irreversible mechanism from consideration. When the more deviant points (the three points which look low in Figure 15) are eliminated, differentiability does not improve although kinetic parameters change significantly. The first four mechanisms become

MECH # 18	HETEROGENEOUS DISPROPORTIONATION
KF1 = 3.78784	$K_2[A]/K_1$ = 0.385866, ERROR = 0.143676E-03
MECH # 12	ECEC, IRREVERSIBLE
KF1 = 4.01528	K_2/K_1 = 1.91578, ERROR = 0.144809E-03
MECH # 1	EC, REVERSIBLE
KF1 = 4.03536	KF/KF = 0.80477E-05, ERROR = 0.148196E-03
MECH # 21	ECC, 1ST AND 2ND ORDER REGENERATIVE
KF1 = 3.96184	$K_2[A]/K_1$ = 1.07697, ERROR = 0.152502E-03

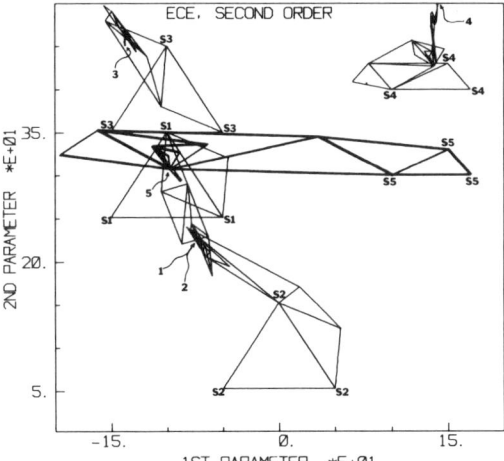

Figure 16. Graphic monitoring of the simplex movement. The simplex designations, Sn (where n is an integer), mark the starting vertices of the respective simplex. The numbered arrows indicate the point to which the correspondingly numbered simplex converged. Simplexes S1, S2, and S3 start from an area centered over the best three points of a 3 by 6 survey mesh. Simplexes S4 and S5 were started from a user specified place. In parentheses following the simplex designation are given k_f, k_b/k_f, and the average square deviation of the fit: S1 (1.90, 4.77, 0.003072), S2 (1.89, 4.78, 0.003072), S3 (0.400, 20.7, 0.003634), S4 (249., 24.6, 0.02597), S5 (1.02, 9.45, 0.003097)

These results emphasize the need for high quality experimental ratios if kinetic information is extracted.

7. The data is fitted to mechanism 1 constrained to be irreversible (the "0." is the fixed value of the second parameter).

8. The experimental points, the interpolated best fitting theoretical curve, and the working curves are plotted. The MISS parameter controls the number of working curves which are plotted; in this case MISS is greater than the number of working curves so none are plotted. Figure 15 shows that the scatter in the data is appreciable and that data at longer switching times would cover the information region better.

9. Another form of the CHARGE directive (there are corresponding CURRENT and ABSORBANCE directives). All members of the mechanism set not excepted are fitted to. Here the prekinetics and the mechanisms which have a concentration ratio as a parameter are eliminated from consideration. The resolution parameters are changed from their defaults.

10. A bar plot is produced where the log (RATIO) axis extends from 0. to 0.5 (similar to Figure 14).

11. The movement of the simplex optionally can be monitored. S1, S2, and S3 in Figure 16 are the three better starting positions of a 3 by 6 survey grid covering mechanism 16. The first two starts converge on what presumably is the absolute minimum, but the third start converges on a local minimum. In Figure 16 the numbered arrows indicate the convergence point of the corresponding starting simplex. This directive specifies where the simplex should be started (the survey feature overridden); simplex S4 also converges on a local minimum.

12. The behavior of simplex S5 shows the complex nature of the surface. Although it moved into the region S1 started from, it was quite small at that point and unlike S1 converged on a local minimum.

Although no single example characterizes the extent of information to be expected from a set of experimental ratios and this program and data set, it is not surprising that more information will in general be needed in a mechanism study. This means employing other techniques as well as a more complete use of double step information. A convenient way to sort out mechanisms containing a second-order kinetic step is to vary the bulk concentration of species A (Table I). Ratios are obtained for a series of switching times and the 1st and 2nd kinetic parameters (Table I) obtained using the program. These parameters depend on the bulk concentration of A if a second-order kinetic step is involved. Mechanisms 15, 18, and 21 differ from mechanisms 1 and 12 in this regard.

Additional information can be obtained from the double step experiments if more than a single ratio from the response curve is utilized. For example, the entire curve can be fitted if a third normalization factor is added to the optimized parameters. This analysis can be restricted to the forward step, or the entire double step response can be used. The degree of fit can be based on a single switching time, or more reliably on the average degree of fit for a number of switching times. To avoid the normalization parameter, the degree of fit could be based on ratios formed at a number of points on the response curve (that is at $n\Delta\tau + \tau/n\Delta\tau$).

Other approaches are to vary temperature or pH in order to change the value of the rate constants. Mechanisms which appear similar for one set of kinetic parameters usually yield distinguishable responses when compared in another region of the kinetic parameters. As mentioned earlier, charge and current ratios in some cases provide complementary information and it seems reasonable to acquire both routinely. The experimental difficulty of obtaining good absorbance data and knowing relative absorptivities in case of overlapping bands reduces the usefulness of this potential source of information, but the possibilities of the approach should not be ignored.

Simplex-Finite Difference Simulator. The mechanism fitting procedure described has two undesirable features. Interpolation error usually affects the result, and the mechanism is restricted to two-parameter flexibility. Both of these deficiencies are eliminated when a simulation routine is coupled to a simplex optimizer so that results are computed on demand from the simplex. As mentioned earlier the only disadvantage with this approach is the amount of computer time required. The best approach is to use a combination of the two. First the experimental data are analyzed using the data set approach and then the results refined if necessary by the coupled simplex-EFDS program. This latter program can optimize more than two parameters and can simulate a more general system, diffusion constants can be unequal, and electron transfer need not be infinitely fast.

Supplementary Material Available: A listing of the charge, current, and absorbance responses (177 pp). Microfiche (105 × 148 mm, 24X reduction, negatives) containing all of the supplementary material for the papers in this issue may be obtained from the Business Operations, Books and Journals Division, American Chemical Society, 1155 16th Street, N.W., Washington, D.C. 20036. Remit check or money order for $2.50 for microfiche, referring to code number ANAL-78-JAN.

LITERATURE CITED

(1) S. W. Feldberg, in "Electroanalytical Chemistry", Vol. 3, A. J. Bard, Ed., Marcel Dekker, New York, N.Y., 1969, pp 199–296.
(2) W. M. Schwarz and I. Shain, J. Phys. Chem., **69**, 30 (1965).
(3) J. H. Christie, J. Electroanal. Chem., **13**, 79 (1967).
(4) T. H. Ridgway, Ph.D. dissertation, University of North Carolina, Chapel Hill, N.C., 1971.
(5) W. V. Childs, J. T. Maloy, C. P. Keszthelyi, and A. J. Bard, J. Electrochem. Soc., **118**, 872 (1971).
(6) R. P. Van Duyne, Ph.D. dissertation, University of North Carolina, Chapel Hill, N.C., 1970.
(7) G. C. Grant and T. Kuwana, J. Electroanal. Chem., **24**, 11 (1970).
(8) L. S. Marcoux, J. Am. Chem. Soc., **93**, 537 (1971).
(9) G. S. Alberts and I. Shain, Anal. Chem., **35**, 1859 (1963).
(10) H. B. Herman and A. J. Bard, J. Phys. Chem., **70**, 396 (1966).
(11) J. H. Harrison and D. W. Shoesmith, J. Electroanal. Chem., **28**, 301 (1970).
(12) R. I. Koopman, Ber. Bunsenges. Phys. Chem., **72**, 32 (1968).

(13) T. H. Ridgway, C. N. Reilley, and R. P. Van Duyne, *J. Electroanal. Chem.*, **67**, 1 (1976).
(14) J. R. Delmastro and G. L. Booman, *Anal. Chem.*, **41**, 1409 (1969).
(15) D. T. Pence, J. R. Delmastro, and G. L. Booman, *Anal. Chem.*, **41**, 737 (1969).
(16) J. R. Delmastro, *Anal. Chem.*, **41**, 747 (1969).
(17) P. Delahay and G. L. Stiehl, *J. Am. Chem. Soc.*, **74**, 3500 (1952).
(18) P. Delahay, "New Instrumental Methods in Electrochemistry", Interscience, New York, N.Y., 1954.
(19) J. Koutecky and J. Koryta, *Collect. Czech. Chem. Commun.*, **19**, 845 (1954).
(20) R. S. Nicholson and I. Shain, *Anal. Chem.*, **36**, 706 (1964).
(21) R. S. Nicholson and I. Shain, *Anal. Chem.*, **37**, 178 (1965).
(22) G. L. Booman and D. T. Pence. *Anal. Chem.*, **37**, 1366 (1965).
(23) R. S. Nicholson, J. M. Wilson, and M. L. Olmstead, *Anal. Chem.*, **38**, 542 (1966).
(24) R. S. Nicholson, *Anal. Chem.*, **38**, 1406 (1966).
(25) D. S. Polcyn and I. Shain, *Anal. Chem.*, **38**, 376 (1966).
(26) R. H. Wopschall and I. Shain, *Anal. Chem.*, **39**, 1514 (1967).
(27) M. D. Hawley and S. W. Feldberg, *J. Phys. Chem.*, **70**, 3459 (1966).
(28) R. H. Wopschall and I. Shain, *Anal. Chem.*, **39**, 1527 (1967).
(29) R. H. Wopschall and I. Shain, *Anal. Chem.*, **39**, 1535 (1967).
(30) J. M. Saveant, *Electrochim. Acta*, **12**, 999 (1967).
(31) J. M. Saveant and E. Vianello, *Electrochim. Acta*, **12**, 1545 (1967).
(32) R. N. Adams, M. D. Hawley, and S. W. Feldberg, *J. Phys. Chem.*, **71**, 851 (1967).
(33) J. H. Christie, R. A. Osteryoung, and F. C. Anson, *J. Electroanal. Chem.*, **13**, 236 (1967).
(34) M. L. Olmstead and R. S. Nicholson, *J. Electroanal. Chem.*, **14**, 133 (1967).
(35) M. L. Olmstead and R. S. Nicholson, *J. Electroanal. Chem.*, **16**, 145 (1968).
(36) J. W. Strojek, T. Kuwana, and S. W. Feldberg, *J. Am. Chem. Soc.*, **90**, 1353 (1968).
(37) M. Mastragostino, L. Nadjo, and J. M. Saveant, *Electrochim. Acta*, **13**, 721 (1968).
(38) M. Mastrogostino and J. M. Saveant, *Electrochim. Acta*, **13**, 751 (1968).
(39) M. L. Olmstead, R. G. Hamilton, and R. S. Nicholson, *Anal. Chem.*, **41**, 260 (1969).
(40) M. L. Olmstead and R. S. Nicholson, *Anal. Chem.*, **41**, 851 (1969).
(41) M. L. Olmstead and R. S. Nicholson, *Anal. Chem.*, **41**, 862 (1969).
(42) M. S. Shuman and I. Shain, *Anal. Chem.*, **41**, 1818 (1969).
(43) S. W. Feldberg, *J. Phys. Chem.*, **73**, 1238 (1969).
(44) R. F. Nelson and S. W. Feldberg, *J. Phys. Chem.*, **73**, 2623 (1969).
(45) G. Kissel and S. W. Feldberg, *J. Phys. Chem.*, **73**, 3082 (1969).
(46) N. Winograd, H. N. Blount, and T. Kuwana, *J. Phys. Chem.*, **73**, 3456 (1969).
(47) H. N. Blount, N. Winograd, and T. Kuwana, *J. Phys. Chem.*, **74**, 3231 (1970).
(48) C. P. Andrieux, L. Nadjo, and J. M. Saveant, *J. Electroanal. Chem.*, **26**, 147 (1970).
(49) M. H. Hulbert and I. Shain, *Anal. Chem.*, **42**, 162 (1970).
(50) M. S. Shuman, *Anal. Chem.*, **42**, 521 (1970).
(51) L. Nadjo and J. M. Saveant, *J. Electroanal. Chem.*, **30**, 41 (1971).
(52) S. W. Feldberg, *J. Phys. Chem.*, **75**, 2377 (1971).
(53) L. Nadjo and J. M. Saveant, *Electrochim. Acta*, **16**, 887 (1971).
(54) T. H. Ridgway, R. P. Van Duyne, and C. N. Reilley, *J. Electroanal. Chem.*, **34**, 267 (1972).
(55) L. Marcoux and T. J. P. O'Brien, *J. Phys. Chem.*, **76**, 1666 (1972).
(56) L. Marcoux, *J. Phys. Chem.*, **76**, 3254 (1972).
(57) S. W. Feldberg and L. Jeftic, *J. Phys. Chem.*, **76**, 2439 (1972).
(58) J. B. Flanagan and L. Marcoux, *J. Phys. Chem.*, **77**, 1051 (1973).
(59) C. Li and G. S. Wilson, *Anal. Chem.*, **45**, 2370 (1973).
(60) A. M. Bond, R. J. O'Halloran, I. Ruzic, and D. E. Smith, *Anal. Chem.*, **48**, 872 (1976).
(61) J. W. Dillard and K. W. Hanck, *Anal. Chem.*, **48**, 218 (1976).
(62) K. Holub and J. Weber, *J. Electroanal. Chem.*, **73**, 129 (1976).
(63) H. Kojima, A. J. Bard, H. N. C. Wong, and F. Sondheimer, *J. Am. Chem. Soc.*, **98**, 5560 (1976).
(64) H. H. Adam and T. A. Joslin, *J. Electroanal. Chem.*, **58**, 393 (1975).
(65) S. Valcher and M. Mastragostino, *Electrochim. Acta*, **17**, 107 (1972).
(66) M. D. Ryan and D. H. Evans, *J. Electroanal. Chem.*, **67**, 333 (1976).
(67) W. Britton and A. J. Fry, *Anal. Chem.*, **47**, 95 (1975).
(68) M. Fleischmann, T. Joslin, and D. Pletcher, *Electrochim. Acta*, **19**, 511 (1974).
(69) K. Holub, *J. Electroanal. Chem.*, **65**, 193 (1975).
(70) J. A. Richards, P. E. Whitson, and D. H. Evans, *J. Electroanal. Chem.*, **63**, 311 (1975).
(71) N. Winograd, *J. Electroanal. Chem.*, **43**, 1 (1973).
(72) T. Joslin and D. Pletcher, *J. Electroanal. Chem.*, **49**, 171 (1974).
(73) S. W. Feldberg, *Electrochemistry*, **2**, 185 (1972).
(74) I. Ruzic and S. W. Feldberg, *J. Electroanal. Chem.*, **50**, 153 (1974).
(75) G. D. Smith, "Numerical Solution of Partial Differential Equations", Oxford University Press, New York, N.Y., 1969, pp 1–97.
(76) J. R. Sandifer and R. P. Buck, *J. Electroanal. Chem.*, **49**, 161 (1974).
(77) S. N. Deming and S. L. Morgan, *Anal. Chem.*, **45**, 278A (1973).
(78) R. W. Hamming, "Numerical Methods for Scientists and Engineers", 2nd ed., McGraw Hill, New York, N.Y., 1973, p 352.
(79) J. H. Ahlberg, E. N. Nilson, and J. L. Walsh, "The Theory of Splines and Their Applications", Academic Press, New York, N.Y., 1967.
(80) M. H. Schultz, "Spline Analysis", Prentice-Hall, Englewood Cliffs, N.J., 1973.

RECEIVED for review May 29, 1975. Resubmitted July 22, 1977. Accepted October 18, 1977. This work was supported by the National Science Foundation.

DATA ENHANCEMENT

Editors' Comments on Papers 19 Through 24

19 **HORLICK and MALMSTADT**
 Basic and Practical Considerations for Sampling and Digitizing Interferograms Generated by a Fourier Transform Spectrometer

20 **COOPER**
 Errors in Computer Data Handling

21 **den HARDER and de GALAN**
 Evaluation of a Method for Real-Time Deconvolution

22 **SKOGERBOE et al.**
 A Dynamic Background Correction System for Direct Reading Spectrometry

23 **NIEMCZYK and ETTINGER**
 A Computer-controlled Photon Counting Spectrometer for Rapidly Scanning Low Light Level Spectra

24 **McLAFFERTY et al.**
 Signal Enhancement in Real-Time for High Resolution Mass Spectra

Data enhancement is an area of increasing concern as scientists continue to search for improvements in methods to extract the maximum information possible from observations. Three of the papers in this section provide useful guidelines to computerized handling of data. Practical considerations are emphasized in the article by Horlick and Malmstadt (Paper 19), while Cooper (Paper 20) shows how to prevent the data from getting any worse than it was when collected. Den Harder and de Galan (Paper 21) discuss deconvolution procedures to recover data degraded by the process of observation.

Horlick and Malmstadt (Paper 19) make the observation that while any waveform must be digitized at a sampling rate at least twice that of the band width of the system, many factors determine the bandwidth. The primary factor is the spectral region under investigation, which will be further limited by components of the Fourier spectrometer. These limitations are important in that it is the bandwidth, and not necessarily the highest frequency of the waveform, that is important for determination of sampling interval.

This paper is a good example of how computer simulation can be used to examine the effect of experimental conditions or errors on the resulting data. The effects of missed or extra points in an interferogram are conveniently simulated and presented by computer to illustrate the distortion and loss of resolution that can result from these problems.

Cooper's paper (Paper 20) on errors in computer data handling appeared as a feature article in *Analytical Chemistry*. We chose this paper rather than two others[1,2] because it includes the ideas of both in a more general treatment. The earlier papers should be consulted for additional details. In these papers Cooper has discussed the nature of digital signal averaging in Fourier transform NMR and the use of a computer in an experimental investigation of noise and dynamic range in the Fourier transform process. He finds that the dynamic range observable following a Fourier transform process is proportional to the computer's word length and inversely proportional to the number of transformed words and the number of memory locations full at the outset of the transform.

Paper 21, by den Harder and de Galan, is a more specialized article on real-time deconvolution. It is a well-illustrated, thorough evaluation of a method to recover a spectrum from a convoluted signal. The method is applied to simulated and real data, and the advantages and disadvantages, as compared to other deconvolution procedures, are clearly pointed out. A particularly significant advantage is that the spectrum is deconvoluted while it is being recorded, allowing a major reduction in computer memory required as compared to the Fourier transform method. The original band shape can be recovered by using only the first and second derivative of the experimental data, provided that the width of the broadening function is less than or equal to the width of the undistorted band. In part one of this volume, a paper was presented[3] that described a method for deconvolution of spectra by "pseudo-deconvolution." This procedure is most useful if the width of the broadening function is less than and does not approach that of the true profile, as this procedure generally does not converge to the correct result if the two functions approach equal width.[4]

The present approach to deconvolution is to expand the Fourier transform of the broadening function and use this expansion with a set of moments of the broadening function. Coefficients can be derived from these moments and an equation for the true profile can be obtained. Deconvolution is thus accomplished by calculating derivatives of the observed data and subtracting a weighted form of the derivation curve from the original data. The derivatives are calculated from a nine-point third-degree polynomial, using the method of Savitzky and Golay.[5,6] The objective is the correct recovery of the complete profile. In this regard, the effect of noise on the deconvolution procedure is examined and the criteria of evaluation clearly stated. Two criteria are used for judging the success of the deconvolution. One is the root mean square deviation of a data point between the original function and the result of deconvolution, and the other is the recovery of peak height. Both of these tests require knowledge of the original function, making computer simulation a necessary method of evaluation.

Paper 21 is thorough in its examination of the method by computer simulation and follows through with real data. Examples are from atomic emission and X-ray diffraction profiles.

Skogerboe et al. (Paper 22) report on a dynamic background correction system for emission spectrometry. They discuss several earlier approaches to improving emission data and conclude that a background correction system should use a single photomultiplier at each wavelength and alternate data collection rapidly between a spectral line and an adjacent background region. A spectrometer to meet these requirements was built, using computer control of photomultiplier selection and signal versus background acquisition. The improvement in standards curves at low concentration is illustrated by Figure 6, in which the described system has a greater linear range than correction by subtraction of blanks.

Niemczyk and Ettinger (Paper 23) describe another method of signal enhancement. In their work on a photon counting spectrometer, the time spent making an observation is varied via a feedback circuit based on the signal-to-noise ratio. In instrument systems the use of feedback in the control circuit to modify the experimental conditions is a step beyond that of conventional data acquisition. In general, when considering feedback one is concerned with negative feedback. Negative feedback corrects the system response. Niemszyk and Ettinger use the signal-to-noise ratio (S/N) of the observations to determine (correct) the scan rate of a spectrometer. Thus more time is spent in regions of low S/N than in regions of high S/N. To prevent the spectrometer from stalling, the time required to produce an acceptable S/N is also determined so that if the time is too long, the scan can be advanced to the next region without wasting time on a hopeless situation. The process is implemented by making a short initial measurement at each point from which the computer advances the scan either because it has a high enough S/N, or because waiting longer is of no use, or it continues to collect more data until an acceptable S/N is reached. The significant point is made that a large gain in efficiency is obtained because a computer is acting as an intelligent controller instead of just a sophisticated data acquisition system.

The final paper in this section, by McLafferty et al. (Paper 24), explains how a computer can be used to enhance the signal generated by a high resolution mass spectrometer. Alternate modes of achieving the same end are discussed and analysis of the results indicates the substantial improvement achieved by the approach used.

Paper 24 sets forth an example of how computer control of the data collection process can enhance the quality of the data. The idea is to rescan effectively each mass by using an offset to counter the continually changing magnetic field back to the point where the ion is again in focus. The time between each peak is thus made use of, which otherwise would be wasted. The offset is accomplished with either the ion accelerating voltage or the electrostatic analyzer voltage. The major advantage of the technique, apart from the signal enhancement, is that it is accomplished with no increase in total scan time.

Other papers on computer data enhancement include those by Mattson, Hirschfeld, Keir et al., and O'Halloran and Smith. Mattson[7] has made a case for the rejuvenation of older analytical techniques in a paper on computer automation of dispersive infrared spectrophotometry. He describes the design of an on-line minicomputer system, including circuitry and flow charts. Several examples are given of the effect of a 21-point, fourth-degree polynomial Savitzky-Golay smoothing operation on infrared data. The effect of computer averaging is also demonstrated. He points out that many of the functions attributed to Fourier transform infrared equipment, other than speed, are essentially operations performed by the computer, not by the interferometer.

Hirschfeld[8] has demonstrated how, by computer manipulation of repeat spectra of partially fractionated samples, recognizable spectra of mathematically separated constituents may be obtained from infrared spectra of unknown mixtures. This is an extension of the "spectrum stripping" technique. It is pointed out that the partial fractionation may be achieved by letting the sample stand a few hours in a heated watch glass, by extraction, or by using the leading or trailing edges of unresolved chromatographic peaks. Hirschfeld states that the procedure will almost always work for two-component mixtures and most of the time for three-component ones.

Keir et al.[9] have investigated Hadamard transform spectroscopy (HTS) in multielement atomic absorption, using a new computation and superposition procedure. They find that HTS offers a slight multiplex advantage in the signl-to-noise ratio over single-slit scanning of the spectrum when the signal is intense, but is disadvantageous for measuring small signals in atomic emission and fluorescence.

A note by O'Halloran and Smith[10] treats fast Fourier transform-based interpolation of sampled electrochemical data. They point out that rather than increasing data density by special sampling techniques, which put significant demands on measurement repeatability, the data density may be enhanced by Fourier domain interpolation. Further, when this technique is combined with fast Fourier transform digital filtering, substantial data enhancement is realized.

REFERENCES

1. J. W. Cooper, "Computers in NMR, I: Signal Averaging in Fourier Transform NMR," *Comput. Chem.*, **1**, 55 (1976).
2. J. W. Cooper, "Computers in NMR. II. Experimental Investigation of Noise and Dynamic Range in the Fourier Transform Process," *J. Magn. Reson.*, **22**, 345 (1976).
3. R. N. Jones, R. Venkataraghaven, and J. W. Hopkins, "The Control of Errors in Infrared Spectrometry, I. The Reduction of Finite Spectral Slit Distortion by the Method of Pseudo-deconvolution," *Spectrochim. Acta*, Part A, **23**, 925 (1967).
4. D. A. Ramsay, "Intensities and Shapes of Infrared Absorption Bands of Substances in the Liquid Phase," *Chem. Soc.*, **74**, 72 (1952).
5. A. Savitzky and M. J. E. Golay, "Smoothing and Differentiation of Data by Simplified Least Squares Procedures," *Anal. Chem.*, **36**, 1627 (1964).
6. J. Steiner, Y. Termonia, and J. Deltour, "Comments on 'Smoothing and Differentiation of Data by Simplified Least Squares Procedures'," *Anal. Chem.*, **44**, 1906 (1972).
7. J. S. Mattson, "Design and Applications of an On-Line Minicomputer System for Dispersive Infrared Spectrophotometry," *Anal. Chem.*, **49**, 470 (1977).
8. T. Hirschfeld, "Computer Resolution of Infrared Spectra of Unknown Mixtures," *Anal. Chem.*, **48**, 721 (1976).
9. M. J. Keir, J. B. Dawson, and D. J. Ellis, "Multielement Atomic Absorption Analysis using Hadamard Transform Spectroscopy with a New Computation and Superposition Procedure," *Spectrochim. Acta*, **32B**, 59 (1977).
10. R. J. O'Halloran and D. E. Smith, "Fast Fourier Transform Based Interpolation of Sampled Electrochemical Data," *Anal. Chem.*, **50**, 1391 (1978).

19

Copyright © 1970 by the American Chemical Society
Reprinted from *Anal. Chem.* **42**:1361–1369 (1970)

Basic and Practical Considerations for Sampling and Digitizing Interferograms Generated by a Fourier Transform Spectrometer

Gary Horlick[1] and Howard V. Malmstadt

Department of Chemistry and Chemical Engineering, University of Illinois, Urbana, Ill. 61801

A number of problems associated with the sampling and digitizing of interferograms and their effect on the resulting spectra are investigated and described. These include studies of the sampling rate, the accuracy of the frequency axis, the effects of missed, extra, and bad points, and the resolution required for the analog-to-digital converter. These problems are all studied using computer simulation on real and synthetic interferograms, and spectra are calculated to demonstrate the erroneous effects that can result. Criteria are established for selecting the sampling interval and a number of examples are given to show how the interval and the spectral region under investigation determine the final labeling of the frequency axis. It is shown that the occurrence of a missed, extra, or bad point during the sampling and digitizing steps can distort the final spectrum. Thus these types of errors must be avoided if accurate spectra are to be calculated. Finally, it is shown that a lack of resolution in the analog-to-digital conversion step results in a loss of spectral resolution.

WITH THE AVAILABILITY of commercial instrumentation Fourier transform spectrometers are becoming more common. The main distinguishing feature of this spectrometric technique is that the spectrum is obtained by taking the Fourier transformation of an interferogram which is generated by a two-beam interferometer (*1, 2*). Considerable data handling is necessary with Fourier transform spectrometry and many

---

[1] Present address, Department of Chemistry, University of Alberta, Edmonton, Alberta, Canada. To whom requests for reprints should be sent.

(1) Gary Horlick, *Appl. Spectry.*, **22**, 617 (1968).
(2) H. A. Gebbie, *Appl. Opt.*, **8**, 501 (1969).

1362 G. Horlick and H. V. Malmstadt

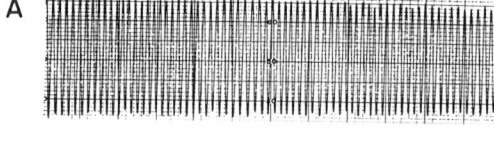

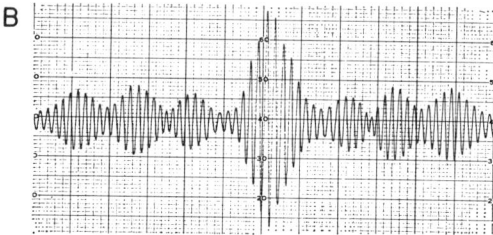

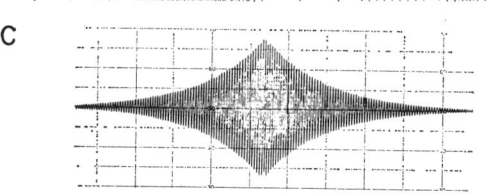

Figure 1. Partial interferograms for a He–Ne Laser (A), an iron hollow cathode lamp (B), and an 8400-Å interference filter (C)

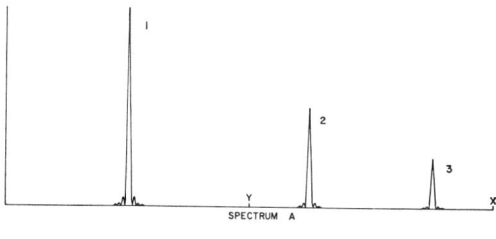

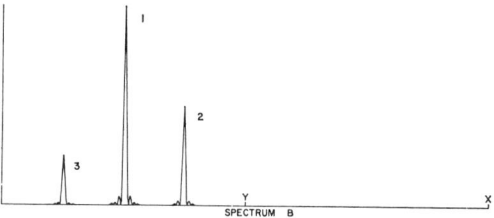

Figure 2. Simulated aliasing

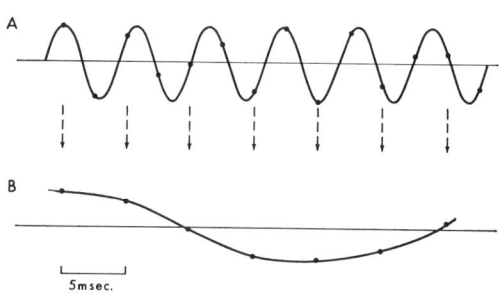

Figure 3. Aliasing of a 175-Hz sine wave to a 25-Hz sine wave by undersampling

apparently minor errors can result in completely erroneous spectra. The interferogram must, in general, be digitized because the data handling steps necessary for the reduction of an interferogram to a spectrum normally require a digitized interferogram. An interferogram can only be digitized at a certain finite number of sampling points. It is necessary that the resulting set of discrete digital values be an accurate representation of the original continuous interferogram. A number of questions then arise. How often must the interferogram be sampled? What is the effect of the sampling rate on the frequency axis of the final spectrum? What is the effect of a missed or additional sampling point? What is the effect of an erroneous value being assigned to a sampling point? What is the effect of the resolution of the analog-to-digital converter making the conversion?

In order to obtain a practical understanding of the erroneous effects that can result from the improper sampling and digitization of an interferogram, the problems posed by the above questions were investigated using experimental interferograms in conjunction with computer simulation.

It should be noted that the considerations investigated and discussed here are directly applicable, with little modification, to the sampling and digitizing of the free induction decay signal obtained in the Fourier transform NMR experiment (3, 4).

EXPERIMENTAL

Interferogram Measurements. The experimental interferograms were measured with a Fourier transform spectrometer system constructed at the University of Illinois (5). Interferograms of a He–Ne laser, an argon-filled iron hollow cathode lamp, and the transmission of an interference filter were measured in the visible and near infrared spectral regions.

Portions of the continuous interferograms for each of these three sources [He–Ne laser (A), iron hallow cathode lamp (B), 8400-Å interference filter transmission (C)] are shown in Figure 1. These illustrate typical interferograms to be expected from these sources. Only very small portions of the interferograms for the He–Ne laser and the iron hollow cathode lamp are shown in Figure 1, representing a change in retardation of about 44 μ in both cases. The complete interferograms were measured with a change in retardation of about 1120 μ.

In the case of the 8400-Å interference filter, the complete interferogram is shown (120-μ retardation change). The line shape of an interference filter transmission approximates a Lorentzian. Thus, the interferogram would be expected to be an exponentially damped cosine wave as observed.

Computer Calculations. All spectra were calculated on an IBM 360 computer utilizing a software system developed for the reduction of interferograms to spectra (5). This included apodization, Fourier transformation using the Cooley–Tukey algorithm, and the calculation and utilization of phase information. All spectra were plotted using the standard CALCOMP plotting system.

SAMPLING RATE AND THE EFFECTS OF UNDERSAMPLING

How often must the interferogram by sampled? This is one of the first questions that must be asked when making a spectral measurement with a Fourier transform spectrometer. The answer to this question is found in information theory

---

(3) R. R. Ernst and W. A. Anderson, *Rev. Sci. Instrum.*, **37**, 93 (1966).
(4) R. R. Ernst, "Advances in Magnetic Resonance, Vol. 2," J. S. Waugh, Ed., Academic Press, New York, N. Y., 1966, p 1.
(5) Gary Horlick, Ph.D. Thesis, University of Illinois, Urbana, Ill. 61801.

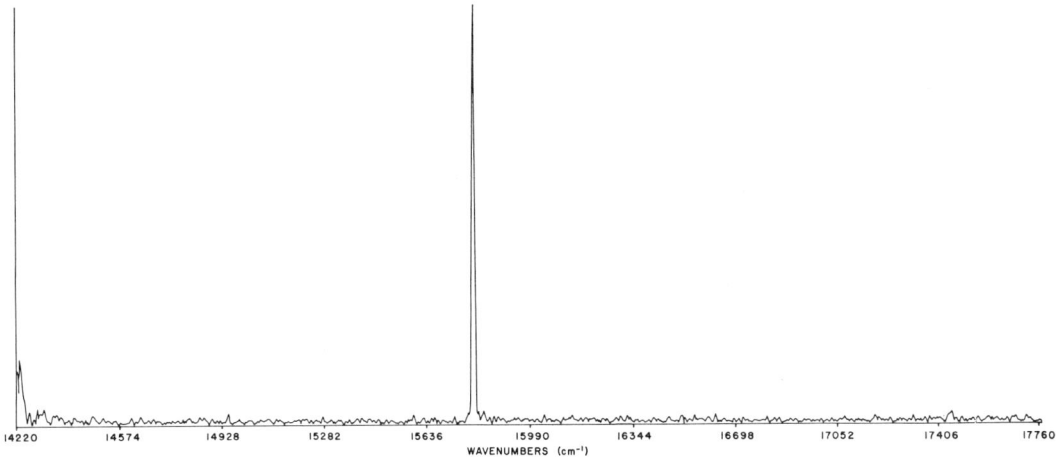

Figure 4. Spectrum of a He–Ne laser

(6, 7). An interferogram, or any waveform that is a function of time or distance, must be digitized at a sampling rate that is twice the bandwidth of the system in order that the spectrum can be accurately recovered. A number of experimental parameters and components in the Fourier transform spectrometer system determine this bandwidth.

The spectral region under investigation is the primary factor determining the range of frequencies that constitute the bandwidth of the system. In the case of a rapid scan instrument (1 sec) in the middle infrared (40–4 μ) the frequency content is in the low audio range (25–250 Hz). For slow scan instruments, such as the one constructed for this investigation, the frequency content is sub-audio. However, in each case the physical distance that the mirror must be moved to modulate any particular wavelength is the same. Thus, to determine the sampling interval, the amount the retardation must change in order to modulate the highest optical frequency (shortest wavelength) passed by the system

(6) R. Bracewell, "The Fourier Transform and Its Application," McGraw-Hill Book Co., New York, N. Y., 1965.
(7) T. Kobylarz, *Electronics*, **41** (8), 124 (1968).

through one-half period is calculated and then the interferogram is sampled at this interval. This becomes a sampling rate if the velocity of the mirror is taken into consideration. Note that the sampling interval calculated on this basis (proper sampling of highest frequency) involved the simplifying assumption that the bandwidth of the interferogram extended from some maximum frequency to zero frequency. This is, of course, not true in real practice.

The spectral region actually measured, *i.e.*, the bandwidth, will be limited by the components in the Fourier transform spectrometer system. These will include the optical transmission of the interferometer as determined by the beamsplitter material, the beamsplitter substrate, the compensator, the reflectivity of the Michelson mirrors, and any input and output mirrors or lenses; the spectral and frequency response of the detector; the frequency response of the detector amplifier; and the time constant of the final readout device.

The optical and electronic filters used in the instrument never have perfectly sharp bandwidth cutoffs. As a result there is the possibility of aliasing (6, 7) of high frequency noise and unwanted signal. Aliasing refers to the improper

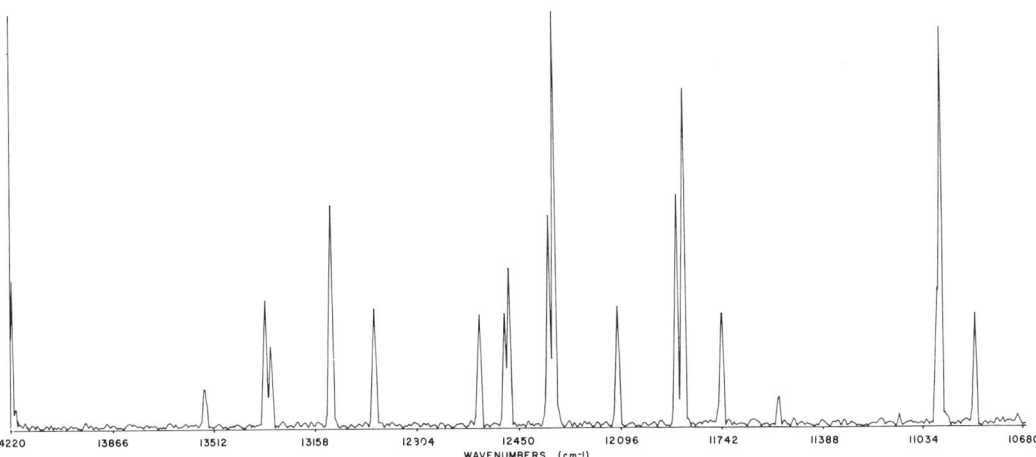

Figure 5. Spectrum of an iron hollow cathode lamp

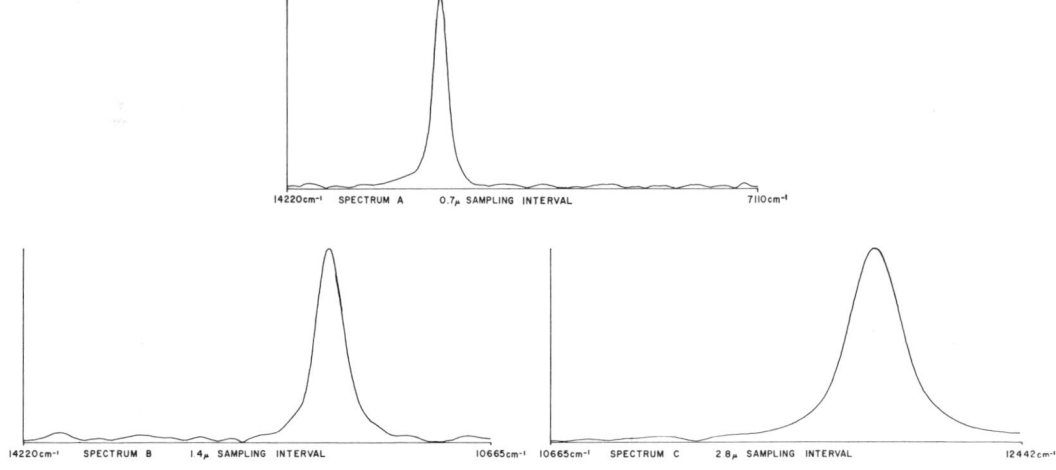

Figure 6. Absorption spectra of an 8400-Å interference filter

sampling of frequencies greater than those sampled correctly by the sampling interval used. These improperly sampled high frequencies show up as spurious low frequencies in the final spectrum. This is shown in Figure 2. Spectrum A resulted from the Fourier transformation of a synthetic interferogram which contained three frequencies each with a different amplitude. The interferogram was sampled at a rate such that all frequencies from zero to point X were sampled correctly. Then every other sampling point was dropped, the spectrum was calculated again and Spectrum B resulted. Now the interferogram is sampled so that frequencies out to point Y are sampled correctly. Peaks 2 and 3 now appear as completely spurious low frequencies and their position can be predicted as a folding over about the central point of the original frequency axis, when, as was done in this case, the number of sampling points is exactly halved and their spacing doubled. Peak 1 remains at its proper position on the frequency axis as it is still sampled correctly in the interferogram at this lower rate. Thus if an interferogram signal has a maximum frequency of 200 Hz (point X) it must be sampled at a rate of at least 400 Hz. In the case of undersampling as described above a signal of 175 Hz (peak 3) appears as a signal of 25 Hz; i.e., it has an alias of 25 Hz. A 125-Hz signal (peak 2) has an alias of 75 Hz. This can readily be appreciated by a simple exercise. A 175-Hz sine wave sampled at a 400-Hz sampling rate is shown in Figure 3A. If every other point is dropped out simulating the undersampling of Figure 2 and the remaining points are connected (Figure 3B), a 25-Hz sine wave is the result. If this same exercise is repeated for a 125-Hz sine wave, a 75-Hz sine wave will be the result.

It should be noted that if Peak 1 (Figure 2) did not exist, the fold over of Peaks 2 and 3 would not be serious as it occurs in an accurately predictable manner. This simply illustrates that it is the bandwidth and not necessarily the highest frequency of the waveform that is important when determining the sampling interval. If no low signal or noise frequencies exist, then the high frequencies of interest can be allowed to fold over without introducing any error. However, care must be exercised in ensuring the absence of spurious fold over and also in the proper labeling of the frequency axis in order that the interferogram can be transformed to an accurate spectrum.

EFFECT OF SAMPLING INTERVAL ON FREQUENCY AXIS PRESENTATION

It is important to see how the choice of sampling interval determines the labeling of the frequency axis because the validity and accuracy of the final spectrum is only as good as this knowledge of the frequency axis. With most Fourier transform spectrometers, it is not possible to have complete freedom of choice of the sampling interval. It is usually set by the angular size of step for a stepping motor drive, or the wavelength of a monochromatic line chosen as a reference wavelength. The reference line used in the interferometer constructed for this work was a neon line at 7032.4 Å (14220 cm^{-1}). If one sampling point is taken for each cycle of the reference line, that is for every 0.7-μ change in retardation, wavelengths up to 1.4 microns (7110 cm^{-1}) can be sampled correctly. Thus a bandwidth of 7110 cm^{-1} can be covered. If modulation frequencies resulting from wavelengths shorter than 1.4 microns are optically or electronically removed, then the frequency axis of the final spectrum will extend from zero to 7110 cm^{-1}. With appropriate filtering, this bandwidth could equally well cover the regions of 14220 to 7110 cm^{-1}, 28440 to 21330 cm^{-1}, or 14220 to 21330 cm^{-1}. These constraints are rigorous with respect to this specific sampling interval (0.7 μ) and the 7110 cm^{-1} bandwidth cannot be centered at any other frequency positions.

If the frequency of the reference line is divided by two, the sampling interval becomes 1.4 μ and the bandwidth that can now be covered is 3555 cm^{-1}. The regions that can be covered in the infrared and visible are listed in Table I. Again it should be emphasized that these are the only regions that can be covered starting with the 0.7-μ reference line, dividing its frequency by two and thus sampling at 1.4-μ intervals. To utilize any specific bandwidth, all wavelengths outside of it must be filtered out in some fashion in order to avoid aliasing. Thus starting with a certain fixed sampling interval, a specific bandwidth can be covered and that bandwidth is centered at certain specific spectral regions. Spectra of a He–Ne laser, an iron hollow cathode lamp, and the transmission of an interference filter have been measured and calculated in order to illustrate the above points.

The spectrum that was calculated from an interferogram measured using a small He–Ne laser as a source is shown

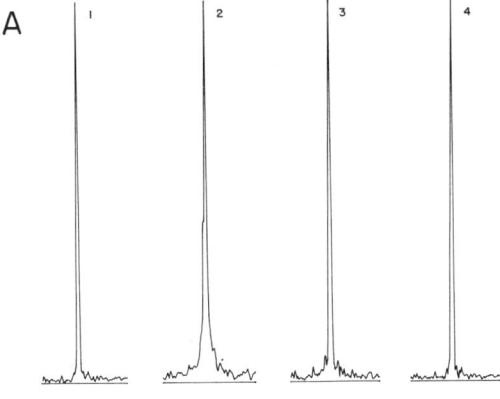

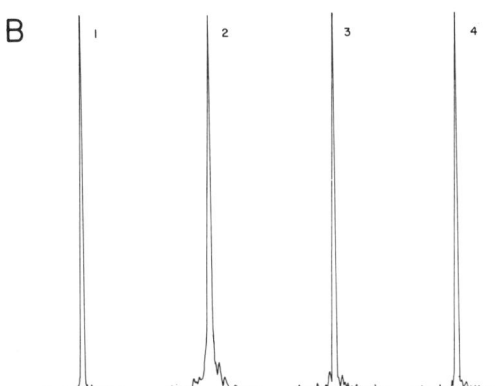

Figure 7. Effect of missed (A) and extra (B) points on the spectrum of a He–Ne laser

Table I. Bandwidths Covered with a 1.4-μ Sampling Interval

Region No.	Region, cm⁻¹	Region, μ
1	0– 3,555	∞ –2.8
2	7,110– 3,555	1.4 –2.8
3	7,110–10,665	1.4 –0.94
4	14,220–10,665	0.7 –0.94
5	14,220–17,775	0.7 –0.56
6	21,330–17,775	0.47–0.56
7	21,330–24,885	0.47–0.4
8	28,440–24,885	0.35–0.4

Table II. Argon Emission Lines

Peak No.	Measured values cm⁻¹	Measured values Å	Listed values, Å
1	13324	7505	7505.1
2	13307	7515	7514.6
3	13095	7636	7635.1
4	12946	7724	7723.7, 7724.2
5	12578	7950	7948.2
6	12488	8008	8006.2
7	12474	8017	8014.8
8	12338	8105	8103.7
9	12321	8116	8115.3
10	12099	8265	8264.5
11	11866	8410	8408.2
12	11734	8522	8521.4
14	10971	9115	Fold over
15	10960	9124	9123.0
16	10839	9226	9224.5

in Figure 4. The interferogram (See Figure 1) was a slowly damped sinusoidal waveform and was sampled at 1.4-μ intervals, thus the plot covers a bandwidth of 3555 cm⁻¹. The resolution was approximately 8.4 cm⁻¹ (3 Å) as the interferogram was double sided and 1710 points long. Triangular apodization was used. The wavelength of the laser source is 6328 Å (15,803 cm⁻¹) which falls in region 5 of Table I. The simplicity of the laser source provides the bandwidth limiting in this case. A check on the validity of this choice of axis labeling is provided by calculating the wavelength of the laser line from the plot parameters. The plot is 1024 points long and the maximum of the laser line is at point 456. Assuming that the plot starts at 14220 cm⁻¹ and is 3555 cm⁻¹ long, the wavelength of the laser line is calculated to be 6328 Å (15804 cm⁻¹).

The spectrum of an iron hollow cathode lamp (Argon filler gas) in the near infrared is shown in Figure 5. The resolution was approximately 10 cm⁻¹ (4 Å) as the interferogram (See Figure 1) was double sided and 1540 points long. The interferogram was sampled at 1.4-μ intervals, and thus the bandwidth is again 3555 cm⁻¹. However, the bandwidth is now limited by a Corning glass filter (No. 7-69) and the spectral response of the Si cell detector (8)

(8) W. L. Wolfe, Ed., "Handbook of Military Infrared Technology," Office of Naval Research, Department of the Navy, Washington, D. C., 1965.

to region 4 of Table I, 14220 to 10665 cm⁻¹ (0.7 to 0.94 μ). Again a check of the validity of this axis choice can be provided by calculating the wavelengths of the major lines present in the measured spectrum and comparing these with literature values. A listing of the wavelengths of the major lines as calculated by the computer when the interferogram was transformed to the spectrum is contained in Table II. The maximum of each peak was found, and using this value, the wavelength of the line was calculated as for the laser line. A more accurate determination of the peak maximum could be made using a program that fits the region around the peak with a cubic equation and then differentiates the equation to find that maximum (9). Also contained in Table II is a listing of major Argon emission lines in the near infrared as tabulated by Zaidel (10). All the lines are argon lines and essentially all the wavelengths match within the limit of the plot accuracy, which is one part in a thousand (1024 words) or approximately 2.2 Å. The side peak at 9115 Å (10970 cm⁻¹) is the result of the fold over (aliasing) of a strong argon emission line at 9657.8 Å.

Three absorption spectra of an interference filter are shown in Figure 6. The peak of the transmission is at 8400 Å as measured with a Cary 14. The source was a tungsten bulb. The bandwidth of the system is limited by the spectral simplicity of the interference filter transmission and the spectral response of the Si cell detector. Each spectrum resulted from the transformation of an interferogram which was sampled at three different intervals. The interferogram (See Figure 1) was about 120 μ long and the same length was used in each case. Therefore the resolution (about 170 cm⁻¹) is the same for each spectrum. Pertinent information about the plots and a list of the wavelengths of the

(9) R. N. Jones, *Appl. Opt.*, **8**, 597 (1969).
(10) A. N. Zaidel', V. K. Prokof'ev, and S. M. Raisku, "Tables of Spectrum Lines," Pergamon Press, Ltd., Oxford, England, 1961.

Table III. Summary of Parameters from Figure 6

Spectrum	Sampling interval, μ	Bandwidth, cm^{-1}	Axis, cm^{-1}	Transmission peak, Å
A	0.7	7110	14220–7110	8412
B	1.4	3555	14220–10665	8407
C	2.8	1777	10665–12442	8405

transmission peak as calculated by the computer are summarized in Table III.

This last example clearly shows how the choice of sampling interval and the resulting frequency axis presentation are intimately related. In addition all the examples illustrate that some *a priori* knowledge of the spectral bandwidth under investigation is necessary in order to accurately calculate the resulting spectrum.

EFFECT OF MISSED AND EXTRA SAMPLING POINTS

It is possible in some experimental systems for a sampling point to be missed or for an extra sampling point to be taken. Thus it was necessary to determine what effect such occurrences would have on the resulting spectra. It was felt that computer simulation of a missed or extra point in a normal interferogram would provide the best practical indication of the effects to be expected. Interferograms of a He–Ne laser, an iron hollow cathode lamp, and the transmission of polystyrene were used.

The effect of a missed point was simulated in an interferogram that resulted from the measurement of a He–Ne laser source. The interferogram was double sided, 1710 points long, and was triangularly apodized before the dropped points were simulated. A single point was dropped out at three different locations from the center of the interferogram and then the rest of the points were appropriately indexed. For example, if point 1000 was dropped, the values of points 1001 to 1710 would be reindexed as 1000 to 1709. This then results in an abrupt phase shift at point 1000 and the

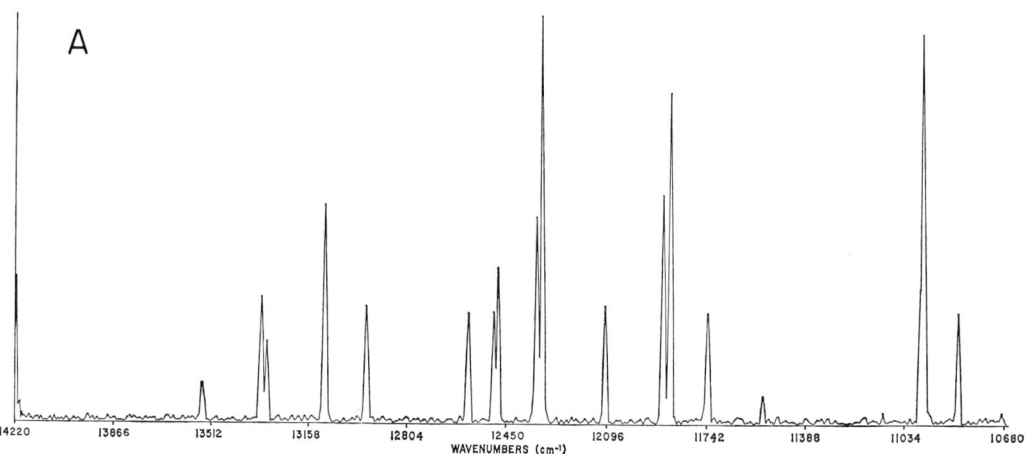

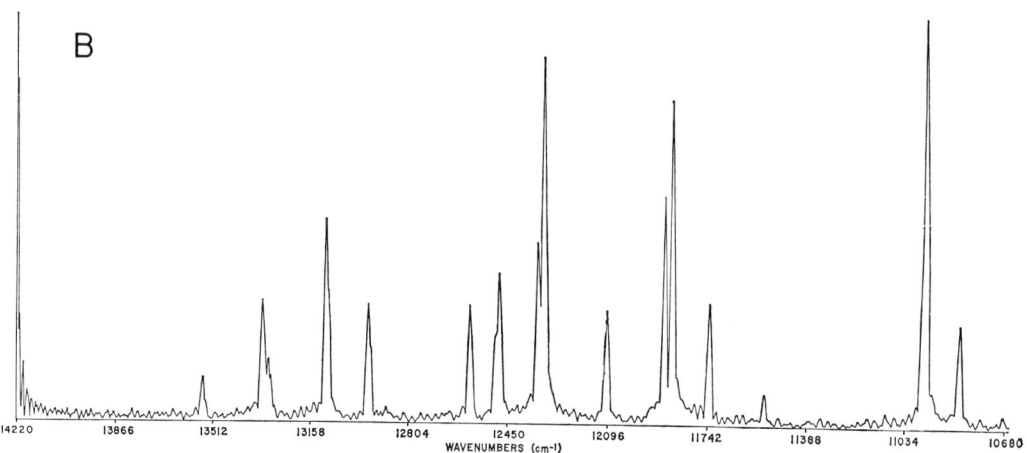

Figure 8. Effect of an extra point on the spectrum of an iron hollow cathode lamp

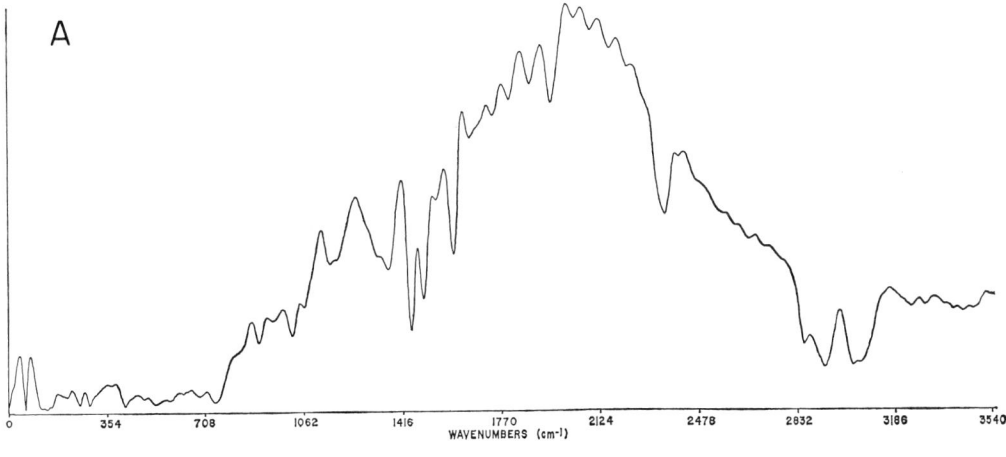

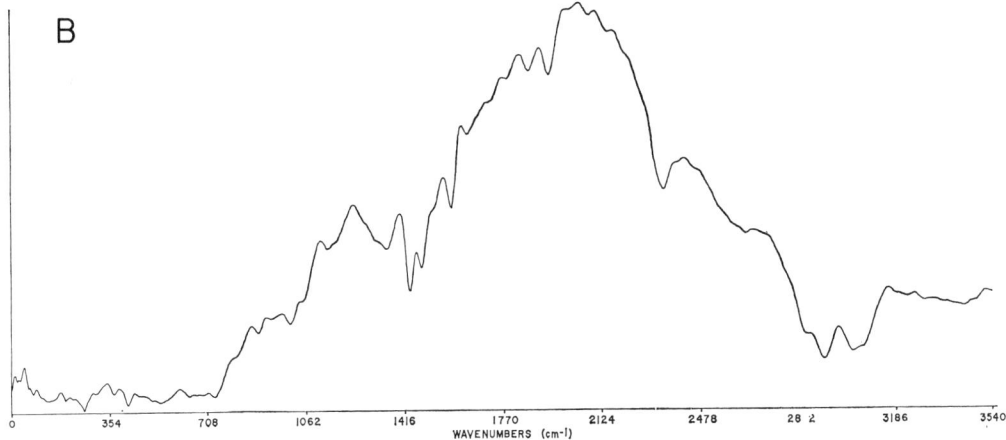

Figure 9. Effect of an extra point on a polystyrene spectrum

Fourier summation will not proceed properly. This would tend to be more serious the closer the dropped point is to the center of the interferogram. The effect this has on the resulting spectrum is shown in Figure 7A. Peak 1 is the normal case. For peak 2 the missed point occurred at point 1000, for Peak 3 at 1300, and for Peak 4 at 1600. Note that Peak 2 is severely distorted and as the missed point is moved further out from the center of the interferogram (Point 855) the distortion becomes less severe as predicted above. Since the source is a laser line, these figures approximate the resolution function of the spectrometer if a data point is missed. It is obvious that this type of error is very serious and must be completely avoided if accurate spectra are to result.

In a similar manner, the effect of an extra point was simulated on the same interferogram. This time a point was added between two present points at three different locations from the center of the interferogram and the series of data points were appropriately reindexed. For example, if a point was to be added between points 1000 and 999, it would be indexed as point 1000 and its value set as the average value of the original point 1000 and point 999. The values of the original points 1000 to 1710 would be reindexed as 1001 to 1711, point 1711 having a value of zero. This would result in a discontinuous phase shift similar to that caused by the missed point and a similar effect would be expected. This is shown in Figure 7B. Peak 1 is again the normal case. For Peak 2, the extra point was set at point 1000, for Peak 3 at 1300, and for Peak 4 at 1600. Again, for the point closest to the center of the interferogram, the peak is severely distorted. Thus, an extra point is also a very serious error and must be completely avoided.

The effects that an extra point has on line and continuous spectra are shown in Figure 8 and Figure 9, respectively. Figure 8A is the normal iron hollow cathode spectrum and Figure 8B shows the distortion that results when an extra point is added at point 1100. The interferogram is 1540 points long, and the center is at point 800. Figure 9A is a normal low resolution polystyrene single beam absorption spectrum, and Figure 9B is the resulting spectrum when an extra point is added at point 270. The interferogram is 490 points long and the center is at 235. Triangular apodization was used in all cases.

The overall effects are a distortion of normal bands and lines and an overall reduction in resolution, both of which are very hard to quantitate. The main point to realize is that both errors are very serious and must be avoided at all costs. This computer simulation study proved effective

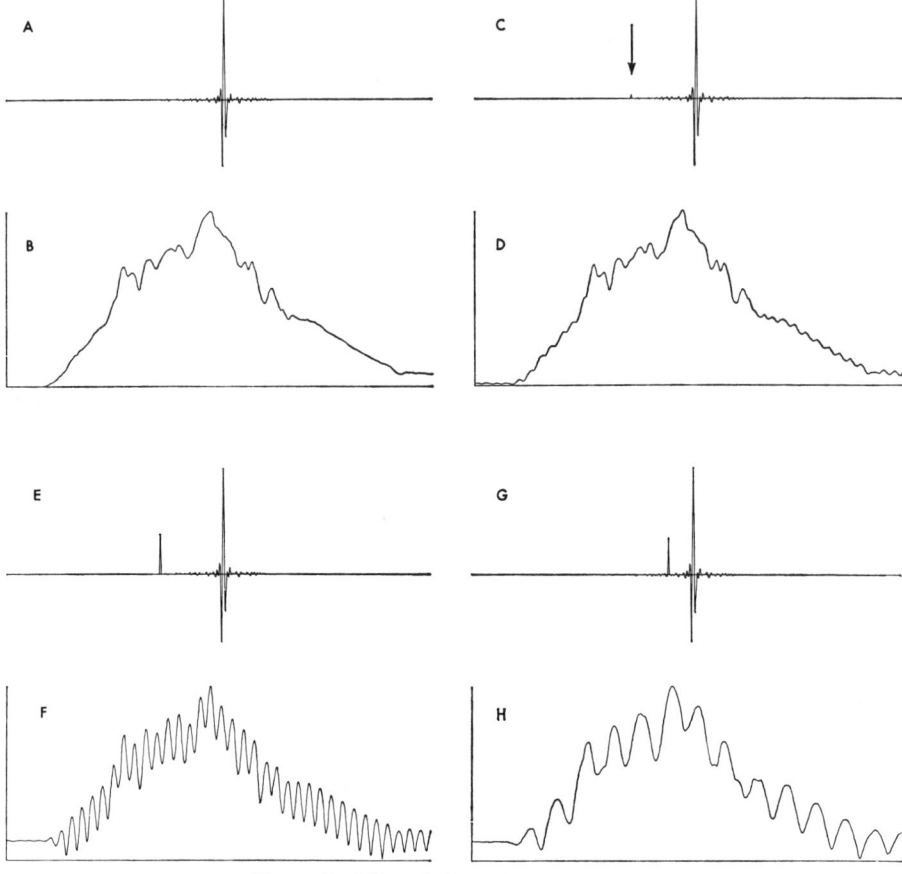

Figure 10. Effect of a bad point on a spectrum

in illustrating their severity and also indicated the types of effects these errors have on a spectrum and thus allows for their recognition and, hence, detection.

EFFECT OF A BAD SAMPLING POINT

A complication that can arise with analog-to-digital converters and readout systems is that a so-called bad point can sometimes occur. In this case an erroneous value is assigned to a sampled point, often orders of magnitude different than what it should have been. This type of error has been simulated in an interferogram and the results are shown in Figure 10. This type of error is completely intolerable in Fourier transform spectrometry as can be seen from the spectra resulting from the transformation of such interferograms. Even the small bad point in interferogram C causes visible fringing in spectrum D. A comparison of spectra F and H calculated from interferograms E and G shows that the frequency of the fringes produced in the spectrum depends on the position of the bad point. Higher frequency fringes are produced by points further from the central fringe. This type of error could also be caused by a noise spike on the wing of an interferogram.

This error is most serious when the interferogram is measured in one scan. If several scans are time averaged, this type of error may be minimized if it occurs in the measurement step rather than during the analysis of the interferogram.

Some of the advantages of rapid and repetitive scanning of the interferogram are discussed by Mertz (11).

RESOLUTION OF THE ANALOG-TO-DIGITAL CONVERTER

An interferogram has a wide dynamic range, especially when a broadband source is observed, and the small fluctuations on the wings are important. This dynamic range, in general, is greater in the interferogram than in the spectrum (12). This can be seen intuitively because the central fringe intensity represents the summation of all the frequencies in the spectrum when the interferometer is fully compensated. Also the process of Fourier summation is capable of detecting small sinusoidal fluctuations buried in the noise on the wings of the interferogram. Thus the analog-to-digital converter must have a high resolution, on the order of 12 to 14 bits in order to avoid introducing digitization noise when high resolution spectra are to be calculated from interferograms. Twelve-bit resolution in an analog-to-digital converter means that the peak input voltage can be divided into a

(11) L. N. Mertz, *J. Phys. (Paris), Suppl.*, **28**, C2-87 (1967).
(12) I. Coleman and L. N. Mertz, "Experimental Study Program to Investigate Limitations in Fourier Spectroscopy," Block Engineering, Inc., Cambridge, Mass. 02139. This publication is available from the Clearing House for Federal Scientific and Technical Information, Springfield, Va., 22151. The order number is Ad 665,890 (1968).

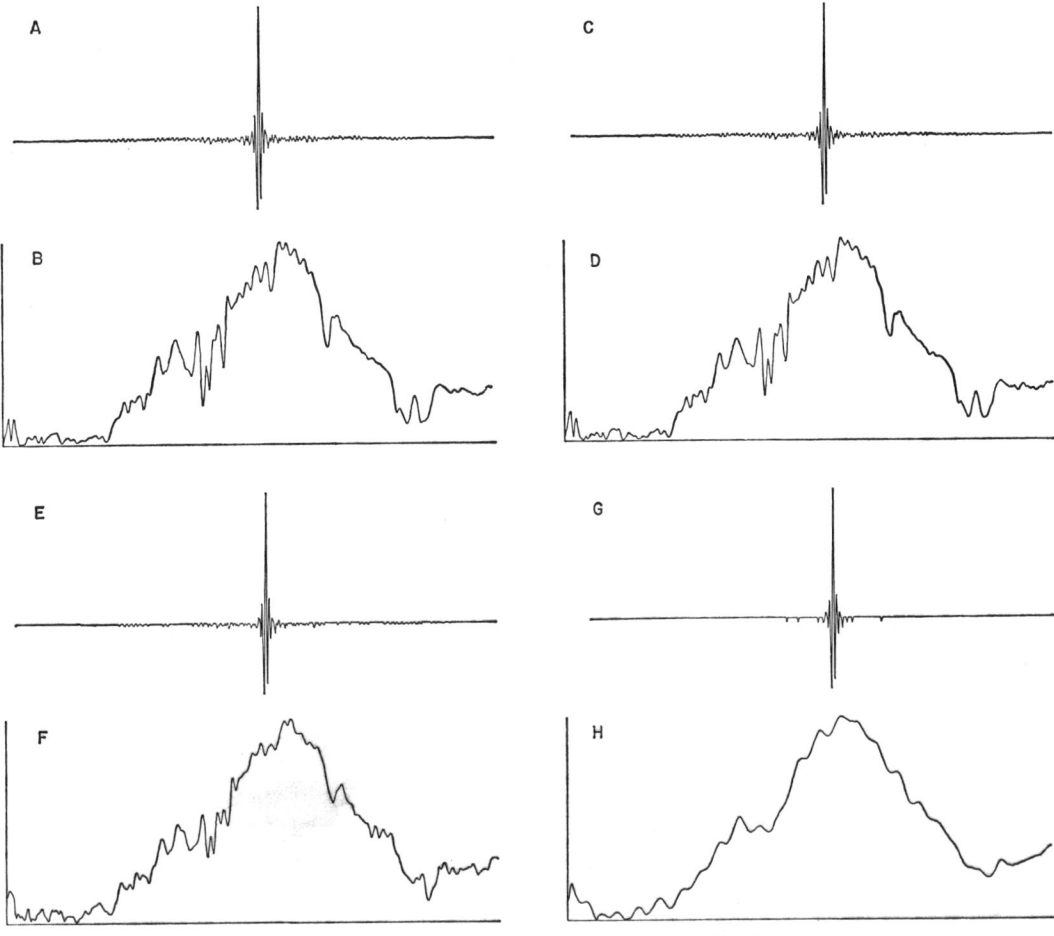

Figure 11. Effect of insufficient resolution in the analog-to-digital converter

maximum of 4096 levels (2^{12}). Thus a signal with a dynamic range of up to 4000 to 1 can be accurately digitized.

The effect of not having enough resolution in the analog-to-digital conversion step was simulated on a computer using a low resolution polystyrene interferogram. The decimal digital value of each point was converted to a thirteen-bit binary number plus a sign bit using a successive approximation method. The most significant bit had a value of 4.096 volts and the least significant bit a value of 0.001 volt. The decimal digital values of the interferogram points could be easily regenerated from the binary numbers by a simple weighted summation on the computer. It is a simple matter to set one or several of the least significant bits equal to zero before regenerating the decimal digital values of the interferogram samples. Thus if the last four least significant bits are set equal to zero, the resolution is only 9 bits and the value of the least significant bit would now be 0.016 volt rather than 0.001 volt. The interferograms and spectra resulting from a series of such truncations are shown in Figure 11. Spectrum B resulted from the transformation of 13-bit interferogram A, spectrum D from 8-bit interferogram C, spectrum F from 6-bit interferogram E, and spectrum H from 4-bit interferogram G. The main effect is a loss in resolution as would be expected when the small fringes on the wings of the interferogram are not properly digitized.

It is interesting to note that the resolution of an analog-to-digital converter can be improved by the addition of random noise (13, 14). If the amplitudes of the signal modulation frequencies are smaller than the size of the least significant bit, as they might be on the wings of the interferogram, they will not be digitized accurately. However, a random noise signal can be added to the interferogram signal. This raises the instantaneous signal level so that it is digitized accurately. A number of scans must be made and averaged in order to reduce the noise level to its original level, or in the case of Fourier transform spectrometry, band limited random noise can be added such that its frequency spectrum does not overlap with that of the signal frequencies.

RECEIVED for review May 13, 1970. Acception July 27, 1970. Taken in part from the Ph.D. Thesis of Gary Horlick, University of Illinois, 1970. This work was supported, in part, by a Division Summer Fellowship awarded to one of the authors (G.H.) by the Analytical Division of the American Chemical Society.

(13) J. Butterworth, P. E. MacLaughlin, and B. C. Moss, *J. Sci. Instrum.*, **44**, 1029 (1967).
(14) I. Coleman, *J. Opt. Soc. Amer.*, **56** (8), IV (1966).

ERRORS IN COMPUTER DATA HANDLING

James A. Cooper
Tufts University

As the laboratory minicomputer and microprocessor become more and more an integral part of laboratory instrumentation, it becomes necessary for the modern analytical chemist to have a greater appreciation of the minicomputer's role (1–3). He must not only be able to give intelligent direction to programmers who may prepare software for specific data acquisition and analysis tasks, but must also be able to appreciate the nature of the tasks performed within commercial data packages. In many cases the programs may provide for certain choices that can only be understood with some knowledge about the nature of the data gathering process. In this brief article we will divide these tasks into three sections: getting data into the computer, processing the data, and getting data out of the computer. We will also take a close look at the type of special problems that may occur in various types of Fourier transform spectroscopy.

Getting Data Into the Computer

Data to be put into the computer is quite often a varying voltage from some sort of chromatograph or spectrometer. It is a plot of some measurement vs. time where time may also run parallel with, say, frequency. These data are converted to a series of binary numbers and placed in successive computer memory locations, through the use of an analog-to-digital converter (ADC) and a real time clock. Conversion is rather like placing a window screen over a plotted spectrum and letting each horizontal space in the screen represent a successive memory location and the vertical location that the spectrum passes through represent the size of the number placed in that memory location.

Obviously, the dimensions of the screen must be suitable for our computer. There must be a sufficient number of memory locations for all of the data points thus acquired, and each word must be large enough to hold the number of counts produced by the analog-to-digital converter, although today, these are seldom problems. Each computer memory location is composed of *bits* that can be either 0 or 1 and represent successive powers of two. Minicomputers may have word lengths from 8 to 24 bits, with the most popular sizes being 12-, 16-, 20-, and 24-bit words. Thus, the common 16-bit word can hold numbers as large as about 2^{16} or 65536 counts. The analog-to-digital converters, on the other hand, are usually shorter, on the order of 6–13 bits, with the longer 15- or 16-bit ADC's reserved for special applications. As long as only one datum is placed in each word, there will always be room for a sufficient number of bits to hold the result.

Often the computer is used to enhance weak signals by adding together successive scans through the data. This is known as *signal averaging*. The signal grows at a rate proportional to the number of scans, and the rms value of the noise increases at a rate proportional to the square root of the number of scans. Thus,

$$S = \frac{k_1 B}{k_2 B^{1/2}} = K B^{1/2} \quad (1)$$

where S represents the signal to noise, B the total number of scans, and k_1, k_2, and K are constants. The signal-to-noise thus grows at a rate proportional to the square root of the number of scans.

There are a number of possibilities for errors in data gathered by signal averaging methods which we summarize here. First, the very small word length computer may quickly fill up memory as successive scans are added together. While it might at first seem that the number of scans is limited to the 2 raised to the power of the number of bits in excess between the word length and the ADC length, the total number of scans is dependent on the initial signal-to-noise S, the word length w, and the ADC length d according to the equation

$$B = \left[\frac{-1 + [1 + 4S(S+1)2^{w-d}]^{1/2}}{2S} \right]^2 \quad (2)$$

Table I. Total Scans Possible for Various Initial S/N and ADC Resolution Parameters

Initial S/N:	1		0.1		0.01	
$w - d$	Max scans	Final S/N	Max scans	Final S/N	Max scans	Final S/N
2	5	2.2	10	0.31	15	0.038
4	26	5.1	84	0.91	200	0.14
6	117	10.8	484	2.20	1996	0.44
8	489	22.1	2332	4.83	14016	1.18
10	2003	44.7	10251	10.10	75878	2.75
12	8101	90.0	42982	20.70	354182	5.95
14	32587	180.5	176028	41.95	1531040	12.37
16	130710	361.5	712455	84.40	6366810	25.23
18	523564	723.5	2866650	169.30	25966900	50.95

This equation is evaluated for various values of $w - d$ and S in Table I. The error that many people make in assuming that only a few scans may be taken before the computer word is full is not justified when the initial S is small.

Since more scans can be taken when $w - d$ is large, the obvious conclusion is that the smaller d is the better. In fact, if all peaks in the data are small, it is possible to do substantial amounts of signal averaging with an ADC as small as 1 or 2 bits. However, there must be at least one-half bit of noise in the signal for averaging to occur. If there is insufficient noise, the data will simply be quantized into little steps that will never average out. This point has been carefully discussed by Horlick (4). This quantizing error can occur, then, when there is a high signal-to-noise and a small ADC.

The number of scans may not be limited by Equation 2 when we do not care if some points in memory overflow. For example, if there is a large, uninteresting peak caused by some solvent, and some small, separated weak peaks of interest, there is no harm in letting the large one overflow memory; the operator "error" would be in stopping when memory overflow is first detected. On the other hand, it might be possible to continue averaging without overflow if all the data are divided down by 2 and the ADC length reduced correspondingly. In this case, more scans can be taken before memory again comes close to overflow, but the ADC may no longer have sufficient resolution to detect the large peak and still return some information about the small peaks. The program "error" is in dividing down the ADC when, in fact, the large dynamic range between the two peaks must be maintained for signal averaging to occur. This rather subtle problem has not been dealt with too effectively in current commercial programs, which generally *always* divide down the data when overflow is imminent. The problem of memory overflow in FT spectroscopy is illustrated in Figure 1, showing that overflow of even a few points can cause substantial decreases in S/N after the transform.

The only real solution to this problem is to use a longer word length for signal averaging. This can be done when buying a new computer by specifying one of the longer word length models or on shorter word length computers by utilizing double precision storage for all data, so that two computer words are used for each data point. Unfortunately, double precision has a number of disadvantages for some applications: notably that it cuts the total number of data points obtainable in half and/or increases the total cost of memory that must be purchased. Since many minicomputers are physically or logically limited as to the maximum available memory, some high resolution applications cannot be accomplished in double precision because of insufficient memory space. With FT spectroscopy, only powers-of-2 numbers of points can be conveniently Fourier transformed. This may further limit the available memory to the next lower power-of-two number of points.

These problems are common in sophisticated limiting cases such as FT–infrared spectroscopy, where they are usually solved by acquiring large arrays of data onto magnetic disks (5). This is a solution to the problem whenever data acquisition rate does not need to be greater than, say, 50 μs per point and when the substantially longer data processing times are not considered a disadvantage.

Coherent Noise

An additional error commonly made in signal averaging is in attenuating the input signal so that the ADC is not filled during each scan. While this is theoretically equivalent to the digital dividing down of the ADC resolution, it is not practically equivalent, because in the laboratory there are a number of noise sources that may add to the signal. If that signal voltage is small when entering the ADC, then external noise sources, some of them coherent, may add to the signal more significantly, thus obscuring the data of interest.

One of the commonest forms of coherent noise is the dc bias that may be added to the signal because of improper grounding or unbalanced amplifiers in the spectrometer. This dc bias is not random noise that will average out during successive scans according to Equation 2, but is a constant that will be added to the data on every scan. Unless this dc bias is removed by some hardware technique such as AC-coupling or by some software technique such as subtraction of a constant, memory will rapidly overflow, making the acquired data useless for many purposes.

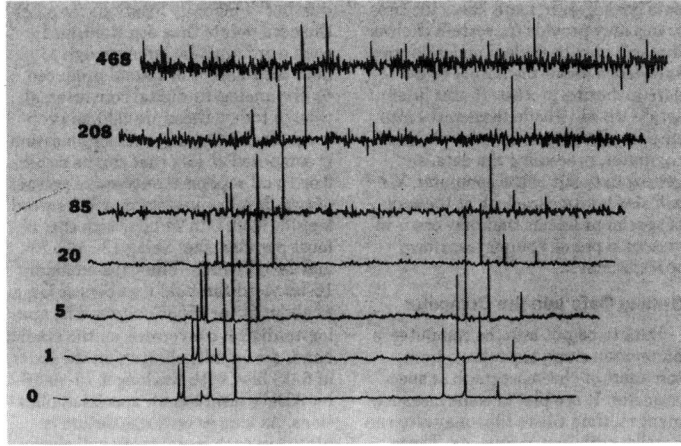

Figure 1. Effect of memory overflow on data after Fourier transformation
Transform of 4096-point FID after indicated number of data points have been allowed to overflow memory

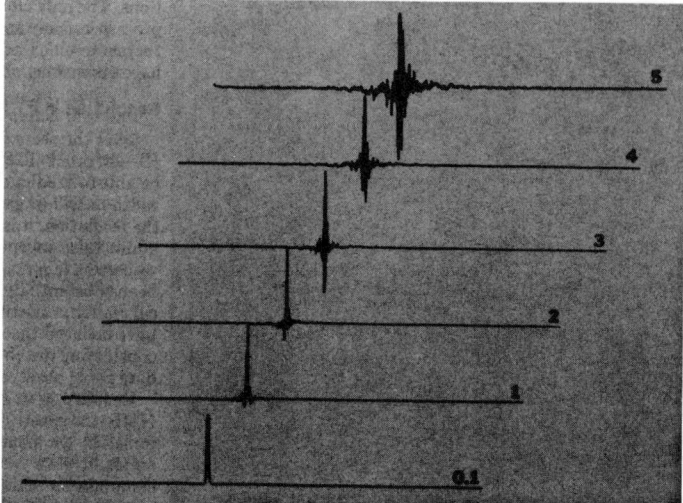

Figure 2. Effect of timing jitter on data acquisition
Fourier transform of 2500 Hz sine wave sampled every 100 µs with timing error randomly selected from the range ±0.1, 1, 2, 3, 4, or 5 µs. Line broadening factor of 1.0 Hz was applied to resulting sine wave before transform

A subtle form of design or programming error can be recognized in some general purpose computers that offer "laboratory peripheral systems", when the designers are not familiar with laboratory requirements. One popular 16-bit computer, for example, has a bipolar input to its ADC, which allows a signal of ±1 V. It would seem reasonable that it should convert these voltages to signed numbers as well, say from −2048 to +2047 for a 12-bit ADC. However, the hardware produces positive numbers after the conversion in the range of 0–4095. This is a serious drawback if signal averaging is planned, since even zero volts produces a digital dc bias of 2048 counts and memory will fill up in a maximum of 16 scans even with a zero-volt input. Unless the programmer and user recognize this design characteristic, little or no averaging will be possible without overflowing memory. Further, the hardware's maximum data rate may not be attainable in such machines because of the necessity of continual subtraction of a constant during acquisition.

Timing in Data Acquisition

One error condition sometimes overlooked in both commercial and home-grown data systems occurs when an inexperienced programmer decides to generate sampling interval timing by counting memory cycles. This may work on one computer, but when transferred to another of slightly newer vintage, the timing may be entirely different. Since many modern computers have variable instruction times that are data dependent and may vary with interrupts, cycle stealing disk transfers and memory refresh timing, this technique is no longer considered acceptable since the "jitter" between points may become fairly large. The effect of this jitter can be imagined as stretching the spectrum out of shape since the points are not uniformly distributed in time across the data. This effect can also be illustrated quite graphically by sampling a 2500 Hz sine wave every 100 µs ± some variable quantity and then Fourier transforming the resulting data. This has the effect of changing a plot of absorption vs. time (the sine wave) to a plot of absorption vs. frequency (6) (a peak). As shown in Figure 2, the effect of this problem is to generate a large number of spurious harmonics as the jitter time increases.

Data Handling in Fourier Transform Spectroscopy

In FT spectroscopy, additional problems regarding sampling rate must be recognized, since we are sampling sine waves and converting them to peaks by the Fourier transform calculation process. If we wish to examine a spectrum that covers a 5000 Hz spectral width, then we are going to be sampling sine waves whose frequencies may vary from 0 to 5000 Hz. According to the sampling theorem (7), we must sample a sine wave at least twice per cycle to represent it accurately to the computer. This highest frequency, which is sampled only twice per cycle, is known as the *Nyquist frequency* (8).

Suppose that we are only interested in 2000 Hz of this spectrum that has

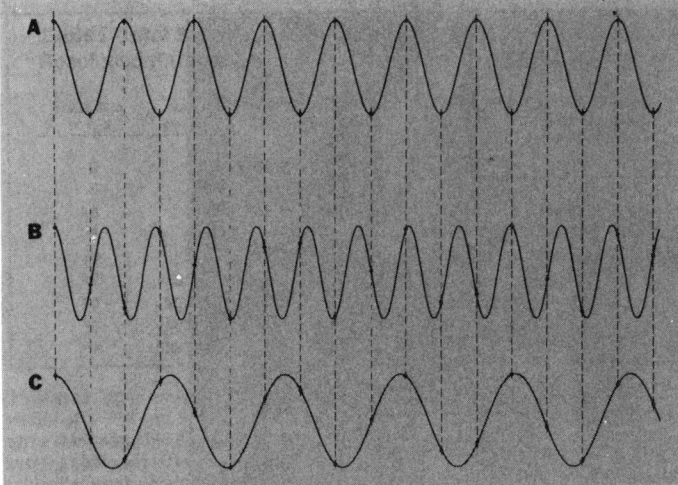

Figure 3. A: Sampling of Nyquist frequency sine wave; B: sampling of frequency $N - \Delta f$; C: sampling of frequency $N + \Delta f$

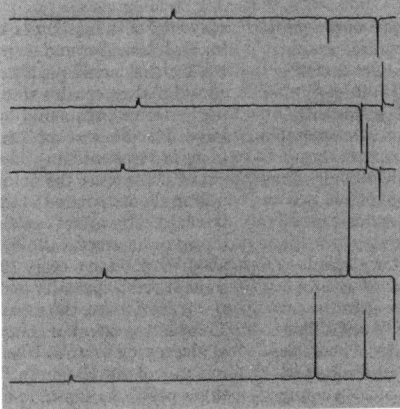

Figure 4. Illustration of foldback of lines in carbon-13 NMR spectrum

rf carrier frequency was shifted downfield for each spectrum above bottom one so that upper lines lie outside spectral width and fold back as shown

peaks covering 5000 Hz. We might be tempted to set the computer's "spectral width" to 2000 Hz and assume that only these peaks will appear in the spectrum. This, unfortunately, is not what occurs. By setting the spectral width to 2000 Hz, we have simply instructed the computer to sample the data less often: at a rate of 4000 Hz instead of 10 000 Hz. The frequencies that lie outside the spectra width are still sampled, however, and their effect can be seen from Figure 3. Figure 3A represents a sine wave having a frequency equal to the Nyquist frequency, sampled at 2 points per cycle, Figure 3B a frequency Δf less than the Nyquist frequency, and Figure 3C the Nyquist frequency plus Δf. The dots represent the instants at which the computer samples the data. Careful examination of Figure 3B and C will show that the computer "sees" the same data in both cases and that the frequency $N + \Delta f$ outside the spectral window or bandwidth will indeed appear in the spectrum having a frequency equal to $N - \Delta f$. This phenomenon is called *fold-back* or *aliasing*; while it is usually to be avoided, the cases where it cannot be avoided must be recognized carefully. In Figure 4 the carbon-13 NMR spectrum of ethyl iodide in chloroform-d is shown. In the lowest trace, the entire spectrum lies within the spectral window created by sampling at the correct rate. In the upper traces, the rf carrier is moved downfield (to the left) so that the spectrum has increasingly higher frequencies relative to the observation position. The lines in the spectrum fold back one by one until the lines due to the methyl and meth-ylene carbons have interchanged positions. The only clue to the foldback process in many spectra may be that there are some lines that appear to have anomalous phase.

Resolution in FT Spectroscopy

Since the user of an FT–NMR, FT–IR, or FT–ICR (9) system has to be able to predict not only the spectral width he will be examining, but also the resolution, it is instructive to examine this concept here. The reported resolution in a number of cases has been substantially at odds with physical reality primarily because users have believed their computers without considering the characteristics of the data being observed.

Resolution in FT–NMR. In FT–NMR, the spectral width is determined by the sampling rate. For example, to observe a 1000 Hz width, the computer must sample at a rate of 2000 Hz, or every 500 µs. More resolution cannot be obtained by sampling more often: this will only change the spectral width. Additional resolution is obtained by sampling for a longer period of time, in other words by taking more data points. For example, if 4096 data points are taken in a 1000 Hz spectral width, the resolution would seem to be about 0.25 Hz/point. If the data were sampled for twice as long to obtain 8192 points, the resolution would seem to be 0.122 Hz. However, because of the usual practice among instrument manufacturers of providing software that performs *in place* transforms (10), the resolution is actually only half that good. This occurs because a Fourier transform must produce both real and imaginary coefficients, even from data that were all real. Transforming a 4096-point real array produces 2048 real and 2048 imaginary points, and the resolution is only about 0.5 Hz/point.

It would appear that the longer one samples the data, the greater the resolution obtained in FT–NMR. In actual fact, the data do not persist forever (11, 12). Although some modern data systems allow the possibility of accumulating as many as 512 000 points onto magnetic disk and Fourier transforming them, the signal does not persist for 512 000 × 500 µs = 256 s, and the calculated resolution of 1000/256 000 = 0.0039 Hz/point is never obtained. Much of the data array is filled with noise that remains after the signal has died out, and this noise is spread throughout the entire spectrum by the transform, leading to reduced sensitivity and resolution compared to the spectrum sampled only to the point at which signal has died out. This common computer-user error often leads to extravagant expectations for resolution that are never achieved since the expected resolution is obscured by noise at the end of the data.

Better resolution can actually be obtained by sampling a smaller number of data points, to the point where the data have nearly died out, and then adding an equal number of zeroes to the data before the transform. This technique, called *zero-filling*, is illustrated in Figure 5. Figure 5A shows two frequencies differing by 0.13 Hz, sampled for 8192 points, at 500 μs/point leading to an expected spectral width of 1000 Hz and an expected resolution of 1000/8192 = 0.122 Hz. Figure 5B shows an expanded portion of their 4096-point Fourier transform, showing that despite our expectations, the two lines are not resolved. However, Figure 5C shows an expanded portion of a 16 384-point transform consisting of the same data and 8192 zeroes. Here the lines are resolved as we expected. Bartholdi and Ernst (*13*) showed that one power-of-2 of zero filling is indeed desirable to extract all the resolution information present in the spectrum, and failure to utilize this technique is also a computer or user error leading to a loss of resolution. Note particularly in FT-NMR that it is more desirable to terminate the data acquisition when the data have died out and then fill with zeroes than to acquire more data points consisting purely of noise and not zero-fill.

Resolution in FT-IR. In FT-IR the data are acquired from an interferometer as a function not specifically of time but of mirror travel distance (*5*). If the data are acquired over a longer mirror path length, more resolution is observed. But if more points are taken during that mirror travel, this only results in a different spectral width being observed. Since the mirror path length is fixed, resolution is fixed for a given mirror speed. Instead, the rate of mirror travel is reduced so that more points can be taken at the proper sampling frequency for the desired spectral width. The native resolution of FT-IR spectrometers is thus more or less controlled by the system designer, and scan rates slower than that will not generally result in more resolution.

In FT-IR many users feel compelled to engage in several orders of zero filling: filling a 32 768-point spectrum not just to 65 536 points before the transform but to 131 072 or even 262 144 points. Although this may produce a darker spectrum upon plotting since the pen moves in smaller increments, it *cannot* produce more resolution than the 65 536-point case and will only require much larger amounts of computer time to process the data. The resolution is determined by the spectrometer configuration, and the mirror travel rate cannot be further

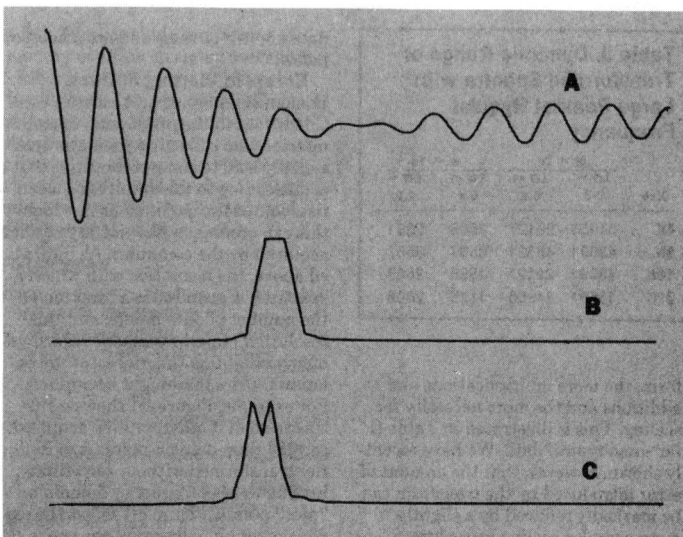

Figure 5. A: 8192-point free-induction decay consisting of two lines separated by 0.12 Hz; B: expansion of in place Fourier transform of A; C: expansion of transform of 16 384-point array created by adding 8192 zeroes to A

enhanced by additional zero filling. The data will essentially amount to a spectrum containing more interpolation between points according to the instrument line shape.

Resolution in FT-ICR. With the advent of Fourier transform ion cyclotron resonance (*9*), extremely rapid acquisition of mass spectral data has become possible. Resolution in FT-ICR is not linear with mass as Comisarow and Marshall have pointed out (*14*) and depends on the applied magnetic field. Substantial resolution improvements over low resolution ICR spectra are possible by sampling the data as long as possible. Unfortunately, since the cyclotron resonance frequencies of masses 16–400 range from 1 MHz to 35 kHz, the computer hardware requirements are prodigious to obtain extremely high resolution spectra. Data must be acquired as data rates as fast as fractions of a microsecond. Because of the extremely high data rates, acquisition onto disk is not always possible. Resolution is limited to the number of data points in the memory of the computer. Resolution can still be enhanced by taking advantage of computer programs to perform one level of zero-filling through a disk-based Fourier transform. This possibility should not be overlooked if truly high resolution ICR spectra are desired.

Errors in Processing of Fourier Transform Data

Many minicomputer programs are employed for the purpose of Fourier transforming data from NMR, IR, or ICR spectrometers. In fields such as Raman spectroscopy, Fourier transforms are used to simplify data smoothing and baseline correction (*15*). These programs are usually written to operate as rapidly as possible and utilize as little storage as possible, even at the occasional sacrifice of some accuracy. One of the principal problems in such Fourier transforms is that they are written in *integer mode*, where the largest number that can be represented is a fairly small integer, for example, ±32 767 in a 16-bit word computer. Since the Fourier transform requires a number of multiplications and additions for each "pass" through the transform (*6*), there is a substantial possibility that the numbers will become too large for the computer to represent as integers. This is particularly true when the computer's memory is full at the outset of the transform, so that a large amount of data scaling will have to be done during the transform.

In cases where small peaks must be detected in the presence of large peaks, particularly in FT-NMR and FT-ICR, there is the possibility that the round-off and scaling errors during the transform will cause these smaller peaks to be lost in the noise. This has been shown (*16*) to be a particular problem in 16-bit word computers and a somewhat lesser problem in longer word length computers and those utilizing double precision. The variation in post-transform dynamic range is affected by the computer word length and by the size of the transform, since the larger the trans-

Table II. Dynamic Range of Transformed Spectra with Large Peak at Nyquist Frequency

Size	w = 20		w = 16	
	LB = 0.4	LB = 0.2	LB = 0.4	LB = 0.2
4K	34953	29127	2608	5891
8K	43691	43691	3567	3567
16K	43691	29127	1556	2608
32K	43691	24966	1725	2608

form, the more multiplications and additions and the more necessity for scaling. This is illustrated in Table II for some typical data. We have recently shown, however, that the amount of error introduced by the transform can be markedly reduced by a slightly more complex scaling procedure, which adds only about 50% to the transform time (17).

In FT–IR, the dynamic range problem that exists because of the beat at zero path difference disappears in the frequency domain and becomes a large dc term in the background. Nonetheless, transform errors in FT–IR will also reduce the accuracy of small peaks. The exact amount of this reduction is the subject of some of our current research.

Getting Data Out of the Computer

The usual ways of observing computer-acquired data are by examining a crt display, plotting out the data, or asking for some sort of summary of the observed peaks. Observing the crt display is useful only for a rough idea of the nature of the data. The errors in making decisions from such a display are obvious. The possible errors in the programs for producing hard copy data are less obvious and are discussed below.

Errors in Plotting of Data. Human engineering and esthetic considerations in the production of commercial data collection packages occasionally lead to computer output that is misleading to the uninitiated scientist. One of the major areas in which this can occur is in the plotting of data acquired by the computer. As indicated above, the resolution with which a spectrum is acquired is a function of the number of data points, and this resolution remains constant regardless of any weighting functions or interpolation factors that might be applied. For example, Figure 6A shows a ^{13}C spectrum of 3-ethylpyridine acquired as 4096 time-domain points and Fourier transformed without zero-filling, leading to 2048 frequency domain or "plot" points. Figure 6B shows the expansion of a small area of the spectrum plotted so that the pen moves only horizontally and vertically giving a histogram of the existing information in the computer's memory. This is an accurate but unesthetic representation of the actual data that have been measured. Figure 6C, on the other hand, is an esthetic but inaccurate representation of the data measured using a simple linear interpolation scheme that draws the best *diagonal* line between points. Plot routines using quadratic and cubic interpolation schemes are not uncommon and unless labeled, these interpolated spectra give the appearance of more data than have actually been measured. They imply certain peak positions, line splittings, and line shapes that may or may not really exist but in any case have not been measured in the actual spectrum. These data can be misleading to the user if he is not properly informed.

A further confusion in the study of published or externally obtained spectra is the possibility of baseline smoothing of data during plotting so that a 3-point or higher smoothing function is applied to all baseline areas not part of peaks and not applied to peak areas. The spectrum has an unusually high apparent signal-to-noise; unless the peak selection algorithm is quite sensitive, some small but important peaks may also be smoothed by the function. This sort of data manipulation is a perfectly acceptable technique when it is understood to have taken place, but some spectra have been published without explaining this point. Worse, some manufacturers are using such techniques to artificially enhance the apparent signal-to-noise capabilities of their instruments in sales situations.

Errors in Peak Detection. The actual method by which a peak is detected can be rather complex (10). However, once the starting, ending, and maximum of a peak are determined by a program, the data printed out may indicate more information than is really available. A number of computer programs use some sort of interpolation process to arrive at the actual center of a peak when the data indicate that the top of a peak falls between the measured data points. This can be a risky process, depending on the signal-to-noise ratio in the spectrum, since a small amount of noise at the crest of a peak can throw this interpolation scheme off completely. Programs that do this usually interpolate the actual x-position of the center of the peak between adjacent channels by a simple calculation of the point of intersection of two lines drawn through the two points on either side of the position where the presumed maximum occurs (10). This approach is fraught with error if

a) there is substantial noise near the crest of the peak

b) there is the shoulder of another peak influencing one of the outer points

c) there is an undetected splitting between the two measured points

d) the peak is actually asymmetrical.

In any of these cases, the data will be incorrect, although in case c the data will be roughly correct, assuming that the splitting did not occur. To handle peak listings correctly, the user must be aware of what the program is doing and be able to suppress this interpolation for comparison in special cases.

Summary and Conclusions

The minicomputer can be an extremely powerful tool in the laboratory if its limitations are well under-

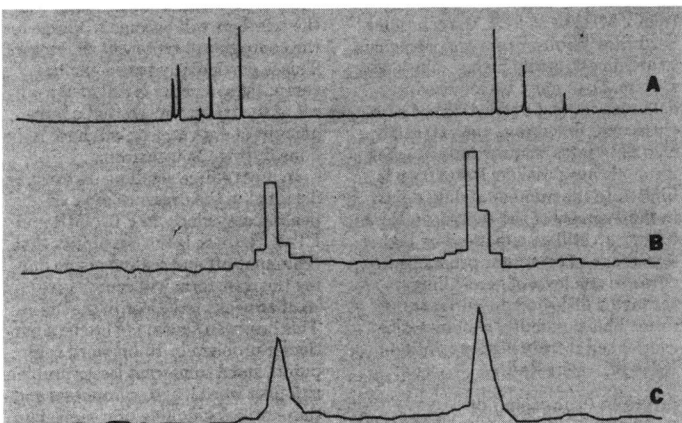

Figure 6. A: Plot of 2048-point spectrum of 3-ethylpyridine; B: plot of segment of spectrum in A without any interpolation between points; C: plot of same segment of spectrum with linear interpolation between points

stood. These include limitations of memory overflow in signal averaging, foldback of high frequency lines in time domain spectra, round-off errors in Fourier transforms, and errors in interpolation during plotting or peak printout. In all cases the knowledgeable analyst should be able to recognize these as potential hazards and circumvent them if necessary.

References

(1) R. E. Dessy and J. Titus, *Chem. Technol.*, **3** (7), 436–8 (1973).
(2) S. Perone, *Comput. Chem. Biochem. Res.*, **1**, 1–35 (1972); S. Perone and D. O. Jones, "Digital Computers in Scientific Instrumentation", McGraw-Hill, New York, N.Y., 1973.
(3) J. W. Cooper, *Comput. Chem.*, **1**, 55–60 (1976).
(4(G. Horlick, *Anal. Chem.*, **47** (2), 352–4 (1975).
(5) P. R. Griffiths, "Chemical Infrared Fourier Transform Spectroscopy", Wiley, New York, N.Y., 1975.
(6) E. Oran Brigham, "The Fast Fourier Transform", Prentice-Hall, Englewood Cliffs, N.J., 1974.
(7) M. Woodward, "Probability and Information Theory", Pergamon Press, New York, N.Y., 1955.
(8) R. Swanson, D. J. Thoennes, R. S. Williams, and C. L. Wilkins, *J. Chem. Educ.*, **52** (8), 530–3 (1975).
(9) M. B. Comisarow and A. G. Marshall, *J. Chem. Phys.*, **64** (1), 110–19 (1976).
(10) J. W. Cooper, "The Minicomputer in the Laboratory: With Examples Using the PDP-11", Wiley-Interscience, New York, N.Y., 1977.
(11) J. A. Pople, W. G. Schneider, and H. J. Bernstein, "High Resolution Nuclear Magnetic Resonance", McGraw-Hill, New York, N.Y., 1965.
(12) T. C. Farrar and E. D. Becker, "Pulse and Fourier Transform Nmr", Academic Press, New York, N.Y., 1971.
(13) E. Bartholdi and R. Ernst, *J. Magn. Reson.*, **11** (1), 9 (1973).
(14) M. B. Comisarow and A. G. Marshall, *J. Chem. Phys.*, **62** (1), 293–5 (1975).
(15) G. Horlick, *Anal. Chem.*, **44** (6), 943–7 (1972).
(16) J. W. Cooper, *J. Magn. Reson.*, **22**, 345–57 (1976).
(17) J. W. Cooper, I. S. Mackay, and G. B. Pawle, *ibid.*, **28**, 405 (1977).

21

Copyright © 1974 by the American Chemical Society
Reprinted from *Anal. Chem.* 46:1464-1470 (1974)

Evaluation of a Method for Real-Time Deconvolution

Arend den Harder[1] and Leo de Galan[2]

Laboratorium voor Instrumentele Analyse, Technische Hogeschool, Delft, Nederland

This paper presents an evaluation of a procedure for the deconvolution of two-dimensional profiles, such as spectra, chromatograms, etc. It is based upon a proposal by Hardy and Young that has apparently never been tested on real spectra. It is shown that by using only the first and second derivative to the experimental spectrum, the original band system can be recovered if the width of the broadening function is less than or equal to the width of the nondistorted bands. The method can be applied to a variety of broadening functions and provides for simple and effective noise suppression. In comparison with other methods for deconvolution, the proposed procedure offers the advantage that a spectrum can be deconvoluted while it is being scanned. This makes it especially attractive for small on-line computers.

Experimentally observed spectra, chromatograms, and the like are generally subject to several, independent broadening processes. Atomic spectral lines are subject to Doppler broadening and collisional broadening (1); atomic and molecular spectra are distorted by the finite resolving power of the monochromator (2-5); electronic filtering introduces asymmetry (6, 7); X-ray diffraction profiles are influenced by structural effects in the specimen (8), etc. All such interactions are mathematically described by the convolution integral

$$T(y) = \int_{-\infty}^{+\infty} t(x) B(y - x) \, dx \quad (1)$$

where $t(x)$ is the nondistorted line profile, $B(y-x)$ is the broadening function, and $T(y)$ is the recorded line profile.

The mathematical properties of this integral have been described by Hsu (9) and Shapiro (10). Whereas the calculation of $T(y)$ from known $t(x)$ and $B(y-x)$ is simple and straightforward, the reverse procedure of calculating $t(x)$ from known $T(y)$ and $B(y-x)$ is rather more difficult. This procedure is known as deconvolution.

Two methods of deconvolution appear to be widely used. The first method is of long standing (11-15) and has been designated by Jones (2) as pseudo-deconvolution. Here the observed profile $T(x)$ is first convoluted once more with the broadening function $B(x)$ to give a convolution result $C(x)$. The true profile is then calculated as $t(x) = T(x)^2/C(x)$. This is an approximation since the inter-

---

[1] Present address, Hoogovens-Estel, IJmuiden, The Netherlands.
[2] Author to whom correspondence should be directed.

(1) A. C. G. Mitchell and M. W. Zemansky, "Resonance Radiation and Excited Atoms," University Press, Cambridge, England, 1961.
(2) R. N. Jones, R. Venkataraghavan, and J. W. Hopkins, *Spectrochim. Acta, Part A*, **23**, 925, 941 (1967).
(3) D. A. Ramsay, *J. Amer. Chem. Soc.*, **74**, 72 (1952).
(4) J. R. Morrey, *Anal. Chem.*, **41**, 719 (1969).
(5) L. de Galan and J. D. Winefordner, *Spectrochim. Acta, Part B*, **23**, 277 (1968).
(6) I. G. McWilliam and H. C. Bolton, *Anal. Chem.*, **41**, 1755, 1762 (1969).
(7) E. Grushka, *Anal. Chem.*, **44**, 1733 (1972).
(8) B. E. Warren and B. L. Averbach, *J. Appl. Phys.*, **21**, 595 (1950); **23**, 497 (1952).
(9) H. D. Hsu, "Outline of Fourier Analysis Including Problems with Step by Step Solutions," Uniteck, New York, N.Y., 1967.
(10) K. S. Shapiro, "Smoothing and Approximation of Functions," Van Nostrand, New York, N.Y., 1969.
(11) H. C. Burger and P. H. van Cittert, *Z. Phys.*, **79**, 722 (1932).
(12) H. C. van Hulst, *Bull. Astron. Inst. Neth.*, **9**, 225 (1941).
(13) A. L. Khidir and J. C. Decius, *Spectrochim. Acta*, **18**, 1629 (1962).
(14) S. Ergun, *J. Appl. Crystallogr.*, **1**, 19 (1968).
(15) J. Szöke, *Chem. Phys. Lett.*, **15**, 404 (1972).

section points of $T(x)$ and $C(x)$ are not modified. Although the procedure can be repeated to give a better approximation, prolonged iteration increases computing time and generally does not converge to the correct result if the width of the broadening function $B(x)$ approaches that of the true profile $t(x)$, (3).

The second popular method of deconvolution uses Fourier transformation (16, 17). The recorded profile and the broadening function are first Fourier transformed to give

$$FT(\omega) = \frac{1}{\sqrt{2\pi}} \int_{-\infty}^{+\infty} T(x) e^{-i\omega x} dx \quad (2)$$

$$FB(\omega) = \frac{1}{\sqrt{2\pi}} \int_{-\infty}^{+\infty} B(x) e^{-i\omega x} dx \quad (3)$$

and the ratio of these transforms gives the Fourier transform of the true profile (18–21), which is then back transformed to

$$t(x) = \frac{1}{\sqrt{2\pi}} \int_{-\infty}^{+\infty} (FT(\omega)/FB(\omega)) e^{i\omega x} d\omega$$

In principle, this method is generally applicable and free of errors, while lengthy calculations are sped up by using a fast Fourier transform (22). In practice, the base line must be determined before the transforms can be executed and the transformed functions must be terminated to keep the noise within limits (17, 23). This introduces errors.

It should finally be noted that in both deconvolution methods, the complete profile $T(x)$ and the broadening function $B(x)$ must be recorded and stored in the computer memory before the deconvolution can be started. Both methods are numerical procedures and can be executed only with digital functions. In the pseudo-deconvolution method, the broadening function and the measured spectrum must have the same sampling interval. For the Fourier method, there must be a known relation between the sampling interval of both functions. These disadvantages are avoided in the procedure presented in this paper.

DESCRIPTION

Several years ago, Hardy and Young (24) proposed a simplification of the Fourier transform method. As has been remarked by Khidir and Decius (13), their suggestion seems to have gone unnoticed, although similar expressions have been derived independently by Allen, Gladney, and Glarum (25) for resolution enhancement of ESR spectra and by Jones and Misell (26) for deconvolution. However, the latter authors give no practical application.

According to Hardy and Young, Equation 3 is expanded to give·

$$FB(\omega) = \frac{1}{\sqrt{2\pi}} \int_{-\infty}^{+\infty} B(x) \left[1 - i\omega x + \frac{(i\omega x)^2}{2!} - \frac{(i\omega x)^3}{3!} + \ldots \right] dx$$

(16) G. Horlick, *Appl. Spectrosc.*, **26**, 395 (1972).
(17) A. Goldman and P. Alon, *Appl. Spectrosc.*, **27**, 50 (1973).
(18) G. Horlick, *Anal. Chem.*, **44**, 943 (1972).
(19) R. Bracewell, "The Fourier Transform and Its Applications," McGraw-Hill, New York, N.Y., 1969.
(20) D. W. Kirmse and A. W. Westerberg, *Anal. Chem.*, **43**, 1035 (1971).
(21) H. Schrijver, *Physica*, **49**, 135 (1970).
(22) J. W. Cooley and J. W. Tukey, *Math. Comp.*, **19**, 297 (1965).
(23) J. D. Bregman and F. F. M. de Mul, *Nucl. Instrum. Methods.*, **93**, 109 (1971).
(24) A. C. Hardy and F. M. Young, *J. Opt. Soc. Amer.*, **39**, 265 (1949).
(25) L. C. Allen, H. M. Gladney, and S. H. Glarum, *J. Chem. Phys.*, **40**, 3135 (1964).
(26) A. F. Jones and D. L. Misell, *Brit. J. Appl. Phys.*, **18**, 1479 (1967).

and we define moments of order n divided by $n!$ as follows:

$$\alpha_n = \frac{1}{n!} \int_{-\infty}^{+\infty} B(x) x^n dx \quad (4)$$

so that, if $B(x)$ is normalized to unit area:

$$FB(\omega) = \left[1 + \sum_{n=1}^{\infty} \alpha_n (-i\omega)^n \right] / \sqrt{2\pi}$$

The Fourier transform of the true profile is then given by:

$$Ft(\omega) = \frac{FT(\omega)}{FB(\omega)} = \frac{\sqrt{2\pi}\, FT(\omega)}{1 + \sum_{n=1}^{\infty} \alpha_n (-i\omega)^n}$$

$$= \sqrt{2\pi}\, FT(\omega) \left\{ 1 - \sum_{n=1}^{\infty} \alpha_n (-i\omega)^n + \left(\sum_{n=1}^{\infty} \alpha_n (-i\omega)^n \right)^2 - \ldots \right\}$$

Collecting equal powers of $(-i\omega)$ gives:

$$Ft(\omega) = \left[FT(\omega) - \sum_{n=1}^{\infty} a_n (+i\omega)^n FT(\omega) \right] \sqrt{2\pi} \quad (5)$$

where the coefficients a_n are related to the α_n, defined by Equation 4, e.g.,

$a_1 = -\alpha_1$
$a_2 = \alpha_2 - \alpha_1^2$
$a_3 = -\alpha_3 + 2\alpha_1\alpha_2 - \alpha_1^3$
$a_4 = \alpha_4 - \alpha_2^2 + 2\alpha_1\alpha_3 - 3\alpha_1^2\alpha_2 + \alpha_1^4$ etc.

It follows from Fourier transform theory that

$$\int_{-\infty}^{+\infty} T^n(x) e^{-i\omega x} dx = (i\omega)^n \int_{-\infty}^{+\infty} T(x) e^{-i\omega x} dx = (i\omega)^n FT(\omega)$$

where $T^n(x)$ is the nth derivative of $T(x)$.

Substitution into Equation 5 and back-transformation then gives:

$$t(x) = T(x) - \sum_{n=1}^{\infty} a_n T^n(x) \quad (6)$$

It is clear that the calculations of the coefficients a_n proceed completely independent from the measurement of the observed spectrum $T(x)$. Specifically, the intervals between the data points in the spectrum $T(x)$ and the broadening function $B(x)$ need not be equal. The precision in calculating a_n-values can thus be adapted to the accuracy with which the broadening function is known, either analytically or numerically. From the examples presented in Table I, it is also clear that a variety of broadening functions can be accommodated, but the method fails for slowly decreasing broadening functions, such as a Lorentzian.

According to Equation 6, the deconvolution procedure has now been reduced to the calculation of the derivatives of the recorded spectrum and appropriate subtractions.· These simple operations can be performed with an on-line computer while the spectrum is being recorded and require only a minimum of computer memory and computing time.

On the other hand, it is clear that a correct recovery of the true profile would require an infinite number of terms, which apart from practical limitations, increases the influence of noise through the inclusion of higher derivatives. In order to evaluate the number of terms required in practice, we have analyzed the deconvolution of a Gaussian peak T of width $\Delta\lambda_T$ for another Gaussian of unit area and width s, representing the broadening function. (Here and in the remainder of this article the width will always refer to the full width at half height). The result of this deconvolution should be the original profile function,

Table I. Coefficients Needed for Deconvolution

		Broadening function		
Coefficient	Triangle $-\|x\|/s^2 + 1/s$ ($\|x\| < s$)	Gaussian[a] $(4\ln2/\pi s^2)^{1/2}\exp[-4x^2\ln2/s^2]$	Lorentzian $\dfrac{1/2\pi s}{(1/2s)^2 + x^2}$	Exponential $(\ln2/s)\exp[-x\ln2/s]$ ($x > 0$)
α_1	0	0	0	$s/\ln2$
α_2	$s^2/12$	$s^2/16\ln2$	∞	$(s/\ln2)^2$
α_3	0	0	0	$(s/\ln2)^3$
α_4	$s^4/360$	$s^4/512(\ln2)^2$	∞	$(s/\ln2)^4$
$\alpha_n{}^b$	$\dfrac{2s^n}{(n+2)!}$	$\dfrac{1,3,5\ldots(n-1)(1/2s)^n}{(2\ln2)^{n/2}n!}$	$\ldots$	$(s/\ln2)^n$
a_1	0	0	0	$s/\ln2$
a_2	$s^2/12$	$s^2/16\ln2$	∞	0
a_3	0	0	0	0
a_4	$-s^4/240$	$-s^4/512(\ln2)^2$	∞	0

[a] The values of a_2 and a_4 for a Gaussian broadening are the same as given by Allen et al. (20). If the formulas of Jones and Missell (21) are corrected, they also give the same results. [b] For triangle and Gaussian, n is even.

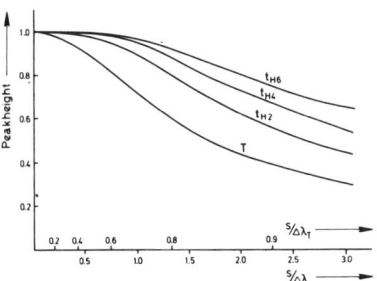

Figure 1. Influence of higher derivatives on the recovery of the peak height after deconvoluting a Gaussian peak of width $\Delta\lambda_T$ for another Guassian profile of width s. The deconvolution should produce a Gaussian of unit height and width $\Delta\lambda$

T = height of broadened band, t_{H_2} = height of band after deconvoluting with the 2nd derivative, t_{H_4} = height of band after deconvoluting with 2nd and 4th derivative, t_{H_6} = height of band after deconvoluting with 2nd, 4th, and 6th derivative

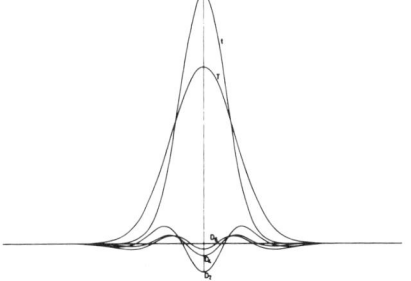

Figure 2. Contribution of higher derivatives to the recovery of a Gaussian band, t, after deconvolution for an equally broad Gaussian ($s/\Delta\lambda = 1$)

T = band after broadening, t = original Gaussian band, D_2 = difference between the deconvoluted curve, and t, using only the second derivative, D_4, D_6 = as D_2 but including the 4th and 6th derivative

which in this case is again a Gaussian, t, with an area equal to the area of T and a width equal to $\Delta\lambda = \sqrt{(\Delta\lambda_T^2 - s^2)}$. This example is chosen, because the derivatives to the Gaussian profile and the deconvolution can be expressed in closed form.

Not surprisingly, the number of terms required in Equation 6 to recover the true profile, t, depends upon the relative widths of the broadening function and the true profile, $s/\Delta\lambda$. Figure 1 shows the recovery of the peak height as a function of this ratio. It is clear that the contribution of the sixth derivative is already very small and that the results for $s/\Delta\lambda \geq 1$ are poor anyhow. (For symmetrical functions the odd derivatives are unimportant, of course.)

Figure 2 presents the complete profiles of the original function, t, and the broadened function, T, for $s/\Delta\lambda = 1$. The difference between the deconvoluted profile and the true original profile is also shown for a successive number of terms. Again, the influence of the sixth derivative is small, but the fourth derivative cannot be ignored. Because of the presence of noise, however, the fourth derivative will already be seriously distorted in any practical observation. Now, Figure 2 also shows that the difference, D_2, between the true profile and the result of deconvolution taking only two terms very closely resembles the shape of the second derivative to the experimental function, T. This suggests that a better fit can be obtained, if the coefficient a_2 is modified.

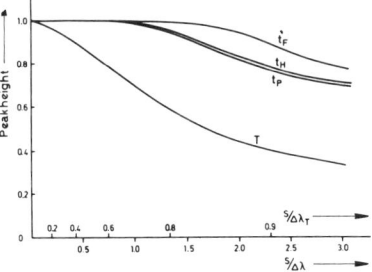

Figure 3. Recovery of peak height from the convolution of two Gaussian bands as in Figure 1

T = height of broadened band, t_H = height of band after deconvolution with Equation 8, t_F = height of band after deconvolution with Fourier method, t_P = height of band after pseudo-deconvolution

For the case of two Gaussian functions, it can be derived mathematically that in the peak maximum ($x = 0$) the fourth term $a_4 T^4(0)$ can be replaced by $ka_2(s/\Delta\lambda_T)^2 T^2(0)$, where the coefficient k depends only slightly upon the width ratio. By analyzing a number of different cases, we have found that good results can be obtained, if the coefficient of the second derivative is changed to

$$a_2' = a_2[1 + 1.3(s/\Delta\lambda_T)^2] \qquad (7)$$

Table II. Results of Deconvolution of a Single Band System

Original profile, t (width $\Delta\lambda$, unit height)	Broadening function, B (width s, unit area)	$s/\Delta\lambda$	Peak heights after Convolution, T	Deconv. Eq. 6, 2 terms, t_{H_2}	Deconv. Eq. 8, t_H	Root mean square deviations in % of original peak height Between t and t_{H_2}	Between t and t_H
Gaussian	Gaussian	0.25	0.97	0.99	1.00	0.05	0.03
		0.5	0.90	0.98	1.00	0.34	0.22
		0.75	0.80	0.94	1.00	1.12	0.78
		1	0.70	0.89	1.00	2.38	1.70
Gaussian	Triangle	0.25	0.96	0.99	1.00	0.14	0.13
		0.5	0.90	0.96	1.00	0.42	0.26
		0.75	0.80	0.92	0.98	1.39	0.94
		1	0.70	0.86	0.94	2.94	2.19
Lorentzian	Triangle	0.25	0.96	0.99	1.00	0.14	0.13
		0.5	0.87	0.95	0.97	0.65	0.53
		0.75	0.79	0.90	0.93	1.56	1.36
		1	0.70	0.83	0.88	2.86	2.42
Lorentzian	Gaussian	0.25	0.95	0.99	1.00	0.13	0.14
		0.5	0.90	0.95	0.97	0.53	0.64
		0.75	0.80	0.91	0.94	1.36	1.19
		1	0.70	0.85	0.92	2.41	2.20

Table III. Deconvolution of an Exponentially Broadened and Shifted Gaussian of Width $\Delta\lambda$ and Unit Height

Relative width of exponential $s/\Delta\lambda$	Recovered peak height	Root mean square dev, %	Shift/$\Delta\lambda$ After broadening	After deconv.
0.25	1.00	0.48	0.30	0.0
0.50	1.00	0.49	0.45	0.0
1.00	1.00	0.49	0.50	0.0
1.25	0.99	0.50	0.65	0.0
1.50	0.99	0.53	0.70	0.0

Substitution into Equation 6 then yields the final expression

$$t(x) = T(x) - a_1 \frac{dT(x)}{dx} - a_2[1 + 1.3(s/\Delta\lambda_T)^2] \frac{d^2T(x)}{dx^2} \quad (8)$$

As is shown in Figure 3, this expression allows a complete recovery of the peak height from a convolution of two Gaussian function up to $s/\Delta\lambda = 1$. Results for other combinations lead to similar conclusions as will be discussed below (Table II). It is also seen that deconvolution by means of Eqaution 8 is fully comparable to the pseudo-deconvolution method (compare curves t_P and t_H). The Fourier transform procedure provides slightly better results for wider broadening functions (however, see Table IV). The reasons for the limitation of the former two methods are obvious. In the pseudo deconvolution method, the points of intersection of the experimental profile, T, and its first convolute remain unchanged as remarked earlier. In the proposed deconvolution method based on Equation 8, the points of inflection of the experimental profile, T, will remain unaltered, because the second derivative is zero. These inflection points of the experimental profile T generally do not coincide with the inflection points of the true profile t.

The recovery of a complete profile is shown in Figure 4A for a combination of two Gaussian profiles with $s/\Delta\lambda = 0.75$. The difference between the original profile, t, and the deconvoluted curve, t_H, is always less than 2% of the peak height, as is shown by the difference curve, D.

EVALUATION

In order to assess the potential and the limitations of the proposed procedure, it was subjected to several tests

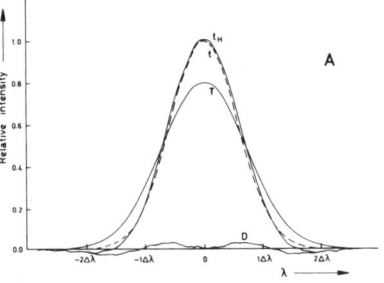
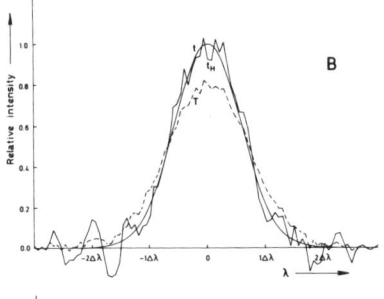
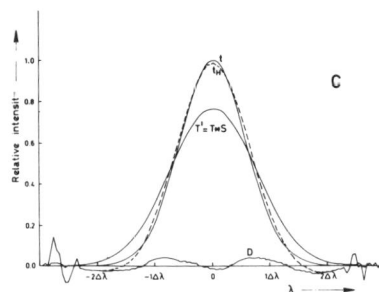

Figure 4. Recovery of a single Gaussian band in the presence of noise

(A) t = non-broadened Gaussian, width $\Delta\lambda$; T = after broadening with Gaussian, width $s = 0.75\ \Delta\lambda$; t_H = deconvolution of T using equation 8; $D = t_H - t$

(B) Same as (A), but with noise superimposed on T; no additional smoothing

(C) Same as (B), but including smoothing of T with a triangle, S, of width $0.5\ s = 0.37\ \Delta\lambda$

Table IV. Comparison between Different Methods of Deconvolution and the Influence of Noise

	Original profile, t (width $\Delta\lambda$ unit height)	Broadening function B (width s unit area)	Present method		Fourier transform		pseudo-deconv.	
			Peak height	rms dev, %	Peak height	rms dev, %	Peak height	rms dev, %
$s/\Delta\lambda = 0.75$,	Gaussian	Gaussian	1.00	0.78	1.00	0.04	0.98	0.33
spectra	Gaussian	Triangle	0.98	0.94	0.98	0.52	0.98	0.38
without	Lorentzian	Gaussian	0.94	1.19	0.99	0.30	0.96	0.71
noise	Lorentzian	Triangle	0.93	1.36	0.98	0.47	0.95	0.82
$s/\Delta\lambda = 0.75$,	Gaussian	Gaussian	0.98	2.29	0.99	2.25	0.96	3.03
spectra	Gaussian	Triangle	0.97	2.13	1.01	3.05	0.96	3.16
with 3%	Lorentzian	Gaussian	0.95	2.43	0.96	3.03	0.91	3.60
noise and smoothed	Lorentzian	Triangle	0.94	2.23	0.96	3.45	0.91	3.71

using computer programs written in FORTRAN IV and run on an IBM 360/65 computer. The program calculates convolutions of any two functions and then branches into deconvolution routines either through Fourier transformation or pseudo-deconvolution or using Equation 8. The derivatives needed in the last option are calculated from a nine-point third degree polynomial after Savitsky and Golay (27, 28). This is the most simple polynomial possible. Indeed, a third degree polynomial is necessary to enable calculation of the third derivative which may be required in the deconvolution of asymmetric functions. A minimum of nine points is required to adequately fit this polynomial over a limited portion of the experimental spectrum, where each band is described by a minimum of 30 points. In the following discussion, it will be assumed that the broadening function $B(x)$ is constant over the spectral range of interest and that it is known or can be independently determined.

Because the deconvolution works only for $s/\Delta\lambda_T < 0.7$, the correction factor in Equation 7 is less than 0.65, so that precise knowledge of the experimental band width, $\Delta\lambda_T$ is not required. In most cases, a sufficiently accurate value will be known beforehand but, if necessary it can be determined from a representative band in the spectrum. Problems arise if the band width varies appreciably from one band in the spectrum to another. In that case, the spectrum must either be divided into different parts that are treated separately or the present method must be abandoned.

A major problem is noise. Although the polynomial fitting of Savitsky and Golay reduces the noise, it does not eliminate it completely and excessive noise was found to deteriorate the deconvolution. This can be seen from a comparison of Figures 4A and 4B where the same curves are shown without noise and with noise, respectively. Although the result of deconvolution, t_H, undulates around the true original profile, t, the noise superimposed on the experimental profile, T, is substantially increased even though polynomial fitting has been used. It is, of course, possible to suppress the noise in t_H by another smoothing routine. However, such smoothing generally distorts the spectrum, and this is unacceptable if the true profile, t, is desired.

However, the approximate nature of the present deconvolution procedure provides an elegant way to reduce the noise further, prior to deconvolution. To this order, the experimental spectrum is smoothed by convoluting it with a triangle, with a width intermediate between the total band width, $\Delta\lambda_T$ and the width of the noise peaks. Because the order of deconvolution can be changed freely, the spectrum obtained after smoothing can be recovered

(27) A. Savitsky and M. J. E. Golay, *Anal. Chem.*, **36**, 1627 (1964).
(28) J. Steiner, Y. Termomia, and J. Deltour, *Anal. Chem.*, **44**, 1906 (1972).

by deconvolution with a function, which is the convolute of the broadening function $B(x)$ and the smoothing triangle. This procedure can also be utilized in the Fourier transform method. The results of this operation are shown in Figure 4C. In this case, the width of the smoothing triangle is 0.5 s. Because the broadening function is expanded by the smoothing triangle, the deconvolution result in Figure 4C is poorer than in Figure 4A, but in comparison with Figure 4B, the noise is greatly reduced.

The true potential of this method of noise reduction becomes apparent, when the broadening function $B(x)$ is not analytically known, but must be determined experimentally (for example, by scanning a narrow band profile). This independent measurement will also be subject to noise and this is reflected in the deconvolution through the calculated coefficients a_n. If, however, the same smoothing triangle is used for smoothing the broadening function and the subsequently measured spectrum, then the result is exactly the function needed for the deconvolution of the spectrum, i.e.,

$$T' = T * S = (t * B) * S = t * (B * S) = t * B'$$

where * denotes convolution and T' and B' are the results of smoothing the experimental spectrum T and the broadening function B with the smoothing triangle S.

One final point is the location of the boundaries of the broadening function. It is clear from Equation 4 that the influence of noise in the wings of the broadening function is greatly magnified in the calculation of a_n through the multiplication by x^n. It is therefore important to analyze the broadening function carefully to establish the proper boundaries for the calculation of a_n. A premature cut-off underestimates the a_n-values but, after a certain point, only noise is sampled. A few trial runs have been found to be very helpful.

RESULTS

The applicability of the proposed deconvolution method was checked by deconvoluting different functions. The results obtained for a single band system are presented in Tables II, III, and IV. It might be mentioned here that the objective of the proposed method of deconvolution is a correct recovery of the complete profile. The realization of this objective is judged from two criteria. The first criterion is the root mean square deviation of all data points between the original function and the result of deconvolution. A second criterion is the recovery of peak height, which is very sensitive to convolution broadening and provides an estimate of the maximum absolute deviation.

Table II presents the deconvolution for either a Gaussian or a triangular broadening function to recover an originally Gaussian or Lorentzian profile. The peak height, which is seriously decreased upon broadening, is adequately recovered up to $s/\Delta\lambda = 1$. The improvement of

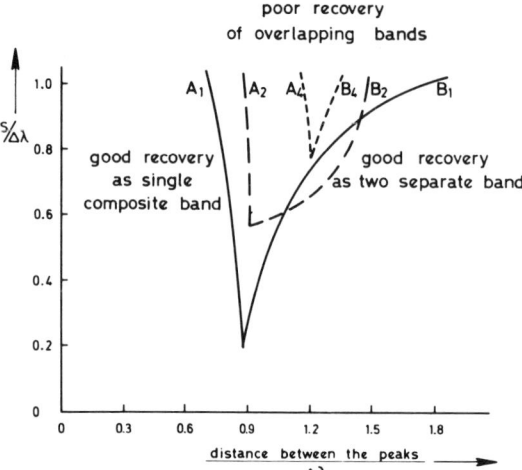

Figure 5. Recovery of a two-band system consisting of two Gaussians of equal width $\Delta\lambda$ and broadened by a triangle of width s; further explanation in the text

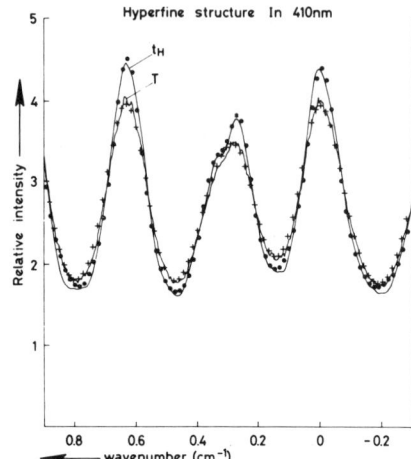

Figure 7. Deconvolution of the atomic emission line profile of In 410 nm

T = experimental spectrum; t_H = after deconvolution; (x-x) indicates convolution of t_H with Gaussian broadening function (s = 0.09 cm^{-1}); (●-●) represents curve fitting with four pure Lorentzian curves, yielding the following data

component	Width, cm^{-1}	Position, cm^{-1}		Intensity	
		Found	Expected	Found	Expected
1	0.16	0	0	31	33
2	0.12	0.27	0.28	21	22
3	0.13	0.37	0.38	14	12
4	0.15	0.65	0.66	34	33

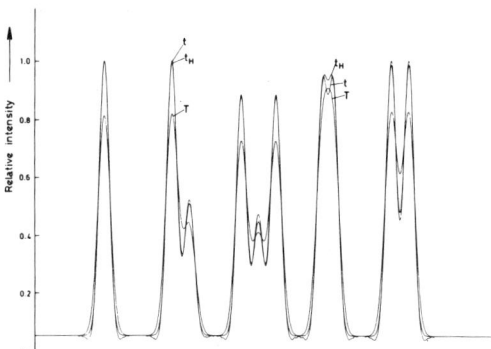

Figure 6. Deconvolution of a composite system of ten Gaussian bands broadened by another Gaussian; $s/\Delta\lambda$ = 0.75.

t = original band system, T = after broadening, t_H = result of deconvolution

using Equation 8 over Equation 6 with only two terms is seen from a comparison of columns 5 and 6. The final two columns present the root mean square deviation between the deconvoluted curves and the original function. Again, the modification implied in Equation 8 offers a substantial improvement for Gaussian profiles and some improvement for Lorentzian bands.

If the broadening function is an exponential, Table I predicts that only the first derivative is needed in Equation 6 to obtain an exact result. This is demonstrated in Table III, from which it is seen that both the peak height and the peak position (which is shifted by the exponential) are excellently recovered up to high values of $s/\Delta\lambda$.

Table IV compares the present method with pseudo-deconvolution and Fourier transform procedures for $s/\Delta\lambda$ = 0.75. In the absence of noise, the present method is fully comparable to the pseudo-deconvolution, but inferior to Fourier transformation. In the presence of noise, however, the present method is superior to either of the other two methods. This is shown by the data in the lower half of the table where, in all deconvolution routines, about 3% peak-to-peak noise has been superimposed and smoothed by convolution with a triangle indicated earlier. Compare also Figures 4B and 4C, where T and T' are the experimental profile before and after smoothing, respectively. Smoothing apparently reduces the noise to less than 1%, but hardly affects the recovery of the original profile through deconvolution.

The recovery of a composite band system depends not only upon the ratio $s/\Delta\lambda$, but also upon the separation between the bands and their intensity ratio. The results for a system of two Gaussian bands of equal width, $\Delta\lambda$, and intensity ratios between 1 and 4 are presented in Figure 5 for a triangular broadening function. For small band separation, the system is deconvoluted into the original composite band; this is the area to the left of the curves A_1 (equal intensity), A_2 (intensity ratio two), and A_4 (intensity ratio four). For large separations, the system is deconvoluted into two separate bands; this is the area to the right of the curves B_1, B_2, and B_4. In the intermediate area, the deconvolution fails. Here two inherently slightly overlapping bands have merged into one because of broadening, and the restriction to only the second derivative (Equation 8) prevents the present deconvolution method from recovering the original unbroadened band system.

Figure 6 presents the deconvolution of a synthetic spectrum consisting of 10 Gaussian bands of equal width, broadened by another Gaussian of width s = 0.75 $\Delta\lambda$. In general, recovery is seen to be good except for the two peaks of equal intensity with a separation equal to the peak width $\Delta\lambda$; this could be expected from Figure 5.

PRACTICAL APPLICATIONS

The first example is the deconvolution of the hyperfine structure of the atomic emission line of indium at 410 nm, emitted by a flame and measured with a Fabry-Perot in-

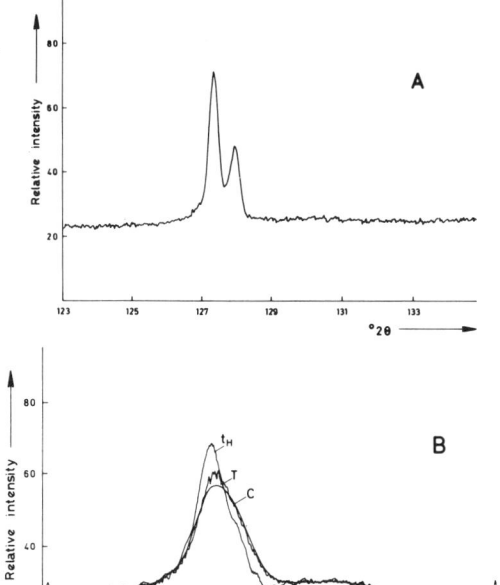

Figure 8. Deconvolution of X-ray diffraction profile

(A) Broadening function; Cu Kα radiation reflected from large magnesium oxide particles

(B) T = experimental curve after reflection from small magnesium oxide particles, t_H = result of smoothing and deconvolution of T, C = convolution of t_H with profile shown in (A)

terferometer (29). Each of the four hyperfine components has a width of 0.19 cm^{-1} and is the convolution of an unknown Lorentzian with a known Gaussian with a width of 0.09 cm^{-1}. Figure 7 shows the experimental profile, T, the spectrum deconvoluted for the Gaussian, t_H, and the result of convoluting this curve with the known Gaussian. The root mean square deviation between the experimental profile and the convoluted curve is 1.1%. If the result of deconvolution is curve-fitted with four Lorentzians to resolve the central, naturally overlapping bands, both the peak positions and the relative intensities are excellently recovered. Contrary to expectation, however, the width of the Lorentzian profiles is found to be unequal. The extreme bands are nearly completely Lorentzian and their width agrees well with previous data (29). The recovery of the central bands is somewhat less than adequate due to the strong overlap.

The second example concerns the correction of the X-ray diffraction profile of the {422} reflection of an MgO-powder sample for broadening caused by the finite width of the Cu-Kα lines of the irradiation source and geometrical factors of the instrument. This complex and highly asymmetrical broadening function has been measured using a sample of relatively large and perfect MgO crystallites (Figure 8A).

Deconvolution of the experimental profile (Figure 8B, curve T) for the broadening function shown in Figure 8A leads to a profile from which structural effects can be derived (8). The result of the deconvolution is shown in Figure 8B (curve t_H). The noise in the experimental spectrum T is smoothed as described above, and deconvolution is executed with first and second derivative. The correction implied in Equation 8 has been included, although it is small because each individual Kα line in Figure 8A is relatively narrow. For the calculation of the coefficients α_1 and α_2 (Equation 4), the zero point for x is chosen at the maximum of the highest peak.

The deconvolution has been checked by convoluting the result with the broadening function, yielding curve C in Figure 8B. The root mean square deviation between curve C and the experimental band is 0.7%. The results obtained were found to be fully comparable to those obtained from Fourier transform analysis.

CONCLUSION

The deconvolution procedure presented in this paper offers several advantages over existing methods. Over the Fourier transform method, it has the advantage that the base line in the observed spectrum, T, need not be subtracted beforehand. Just as in the pseudo-deconvolution method only the broadening function must be corrected for a non-zero base line. However, the pseudo-deconvolution only starts when the broadening function completely covers the recorded spectrum for the first time, and it ends when the broadening function is about to drop from the recorded spectrum. The combined losses in data points at the front and the end of the experimental spectrum are then equal to the total number of points used to define the broadening function. In the present method, only the first (and the last) four points in the experimental spectrum are lost, because a nine-point cubic function is used to calculate the derivatives.

No doubt, the main advantage of the present method is its capability to execute the deconvolution in real-time, while the spectrum is being recorded. This not only reduces the time between the completion of a spectrum scan and the presentation of the deconvoluted result (which is rarely important), but it also reduces substantially the amount of computer memory required. In the absence of noise, deconvolution can begin after a sufficient number of data points have been collected to initiate the calculation of the derivatives to the spectrum. In the presence of noise, the number of data points is slightly larger to execute any smoothing routine desired.

ACKNOWLEDGMENT

The authors wish to express their gratitude to R. Delhez and E. J. Mittemeyer of the Laboratory of Metallurgy for the measurement and the Fourier transform deconvolution of the MgO diffraction profiles, for many helpful discussions, and for careful reading of the manuscript. The authors are indebted to H. C. Wagenaar for the measurement of the indium hyperfine spectrum.

RECEIVED for review October 23, 1973. Accepted January 28, 1974.

(29) H. C. Wagenaar and L. de Galan, *Spectrochim. Acta, Part B*, **28**, 157 (1973).

Copyright © 1976 by the Society for Applied Spectroscopy
Reprinted from *Appl. Spectrosc.* **30**:495–500 (1976)

A Dynamic Background Correction System for Direct Reading Spectrometry

R. K. SKOGERBOE,* P. J. LAMOTHE,† G. J. BASTIAANS,‡ S. J. FREELAND and G. N. COLEMAN

Department of Chemistry, Colorado State University, Fort Collins, Colorado 80523

A direct reading emission spectrometer system is described which permits dynamic background correction measurements at each analytical wavelength. The system utilizes a refractor plate mounted on a tuning fork as a rapid means for square-wave shifting the spectral band pass from each analytical wavelength to the adjacent background wavelength position. A laboratory computer controls the system, receives the data, and performs the background subtraction. Evaluation of the system has shown that photon counting statistics are applicable and that reliable background corrections are obtained. Comparisons of this correction approach with others that have been used have further demonstrated its advantages for multielement analysis by emission spectrometry.

Index Headings: Direct reading emission spectrometer; Background correction; Multielement analysis.

INTRODUCTION

The problem of correction for background emission in direct reading optical spectrometry has been discussed in several publications.[1-8] These have generally shown that the detection capability for direct reading spectrometry is often limited by the magnitude and the reproducibility of the background observed at any analytical wavelength. Some reports have emphasized that the background intensities observed may be highly affected by relatively minor changes in the general sample composition.[1-5,7] These have suggested that the background fluctuations observed may account to an appreciable extent for inaccuracies in the analyses. Consequently, the development of means for monitoring and correcting of background has received considerable attention.

Several approaches to background correction have been used. It has been common to measure the background emitted by blank samples at each analytical line and use this for correction. The utility of this approach depends on how well the background from the blanks simulates that from the samples and how reproducible such measurements are from one sample to another. A second approach has involved movement of an alignment refractor plate behind the entrance slit to permit the measurement of background just off each analytical wavelength with actual samples. This may produce better estimates of the actual sample backgrounds but again depends ultimately on the reproducibility of the background between differing samples. Weekley and Norris[4] made corrections by dedicating one readout channel to background measurement at a selected wavelength. The method relies on the assumption that a change in background intensity at the monitor wavelength is accompanied by a directly proportional change in background at all other analytical wavelengths. An analog computation system was used in conjunction with a multiple regression equation to demonstrate that this assumption was generally valid for spark excitation of samples of quite similar composition.[4] When the assumption was imposed on arc excitation and the major variations in background emission so often associated with it, the approach was found much less useful. In an attempt to improve on it, Thompson and Bankston[5] used several background monitor positions to measure and compute background regression equations of higher order and thence the correction at each analytical wavelength. While the general concept is good, the results obtained were not particularly impressive in terms of precision.

Efforts have also been made to measure photoelectrically the background and the line plus background signals simultaneously at each analytical wavelength. Saunderson et al.[1] described a system which utilized a prism and a mechanical shutter at the focal plane to allow measurement of the line plus background and background adjacent to the line for alternate 1 sec intervals using the same photomultiplier tubes. This system, also used by Carpenter et al.,[2] cannot be easily employed for more than a few readout channels. Leys[6] described a system based on the use of offset exit slits coupled with a mechanical chopper at the entrance slit to alternately examine the analytical and background signals. The principal difficulty encountered derived from the use of different areas of the photocathode surfaces. Calibration of each photomultiplier for the differences in photocathode sensitivity for the different areas used was required. Data were presented indicating that this background correction system was instrumental in extending the linearity of analytical curves by factors of three to ten with corresponding improvements in detection capabilities.

Received 28 April 1976.
* Author to whom requests for reprints should be addressed.
† Present address: Environment Protection Agency, Research Triangle Park, 27711.
‡ Present address: Department of Chemistry, Georgetown University, Washington, DC 20007

Recently, Gordon et al.[8] described a modification of the rotating refractor plate method for short segment optical scanning reported by Snelleman et al.[9] A refractor plate behind the entrance slit rotates through a small angle thereby shifting the spectrum passing the exit slit. A mercury-wetted reed relay phased to the plate rotation was used to switch the line plus background signal to an integrating capacitor and the adjacent background signal to another capacitor. After measuring the two signals over an appropriate integration period, a simple subtractive capacitor discharge network was used to obtain the background corrected signal. The authors[8] claim an increase in measurement precision by a factor of 3 over background correction techniques which rely on separate photomultiplier tubes located at adjacent wavelengths. They also observed an extension of the linearity of analytical curves to lower concentration levels.

In considering the general analytical requirements and the experiences of others, the following operational characteristics for a background correction system may be formulated. The system should permit measurement of the background at reproducibly selectable positions adjacent to the analytical lines in a rapidly alternating time segment mode with a single photomultiplier at each wavelength. The present report describes a system designed for this purpose and presents data indicative of the value of this approach compared to others.

I. DESCRIPTION OF THE SYSTEM

The system designed and tested in the present study is conceptually illustrated in Fig. 1. Light from the excitation medium enters the spectrometer slit, passes through the optical scan refractor plate (SRP) mounted at an angle to the path of travel, and then passes through an alignment refractor plate (ARP) to the grating. The dispersed radiation is then directed to the exit slits and photomultiplier at each analytical wavelength by passage through the exit alignment refractor plates. Thus, the only addition to the optical system has been the scan refractor plate. The SRP is mounted on a tuning fork operating at 100 Hz so it can be moved into, and out of, the optical path in a square wave mode. The function of this plate is to displace the entrance slit image incident on the grating by a distance, d, which (for small angles) can be approximated by[9]:

$$d = t\theta\left[\frac{n-1}{n}\right] \quad (1)$$

where t is the plate thickness, θ is the angle of the optical path from the plate normal in radians, and n is the index of refraction of the quartz plate. The values of t and θ used in the present spectrometer (1.5 m, Jarrell-Ash, model 66-000) were 1.6 mm and 12°, respectively, as dictated by the dispersion characteristics of the spectrometer and the optical scan increment required. The spectrometer has a reciprocal linear dispersion (plate factor) of 0.56 nm/mm. The effective width of an emission line at the focal plane may be approximated by the product of the slit width and the plate factor.[10] Thus, the line width for the present system was 0.014 nm for the 25 μm entrance slit used. For a maximum exit slit width of 100 μm, the spectral band pass is 0.056 nm in the first order or 0.028 nm in the second. The plate thickness and angle were consequently selected to provide wavelength shifts of 0.06 and 0.03 nm in the first and second orders, respectively. Consequently, the photomultipliers are illuminated for alternate 5 msec half-cycles with the line plus background signals over the spectral band pass increment of each readout channel when the scan refractor plate is out of the light path and the background signals over the same wavelength intervals adjacent to each line when the plate is resident in the path. Moreover, the same area of each photocathode is illuminated during each half-cycle so spatial variations in photosensitivity are not encountered. To minimize rise and fall times in the optical signal, the leading edge of the SRP was ground to a nominal 12° so the thickness is constant when the plate enters the optical path.

The computer control and readout system conceptually illustrated in Fig. 2 was utilized. The square wave, on-line and off-line photomultiplier currents from each of n analytical wavelengths (channels) are converted to a voltage and sent to a multiplexor and A/D convertor unit (Zeltex, model 7210-120-025-245). The computer (Data General Corp., Nova 2/4) also receives a reference signal from the tuning fork drive which has been converted to square wave and electronically phased with the channel signals. The reference signal is used under program control to direct the multiplexor to sample the voltage from a given channel and perform the A/D conversion and storage operation. For the present system, a total of approximately 15 μsec are required for these operations. Thus, sampling of 20 channels in sequence can be repeated about 16 times for each 5 msec scan half-cycle. The computer program used is such that preselected readout (measurement) intervals ranging from 10 msec to several minutes may be utilized. The printout provides the background signal (B), the line plus background signal (L + B), and the net signal (L) for each successive measurement interval selected as well as the integral of the net line signal over a selected total measurement period as shown in Table I. Thus, examination of the signals on both time resolved and integral bases is possible.

It should be emphasized that the design of the sys-

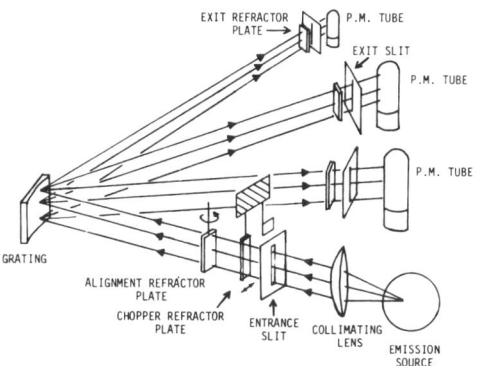

FIG. 1. Schematic diagram of the system.

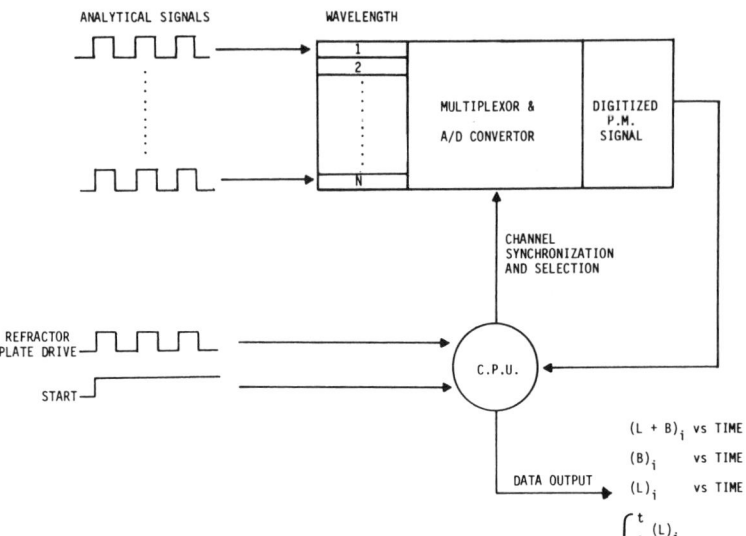

FIG. 2. Schematic of computer control and readout system.

TABLE I. Sample readout data for the plasma source.

Time (sec)	Intensity ($\times 10^{-7}$)			
	B	L + B	(L + B) − B	Integral
0–2	0.0166968	0.4131664	0.3964696	0.3964696
2–4	0.0163126	0.4126018	0.3962892	0.7927587
4–6	0.0166876	0.4152318	0.3985442	1.1913029

tem described in general terms above has resulted from experimental examination of other design possibilities. A more complete description of the system is available from the authors. The advantages of this design relative to others are pointed out in the following section.

II. CHARACTERIZATION OF THE SYSTEM

The present design is a modification of that based on the use of a rotating refractor plate as described by Snelleman et al.[9] and utilized for a polychromator by Gordon et al.[8] Their approach of oscillating a refractor plate continuously resident in the optical path was rejected in the present development because such rotation provides an optical scan that is sine wave in nature ranging from $\lambda - d\lambda$ to $\lambda + d\lambda$ with the analytical wavelength centrally located between these extremes. As a result, the analytical line is centrally located in the spectral band pass only during the peak period of the sine wave scan period while the background signals on either side of the line are measurable only during the trough periods of the sine wave. Consequently, only about 10 to 20% of the time during a scan cycle can be devoted to signal measurement; the signals received during the remainder of the cycle are essentially rejected by phase sensitive detection methods. For sequential sampling of 20 channels at a scan rate of 100 Hz, one would be restricted to sampling each line plus background and background signal 1 to 3 times per scan cycle. Since sequential sampling in this manner is essentially time averaging, one may predict on the basis of the central limit theorem that the measurement precision will improve with the number of measurements per channel per scan cycle. The choice of the square wave optical scan approach was thus justified on the basis of minimizing the dead-time required of the readout system and the anticipated increase in analytical precision. Moreover, less phasing problems could be expected for the square wave scan approach.

Kelley and Horlick[11] have indicated that a prime requisite of a digital data readout system is the ability to represent accurately the analog input signal. As a test of this, the computer was programmed to run at the maximum data acquisition rate while ignoring the optical scan status signal. Hollow cathode lamps were used to obtain variable, but stable, intensities for several elements. The computer-acquired data were compared with the analog signals observed with an oscilloscope. An example of the results is given in Fig. 3 where each point represents one cycle ~15 μsec through the measurement, conversion, and storage routine. The square wave nature of the scan may be readily noted. The observed rise and fall times during which the refractor plate edge is crossing the optical path may also be noted to be approximately 0.15 msec. Consequently, about 94% of the total cycle period may be devoted to actual measurement. These checks demonstrated a direct correspondence between the analog and digital signals in terms of intensity and in terms of reproduction of the fluctuations in the photomultiplier outputs due to shot noise. The digital representation was accordingly judged accurate.

Other measurements were made to assess the ultimate precision attainable with the system. Again, hollow cathode lamps were used as light sources because they are the most stable source available. Typical results are summarized in Fig. 4 where the relative standard deviation was calculated from 10 replicate measurements for each integration time. Two inferences may be drawn from these results. First, the experimental points closely follow the square root of integration time trend predicted by the Central Limit Theorem such that photon counting statistics may be considered applicable.[12] Second, if one accepts the hollow cathode sources as the most stable excitation media likely to be encountered, the precision plateau of about 0.2% for integration times beyond 60 sec may be taken as an estimate of the limiting precision of the total system.

Having established the analog representation validity of the system and its limiting precision characteristics, we ran comparative experiments to determine whether a capacitor storage system would show advantages over the sequential sampling approach described above. To make this assessment, dual capacitor systems were set up for five elements such that the background signal was accumulated in one capacitor while the corresponding line plus background was received by the other.[6,8] Typical analytical calibration experiments were run on a series of aqueous standards using a microwave-induced plasma[13] as an excitation source for both the capacitor storage and computer readout systems using 120 sec integration periods. Analytical curves representative of the general case for the two readout methods are given in Fig. 5. The colinearity of the curves with slopes very near the theoretically expected value of unity is generally indicative of at least equivalency between the two approaches. The replicate data used to determine the analytical curves were also used to compute pooled estimates of the standard deviations for each measurement approach. These clustered closely around 3% (relative standard deviation) for the computer readout approach while values clustered closely around 6% were observed for the capacitor system. Although these differences were not inordinately large, they were significant at the 95% confidence level. It is suspected that the poorer precision for the capacitor system is due primarily to its susceptibility to switching transients and leakage problems. A final comparison of the two systems was based on the detection limits summarized in Table II. These are generally consistent with the precision data since approximately a factor of 2 improvement accrued from the use of the computer readout system. Based on these comparisons, the computer readout approach was judged to be at least equivalent, if not superior, to the capacitor readout system. All subsequent measurements were consequently based on utilization of the computer readout system.

Other experiments designed to demonstrate the utility of the system involved comparison of its capabilities with those derived from more common approaches to background correction. In one such experiment, analytical curves were determined for aqueous solutions introduced into a microwave-induced plasma.[13] The

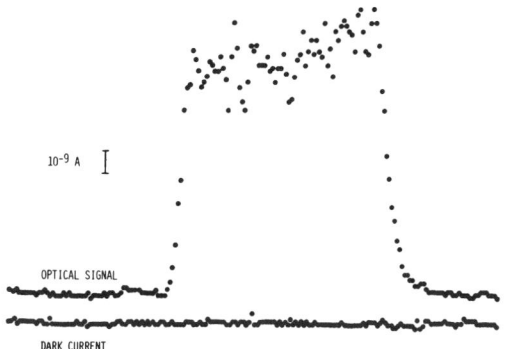

FIG. 3. Example of digital representation of the analog optical scan signal.

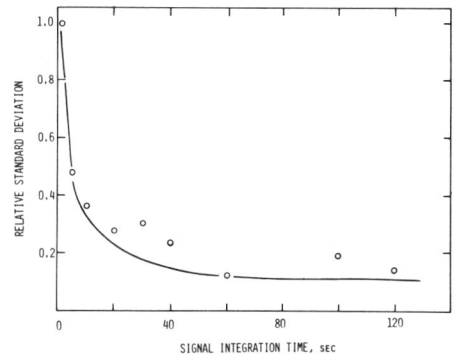

FIG. 4. Effect of measurement time on precision. Points are measured values; solid curve calculated from Central Limit Theorem.

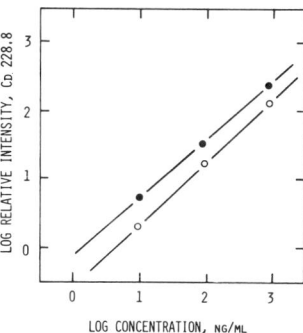

FIG. 5. Typical analytical curve for computer readout (●——●) and capacitor storage (○——○) measurement systems.

TABLE II. Comparison of detection limits obtained with the computer and capacitor readout systems.

Element	λ (nm)	Detection limits (μg/ml)	
		Capacitor readout	Computer readout
Cu	327.4	0.015	0.009
Cd	228.8	0.025	0.005
Fe	372.0	0.01	0.01
Pb	405.8	0.02	0.008
Zn	213.9	0.001	0.0005

standard solutions used were prepared from high purity materials dissolved in 0.09 M nitric acid doubly redistilled from quartz. Each solution and an acid blank were analyzed four times using 2 min integration periods. An example of typical results is given in Fig. 6.

For most elements studied, plots of the line plus background intensity alone indicated significance of background emission at low concentrations as evidenced by the nonlinearity shown in Fig. 6. When the line plus background readings for the blank solution were subtracted from these, as is often done to correct for background, some compensation was obtained which resulted in a decrease in the degree of nonlinearity at low concentrations. In the general case, however, curve tailing similar to that shown was still experienced indicative of the inadequacy of this means of estimating background. In every case, the measurement of the actual background adjacent to each analytical wavelength for each solution resulted in a more reliable correction and a concomitant extension in the linearity of the analytical curves. This is clearly due to the fact that the background emission observed is dependent on the actual sample composition even though all other experimental parameters are the same. In the present experiments, the analytical solutions used contained five or more elements. No two elements were present at the same concentration in any solution and the respective concentrations ranged in steps from 0.01 to 3 µg/ml. Under these circumstances, it was not unusual to observe background intensities for an analytical line that varied 20 to 50% or more from one solution composition to another.

It should be emphasized that the background intensities are measured while the quartz scan plate is resident in the optical path. Since the quartz has a finite transmission value less than unity, the background intensity measured is less than its actual value. Correction for this is obtained simply by dividing the measured background intensity value by the transmittance of the quartz plate at each respective wavelength. These corrected values are subsequently subtracted from the line plus background intensities measured with the SRP out of the optical path.

Background intensities measured with the present system were also used as means of evaluating the approach in which several readout channels are dedicated to background measurement and used to fit a regression equation from which corrections are calculated for the analysis wavelengths.[4,5] Examples of regression equations obtained by using seven and eight background measurement wavelengths spaced at about 30 to 40 nm increments throughout the range are shown in Fig. 7. A polynomial least-squares fit was made for each set of data in which the order of the equation for the best fit was chosen, i.e., for seven background channels a third-order equation provided the best fit while a fourth-order equation for the eight channel data best. Examination of the figures indicates major differences in the background intensities predicted by the equations even though data for seven of the wavelengths used for the computations were the same in both instances. These equations were used to compute background intensities for five analytical wavelengths and these were compared with the actual measured values. The results given in Table III are indicative of the general insufficiency of this approach. As expected, the discrepancies, indicated by the relative differences between the measured and computed values, depend on the wavelength. The occurrence of

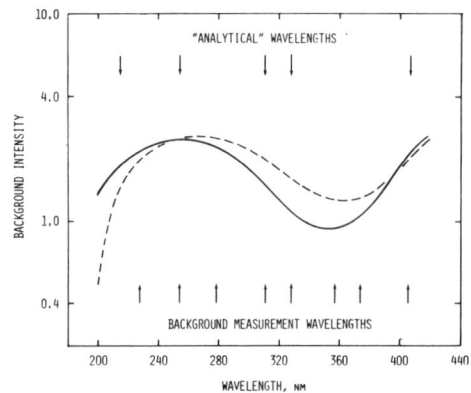

FIG. 7. Background intensity vs wavelength curves obtained from regression equations fit to measured intensities. ———, measurement of background at seven wavelengths; third-order equation for best fit; - - -, measurement of background at eight wavelengths; fourth-order equation for best fit.

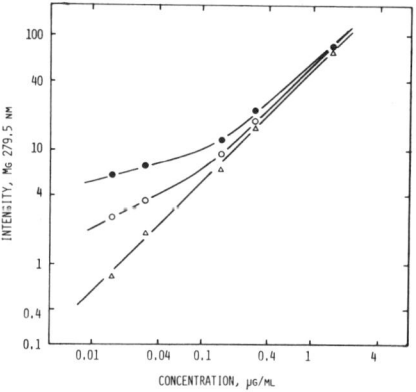

FIG. 6. Typical effect of different methods of background correction on analytical curves. ●———●, line + background intensity, ○———○, [(line + background) − blank intensity], △———△, [(line + background) − background intensity].

TABLE III. Comparison of background intensities estimated from regression equations with measured values.

No. of background measurement wavelengths	Order of "best-fit" equation	% Error = 100 (measured intensity − computed intensity)/measured intensity				
		Zn 213.8 nm	Hg 253.6 nm	Al 308.2 nm	Ag 328.1 nm	Pb 405.8 nm
5	3	−118	−22.1	97.6	−76.8	31.2
7	3	70.2	−50.9	−8.8	−88.9	28.9
8	4	33.4	−46.4	17.1	−85.3	18.7
Average		73.9	39.8	50.9	83.7	26.3

discrepancies of nearly 100% or more, even though the computations are based on the use of many more background measurement wavelengths than most users would be willing to commit, strikingly demonstrates that this approach is grossly inadequate.

In view of these comparisons, the present system shows several advantages over other approaches to background correction. These are more completely delineated in other reports by Skogerboe et al.[14]

ACKNOWLEDGMENT

This research was supported by National Science Foundation Grant GI-44423.

1. J. L. Saunderson, V. J. Caldecourt, and E. W. Peterson, J. Opt. Soc. Am. 35, 681 (1945).
2. R. O'B. Carpenter, E. DuBois, and J. Sterner, J. Opt. Soc. Am. 37, 707 (1947).
3. M. Slavin, Appl. Spectrosc. 16, 173 (1962).
4. B. E. Weekley and J. A. Norris, Appl. Spectrosc. 18, 21 (1964).
5. G. Thompson and D. C. Bankston, Spectrochim. Acta 24B, 335 (1969).
6. J. A. Leys, Anal. Chem. 41, 396 (1969).
7. P. Frank, P. Hahn-Weinheimer, and G. Markl, Appl. Spectrosc. 25, 529 (1971).
8. W. A. Gordon, K. M. Hambidge, and M. L. Franklin, in *Aplications of Newer Techniques of Analysis*, I. L. Simmons and G. W. Ewing, Eds. (Plenum, New York, 1973).
9. W. Snelleman, T. C. Rains, K. W. Yee, H. D. Cook, and O. Menis, Anal. Chem. 42, 394 (1970).
10. F. A. Jenkins and H. E. White, *Fundamentals of Optics* (McGraw-Hill, New York, 1957).
11. P. C. Kelley and G. Horlick, Anal. Chem. 45, 518 (1973).
12. H. V. Malmstadt, M. L. Franklin, and G. Horlick, Anal. Chem. 44, 63A (1972).
13. F. E. Lichte and R. K. Skogerboe, Anal. Chem. 45, 399 (1973).
14. R. K. Skogerboe and G. N. Coleman, Appl. Spectrosc. 30, 504 (1976).

A Computer-controlled Photon Counting Spectrometer for Rapidly Scanning Low Light Level Spectra

T. M. NIEMCZYK and D. G. ETTINGER
Department of Chemistry, University of New Mexico, Albuquerque, New Mexico 87131

A method of optimizing the scan time of a spectrometer based on the signal level is discussed and its implementation in a computer-controlled photon counting spectrometer is presented. Previous implementations of signal-dependent methods have been useful when the spectra being measured contain only regions where a measurable signal exists. This problem has been eliminated by taking advantage of the ability of the computer to make a decision. Data are presented showing how the use of an "intelligent" instrument, such as that described here, can increase the efficiency of any experiment.

Index Headings: Photon counting; Computer-controlled spectrometer; Luminescence spectroscopy.

INTRODUCTION

There are many instances in chemistry, physics, astronomy, etc., where the spectrum of a very faint source must be measured. In general, a photomultiplier tube is used as the photon flux to electronic signal transducer and the electronic signal is then detected by one of several different techniques. The techniques include dc current or voltage measurement, charge integration, synchronous detection, photon counting, and the shot-noise method.[1,2] The instrument described in this paper was designed to measure visible luminescence spectra obtained when certain materials are exposed to an infrared excitation source.[3] The very nature of the process makes most of these phosphors very inefficient and thus some of the spectra measured are exceedingly weak. Because of this application, the overriding consideration when the detection technique was chosen was that it be the optimum system for very low level signals. Photon counting has been shown both theoretically and experimentally to be superior to other detection techniques when low light levels are encountered.[4-8] On this basis photon counting was selected for use in the instrument described here.

Photon counting in spectrometric systems has been discussed and compared to other detection techniques by many authors.[4-13] These discussions point out many inherent advantages for photon counting when compared to other current measuring techniques. The signal is processed in a discrete manner, reducing the number of domain conversions,[14] leaving the information in a form directly compatible with the computer needed in thi

Received 28 April 1978.

application. The processing of information by digital circuitry makes photon counting detection less susceptible to long term drift and $1/f$ noise which often limits analog circuitry. This feature allows the spectroscopist to utilize much longer averaging times than would be feasible with analog techniques. Photon counting systems also include the circuitry to discriminate against photomultiplier dark current originating down the dynode chain and because the output is in digital form the reading error inherent to analog systems is virtually eliminated. The overall result is that at low light levels photon counting yields a higher signal/noise ratio (SNR) when compared to other techniques.

The spectrum shown in Fig. 1 is typical of the spectra being measured in our laboratory. In order to resolve the fine structure in the spectrum the slits of the monochromator must be kept narrow so that the resolution of the monochromator is high. This reduces the light flux through the monochromator and in the situation where a weak source is encountered, exceedingly long measurement times are needed to produce an acceptable SNR. Note that when a spectrum, such as that shown in Fig. 1, is being scanned much time is wasted measuring points where there is no useful information. Also, at the points in the spectrum where the signal level is high the measurement time can be significantly reduced and still produce an acceptable SNR. The spectrometer system described here takes advantage of these facts to scan a spectrum in the fastest possible time and still achieve the minimum acceptable SNR required.

I. SIGNAL-DEPENDENT TIME INTERVAL SCANNING

Most conventional spectrometers scan through a spectrum at a constant scan rate. The scan rate is selected so that a point, or resolution element, in the spectrum where the signal is small is determined to the desired precision. If the spectrum is very weak the total scan time can be prohibitively long. In order to speed up the measurement time the concept of signal-dependent time interval (SDTI) scanning has been introduced. This concept has been applied to analog systems[15] as well as to photon counting systems.[16,17] In these systems the scan time is controlled by feeding back SNR information such that more time is spent in areas of low SNR and less time is spent in regions of high SNR. Thus, as implemented, these systems reduce the scan time significantly only when the spectrum consists mostly of regions where the signal level is above the baseline noise level. In a spectrum such as that shown in Fig. 1, a great deal of time would be spent recording the regions between the phosphor emission bands because the SNR in these regions is at its lowest. In the system described here we take advantage of the decision making ability of the computer to determine how much time is needed at each point to produce an acceptable SNR in the same manner as discussed previously.[16,17] The computer, however, has the ability to determine that no measurable signal exists at a point so that no more time will be wasted on this point even though it is a point of very low SNR.

To see how the SDTI method is applied it is useful to consider the factors that determine the SNR in a photon counting experiment. The dark current pulses do not in general constitute noise, but it is the fluctuation in the dark current signal and the desired signal that constitute noise. The fundamental noise in a photon counting system is the fluctuation in the signal count as determined by counting statistics. The arrival of photons at the photocathode is essentially random; thus, the probability of fluctuations about the rate of arrival is determined by taking the square root of the number of counts.[18] If the assumption is made that a Gaussian distribution describes the arrival, the square root of the number of counts is equivalent to the standard deviation. Thus, one may estimate the SNR for a specific measurement by dividing the count by the square root of the count.

To be more analytical one must take into account the fluctuations in the dark current as well. The SNR expression becomes slightly more complicated.[19] The error, noise, in the signal count is calculated using the standard equation for the counting statistics of a difference. The equation is:

$$E_s = (R_s + nR_d)^{1/2} t^{1/2} \qquad (1)$$

where E_s is the error in the signal count, R_s is the signal count rate, and R_d is the dark count rate. The coefficient, n, is 2 when R_d is measured in the same time period, t, as the signal or is unity if R_d is measured separately to arbitrary accuracy using a long measurement period.

The observation time, t, was the same for the signal and dark count determinations described here so n will be treated as having a value of 2. The signal is $R_s t$; thus, the signal SNR can be expressed as:

$$\frac{S}{N} = \frac{R_s^{1/2} t^{1/2}}{(1 + 2R_d/R_s)^{1/2}} \qquad (2)$$

It can be seen from Eq. (2) that three factors determine whether a measurement is desirable: the minimum acceptable SNR, $(S/N)_{min}$; the weakest signal of interest, R_o; and the dark count rate, R_d. The minimum acceptable SNR will occur when the signal rate is equal to R_o, or:

$$\left(\frac{S}{N}\right)_{min} = \frac{R_o^{1/2} t^{1/2}}{(1 + 2R_d/R_o)^{1/2}} \qquad (3)$$

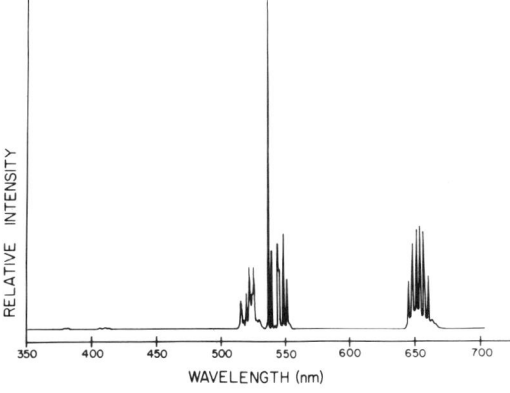

FIG. 1. Visible emission spectrum of $(Y_{0.80}Yb_{0.19}Er_{0.01})F_3$ obtained with excitation at 970 nm.

If the signal rate is large with respect to the dark count rate ($R_s \gg R_d$) the SNR reduces to:

$$\frac{S}{N} = R_s^{1/2} t^{1/2} \quad (4)$$

To compare scanning speeds these equations can be solved for the time needed to measure a point to the desired SNR. When $R_s = R_o$ the maximum amount of time, t_0, will be needed to reach $(S/N)_{min}$ and is given by:

$$t_0 = \left(\frac{S}{N}\right)_{min}^2 (1 + 2R_d/R_o)/R_o. \quad (5)$$

In the conventional fixed time interval methods this is the amount of time spent on every point so that the total scan time, T_s, is given by

$$T_s = Mt_0, \quad (6)$$

where M is the number of points in the spectrum. In the work of Pruett[16] and Wenke[17] the measurement time at each point, t, was calculated by replacing R_o with R_s in an equation similar to Eq. (5):

$$t = \left(\frac{S}{N}\right)_{min}^2 (1 + 2R_d/R_s)/R_s. \quad (7)$$

Thus, the measurement time needed to produce the desired SNR is determined at each point in the spectrum. At points in the spectrum where $R_s > R_o$ a considerable time savings is achieved and the net result is an overall increase in scanning speed. In this case T_s can be given by

$$T_s = Xt_0 + Y\bar{t} \quad (8)$$

where there are X points of $R_s \le R_o$, Y points of $R_s > R_o (X + Y = M)$ and $\bar{t}$ is the average time used to measure the points where $R_s > R_o$. In a spectrum of wide dynamic range $\bar{t}$ can be considerably less than t_0 so a substantial time savings is achieved.

The SDTI technique as implemented by Pruett[16] and Wenke[17] works well if the spectrum being measured consists entirely of a large band of varying intensity. It fails to improve the scan time significantly when the spectrum consists of widely separated peaks, between which the signal is of no interest. The photon counting spectrometer discussed here incorporates a laboratory computer for experimental control and data analysis. Thus, the decision-making capability of the computer can be used to alleviate the problem of wasting a great deal of time scanning spectral regions containing no information.

In the system utilized in our laboratory a short initial measurement is made at each point and the computer then makes a decision based on the count rate at that point. Three cases arise. In regions where R_s is large and $(S/N)_{min}$ is already realized no further measurement is needed. When $R_s \ge R_o$ but $(S/N)_{min}$ has not been realized the computer determines how long the measurement time need be and the measurement at that point is continued. In the third case $R_s < R_o$ and the point is of no interest; therefore, no additional time is spent measuring this point. The result is that a considerable savings in scanning time is achieved regardless of the structure of the spectrum being measured.

It should be pointed out that several experimental conditions must be realized for the SDTI method to be advantageous. The time required for the computer to make a decision and the time needed to step the monochromator to the next wavelength must be small compared to the measurement time. The time needed to perform the calculations and change the monochromator wavelength setting takes much less than a second, and the measurement time for the low level spectra for which this instrument was designed is often on the order of hours for a several hundred point spectrum. Thus, these requirements are easily met. In addition the desired SNR should be constant over the spectrum of interest.

II. EXPERIMENTAL

A block diagram of the computer-controlled spectrometer is shown in Fig. 2. The computer used in the experiments described here was a Digital Equipment Corp. PDP8/e equipped with 16K memory, a dual floppy disk system, a CalComp plotter and a real time clock. The monochromator used was a Heath model EU-700. The photomultiplier, several different kinds of which have been used, was housed in Products for Research model 1405 RF housing.

A mechanical shutter is available and was used when needed. For most of the measurements of the upconverting phosphors it was convenient to utilize a gallium arsenide infrared emitting diode as the source. The power supply designed to operate this diode had the feature of being modulated with a TTL level signal. When the diode source was used the shutter control was used to modulate the diode output. The ability to modulate the light is important in that the measurement has the advantages associated with synchronous detection. The frequency of modulation used in the instrument described here was quite low, but the effects of $1/f$ noise, or drift, is greatly reduced by using the synchronous mode of operation.

The amplifier/discriminator is a PAR model 1120. The counter is a 13-bit synchronous counter. The thirteenth bit is used as an overflow bit. When this bit changes state the computer increments a location in core so that no information is lost. The first two bits of the counter are made of high speed Shockty TTL flip-flops so that the counter is capable of operating as fast as the amplifier/discriminator. The remainder of the counter

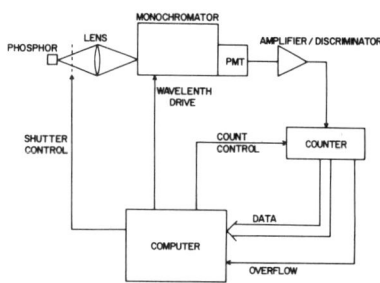

FIG. 2. Block diagram of the computer-controlled photon counting spectrometer.

was built from conventional TTL counter chips. An ECL-to-TTL converter was used to make the output pulses from the amplifier/discriminator compatible with the TTL counter. Upon completion of the count period the output of the 12-bit counter was gated into the accumulator of the PDP8/e and stored for further use. More specific details of the counter construction will be provided to any reader who requests them.

III. RESULTS

The spectrum shown in Fig. 3 is typical of the spectra being measured in this laboratory. This is the green band produced when a $(Y_{0.80}Yb_{0.19}Ho_{0.01})F_3$ phosphor is held at 0 °C and excited at 970 nm. This sample was chosen because it illustrates a typical example of a spectrum with regions of no signal and a wide dynamic range overall. The software normally used was modified to count the number of points in the spectrum that fell into each class.

The spectrum was scanned from 500 to 600 nm and a data point was recorded every 0.1 nm for a total of 1000 data points. An upper time limit for each point was set at 100 s. Thus, in fixed time interval photon counting the spectrum would take about 38 h to record if 100 s were required to count each point. In the methods of Pruett[16] and Wenke[17] a considerable amount of time would be saved recording the data points between 530 and 560 nm, but the remaining portion of the spectrum contains no signal and the SDTI methods they used would require the maximum amount of time measuring these points. Thus, about 20 h would be used to measure the spectrum.

In the method used here 654 points fell into the class of no information and thus only 2 s were spent at each of these points. At 11 points the SNR was at the desired level after the initial 2-s measurement period. The SNR at the remaining points was found to be improvable by increasing the measurement time.

The overall measurement time used to record this spectrum was slightly less than 7 h, which included the wavelength setting and calculation time. Note that only about 22 min of this time was used to record the spectrum in regions of no signal, and the balance, about 6.5 h, was spent recording the emission band. Thus, most of the measurement time is spent producing an acceptable SNR in regions of the spectrum where there is useful information. This is in direct opposition to the previously employed SDTI methods.

IV. CONCLUSION

The system described here has been in operation for several months and has proved to be a very powerful tool for the recording of very low light level spectra. The system has proved itself to be very stable, even over very long runs, and very convenient due to the lack of attention required from the experimenter after the initial set up. The advantage of having spectra stored in a digital form ready for processing, i.e., smoothing or plotting, is also not small.

Many experimental apparatuses have been improved via a feedback system. Generally, information about some measurable parameter is fed back to the controller to add stability or flexibility to the system. The instrument described here illustrates the power of having a computer to control an experiment, where the control is based on information fed from the experiment to the computer. Thus, when an intelligent controller, such as a computer, is employed in an instrument the experimental parameters can be automatically optimized to produce the maximum amount of information from any particular experiment. Although the system described here is far from complex, it does illustrate a large gain in efficiency due to the computer acting as an intelligent controller as opposed to merely a sophisticated data acquisition system.

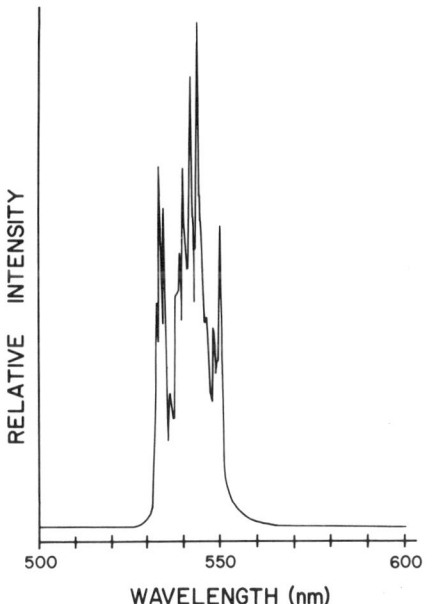

FIG. 3. Green emission band of $(Y_{0.80}Yb_{0.19}Ho_{0.01})F_3$ obtained with excitation at 970 nm.

1. Y.-H. Pao, R. N. Zitter, and J. E. Griffiths, J. Opt. Soc. Am. 56, 1133 (1966).
2. Y.-H. Pao and J. E. Griffiths, J. Chem. Phys. 46, 1671 (1967).
3. D. G. Ettinger and T. M. Niemczyk, J. Chem. Phys. 68, 872 (1978).
4. R. G. Tull, Appl. Opt. 7, 2023 (1968).
5. J. D. Ingle, Jr. and S. R. Crouch, Anal. Chem. 44, 785 (1972).
6. J. K. Nakamura and S. E. Schwarz, Appl. Opt. 7, 1073 (1968).
7. M. K. Murphy, S. A. Clyburn, and C. Veillon, Anal. Chem. 45, 1468 (1973).
8. R. Jones, C. J. Oliver, and E. R. Pike, Appl. Opt. 10, 1673 (1971).
9. K. C. Ash and E. H. Piepmeier, Anal. Chem. 43, 26 (1971).
10. F. Robben, Appl. Opt. 10, 776 (1971).
11. M. L. Franklin, G. Horlick, and H. V. Malmstadt, Anal. Chem. 41, 2 (1969).
12. J. Rolfe and S. E. Moore, Appl. Opt. 9, 63 (1970).
13. A. T. Young, Appl. Opt. 8, 2431 (1969).
14. C. G. Enke, Anal. Chem. 43(1), 69A (1971).
15. Astronomische Nachrichten 196, 356 (1913).
16. H. D. Pruett, Appl. Opt. 11, 2529 (1972).
17. D. C. Wenke, Ph.D. thesis, Michigan State University, 1972.
18. S. Goldman, *Frequency Analysis, Modulation and Noise* (McGraw-Hill, New York, 1948), p. 306.
19. G. A. Morton, Appl. Opt. 7, 1 (1968).

24

Copyright © 1972 by the American Chemical Society
Reprinted from *Anal. Chem.* 44:2282–2287 (1972)

Signal Enhancement in Real-Time for High-Resolution Mass Spectra

F. W. McLafferty, John A. Michnowicz, Rengachari Venkataraghavan, Peter Rogerson, and B. G. Giessner
Department of Chemistry, Cornell University, Ithaca, N.Y. 14850

An on-line, real-time computerized method for effectively increasing the sensitivity, resolution, and mass measuring precision of a high-resolution mass spectrometer has been developed. This method for Signal Enhancement in Real Time (SERT) utilizes the relatively large vacant areas between peaks to rescan peaks in real-time under direct computer feedback control. The ensemble-averaged rescans have an increased signal/noise ratio when compared to the single scans and significantly increase the effective sensitivity, resolution, and mass measuring precision of the instrument without increasing the scanning time, in contrast to most methods for ensemble-averaging of spectral data.

THE DIRECT APPLICATION of computers to mass spectrometry has made possible many advances in the field of mass spectrometry. The power of the computer in acquiring, reducing, and displaying data has been well documented. High-resolution mass spectrometry especially has benefitted by the tremendous speed and efficiency of the computer in collecting data, computing exact masses, and relating these masses to elemental compositions (*1–6*). The computer, however, has not, up to this time, been used in improving the rather inefficient process of magnetically scanning a high-resolution mass spectrum. When compared to photoplate detection which is a time-integrated process, magnetically scanned mass spectrometric data have a low information content per unit time (*2*). These methods have approximately comparable sensitivities only when a high-gain electron multiplier is used with the latter. At a resolution of 10,000, a *m/e* 100 peak in a magnetically scanned spectrum will be only 0.01 amu wide, and therefore a spectrum will exhibit peaks in less than 1% of

(1) "Biochemical Applications of Mass Spectrometry," G. Waller, Ed., John Wiley & Sons, New York, N.Y., 1972.
(2) John Roboz, "Introduction to Mass Spectrometry, Instrumentation and Techniques," Interscience Publishers, New York, N.Y., 1968, pp 361–4.
(3) J. S. Halliday, in "Advances in Mass Spectrometry," E. Kendrick, Ed., Vol. 4, Institute of Petroleum, London, England, 1967, p 239.
(4) F. W. McLafferty, R. Venkataraghavan, J. E. Coutant, and B. G. Giessner, ANAL. CHEM., **43**, 967 (1971).
(5) R. J. Klimowski, R. Venkataraghavan, F. W. McLafferty, and E. B. Delany, *Org. Mass Spectrom.*, **4**, 17 (1970).
(6) D. H. Smith, R. W. Olsen, F. C. Walls, and A. L. Burlingame, ANAL. CHEM., **43**, 1796 (1971).

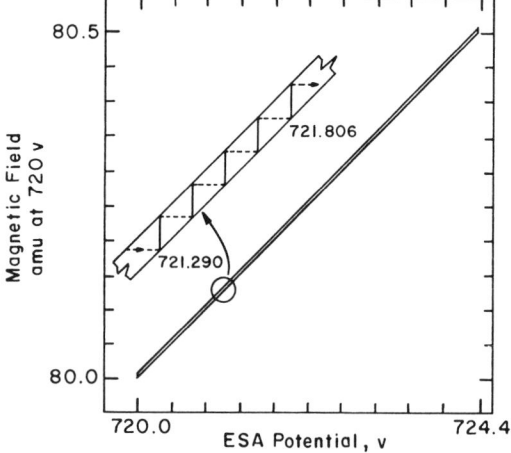

Figure 1. Effect of ESA potential on magnetic field required to focus the *m/e* 80 peak. The insert illustrates the incremental shifts in ESA potential which make possible the rescanning of the peak at a higher magnetic field

the mass axis (*3*). The other 99% of the time is used waiting for a peak to appear on the detector. At a resolution of 100,000, 99.9% of the mass axis is "vacant" time. In a preliminary communication, it was suggested that these vacant regions between peaks can be used to collect useful data (*4*). By placing the mass spectrometer under computer control and rescanning each peak repeatedly after it originally passes the detector, the time between peaks can be used to increase the information content of a spectrum. We report here how this method of signal enhancement in real-time (SERT) can effectively increase the sensitivity, resolution, and mass measuring precision of the mass spectrometer. A multiscan averaging technique has been used to improve mass measuring accuracy (*5, 6*), and other spectrometric techniques use similar methods of multiple scanning (*7*), but all require that the time for data

(7) M. Silverstein and G. Bassler, "Spectrometric Identification of Organic Compounds," J. Wiley & Sons, New York, N.Y., 1968, p 114.

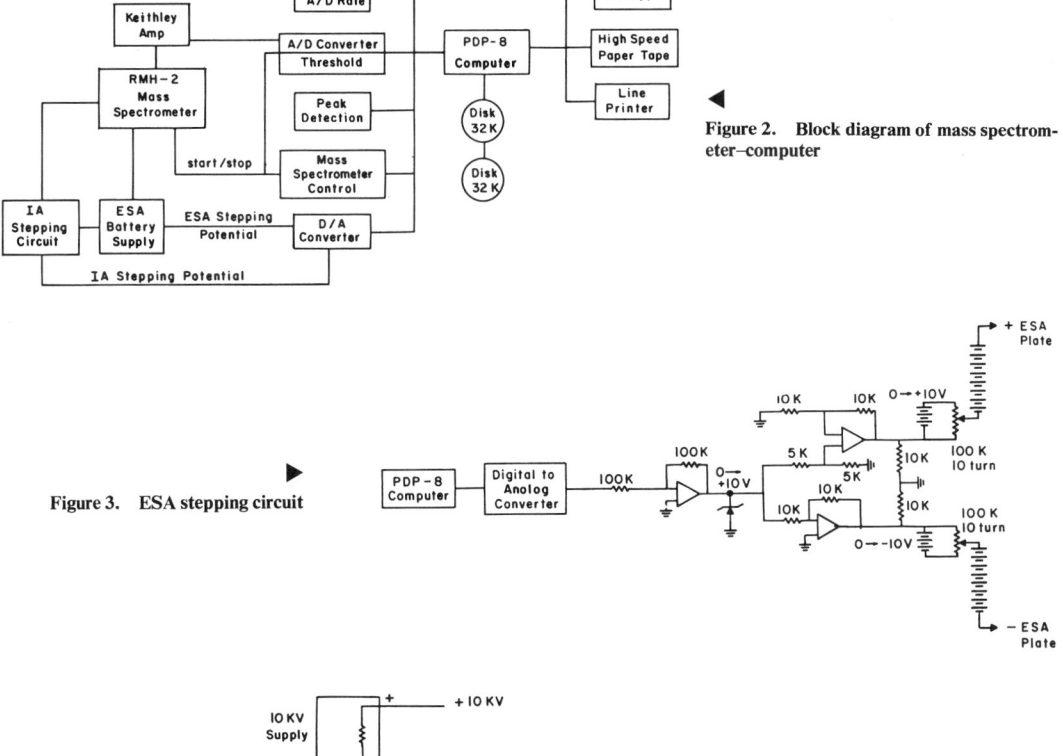

Figure 2. Block diagram of mass spectrometer–computer

Figure 3. ESA stepping circuit

Figure 4. IA stepping circuit

collection be increased in proportion to the signal enhancement achieved.

Three methods were considered for rescanning a peak once it has passed the detector in a magnetically scanned, double-focusing, high-resolution mass spectrometer. A small opposing magnetic field may be introduced into the system, but the large time constants needed to change magnetic fields makes this method impractical. A change in either the ion accelerating (IA) or electrostatic analyzer (ESA) voltage can also be used to rescan a peak. We initially investigated the ESA potential change method because of the smaller voltages present in the ESA circuit and also our laboratory had previous experience in computer control of the ESA potential, developed for defocused metastable analysis (8).

The finite width of the β-slit between the ESA and magnetic sectors allows the ESA potential to be varied over a small range (720 ± 4.4 V at maximum slit width in the Hitachi RMH-2) without defocusing the main ion beam (9). Thus as the magnet is continuously scanning, the ESA potential can be varied to keep a peak on the detector. The possible combinations of ESA and magnetic field values produce the "ion ridge" illustrated in Figure 1; as the magnetic field is increasing, a peak at m/e 80 will be in focus at an ESA potential range of 720.0 to 724.4 volts, making it possible to rescan the peak many times by incrementing the ESA voltage. At mass 80, the ion ridge extends for a half mass unit, and at mass 600, 3.3 mass units. There are major disadvantages to this method; maximum instrument resolution cannot be achieved with the β-slit at its maximum setting, the resolution changes across the ESA potential range, and only a limited degree of offset is possible.

In a similar fashion, the same mass can be brought back into focus at a higher magnetic field by increasing the IA relative to the ESA potential. The IA potential can be changed over a

(8) J. E. Coutant and F. W. McLafferty, *Int. J. Mass Spectrom. Ion Phys.*, **8**, 323 (1972).

(9) John Roboz, "Introduction to Mass Spectrometry," Interscience Publishers, New York, N.Y., 1968, p 61.

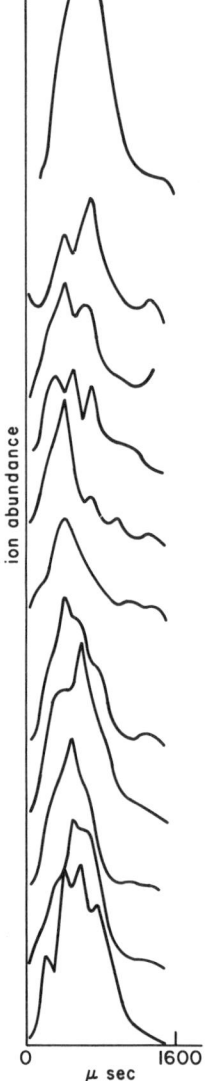

Figure 5. Rescanned m/e 77 peaks; abscissa is time (mass), ordinate is ion current. Bottom: first five and last five rescans. Top: ensemble-average of 45 rescans

10,000. For the IA/ESA stepping the β-slit was closed to approximately 0.90 mm and adjustment of other slits gave a resolution of 20,000.

ESA Modifications. The existing ESA power supply was replaced by two battery stacks, one for each ESA plate. The low potential side of each stack is driven by an operational amplifier as shown in Figure 3. The voltage supplied by a digital-to-analog converter (D/A) is applied to the bottom of both battery stacks. An amplifier is used as a line driver to enhance the low output current of the D/A. The potential of the ESA plates can be changed in approximately 1 μsec; a slight modulation extends the effective response time to 5 μsec.

IA Modifications. Stepping of the 9.2-kV accelerating voltage is achieved by adding the incremented D/A voltage to the bottom of the RMH-2 Hitachi power supply. The normal ground of the power supply is disconnected, thus allowing the 10-kV supply to be incremented in voltage steps supplied by the D/A. Grounding of the power supply is then made through the load resistor network of the stepping circuit, as seen in Figure 4. When used in the IA/ESA stepping mode, the voltage increments required for the ESA plates are obtained from the same circuit through a resistor network which uses a 500-ohm potentiometer for fine adjustment of the IA/ESA ratio. The IA can presently be stepped a maximum of 100 V, although a 1000-V stepping circuit is being planned for our future system. Rise time of the IA step varies from a minimum of 5 μsec for steps of less than 5 V to a maximum of 13 μsec for a 100-V step.

Additional Modifications. The existing electron multiplier and signal amplifier were replaced with a Bendix Spiraltron Model 4219-X in-line electron multiplier whose output is connected to a Keithley 421 current amplifier.

Hardware. The mass spectrometer was coupled to a Digital Equipment Corporation PDP-8 computer with 4K of core, two 32K random access disk units, a high speed paper tape punch and reader, and a Shepard 880 high speed line printer. The output of the Keithley amplifier is connected to the computer via a 12-bit unsigned analog-to-digital converter (A/D) with a sample-and-hold amplifier, a 24-bit binary counter driven by a 20-kHz crystal oscillator, a 24-bit buffer register, and logic for automatic sequencing of all interface functions; operation of this system for normal high-resolution data collection has been described previously (5). Output from the computer to the stepping circuits is made through a 12-bit D/A developed for automated metastable ion analysis (8). All data were collected at 10 kHz, which with a 60 sec/decade scan speed gave approximately 20 data points per peak at 10,000 resolution.

Software and System Operation. The system operation and programming techniques for SERT will be demonstrated by outlining the operations required to rescan selected peaks in a spectrum by incrementing the ESA voltage. A scheme for applying SERT to an entire spectrum will then be discussed.

At a resolution of 10,000, a peak at m/e 80 will have a width of 8 millimass units (mmu) and contain approximately 25 data points at a 10-kHz digitization rate. (Thus if a 5% valley definition of resolution is used, the peak width will be 25 data points at 5% of its height.) With these conditions, the computer calculates that a shift of 0.129 V in the ESA potential will cause a displacement in the mass scale equivalent to one peak width, permitting the m/e 80 ion to be rescanned. With the total ESA potential range of 4.4 V available in the RMH-2, 4.4 V /0.129 V or 34 rescans of the m/e 80 ion are possible. A graphical representation of these rescans is shown in Figure 1. At a resolution of 50,000, the peak width is reduced to 0.0016 amu and 170 rescans could be obtained. However, the peak would contain only four data points at a 10-KHz digitization rate. If 20 data points are desired for adequate identification of the peak profile (3), a 50-kHz digitization rate would be required. By stepping the IA/

much wider range of values (at least ±10%), permitting a longer mass range to be rescanned and making dynamic peak matching possible. Also the β-slit can be closed to a setting which allows optimum instrumental operation despite the changing IA values. Since the IA/ESA stepping method appears to be preferable, we have recently developed the circuitry to step the IA/ESA potential.

EXPERIMENTAL

Instrumentation. A block diagram of the instrumentation is shown in Figure 2. The mass spectrometer is a Hitachi RMH-2 modified to allow computer control of the ESA or IA potential. It operated at a 9.2-kV accelerating potential and a scan speed of 60 seconds/decade.

In the ESA stepping experiments the β-slit was opened to its maximum, 3.7 mm, and the resolution was approximately

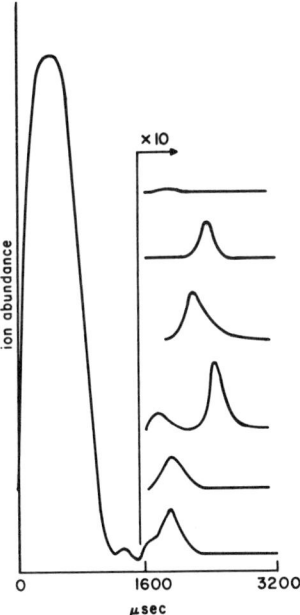

Figure 6. Left: $C_5H_5N^{.+}$ peak. Right: 5 rescans of base line and (Top) ensemble-average of 40 baseline rescans

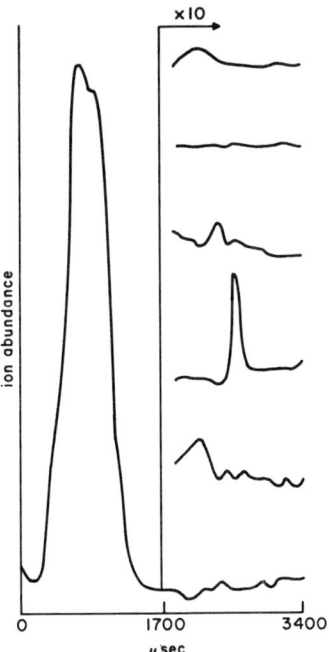

Figure 7. Left: $C_5H_5N^{.+}$ peak. Right: 5 rescans of $^{13}CC_5H_6^{.+}$ ion and (Top) ensemble-average of 40 rescans

ESA voltage, the same m/e 80 peak at a 10,000 resolution can be rescanned by changing the IA 1.25 volts. The present IA voltage range of 100 V permits 80 rescans to be performed.

The limited 4K of core in our PDP-8 necessitates that the program be divided into two phases. The first phase calculates the necessary IA/ESA or ESA voltage increments, collects data, and, when time becomes available, transfers it to the disk. Phase two retrieves the data from the disk, sums all the individual rescans of an ion ridge, and outputs the composite profile and centroid time.

RESULTS AND DISCUSSION

Alignment and Signal Enhancement. The exact mass range covered in each rescan must be invariant for proper averaging of the scans; for this the computer must calculate the exact offset potential required to rescan a peak. Figure 5 shows rescanned peaks containing 15 data points (approximately 5 mmu) from the m/e 77 ion of benzene produced by stepping the ESA potential during magnetic scanning. The first five and the last five of 45 rescans show similar peak shapes, indicating that alignment has been achieved. The high frequency response of the amplifier produces peaks which are not electrically smoothed, but the random noise produced in the signal is almost completely eliminated in the composite of the 45 rescans, resulting in a smooth, Gaussian-shaped peak profile.

Sensitivity. By ensemble-averaging the data of multiple rescans of a particular portion of a spectrum and thereby increasing the signal/noise ratio, it should be possible to improve the sensitivity of the mass spectrometer and detect peaks previously lost in base-line noise. To test this concept, mixtures with varying concentrations of benzene and pyridine were made and the doublet at m/e 79 ($m/\Delta m = 9{,}745$) was rescanned.

The averaged base-line noise that could be expected was first determined using a sample of pure pyridine. Figure 6 shows the normal $C_5H_5N^{.+}$ ion and an 8.8 mmu, portion of base line where the $^{13}C^{12}C_5H_6^+$ ion of benzene would appear if it were present. Time requirements allow approximately 40 rescans, 5 of which are illustrated along with the ensemble average of the 40 rescans. A numerical analysis indicates a 3.5-fold increase in the S/N ratio of the composite. The theoretical increase should be proportional to the square root of the number of rescans, or 6.3. The occurrence of 60 Hz noise, clearly present in the base-line computer printouts, is most likely responsible for the discrepancy.

Figure 7 shows 5 of the 40 base-line rescans of a pyridine: benzene mixture which should produce a 125/1 peak-height ratio. Although some rescans show apparent peaks, these can be misleading; however, from the composite of the 40 scans, there is no doubt as to the existence and location of the second component. The observed mass difference between the two peaks is 8.7 mmu; theoretical separation is 8.3 mmu. Experimental peak height ratio is lower than theoretical, but well within our experimental limits. A 1000/1 peak-height ratio was found to be the lower limit of detectability under these conditions using 40 rescans.

Resolution. Figure 8 shows five ESA rescans of the $^{13}CC_6H_7^+ : C_7H_8^+$ doublet ($m/\Delta m = 20{,}595$) scanned at a resolution of approximately 5,000. The poor resolution and noise present in the signal obviously make recognition of the doublet and subsequent attempts to deconvolute (10, 11) any single scan impossible. However, the addition of 27 of these

---

(10) R. Venkataraghavan, F. W. McLafferty, and J. W. Amy, ANAL. CHEM., **39**, 178 (1967).
(11) D. D. Tunnicliff and P. A. Wadsworth, *ibid.*, **40**, 1826 (1968).

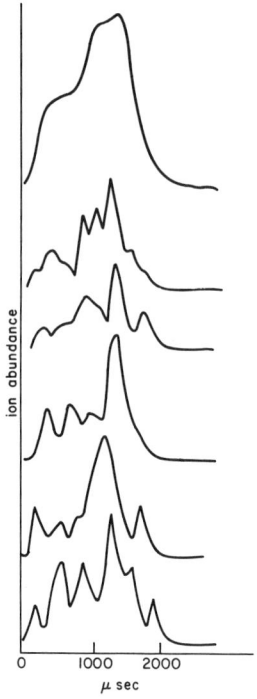

Figure 8. Bottom: 5 rescans of $C_6{}^{13}CH_7{}^+ : C_7H_8{}^+$ doublet. Top: Composite of 27 rescans

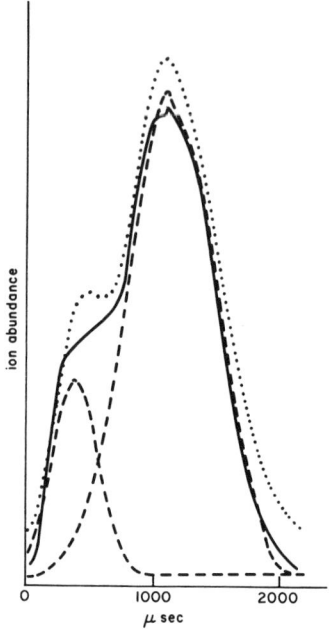

Figure 9. Deconvolution of $^{13}CC_6CH_7{}^+ : C_7H_8{}^+$ doublet

—— Composite
- - - Deconvolution of composite
.... Summation of deconvoluted composite (offset for clarity)

rescans produces a composite (Figure 8) which clearly indicates the presence of a doublet. This composite profile was deconvoluted (10) to the doublet shown in Figure 9. The experimental separation measured between the deconvoluted peaks is 4.56 mmu; the theoretical value is 4.47 mmu. Similar results were obtained from rescans obtained at 7,000 and 10,000 resolution.

Mass Measuring Precision. The increase in mass measuring precision utilizing a composite peak was investigated by rescanning the m/e 77, 78, and 79 ions of benzene. Centroid times were then computed for the composites of the rescanned 77, 78, and 79 peaks and also for a randomly-picked single scan from each of the rescanned peaks. A linear relationship between the centroid times and exact masses of the 77 and 78 ions was then used to calculate the mass of the 79 ion. Separate experiments in both the ESA and IA modes were performed 10 times with the minimum number of rescans for a peak being approximately 50. The results are shown in Table I. The theoretical decrease in the deviation should be $\sqrt{50}$ or 7.1. Our results are high by a factor of 2.5. Again it is believed that the discrepancy is due to a periodic noise signal, which unlike random noise will not be averaged in the composite. This hypothesis is supported by the fact that when 80% of the rescans in the ESA experiments are omitted, the deviation increases to only ±0.00027 amu.

It should be possible to effect further improvements in mass accuracy by dynamic peak matching. Thus if the unknown peak and a known reference peak are rescanned alternatively by proper offsetting of the IA potential, periodic fluctuations which would have occurred during the time necessary to scan between them magnetically will be reduced by averaging. Preliminary experiments support the feasibility of this concept.

Proposed System. The results obtained on narrow regions of a spectrum indicate that SERT can be applied to scan a complete high resolution spectrum. Experience with the present hardware indicates that the software can be written to scan a mass range of 60–600 in 60 sec utilizing a small computer system like the PDP-8 with 8K of memory, a bulk storage device like a 32-K disk, and an appropriate mass spectrometer/computer interface.

In our proposed general system, a low-resolution, high sensitivity scan is first taken from which the computer finds the time at which each peak appears in the spectrum and locates broad peaks which will require a higher number of data points per scan. The computer checks to see if any adjacent peaks in the spectrum are separated by a greater mass ratio than can be accommodated by the IA stepping range; if they are not, the full 60-sec scan time can be utilized in rescanning. The number of rescans possible per peak is then calculated from the time available for rescanning, taking into account the number of peaks and the time required to manipulate the data between peaks.

The high-resolution SERT mass spectrum is then run. Before each peak is scanned, the computer calculates the IA offset required by dividing the total IA range by the number of rescans. Each peak scan is directly ensemble-averaged by setting up buffer tables in the computer memory. If a peak is 15 data points wide, 15 core locations are set up, and the corresponding points from each rescan are added using multiple precision arithmetic to the previous totals to prepare an averaged peak envelope. The centroid and area of the envelope will be computed and stored if the peak is a singlet, or the entire profile will be saved for later deconvolution if the peak is an unresolved multiplet. Since the time required to move information from core to a bulk storage device like a disk is relatively high (~40 msec), all of the data will be held in core until the spectral scan is complete; for a normal spectrum, a core of 8K should be sufficient to accomplish this.

It also appears possible to utilize SERT under more demanding conditions, such as a scan speed of 6 sec/decade at 10,000 resolution or 60 sec/decade at 100,000 resolution. Such a system has been designed based on a computer with a 400-nsec memory cycle time (DEC PDP 11/45) and an interface with a high speed A/D or a pulse counter and digital logic for various control operations such as thresholding, multiple sampling of the A/D, and automatic updating of the D/A. Data transfers between the interface and the computer will be accomplished through high speed channels without any program supervision.

It may also be possible to obtain the necessary peak information without the preliminary low-resolution scan. If only a limited combination of elements is possible, substantial regions of the spectrum (especially at lower masses) cannot contain peaks and can thus be used for rescanning. Further, it may be feasible to predict in real time the elemental compositions possible at higher masses from those found to be present at lower masses. Special problems, such as the sequencing of peptide mixtures (12), required the detection and exact mass determination of peaks at only a limited number of possible masses, so that a much larger number of rescans and a concomitant increase in sensitivity should be possible for each peak.

Table I. Precision of 10 Separate Mass Determinations from Single Scan vs. Rescan Measurements

Mode	Single scan	Rescan
ESA	79.04323 ± 0.00062	79.04402 ± 0.00024
IA	79.04296 ± 0.00080	79.04308 ± 0.00030

CONCLUSION

The signal enhancement of high-resolution mass spectral data using SERT is unique in that it is performed on-line, in real time, and does not increase the time requirements for a spectrum. Previously mentioned methods for increasing S/N ratios (5-7) require many spectra to be obtained, an obvious disadvantage when dealing with the small sample sizes often required in mass spectral studies. This technique should be applicable to the recording of other spectra and chromatograms in which a substantial proportion of the base line does not contain any real data.

RECEIVED for review June 6, 1972. Accepted July 27, 1972. Financial support for this work was provided by National Institutes of Health Grant GM 16609.

(12) F. W. McLafferty, R. Venkataraghavan, and P. Irving, *Biochem. Biophys. Res. Commun.*, **39**, 274 (1970).

INFORMATION RETRIEVAL

Editors' Comments on Papers 25 Through 29

25   LYTLE
Computerized Searching of Inverted Files

26   GROTCH
Matching of Mass Spectra When Peak Height Is Encoded to One Bit

27   HERTZ, HITES, and BIEMANN
Identification of Mass Spectra by Computer-Searching a File of Known Spectra

28   WANGEN, WOODWARD, and ISENHOUR
Small Computer, Magnetic Tape Oriented, Rapid Search System Applied to Mass Spectrometry

29   FELDMANN et al.
An Interactive Substructure Search System

Searching files of chemical data is an important aspect of many branches of chemistry. It is particularly important in analytical chemistry for the identification of unknowns through comparison with known spectra. The five papers in this section present different approaches to the problems of searching files of analytical data.

As collections of spectra continue to grow, the ability to eliminate as much as possible of the file from consideration becomes increasingly important. In Paper 25 Lytle presents the use of inverted files for rapid searching of large spectral collections on a small computer. In an inverted file, data is stored by spectral characteristics rather than by compound. Instead of putting all the spectral data of one compound together as in a spectrum, one puts together the data for one spectral interval for all the compounds. The advantage in search characteristics is that one need only search a subset of the entire file, which is defined by the spectral characteristics of the unknown.

Hertz, Hites, and Biemann (Paper 27) report on an algorithm for the automatic comparison of unknown mass spectra generated by a combined gas chromatograph–mass spectrometer. The GC/MS combination is capable of producing far more information than can be manually processed. The authors have solved this problem by first reducing the data and then comparing the reduced mass spectra to a similarly reduced reference file. A similarity measure is calculated and the results of the search of the reference file reported to the operator.

Their algorithm uses an abbreviated reference spectra and a pre-search to eliminate spectra that are very dissimilar to the unknown spectrum. Search comparisons are reduced by using only the two largest peaks in each fourteen mass unit interval. The algorithm is explained very clearly and examples illustrate the method.

An important aspect of searching is the degree of data compression that can be achieved without significant loss of information. Grotch (Paper 26) uses statistical and information theory to show that one can reduce intensity information in mass spectra to the level of peak presence or absence and still retain most of the information in the mass spectra. This reduction to a single bit for each mass results in considerable saving of space, since sixteen masses can be coded in a sixteen-bit word. Computation time is also reduced because a fast "EXCLUSIVE OR" instruction can be used to make comparisons. When a data set described in the paper was encoded in this form, the number of times different compounds produced the same binary spectrum occurred only three times per 10^6 attempted matches, indicating that almost all the information necessary for search purposes is retained in the binary spectra.

Wangen, Woodward, and Isenhour (Paper 28) present a fast search system that does not require the library spectra to be held in memory. The spectra are coded in binary form and compressed so that 6652 low resolution mass spectra can be searched in fifteen seconds on a minicomputer. The paper is interesting in its use of information theory as a guide to reduce the data. The information content of each mass position can be calculated by considering each mass as a channel for carrying information. The information content will be maximized when each channel is on (has a peak present) half the time. For a collection of mass spectra, most mass positions have a probability of peak presence far less than half the time. When mass positions are combined, this probability can be brought closer to the maximum information content value of one-half. In general, positions differing by 1, 2, 13, 14, and 15 mass units are combined to create an average probability of 0.45 for the resulting "combined mass" positions. Comparisons between the unknown stored in memory and the library spectra on magnetic tape are made by the "EXCLUSIVE OR" instruction. This fast operation is combined with a look-up table of possible results to allow searches at a rate limited only by the speed of the tape drive.

One significant development in computer technology has been the growth of chemical information systems available to many users over networks of computers. Chemical information in the form of the *Chemical Abstracts* data base is available to be searched. Choosing the method of posing questions for these systems is of more than trivial importance. The method of encoding chemical information—in particular, chemical structures—is a topic of considerable interest.[1,2] Codes such as the Wiswesser Line Notation[3] are one method of encoding structures in a canonical, or unambiguous, manner. A linear notation has the desirable property of being in computer compatible form and therefore easily stored. Such codes have a problem in that the priority of structural representation often makes it difficult to search for substructures that are not easily decoded from the representation. Several well-known systems are discussed in the paper by Feldmann et al. (Paper 29). The ability to perform substructure searches is of particular use in fields such as synthetic organic chemistry. In an example from Feldmann et al., a search was defined for all structures with 5 and 6 atom fused rings that were not imbedded in a larger molecule. The structure was allowed to have a noncarbon atom at one specified node of the structure and a substituent at another node of the structure. A search of the NIH-EPA Mass Spectral Search System[4] yielded eighteen compounds with these structural features.

Heller et al.[5] have reported on the conversational mass spectral search system at NIH. They describe the dissimilarity index comparison option and the use of microfiche and graphics terminals.

REFERENCES

1. W. T. Wipke, S. R. Heller, R. J. Feldman, and E. Hyde, eds., *Computer Representation and Manipulation of Chemical Information*, Wiley, New York, 1974.
2. J. Ash and E. Hyde, eds., *Chemical Information Systems*, Wiley, New York, 1975.
3. E. G. Smith, *The Wiswesser Line-Formula Chemical Notation*, McGraw-Hill, New York, 1968.
4. S. R. Heller, G. W. A. Milne, R. J. Feldmann, and S. R. Heller, "An International Mass Spectral Search System (MSSS) V. A Status Report," *J. Chem. Inf. Comput. Sci.*, **16**, 176 (1976).
5. S. R. Heller, D. A. Koniver, H. M. Fales, and G. W. A. Milne, "Conversational Mass Spectral Search System," *Anal. Chem.*, **46**, 947 (1974).

Computerized Searching of Inverted Files

F. E. Lytle
Department of Chemistry, Purdue University, Lafayette, Ind. 47907

INFRARED SPECTRAL interpretation has been widely used as an analytical tool for both research and quality control applications. These absorption bands can indicate various functional groups and may ultimately lead to an identification of a sample's structure. The analysis is, however, only completely certain when a point-to-point spectral comparison has been made with a known sample. To this end an indexed collection of spectra is invaluable. As the volume of cataloged data has grown, various attempts have been made toward rapid retrieval of specific reference curves. Kenyon (*1*), in his excellent review article, discusses five classes of indexes used for such data retrieval. These classes and some of their characteristics are summarized in Table I. Curry, Read, and Brown (*2*), discuss the same type of indexes and arrive at the conclusion that edge notched cards are probably best for searching small collections of spectra (<200), inverted files (optical coincidence cards) for intermediate collections (~10,000) and computers for large collections. Kaiser (*3*) also favors inverted files for medium sized collections and discusses in detail their use with a Document of Molecular Spectroscopy literature index. Recently, several computer-oriented searches have been described. Anderson and Covert (*4*) have developed a sophisticated large scale routine for the IBM 7080 computer system. These authors also reduced the time involved for a typical search by decreasing the length of the data records corresponding to one compound. Erley (*5*) has gone several steps further by developing a computerized search for the smaller IBM 1130 utilizing a disk storage device. Although his search was based on a normal file, he went to some depth in reducing its size by packing the data within computer words and subsequently using logical masking operations to extract the necessary data. This report describes a computerized method for inverted file searching of infrared spectra. By combining the benefits of both the optical-coincidence and computer methods, it is possible to design a search for large collections of data having optimum characteristics.

The Inverted File. The particular collection of spectra described in this report were assembled from numerical infrared data available in Sadtler's Spec Finder (Sadtler Research Laboratories, Inc., Philadelphia, Pa. 19104). This spectral index is an organized listing of compounds according to their strongest absorption band in each micron interval from 2 to 14 microns. In a normal file, this data would be punched on cards, one spectrum per card. Then to search the file, all of the spectra in the collection would be examined to see if they matched the unknown. In the inverted file,

Table I. Characteristics of Data Retrieval Indexes

Class	Advantages	Disadvantages
Manual card or book indexes	Low cost; availability to user	Searching restricted to a few items at a time; difficult to update
Edge notch cards	Some of the data on same card; easy to update; availability to user	Handling difficult for large collections; new characteristics cannot be readily added
Punched cards for machine sorting	Automatic retrieval; easy updating	High cost of equipment and storage space
Optical coincidence cards	Small storage space; easy updating; availability to user	Cannot have a near hit (go-no-go)
Computers	High speed; low cost for multiple searches	Lack of availability to the user for non-time sharing systems

each spectral characteristic is considered a set of data. Each micron interval has ten possible characteristics corresponding to absorption in one-tenth micron intervals and one "no absorption" characteristic. To encode the Spec Finder in this fashion thus requires 143 sets of data. If the resultant files were in computer card form, each set would resemble that of Figure 1. Kaiser (*3*) actually shows an example of punched computer cards being used as optical-coincidence cards. The deck of cards can be thought of as a three dimensional data array A_{ijk}, where i represents the card sequence number, j represents the column, and k represents the row. Every compound in the collection corresponds to some element in the data array. Any one compound corresponds to the same position in all of the data decks. If any given card position contains a punch, the compound associated with that position has the absorption characteristic represented by that deck as the strongest band in the appropriate micron interval. To search a collection of spectra, several decks of cards corresponding to the strongest absorption bands in various micron regions of the sample are read into the computer and the corresponding arrays are numerically added together. If seven bands were used in the search, seven decks would be read in

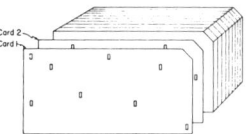

Figure 1. An inverted file in computer card form. Each card has 80 columns and 12 rows. Each deck would represent the strongest absorption in a given micron interval (corresponding to one optical-coincidence card)

---

(1) W. C. Kenyon, ANAL. CHEM., **35** (12), 27A (1963).
(2) A. S. Curry, J. F. Read, and C. Brown, *J. Pharm. Pharmacol.*, **21**, 224 (1969).
(3) H. Kaiser, *Hilger Journal*, November, 64 (1965).
(4) D. H. Anderson and G. L. Covert, ANAL. CHEM., **39**, 1288 (1967).
(5) D. S. Erley, *ibid.*, **40**, 894 (1968).

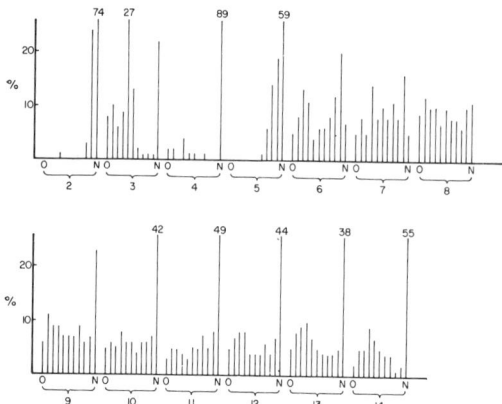

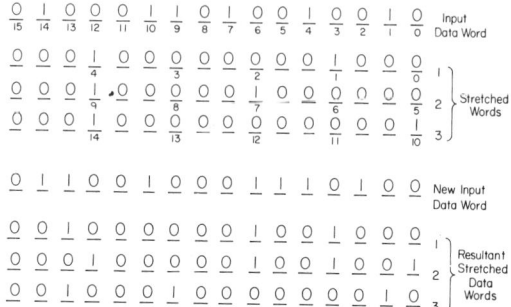

Figure 2. Distribution of strongest absorption band data among the eleven possibilities (.0 through .9 and N- "no-absorption") for each of the thirteen micron regions (2 through 14)

Figure 3. Bit manipulations
(a) Reading first characteristic and stretching
(b) Reading second characteristic, stretching and adding

and summed. Those compounds having a perfect match with the unknown would contain a seven in their corresponding array elements. Those compounds having a near hit would contain a six in their element, etc. This is simply a mathematical equivalent to the optical-coincidence cards, except that near hits are just as easy to obtain as perfect matches. A near hit is often the desired compound, the spectral differences being ascribed to several factors [see reference (2)]. The level of mismatch acceptable can be chosen at will.

Search Time and File Size Considerations. With standard computer punched card systems, each card represents one compound, thus a collection of 96,000 spectra requires an equal number of cards (48 boxes). A typical card reader has a speed of 500 cards per minute, requiring $3\frac{1}{4}$ hours to process the data. With an inverted file, 960 compounds can be coded onto one card, thus 96,000 compounds would yield a deck of 100 cards. This so compresses the data that it requires only 14,300 cards ($\sim$7 boxes) to represent the 143 possible absorption characteristics. A search involving seven characteristics would then require reading 700 cards and take 84 seconds of input time. It becomes obvious that searching operations of this type have input–output time as their limiting factor and *not* central processor power. This means that users of small computers limited to card input can perform a high quality search on very large collections of spectra previously relegated to computers with magnetic tape or disk input devices.

The same time saving described above for cards is also available with papertape or disk. There are two reasons why the inverted file is faster in all of these cases. First, the data are compressed as tightly as possible on the input stream, eliminating any wasted space; and second, only the data known *a priori* to be pertinent to the problem must be read into the computer. This last point cannot be over-emphasized.

Data Characteristics. To examine the characteristics of the Spec Finder method of coding, several studies were performed on a large computer. These operations were done with Fortran IV programs on the CDC 6500 computer. The original data randomly selected from Sadtler's Spec Finder was hand punched onto cards. The first involved a frequency distribution study of the data among the eleven possible choices for each micron interval. These data are shown in Figure 2. It is readily noted that the 2, 3, 4, and 5 micron intervals contain unevenly distributed information. On this basis, it was decided to retain only the 6 through 14 micron regions. This further reduces the total amount of punched data to 99 characteristics or $\sim$10,000 cards for 96,000 compounds.

A sifting factor for each characteristic can be defined at this point as the percentage of compounds in the collection having that particular characteristic. The sifting factor simply indicates how much the collection will be reduced utilizing that set of data. The total sifting factor for a search is the product of the sifting factor for each band utilized in the search. Thus using the 6.3, 7.0, and 8.4 bands, the collection should be reduced by a total sifting factor of 0.11 $\times$ 0.05 $\times$ 0.07 = 0.00039. For most characteristics, excepting the "no-absorption" possibilities, the sifting factor is approximately 0.1. Since the computer program allows seven characteristics to be used, the total sifting factor should be in the vicinity of 10^{-7}—*i.e.*, it would on the average reduce 10 million compounds to one. However, if the no-absorption characteristics are utilized in a search, the efficiency is reduced. For example, if such characteristics for the 11, 12, and 14 micron regions were used, the total sifting factor would only be 0.49 $\times$ 0.44 $\times$ 0.55 = 0.12. This simply indicates that within this collection of data "no-absorption" characteristics should not be utilized unless absolutely necessary.

A check for inter-characteristic correlations was made on the data. This is again related to the sifting factors. For example, if the 6.2 characteristic sifted out the same compounds as the 8.4 characteristic, the search sifting factor would no longer be a simple product of the characteristic sifting factors. The two pieces of data would be redundant and cause an inefficient search. Surprisingly, the Spec Finder method of coding the data is exceedingly free of any correlations. There were no band-band correlations at the 50% or higher level of correlation. There were many band-no-absorption and no-absorption-no-absorption correlations at this level, but the statistical validity of concluding an actual relationship in these cases is doubtful. Also, since the no-absorption characteristics were not used in the searches, the point became academic in nature, and was not pursued further.

Small Computer Aspects. The actual search results described in this report were obtained on a sixteen bit, 8K core memory, Hewlett-Packard 2115 laboratory computer utilizing papertape input. The data files are fed into the computer as bit strings packed 15 bits to the word. Once in the computer each word of the bit string is stretched into three new

Computerized Searching of Inverted Files

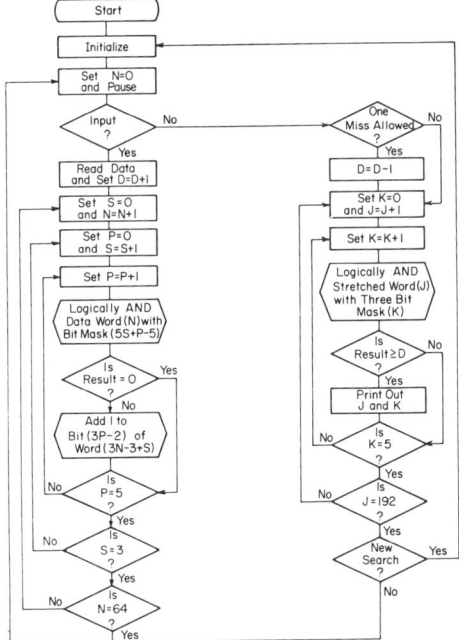

Figure 4. Search program designed to read 64 words of data (960 compounds). S, P, N, D, K, and J are constants used to control the flow of the program

Table II. Retrieval Patterns for 45 Searches Involving Compounds Known to Be in the Collection

No. of characteristics used	No. of compounds retrieved							Av. No. retrieved
	1	2	3	4	5	6	7	
3	18	12	7	5	2	0	1	2
4	43	2	0	0	0	0	0	1
5	45	0	0	0	0	0	0	1

words by packing two empty bits between each data bit as shown in Figure 3a. For the file of 960 compounds used, this required reading 64 words of data and storing the stretched results in a 192 word array. Then the data for the next characteristic absorption band are read and stretched in a similar manner. This time, however, the stretched bits are added to the previously stretched data words. This is shown in Figure 3b. A maximum of seven bands can be utilized as input for the search, since any compound having a perfect match with the unknown would have the binary number 111 in the appropriate bit positions of the stretched words. When all of the characteristics have been read in and the search is performed, each set of three bits is examined for a perfect match or near hit. Usable cross reference data are outputed on the teletype for such occurrences.

The computer program was written in normal Hewlett-Packard Fortran, and is shown in Figure 4. The program required a surprisingly small amount of memory. This, in conjunction with the method of packing the data, allowed a search of 15,000 compounds. With minor changes in the input-output formater, this could easily be raised to 20,000 compounds or the program could be written in assembly language and possibly raised to the vicinity of 30,000 compounds.

To determine searching time parameters, a "dummy" file of 14,040 compounds was added to the 960 compound working file. The dummy compounds were always initialized to zero before each search so that they could not match an unknown. The machine time involved in searching a dummy collection is equivalent to that of an equal sized real collection. Actual times involved in the prototype system were encouraging. If no matches were encountered by the computer, the machine time necessary to search 15,000 spectra was 11 seconds (not including input and set-up time). The average time required to do a complete search, including selecting files and input/output time was three minutes. Most of this does not involve computation; thus, searching 30,000 spectra would require less than 60 seconds of additional time. A search involving 90,000 compounds would require three separate runs. However, gross cataloging of the data such as organic–inorganic or liquid–solid could reduce the input file size. For very large collections it would be best to use card or disk files.

To test the program and the validity of the sifting factors, the ability to recover compounds known to be in the collection was checked. Since the file was composed of 960 compounds, three characteristics should reduce the data to the correct answer. Table II shows the results of this study and indicates that on the average three characteristics reduced the collection to two possibilities. This is in good agreement with the idealized value of the sifting factor. The same degree of reduction occurred for search compounds known not to be in the collection. It should also be noted that in every case, the program arrived at the correct answer.

CONCLUSIONS

The computerized inverted file represents a search for large collections of analytical data that has many good characteristics:

1. It allows the users of small computers restricted to card or papertape input to perform a high quality search on large data collections.
2. The availability is enhanced since it is possible to perform the search on a laboratory computer and no longer necessary to make arrangements with a central computer center.
3. The cost is minimal, since hourly rates on the necessary sized computer are usually small.
4. New characteristics are easily added to the collection.
5. It has the advantage over optical-coincidence cards of being able to recognize near hits.
6. Since data compression is utilized to its fullest extent, the necessary shelf storage space is minimal.
7. The most important advantage is that it easily lends itself to automation and can allow "on-line" searches of collections of analytical data while the spectrum is being determined.

Two possible drawbacks should be mentioned. First, new compounds are easily added to card or disk files; however, entire new paper tapes have to be punched each time the collection is updated. Second, only one search can be done at a time.

ACKNOWLEDGMENT

The author thanks Purdue University's Computer Reserve Fund for purchasing computer time.

Received for review October 31, 1969. Accepted December 22, 1969. Work partially supported by National Science Foundation Departmental Grant GP-8528.

Matching of Mass Spectra When Peak Height Is Encoded to One Bit

S. L. Grotch

Jet Propulsion Laboratory, California Institute of Technology, Pasadena, Calif.

RECENT ADVANCES in analytical instrumentation have inundated the chemist with a deluge of data. For example, the coupled gas chromatograph/mass spectrometer (GC/MS) can produce a mass spectrum every few seconds for periods of many minutes in a single sample analysis. A commercially available infrared Fourier transform spectrometer can be scanned in a few seconds to produce a complete infrared spectrum. To rephrase a familiar quotation of Sir Winston Churchill, "Never have so many, owed so much, to so few instruments."

To even log this flood of data the chemist has called on the assistance of computers. In the last few years a number of computer systems for the real-time acquisition and processing of mass spectral data have been reviewed in the literature (*1, 2*). Since a detailed interpretation of these data is generally time-consuming, it is also natural to explore methods for automating this interpretation.

In the past, it was felt that spectral interpretation was essentially different for each class of spectral information. With the increasing use of automated equipment and computers, it is becoming more apparent that spectral interpretation has many common features. In the work with mass spectra reported here, for example, many common features can be found which are related to the interpretation of infrared spectra.

Simply stated, the primary question raised in this study is, "How specific a signature is a mass spectrum when peak height is quantized to only two levels?" The answer to this question will be sought using statistics and information theory.

As a simple introduction, assume that low resolution mass spectra are available over a mass range of 200 amu. Assume that at each mass the only information known is the absence or presence of a peak. Peak height is thus quantized to two levels, or one bit; a "zero" denoting no peak, a "one" denoting a peak. Since 200 mass units are covered, the maximum information in a spectrum is 200 bits.

If all possible combinations of these 200 bits occurred, a total of 2^{200}, or approximately 2×10^{60} distinct combinations would result. Since less than 10^7 compounds are now referenced by *Chemical Abstracts* (*3*), a mass spectrum of only 200 bits would, in theory, contain more than enough information to provide for the unambiguous identification of all known compounds.

Obviously, these arguments are oversimplified. All 2^{200} possible combinations of patterns do not, in fact, arise.

Many bit combinations occur more frequently than others, while still others do not occur at all. Nevertheless, the viewpoint posed above is useful because it provides a basis for quantitatively calculating the "information" in spectra and for relating this information to the ambiguity or uniqueness of spectra in general.

The approach of the present study was a statistical examination of the matching of large groups of mass spectra. The spectrum of each member of a group was compared with that of every other member in the group and histograms were obtained for the number of mismatches observed. These histograms are a quantitative measure of the degree to which the spectra of the group differ from one another. A theoretical attempt has been made to predict this histogram from the statistics of the spectral groups.

MOTIVATION FOR STUDY

The present study is an extension of work carried out in support of a pyrolysis gas chromatograph–mass spectrometer (GC/MS) experiment for the Mars lander of Project Viking (*4*). In this experiment a sample of Martian soil is to be pyrolyzed in an oven and the effluent passed first through a gas chromatograph for separation, and then into a mass spectrometer for compound identification. If the pyrolysis products can be identified from their mass spectra, it should be possible to infer in some detail the compounds introduced into the pyrolysis oven.

At planetary distances, the data produced by the mass spectrometer must be digitally encoded before transmission to earth. The encoding of these data is important particularly with respect to compound identification when spectral peak height is encoded to only a few bits (≤ 4). Another question of importance is, "For a specified number of levels, where should the transitions between levels be set?"

It is believed that by approaching the problem of identification from information theory, it may be possible to develop matching criteria which have a firmer theoretical basis and are not merely the results of intuitive reasoning. This approach may also provide a better understanding of other spectral identification schemes. Additionally, it becomes possible to answer in a more definitive fashion, at least for the binary case, the question raised by Crawford and Morrison (*5*), "How specific is a mass spectrum?"

---

(1) C. W. Childs, P. S. Hallman, and D. D. Perrin, *Talanta*, **16**, 629 (1969).
(2) E. Kendrick, Ed., "Advances in Mass Spectrometry," Vol. 4, The Institute of Petroleum, London, 1968, pp 3–122.
(3) D. P. Leiter, H. L. Morgan, and R. E. Stobaugh, *J. Chem. Doc.*, **5**, 238 (1965).
(4) P. G. Simmonds, G. P. Shulman, and C. H. Stembridge, *J. Chromatogr. Sci.*, **7**, 36 (1969).
(5) L. R. Crawford and J. D. Morrison, ANAL. CHEM., **40**, 1464 (1968).

Table I. Statistics of the Mass Spectral Library

	Group of 1000 spectra			
Molecular weight	1	2	3	Total
Minimum	16.0	18.0	32.0	16.0
Average	123.8	143.5	210.0	159.1
Maximum	164.0	536.0	536.0	536.0
No. of Peaks $\geq 0.01\%$				
Minimum	3	2	1	1
Average	65.6	76.2	93.6	78.4
Maximum	158	165	177	177
No. of Atoms				
Minimum	2	1	6	1
Average	20.5	18.7	29.2	22.8
Maximum	34	105	110	110
No. of Compounds Containing only				
C, H	134	54	45	233
C, H, O	625	375	481	1481
C, H, N	131	12	43	186
Contain Halogen	0	516	336	852

INTERPRETATION OF MASS SPECTRA

The fundamental problem of mass spectrometry is the deduction of chemical structure from a mass spectrum. Because this problem lies at the heart of mass spectrometry, it has received very extensive attention with a resultant vast literature devoted to the problem (*6–13*).

As is often the case in problems of this sort, no single approach is "best" in all situations since each technique has certain features which can make it well suited to a particular application, while poorly suited to yet another.

The approach investigated here is the library search technique; the comparison of an unknown spectrum against a library of previously measured known compounds. This technique readily lends itself to digital computation with a minimum of programming effort. It is of great generality, since in principle, it requires no additional information other than a library of low resolution mass spectra.

The criteria used for "best fit" are generally chosen arbitrarily. This is due in large part to the observation of workers (*5*) that any number of different criteria will correctly identify an unknown if it is present in a library.

Much work remains in the selection of matching criteria which have more theoretical justification. More attention must be directed to the problems of identification in the presence of spectral differences due to both instrumental factors and operating conditions. Also worthy of greater attention is the proper use of a library search even when the unknown spectrum is not present in the library.

Low resolution mass spectral libraries now available in computer-compatible format contain about 5000 nonredundant spectra. The existence of organizations such as the Mass Spectrometry Data Centre in Great Britain should facilitate the continued accumulation and dissemination of these collections.

MASS SPECTRAL LIBRARY

For this study, a large library of experimentally-measured low resolution (unit mass) mass spectra was used. These spectra were normalized in the conventional manner (*i.e.*, the maximum intensity peak in each spectrum was 100). Two sources of spectral data were utilized: The Dow Chemical Company uncertified collection (1967 spectra) (*14*) and a collection provided by Prof. K. Biemann of MIT, consisting of the ASTM collection and other spectra measured in this

---

(6) K. Biemann, P. Bommer, and D. M. Desiderio, *Tetrahedron Lett.*, **26**, 1725 (1964).
(7) A. L. Burlingame and D. H. Smith, *Tetrahedron*, **24**, 5749 (1968).
(8) J. Lederberg and E. Feigenbaum, "Formal Representation of Human Judgement," B. Kleinmuntz, Ed., Wiley & Son, New York, 1968, Chap. 7.
(9) B. Petterson and R. Ryhage, *Ark. Kemi*, **26**, 293 (1967).
(10) V. V. Raznikov and V. L. Talroze, *Dokl. Akad. Nauk SSSR*, **170**, 379 (1966).
(11) B. Petterson and R. Ryhage, ANAL. CHEM., **39**, 790 (1967).
(12) P. C. Jurs, B. R. Kowalski, T. L. Isenhour, and C. N. Reilley, *ibid.*, **41**, 690 (1969).
(13) L. R. Crawford and J. D. Morrison, *ibid.*, **40**, 1469 (1968).

(14) "Uncertified Mass Spectral Data," R. S. Gohlke, Ed., Eastern Research Laboratory, Dow Chemical Co., Framingham, Mass., Oct. 1963.

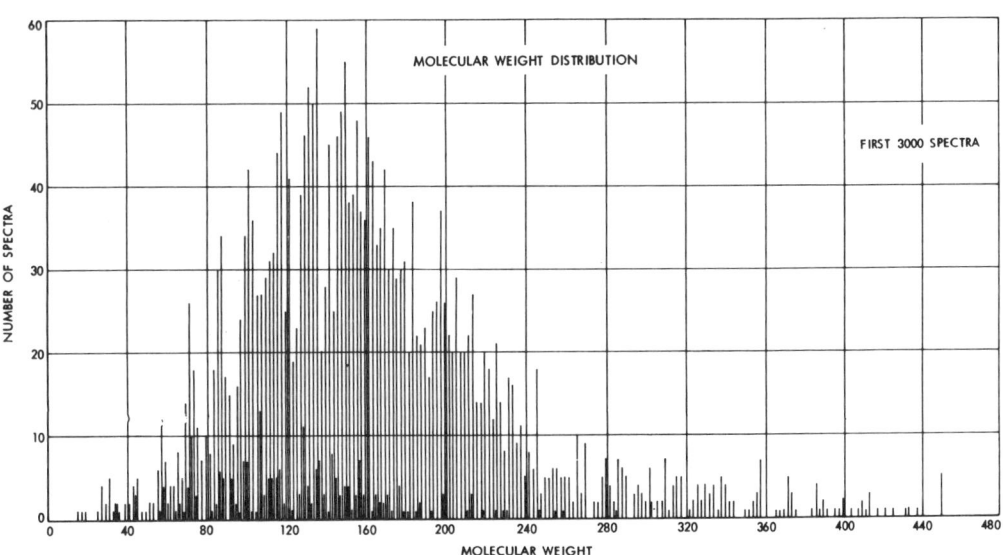

Figure 1. Molecular weight distribution of 3000 spectra

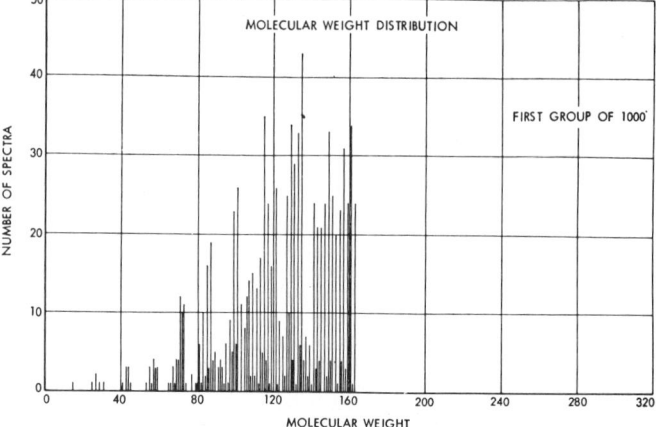

Figure 2. Molecular weight distribution of first group of 1000 spectra

laboratory (15) (2004 spectra). Approximately 3000 additional spectra (primarily the API and MCA collections) were available but were only partially used in the work reported here.

The Dow and MIT collections were screened to remove multiple spectra and a library of 3000 nonredundant spectra (1903 Dow and 1097 MIT) was finally obtained. Characteristics of this library are presented in Table I, and the molecular weight distribution is plotted in Figure 1.

To assess the sensitivity of results to the choice of library, the larger group of 3000 spectra was arbitrarily divided (by accession number) into three subgroups of 1000 spectra each. The characteristics of these groupings (which remained the same in all calculations) are also given in Table I and Figures 2–4. It should be noted that a few of the spectra used in this work may be incomplete or even incorrect. It is felt, however, that these few spectra will only influence the statistical results of a group of 1000 very slightly.

MATCHING OF LARGE GROUPS OF MASS SPECTRA

To quantitatively determine how different a group of N spectra are from one another, each member of the set should be compared against every other member using some criterion of disagreement. The statistical distribution of this criterion is a quantitative measure of how different these spectra are, on the average. This is the approach adopted here. Previous workers (5, 9) have done similar studies but have presented matching results for only a few specific compounds, rather than for a group in the statistical sense. Unlike previous work, however, spectral peak height is quantized here to only one bit at each mass position. In effect, for each spectrum one only knows whether or not a peak is present above a specified transition at a particular mass over a specified mass range.

One-bit encoding has been adopted for several reasons:

In order to more clearly understand the principles underlying spectral matching, the simplest case should be understood first.

In space applications, the number of bits available for storage or transmission is often severely limited. It is, therefore, important to investigate the uniqueness of spectra when encoded to only a few bits. One-bit encoding is the simplest case and provides a very useful datum point.

If very large groups of spectra are to be compared in detail, computer costs become prohibitive if traditional comparison techniques are employed. For example, Crawford and Morrison (5) cite search times of approximately 100 comparisons per second (on a CDC 3200 computer) for comparing only the 6 major peaks of a spectrum. At this rate, approxi-

(15) K. Biemann, Massachusetts Institute of Technology, Cambridge, Mass., personal communication, June 1968.

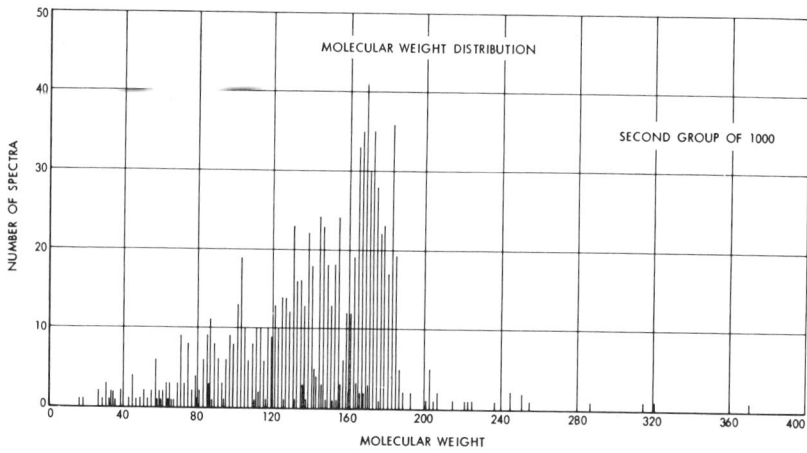

Figure 3. Molecular weight distribution of second group of 1000 spectra

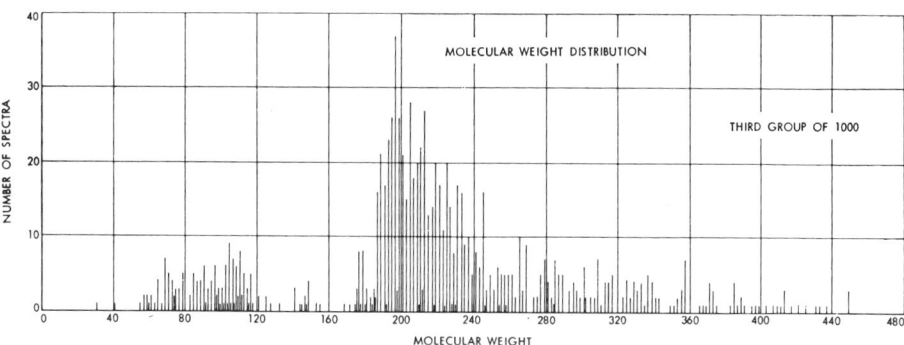

Figure 4. Molecular weight distribution of third group of 1000 spectra

mately one hour of computer time is required for a complete pair-wise comparison of only a 1000-spectrum library. Since computation time increases as the square of the size of library, such investigations soon become prohibitively expensive. As will be seen, a one-bit quantization results in an order of magnitude increase in comparison rates.

One-bit encoding results in a very efficient utilization of computer core storage since 32–36 masses can be stored in each computer word instead of a single channel as in conventional techniques. It becomes possible to increase significantly the number of spectra which can be simultaneously stored in core or alternatively, for a given number of spectra, to reduce core requirements.

CALCULATION PROCEDURE

Prior to the matching calculations, the original spectral data were first converted into a new set in which all peak heights above a specified threshold were represented by a "1," all others by an "0." These bit strings were then packed into computer words. In these calculations three different computers were utilized: an IBM-7094, a UNIVAC 1108, and an IBM 360/44.

The preparatory calculations required a total of about five minutes of IBM-7094 time to pack 3000 spectra for each transition. Three factors were fixed for each set of matching calculations: the group of spectra used; the mass range examined (maximum 12–200); and the transition in peak height between a level of 0 and a level of 1. (Assumed to be a constant, independent of mass.)

The matching calculations were performed in the following manner:

Every member of a selected set of N spectra was compared with every other member of the group. For N spectra, $N(N - 1)/2$ comparisons were required. (The number of combinations of N things taken 2 at a time.)

In the comparison of two spectra A and B, the appropriate computer words were combined using a logical "exclusive or" instruction. A bit in the resultant word $C = \text{XOR}(A,B)$ contains a "0" where the two spectra agree (both 0 or both 1) and a "1" where they disagree (one 0 and the other 1). (See Table II for an example.)

The number of "1" bits in word C is the number of channels which disagree between spectrum A and spectrum B. These "1"s are counted and a histogram of disagreements is maintained for each spectral comparison.

For a search over the mass range 12–140, approximately 1000 spectral comparisons per second could be performed using the IBM-7094. With the Univac 1108, the rate was about 3000 comparisons per second. It should be noted that all these programs were written in Fortran IV and it should be possible

Table II. Example of Logical Exclusive or (XOR) Operation C = XOR(A,B)

WORD	BIT POSITION							M
	1	2	3	4	5	6	7	M
A	0	1	1	0	1	1	0	....	1
B	1	1	0	0	1	0	1	....	1
C	1	0	1	0	0	1	1	....	0

M = Number of bits in a computer word
M = 36 for Univac-1108 or IBM-7094; 32 for IBM-360

to increase even further these processing speeds through the use of assembly language.

MATCHING STATISTICS FOR ONE-BIT ENCODED SPECTRA

Using the procedure outlined above, calculations were performed on a Univac-1108 computer on three different groups of 1000 mass spectra covering the mass range 12–200. In this section, these numerical results will be presented and in the following section, a theoretical analysis will be given.

The observed matching statistics for the first group of 1000 spectra is summarized in Figure 5 for level transitions of 0.01, 1, and 10% of the base peak. For 1000 spectra, $1000 \times 999/2 = 499500$ matches were performed. This histogram represents the distribution of the number of peak mismatches when all binary spectra in the group were matched against one another, mass by mass, for the amu range 12–200. Although this curve appears continuous, it is in fact discretized since only integral numbers of disagreements can arise. The probability that a given number of disagreements M occurs, is given by the ratio of the ordinate corresponding to the abscissa M, divided by the total number of matches attempted. This histogram statistically represents how different the mass spectra are in the grouping when quantized to only one bit.

Although quantitatively different, the matching results for the other two groups of 1000 spectra are typified by Figure 5.

Several conclusions are apparent from these results:

As the transition between level "0" and level "1" increases, the maximum of the frequency distribution shifts to fewer disagreements, i.e., the spectral patterns become more similar. One would intuitively expect this behavior, since as the transi-

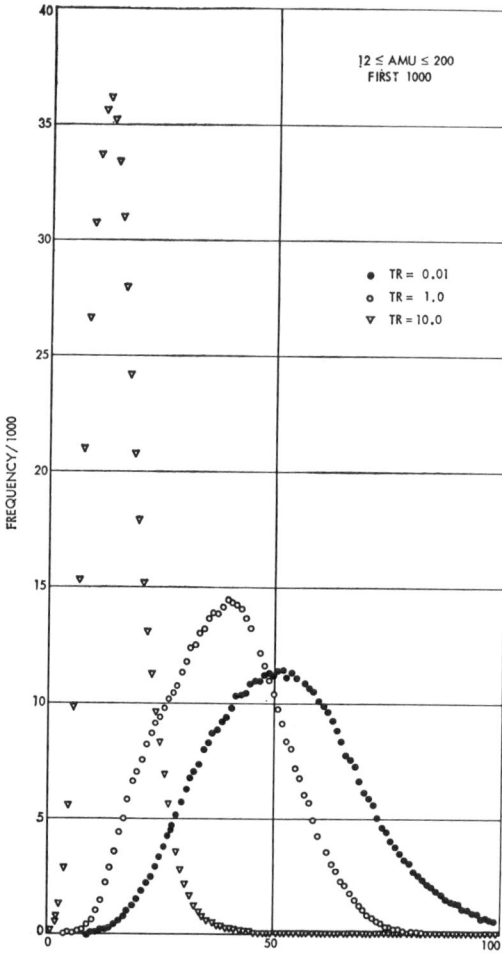

Figure 5. Matching histogram for first group of 1000 spectra

Table III. Perfect Matches

Perfect matches observed in the matching of 3246 compounds, encoded to one bit with a level transition at 1.0% base peak for the mass range 13 to 140

1. 1-Butene
   2-Butene
2. , m-Xylene
   p-Xylene
3. 2,3-Dimethylpyridine
   3,4-Dimethylpyridine
4. 1,4-Dicyanobenzene
   1,2-Dicyanobenzene
5. 2,4-Dimethylbenzyl alcohol
   3,5-Dimethylbenzyl alcohol
6. o-Chloro-toluene
   p-Chloro-toluene
7. t-Dichloroethylene
   c-Dichloroethylene

Additionally, eight other perfect matches were found for high molecular weight compounds (generally, natural products) which had peaks in excess of 1.0% of the base peak at all masses above 50–70 to 137.

tion increases, fewer peaks occur, and, hence, fewer disagreements will arise, on the average.

As the transition increases, the spread in the histograms becomes smaller. Since the total number of matches is constant, the frequency of the maximum increases as the spread is reduced. The result will be discussed in the following section.

From the viewpoint of spectral uniqueness, it is more informative to consider the matching distribution as a cumulative plot; *e.g.*, the percentage of matches with less than a given number of disagreements *vs.* the number of disagreements. The results of Figure 5 are presented in this manner in Figure 6 using a nonlinear scale for the abscissa. (On this scale, a true normal distribution would be a straight line.)

For a level transition of 1.0% of the base peak, typically less than one match in two thousand yields fewer than five disagreements. Furthermore, of the 189 channels examined, approximately 40 disagree on the average for this transition.

These results show that the mass spectra of pure compounds are indeed quite unique even when encoded to only one bit. A similar conclusion was reached by Jurs *et al.* (*12*) who found in their pattern classification work with mass spectra that one-bit encoding still provided adequate information for good pattern separations. Furthermore, such results provide assurance that one-bit encoding can be an important aid in the identification of mass spectra [A spectral matching computer program based on these principles has been developed (*16*)].

It is interesting to examine those cases in which perfect matches resulted when the spectra of two different compounds were compared. In these cases, ambiguity of identification would result with one-bit encoding.

A library of 3246 mass spectra (including the 3000 previously discussed) was compared using an IBM-360/44. All spectra were encoded to one bit, with the level transition set at 1.0% of the base peak. Each encoded spectrum of the group was compared with every other spectrum over the mass range 13–140 for a total of 5.2×10^6 matches.

Perfect matches between spectra of different compounds were found in only 15 cases (approximately 3 perfect matches per 10^6 attempted). Seven of these compound pairs are listed in Table III. The remaining eight pairs were all high molecular weight natural products with a peak at every mass in the mass range of 50–60 to 140. In all cases in Table III, both compounds in the pair are close isomers. This result again indicates that a binary mass spectrum is a very specific chemical signature.

In the following section a theoretical analysis of these results will be presented with a view to the *a priori* prediction of these matching statistics.

PREDICTION OF MATCHING RESULTS FROM STATISTICS OF SPECTRA

Although the frequency distributions for the matching of large groups of spectra can be obtained by actually performing the matching calculations, it is important to have alternative methods available.

If the matching histograms could be predicted, a deeper understanding of the principles underlying the matching process might become apparent. This understanding would serve as a foundation for further studies of the more commonly encountered, but more complex, multibit situation.

---

(16) S. L. Grotch, Eighteenth Annual Conference on Mass Spectrometry and Allied Topics, San Francisco, Calif., June 1970.

Even though the calculations in the matching process can be made rapidly, the time required increases as the square of the number of spectra considered, and for very large groups of spectra (say, $N > 5000$) it becomes impractical to perform such calculations. If a faster means of prediction could be developed, the matching calculations might prove unnecessary.

In this section the matching process is discussed from the viewpoint of statistics for the one-bit situation. The generalization to the multibit case is simple and is discussed after the one-bit case is explored.

It is obvious that the mathematical results derived here are applicable to spectral classes other than mass spectra (for example, infrared spectra), but primary attention will be focused on the mass spectral example.

Consider a group of N spectra in which peak height has been quantized to one bit over a total of L channels (atomic mass units in the case of mass spectra, wave number or wavelength increments in the case of infrared spectra). Note that the transition between level "0" and level "1" need not be a constant, but could be a function of channel number.

Consider the matching of two spectra, X and Y, each represented by a vector of L elements ("0's" or "1's"). The individual elements x_i and y_i denote the presence ("1") or absence ("0") of a peak above the specified level transition in channel i. Define functions $F(x_i, y_i)$ and D_L in the following manner:

$$F(x_i, y_i) = 0 \quad \text{if } x_i = y_i \quad (1)$$

$$F(x_i, y_i) = 1 \quad \text{if } x_i \neq y_i \quad (2)$$

$$D_L = \sum_{i=1}^{i=L} F(x_i, y_i) \quad (3)$$

In the context of spectral matching, if two spectra (X and Y) are compared, channel by channel, a peak mismatch [$F(x_i, y_i) = 1$] results only when one spectrum has a peak in a given channel i and the other does not ($x_i \neq y_i$). The number of disagreements over a range of L channels is the sum of the $F(x_i, y_i)$, or D_L.

The frequency histograms presented are the observed distributions of the variable D_L. The problem, therefore, is the prediction of the distribution of D_L knowing the statistics of X and Y. To this end, the mean and variance of D_L will be calculated.

MEAN OF D_L

Define $\bar{D}_L$ as the expected value or mean of the variable D_L and let the operator $E[\ ]$ denote expected value. Assume that X and Y come from the same probability distribution and let:

p_i = probability of a "1" in channel i (i.e., Prob. ($x_i = 1$))
q_i = probability of an "0" in channel i ($=1-p_i$) (Prob. ($x_i = 0$))

From the definition of expected value it follows that:

$$E[F(x_i, y_i)] = 2p_i q_i \quad (4)$$

Since the expected value of a sum is the sum of the expected values:

$$\bar{D}_L = 2 \sum_{i=1}^{i=L} p_i q_i \quad (5)$$

Using Equation 5, it is a simple matter to predict the average number of disagreements for the matching process from the statistics of the spectral group. This calculation is extremely rapid and the time required increases only linearly with N, the

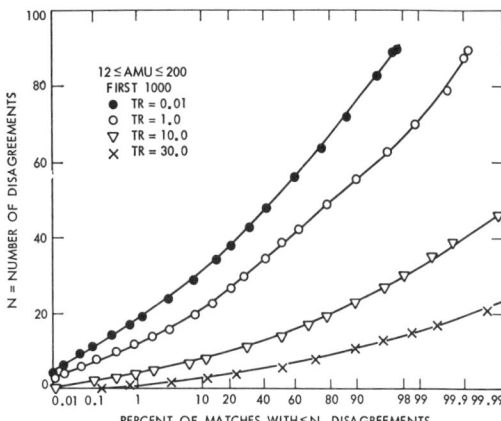

Figure 6. Cumulative matching distribution for first group of 1000 spectra

Table IV. Average Number of Disagreements ($\bar{D}_L$) for Matching 3 Groups of 1000 Mass Spectra over the AMU Range 12–137

	Level transition		
	0.1	1.0	10.0
1st 1000			
Observed	47.56	37.41	14.62
Predicted	47.51	37.38	14.61
2nd 1000			
Observed	53.21	40.60	13.29
Predicted	53.16	40.56	13.10
3rd 1000			
Observed	54.35	49.20	18.76
Predicted	54.30	49.16	18.75

number of spectra in the group, rather than as N^2, for the acutal matching.

It can easily be shown that $\bar{D}_L$ is maximized when for all i, $p_i = 1/2$. That is, the average number of disagreements is maximized when it is equally likely that in each channel a "0" or "1" will occur. In this case, the average number of disagreements is one-half the number of channels considered.

For identification purposes, it is desirable to maximize the differences between spectra. To maximize $\bar{D}_L$, the transitions between "0" and "1" should be chosen such that all p_i approach 0.5.

A comparison of the predicted average number of disagreements (from Equation 5) and that actually found in the matching of three different groups of 1000 spectra is presented in Table IV for level transitions: 0.1, 1, and 10% of the base peak, over the mass range 12–137. The observed values agree very closely with those calculated from Equation 5, as they should, since both calculations should yield identical results.

VARIANCE OF D_L

The calculation of the variance of D_L is more complex since in general cross-correlations between channels must be considered. Define σ^2 as the variance of D_L. By definition:

$$\sigma^2 = E(D_L^2) - [E(D_L)]^2 \quad (6)$$

Table V. Variance (σ^2) of Distribution of Disagreements for the Matching of Three Groups of 1000 Mass Spectra over the AMU Range 12–137

	Level transition		
	0.1	1.0	10.0
1st 1000			
Observed	206.6	153.7	36.92
Predicted			
Equation 15	28.0	23.8	11.23
Equation 14	208.7	154.9	37.09
2nd 1000			
Observed	197.6	149.4	42.49
Predicted			
Equation 15	29.9	25.7	10.80
Equation 14	200.3	151.0	42.83
3rd 1000			
Observed	261.4	241.2	141.0
Predicted			
Equation 15	30.0	28.3	14.7
Equation 14	265.0	243.4	141.3

Substituting Equation 3 into Equation 6 and recalling that $E[D_L] = \bar{D}_L$ yields:

$$\sigma^2 = -\bar{D}_L^2 + E\left[\sum_i F^2(x_i, y_i) + 2\sum_i \sum_{j>i} F(x_i, y_i) F(x_j, y_j)\right] \quad (7)$$

Since $F(x_i, y_i) = 0$ or 1 only:

$$E[F^2(x_i, y_i)] = E[F(x_i, y_i)] \quad (8)$$

Equation 7 becomes:

$$\sigma^2 = \bar{D}_L(1 - \bar{D}_L) + 2E\left[\sum_i \sum_{j>i} F(x_i, y_i) F(x_j, y_j)\right] \quad (9)$$

The expected value of the second term may be computed from the joint probability distribution of the peaks in a spectrum. For any pair of channels i and j in a spectrum define:

$\pi_{x_i x_j}$ = Probability that channel i is x_i and channel j is x_j (where $x_i, x_j = 0, 1$)

For example, $\pi_{1,0}$, is the probability that channel i contains a "1" and channel j contains an "0".

For each channel pair i,j the contribution to the sum in Equation 9 is:

$$E[F(x_i, y_i) F(x_j, y_j)] = 2[\pi_{0,0_j} \pi_{1_i 1_j} + \pi_{1_i 0_j} \pi_{0_i 1_j}] \quad (10)$$

The π may be readily calculated from a set of spectra by tabulating the number of times a particular peak configuration (00, 11, 01, 10) occurs for each channel pair $i, j > i$. For each channel pair in a set of N spectra define the mean values:

$$\mu_{ij} = \frac{1}{N} \sum_{k=1}^{k=N} x_i x_j \quad (11)$$

and

$$p_i = \frac{1}{N} \sum_{k=1}^{k=N} x_i \quad (12)$$

The summations are taken over all N spectral members in a given set. Again, p_i denotes the probability of finding a "1" in a single channel i, whereas μ_{ij} is the probability of simultaneously finding a "1" in both channels i and j. Equation 10 becomes:

$$E[F(x_i, y_i) F(x_j, y_j)] = 2[2(\mu_{ij} - p_i)(\mu_{ij} - p_j) + (\mu_{ij} - p_i p_j)] \quad (13)$$

The final result for σ^2 is obtained by substituting Equation 13 into Equation 9.

$$\sigma^2 = \bar{D}_L(1 - \bar{D}_L) + 4 \sum_i \sum_{j>i} [2(\mu_{ij} - p_i)(\mu_{ij} - p_j) + (\mu_{ij} - p_i p_j)] \quad (14)$$

Equation 14 is simplified considerably if the assumption is made that all channels are independent. In this case, Equation 14 reduces to:

$$\sigma^2 = \sum_{i=1}^{i=L} 2p_i(1 - p_i)[1 - 2p_i(1 - p_i)] \quad (15)$$

In Table V are presented the observed and predicted variances as calculated from Equations 14 and 15 for three different groups of 1000 spectra. If the channels are assumed independent, the predicted variances are about an order of magnitude lower than those found in the actual matching calculations. When the correlation between channels is included, Equation 14 predicts the variance very well.

These results show that in mass spectra, channels are generally not independent, but are in fact, highly correlated. Other calculations have shown, for example, that on the average, if a peak is observed at a given mass, it is much more likely that a second peak will also be present at an adjacent mass rather than no peak occurring. (This is due to isotopic effects and the common loss of a hydrogen atom.) Calculations indicate that this correlation extends well beyond adjacent masses.

It can be seen in Table V that as the transition between "0" and "1" increases, the discrepancy between the observed and the predicted variance assuming channel independence (Equation 15) tends to become smaller. This is due to a reduced correlation between channels (e.g., the covariance contribution) when only the larger peaks are considered.

Although Equation 14 appears formidable, it is not difficult to program and the calculations can be performed very rapidly (typically, less than one minute is required on the Univac 1108 for 1000 spectra for the mass range 12–200). By using Equation 5 to predict the mean, and Equation 14 to calculate the variance, it is possible to accurately predict, *a priori*, the first two moments of the matching distribution. With these equations, the matching distribution may be calculated about an order of magnitude more rapidly than by actually performing the matching calculations.

FORM OF MATCHING DISTRIBUTION

The process giving rise to the distribution of D_L is analogous to that described by Feller (Ref. 17, p 205) as "Bernoulli trials with variable probabilities." In the present context, each trial consists of the matching of a given channel in two spectra with a probability of success (i.e., a match) that varies as a function of channel number. Feller calculates the mean and variance for the distribution of the number of successes under the assumption that each trial represents an independent event, i.e., there is no correlation between the probabilities of success in different trials. The mean and variance calculated by Feller are equivalent to Equations 5 and 15. A more de-

---

(17) W. Feller, "An Introduction to Probability Theory and its Applications" Vol. I, John Wiley & Sons, New York, N. Y., 1950.

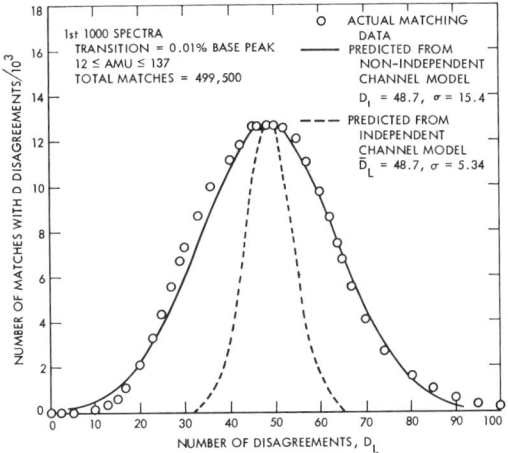

Figure 7. Predictions of matching model with observed data

tailed discussion of these matching distributions may be found in Ref. *18* where a Poisson limit was obtained.

Feller (Ref. *17*, p 217) points out a seemingly paradoxical feature of these distributions; namely, for a given mean, as the probability of success at each trial becomes more uniform (*i.e.*, all $p_i \to$ constant) the magnitude of the variance of the distribution of successes, σ^2, *increases*, *i.e.*, the distribution of successes becomes *less uniform*. In the case of all $p_i = 1/2$, the variance is maximized and equal to $\bar{D}_L$. If separability is to be measured by a parameter such as $\bar{D}_L/\sigma$, rather than $\bar{D}_L$, optimum separability may not be achieved by choosing all $p_i = 0.5$. The matching histograms of Figure 5 do not appear to be statistically normal, although they may be approximated by normal distributions with the variance calculated from Equation 14 (see Figure 7).

GENERALIZATION OF THE ONE-BIT RESULTS TO MULTIPLE LEVELS

To generalize the results derived for the one-bit case define:

p_{ik} = probability that a peak in channel i will fall in level k

$$F(x_i, y_i) = 0 \text{ when } x_i = y_i \quad (16)$$
$$= C(x_i, y_i) \text{ for } x_i \neq y_i \quad (17)$$

The function $C(x_i, y_i)$ depends upon the criterion which is applied to a disagreement between levels. If, for example, the measure of disagreement is the absolute value of the difference between the levels in channel i

$$C(x_i, y_i) = |k_1 - k_2| \text{ for } x_i = k_1 \text{ and } y_i = k_2$$
$$\text{or } x_i = k_2 \text{ and } y_i = k_1 \quad (18)$$

The expected value of F in channel i for M levels is:

$$E[F(x_i, y_i)] = 2 \sum_{k_1=1}^{M} p_{ik_1} \sum_{k_2>k_1}^{M} p_{ik_2} C(k_1, k_2) \quad (19)$$

If D_L is defined by Equation 3, then the mean of D for L channels is:

$$\bar{D}_L = 2 \sum_{i=1}^{L} \left\{ \sum_{k_1=1}^{M} p_{ik_1} \sum_{k_2>k_1}^{M} p_{ik_2} C(k_1, k_2) \right\} \quad (20)$$

The variance of each individual channel is:

---

(18) D. E. Barton, *J. Roy. Statist. Soc.*, **20**, 73 (1958).

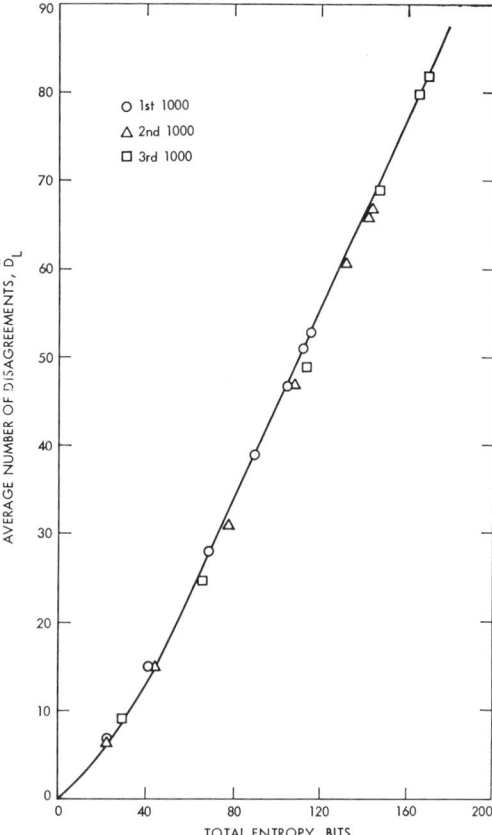

Figure 8. Average number of disagreements *vs.* total entropy for three groups of 1000 spectra

$$\sigma_i^2 = 2 \sum_{k_1=1}^{M} \sum_{k_2>k_1} p_{ik_1} p_{ik_2} C^2(k_1, k_2) -$$
$$\left[2 \sum_{k_1=1}^{M} \sum_{k_2>k_2} p_{ik_1} p_{ik_2} C(k_1, k_2) \right]^2 \quad (21)$$

The variance of D_L, assuming channel independence, is obtained by summing the individual σ_i^2 over L channels. For non-independent channels, the result is more complex and the form depends on the "best fit" criterion, C.

INFORMATION CONTENT OF MASS SPECTRA

In space applications, it is frequently important to determine the information content of a data source for encoding purposes. This content can be expressed quantitatively in terms of the information "entropy" measured in "bits" (not to be confused with the thermodynamic quantity of the same name). Since entropy is a quantitative measure of the inherent information in a source, for spectral identification it is important to encode spectra so as to maximize this entropy, For more detailed descriptions of these concepts see Ref. *19*. or any of the many texts on information theory.

For M levels, the single channel entropy h_i of a spectrum is

---

(19) S. Goldman, "Information Theory," Prentice Hall, Englewood Cliffs, N. J., 1953.

1222   S. L. Grotch

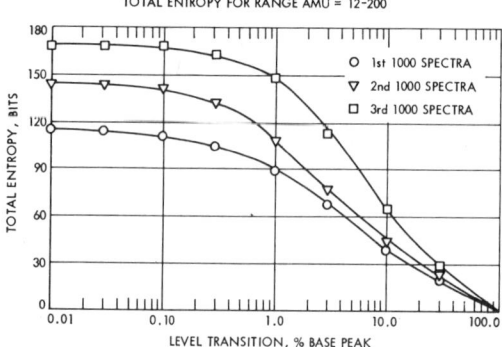

Figure 9. Total information entropy as a function of level transition

defined as:

$$h_i = - \sum_{k=1}^{M} p_{ik} \log_2 p_{ik} \quad (22)$$

where p_{ik} is the probability of a given peak in channel i falling in level k. The total entropy for L channels is:

$$H_L = \sum_{i=1}^{L} h_i = - \sum_{i=1}^{L} \sum_{k=1}^{M} p_{ik} \log_2 p_{ik} \quad (23)$$

For simplicity, consider the one-bit case, $M = 2$. Using the definitions given above, Equation 23 becomes:

$$H_L = - \sum_{i=1}^{L} [p_i \log_2 p_i + (1 - p_i) \log_2 (1 - p_i)] \quad (24)$$

It is easily shown that H_L is maximized and equal to L bits if, and only if, for all i, $p_i = 1/2$. Stated in another way, the total information of a class of spectra is maximized if at each mass the transition level is set so that it is equally likely that a peak will or will not occur.

In terms of the binary matching studies given above, a comparison of Equation 5 for the average number of disagreements, $\bar{D}_L$, with the entropy function of Equation 24 shows that as the entropy of the source increases, $\bar{D}_L$ also increases and that both are maximized when for all channels, $p_i = 1/2$. Thus, to achieve maximum average separability (as measured by $\bar{D}_L$), the level transitions should be set so as to obtain maximum entropy. (See Figure 8.)

The total entropy as calculated from Equation 24 for the three groups of 1000 spectra as a function of a constant level transition is given in Figure 9. For each group the total entropy is substantially constant for transitions over the range 0.01% to approximately 0.3% of the base peak, and the entropy falls rapidly above about 1.0% of the base peak. In effect, the peaks below about 0.3% contribute relatively little to the information content. This is also reflected in $\bar{D}_L$ which increases by less than 13% in going from a transition of 0.3 down to 0.01%.

In a practical situation, the setting of the optimum transition is also strongly influenced by the noise level of the system which is not taken into account in these arguments. The calculations above indicate that for the spectral groups examined, the entropy increases continuously as the transition level drops, at least down to 0.01% of the base peak. In many systems, however, noise may preclude going to such low levels and a compromise has to be made. Figure 9 indicates quantitatively how much information is lost in achieving this compromise.

The generalization of the above arguments to the multilevel case is simple. It can be shown that H_L is maximized and equal to $L \log_2 M$ bits only when the probability that a peak will fall in any level is the same for all levels in all channels (all $p_{ik} = 1/M$).

In the absence of noise, the optimal transitions between levels are dictated by the peak height distribution of mass spectra. This distribution has been found to be approximately lognormal [20]. Thus, nearly optimal levels can be achieved by choosing equal increments on a logarithmic height scale.

Ideas derived from information theory should prove useful in providing quantitative insights into problems of spectral matching. A recent paper [21] explores some of these concepts in general terms for spectroscopic problems. Hopefully, these notions will find application in future studies in this area.

ACKNOWLEDGMENT

The author acknowledges the statistical insights provided by his colleagues H. Lass and C. Solloway.

RECEIVED for review March 4, 1970. Accepted July 2, 1970. Material presented at the Joint Conference of the Chemical Institute of Canada with the American Chemical Society, Toronto, Canada, May 24–29, 1970. This paper presents the results of one phase of research carried out at the Jet Propulsion Laboratory, California Institute of Technology, under Contract No. NAS 7-100, sponsored by the National Aeronautics and Space Administration.

(20) S. Grotch, ASTM E-14, Conference on Mass Spectrometry, Dallas, Texas, May 1969.
(21) H. Kaiser, ANAL. CHEM., **42**, No. 2, 24A (1970).

Identification of Mass Spectra by Computer-Searching a File of Known Spectra

H. S. Hertz, Ronald A. Hites, and K. Biemann

Department of Chemistry, Massachusetts Institute of Technology, Cambridge, Mass. 02139

To relieve the chemist from the tedious task of manually interpreting the large number of mass spectra obtained from gas chromatographic effluents, an automatic technique has been developed which compares the spectrum of an unknown compound to a large file of reference spectra. Both the unknown and reference spectra are abbreviated, before comparison, by selecting the two largest peaks in each fourteen mass unit interval throughout the entire spectrum. After the computer preselects the most similar mass spectra, a similarity index is calculated, which represents the weighted average ratio of the two spectra and is an absolute measure of the degree of match between the unknown and a particular reference spectrum. The algorithm used is described and evaluated, and applications are presented.

THE COMBINATION of a gas chromatograph with a mass spectrometer is an important and powerful analytical tool that facilitates work in diverse fields concerned with complex mixtures of organic compounds (1, 2). The large amount of mass spectrometric data that can be produced during a gas chromatographic run has led to the development of automated recording techniques (3–5) that convert the spectra directly to digital form and present them as mass vs. intensity tables or plots. Nevertheless, even with these modern, automated recording techniques, it is still necessary for the chemist to identify the material in question by a detailed examination of the mass spectrum.

In most complex gas chromatograms, an appreciable fraction of the components is otherwise known and are previously encountered substances. Their identification by interpretation of the corresponding mass spectra takes time and such interpretations are always confirmed by comparing the spectrum in question with the authentic spectrum of the compound, if available. If they are indeed identical, the time-consuming interpretive step was necessary only for the selection of the spectrum for comparison. It is thus only logical to shorten this process by a direct comparison of the mass spectrum in question with a collection of all known spectra, eliminating prior manual interpretation. While a few spectra can be compared with a limited collection using semimechanical sorting procedures (6), the more widespread use of combined gas chromatograph-mass spectrometer systems and the present availability of almost 10,000 reference mass spectra require a relatively powerful computer to perform the large number of comparisons necessary. The availability and routine use in this laboratory of the above mentioned GC-MS-computer system, which is capable of producing 400 mass spectra within a half-hour gas chromatogram, not only vividly presented an acute identification problem, but also provided an excellent opportunity to continuously test the comparison method on real-life data. The approach chosen has been described in preliminary form some time ago (7). In the intervening two years of use, the algorithm has

---

(1) F. A. J. M. Leemans and J. A. McCloskey, *J. Amer. Oil Chem. Soc.*, **44**, 11 (1967).
(2) W. H. McFadden, *Separ. Sci.*, **1**, 723 (1966).
(3) R. A. Hites and K. Biemann, ANAL. CHEM., **40**, 1217 (1968).
(4) W. E. Reynolds, V. A. Bacon, J. C. Bridges, T. C. Coburn, B. Halpern, J. Lederberg, E. C. Levinthal, E. Steed, and R. B. Tucker, *ibid.*, **42**, 1122 (1970).
(5) C. C. Sweeley, B. D. Ray, W. I. Wood, J. F. Holland, and M. I. Krichevsky, *ibid.*, p 1505.

(6) P. D. Zemany, ANAL. CHEM., **22**, 920 (1950).
(7) R. A. Hites and K. Biemann in "Advances in Mass Spectrometry," Vol. 4, E. Kendrick, Ed., The Institute of Petroleum, London, 1968, p 37; presented at the International Mass Spectrometry Conference, Berlin, September 1967.

been continuously refined and made more efficient; the resulting improved algorithm is discussed in this detailed paper.

The possibility of identifying low resolution mass spectra by computer techniques has been explored in several laboratories. Most of these attempts have dealt with low molecular weight compounds and have used only a small number of the largest peaks in the spectrum.

The earliest of these techniques (8) identified compounds by a comparison of the five most intense peaks in the unknown spectrum with the five most intense peaks in each of the reference spectra. If necessary, a "disagreement index," which is the sum of the absolute differences of the intensities taken over all masses, was calculated. No data were presented to demonstrate the use of this technique to solve an actual problem.

Other proposed searching techniques (9) were somewhat similar to the above (8) except that no quantitative disagreement indices were calculated. Results using unknown data indicated that these techniques failed to retrieve the correct compound if the molecular weight was much above 150. Similar conclusions were also reached by other authors (10).

Another searching technique for mass spectra has been reported that makes use of the six most intense peaks in the spectra and several alternate normalization methods (11). For comparing spectra a "dissimilarity index," also based on the sum of absolute intensity differences, was calculated but first the intensities were transformed by using their squares or square roots.

A recently reported searching technique used a similarity index which was based on the n strongest peaks in the spectrum or the n strongest peaks in each interval of m amu (12). This comparison depended on the position i of each selected peak in the unknown relative to the position j of the same peak in the spectrum from the collection, positions i and j being determined by decreasing ion intensity. The examples of searches generally utilized spectra contained in the collection, but data obtained in the course of a gas chromatogram of a mixture of terpenes were also tested. All test searches used a very limited collection of mass spectra (300 spectra or less).

In addition, programs have been written for the interpretation (rather than matching with authentic spectra) of the mass spectra of normal and mono-methyl substituted hydrocarbons and normal and mono-methyl substituted methyl esters of fatty acids (9, 13). The DENDRAL algorithm has been applied to the interpretation of mass spectra of aliphatic ketones, ethers, and amines (14–16). Techniques for the automatic recognition of compound classes from mass spectra have been suggested (17–20). Also, the computer-interpretation of the high resolution mass spectra of simple ketones, esters, amines, and amides has been reported (21, 22). Other investigators utilized high resolution mass spectra, proton magnetic resonance spectra, infrared spectra, and ultraviolet spectra for the computer identification of compounds with less than fifteen carbons and one oxygen (23). Computerized learning machines have been applied to mass spectrometry (24) and to the interpretation of combined mass spectrometric, infrared, melting point, and boiling point data (25).

All of these approaches (13–25) are aimed at aiding in the interpretation of the spectrum of a new compound or a compound whose mass spectrum is not already available. This problem is, however, entirely different from the subject of this paper. Nevertheless, it will be seen that the comparison approach outlined here also leads to very useful results if the spectrum of the compound in question is not available in the collection, but related ones are. From a philosophical point of view the two approaches, interpretation and matching, differ in the uniqueness of the answer obtainable for compounds other than very simple ones. An efficient matching technique will positively identify any compound whose mass spectrum is known (assuming it is in the collection to be searched). Only an ideal interpretation technique will be able to do this, and this event is still years if not decades off. Even then it will probably pay to eliminate known compounds by a matching procedure and reserve the much more costly computer interpretation for the remaining unknown compounds.

For practical reasons, any collection of reasonable size will consist of spectra taken in different laboratories on different types of instruments and under different experimental conditions. Therefore, a generally applicable technique for the comparison of mass spectra should utilize the gross features of an *entire* spectrum rather than depend on a quite exact match of a few intense peaks. In addition, the comparison must provide a quantitative measure of similarity, rather than a "yes" or "no" answer. Thus, the search results will often reveal at least the type of compound, if the authentic spectrum is not in the reference file but related ones are. A high degree of similarity implies that a spectrum of the unknown is present in the reference file. A lower degree of similarity indicates that, although a spectrum of the unknown is not in the file, the unknown spectrum has some features in common with the

---

(8) S. Abrahamsson, G. Häggström, and E. Stenhagen, presented at the Fourteenth Annual Conference on Mass Spectrometry and Allied Topics, Dallas, Texas, May 1966, p 522; S. Abrahamsson, *Sci. Tools*, **14**(3), 129 (1967).
(9) B. Pettersson and R. Ryhage, *Ark. Kemi*, **26**, 293 (1967).
(10) I. C. Smith, W. Kelly, A. Brickstock, and R. G. Ridley, presented at the Fifteenth Annual Conference on Mass Spectrometry and Allied Topics, Denver, Colorado, May 1967, p 102.
(11) L. R. Crawford and J. D. Morrison, ANAL. CHEM., **40**, 1464 (1968).
(12) B. A. Knock, I. C. Smith, D. E. Wright, R. G. Ridley, and W. Kelly, *ibid.*, **42**, 1516 (1970).
(13) B. Pettersson and R. Ryhage, *ibid.*, **39**, 790 (1967).
(14) A. M. Duffield, A. V. Robertson, C. Djerassi, B. G. Buchanan, G. L. Sutherland, E. A. Feigenbaum, and J. Lederberg, *J. Amer. Chem. Soc.*, **91**, 2977 (1969).
(15) G. Schroll, A. M. Duffield, C. Djerassi, B. G. Buchanan, G. L. Sutherland, E. A. Feigenbaum, and J. Lederberg, *ibid.*, p 7440.
(16) A. Buchs, A. M. Duffield, G. Schroll, C. Djerassi, A. B. Delfino, B. G. Buchanan, G. L. Sutherland, E. A. Feigenbaum, and J. Lederberg, *J. Amer. Chem. Soc.*, **92**, 6831 (1970).
(17) V. L. Tal'roze, V. V. Raznikov, and G. D. Tantsyrev, *Dokl. Akad. Nauk SSSR*, **159**, 182 (1964).
(18) V. V. Raznikov and V. L. Tal'roze, *ibid.*, **170**, 379 (1966).
(19) L. R. Crawford and J. D. Morrison, ANAL. CHEM., **40**, 1469 (1968).
(20) M. C. Hamming and R. D. Grigsby, presented at the Fifteenth Annual Conference on Mass Spectrometry and Allied Topics, Denver, Colo., May 1967, p 107.
(21) A. Mandelbaum, P. V. Fennessey, and K. Biemann, presented at the Fifteenth Annual Conference on Mass Spectrometry and Allied Topics, Denver, Colo., May 1967, p 111.
(22) R. Venkataraghavan, F. W. McLafferty, and G. E. Van Lear, *Org. Mass Spectrom.*, **2**, 1 (1969).
(23) S-I. Sasaki, H. Abe, T. Ouki, M. Sakamoto, and S. Ochiai, ANAL. CHEM., **40**, 2220 (1968).
(24) P. C. Jurs, B. R. Kowalski, T. L. Isenhour, and C. N. Reilley, *ibid.*, **42**, 1387 (1970); and earlier papers cited therein.
(25) P. C. Jurs, B. R. Kowalski, T. L. Isenhour, and C. N. Reilley, *ibid.*, **41**, 1949 (1969).

spectra retrieved from the reference file. It should be noted that this fact contradicts the recent statement by Jurs et al. (24) who must have assumed that such search systems aim at a simple and unqualified "yes-no" algorithm, resulting only in the correct answer or failure.

EXPERIMENTAL

Abbreviation of the Mass Spectrum. The most general comparison would involve the use of all peaks. However, a mass spectrum could consist of several hundred peaks and a collection of reference spectra must consist of several thousand spectra, if it is to be useful. Thus storage limitations as well as the speed of the comparison dictate that the spectra must be condensed in some standardized manner. In order to shorten the data but delete as little of structural significance as possible, the spectrum is abbreviated by selecting the two largest peaks in each 14 mass unit interval throughout the spectrum, fourteen being the mass of a methylene group. At the outset of this work the intervals 1–14, 15–28, 29–42, etc. were used (7); they were changed, however, to 6–19, 20–33, 34–47, etc., when it was realized that difficulties occur if the boundaries split common peak clusters (i.e., fall between $m/e = 42$ and 43 for example).

In contrast to other condensation techniques (9, 12), this one selects the *interpretively* significant peaks in a standardized manner. Other condensation techniques generally make use of the n strongest peaks in the mass spectrum, where n is usually five or more (8, 9, 11). Careful consideration of this technique shows that it will omit interpretively significant peaks in many instances. For example, the spectrum of 10-ethyl-10-n-propyldocosane (Figure 1), shows that the six most intense peaks ($m/e = 41, 43, 55, 57, 71, 85$) do not include *any* of the structurally significant peaks; i.e., those indicating the branching at C-10 ($m/e = 351$ M-ethyl; $m/e = 337$ M-propyl; $m/e = 253$ M-nonyl; $m/e = 211$ M-dodecyl). Furthermore, these six most intense peaks are common to all saturated hydrocarbons and would probably be helpful only for eliminating nonhydrocarbon spectra. Comparison of various mass spectra of farnesol by Knock et al. (12) further demonstrates the unreliability of using a small number of intense peaks. In agreement with our earlier paper (7), comparison of these farnesol spectra with a limited reference collection, showed much improved results when a spectrum abbreviation technique (such as two largest peaks every fourteen amu) was used rather than the n most intense peaks (where n was chosen to be roughly equal to the number of peaks in the abbreviated spectrum).

The concept of abbreviated spectra was based originally on a consideration of the steps involved in the individual interpretation of mass spectra. It was realized that the chemist primarily considers the most abundant peaks within peak clusters rather than the absolutely most abundant peaks. Selecting peaks in consecutive regions of 14 mass units assures that the significant peaks belonging to a homologous series of ions are retained if they should happen to be relatively abundant.

Furthermore, this abbreviation technique by definition assures that the molecular ion (if it is present in the complete spectrum) is not deleted, because the heaviest fragment ion cannot possibly be in the same group as the molecular ion (except ions due to loss of hydrogen). To further ensure comparability, the two most intense ions in each region, rather than only one, are retained in the abbreviated spectrum.

This approach has the apparent disadvantage, in comparison with previously mentioned methods, that the abbreviated spectrum of a compound of high molecular weight will contain more peaks than one of low molecular weight. For a computer, however, it is irrelevant whether the number of peaks to be compared is fixed or variable; also it is quite justifiable to characterize a more complex molecule by a larger number of parameters. As an example, the abbreviated spectrum of 10-ethyl-10-n-propyldocosane is shown in Figure 1 together with its complete spectrum. It can be seen that the structurally insignificant peaks are eliminated leaving an abbreviated spectrum that retains almost all of the characteristic features of the original spectrum.

It should be noted that this well-defined method of abbreviation is not only useful for searching purposes but may also solve the problem of documentation in publications, where mass spectra of compounds represent an important aspect of the subject or the experimental evidence. Such abbreviated spectra, if printed in numerical form (see top of Figure 2), take little space and are much better defined and more complete than a small number of peaks subjectively selected by the author.

The Reference File of Mass Spectra. In order to test any concept of computer searching, it is necessary to assemble a file of reference spectra. The next few paragraphs will describe the origin of the spectra included in the reference file and the format of these spectra on magnetic disks.

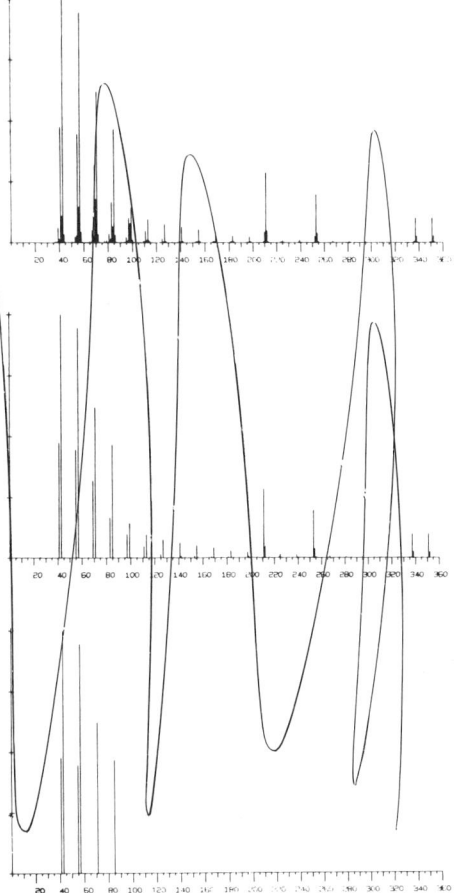

Figure 1. Complete (top), "abbreviated" (middle), and "six most intense peak" (bottom) spectra of 10-ethyl-10-n-propyldocosane

```
SYNTHETIC TEST MIX                              5   26   70   46
ABBREVIATED SPECTRUM

MASS  INT     MASS  INT     MASS  INT     MASS  INT     MASS  INT     MASS  INT     MASS  INT     MASS  INT
 31    7      41   168      43   995      55   101      57   983      71   599      72   850      85   101
 86   47      99   760     100    39     107     3     110    15     128   223     129    23     132     3

RECTANGULAR ARRAY ELEMENTS
   1    2   13    3   11   14    6    9   12    4    5    8   10    7
```

RESULTS

```
                                                    ID.NO.    SIM.
*OCTANONE  3                                          16      0.509
2-ETHYL  BUTYRALDEHYDE     D00249                   3708      0.290
3-HEXANONE    AP0662                                5867      0.286
2-ETHYLHEXANAL                                      1794      0.282
2,5-DIMETHYLOCTANE     AP1942                       7063      0.266
4-ETHYL-1-OCTYN-3-OL    MSC  437                    7964      0.256
2-ETHYL  BUTYRALDEHYDE                              2014      0.255
2-IODOHEPTANE    D01759                             5172      0.247
NOR-DODECANE     AP1028                             6215      0.243
2-ETHYL-HEXANAL     D00538                          3989      0.229

BASE PEAK= 852, NUM PEAKS= 690, REC ARRAYS= 240, 250 PEAKS= 194
```

Figure 2. SERCH output for spectrum No. 46 of the test mixture (see Figure 3)

Names of the compounds do not always conform to the IUPAC notation because the spectrum collection is derived from many different laboratories and no attempt was made to standardize the compound names. (The prefix "nor" has been used for "normal" by some laboratories.) For a discussion of the other entries see text

At present, there are approximately 7600 spectra in the reference file. Of these, approximately 2400 originate with the ASTM Committee E-14 Subcommittee IV spectrum collection (26), approximately 2000 come from the collection of The Dow Chemical Company (27), and approximately 1800 from the files of the American Petroleum Institute (28). The remaining 1400 spectra in the collection are taken from the literature, from the Mass Spectrometry Data Centre collection (29), and from spectra measured in this laboratory. The classes of compounds in this reference collection cover a wide range, and the molecular weights range up to $m/e = 1000$.

For each compound in the reference file, the complete spectrum is stored on magnetic tape and can be used for a variety of purposes. For searching, the name, identification number, and abbreviated spectrum are stored on an IBM 2315 magnetic disk pack (capacity: 512,000 16-bit words). To save disk space and conserve core storage requirements, a mass and intensity pair are packed in one computer word and the data are grouped into convenient units (2560 words) for disk input-output.

The data are transferred to the magnetic disk by a program which reads the data from cards or magnetic tape; tests for obvious errors in the data (such as keypunch errors which interfere with the ascending order of the masses); abbreviates the mass spectrum; packs each mass and intensity pair into one 16-bit word; and, adds the abbreviated spectrum to a 2560-word data array. When the data array is full, it is transferred to the magnetic disk.

This same program can modify or output the data already on the disk. The input to this section of the program is the identification numbers of the spectra to be edited, plotted, deleted, or printed. In addition, data can be added to the disk at any time. Thus, the reference collection can be continuously modified and updated on the disk. Of course, as mentioned before, the complete spectra are stored on magnetic tape and are available for various purposes, such as plotting, displaying on a storage oscilloscope, or reformating the reference collection on disk.

Comparison of Mass Spectra. The abbreviated spectra of unknown compounds are compared with the reference file of abbreviated spectra by a program named SERCH. From the outset, it was realized that this program must be applicable to automatically recorded spectra from a gas chromatograph–mass spectrometer–computer system, because this is an area where manual interpretation methods become extremely inefficient and impractical because of the vast amount of useful information generated within a short time (i.e., a gas chromatogram of less than one-hour duration). It is this principle which has served as a guideline throughout the development and testing of the SERCH algorithm. To begin, the unknown spectrum is read in, abbreviated, and printed (to inform the user of the data which are actually being used for comparison). The largest peak of the unknown spectrum is identified and stored separately for reasons outlined later. The remainder of the SERCH algorithm can be divided into two sections: the presearch and the detailed comparison of mass spectra.

THE PRESEARCH. To eliminate waste of time by comparing spectra that are very unlikely to be identical, several presearches are performed. These presearches are intended to eliminate only obviously dissimilar spectra. The requirements are such that the correct spectrum should *never* be eliminated even if instrumental conditions are very different or if the unknown is not completely pure. Basically the presearch requires: the largest peak in the known spectrum to be at least 25% relative intensity in the unknown spectrum and *vice versa*; the mass range covered by the known and unknown spectra to differ by not more than a factor of approximately three; and the total abundance of homologous series of ions to be similar in the known and unknown spectra.

The rationale behind the presearches employed to accomplish the goals outlined above are as follows: First, the most intense peak in the known spectrum must be at least 25% relative intensity in the unknown spectrum. There is obviously not much gained if this most intense peak in the known is one of several very common peaks ($m/e = 41, 43, 55, 57, 91, 105$) and, if this is the case, it is therefore automatically replaced by the second most intense peak for the purpose of this comparison, but only if this second peak is greater than 50% relative intensity. Second, the number of peaks in the unknown abbreviated spectrum must agree with the number

---

(26) "Uncertified Mass Spectral Data," ASTM Committee E-14 Subcommittee IV, 1960.
(27) R. S. Gohlke, Ed., "Uncertified Mass Spectral Data," the Dow Chemical Co., Midland, Mich., 1963.
(28) "Catalog of Mass Spectral Data," American Petroleum Institute Research Project 44.
(29) Mass Spectrometry Data Centre, AWRE, Aldermaston, Berks., England.

of peaks in the known abbreviated spectrum within ±75% of the number of peaks in the unknown spectrum. This test eliminates the comparison of the spectra of compounds which differ greatly in molecular weight (*i.e.*, more than a factor of approximately three) and thus can not possibly be identical or even similar. Third, a "rectangular array" presearch (*20*) is used as a rough screen for correct compound type. The elements of the rectangular array are obtained by summing the intensities of the ions at m/e $1 + 14n$, $2 + 14n$, ... $14 + 14n$, where $n = 0, 1, 2, ...$ These fourteen sums are then arranged in decreasing order and the mass sequences represented by the five highest sums are used for comparison. For example, if the five highest sums were the homologous series starting with masses 1, 13, 14, 2, 11, respectively, as could be the case for a saturated hydrocarbon, this five-membered ordered array would be used for the comparison. This array is compared with the rectangular arrays of those authentic spectra passing the previous two presearches. The comparison is based upon the position of each of these five highest rectangular array sums of the unknown relative to the presence and position of the same sums in the five highest rectangular array elements of the known. Again, in order to retain similar compounds and varying spectra of the correct compound, similarity but not identity of these rectangular arrays is required. Fourth, the most intense peak in the unknown spectrum must be at least 25% relative intensity in the known spectrum. (This 25% relative intensity limit seems to be a very safe one taking into account pattern distortion due to instrumental differences or due to changes in sample concentration during the emergence of a gas chromatographic fraction.)

An alternate method of preselection involving comparison of spectra reduced to binary intensity data (present or not present) (*30*) was tried in this laboratory and rejected since preliminary results were not promising. Using spectra obtained on a gas chromatograph–mass spectrometer system as test data, the variable intensity "grass" (spurious, small peaks) always present in spectra corresponding to small gas chromatographic peaks presented serious problems when these data were reduced to a binary spectrum. Furthermore, the time involved in performing this comparison on an IBM 1800 computer was prohibitive. On other computers the time requirements for a binary presearch may be reduced to a tolerable level (*30*).

It has been suggested (*9, 12*) that the molecular weight should be used as the basis for narrowing the number of spectra to be compared in detail. This is a very unrealistic suggestion, unless one deals with a group of compounds the molecular weight range of which is well known in advance, which is seldom the case. To interrupt an automated sequence for visual inspection of the data is illogical even if it were simple to deduce the molecular weight of a compound by inspection of its mass spectrum. In addition, spectra from a gas chromatograph–mass spectrometer system often contain column bleed or low intensity noise at a higher m/e ratio than the molecular ion, which makes an automatic computer identification of the molecular ion extremely difficult. Finally, using the molecular weight as a presearch criterion makes it less likely to find structurally similar compounds if the actual one is not part of the collection, a feature which must be part of any useful search technique.

The initial filters discussed above, implemented by simple table look-up procedures, result in a list of identification numbers of known spectra to be compared further. The relatively few known spectra selected in this manner are read from the disk in turn and compared in detail, *quantitatively*, to the unknown spectrum. The numerical result of this comparison is a "similarity index."

DETAILED COMPARISON OF MASS SPECTRA. It was felt that the similarity index should indicate the correct spectrum even if the unknown spectrum were greatly distorted because of changing sample concentration during the emergence of a gas chromatographic fraction, because of instrumental differences (*i.e.*, reference spectrum determined with another type of spectrometer), or because of the inclusion of peaks from impurities such as column bleed. Furthermore, an effective search should indicate compounds similar to the unknown if an exactly matching spectrum of the authentic material were not part of the file. For these variations between the unknown and the reference spectra, it seemed that the intensity ratio of the peaks at matching mass numbers in the two abbreviated spectra would give a better indication of similarity than would a sum of the differences between them. In addition, since ratios would range between 1.00 for complete agreement to 0.00 for complete disagreement, they are preferred to the sum of differences which has no unique upper limit (unless the spectra are normalized such that the sum of intensities equals a constant). Thus, a ratio is equivalent to the probability of agreement and gives an absolute degree of match which is useful for comparisons among various unknown spectra.

In essence, the similarity index is the *weighted* ratio of the known to unknown abbreviated spectrum taken mass for mass. The ratios of intensities of peaks at the same mass in the known and unknown spectra are calculated starting at the lowest mass that is common to both spectra. This precaution is necessary to avoid comparison of mass regions where one of the two spectra may not have been recorded. If a given mass is present in only one spectrum, the corresponding ratio is set to zero.

Ratios much lower or much higher than unity can be caused either by great differences in mass discrimination (caused by different instruments or smooth changes in sample concentration during the emergence of a gas chromatographic peak), or by similarly extreme ratios due to the actual non-identity of two spectra. To distinguish between these two cases, all ratios are divided by the average of only those ratios due to large peaks (>10% relative intensity). In the case of actual non-identity, the average ratio of large peaks would be far from unity and division by this value generates even greater differences in the ratios. For similar spectra, however, the average ratio is near unity and division by this value causes little change in the ratios. An earlier approach involving scatter from a curve fitted to ratio *vs.* mass data (*7*) was abandoned in favor of the simpler and less time consuming technique outlined here.

At this point, in order to give meaningful averages, any of the calculated ratios which are greater than unity are replaced by their reciprocals.

Reasoning that agreement as well as disagreement of the intensities is more significant for abundant ions than for minor ones, the corrected ratios are weighted by the factors 12, 4, and 1, depending on whether the larger intensity making up the ratio is >10%, 1% to 10%, or <1% relative intensity, respectively. (These intervals and weights were empirically determined.) The individual corrected ratios are multiplied by the corresponding weighting factor and the average weighted ratio is calculated.

A further indication of the similarity of the spectra is the fraction of the total ionization (the sum of all intensities of both spectra) due to peaks that do not have corresponding masses in the other spectrum. This "fraction of unmatched intensities" plus one is divided into the average weighted ratio to give the final similarity index:

$$\text{similarity index} = \frac{\text{average weighted ratio}}{\text{fraction of unmatched intensities} + 1}$$

The reason for this last division is merely to combine these two measures of agreement into one number which has a theoretical maximum of 1.00.

In this manner, a similarity index is calculated for each of

---

(30) S. L. Grotch, ANAL. CHEM., **42**, 1214 (1970).

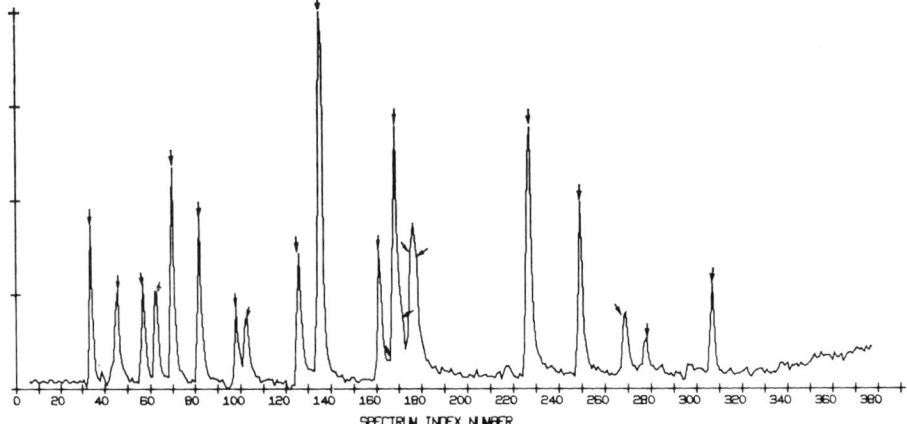

Figure 3. Computer-generated total ionization plot for the synthetic test mixture

Arrows designate the positions on the total ionization plot of the spectra which were compared to the computer-searchable collection of mass spectra

the pre-selected reference spectra to which the unknown is compared. These indices are stored in core memory along with the corresponding reference compound name in a list ordered by decreasing similarity index.

A high and unique similarity index indicates a particular match that is much better than all others and, therefore, most probably is a correct identification. For this reason, after all of the indicated reference spectra have been examined, the difference between the first and second highest similarity index is calculated. If this difference is greater than 0.05 unit and the highest similarity index is greater than 0.350 unit, an asterisk is printed before the name of the first member of this ordered list as an indication that there is one spectrum that is much more similar to the unknown than any of the others. (These two numbers were empirically determined and are the ones usually used. However, they can be varied depending upon the quality of the data and the experimental conditions.)

In any case, the first several members of this ordered list are printed (usually ten, but their number can be limited at the discretion of the investigator). As can be seen from Figure 2, this list contains matches of the unknown spectrum to the reference collection in decreasing order of similarity. Since the first member is preceded by an asterisk in Figure 2, the probability is very high that this is the correct result, and indeed it is. SERCH now re-initializes and reads in the next unknown spectrum.

Some notes on the amount of computer time required are in order. Using an IBM 1800 computer and searching the complete collection of 7600 spectra, the search for most unknowns requires under two minutes. For the SERCH comparisons listed in Table I, the average time was 60.5 seconds. The lowest SERCH time was 28 seconds (spectrum No. 126, quinoline) and the longest SERCH time was 2 minutes and 59 seconds (spectrum No. 70, dodecane). Eighty-five per cent of the searches required less than 90 seconds. The variation in time for these comparisons is dependent on the number of spectra selected in the presearch. Hence a compound with mass spectrometric characteristics representative of a large group of compounds will have many presearch finds and require a longer search time than a compound with a very unique mass spectrum. For example, a hydrocarbon unknown such as dodecane will have many presearch finds (249 for spectrum No. 70) because the pre-search is not intended to eliminate similar hydrocarbon spectra, such as the many contained in the API collection; on the other hand, quinoline has an uncommon, unique spectrum and hence few compounds will be retrieved in the presearch (6 for spectrum No. 126). It currently takes somewhat less than $1/3$ of a second to calculate one similarity index. It is anticipated that this time will be considerably reduced when the present FORTRAN program is rewritten into IBM 1800 assembler language and hardware for floating point arithmetic is installed. Minimizing the time required to perform a search is important, because one eventually wants to routinely search most of the spectra recorded during a gas chromatographic run and not only the few recorded at the tops of gas chromatographic peaks.

There are several additional options in the SERCH algorithm, some of which should be mentioned. The manner in which the reference file is currently formated allows one to compare an unknown with a particular data collection; for example, one could compare an unknown hydrocarbon to only the API collection. Provisions are being made to allow comparison with a particular type of compound (e.g., esters), if one knows the nature of an unknown sample. Of course, in the usual mode of operation the entire collection is searched. The user may also input masses which are to be deleted from the unknown spectrum, for example $m/e = 207$ and 281 (due to silicon column bleed) in the case of a gas chromatographic fraction. Another option allows the user to compare a whole gas chromatograph–mass spectrometer run, printing the most probable SERCH finds along the contours of the total ionization plot. Programs are also available which allow comparisons of unknowns with smaller reference collections using the same comparison routine as described above. Examples of such sub-collections are: spectra pertaining to compounds derived from a specific source or for the identification of a limited type of compounds, such as a collection of drug and drug metabolite mass spectra (31); and the use of mass spectra from one gas chromatograph–mass spectrometer run as reference data for comparison with another one, in situations where one wishes to recognize the variations in

---

(31) T. Sakai, H. S. Hertz, J. R. Althaus, and K. Biemann, Massachusetts Institute of Technology, Cambridge, Mass., unpublished data, 1971.

Table I. SERCH Results on Synthetic Test Mixtures[a]

Spectrum Index No.	SERCH Find No. 1 Name	Sim. Ind.	SERCH Find No. 2 Name	Sim. Ind.	Compound in mixture
34	Isopropylbenzene Dow 444	0.695	Isopropylbenzene (Cumene) API 311	0.678	Isopropylbenzene
46	Octanone 3	0.509	2-Ethylbutyraldehyde Dow 249	0.290	3-Octanone
57	n-Butylbenzene Dow 623	0.590	nor-Butylbenzene API 494	0.463	n-Butylbenzene
63	Cycloheptanone Dow 364	0.817	Cycloheptanone ASTM 1421	0.713	Cycloheptanone
70	nor-Dodecane API 1598	0.770	n-Dodecane API 23	0.575	n-Dodecane
82	Methyl benzoate Dow 669	0.647	Methyl benzoate MCA 88	0.469	Methyl benzoate
98	3-Acetylpyridine	0.460	2-Ethylpyridine	0.306	3-Acetylpyridine
103	p-Hydroxyacetophenone Dow 685	0.428	m-Hydroxyacetophenone Dow 684	0.367	o-Hydroxyacetophenone[c]
126	Quinoline MCA 105	0.499	Quinoline API 625	0.497	Quinoline
135	1-Methylnaphthalene Dow 738	0.658	2-Methylnaphthalene Dow 753	0.581	1-Methylnaphthalene
161	ar-Methoxybenzaldehyde Dow 675	0.242	p-Methoxybenzaldehyde	0.241	p-Methoxy methyl benzoate[c]
166[b]	o-Fluorochlorobenzene	0.064	1-Chloro-2-fluoro-benzene Dow 562	0.051	3-Methylindole
168	Methyl dodecanoate	0.726	Methyl tridecanoate ASTM 1905	0.441	Methyl dodecanoate
172[b]	Propiophenone Dow 633	0.268	Phthalide Dow 656	0.255	Phthalide
175	Dihydrocoumarin Dow 826	0.514	Anethole MSDC 61	0.096	Dihydrocoumarin
178	Vanillin ASTM 1045	0.258	p-Methoxybenzoic acid ASTM 1188	0.117	Vanillin
227	Ethyl myristate	0.602	Ethyl caprylate ASTM 1068	0.211	Ethyl tetradecanoate (myristate)
249	9-Fluorenone Dow 1297	0.664	Fluorenone ASTM 1435	0.583	9-Fluorenone
268	Carbazole MSDC 62	0.197	Carbazole MCA 106	0.178	Carbazole
278	Caffein MSDC 63	0.091	10,10-Dimethylacridane ASTM 1367	0.027	Caffein
307	4-Bromobenzophenone Dow 1872	0.053	ar-Methyl-ar-chlorophenyl phenyl ether Dow 1693	0.039	Triphenylcarbinol[c]

[a] It will be noted that the names of the compounds do not always conform to the IUPAC notation. This is because the spectrum collection is derived from many different laboratories and no attempt was made to standardize the compound names. (The prefix "nor" has been used for "normal" by some laboratories.)

[b] This compound represented a small shoulder on a previous peak and the spectrum searched was obtained by subtracting the spectrum of the previous component (see text).

[c] The spectrum of this compound is not in the reference collection.

---

mixtures from different but related sources, even if the components of the mixtures are not yet identified.

DISCUSSION AND RESULTS

The computer searching algorithm was tested as it evolved over a period of years almost exclusively on data taken from gas chromatographic effluents. Such spectra were interpreted in the normal manner and the results compared to the results of SERCH. Many of these comparisons led to program changes and thus to a continuous improvement of the algorithm.

To illustrate the use and capabilities of the SERCH routine two sets of data which were generated on the gas chromatograph-mass spectrometer-computer system (3) will be discussed. It was felt that a valid example must utilize data obtained in actual experiments rather than spectra that are part of the reference collection. There is obviously no point in testing whether a program can find an identical set of data and the number of different spectra of the same compound present in the collection is too limited to provide a valid test. The first set of data to be discussed is a synthetic test mixture, designed to evaluate the capabilities of the SERCH routine on a wide class of compounds, some of which are present and others absent in the spectrum collection, and also involving some instances where more than one compound occurs in a single gas chromatographic peak. The second mixture represents a series of methylated acids obtained from a methanolic potassium hydroxide hydrolysis of kerogen from the Green River Formation (32).

The computer-generated total ionization plot for the synthetic test mixture is shown in Figure 3. Table I lists the first and second SERCH finds for the compounds in the test mixture, as well as the components actually present in the mixture. It will be recalled that the computer lists the ten most similar spectra, but only the first two are presented in Table I to conserve space.

As is often the case in a gas chromatographic experiment, the first few components are well resolved and free from column bleed. Hence, the similarity indices for the compounds retrieved are generally high if a similar compound is found in the collection of mass spectra. The very high similarity indices for the primary SERCH finds coupled with the

(32) R. C. Murphy, M. V. Djuricic, D. and Vitorovic, Massachusetts Institute of Technology, Cambridge, Mass., unpublished data, 1968.

688 H. S. Hertz, R. A. Hites, and K. Biemann

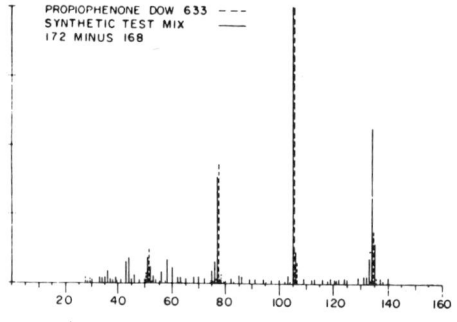

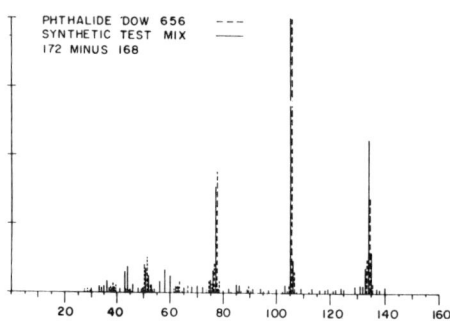

Figure 4. "Overplot" which compares unknown "172 minus 168" of the test mixture (see Figure 3) to propiophenone (top) and phthalide (bottom)

difference in similarity indices between the first and second compound retrieved for each of the first ten components in the synthetic test mixture suggest that the primary SERCH finds are correct. Indeed, they are correct in every instance except for o-hydroxyacetophenone (spectrum No. 103). The spectrum of this compound is not contained in the collection of mass spectra. The spectra of the p- and m-isomers which are in the collection are thus retrieved by SERCH as the most similar spectra. Because of the known similarity of the mass spectra of aromatic isomers, one always has to be careful with assignment of aromatic substitution patterns. This result is thus just as good as if the output would have listed o-hydroxyacetophenone first and the p- or m-isomer second. Final identification in the case of such aromatic isomers always requires comparison with authentic spectra determined under identical conditions or other additional information.

Figure 2 shows an actual SERCH output for the second component of the mixture, 3-octanone. The series of numbers below the abbreviated spectrum represent the ordered rectangular array elements for the unknown. The numbers below the results are the number of spectra in the collection still being considered after successive presearches. For 3-octanone, there were 194 presearch finds and, hence, 194 similarity index calculations were performed.

Inspection of Table I reveals the similarity index for the primary SERCH find for spectrum 161 is considerably lower than the similarity indices for the preceding components in the mixture. Furthermore, two more p-methoxy carbonyl compounds (p-methoxyacetophenone and p-methoxypropio-phenone) are found third and fourth with similarity indices very close to the first two (as indicated earlier the ten most similar spectra are listed by the computer but only the first two are shown here in Table I to keep it within a manageable size). Obviously, the compound is very likely at least a methoxybenzoyl derivative, if it is not indeed a methoxybenzaldehyde. In such instances comparison of the two spectra (scan 161 and Dow spectrum 675) is the logical next step, which can be taken using a program which overplots any two spectra on the same grid, in two different colors (black and red) and slightly offset on the x-axis (e.g., see Figure 4, in which the red trace is shown as dotted lines). In this particular instance, it is immediately obvious that the unknown differs from the known merely in the molecular weight, which is 30 amu higher, and the loss of 31 amu instead of 1 amu (hydrogen) from the molecular weight. A methyl methoxybenzoate is the logical interpretation. Indeed, methyl p-methoxybenzoate had been added to the test mixture to evaluate the results for a compound whose spectrum is not present in the collection.

The spectrum of 3-methylindole would probably not have been found without the use of some of the computer algorithms for finding significant masses in a gas chromatograph–mass spectrometer run (33). One of these algorithms indicated that m/e 130 and 131 are of relatively high abundance in only a few spectra. Mass chromatograms (33) of m/e 130 and 131 revealed that these spectra were centered around scan 166. Inspection of the total ionization plot (Figure 3) reveals, however, that this scan is in the valley between two gas chromatographic peaks. Utilizing the algorithm described earlier (7), the spectrum of scan 161 was subtracted from that of scan 166. The resulting spectrum was of poor quality but when compared with the spectrum collection the first two finds were those listed in Table I, followed by a number of mono-methyl indoles for which the similarity indices were only 0.008 to 0.009. Hence, even these extremely poor similarity indices can provide useful hints for the identification of a minor, poorly resolved component. In this instance it was particularly fortuitous that this unknown spectrum was that of 3-methylindole. Because this compound has essentially a two-peak spectrum (m/e = 130 and 131), it was identified without trouble by the computer because the intense peaks were well above the constant low intensity noise.

Returning again to Figure 3, spectrum 172 is a shoulder on the methyl dodecanoate peak (spectrum No. 168). Subtracting spectrum 168 from spectrum 172 yielded a reasonably clean spectrum of the component on the shoulder of the peak. Since the first SERCH find was too low to be a "starred" fit and since it differed from the second find by only 0.013, a manual examination of the spectra was required. Figure 4 represents overplots of the subtracted spectrum and the reference spectra of propiophenone and phthalide, the first two SERCH finds. Comparison of these overplots, especially in the region of the molecular ion, shows greater similarity between phthalide and the unknown than between propiophenone and the unknown.

The next peak in the total ionization plot (centered at spectrum index No. 176) contains different components on either side. Nevertheless, the compounds were separated well enough before entering the mass spectrometer to allow both compounds to be retrieved, each as the first SERCH find. Vanillin (spectrum No. 178) was retrieved with a similarity index of only 0.258, but this is reasonable since the vanillin

---

(33) R. A. Hites and K. Biemann, ANAL. CHEM., **42**, 855 (1970).

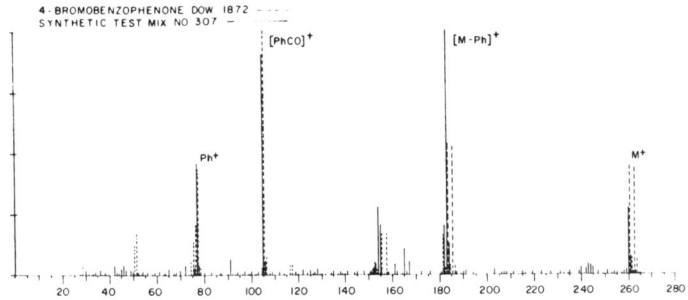

Figure 5. "Overplot" of spectrum No. 307 of the test mixture (see Figure 3) and 4-bromobenzophenone

was not perfectly separated from the previous component in the mixture (spectrum No. 175).

It can be seen (Table I) that similarity indices decrease as a gas chromatogram proceeds; the results are, however, still good. The component of longest retention time, triphenylcarbinol (spectrum No. 307), was also not in the collection of mass spectra. The spectrum of 4-bromobenzophenone, the best SERCH find, does show some similarity to that of triphenylcarbinol (Figure 5), and indeed, there are some similarities in the structures of these two compounds. The similarities are enhanced in the mass spectrometric data because of the fortuitous coincidence of the nominal mass of C_6H_6 and ^{79}Br minus H, which, when added to benzophenone, give triphenylcarbinol and bromobenzophenone, respectively. The very low similarity index shows that the SERCH find probably is incorrect, but that some structural similarities exist.

In summary, the results of the SERCH retrievals on this test mixture indicate that the routine appears to perform quite well. Of the eighteen components contained in the collection of spectra, sixteen were retrieved as the first SERCH find and the two spectra obtained by subtraction were retrieved fourth and second, respectively. For the three compounds not contained in the collection, very similar compounds were retrieved for two of them, and a compound with a related structure for the third. This information naturally aids the quick identification of the compound by manual interpretation. Furthermore, these results demonstrate that the presearch is very effective. Out of 7600 possible spectra, an average of 46 spectra were preselected for each unknown in the test mixture; and in no case was the correct spectrum discarded by the presearch. In addition, for the compounds not contained in the reference collection, the SERCH results reveal that similar compounds are also retained by the presearch.

The second example represents the methanolic hydrolysis product of Green River Shale kerogen (32). The resulting mixture contains a series of straight chain fatty acids, a series of α,ω-diacids, and a series of terpenoid acids. Although not all the components of the mixture were in the collection, a sufficient number were present to allow rapid identification of these series of compounds from SERCH finds, manual inspection of the mass spectrometric data, and gas chromatographic regularities revealing homologous series of compounds. The computer-generated total ionization plot for this mixture is shown in Figure 6. Table II shows the first and second SERCH finds for each component.

The correct compound or a very similar compound was

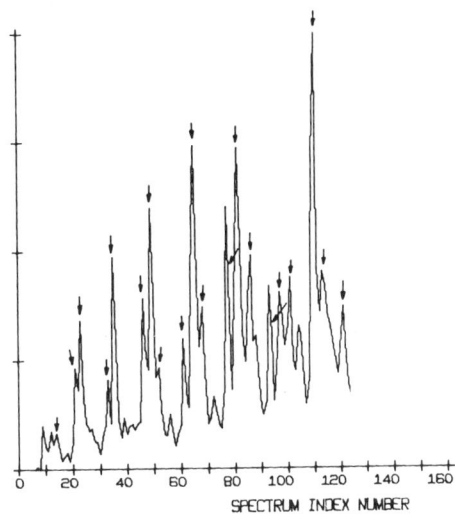

Figure 6. Computer-generated total ionization plot for Green River Shale kerogen hydrolyzate

Arrows designate the positions on the total ionization plot of the spectra which were compared to the computer-searchable collection of mass spectra

retrieved in all cases except that of methyl heptane-1,7-dioate (spectrum No. 23). In this case the lone compound retrieved (a highly branched alcohol) showed a low similarity index and was somewhat suspect since the mixture represents an acid fraction converted to methyl esters. Manual comparison of the retrieved spectrum to that of the unknown shows that the two spectra are indeed very different. Nevertheless, the retention times of the other, quite unambiguously identified components indicate that spectrum No. 23 is the C_7 member of the diester series. Examination of the spectrum confirms this hypothesis. It should be noted that the mass spectra of diesters of less than nine carbon atoms in the

Table II. SERCH Results on DV V-2m

Spectrum Index No.	SERCH Find No. 1 Name	Sim. Ind.	SERCH Find No. 2 Name	Sim. Ind.	Compound in mixture
14	Methyl hexane-1,6-dioate	0.109	Dimethyl-2-ethyl suberate ASTM 1906	0.093	Methyl hexane-1,6-dioate
21	Methyl caprate Dow 1381	0.250	Methyl decanoate ASTM 1834	0.243	Methyl decanoate (caprate)
23	4-Ethyl-2,6-dimethyl-4-heptanol Dow 1213	0.085			Methyl heptane-1,7-dioate[a]
33	Methyl heptadecanoate	0.246	Methyl undecanoate	0.244	Methyl undecanoate
35	Methyl octane-1,8-dioate	0.393	Dimethyl suberate ASTM 1903	0.366	Methyl octane-1,8-dioate (suberate)
46	Methyl dodecanoate	0.461	Methyl laurate ASTM 1081	0.415	Methyl dodecanoate (laurate)
49	Methyl nonane-1,9-dioate	0.498	Trimethyl aconitate Dow 1682	0.042	Methyl nonane-1,9-dioate
52	Methyl 2,6,10,14-tetramethylpentadecanoate	0.282	Ethyl 9-methyloctadecanoate	0.171	Methyl 2,6,10-trimethylundecanoate[a]
61	Methyl tridecanoate ASTM 1905	0.446	Methyl tridecanoate	0.400	Methyl tridecanoate
65	Methyl 10-undecenoate ASTM 1786	0.157	Dimethyl suberate ASTM 1903	0.130	Methyl decane-1,10-dioate[a]
68	Methyl 3,7,11,15-tetramethylhexadecanoate	0.306	Methyl 2,4,6-trimethyltetracosanoate	0.151	Methyl 3,7,11-trimethyldodecanoate[a]
78	Methyl tetradecanoate	0.449	Methyl tridecanoate ASTM 1905	0.386	Methyl tetradecanoate
81	Methyl docosane-1,22-dioate	0.107	Methyl 3-methylhexadecane-1,16-dioate	0.102	Methyl undecane-1,11-dioate[a]
86	Methyl 4,8,12-trimethyltridecanoate	0.436	Methyl 4,8,12,16-tetramethylheptadecanoate	0.305	Methyl 4,8,12-trimethyltridecanoate
94	Methyl pentadecanoate	0.334	Methyl palmitate ASTM 1907	0.308	Methyl pentadecanoate
97	Methyl docosane-1,22-dioate	0.145	Methyl octadecane-1,18-dioate	0.145	Methyl dodecane-1,12-dioate[a]
101	Methyl 5,9,13-trimethyltetradecanoate	0.355	Methyl 18-n-propylheneicosanoate	0.316	Methyl 5,9,13-trimethyltetradecanoate
110	Methyl hexadecanoate	0.589	Methyl palmitate ASTM 1907	0.574	Methyl hexadecanoate (palmitate)
113	Dioctyl phthalate Dow 1960	0.060	Dioctyl phthalate	0.053	Dioctyl phthalate
121	Methyl 2,6,10,14-tetramethylpentadecanoate	0.344	Ethyl 9-methyloctadecanoate	0.186	Methyl 2,6,10,14-tetramethylpentadecanoate

[a] The spectrum of this compound is not in the reference collection.

acid moiety are quite different from the higher homologs and also differ appreciably from homolog to homolog (34). Therefore no other diester was retrieved.

The mass spectra of three of the other methyl esters of dicarboxylic acids are also not contained in the collection. In all three cases similar compounds were retrieved (but with clearly lower similarity indices), which enabled rapid identification of the correct compound based on its retention time and inspection of the mass spectrum.

The mass spectra of two of the terpenoid acid methyl esters (spectrum No. 52 and 68) in the mixture are also not contained in the reference collection. In both cases SERCH retrieved the next higher "isoprenolog," i.e., the one containing one more isoprene unit. By glancing at the spectra, one realizes that the unknowns were the lower homologs of the compounds retrieved.

One compound of interest in this mixture is dioctyl phthalate. Knowing the overall composition of the naturally occurring mixture, one would not have expected to find an aromatic compound. The computer, however, has no bias based on origin of the sample and hence retrieves unexpected compounds as easily as predicted components. The phthalate is most probably an artifact introduced during the chemical treatment of the kerogen.

(34) R. Ryhage and E. Stenhagen, Ark. Kemi, 14, 497 (1959).

The mass spectra of fourteen components in this sample are contained in the collection of mass spectra. Of these, thirteen correct spectra were retrieved as the first SERCH find and one was retrieved as the second SERCH find (methyl undecanoate, No. 33). However, for methyl undecanoate the difference in similarity index between the first and second SERCH finds was only 0.002, a case which calls for detailed comparison of the unknown spectrum with both SERCH finds.

The results discussed above demonstrate that the identification of unknown mass spectra by comparison with the reference file can be accomplished rapidly in either one or two steps. In cases where a high and unique similarity index is obtained and the nature of the sample is known, the SERCH result plus inspection of the unknown spectrum will allow identification of the compound. In most other cases, a second step involving comparison of the unknown spectrum to the spectrum retrieved will enable the investigator to identify the unknown.

Over the past few years, this retrieval system has been employed in this laboratory for the identification of the components of many complex mixtures. A recent application of the system involving the search for drugs and their metabolites in body fluids is described in another publication (35).

(35) J. R. Althaus, K. Biemann, J. Biller, P. F. Donaghue, D. A. Evans, H.-J. Förster, H. S. Hertz, C. E. Hignite, R. C. Murphy, G. Preti, and V. N. Reinhold, Experientia, 26, 714 (1970).

ACKNOWLEDGMENT

The authors thank Edward Ruiz and Mrs. Vivian Zoller for programming assistance, and R. G. Ridley, Mass Spectrometry Data Centre at Aldermaston, and H. G. Boettger, Jet Propulsion Laboratory, for their collaboration in compiling the spectrum collection. Robert C. Murphy kindly donated the sample of Green River Shale extract discussed in this paper.

RECEIVED for review October 23, 1970. Accepted January 28, 1971. This work was supported by a National Institutes of Health Research Grant (No. RR00317 from the Division of Research Resources), a National Institutes of Health Training Grant (No. GM 01523), and a National Aeronautics and Space Administration Research Grant (No. NGR 22-009-005).

Small Computer, Magnetic Tape Oriented, Rapid Search System Applied to Mass Spectrometry

L. E. Wangen, W. S. Woodward, and T. L. Isenhour

Department of Chemistry, University of North Carolina, Chapel Hill, N. C. 27514

A fast search procedure capable of searching a spectral library at a rate of 10000 16-bit words per second directly from magnetic tape has been developed for a small computer. No computer memory is devoted to library spectra. A library of 6652 low resolution mass spectra with 352 mass positions coded to peak/no peak information can be completely searched for nearest as well as perfect matches in 15 seconds. Statistical considerations and some principles of information theory are used to reduce to 48 the number of bits necessary to code a mass spectrum with minimal loss of pertinent information. Mass positions that consistantly correlate throughout the data set are combined such that all spectra are reduced in dimensionality by the same procedure. This makes it unnecessary to perform any decoding operations on library spectra prior to or during the search. Results are presented for searching 352 dimensional spectra as well as the same spectra reduced to 80 and 48 dimensions.

THE AVAILABILITY of large libraries of spectrometric data in computer compatible form has led to an increasing use of spectral comparison as an aid to structure determination and compound identification. Powder diffraction files have been utilized in mineral identification while the use of infrared and mass spectra in compound identification has become of increasing import as spectra libraries (1–5) are made available. Several recent papers have dealt with the numeric representation of the data and methods for efficient search and comparison.

Anderson and Covert (1) reported a system developed for infrared data on the IBM 7080 computer. As many as 20 spectral terms (adsorption maxima or no absorption) and 15 chemical classification terms together with melting or boiling point information could be compared with the library spectra to identify a compound. This system allowed for a ±0.1-μm ambiguity in the wavelength of adsorption peaks and could search for five unknowns at a time giving up to the 100 best matches for each unknown. They achieved a rate of 167 spectral comparisons per second. Erley (2) compacted the ASTM infrared file into 10 16-bit words per spectrum and used logical operations to perform comparisons. The data coding included chemical group and elemental

---

(1) D. H. Anderson and G. L. Covert, ANAL. CHEM., **39**, 1288 (1967).
(2) D. S. Erley, *ibid.*, **40**, 894 (1968).
(3) D. S. Erley, *Appl. Spectros.*, **25**, 200 (1971).
(4) F. E. Lytle, ANAL. CHEM., **42**, 355 (1970).
(5) F. E. Lytle and T. L. Brazie, *ibid.*, p 1532.

information in addition to the spectra. His searching procedure requires that the library and the unknown spectra have exactly the same peaks present although there may be a ±0.1-μm ambiguity in peak position. This assembler language search program required about 7000 16-bit words of memory in the IBM-1130 computer. It could search 1000 spectra (10000 16-bit words) per second from disk data files. [Where possible, for comparative purposes, search speeds are given in number of 16-bit words (a common computer word size) searched per second.] More recently Erley (3) mentions, but doesn't elaborate on, a more flexible extension of the above searching procedure that allows one or more complete mismatches between the unknown and standard spectra.

Lytle (4) reported a computerized search developed on portions of the infrared data in Sadtler's Spec-Finder. The Spec-Finder library contains the position of the most intense absorption in each of the 1.0-μm intervals from 2 through 14 where each micrometer interval is divided into ten subintervals of 0.1 μm width or, in case a given micrometer interval contains no maximum, a no absorption is recorded. Thus of eleven possibilities for each micrometer interval corresponding to a given compound, only one is recorded. Lytle organized the data into 943 separate files according to wavelength subintervals such that there was a file corresponding to every 0.1-μm interval or no absorption possibility. Each file contained a position (bit) for every compound in the collection. The bit was *on* if the position corresponded to the compounds most intense absorption in the appropriate 1.0-μm interval, otherwise it was *off*. Thus only $1/_{11}$ of the total bits are *on*. A major advantage of this method is that only those files corresponding to the unknowns' absorption maxima are searched. In addition, these files are input to the computer sequentially, thus reducing computer memory requirements. This system, although not allowing any ambiguity in peak position, does give closest matches as well as perfect matches. The method searches for only one unknown at a time but can search 20000 to 30000 spectra at a rate of approximately 1000 per second in a computer with 8000 16-bit words of memory using paper tape input.

More recently Lytle and Brazie (5) reported a method using statistical compression of the same infrared compilation. They used statistical considerations and principles of information theory in an attempt to maximize information content relative to the number of bits used to represent the data. The idea is to reduce the number of bits needed to store the data with a minimal loss of information. Each spectrum is combined and packed into 1 16-bit word and the data are input to the computer in blocks from disk. The number of spectra that can be searched per second depends on the amount of memory available as an input buffer. In this case an XDS Sigma 5 computer with 16000 32-bit words of memory was used allowing a maximum search speed of 18000 16-bit spectra per second for a single unknown. The method as presented requires perfect matches and doesn't allow any ambiguity in peak position. However perfect matching is undoubtedly a program imposed limitation rather than a characteristic of the general method.

Jurs (6) used "hash coding" to obtain near optimum speeds for retrieval from computer memory of perfect matches of simulated infrared spectra. This method gives fast retrieval with only a fraction of excess memory, however it suffers from the inherent disadvantage of allowing no room for error in experimental data acquisition.

Of the above search systems the first two have been extensively tested. Erley (3) makes some practical suggestions for quantitative comparison of infrared search systems and presents data comparing three different search algorithms.

A further area of current interest is the coding and searching of mass spectra. Knock *et al*. (7), compared various matching methods using the most intense peaks for compound identification from low resolution mass spectra. Attempts were made to compensate for instrumental variations by dividing the spectrum into equal ranges, each containing the same number of mass units. In each mass range, the *n* most intense peaks are retained in order of decreasing intensity. Corresponding mass positions are then compared as usual. A success rate of 97% was reported, and similar compounds were generally retrieved in addition to the unknown. Therefore the method appears useful even when the unknown is not included in the library. A limited range of molecular weights is searched to reduce time and memory requirements. Search time depends on several variables and ranges from 3 to 30 seconds per unknown for a file of 8000 spectra on direct access disk with an IBM 360/50 computer. Hertz *et al*. (8) report a search procedure designed for use with a GLC-MS system. The method of representing the mass spectra is similar to one described more generally in reference 7. In the work by Hertz *et al*., the two largest peaks in each 14-mass unit interval beginning with mass position 6 are retained to form an "abbreviated" mass spectrum, This method is intended to allow reduction of the spectrum while retaining "interpretively" significant peaks. The search procedure involves extensive preprocessing to eliminate the necessity of making exhaustive comparisons of dissimilar spectra. After preprocessing detailed comparisons are made with the selected standard spectra by a ratio method. The system has been extensively tested and a high success rate is reported. Approximate search time for processing a library of 7600 abbreviated low resolution mass spectra from disk with an IBM 1800 computer is 2 minutes per unknown.

In related work, Grotch (9) did an extensive study designed to ascertain the feasibility of using low resolution mass spectra coded as peak/no peak information for compound identification. Each mass position in a spectrum is coded to one bit, if the peak intensity in that position is greater than a given fraction of the base peak, the bit is *on;* if not, the bit is *off*. Exhaustive comparisons were made using various fractions of the base peak (transition levels) as cut offs and this method of encoding was shown to retain much useful information. His results indicate that a compound could be identified by searching a library of mass spectra encoded in this way. In addition this method of coding offers a substantial savings in storage and, as will be shown below, lends itself nicely to efficient searching.

This work presents a search procedure developed on (but not restricted to) the Atlas of Mass Spectral Data compilation (*10*); the procedure incorporates many of the desirable features of earlier work including speed, minimal memory requirement, and applicability to real data and is designed to operate on a small computer system.

If a search system is to be practical for spectrometry data, it should have the following two attributes. The data should

---

(6) P. C. Jurs, ANAL. CHEM., **43**, 364 (1971).

(7) B. A. Knock, I. C. Smith, D. E. Wright, and R. G. Ridley, ANAL. CHEM., **42**, 1516 (1970).
(8) H. S. Hertz, R. A. Hites, and K. Biemann, *ibid.*, **43**, 681 (1971).
(9) S. L. Grotch, *ibid.*, **42**, 1214 (1970).
(10) "Atlas of Mass Spectral Data," E. Stenhagen, S. Abrahamsson, and F. W. McLafferty, Ed., John Wiley and Sons, New York, N. Y., 1969.

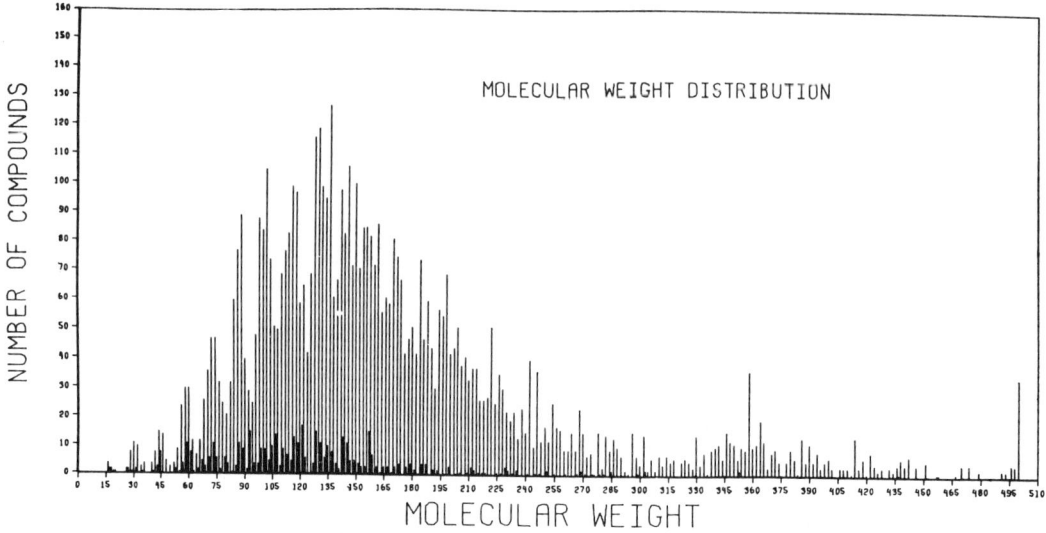

Figure 1. Molecular weight distribution of the 6652 low resolution mass spectra

be coded and the search performed so as to allow successful retrieval given the nearly unavoidable experimental and human errors resulting from data collection—*i.e.*, the coding should retain sufficient redundancy such that close matches are meaningful and the search should find such near misses. The results of searching for unknowns not contained in the library should be intuitively meaningful. These two factors are closely related.

The development of a search and retrieval system logically concerns two related problems—representation of the data and the search algorithm. These will be discussed separately.

SPECTRUM REPRESENTATION AND CODING

The usual approach for search and retrieval of mass spectra is to select a given number of the most intense peaks, assuming that these are most important for compound identification. This selection reduces both the time and storage requirements of a search system; retaining the complete spectrum including intensities presents an inordinate amount of data for storage and searching. Additional savings can be realized if peaks are coded in order of decreasing intensity rather than as actual intensity. This is the basic approach successfully taken by Knoch *et al.* (7). However the work of Grotch (9) indicates that there may be sufficient information for compound identification when the entire spectrum is retained with peak height encoded to only one bit. Because this representation lends itself nicely to computer coding and searching, it was decided to investigate its utility with the Atlas library.

The Atlas of Mass Spectral Data used in this study consists of 6652 low resolution mass spectra including all peaks present at intensities greater than 0.1% of the base peak (transition level of 0.1%). The spectra were originally measured by a variety of laboratories using different instruments, but were all obtained by electron-bombardment ionization, generally between 30 and 100 eV. There are about 5000 different compounds represented in the compilation; however, all 6652 spectra were retained because duplications resulting from different laboratories allow an estimation of a search methods applicability under actual experimental conditions. The elemental statistics, molecular weight distribution, and mass fragment populations of the original spectra are shown in Table I and Figures 1 and 2, respectively.

Three transition levels (0.1, 1.0, and 5.0%) were tested to ascertain the utility of each for spectral representation. To completely perform this task would require matching each spectrum with every other spectrum at each transition level for all 6652 spectra—a large expenditure of computer time. However, it is a relatively fast computer task to sort the spectra after coding and packing the data into computer words. Once sorted, a single iteration through the 6652 spectra, writing out neighbors that match perfectly, will provide all identical spectra. In this way, the perfect matches can be obtained although not the close matches or the average number of disagreements. The number of perfect matches at each transition level together with the type of compounds involved in such matches is shown in Table II. The table shows that the number of perfect matches increases with transition level and indicates that the information loss is going from a transi-

Table I. Statistics of the 6652 Mass Spectra in the Atlas Compilation

No. of compounds containing[a]

C	6607
H	6216
O	3521
Cl	792
N	536
S	467
F	295
Br	259
D	90
Si	48
I	42
P	10
C^{12}	1
C^{13}	1
Av no. peaks $\geq 0.1\%$	86
Av molecular weight	167
Av no. of elements	3

[a] Each elemental symbol implies the normal isotopic distribution unless indicated otherwise.

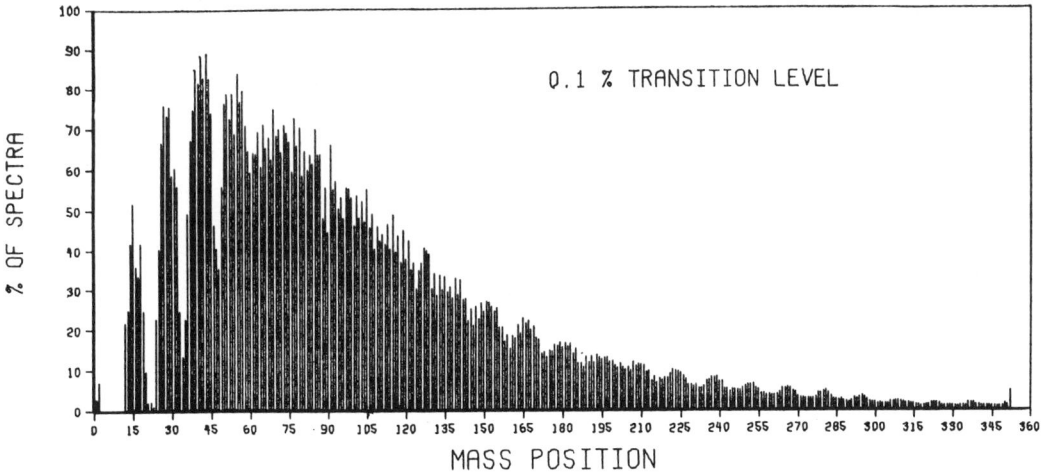

Figure 2. Percentage of spectra having a given mass fragment at a 0.1% transition level

Table II. Comparison of Number of Perfect Matches at Transition Levels of 0.1%, 1.0%, and 5.0%

Transition level, %	No. perfectly matching groups[a]	No. spectra involved	Chemical characteristics of matching spectra		
			Same Compound	Similar isomer	Related structure or miscellaneous
0.1	167	370	112	42	14
1.0	209	451	156	38	15
5.0	376	836	335	98	43

[a] Matching groups in some cases contain several spectra meaning that the same compounds, similar isomers and related structure, or miscellaneous can all occur in a given matching group.

tion level of 0.1 to 1.0% is small relative to the comparable loss at 5.0%. Most of the perfect matches derive from identical compounds while those remaining are predominantly similar isomers. Thus it appears that using a transition level of 1.0% as opposed to 0.1% would be quite acceptable.

In addition to showing that peak/no peak coding of mass spectra is useful for compound identification, the tests at levels other than the original of 0.1% were prompted by noise and/or contamination considerations. Higher transition levels are desirable for minimizing the effect of spectral noise, *i.e.*, an extremely low transition level would result in a peak at every position. These considerations (noise *vs.* information loss) prompted a compromise in favor of a 1.0% transition level and further discussion is with respect to this level. It should be noted that the representation of the mass spectra by 352 positions was purely a matter of convenience. It is believed that the spectra could be equally well represented by about 300 positions with all fragments greater than 300 assigned to one of the other positions.

REDUCTION OF DIMENSIONALITY

The average number of peaks per spectrum at a 1.0% transition level is 55 and as shown in Figure 3, the majority of mass

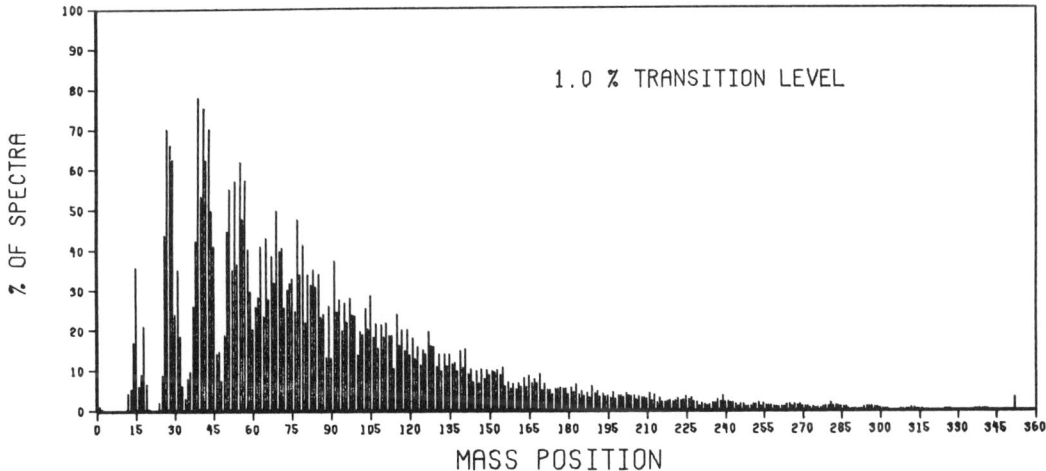

Figure 3. Percentage of spectra having a given mass fragment at a 1.0% transition level

positions are sparsely populated, hence an average mass spectrum will contain predominantly zeros. Thus when the entire mass spectrum is coded, the information content of many mass positions is very low. In information theory, information entropy provides a quantitative measure of the inherent information of a source. If, as in the case under discussion, there are only two possible states per channel (*e.g.* peak/no peak), the information entropy per channel (mass position) is given by

$$h = -[p \log_2 p + (1 - p) \log_2 (1 - p)]; 0 \leq p \leq 1 \quad (1)$$

where p is the probability of the given mass position containing a peak (the fraction of spectra having the given mass fragment). The total entropy is obtained by summing over all channels. Equation 1 is a maximum at $p = 1/2$ implying that the information content will be maximized when all bits (mass positions) are *on* for exactly half the compounds in the library. If the actual number of bits of an information source are most efficiently utilized, the entropy will have its maximum possible value—*i.e.*, it will be numerically equal to the number of real bits. This efficiency concept is quantified by the relative entropy defined as the actual (measured) entropy divided by the maximum possible entropy of the information source. Thus for ten data bits with an entropy of 8.5, the relative entropy is 0.85; and the efficiency of the representation is 85%.

If possible, different transition levels might be chosen for each mass position such that the bit corresponding to that position would be *on* half the time. However as shown by Figure 2 most positions are *on* for less than 50% of the library spectra making this impossible. Even if this procedure were possible, noise considerations would make it less useful. An alternative approach, similar to one taken by Lytle and Brazie (5), is to combine various mass positions such that the resultant bit position will have 50% occupancy. For example Figure 3 shows that if mass positions 86 and 87 (p's of 0.236 and 0.246, respectively) are combined, the resultant bit would be *on* for about 48% of the compounds assuming few correlations between the two mass positions. If such a procedure is carried out for the entire spectrum, a reduction in dimensionality will result. Hopefully, the accompanying loss of total information will be less significant. Thus for cases where the data contain many zeros, it might be possible to drastically reduce the number of bits required to represent a spectrum with a minimum loss in total information content. (The lost information can be calculated using Equation 1.) As always a compromise must be made, in this case between dimensionality reduction and information content (efficacy of the resulting spectral representation).

One procedure of reduction involves arbitrarily combining various mass positions, the sole criterion being that the individual occupancy levels sum to ~50%. This procedure reduced the number of bits used to represent a spectrum from 352 to 80. The perfect matches in addition to those originally present at a 1.0% transition caused by this reduction are shown in Table III as are the entropy calculations (Equation 1). It is apparent that there has been a significant decrease in entropy compared to the original spectra; however the amount of important information lost as measured by additional matching groups seems quite acceptable, considering the large reduction in dimensionality. Of 51 additional matching groups, 35 are either identical or similar isomers while the majority of the remainder are structurally very similar, *e.g.*, differing only in length of carbon chain.

Further dimensionality reduction can be accomplished by using the correlations between various mass positions. For example when mass positions 86 and 87 are combined, the resultant bit is turned *on* if either 86 or 87 or both contain a peak giving an apparent p of 0.48 for the bit. Actually this value is an upper limit for p. In reality if position 86 contains a peak, there is a 56% probability that position 87 also contains a peak. When the two are combined, the resultant bit will be turned *on* by peak 86 only 44% of the time that 86 actually contains a peak because, for the other 56% of the spectra that have a fragment at position 86, the bit will have already been turned *on* by position 87. Thus the resultant bit will be on for 35% rather than 48% of the compounds or $p = 0.35$.

A measure of how well two mass positions correlate can be obtained by logically *anding* the corresponding 6652-dimensional vectors and summing the result. These operations are identical to the dot product of two mass positions treated as vectors, each vector containing an entry for every spectrum. Not surprisingly the best correlations were obtained for positions separated by 1, 2, 13, 14, and 15 mass units corresponding to 1 H, 2 H's, CH, CH$_2$, and CH$_3$ fragments, respectively. No other masses gave significant correlation. This does not suggest that strong correlations do not occur between various mass positions for specific classes of compounds as these would not necessarily be discovered by performing the calculation with all classes of spectra.

This correlation information was used as an aid in combining mass positions that correlate well in an attempt to obtain maximum reduction of dimensionality consistent with maximum average entropy per channel and minimum loss of important information. Thus positions differing by 1, 2, 13, 14, and 15 mass units were most commonly combined. The spectra were reduced to 48 dimensions by this method. The average p was 0.45. The additional matching groups together with the entropy calculations are shown in Table III. The additional perfect matches again consist primarily of identical compounds, similar isomers, or structurally related compounds of a different molecular weight. As might be expected, the result of the large reduction in dimensionality is to make the near neighbors closer to the unknown. Another perhaps unexpected result is that the near neighbors are often more intuitively meaningful. This is undoubtedly an effect of combining highly correlating mass positions.

CODING LOGIC AND SEARCH ALGORITHM

Computer hardware is most efficiently utilized by performing operations on "word" size bit strings. The search system

Table III. Information Loss at 1.0% Transition Level Due to Reduction of the Mass Spectra from 352 to 80 and 48 Dimensions ("Mass" Positions)

		80-Dimensions	48-Dimensions
No. additional perfectly matching groups[a]		51	230
Chemical characterization of additional perfectly matching groups	Same	22	76
	Similar isomers	13	115
	Related structure or miscellaneous	16	60
No. spectra involved in matching groups		115	555
Total entropy (Equation 1)		76.5	45.8
Relative entropy (coding efficiency), %		95.8	95.5
Decrease in entropy with respect to 352-dimensions at a 1.0% transition level		55.2	85.9

[a] See footnote to Table II.

Table IV. Logical Operations *and* and *exclusive or*

A	0 1 1 0 1 0 1 1 1 0 0 0 0 1 1 0	
B	1 1 1 0 1 1 0 0 1 0 1 0 0 1 0 0	
AND(A,B)	0 1 1 0 1 0 0 0 1 0 0 0 0 1 0 0	Sum bits *on* = 5
EXCLUSIVE OR(A,B)	1 0 0 0 0 1 1 1 0 0 1 0 0 0 1 0	Sum bits *on* = 6
$(a_i - b_i)^{2a}$	1 0 0 0 0 1 1 1 0 0 1 0 0 0 1 0	Sum $(a_i - b_i)^2 = 6$

^{a} a_i, b_i are the individual components of A and B considered as 16-dimensional vectors.

reported here was developed on a Raytheon 706 computer with 16000 16-bit words of memory; hence, the logic is developed in terms of this word size but is in no way limited to it. If peak height is encoded to one bit, one computer word can contain 16 "mass" positions. Thus, for example, the 80 dimensional spectra will require 5 words per spectrum for a total of 33,260 words to encode all 6652 spectra in the Atlas Library. If it is desired to have more than two levels of intensity, additional bits per mass position are used—*i.e.*, the levels of intensity that can be represented by n bits is 2^n. Given the uncertainty in peak height, it is probably realistic to say that at most 16 levels of intensity, 4 bits, should provide adequate representation.

Given coded spectra there remains the matching problem. If it is desired to find best matches rather than perfect matches, most higher languages such as Fortran are highly inefficient for performing the necessary bit operations. However, this is a limitation of the language not the machine as most computers are hardwired to perform bit operations on computer words. There are two useful logical operators for comparing corresponding spectral bits. The logical *and* is used to obtain the dot product (a measure of correlation) whereas the *exclusive or* gives the number of mismatches and geometrically provides a measure of the distance between the two spectra. (This distance is the square of the euclidian distance for the one bit situation, because if a position contains a peak it is coded as a one, otherwise as a zero. The spectra can be thought of as multidimensioned points consisting of ones and zeros such that the sum of the *on* bits after the *exclusive or* of two spectra is identical to the sum of squares of the differences between corresponding mass positions.) Table IV clairfies these operations. In this work the *exclusive or* was used to perform the comparisons because it is believed that number of mismatches is superior to number of matching peaks as a measure of mass spectral similarity.

After the *exclusive or* operation is performed on corresponding spectral words, the 16-bit result is contained in the computer's accumulator as a binary number between zero and $2^{16} - 1$. However, for our purpose, it is necessary to know the number of *on* bits or mismatches; hence, the binary number must be translated in some way. One efficient way to accomplish this is via a translation table containing an entry corresponding to the number of *on* bits for every possible binary number. There are 2^{16} or 65,536 possible states for 16 bits; hence, a table to translate a 16-bit binary number into number of *on* bits would require precisely 2^{16} entries—an unacceptable and extremely inefficient use of computer memory. This problem can be suitably circumvented by breaking the 16-bit result into 2 8-bit halves (bytes). Thus the translation can be done successively on each 8-bit half of the word necessitating only 2^8 or 256 entries in the table. This process can be carried further when made necessary by memory restrictions. As an example consider the results of the *exclusive or* shown in Table IV. The 16-bit binary number (1000011100100010) = 34594 decimal so if a translation table with 2^{16} entries is used, the 34595th entry (0 is the 1st entry) would contain a 6 corresponding to 6 ones or *on* bits in the binary number. If this word is divided into bytes, the right byte (00100010) equals 34 decimal hence the 35th entry in a 256 entry translation table is 2. The left byte (10000111) equals 135 decimal so that the 136th table entry is 4. These are summed (2 + 4 = 6) to give the correct result for the distance between spectra A and B. The sum is accumulated in this way over all words in a spectrum.

It is worth noting that by concatenating segments of reference and unknown spectra, the method can be generalized to the case of encoding peak height to more than one bit. However concatenation necessitates twice as many table "look up" operations per bit because the *exclusive or* can no longer be used to reduce two corresponding words to a meaningful one-word result. Naturally the contents of the translation table are dependent on the number of encoded intensity levels.

SEARCHING PROGRAMS

The search programs were written as Fortran called assembler subroutines to enable most efficient use of hardware. The packed spectra are passed from magnetic tape directly to the computer accumulator one word at a time at a speed determined by the tape drive. While in the accumulator the word is matched (*exclusive or*) with the corresponding word of the "unknown" spectrum residing in memory. The 9-track magnetic tape contains 400 words per inch and the tape drive speed is 25 inches per second; hence 10000 16-bit words per second can be processed by the computer. Each word resides in the accumulator for about 100 μsec before the next word *must* be loaded. This time is sufficient to allow comparison with at most two unknowns in memory and to look up and sum the number of mismatches for each. It is apparent that the tape drive limits maximum obtainable search speed to 10000 library words per second for two unknowns. Thus a maximum of 10000 one-word spectra per second can be searched from magnetic tape with no memory devoted to library spectra, thereby requiring only enough memory to store the searching program (a few hundred words at most).

Two search routines were programmed—a constant radius (R) search and a constant number (N) search. The constant number search outputs the N closest matches ($N \leq 10$) together with number of mismatches whereas the constant radius search outputs all library members (ordered as to distance from unknown) within a given distance (R) of the unknown. Each method has particular advantages. With a constant number search, the results of a particular spectral match must be compared to the N previous closest matches for possible inclusion as a new closest match. This procedure for the case $N = 10$ and two unknowns requires about 200 μsec, a time equivalent to two words of tape motion. Thus the library tape for the constant number search is lengthened by the insertion of two dummy words between each spectrum.

Table V. Chemical Characterization Relative to the Unknown of the Best Matches Resulting from Searching for Randomly Chosen "Unknowns" at a 1.0% Transition Level

Spectrum coding		Percentage of closest matches found in various chemical classifications		
No. of dimensions	No. of words	Identical or close isomer	Structurally similar	Little or no apparent relation
352	22	84	13	3
80	5	76	19	5
48	3	74	15	11

The constant radius method has no similar requirements, hence is faster by 200 μsec per library spectrum. For ten word spectra, this means that the constant number search requires 20% more time than the constant radius method while for one word spectra there is a 200% increase in search time. The major disadvantage of a constant radius search is uncertainty in assigning an R value for different unknowns. A practical limit on the number of neighbors retained must be set if it is necessary to minimize memory requirements. Thus if R is too large, the library may not be completely searched before filling available memory with matches. On the other hand if R is too small, there may be no recorded matches at all. However R can be varied and if a reasonable maximum is set on number of near neighbors, it should be possible with experience to overcome these disadvantages and therefore obtain maximum search speed without use of excessive memory.

SEARCH RESULTS

Library tapes of the 6652 Atlas Mass Spectra were prepared at 352, 80, and 48 dimensions corresponding to 22, 5, and 3 16-bit words per spectrum, respectively. (These tapes were prepared on the Triangle Universities Computation Centers IBM 360/75 computer.) A set of 100 unknowns was chosen at random from the complete set for purposes of testing and comparing different coding methods. These results are summarized in Table V for the 80 cases where the "unknown" had either duplications or similar isomers present in the library. Of course, the "unknown" is retrieved as an exact match in every case. The results at various codings are comparable and the 3-word spectra generally gave good results, thereby allowing considerable reduction in storage requirements in addition to greatly increasing search speeds.

To enable intuitive feeling for search output, some illustrative results are shown in Tables VI–VIII for the 22 and 3 word cases. (Again in these as in all other cases, the "unknown" is always retrieved as an exact match but is not shown.) Two of these (Table VI, VII) illustrate a better as well as one of the poorer search results in addition to a case (Table VIII) where the library contains no similar compounds. Search time for the 3-word library is approximately 2 seconds with the constant radius search and 3½ seconds for a constant number search whereas with 22 word spectra search time is 16 seconds for a constant number search of all 6652 spectra.

Although conclusive comparisons with the work of Knock et al. (7) are difficult because of the limited molecular weight ranges and to some extent the different set of spectra searched in that work, it was felt desirable to do so. They pointed out that the spectra of Farnesol originating from different instruments show large variations. The Atlas library contains Farnesol seven times representing 4 different sources, thus seven searches were made, one for each "unknown" Farnesol spectrum. The results including five closest matches are shown in Table IX for what are considered the best and worst cases. Of the seven searches, Farnesol showed up first in 4 cases and in the nearest 5 for all cases. Furthermore, if compounds were eliminated because of molecular weight information, Farnesol was 1st in every instance, thus compar-

Table VI. Search Results at 352 and 48 Dimensions for 2,6-Dimethyl-4-Thiaheptane

Mol wt = 146, mol form. = $C_8H_{18}S$, ref no. = BAR 0029, struct. form. = $(CH_3)_2CHCH_2SCH_2CH(CH_3)_2$

No. dim.	Ref no.	Mol wt	Mol form.	No. mismatches	Compound name	Structural formula
3 5 2	SIK 2269	146	$C_8H_{18}S$	9	2,6-Dimethyl-4-thiaheptane	$(CH_3)_2CHCH_2SCH_2CH(CH_3)_2$
	SIK 2267	146	$C_8H_{18}S$	12	5-Thianonane	$CH_3(CH_2)_3S(CH_2)_3CH_3$
	API 1391	146	$C_8H_{18}S$	13	5-Thianonane	$CH_3(CH_2)_3S(CH_2)_3CH_3$
	WUR 1770	132	$C_7H_{16}S$	17	3-Methyl-4-thiaheptane	$C_2H_5CHCH_3S(CH_2)_2CH_3$
	API 1409	132	$C_7H_{16}S$	18	2,4-Dimethyl-4-thiahexane	$CH_3CHCH_3SCHCH_3C_2H_5$
	SIK 2264	132	$C_7H_{16}S$	18	2,4-Dimethyl-4-thiahexane	$CH_3CHCH_3SCHCH_3C_2H_5$
4 8	SIK 2269	146	$C_8H_{18}S$	3	2,6-Dimethyl-4-thiaheptane	$(CH_3)_2CHCH_2SCH_2CH(CH_3)_2$
	SIK 2267	146	$C_8H_{18}S$	5	5-Thianonane	$CH_3(CH_2)_3S(CH_2)_3CH_3$
	API 1391	146	$C_8H_{18}S$	5	5-Thianonane	$CH_3(CH_2)_3S(CH_2)_3CH_3$
	WUR 1770	132	$C_7H_{16}S$	5	3-Methyl-4-thiaheptane	$C_2H_5CHCH_3S(CH_2)_2CH_3$
	API 1409	132	$C_7H_{16}S$	7	2,4-Dimethyl-3-thiahexane	$CH_3CHCH_3SCHCH_3C_2H_5$
	SIK 2264	132	$C_7H_{16}S$	7	2,4-Dimethyl-3-thiahexane	$CH_3CHCH_3SCHCH_3C_2H_5$
	SIK 2263	132	$C_7H_{16}S$	7	2-Methyl-3-thiaheptane	$CH_3CHCH_3S(CH_2)_3CH_3$
	API 0573	116	$C_6H_{12}S$	7	2,5-Dimethyl-thiacyclopentane	$CH_2CHCH_3SCHCH_3CH_2$

Table VII. Search Results at 352 and 48 Dimensions for 2-Myristo-1,3-Diacetin

Mol wt = 386, mol form. = $C_{21}H_{38}O_6$, ref no. = HOL 9273, struct form. =
$$\begin{array}{c} H_2COCOCH_3 \\ | \\ HCOCO(CH_2)_{12}CH_3 \\ | \\ H_2COCOCH_3 \end{array}$$

No. dim.	Ref no.	Mol wt	Mol form.	No. mismatches	Compound name	Structural formula
	HOL 9437	414	$C_{23}H_{42}O_6$	52	1-Palmityl-2,3-diacetin	$H_2COCO(CH_2)_{14}CH_3$ / $HCOCOCH_3$ / $H_2COCOCH_3$
	HOL 2652	442	$C_{25}H_{46}O_6$	55	2-Stearo-1,3-diacetin	$H_2COCOCH_3$ / $HCOCO(CH_2)_{16}CH_3$ / $H_2COCOCH_3$
352	DOW 2550	210	$C_{15}H_{14}O$	69	2,3-Dihydro-2-methyl-7-phenylbenzofuran	(structure)
	DOW 3562	210	$C_{15}H_{14}O$	70	2,3-Dihydro-2-methyl-7-phenylbenzofuran	(structure)
	DOW 0520	200	C_8H_9BrO	70	(1-Bromo-2-hydroxyethyl)benzene	$\phi\text{-CHOHCH}_2Br$
	HOL 9283	358	$C_{21}H_{42}O_4$	7	2-Monostearin	H_2COH / $HCOCO(CH_2)_{16}CH_3$ / H_2COH
	HOL 9272	386	$C_{21}H_{38}O_6$	7	1-Myristo-2,3-diacetin	$H_2COCO(CH_2)_{12}CH_3$ / $HCOCOCH_3$ / $H_2COCOCH_3$
48	HOL 2652	442	$C_{25}H_{46}O_6$	7	2-Stearo-1,3-diacetin	$H_2COCOCH_3$ / $HCOCO(CH_2)_{16}CH_3$ / $H_2COCOCH_3$
	SHH 0057	410	$C_{30}H_{50}$	9	Squalene	$[(CH_3)_2C=CH(CH_2)_2\underset{CH_3}{C}=CH(CH_2)_2\underset{CH_3}{C}=CHCH_2-]_2$
	HOL 9263	358	$C_{19}H_{34}O_6$	9	1-Lauro-2,3-diacetin	$H_2COCO(CH_2)_{10}CH_3$ / $HCOCOCH_3$ / $H_2COCOCH_3$

Table VIII. Search Results at 352 and 48 Dimensions for Benzoyl Fluoride

Mol wt = 124, mol form. = C_7H_5FO, ref no. = DOW 4693, struct form. = C_6H_5COF

No. dim	Ref no.	Mol wt	Mol form.	No. mismatches	Compound name	Structural formula
	DOW 1230	106	C_7H_6O	17	Benzaldehyde	C_6H_5CHO
	DOW 1042	122	$C_7H_6O_2$	17	Benzoic acid	$C_6H_5CO_2H$
352	DOW 5152	134	$C_8H_6O_2$	18	Phthalide	(structure)
	DOW 1586	226	$C_{14}H_{10}O_3$	18	Benzoic anhydride	$(C_6H_5CO)_2O$
	MOR 0077	106	C_7H_6O	19	Benzaldehyde	C_6H_5CHO

Continued

Table VIII. Continued

Mol wt = 124, mol form. = C_7H_5FO, ref no. = DOW 4693, struct form. C_6H_5COF

No. dim	Ref no.	Mol wt	Mol form.	No. mismatches	Compound name	Structural formula
	DOW 1586	226	$C_{14}H_{10}O_3$	7	Benzoic anhydride	$(C_6H_5CO)_2O$
	DOW 2050	96	C_6H_5F	7	Fluorobenzene	C_6H_5F
	DOW 2456	130	C_6H_4ClF	8	1-Chloro-2-fluorobenzene	
4						
8	DOW 5304	138	C_8H_7Cl	9	P-Chlorostyrene	
	DOW 2523	137	C_7H_4NCl	9	P-Chlorobenzonitrile	
	DOW 2823	140	C_8H_9Cl	9	4-Chloro-M-xylene	

Table IX. Search Results at 352 Dimensions for Farnesol

Mol wt = 222, mol form. = $C_{15}H_{26}O$, struct form = $(CH_3)_2CCH(CH_2)_2C(CH_3)CH(CH_2)_2C(CH_3)CHCH_2OH$

	Ref. no.	Mol wt	Mol form.	No. mismatches	Compound name	Structural formula
"Best Result," Ref no. = SSA 0033	SSA 0032	222	$C_{15}H_{26}O$	17	Farnesol	Same as unknown
	WUR 0043	222	$C_{15}H_{26}O$	18	Farnesol	Same as unknown
	SSA 0034	264	$C_{17}H_{28}O_2$	20	Farnesyl acetate	Replace OH by $OCOCH_3$ in unknown
	API 1516	386	$C_{28}H_{50}$	23	1,10-Di(5-primhexahydroindanyl)decane	
	API 1267	414	$C_{30}H_{54}$	24	1,10-Di(α-decalyl)decane	
"Poorest Result," Ref no. = SIK 2486	SSA 0009	204	$C_{15}H_{24}$	18	β-Caryophyllene	
	SIK 1632	204	$C_{15}H_{24}$	19	γ-Cadinene	
	SSA 0007	204	$C_{15}H_{24}$	21	α-Cedrene	
	SIK 2484	204	$C_{15}H_{24}$	21	Thujopsene	
	SIK 2485	222	$C_{15}H_{26}O$	21	Farnesol	Same as unknown

ing favorably with (7). To allow further comparison, search results for 1,3-dimethyl benzene and N-dodecane are shown in Table X. In all these cases, the results at 80 and 48 dimensions were comparable to those at 352.

CONCLUSIONS AND SUMMARY

It is felt that any search and retrieval procedure that finds only perfect matches is unrealistic for dealing with experimentally derived spectra. Although there exist rather elegant procedures for finding perfect matches, there are no known useful alternatives to exhaustive searches for finding a best match within a library (11). Although certain advantages

(11) M. Minsky and S. Papert, "Preceptrons," M.I.T. Press, Cambridge, Mass., 1969, p 215ff.

may derive from searching only selected parts of the library (such as compounds of a certain molecular weight range), the entire library must be searched if it is desired to be certain that the best match is obtained. As pointed out in (7), it may be necessary to consider compounds of different molecular weight for structural clues if the "unknown" is not contained in the library. A case in point is recorded in Table VIII.

This study was undertaken to develop a practical search and retrieval system for low resolution mass spectra. The primary considerations were optimal use of available hardware and a minimization of search time consistent with maintaining the two critera felt necessary for a useful search stated in the introduction. For most small computer systems, the most practical means of storing vast amounts of data is on magnetic

Table X. Search Results at 352 Dimensions for M-Xylene and N-Dodecane

M-Xylene, mol wt = 106, mol form. = C_8H_{10}, ref no. = API 0254, struct form. = [benzene ring with two CH_3 groups]

Ref no.	Mol wt	Mol form.	Mismatches	Compound name
API 0253	106	C_8H_{10}	1	O-Xylene
API 0255	106	C_8H_{10}	1	P-Xylene
API 0178	106	C_8H_{10}	2	O-Xylene
API 0422	106	C_8H_{10}	2	P-Xylene
API 0179	106	C_8H_{10}	2	M-Xylene

N-Dodecane, mol wt = 170, mol form. = $C_{12}H_{26}$, ref no. = API 0404

Ref no.	Mol wt	Mol form.	Mismatches	Compound name
API 0403	156	$C_{11}H_{24}$	5	N-Undecane
AST 2004	156	$C_{11}H_{24}$	5	5-Methyldecane
AST 2003	156	$C_{11}H_{24}$	5	4-Methyldecane
API 1028	170	$C_{12}H_{26}$	6	N-Dodecane
AST 2013	170	$C_{12}H_{26}$	6	4-Methylundecane

tape. Therefore most conventional search systems would require transferral of the data from tape to memory or disk prior to searching, a process requiring an appreciable amount of time given the limited buffer size of small machines. The system reported here uses this otherwise wasted time to actually perform the search.

It has been shown that peak height encoded to one bit retains sufficient information to allow useful characterization of mass spectra. The spectra can be further reduced in dimensionality by combining correlating mass positions with minimal loss of important information. Encouraging results were obtained with 48-dimensional spectra requiring only 3 16-bit words to completely encode a reduced mass spectrum. A magnetic tape library of the 6652 mass spectra encoded in this way can be searched in 2 seconds for two unknowns. On the other hand by using the techniques of this study, it would be possible to search an entire library of 10000 low resolution mass spectra with peak height encoded to 16 intensity levels (4 bits per mass position) in 60 seconds from magnetic tape using minimal memory. Minimum required hardware consists of a moderate speed CPU with 4K words of core memory, one 25 IPS, 9-track tape drive with accumulator I/O, and some means of inputting the unknown spectra and outputting search results.

RECEIVED for review April 30, 1971. Accepted June 29, 1971. Research supported by the National Science Foundation.

29

Copyright © 1977 by the American Chemical Society
Reprinted from *J. Chem. Inf. Comput. Sci.* 17:157-163 (1977)

An Interactive Substructure Search System

R. J. FELDMANN and G. W. A. MILNE

National Institutes of Health, Bethesda, Maryland 20014

S. R. HELLER*

Environmental Protection Agency, Washington, D.C. 20460

A. FEIN, J. A. MILLER, and B. KOCH

Fein-Marquart Associates Inc., Towson, Maryland 21212

Received April 29, 1977

A family of programs for searching on the basis of chemical structure through data bases of chemical information has been assembled and is now publically available on a commercial computer network. The design of and results obtained with these programs are reported, and the status of the system is described and discussed with particular reference to the NIH-EPA Chemical Information System (CIS) and the Toxic Substances Control Act (TSCA).

INTRODUCTION

The ability to use a computer to search for a particular chemical structure or substructure in files of chemical data has for some time been sought after by chemists, and the need for such capability is currently becoming very pressing. A widening interest in the relationships between chemical structure, on the one hand, and various properties, such as toxicity, pharmacological activity, and mutagenicity, on the other, has led in recent years to considerable efforts to generate computer programs which will enable the scientist to locate all occurrences of a given structure or substructure in chemical databases. Further decisive pressure behind these developments has been provided by the enactment, in November 1976, of the Toxic Substances Control Act (Public Law 94-469). This law will require that chemical compounds whose use in commerce is envisaged must first be located within Governmental regulatory files. If they are not in these files, they are deemed "new" and their use in commerce becomes subject to a series of regulations, depending upon their respective toxicities.

In this paper, we describe the NIH-EPA substructure search system, a family of interactive computer programs which allow the user to define a chemical structure or substructure and then to search for occurrences of the structure or substructure in the various databases of the NIH-EPA Chemical Information System.

During the past 25 years, a considerable number of methods of machine representation and handling of chemical structure have been proposed and studied for their utility in manual and automatic data retrieval methods. Some of the better known among these include the German GREMAS system,[1] the British CROSSBOW system,[2] and, in the U.S., the programs developed at the National Cancer Institute,[3] Walter Reed Army Institute of Research,[4] Chemical Abstracts Service,[5] and the Army's Chemical Information Data System.[6]

With the resulting progress in the area of computer-handling of chemical structures, it has become clear that structure records in the form of two-dimensional connection tables are absolutely necessary for structural representation and that both linear notations and chemical nomenclature are at a serious disadvantage vis-à-vis connection tables as far as unambiguity and completeness are concerned. For many years, however, there was no adequate means of screening such connection tables and so, in spite of their intrinsic value, they were not used in any retrieval system.

In the area of structure retrieval, most effort was expended in the development of systems that were designed to fulfill a specific local need. A system of this type that is currently perhaps the most widely used by the chemical industry is the CROSSBOW program,[2] a dozen or so versions of which are in operation around the world. Other systems, such as the GREMAS system[1] or the Walter Reed system[4] require special equipment that is not generally available. While the larger industrial organizations can often afford such luxuries and often also demand in-house facilities of this sort, such systems are of little value to the general chemist. It is this dilemma which led to the development of the NIH nested tree structure searching system, which can operate on a connection table database and which is susceptible to wide dissemination and

Table I. Databases That Can Be Searched by the Substructure Search System.

Cambridge (Xray) Crystal File.
CPSC Chemicals in Consumer Products.
EPA AEROS SOTDAT File.
EPA Las Vegas Chemical Spill File.
EPA Storage and Retrieval of Air Data.
EPA Pesticide Standards.
EPA STORET Water Data Base.
EPA-FDA Pesticide Repository Standards.
EPA Inactive Ingradients in Pesticides.
EPA Oil and Hazardous Materials File.
EPA Pollutants in Drinking Water.
EPA Pesticides File.
EROICA Thermodynamics Data File.
Merck Index.
NBS Gas Phase Proton Affinities.
NBS Heats of Formation of Gaseous Ions.
NBS Single Crystal File.
NCI-SRI Industrial Chemicals File.
NCI PHS-149 File of Carcinogens.
NIMH File of Psychotropic Drugs.
NIH-EPA Carbon-13 Nuclear Magnetic Resonance Search System.
NIH-EPA Mass Spectral Search System.
WHO International Non-proprietary Name File of Drugs.

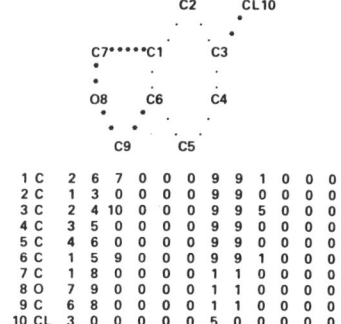

Figure 1. A connection table. The structure is not normally included but is given here for the sake of clarity.

use on a time-shared computer. This system was originally conceived by Feldmann in 1971,[7] and since that time there has been considerable use, testing, and further development. These developments, described below, have been in various different directions, and the more significant features are the following. (1) The programs have been extensively rewritten to reduce CPU demand and improve the user-machine dialog; the new software has been completely documented. (2) A comprehensive User's Manual has been written. (3) The Chemical Information Data System (CIDS) structure codes have been introduced into the substructure search programs where they can be used for searching, or, more importantly, as the basis of fragment screen procedures. (4) While the interactive programs run on a DEC PDP-10, all the database preprocessing is now accomplished on an IBM 370/168. (5) Other additional search features based upon user feedback and on successful features of other systems have been added.

DATABASES

The substructure search programs are designed to support the searching of a number of separate and independent databases. In the current version of the software, a decision must be made by the user as to the identity of the database that will be searched. There is, however, considerable overlap between separate databases and so work is now in progress to merge all the databases, remove duplicates, and arrive at a single consolidated file.

There are currently 23 distinct databases associated with the substructure search system. These are listed in Table I. Many of these files of chemicals have been derived from the files of the EPA. This agency, alone in the U.S. government, has an internal regulation[8] requiring registration of all agency databases containing chemical information. A second major source of files of connection tables is the NIH/EPA Chemical Information System.[9] This is a collection of databases containing spectroscopic and other data that relate to chemical compounds. A decision fundamental to the development of the CIS has been to register all chemicals in the component files of the system. Finally, other databases that are used by the substructure search system have been obtained from other U.S. government agencies. If necessary, they have then been registered for inclusion into the substructure search system.

The process of registration of a compound by CAS takes place in three steps which have been described in detail elsewhere.[10] When a database is obtained for registration and inclusion into the CIS and the substructure search system, the names of all chemicals in the database, together with the appropriate accession numbers, are delivered to CAS. The first step of registration is an attempt to find the compound name in the master CAS nomenclature files. If this name is found, then the "correct" name and the CAS registry number can be extracted and labeled with the accession number. If the name is not found, then CAS, in the second step of the registration process, establishes a structure for the compound and seeks to locate that structure in its structure files, and so arrive at the correct name and registry number. If this step fails, the compound is assumed to be absent from the CAS master authority files and is then given a name and registry number and incorporated into the files.

Ultimately, all the identifiable chemicals in the NIH/EPA CIS files are registered by one or another of these methods and, at that point, a database of accession number, CAS registry number, the name under which it is listed in the CAS (8th or 9th) Collective Index, and the connection table is returned to NIH/EPA for merging into the CIS. A separate file of registry number and synonyms is also obtained from CAS and used in the CIS.

Finally, as a general procedure, all the information that is derived from the CAS files is subject to annual updating. In this way, errors and refinements such as additional synonym information may be incorporated in the substructure search files.

The connection table that is supplied by CAS is not used directly by the substructure search system. Rather, it is translated into a derived connection table which is itself merged into the substructure search system files. This articulation leads to an important advantage in that any changes made by CAS in the format of their connection tables can be handled simply by an adjustment in the program that carries out the translation. The more complex programs that actually handle the substructure searching do not have to be changed.

The basic connectivity information for a compound is supplied by CAS in the form of three separate data elements in a single structure record. These data elements, which are designated the "graph", "nodes", and "bonds" elements, define the element type of each of the nodes, the connections between nodes, and the bond type of each of the connections.

The basic connectivity information present in the structure records supplied by CAS is used to generate the tabular connection table that is shown in Figure 1 and which is of the type that is employed by the substructure search system. The main feature of this derived connection table is that it permits rapid access to all connectivity information associated with a particular node or atom. Generation of the derived connection table is relatively straightforward, although a few minor problems arise. For example, dot-disconnected structures (e.g., of anion–cation pairs) are independently

Table II. Structure Generation Commands.

COMMAND	EFFECT
AATOM n1 m1	Insert an atom between atom n1 and atom m1
ABOND n1 m1	Insert a bond between n1 and m1.
ABRAN l1 at n1	Add a branch of length l1 at atom n1.
ALINK n1 l1 m1	Insert a chain of length l1 between n1 and m1.
ALTBD n1 m1	Define alternate bonds in the smallest ring containing n1 and m1 as aromatic bonds.
ARING n1 m1 l1	Create a ring of l1 atoms between n1 and m1.
CHAIN l	Create a chain of l atoms.
CLEAR	Erase the existing query structure.
CRING n1 l1	Create a ring of l1 atoms including atom n1.
DATOM n1	Delete atom n1.
DBOND n1 m1	Delete the bond joining n1 and m1.
MORGA	Renumber the query structure by the Morgan algorithm.
NUC 66	Create a structure of two fused six-membered rings.
REG	Retrieve the structure corresponding to a specific registry number.
REST	Negate the effect of the previous command.
RING l	Create a ring of l atoms.
SATOM n1	Define the elemental nature of atom n1.
SBOND n1 m1	Define the nature of the bond joining n1 and m1.
SPIRO n1 l1	Create a spiro-attached ring of (l1 +1) atoms at n1.
WISBD n1 m1	Define alternate bonds in the smallest ring containing n1 and m1 as double bonds.

```
OPTION ? NUC
SPECIFY NUCLEUS LINE CODE

LINE CODE  =  65
OPTION ? ABRAN  1 AT  3
OPTION ? ALTBD  1 2
OPTION ? SBOND  3 10
BOND TYPE (H FOR HELP)     = CS
OPTION ? SBOND  1 7 7 8 8 9 6 9
BOND TYPE (H FOR HELP)     = RS
OPTION ? SATOM  8
SPECIFY ELEMENT SYMBOL     = O
OPTION ? SATOM  10
SPECIFY ELEMENT SYMBOL     = CL
OPTION ? D
            2       10CL
             .       .
             .       .
     7******1       3
     *          .
     *          .
     80     6       4
       *  .  .
         .  .
         9    5
```

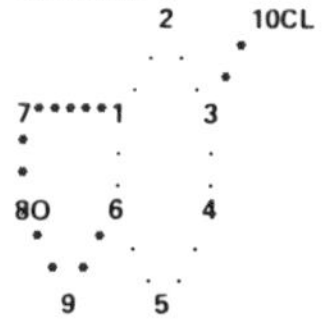

Figure 2. Use of the structure generation commands to define a structure.

numbered and stored by CAS but must be assembled and renumbered for processing within the substructure search system.

In addition to the basic connectivity information described above, CAS also provides information describing a large number of more unusual structural features, such as charge, abnormal mass, abnormal valency, stereochemistry, and so on. These additional features are not currently handled by the substructure search system.

STRUCTURE SEARCHING

A search through a database of connection tables for a specific complete structure, as opposed to an imbedded partial structure, can take advantage of different design techniques and is considered here independently of substructure searching.

It is vital to the implementation of the Toxic Substances Control Act (PL 94-469) that the government-maintained inventory of chemicals used in commerce can be examined rapidly and accurately for the presence or absence of specific compounds. This requires a search for a full structure and is performed effectively by searching through the database of connection tables for a specific connection table, defined by the user.

To do this, the system permits the user to generate a "query structure", as described in the next section. Once the query structure is complete, an identity search can be requested. At that time, a modified form of the connection table corresponding to the query structure is converted to a hash-coded form. The process of hash-encoding involves conversion of the connection table to a single number which is a probable, but not guaranteed, unique representation of the compound's structure. The searching programs then scan a set of searchable files which are indexed by that number for an exact match.[11] If the match is found, the registry number of the matched entry is reported.

SUBSTRUCTURE SEARCHING

The substructure search process involves three distinct steps. The first of these is the generation of a query structure. This is followed by the actual search, and the final step is the display of the structures that have been retrieved from the database by the search. Each of these steps will be treated separately below.

1. Query Structure Generation. One of the more difficult problems in the design of an interactive substructure search system is how the chemist can enter a chemical structure or partial structure into the machine. In the present case, this is accomplished by means of a family of programs that permit the generation of a structure at the computer terminal. These programs have been designed in such a way that the simplest of computer terminals is adequate for their use and so the substructure search system can be accessed by a simple teletype terminal as well as by an advanced graphics terminal. In either case, the user must, with the commands that are given in Table II, generate the query structure of interest. As each command is received, the system generates or modifies a connection table so that it reflects the current structure. The connection table is invisible to the user, although a version of it can be printed out at a command. A more useful option, however, is the DISPLAY command (D), which, working from the current connection table, produces a drawing of the corresponding structure, using procedures designed to produce an unambiguous two-dimensional representation. With the commands given in Table II, the chemist can generate rings of specified sizes, add branches to existing atoms, and specify the identity of specific atoms or bonds. Various other, more powerful, commands include the NUCLEUS command, with which one can generate a fused multi-ring system, such as that common to steroids, in one step. As a query structure is being developed, it is often necessary to be able to inspect it so as to identify the numbers assigned by the program to various nodes (atoms), and it is here that the display command is very useful. When certain modifications are made to the query structure, the program will renumber the atoms, and so it is necessary to examine the structure in order to proceed. An example of the dialog involved in the generation of a query structure is shown in Figure 2. The basic ring system is provided by the NUC command, a branch is added by the ABRAN option, bonds are all specified by the ALTBD and SBOND options, and noncarbon atoms are defined by the SATOM command. The result is the query structure that is finally examined with the help of the display command "D".

Once the query structure has been generated, the second step, a search through the database, may be undertaken. This searching may be accomplished in a variety of different ways. The most trivial of these are the special property searches which scan the database for all compounds whose molecular formula corresponds to that of the query structure, or which have a given molecular weight, and so on. The most used of the structural searches are those in which a specific atom-centered fragment from the query structure is searched for in the database (FPROB) and those (RPROB) in which a particular ring or rings from the query structure is the object of the search. Finally, the SUBSTRUCTURE SEARCH, which is the most exhaustive of all the searches, examines every connection table in subsets of the database on an atom-by-atom, bond-by-bond basis in a search for an exact or imbedded match for the query structure. These options are described in more detail below.

2. Molecular Formula Search. Historically, the molecular formula of each compound was derived from the connection table by summing the entries in column 2, the element column. The molecular formula so obtained contains no hydrogen atoms because hydrogen atoms are not explicitly described in the connection table.

This method has now been supplanted by a simpler process which uses the molecular formulas supplied by CAS as a part of the compound identification. These molecular formulas do, of course, include the number of hydrogens in the compound.

The molecular formulas are hash-encoded as has been described previously,[12] and the file of hash-encoded formulas vs. registry numbers is sorted, primarily upon the hash-encoded formulas, and secondarily upon the registry number. Pointers to the sorted file are generated, and the file of pointers together with the file of encoded formulas and registry numbers become the basis of the molecular formula search.

3. Special Properties Searches. A number of different items are included in the general category of special properties searches. All of these items are organized into a single set of hierarchically ordered files for searching purposes, but the different items are separately identified by property type. Currently, five different property types are included as follows.

a. Molecular Weight. The molecular weight of each compound is derived from the molecular formula provided by CAS as part of the compound identification. Searches may be conducted for a specific molecular weight or for all compounds whose molecular weights fall within a specified range of molecular weights.

b. Total Atom Count. The total atom count for a compound, as used here, is simply the total number of nonhydrogen atoms in the compound, which is the total number of atoms defined in the connection table. The ACOUN search may be used to identify all compounds having a specific total number of nonhydrogen atoms or with a total atom count that falls within a specified range.

c. Atom Population. The atom count for each element in the compound is also extracted from the molecular formula provided by CAS. This information forms the basis of the partial and ranged molecular formula search options which can be used to identify compounds having a specified partial molecular formula. In such a case, specific requirements are defined for some elements, but any number of other elements is also permitted. Alternatively, a partial formula can be defined as the permissible ranges of the appropriate elements.

d. Total Ring Count and Ring Population. These last two types of special property are concerned with the smallest set of smallest rings (SSSR) present in the compound. Both the searchable files and the query structure are examined by algorithms which can identify an SSSR correctly. These algorithms trace pathways through the connectivity section of the connection table (columns 3–8 of Figure 1) and locate the different smallest rings. Standard techniques for starting the tracing and continuing it once a ring has been closed are used in order to ensure that the rings located do indeed constitute an SSSR. Thus, in the example given in Figure 1,

4	C	C	9	C	9	C	1	0	0	2
3	C	C	9	C	1	0	0	0	0	4
3	C	C	9	C	9	0	0	0	0	6
4	C	CL	5	C	9	C	9	0	0	1
3	C	CL	5	C	9	0	0	0	0	2
3	C	O	1	C	1	0	0	0	0	2
3	O	C	1	C	1	0	0	0	0	1

Figure 3. Fragment table corresponding to the connection given in Figure 1.

the algorithms will discover that node 8 is connected to node 9. Node 9, in turn, is connected to node 6 which is connected to node 1, which is connected to node 7 which is connected back to node 8. In this way, the five-membered ring is traced, and in the same way the six-membered ring in the structure can be identified, but the larger nine-membered ring is ignored. The total ring count in this case, then, is two, the total number of rings constituting the SSSR. Similarly, the ring population information provides the numbers of rings of different sizes in the SSSR (one five-membered ring and one six-membered ring in the example of Figure 1). The RCOUN search permits location in the database of all compounds containing either a specified total number of rings or a given number of rings of a specific size. The same command can also be used to impose range requirements upon either of these criteria.

In the generation of the special properties file, the property type, property value, and registry number for each compound are combined into a file in that form. This file is then sorted, first upon property type, secondarily upon the property value, and finally upon the registry number. For search purposes, this inverted file is reorganized into a set of hierarchically ordered files which are accessed by the various commands described above (MW, ACOUN, partial MF, ranged MF, and RCOUN) to identify compounds in the database that fulfill the criteria described by the user.

4. Fragment Probe Search. The fragment probe permits a search through the database for all occurrences of a specific fragment defined by the user. This search operates on a file of fragment properties vs. registry number which is derived from the connection table in the following way. The connection table is scanned, an atom at a time, and a corresponding fragment table is built by a process involving an analysis of that atom and all its immediate neighbors. Each line in this derived fragment table contains (1) the size of the fragment, i.e., the total number of nonhydrogen atoms in the fragment, including the central atom; (2) the nature of the central atom; (3) the nature of the first neighbor; (4) the nature of the bond joining the central atom to the first neighbor; (5) the nature of the second neighbor, and so on until all neighbors have been described. The first neighboring atom to be considered is the one representing the least common element. In the substructure search files, carbon is clearly the most common element, followed by oxygen and nitrogen in that order. In the event that two neighbors are found to represent the same element, the one joined to the central atom by the less common type of bond is considered first. Thus if all the neighbors are carbon, then that carbon that is doubly bonded to the central atom is taken before any that are singly bonded to the central atom. Ten fields are available for this description, unused fields are filled with zeroes, and an eleventh field contains the number of times the fragment occurs in the molecule. When a fragment table entry is generated, the first ten fields of that entry are compared with all existing entries in the part of the table that has already been generated. If a complete match is found, the occurrence count for that entry (field 11) is incremented by one rather than creating a new entry in the table.

Storage of the fragments in this manner permits the

identification, during a search, of any compounds containing a sufficient number of occurrences of the fragment in question. One additional possibility must, however, also be considered. When a search is performed for a particular fragment, the occurrence of that fragment in a compound as a subset of a larger fragment as well as in a stand-alone manner, must be detected. To accomplish this, every node in the structure will be described in the new fragment table several times when the compound is entered into the database. It will first be entered with all its neighbors described. Then the node with one neighbor removed will be described, then a different neighbor will be removed, and so on, until all fragments from which one neighbor has been removed have been entered. If the atom, as it appears in the molecule, has four neighbors, then all the possibilities that result from the stripping of two neighbors will be computed and entered into this new fragment table. Every nonhydrogen atom is therefore described in this table at least once and possibly as many as ten times, depending upon the number and nature of its neighbors. Many of the multiple entries will, of course, be the same and can be merged by incrementing the counter in field 11. For this reason, the fragment table shown in Figure 3, which is derived in this way from the original connection table in Figure 1, has only seven lines.

Once the derived fragment table shown in Figure 3 has been generated, each line, with the appropriate registry number appended to it, is entered into the master database. When this database has been completely assembled, it is sorted on all fields, and the resulting inverted file is hierarchically organized for searching by the FPROBE option. Each level in the hierarchy is associated with one of the eleven fields described above and contains pointers for progressing to the next level of the search. At each level, the appropriate property in the query structure is sought in the database. If it cannot be located, the search fails, but if it is found, the search proceeds to the next level. The eleventh, and final, level contains a pointer to a list of registry numbers of compounds that satisfy the set of properties that were satisfied individually at each level.

If the bonds in the query structure are not all specified, it is possible for more than one database property to satisfy the query requirement. Multiple matches also normally occur at the eleventh level, since any number of occurrences of a fragment in a database structure in excess of the number of occurrences in the query structure will also satisfy the requirement. In these cases, multiple paths are traced to the bottom of the hierarchical structure, and each of the resulting registry number lists is merged to provide the composite set of responding compounds.

5. Ring Probe Search. The ring probe search permits the locating of structures containing a specific ring or rings, with or without heteroatoms, whose identity and position may be specified or not, and having a given substitution pattern. The search is hierarchical and properties are sought in the order given above. Compounds may be retrieved either because the query structure feature is imbedded in them or because they represent an exact match for the query structure.

As in the case of all the other search options, RPROB uses a file derived directly from the connection table shown in Figure 1 by the following series of steps. First the connection table is scanned to identify each nucleus (i.e., set of contiguous rings) in the compound. This is accomplished through a simple determination of which atoms are connected to which other atoms by ring bonds rather than chain bonds, for example. The numbering of the nodes is then reorganized according to a modified Morgan algorithm,[13] which tends to cause the numbering to radiate from the most central node. Figure 4 shows the nucleus from Figure 1, numbered arbitrarily (left) and according to the Morgan algorithm (right).

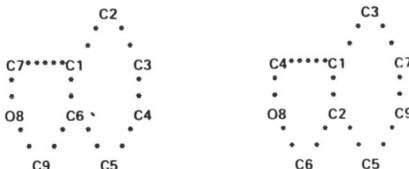

Figure 4. Query structure before (left) and after (right) application of the Morgan algorithm.

In the third step, the connection table for the renumbered nucleus is generated and fields 3–8, which describe the connectivities, are hash-encoded. It is this hash-encoded information that constitutes the first level in the hierarchical search. Next, the heteroatom positions are noted, then the heteroatom types, and finally the substitution pattern around the rings. These four levels of information for each compound are then assembled into a single table entry containing the hash code, a list of 12 heteroatom positions, a list of 12 heteroatom types, a list of 12 substituent positions, and, finally, the appropriate registry number.

The information described above is sufficient to permit the identification of complete nucleus structures within a database. However, to provide the information necessary to permit the identification of imbedded ring structures, i.e., a ring or a set of contiguous rings that is imbedded within a larger ring structure, additional processing is necessary. Path-tracing algorithms are used to identify the individual rings present in the nucleus. All possible combinations of contiguous rings are then formed, and each of these combinations is treated as described above. Thus a unique numbering is assigned using the modified Morgan algorithm, the connectivity information is hash-encoded, and heteroatom and substituent information is noted (bonds to other cyclic nodes which are not included in the set of rings being processed are treated as substituents). This information is then assembled and entered into the table as before.

When the entire database has been formatted in this way, it is sorted. The primary sort fields are those devoted to the hash code, and the subsorts are on heteroatom positions, heteroatom types, substituent positions, and registry numbers. This file is then hierarchically organized in a logically equivalent manner to the fragment files described above, and the resulting files are those that serve as the basis of the RPROBE search.

6. Substructure Search. While it is very useful to be able to learn that a specific fragment or ring is present in various compounds in the database, a more demanding query is whether or not a given complete structure is present, either per se, or imbedded in a larger structure. This is accomplished using the option SUBSTRUCTURE SEARCH.

This program conducts an atom-by-atom, bond-by-bond comparison between the connection table corresponding to the query structure and each of the connection tables in a selected subset of the database. This comparison is done without any subtlety and, in the case of the connection table shown in Figure 1, would proceed as follows. First, atom 1 in the table, i.e., C1 in the structure, is compared to node 1 in the first database connection table that is to be examined. The first check is that both C1 and node 1 represent carbons. If this is not so, then C1 of the query connection table is compared to node 2 of the database connection table. Once two identical atom types, one from each connection table, are found, their neighbors are compared to each other. If the neighbors are not the same, then that node of the database connection table is dropped and a new node is examined. If the neighbors do match, then the various bonds between the central atoms and their respective neighbors are checked. Again, lack of correspondence would cause this pair of atoms to be abandoned, but if the match is perfect, the program proceeds to the next atom in the query structure and repeats the entire process. In this way, every atom in the query structure will, if necessary, be compared to every atom in the database structure, and all bonds will also be compared. Only if the query structure is exactly the same as the database structure or, if an exact copy of the former is located within the latter, will the structure be retrieved from the database as a positive response to the user's question.

The ability of this program to locate and produce compounds in which a specified partial structure is imbedded is very powerful because this is precisely the type of query that chemists are prone to pose. A very typical request, for example, is for all compounds containing a 1-fluoro-3-bromophenyl ring. The fragment probe will allow retrieval of all structures containing an aromatic fluorine and an aromatic bromine, but this search will not guarantee that the fluorine and bromine be in the same ring. The ring probe will permit retrieval of all meta-substituted aromatic ring compounds, but again, the halogen atoms need not be in the same ring. Only the substructure search can limit this list to just those compounds containing the 1-fluoro-3-bromophenyl moiety.

Because of the bluntness of the substructure search program, it can use considerable amounts of processor time, and the most sensible way in which to use this program appears to be to anticipate its application by performing the appropriate fragment or ring probes. In this way, the large database can be reduced to a smaller file of candidate structures known to contain the desired fragments and/or rings. The substructure search program can operate on such a file without incurring intolerable expense, and it can extract from this file just the compounds which represent correct responses to the original query.

7. **Structural Feature Code Search.** As a somewhat different approach to the problem of identifying compounds having various combinations of structural properties, the substructure search system also contains a structural feature code search capability. The structural feature codes, extracted from the CIDS chemical search keys of the U.S. Army CIDS System,[14] consist of a large, somewhat open-ended set of both generic and specific predefined structural characteristics. Generally, several thousand unique codes will be found in a database of a reasonable size. The codes applicable to each compound in the database are automatically assigned. An exhaustive analysis and comparison of the connection table for the compound with the set of predefined structural feature codes is necessary to accomplish this assignment. As in the other types of searches, these codes, along with the registry numbers to which they apply, are sorted to create an index with respect to code. From that index, a set of hierarchically organized files are created for searching purposes.

To use the structural feature code search capability, one must first establish which of the codes are appropriate to the question at hand. The codes can then simply be entered into the search system, which will in turn identify the compounds to which these codes have been assigned. Through the intersection and merging of the results associated with the searches for individual codes, compounds possessing any arbitrarily complex logical combination of the codes can be identified.

RESULTS

In this section, a number of examples of use of the substructure search system are given. All the examples given here are of searches through the structural data base that corresponds to the NIH-EPA Mass Spectral Search System.[15] This

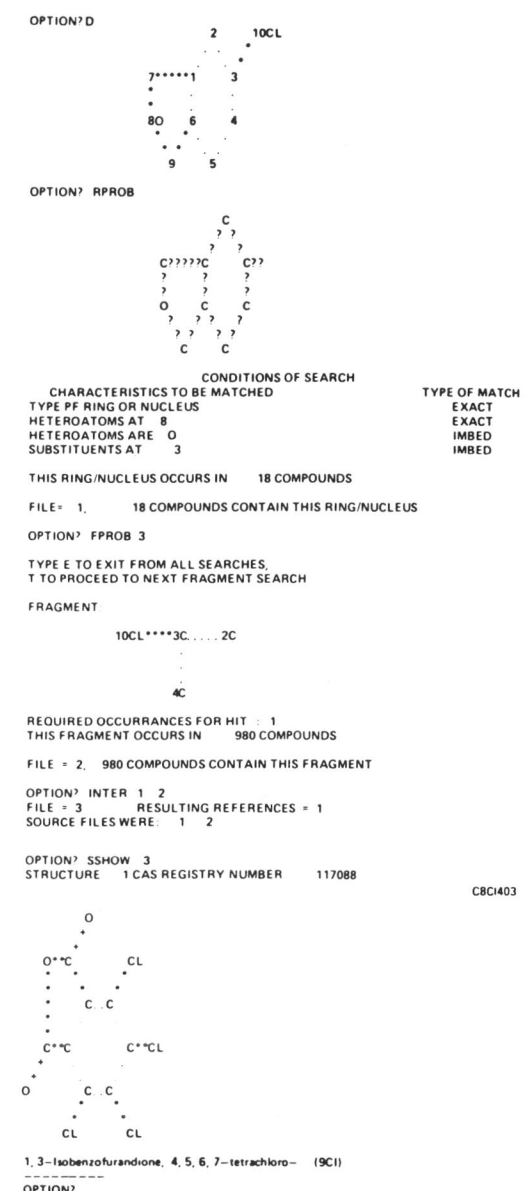

Figure 5. FPROB and RPROB options used with the query structure.

database currently contains just under 30 000 distinct compounds and their respective low-resolution mass spectra.

In the first case, which is shown in Figure 5, several searches were carried out using the query structure given in Figure 1. The first search used RPROB with the following match conditions implied: the ring systems retrieved should be identical with that in the query structure, no imbedment was to be permitted, node 8 and no other node may be a noncarbon, and, finally, there should be substituents at least at node 3. The search for this nucleus produced scratch file number 1 with 18 candidate structures. Next, a fragment probe for all compounds containing a node identical with C3 of the query structure (i.e., the chlorine-bearing carbon) was carried out. This resulted in scratch file number 2, with 980 structures.

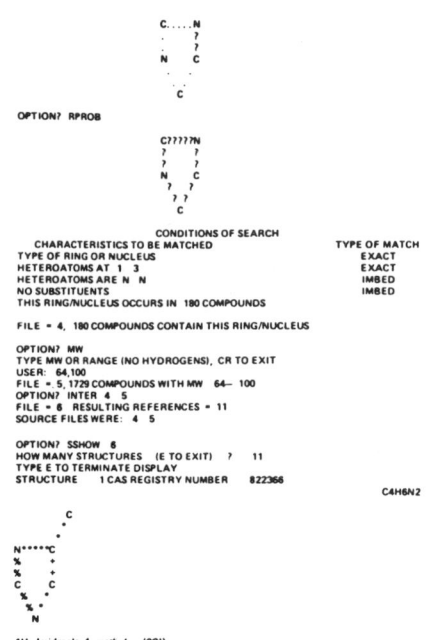

Figure 6. Use of RPROB in conjunction with a molecular weight range specification.

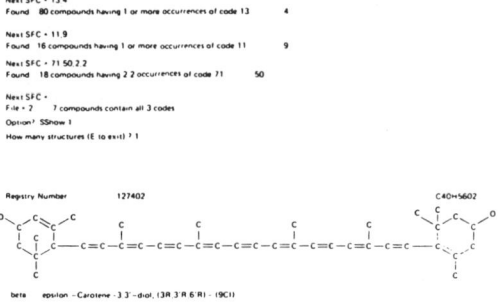

Figure 7. Search using structural feature codes.

Intersection of these two files resulted in a third file containing a single compound, 1,3-isobenzofurandione, 4,5,6,7-tetrachloro-, that satisfied all the criteria defined. In a final command, SSHOW, this structure, its name, registry number, and molecular formula were displayed.

In a different method of searching, shown in Figure 6, the intention was to locate all low molecular weight derivatives of imidazole. An RPROB search led to the retrieval of 180 compounds containing a five-membered ring with two non-carbon atoms in a 1,3 disposition. This was followed by a command to locate all compounds in the database that have molecular weights between 64 and 100. There are 1729 such compounds, but the subsequent intersection showed that only 11 of these were also in the file that resulted from the RPROB search. The first of these, 1H-imidazole, 4-methyl-, registry number 822366, was printed out with the SSHOW command.

Finally, use of the powerful structural feature codes is shown in Figure 7. In this search, the objective was to find all the carotenes in the database. These are compounds that contain two cyclohexene rings joined together by a long (e.g., C18) olefinic chain. The first code that was used, 13,4, implies at least four occurrences of chain branching, i.e., a carbon bonded to at least three other carbons. Eighty compounds fulfilled this criterion. Next, a much more stringent requirement, the occurrence of 9 olefinic bonds was defined with code 11,9, and this produced only 16 hits. In a final criterion, code 71,50,2,2 two, and no more nor less than two, cyclohexene rings were requested. This gave 18 hits. The automatic intersection led to just seven compounds, all carotenes, that met all three requirements, and the first of these, β,ϵ-Carotene-3,3'-diol, (3R,3'R,6'R)-, registry number 127402, was listed with the SSHOW command. In this case, the structure was drawn on a CRT terminal, which permits a better picture.

As can be seen from the foregoing examples, the ways in which queries can be posed to a chemical structure searching system are varied; the chemist can often use nonstructural information, such as an upper limit to molecular weight, to aid in the convergence of a search. This substructure search system has been developed with this in mind. The various search options that are available permit the user to make the greatest use of all the information at his disposal and so complete searches rapidly and efficiently.

SUMMARY

The system that has been described is interactive, and it is this property that is one of its most important features. It is not difficult to learn to use, and once some experience has been gained, queries can be framed very rapidly. At this point, the speed with which answers to the queries are provided becomes extremely valuable because the answers often form the basis of the subsequent query.

As is clear, this is a fairly large program package, and it has not been designed for facile export or transfer from one computer to another. Rather, it is expected that the substructure search system will be most accessible via a networked time-shared computer system, and it is, in fact, already available in this form.[16]

REFERENCES AND NOTES

(1) R. Fugmann, W. Braun, and W. Vaupel, *Angew. Chem.*, **73**, 745 (1961); R. Fugmann in "Chemical Information Systems", J. E. Ash and E. Hyde, Ed., Wiley, New York, N.Y., 1975, Chapter 13.
(2) D. R. Eakin and E. Hyde in "Computer Representation and Manipulation of Chemical Information", W. T. Wipke, S. R. Heller, R. J. Feldmann, and E. Hyde, Ed., Wiley, New York, N.Y., 1974, pp 1–30; D. R. Eakin in "Chemical Information Systems", J. E. Ash and E. Hyde, Ed., Wiley, New York, N.Y., 1975, Chapter 14.
(3) G. F. Hazard and S. Richman, paper presented at the 176th National Meeting of the American Chemical Society, San Francisco, Calif., Sept 1976.
(4) D. P. Jacobus, D. E. Davidson, A. P. Feldman, and J. A. Schafer, *J. Chem. Doc.*, **10**, 135 (1970).
(5) R. J. Rowlett and F. A. Tate, *J. Chem. Doc.*, **12**, 125 (1972).
(6) M. Milne, D. Lefkovitz, H. Hill, and R. Power, *J. Chem. Doc.*, **12**, 183 (1972).
(7) R. J. Feldmann and S. R. Heller, *J. Chem. Doc.*, **12**, 48 (1972).
(8) EPA Internal Regulation No. 2800.2, 1976.
(9) S. R. Heller, G. W. A. Milne, and R. J. Feldmann, *Science*, **195**, 253 (1977).
(10) G. W. A. Milne and S. R. Heller, *J. Chem. Inf. Comput. Sci.*, **16**, 232 (1976).
(11) R. J. Feldmann in "Computer Representation and Manipulation of Chemical Information", W. T. Wipke, S. R. Heller, R. J. Feldmann, and E. Hyde, Ed., Wiley, New York, N.Y., 1974, pp 55–81.
(12) S. R. Heller, *Anal. Chem.*, **44**, 1951 (1972).
(13) H. L. Morgan, *J. Chem. Doc.*, **5**, 107 (1965).
(14) "Handbook of CIDS Chemical Search Keys", Fein-Marquart Associates, Inc., Towson, Md., Nov 1973.
(15) S. R. Heller, H. M. Fales, and G. W. A. Milne, *Org. Mass Spectrom.*, **7**, 107 (1973); S. R. Heller, D. A. Koniver, H. M. Fales, and G. W. A. Milne, *Anal. Chem.*, **46**, 947 (1974); S. R. Heller, R. J. Feldmann, H. M. Fales, and G. W. A. Milne, *J. Chem. Doc.*, **13**, 130 (1973); R. S. Heller, G. W. A. Milne, R. J. Feldmann, and S. R. Heller, *J. Chem. Inf. Comput. Sci.*, **16**, 176 (1976).
(16) The system is available for general use via the TYMSHARE computer network. For further details, please contact AEF.

INTERPRETATION

Editors' Comments
on Papers 30 Through 35

30 ISENHOUR and JURS
    Some Chemical Applications of Machine Intelligence

31 ACZEL et al.
    Computer Techniques for Quantitative High Resolution Mass Spectral Analyses of Complex Hydrocarbon Mixtures

32 CRAWFORD and MORRISON
    Computer Methods in Analytical Mass Spectrometry: Development of Programs for Analysis of Low Resolution Mass Spectra

33 BUCHS et al.
    Applications of Artificial Intelligence for Chemical Inference. IV. Saturated Amines Diagnosed by Their Low Resolution Mass Spectra and Nuclear Magnetic Resonance Spectra

34 DELFINO and BUCHS
    Heuristic Programming as an Ion Generator in Mass Spectrometry. I. Generation of Primary Ions with Charge Localization

35 COREY
    Computer-assisted Analysis of Complex Synthetic Problems

In Paper 30 Isenhour and Jurs present the concept of using the data itself to define pattern recognizers for the interpretation of chemical data. The method is entirely empirical and is based on the concept of training decision functions to recognize patterns in the data. The training is accomplished by an error correction procedure using negative feedback on the decision function to correct itself when its decision is wrong. Once the decision function has been "trained" on a set of known patterns such as mass spectra, it may be used to predict structural properties of unknown spectra. The use of empirical techniques of data interpretation seems particularly advantageous for data of high dimensionality where the theory is incomplete or not applicable due to its complexity.

Paper 31, by Aczel, Allen, Harding and Knipp, provides an example of computer use in an industrial environment. A high resolution mass spectrometer generates data on samples containing approximately 300 components. An on-line 16K computer is used in real time to collect the data, which is later processed off-line by an IBM 360/50 to produce summary tables for each sample, which would be impossible to generate manually.

The data is interpreted in the sense that the program looks for compound types such as $C_n H_{2n}$ through $C_n H_{2n-44}$, $C_n H_{2n-2}S$ through $C_n H_{n-34}S$, and $C_n H_{2n-2}O$ through $C_n H_{n-36}O$. The data is further interpreted on the basis of distribution of aromatic rings and carbon number distribution.

Crawford and Morrison (Paper 32) report on a Fortran IV program for the interpretation of low resolution mass spectra. The program consists of a main program and a series of subprograms, which form probability tables for the unknown to belong to twelve functional group classes. Molecular weight and empirical formula routines are also used in the interpretation procedure. The authors state that "it has not yet proved possible to organize in a systematic form much of the empirical information already known about breakup patterns." It is interesting that the interpretive ability of the program was approximately equal to that of a group of third- and fourth-year chemistry students.

The paper by Buchs et al. (Paper 33) is one of a series from the DENDRAL project at Stanford University. It is concerned with computer interpretation of saturated amines through the incorporation of mass spectrometric theory and exhaustive generation of structures compatible with the available information.

The original conception of the DENDRAL program was to generate a hypothesis or list of hypotheses (chemical structures) to best explain some given spectral data. The algorithms to accomplish this task are written in LISP, a high-level list processing language, and, as stated in the paper, are divided into three main parts: preliminary inference, structure generation, and prediction.

Paper 33 is significant in that the DENDRAL project is a major attempt to incorporate chemical knowledge in an algorithmic form for interpretation of chemical data. The project has pointed to several problems in computer interpretation of chemical data such as mass spectra. One problem is the extraction of mass spectrometry knowledge from chemists; the theory is just not defined well enough, and the individual chemist's method of interpretation is often extremely subjective. The second problem is the modification of the program to include new knowledge. The chemical theory must be separated from the routines that work on the theory if program development is to be accomplished easily.

Another article concerned with interpretation of chemical data is paper XVII[1] in the series of papers on artificial intelligence from Stanford. This paper describes a program that generates all structural isomers, without duplication, consistent with structural constraints supplied interactively by a user. That the program is successful is due in part to its restriction to the part of the structure elucidation problem that allows formal mathematical treatment. The authors point out that it is the interactive capability of the program that prevents the problem of combinatorial explosion of structures, which can occur in the plausible structure generation process. As in any programming effort concerned with a search process through a space of solutions, the programming goal is not to search faster, but to search less. By "pruning" the search tree, it is possible to examine the remaining leads by expansion of the "embedded" structural features. Programs such as this, while not performing the complete interpretation process, greatly aid in the structure elucidation by exhaustively and, importantly, nonredundantly, generating structures that fit the information available on a given unknown's structure.

The program described by Delfino and Buchs (Paper 34) is one of the more innovative approaches in the application of computers to analytical chemistry. It is designed to simulate the formation of ions in the ion source of a mass spectrometer. The use of such a simulation program is twofold. One is to generate exhaustively a nonredundant set of plausible fragmentations for a given structure. The second utility is of a different nature; explicit coding of the problem forces generalization and clarification of fragmentation rules.

The authors have designed the program around the possibility of multiple ionization sites in a molecule, generating molecular ions with different charge loci. Functional groups containing heteroatoms are considered to be the best loci for positive charge. Loss of nonbonding electrons and transfer of hydrogen atoms are also included. This paper is an example of the imaginative use of computers for solving problems in the interpretation of chemical data.

Mass spectra are not the only form of chemical data subject to computer interpretation. Woodruff et al.[2] have applied computer assisted interpretation to carbon-13 nuclear magnetic resonance spectra of natural products. They observe that when dealing with binary data, the Tanimoto similarity measure correctly classifies unknown compounds more often than does the conventional distance measure. One implication of this fact is that one of the fastest search techniques is also one of the best. Such a search is fast because it can be done with a few simple assembly language instructions. This puts the operation of comparing spectra on a binary level immediately rather than through a series of more complex operations such as calculating a Euclidean distance, which even when coded in assembly would require more time.

These operations require not only more time, but usually more processor and software than would be necessary to implement the Tanimoto distance operation. It is useful to develop search procedures that can be programmed into ROM (ready-only memory) in microprocessor controlled spectrometers for comparison against a cassette tape or floppy disk of recorded spectra. Even storing collections of spectra on PROM chips or bubble memories may be possible in the future. The paper has a good discussion on the development of classifier use. By examining the classes of incorrectly classified compounds, a priority of classifier action can be established for automatic compound identification.

In other work on pattern recognition, Mattson et al.[3] have used linear discriminant function analysis of infrared data to classify petroleum pollutants. Infrared spectra of 194 oils were used in the study, which developed a decision tree with high predictive capability.

If one considers analytical chemistry to be an analytic approach to the study of chemistry, then the work by Corey (Paper 35) and co-workers on the computer assisted analysis of complex synthetic problems must surely be included in any collection of significant applications of the computer to chemical problems. No doubt there are those who will claim that such programs will put chemists out of work, but we believe that the use of programs like these can only amplify the chemist's ability to solve problems.

REFERENCES

1. R. E. Carhart, D. H. Smith, H. Brown and C. Djerassi, "Applications of Artificial Intelligence for Chemical Inference XVII. An approach to Computer-Assisted Elucidation of Molecular Structure," *J. Am. Chem. Soc.*, **97**, 5755 (1975).
2. H. B. Woodruff, C. R. Snelling, Jr., C. A. Shelley, and M. E. Munk, "Computer-Assisted Interpretation of Carbon-13 Nuclear Magnetic Resonance Spectra Applied to Structure Elucidation of Natural Products," *Anal. Chem.*, **49**, 2075 (1977).
3. J. S. Mattson, C. S. Mattson, M. J. Spencer, and F. W. Spencer, "Classification of Petroleum Pollutants by Linear Discriminant Function Analysis of Infrared Spectral Patterns," *Anal. Chem.*, **49**, 500 (1977).

Some Chemical Applications of Machine Intelligence

THOMAS L. ISENHOUR
Department of Chemistry
University of North Carolina
Chapel Hill, N.C. 27514

PETER C. JURS
Department of Chemistry
The Pennsylvania State University
University Park, Pa. 16802

THE INTERPRETATION of experimental data and the corresponding establishment of cause and effect relationships are essential aspects of experimental chemistry. In general, the investigator has data which he wishes to place into certain categories. For example, infrared spectra can be used to place compounds into categories defined by functional groups, or pK_a values can be used to define the degradation products of certain protein reactions. Placing data into specific categories, then, is often the basis of interpretation of experimental results.

Two approaches can be used to relate data to categories—theoretical or empirical. Theoretical data interpretation is usually preferred because it is based on explicit causal relationships derived from earlier observations or from logically constructed models. That is, scientists normally prefer interpretations based on theory because they feel they understand the measurement process in some or even all aspects. However, not even the most ardent theoretician would be likely to attempt the interpretation of the dc arc emission spectra of an iron alloy starting from first principles. Empirical methods are, however, readily applied in many common analytical situations; and, most frequently, some combination of the theoretical and empirical approaches is used. For example, while most scientists are satisfied with current theories of light absorption by molecules, it is standard procedure to measure the spectrum of a new compound and select a desirable absorption wavelength empirically in order to develop a colorimetric method.

The learning machine method, presented here, is a totally empirical method of data interpretation. The sole assumption is that a relationship between the data and the defined categories exists—i.e., the experiment measured something related to the property of interest. Even this assumption will be investigated by the empirical method itself. Hence, the learning machine method does not depend upon established theory and, while it is disadvantageous in that accepted hypotheses may not be used, it is simultaneously advantageous in that interpretation will not be restricted to current accepted schools of thought.

The term "learning" used in this context refers to a decision process which improves performance of a task as its experience at performing the task increases. The application of negative feedback causes the decision process to be modified to discriminate against wrong answers, therefore improving its performance with time. In general, empirical relationships are established between available inputs and desired outputs. In this article the inputs will be chemical measurements and the outputs will be the previously mentioned data categories.

Pattern Recognition

Starting in the late 1940's a great many books, papers, and conference reports have dealt with the various phases of the theory, design, development, and use of learning machines ([1-13]). Such studies have been the province of applied mathematicians, statisticians, computer-oriented engineers, and others in several disciplines investigating biological behavior on the neural level. A recent review by Nagy ([14]) demonstrates the amorphous nature of the subject. Applications have appeared in such divergent scientific areas as character recognition (alphabetic and numeric),

Some Chemical Applications of Machine Intelligence

The learning machine method, a totally empirical method of data interpretation, is based on a system whereby a decision process improves its performance as its experience at performing the task increases. Negative feedback causes the decision process to be modified to discriminate against wrong answers, and thus its performance improves with time. One can envision the centralization of such calculations relating to chemical measurements so that small machines in individual laboratories could make decisions on data. Other possibilities, likely and bizarre, can be considered in artificial intelligence

particle tracking (cloud, bubble, spark), fingerprint identification, speech analysis, weather prediction, medical diagnosis, and photographic processing (cell images and aerial photography). Recently, chemical applications have started to appear in a number of areas of spectroscopy (15–29).

The pattern recognition process will be described as four stages: measurement, feature selection and preprocessing, discriminant training, and generalization.

Measurement. The measurement process is generally not a problem in chemical applications. Indeed it has been said that the modern problems of data interpretation have been generated by the incredible rate at which modern instruments can produce data. The quality of data is generally excellent in the physical sciences, and, in most cases, meaningful limits can be placed on accuracy and precision, and experiments can be repeated to check reproducibility.

Data to be used in pattern recognition studies are represented as vectors $\mathbf{X} \equiv (x_1, x_2, \ldots, x_d)$. The utility of this representation is demonstrated by the following example. Figure 1 shows a two-dimensional plot of the melting points and boiling points of several organic compounds. A's represent organic acids and K's represent ketones. Note that each point in the two-dimensional space completely defines the two pieces of information, and, furthermore, the points could be represented as two-dimensional vectors from the origin. It is clear that acids are high boiling and high melting, while the ketones are low boiling and low melting. Hence, from this figure an investigator who knew nothing of chemistry would immediately recognize that the acids and ketones cluster on this plot, and furthermore, that the experimental data available suggest a good method of distinguishing between the two categories, acids and ketones. Figure 2 makes clear the notion of linear separability, meaning the pattern points can be placed into their two classes by a linear decision surface—a line for this two-dimensional case.

Many types of data can be represented in vector form by providing a sufficient number of dimensions. Chemical spectroscopy data, for example, is usually a spectrum of intensities vs. frequency or wavelength. (Mass spectrometry is a notable exception where the abscissa is mass-to-charge ratio.) If the abscissa is quantized, the number of dimensions can simply be the range of the abscissa divided by the resolution. For example, an infrared spectrum recorded from 2.0 to 14.9 μm with data collected every 0.1 μm can be represented by a 130-component vector or a single point in 130-dimensional space without information loss. Approximately 150 dimensions are sufficient to represent low resolution mass spectra of simple organic compounds with up to 10 carbons. The learning machine method is one approach to finding decision surfaces in such a multidimensional space where the experimenter can no longer plot the data and simply look for clusters or other trends.

Feature Selection and Preprocessing (Transformation). The basic overall objective of the pattern recognition method is to classify the patterns into the desired categories. Preprocessing of the data includes algebraic transformations such as the extraction of roots and taking of logarithms, feature selection, or changes in variables through transforms such as the Fourier transform. Such preprocessing can be useful for

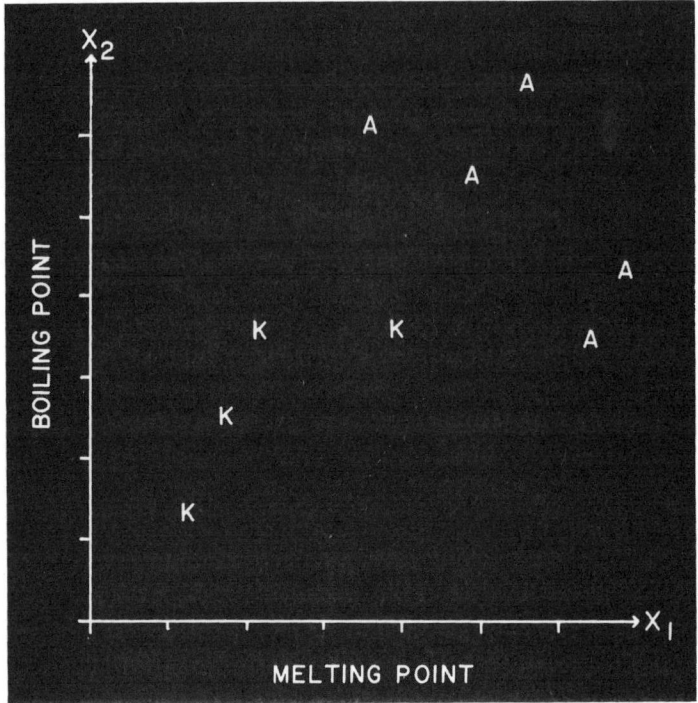

Figure 1. Melting and boiling points of organic acids and ketones plotted in two-dimensional space

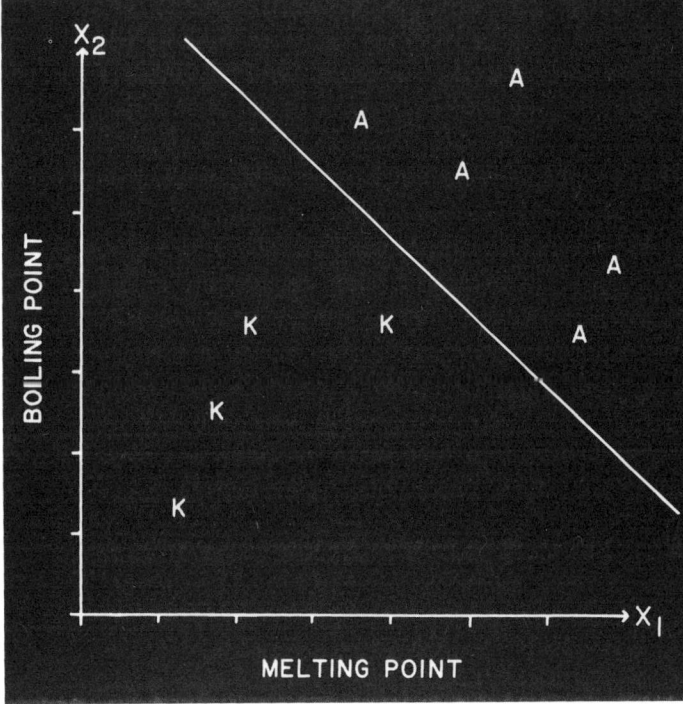

Figure 2. A linear decision surface for organic acids and ketones

the following two reasons. First, some transformations can spread the clusters of the patterns in the two categories further apart in the pattern space, making discrimination easier. Second, preprocessing can reduce the dimensionality of the pattern space, either by discarding dimensions deemed expendable or by combining dimensions (possibly in very complex ways). The advantages gained by reducing the dimensionality will be discussed later. In general, these two goals will not be served simultaneously by a given preprocessing method, and a compromise must be made.

Unfortunately, it is usually not possible to separate the preprocessing stage from the decision stage in pattern recognition systems. This adds to the complexity of studying the overall process. In addition, any preprocessing necessarily imposes some bias on the method of interpretation and some of the advantages of the empirical method are thereby diminished.

Preprocessing of data before classification has received considerable attention (30, 31). In specific studies of chemical data, several preprocessing operations have been investigated, including converting spectroscopic peak intensities to their square roots (16) or their logarithms (23), generating cross terms (26), and using Fourier transforms (25).

Decision Development (Discriminant Training). The widely accepted optimum method for making pattern classification decisions, known as Bayes strategies, depends on having the probability density functions for the classes. Suppose that it is desired to classify patterns represented by d-dimensional vectors $\mathbf{X} \equiv (x_1, x_2, \ldots, x_d)$ into one of two possible categories. Let $F_1(\mathbf{X})$ and $F_2(\mathbf{X})$ be the probability density functions for the two categories, let L_1 and L_2 be the losses associated with misclassifying a member of category one or category two, and let P_1 and $P_2 = 1 - P_1$ be the a priori probabilities of occurrence of patterns in categories one and two. Then it can be shown (2) that the Bayes strategy says to make the decisions as follows:
If
$$P_1 L_1 F_1(\mathbf{X}) > P_2 L_2 F_2(\mathbf{X})$$

then classify **X** in category 1

If

$$P_2 L_2 F_2(\mathbf{X}) > P_1 L_1 F_1(\mathbf{X}) \quad (1)$$

then classify **X** in category 2

This procedure can be generalized to allow decisions among more than two classes. To use the optimum Bayes strategy, the probability density functions, loss functions, and a priori probabilities of each class must be either known or estimated.

If the distribution is not known, or cannot be approximated accurately, then one must either estimate the distribution and proceed accordingly or apply some nonparametric method. For the data which is produced by most chemical experimentation, systems are so complex that rarely is the distribution function known or easily estimable.

In most chemical experimentation, particularly that of chemical analysis, it is rare that any appreciable fraction of the universal set of data is collected under controlled conditions. In mass spectrometry, one of the areas where considerable attention has been paid to the collection of data, large files typically contain several thousand entries, far short of the more than a million known chemical compounds, and even smaller in comparison to the imaginable number of compounds. Hence, at most times we are working with a very small subset of the universal set. Any direct assumption of the universal set from such a small subset could be misleading. For these reasons we resort to an empirical method for developing decision-makers.

The principal decision process to be described here is the threshold logic unit (TLU) (5). We will be concerned with TLU's which are binary pattern classifiers capable of placing a pattern in one of two categories. (This can, however, be made into a complete solution because a series of binary pattern classifiers may be used to subdivide data to any desired degree.)

The original data pattern is denoted by the vector **X**. The TLU implements a plane of the same dimensionality as the patterns which will separate the data into the desired two classes. The two-dimensional data shown in Figure 1 may be separated into the desired categories by any of a family of straight lines (planes in two dimensions), one of which is shown in Figure 2. To cause the decision plane to pass through the origin, an extra degree of freedom is added by augmenting the original d-dimensional pattern vector **X** by a $(d+1) - st$ dimension (which has the same value for every pattern) to give a new vector, **Y**. Usually an arbitrary value of 1 is given for the $d+1$ component of each pattern. [This value can, however, have some effect on the development of decision-makers (20), although it does not affect the separability of pattern sets.] Hence

$$\mathbf{X} \equiv (x_1, x_2, \ldots, x_d)$$

and

$$\mathbf{Y} \equiv (y_1, y_2, \ldots, y_d, y_{d+1}) \quad (2)$$

Figure 3 shows the effect for the two-dimensional case given in Figure 2. Now a three-dimensional plane which passes through the origin may be used to separate the pattern sets.

A convenient way to determine whether a point lies on one side of the plane or the other is to use a vector normal to the plane at the origin. This vector (called a weight vector, **W**) may be thought of as defining the locus of points which constitutes the plane separating the data classes (Figure 3). Because **W** is perpendicular to the plane, the dot product of **W** with any pattern vector (**Y**) will determine whether the pattern lies on one side or the other of the plane.

$$s = \mathbf{W} \cdot \mathbf{Y} = |\mathbf{W}| |\mathbf{Y}| \cos \theta \quad (3)$$

where θ is the angle between the two vectors. $|\mathbf{W}|$ and $|\mathbf{Y}|$ are always positive, and thus

$$-90° < \theta < 90° \quad \cos \theta > 0$$
$$\text{and } s > 0$$
$$90° < \theta < 270° \quad \cos \theta < 0$$
$$\text{and } s < 0 \quad (4)$$

Hence for patterns less than 90° from **W** (and thereby on one side of the plane) the dot product is always

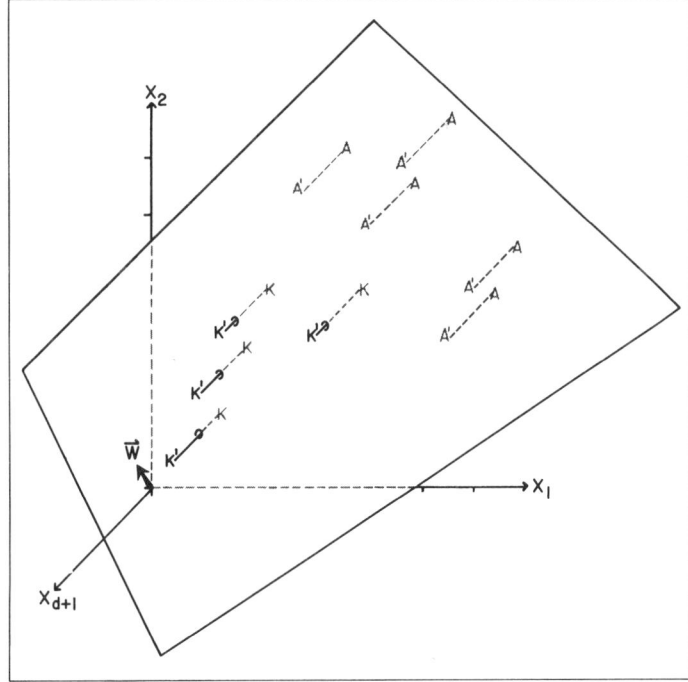

Figure 3. Addition of a $d+1$ dimension to allow the decision surface to pass through the origin

positive, while for patterns on the other side of the plane the dot product is always negative. A further computational convenience is realized in another form of the dot product

$$s = \mathbf{W} \cdot \mathbf{Y} = w_1 y_1 + w_2 y_2 + \ldots$$
$$w_d y_d + w_{d+1} y_{d+1} \quad (5)$$

This sum of the products of components of two arrays is a very easy operation to carry out on a digital computer.

The above derivation generalizes to any number of dimensions and provides a method for determining on which side of a hyperplane a given point lies in hyperspace. Arbitrarily, category 1 can be defined as the positive side of the hyperplane and category 2 as the negative side.

To develop a decision-maker for a given classification, a training set of patterns, for which the correct categories are known, must be used. The members of the training set are presented to the classifier one at a time, and whenever a misclassification occurs, a correction process (negative feedback) is applied to the weight vector. This process continues until all patterns of the training set are correctly classified. If convergence is not obtained, training is arbitrarily terminated after some preset number of feedbacks to conserve computer time.

Several feedback methods have been used. One of the simplest and most effective to date is to move the decision hyperplane along the perpendicular axis between the misclassified point and the plane, so that after the correction it is the same distance on the correct side of the point as it was previously on the incorrect side. This movement is accomplished by adding an appropriate multiple of the pattern vector $\mathbf{Y}$ to the weight vector.

Thus

$$\mathbf{W} \cdot \mathbf{Y}_i = s \quad (6)$$

where s has the incorrect sign for classifying $\mathbf{Y}_i$. Therefore, weight vector, $\mathbf{W}'$, is desired, such that

$$\mathbf{W}' \cdot \mathbf{Y}_i = -s \quad (7)$$

by combining a fraction c of $\mathbf{Y}_i$ with $\mathbf{W}$

$$\mathbf{W}' = \mathbf{W} + c\mathbf{Y}_i \quad (8)$$

Combining Equations 7 and 8 gives

$$s' = \mathbf{W}' \cdot \mathbf{Y}_i = (\mathbf{W} + c\mathbf{Y}_i)\mathbf{Y}_i \quad (9)$$

which can be solved for c to give

$$c = \frac{s' - s}{\mathbf{Y}_i \cdot \mathbf{Y}_i} \quad (10)$$

If, as described above, it is desired that $s' = -s$, then

$$c = \frac{-2s}{\mathbf{Y}_i \cdot \mathbf{Y}_i} \quad (11)$$

and the new weight vector $\mathbf{W}'$ is calculated from the equation

$$\mathbf{W}' = \mathbf{W} - \left(\frac{2s}{\mathbf{Y}_i \cdot \mathbf{Y}_i}\right)\mathbf{Y}_i \quad (12)$$

Of course, other methods can easily be derived which could also be properly termed negative error-correction feedback.

Generalization. The application of an error-correction feedback process to a set of data which is not known a priori to be linear-separable has some interesting aspects. First, successful convergence demonstrates that such a linear classification is possible, even though it might not have been predicted by theory. On the other hand, lack of convergence in less than an infinite number of feedbacks proves nothing. Programs have been written, however, that can accomplish a large number of investigations for a fairly low expenditure in computer time. It has been the experience of the authors, while using an error-forcing computational algorithm, that convergence often occurs within a number of feedbacks that is about twice the number of patterns in the training set. Hence, while inseparability is never proved, it may be reasonably suspected without an inordinate expenditure of computer time.

The use of such empirically developed decision-makers can be applied to the routine classification of data. Furthermore, it may be possible to learn something about the chemistry involved in the experimentation process. Once a relationship is proved to exist, there is encouragement to try to determine the basis of that relationship; and, the constitution of the successful decision-maker may give hints as to its nature. Finally, there is the possibility of learning something about learning itself.

Features of Learning Machines

Four parameters which help evaluate the performance of learning machines, and in the particular example used here, threshold logic units, are called recognition, reliability, convergence rate, and prediction.

Recognition is the ability of the trained pattern classifier to classify correctly the members of the training set. Needless to say, recognition is always 100% for a decision surface which converged to the decision region of a separable set. Once complete training has occurred, the decision surface will always be able to classify correctly any member of the training set. This has potential application as a library. Twenty-six weight vectors were developed (15) to dichotomize correctly the formula subscripts for the mass spectra of 346 compounds from $C_{1-7}H_{1-16}O_{0-3}N_{0-2}$. By storing these weight vectors, each requiring the same dimensionality as the original patterns, the molecular formula could be computed using the binary decision tree shown in Figure 4. This constitutes a storage savings of better than a factor of 10 over retaining all the 346 spectra. It is not meant to imply, however, that the entire information content of the mass spectra has been compressed into the weight vectors, but rather just the necessary set of information to determine molecular formulas. That is, by specifying the question to be answered, it was possible to reduce the amount of data to be retained. Furthermore, the molecular formula may be "computed" by a series of dot product calculations rather than a library search. Such computation takes about 50 msec in a second generation computer or 30 min at a desk calculator.

It is possible that the future may see calculations of weight vectors for large data banks with only the resultant weight vectors supplied to individual laboratories, thereby eliminating the need for repeated

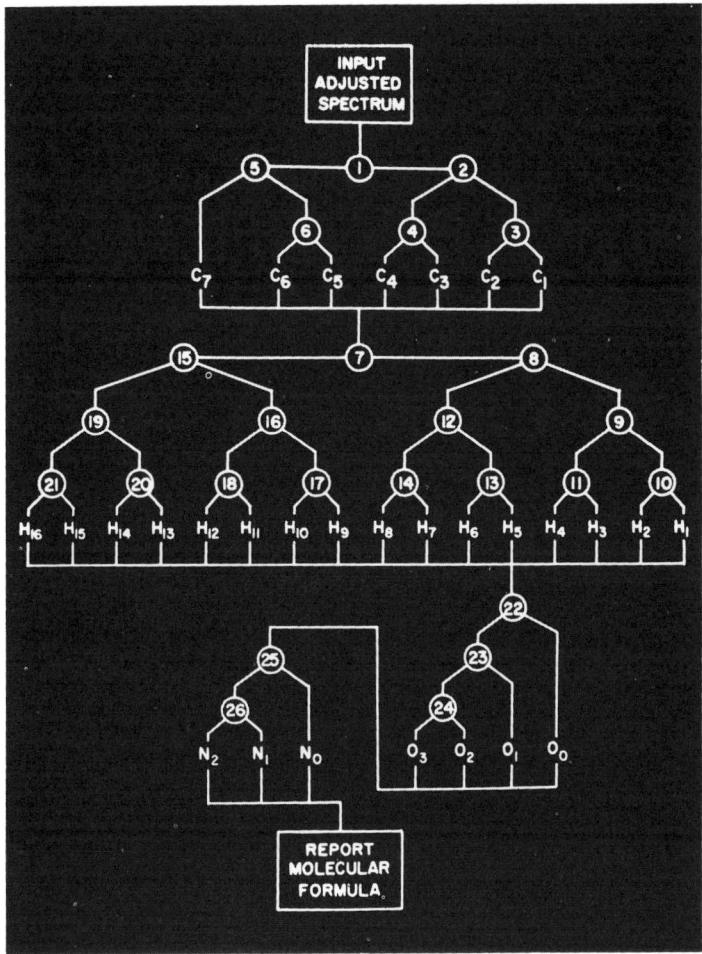

Figure 4. Branching tree of binary decision-makers for determining molecular formula

development of the same decision-makers.

Reliability refers to the ability of the decision-makers to classify members of the training set which have been distorted. Any data collection process has some noise level. In the case of chemical experimentation it is normal to collect slightly different data each time an experiment is run, even if the source of the data is the same compound or system. If a decision-maker is radically affected by small changes in the patterns it classifies, then it is of limited use. To test the procedure, its reliability must be known. To determine the reliability of the weight vectors of Figure 4, the machine was tested on spectra which had been randomly varied to simulate data as it might emerge from a low-resolution mass spectrometer. The error in the peak intensities of the spectra was assumed to be Gaussian. After the intensity of each peak was independently varied using a Gaussian-distributed random number generator, the molecular formula was determined by the master program. A tabulation was kept of the number of times the master program erred in 1000 such trials for various standard deviations and is presented in Table I. The stringency of the random variations imposed upon the spectra should be emphasized, because most instrumental variations in a mass spectrometer cause peak intensities to shift in a related fashion rather than randomly. For example, the University of Washington chemistry department CEC 21-103 had a relative standard deviation in the ratio of the relative intensities of the masses 45/58 of n-butane of 6.9% over a period of six years during which no particular attention was paid to maintaining long-range stability (*32*). For a standard deviation of this magnitude, the learning machine can determine molecular formulas with a reliability greater than 98%. It should be noted that the original data are essentially randomly varied because they come from various mass spectrometers in different laboratories; however, for this learning machine the spectra are standards and thus the reliability testing was in order.

Convergence rate is an important component in determining the expense required to develop decision-makers. Schemes have been developed for minimizing convergence time, thereby maximizing effective convergence rates, the most promising of which is reducing the dimensionality of the data without losing the important information. This will also be important in the prediction. Examples of convergence rate will be given along with those of prediction.

Prediction refers to the ability of the pattern classifier to classify correctly the patterns which were not members of the training set. Prediction is unquestionably the most interesting and exciting aspect of pattern recognition. If the decision-maker can correctly classify patterns that are unknowns, then it is established, or at least strongly implied, that the decision process has learned something of the chemistry of the experiment. Prediction is typically tested by dividing the available known patterns into two sets using a random selection process. One set is used to train the decision vector and the other to test it for prediction. During the prediction test, no further feedback is allowed so that the test is fully one of unknown patterns. For a binary decision-maker, 50% prediction would be expected from random guessing.

Table II shows the results of training weight vectors to determine oxygen presence while using differ-

Table I. Errors in Statistically Varied Data

σ, %	No. incorrect out of 1000	% Correct	Error occurrence			
			C	H	O	N
1	0	100.0	0	0	0	0
5	7	99.3	2	4	0	1
10	24	97.6	12	9	2	1
15	30	97.0	13	10	2	6
20	42	95.8	19	16	4	5

Table II. Convergence Rate and Prediction of Oxygen Presence as a Function of Training Set Size

Training set size	Spectra tested[a]	% predicted	Av. % predicted
50	226	88.3	
	180	84.3	86.2
	255	85.9	
100	654	90.8	
	561	87.7	88.2
	574	86.0	
200	1528	86.6	
	1525	88.6	88.4
	2189	88.1	
300	2590	92.7	
	1959	90.6	90.6
	2986	88.5	

[a] Number of spectra required to produce complete training.

ent-sized training sets chosen randomly from an overall set of 630 low-resolution mass spectra. Normally the convergence rate decreases as the training set becomes larger. This trend is seen in Table II, where all weight vectors were trained to complete recognition. It is interesting to note that predictive ability is high even for training set sizes of 50 and 100 where there are more adjustable parameters (weight vector components corresponding to mass positions) than there are patterns in the training set. Predictive ability is expected to increase as the training set size increases, and Table II demonstrates this, although the noise level is considerable.

Other studies used a set of 387 CH compounds, from which a subset of 200 was chosen randomly to serve as a training set for developing weight vectors. The other 187 were used to test the predictive ability of the weight vectors. Table III shows the results of training and testing the 43 weight vectors developed for the determination of structural parameters of hydrocarbons. Each vector was trained to give a binary decision. In all cases, except the carbon:hydrogen ratio, a positive answer indicated the value was greater than the cutoff number, and a negative value indicated that the value was less than or equal to the cutoff. For example, the first vector (carbon number 9) was trained to give a positive dot product with a mass spectrum of a compound containing 10 carbons and a negative dot product with one containing nine or fewer carbons. The carbon:hydrogen ratio was trained to give a positive result for that particular ratio and a negative result for any other. For example, n-hexane gives a positive dot product for carbon:hydrogen ratio $2n + 2$ and a negative dot product for all other ratios. The categories appearing in Table III are defined as follows: Methyl, ethyl, and n-propyl numbers are the numbers of each group which can be produced by a single-bond rupture—i.e., 3-methylhexane has 3 methyls, 2 ethyls, and 1 n-propyl by this definition; the largest ring classification includes saturated, unsaturated, or aromatic rings; and, branch point carbon number is the number of carbon atoms in the compound which are bonded directly to at least three other carbon atoms. For number of carbon-carbon double bonds, benzene has been classed as three, and the "carbon w/o hydrogen" category refers to carbons which are not bonded to any hydrogens. The final three weight vectors detect presence or absence of benzene rings, acetylenic bonds, and vinyl structural features.

The third and fourth columns of Table III indicate the number of compounds of the training set which fell into each class. The fifth column gives the number of feedbacks necessary for convergence with >2000 indicated for those that had not been completely trained after 2000 feedbacks. (As indicated earlier, failure to train in some arbitrary number of feedbacks does not prove the given training set to be linearly inseparable, and, indeed, incompletely trained vectors may still have considerable recognition and prediction ability.) The sixth and seventh columns indicate the number of compounds in the prediction set which fell into each class. The final column gives the prediction success for the 187 compounds which were not part of the training set. It should be stressed that these compounds were treated by the machine in every way as complete unknowns. The only difference from real unknowns is that the results of these computations may be evaluated. Predictive ability ranged from 61.5% to 98.9% with an average of 90.3%. The prediction percentage can be used as a gauge of the credibility of an answer when produced for an unknown spectrum. Random guessing would give 50% success. In the CH class, considerable structural information may be derived with a high confidence level from a completely empirical calculation method.

Various methods have been demonstrated to improve each of the four features. Undoubtedly the choice of method, particularly in preprocessing and feature selection, will be strongly dictated by the type of data. For example, multicategory systems can be developed to

Table III. CH Class

	Cutoff	Training set			Prediction set		
		Negative category	Positive category	Spectra feedback	Negative category	Positive category	% prediction
Carbon number	9	163	37	227	154	33	89.3
	8	121	79	167	113	74	92.5
	7	80	120	185	77	110	93.6
	6	53	147	99	44	143	94.1
	5	30	170	71	21	166	97.9
	4	15	185	42	7	180	97.3
Hydrogen number	20	196	4	53	182	5	97.3
	18	168	32	170	154	33	95.7
	16	143	57	202	132	55	97.3
	14	110	90	58	100	87	94.1
	12	72	138	51	55	132	95.2
	10	47	153	59	34	153	96.8
	8	28	172	31	19	168	96.8
	6	9	191	34	10	177	97.3
Carbon : hydrogen ratio	$2n + 2$	156	44	25	143	44	96.8
	$2n$	125	75	28	107	80	96.8
	$2n - 2$	153	47	36	154	33	96.8
	$2n - 4$	191	9	39	180	7	98.9
	$2n - 6$	185	15	13	170	17	98.9
Methyl	4	191	9	177	165	22	90.9
	3	160	40	800	136	51	86.1
	2	114	86	859	97	90	86.6
	1	62	138	648	47	140	86.6
	0	29	171	328	24	163	89.3
Ethyl	1	166	34	1518	153	34	80.7
	0	104	96	>2000	97	90	73.3
n-Propyl	1	191	9	211	177	10	90.4
	0	145	55	>2000	148	39	71.7
Largest ring	6	199	1	11	184	3	97.9
	5	142	58	141	137	50	89.8
	4	121	79	149	114	73	90.9
	3	120	80	194	112	75	91.4
Branch point carbons	2	174	26	356	163	24	90.9
	1	108	92	>2000	90	97	61.5
	0	42	158	1486	36	151	87.2
Number of —C=C—	2	171	29	11	165	22	98.4
	1	159	41	153	158	29	92.5
	0	102	98	>2000	98	89	77.5
Carbons w/o hydrogens	1	167	33	432	156	31	83.4
	0	111	89	1327	97	90	70.6
Benzene ring	0	179	21	37	168	19	96.8
—C≡C—	0	184	16	163	174	13	93.6
Vinyl	0	166	34	>2000	158	29	80.2

treat nonlinear systems (28). TLU's can be trained with nonzero thresholds (23) and layers of TLU's can be used (23) to increase predictive ability and reliability. Combining of data from various sources has been demonstrated to be feasible (19). There is no restriction by the pattern recognition method that the data must come all from one source. Indeed, one might go so far as to envision an automated laboratory data analyzer as a simple computer with a minimal central processor capable of calling on a mass storage device containing a variety of weight vectors and carrying out the necessary vector multiplications and binary branches to arrive at the desired results. Such a system could be instructed by the decision vectors developed at a large central system.

Blue Skies—The Future

The suggestion of the centralization of such calculations so that small machines in individual laboratories could make decisions on data is a real possibility. There are other considerations, both likely and bizarre, that remain possibilities in the realm of artificial intelligence.

The basic speed of operations within present electronic computers already far exceed those of the human brain. That is, the rate of flip-flop operations is several orders of magnitude faster than the propagation of signals through the neural network of the human brain. However, these systems appear to make decisions on different bases. For example, the computations involved in catching a baseball thrown with a curve on it seem far too complex for neural speeds to be of any use. Several sightings of the ball would have to be made, and then the optical data fitted to some sort of polynomial or differential equation and finally the equation solved for the time and coordinates of the ball reaching the catcher. Additional problems, such as wind and eyeball error, would make the problem much more complicated.

However it is observed that baseballs are caught by human beings with great regularity. Obviously some sort of learning process employing feedback corrections is involved. It is also likely that the human brain solves problems by some sort of parallel calculations that involve various estimations and extrapolations and even voting machines to make a final decision. The same sort of parallel logic might be introduced to digital computer pattern recognition systems and might give manyfold increases in speed and complexity of problem to be handled.

While digital computers based on electronic phenomena seem to be rapidly approaching certain fundamental limits of speed, these limitations may be overcome by higher component density and even other phenomenological approaches such as using light rather than electronic processes.

Software advances seem almost theoretically unlimited. There are parallels to biological evolution. Biological systems can be placed under stress to encourage evolution toward given goals. The question is, can machines be made to evolve purposefully? If the answer is yes, then we may be looking toward an era of cybernetic chemistry.

In conclusion, it may be said that the problem is "how to solve problems." Perhaps computer systems can eventually answer that question with respect to specific issues and, more excitingly, with respect to general issues.

References

(1) J. Von Neumann, "The Computer and the Brain," Yale Univ. Press, New Haven, Conn., 1958.
(2) G. S. Sebestyen, "Decision-Making Processes in Pattern Recognition," Macmillan, New York, N. Y., 1962.
(3) E. Feigenbaum and J. Feldman, Eds., "Computers and Thought," McGraw-Hill, New York, N. Y., 1963.
(4) J. T. Tou and R. H. Wilcox, Eds., "Computer and Information Sciences," Spartan, Washington, D. C., 1964.
(5) N. J. Nilsson, "Learning Machines," McGraw-Hill, New York, N. Y., 1965.
(6) J. T. Tou, Ed., "Computer and Information Sciences—II," Academic Press, New York, N. Y., 1967.
(7) S. Watanabe, Ed., "Methodologies of Pattern Recognition," Academic Press, New York, N. Y., 1969.
(8) M. Minsky and S. Papert, "Perceptrons," MIT Press, Cambridge, Mass., 1969.
(9) J. M. Mendel and K. S. Fu, Eds., "Adaptive, Learning, and Pattern Recognition Systems," Academic Press, New York, N. Y., 1970.
(10) M. Minsky, *Proc. IRE*, **49**, 8 (1961).
(11) R. J. Solomonoff, *Proc. IEEE*, **54**, 1687 (1966).
(12) C. A. Rosen, *Science*, **156**, 38 (1967).
(13) R. G. Casey and G. Nagy, *Sci. Amer.*, **224**, 56 (1971).
(14) G. Nagy, *Proc. IEEE*, **56**, 836 (1968).
(15) P. C. Jurs, B. R. Kowalski, and T. L. Isenhour, ANAL. CHEM., **41**, 21 (1969).
(16) P. C. Jurs, B. R. Kowalski, T. L. Isenhour, and C. N. Reilley, *ibid.*, p 690.
(17) B. R. Kowalski, P. C. Jurs, T. L. Isenhour, and C. N. Reilley, *ibid.*, p 695.
(18) B. R. Kowalski, P. C. Jurs, T. L. Isenhour, and C. N. Reilley, *ibid.*, p 1945.
(19) P. C. Jurs, B. R. Kowalski, T. L. Isenhour, and C. N. Reilley, *ibid.*, p 1949.
(20) L. E. Wangen and T. L. Isenhour, *ibid.*, **42**, 737 (1970).
(21) P. C. Jurs, B. R. Kowalski, T. L. Isenhour, and C. N. Reilley, *ibid.*, p 3187.
(22) P. C. Jurs, *ibid.*, p 1633.
(23) P. C. Jurs, *ibid.*, **43**, 22 (1971).
(24) L. B. Sybrandt and S. P. Perone, *ibid.*, p 382.
(25) L. E. Wangen, N. M. Frew, T. L. Isenhour, and P. C. Jurs, *Appl. Spectros.*, **25**, 203 (1971).
(26) P. C. Jurs, *ibid.*, in press.
(27) L. E. Wangen, N. M. Frew, and T. L. Isenhour, ANAL. CHEM., **43**, 845 (1971).
(28) N. M. Frew, L. E. Wangen, and T. L. Isenhour, *Pattern Recog.*, **3**, in press.
(29) B. R. Kowalski and C. A. Reilly, *J. Phys. Chem.* **75**, 1402 (1971).
(30) J. T. Tou, *Pattern Recog.*, **1**, 8 (1968).
(31) M. D. Levine, *Proc. IEEE*, **57**, 1391 (1969).
(32) A. L. Crittenden, Univ. of Washington, Seattle, Wash., personal communication, 1968.

31

Copyright © 1970 by the American Chemical Society

Reprinted from *Anal. Chem.* 42:341–347 (1970)

Computer Techniques for Quantitative High Resolution Mass Spectral Analyses of Complex Hydrocarbon Mixtures

Thomas Aczel, D. E. Allan,[1] J. H. Harding, and E. A. Knipp

Baytown Research and Development Division, Esso Research and Engineering Company, Baytown, Texas

HIGH RESOLUTION–LOW VOLTAGE analyses (*1, 2*) are being used extensively in these laboratories for the characterization of complex aromatic mixtures derived from or related to petroleum. This type of analysis determines at the present up to 58 compound types and up to 2900 components per sample. The economic handling of such a large amount of data requires the use of computer systems both for data acquisition and computations. This paper deals with the computer techniques developed for this purpose. Emphasis will be placed on the novel data acquisition techniques, but the data handling techniques will also be reviewed briefly in order to illustrate the application of the overall method.

Quantitative low-voltage analysis requires the identification of the formulas and the determination of the concentrations of all the peaks in the spectrum. The use of a computer system for data acquisition is the most suitable technique for this purpose, as both peak formulas and intensities can be obtained rapidly in a format suitable for further computerized handling and quantitative analysis. Alternate approaches, on the other hand, have major disadvantages. The peak-matching method is extremely slow, does not yield peak intensities, and cannot be used for small or very complex multiplets. The chart-reading method (*1*) is also slower than the computer and cannot be applied well to very complex multiplets.

Automatic data acquisition systems for high resolution spectra of pure compounds at high ionizing voltages have been described in the literature (*3–6*). These systems need to be considerably modified and extended for application to the analyses of very complex mixtures at low ionizing voltages. All the 500–1000 peaks in the spectra need to be correctly and uniquely identified, and this requirement places considerable stress on the accuracy of mass measurement, as most of these peaks are small, 50 to 100 mV, on the average. Very large and very small peaks also occur, so that the dynamic ratio of the peaks to be measured is up to 1/10,000 on an area basis. A novel system of reference standards is required, as none of the conventional compounds used for this purpose (perfluorokerosene, heptacosaperfluorobutylamine) yield peaks at low ionization voltages. Finally, the identified components must be sorted according to homologous series and molecular weight, and the quantitative calculations have to be executed.

EXPERIMENTAL

High resolution–low voltage spectra are obtained on an Associated Electric Industries Model MS9 mass spectrometer, provided with a magnetic scanning system. The resolving power used is 1/10,000.

The instrument is connected to an IBM Model 1802 computer, through an interface consisting of Redcor Model 391030 differential amplifier, and an external clock used to synchronize the analog to digital conversion rate.

The differential amplifier is used to avoid grounding problems in matching to the single-ended input of the IBM 1851 solid-state multiplexer. Other equipment includes an IBM Model 2 analog to digital converter with sample-and-hold amplifier, and an IBM 2401 Model I tape unit for recording the readings. A 16K, 4 μsec Central Processing Unit (CPU) with one disk drive was used in initial operation; it has since been expanded to accommodate other unrelated programs. The initial setup was tested at conversion rates up to 10 KHz and was apparently operational at this rate, though the amplifiers on the mass spectrometer do not allow satisfactory operation at very high scan rates. The tape unit is the rate-limiting part of the computer. A 15 KVA transformer is used to supply power to the computer. Data obtained show that this system has a noise level of about 0.3 mV, or equivalent to the least significant bit of the 14-bit analog-to-digital converter. The interface logic developed includes provisions for communication between the instrument operator and the computer, and for the queuing of the various analytical instruments tied to the computer. The scanning circuit of the MS9 is also controlled by the computer and it is started at the same time as the computer scan.

The IBM Model 1802 computer is used for converting analog to digital data, and for writing these on magnetic tape. About 500,000 readings are collected per spectrum. Variable conversion rates, ranging from 250 to 2000 Hz, are used to maintain an average of approximately 20 conversions per peak at a resolving power of 1/10,000 at various scanning speeds.

The digital signals are temporarily stored as acquired within the IBM 1802 computer in two data tables. These data tables contain 1018 data measurements and are chained

---

[1] Present address, Louisiana State University, Baton Rouge, La.

(1) Thomas Aczel and B. H. Johnson, ANAL. CHEM., **39**, 682 (1967).
(2) Thomas Aczel and B. H. Johnson, 153rd National Meeting of the American Chemical Society, Miami Beach, Fla., April 9–13, 1967, Preprints of Symposia, Volume 12, No. 2, p B-83.
(3) W. J. McMurray, B. N. Greene, and S. R. Lipsky, ANAL. CHEM., **38**, 1194 (1966).
(4) D. Desiderio, P. Bommer, and K. Biemann, *Tetrahedron Lett.*, **1964**, 1725.
(5) D. D. Tunnicluff and P. A. Wadsworth, ANAL. CHEM., **40**, 1826 (1968).
(6) A. L. Burlingame, D. H. Smith, and R. W. Olsen, *ibid.*, p 13.

Table I. Reference Blend for Precise Mass Measurement at Low Voltages

Component	\multicolumn{4}{c}{Peaks used}			
	1	2	3	4
Pyrrole	67.042197			
Fluorobenzene	96.037525			
Chlorobenzene	112.007976	114.005026		
Chlorofluorobenzene	129.998554	131.995604		
Dichlorobenzene	145.969005	147.966055	149.963105	
Bromobenzene	155.957513	157.955543		
Chloronaphthalene	162.023625	164.020675		
Trichlorobenzene	179.930033	181.927083	183.924133	
Chlorobromobenzene	189.918543		193.912643	
Bromonaphthalene	205.973162	207.971192		
Tetrachlorothiophene	219.847485	221.844396	223.841586	225.838636
Iodochlorobenzene	237.904812	239.901862		
Iodonaphthalene	253.959432			
Perfluoronaphthalene	271.987218			
Perfluoroxylene	285.984022			
Dibromotetrafluorobenzene	305.830389	307.828419	309.826449	
Perfluorodiphenyl	333.984022			
Perfluoroacetophenone	361.978930			
Di-iodotetrafluorobenzene	401.802929			
Tetrabromomonofluorophenol	425.672731	427.670761	429.668791	
Octafluorodibromodiphenyl	453.823998	455.822028	457.820058	
Hexabromobenzene	549.506400	551.504430	553.502460	

to each other. After one table has been filled, data are put in the other table. The data from the first table are stored simultaneously on the magnetic tape. This part of the operation is in real time.

The digital tape thus obtained is manually transferred to an IBM Model 360/50 computer, which performs all subsequent operations. A maximum of 18 spectra can be stored on one 2400-foot reel of tape.

DISCUSSION

The programs developed to retrieve and reduce the information contained in the spectra include the integration of peak areas and determination of centroids, the recognition of the externally added reference standard peaks, the calculation of the precise masses and formulas for all the peaks in the spectrum, the determination of the compound type series, and the quantitative analysis. High voltage spectra also can be handled by the system. The final output in this case is a list of the precise masses, formulas, and intensities obtained in a manner similar to that described in the literature (3, 4).

Signals are defined as a peak if more than three data samples are more intense than a threshold. This threshold is a constant 3 mV above the average base line, usually 0.0 V, determined from the first 500 readings. Peak areas and positions are calculated using the centroid model. The integrated areas include the portion of the peak below the previously mentioned threshold. Peak areas are expressed in percent of total ionization, using seven figures which are needed to cover the wide dynamic range in peak intensities. The positions and intensities of all peaks are both printed on computer sheets and punched on cards.

The reference standard used at low ionizing voltages consists of a blend of halogenated aromatics, listed in Table I. The use of this blend is required by the fact that the conventional reference standards used at high voltages, such as perfluorokerosene, do not yield peaks at low ionizing voltages. The halogenated aromatic compounds used are, on the other hand, quite suitable as standards for hydrocarbon samples, as they yield masses with considerable negative mass defects and characteristic isotopes, and thus can easily be recognized either by examination of the printed output, or automatically by the computer. These halogenated aromatics are thought to be suitable also as reference standards for field ionization and chemical ionization spectra. Several programs have been used for standard recognition. The logic starts with the recognition of the first standard, pyrrole, at m/e 67, by its size and its distance from the start of the scan, which is initiated at that value of magnetic field and accelerating voltage at which m/e 614 is foscused in the instrument. Subsequent reference peaks are then identified from their relative positions with respect to m/e 67 and each other. Additional criteria, such as sensitivity, position with respect to other peaks in the multiplet and peakwidth are used if more than one peak is within the time limits set *a priori* around the expected position of the peak. As because of instrument instability none of these programs are entirely and continuously reliable, provisions are made for the input of manually recognized reference times.

Three procedures have been evaluated for mass measurement. The first of these uses the exponential scan law

$$\ln Mx = \ln Ma + \frac{tx - ta}{\tau(a, b)} \quad (1)$$

$$\tau(a, b) = \frac{ta - tb}{\ln [Mb/Ma]} \quad (2)$$

where Ma, Mb are the masses of the reference peaks bracketing a sample peak of mass Mx; ta, tb, tx are the respective occurrence times; and $\tau(a, b)$ is the time constant calculated between the reference peaks a and b.

This approach assumes that τ is constant in the mass interval Ma–Mb. As shown by Lipsky and coworkers (3) this assumption is not perfectly valid, and the value of τ changes during the scan. To compensate for this deviation a second approach was tested, in which the value of τ used was that extrapolated to the mass interval between each sample peak and the standard peak from which it is measured. In this procedure, approximate masses were calculated for the sample masses using Equations 1 and 2. The τ function was then extrapolated linearly to the mass interval between the closest reference peak and the sample peak being measured, and a final value for this mass was obtained by substituting in Equation 1 the extrapolated value of τ. Although somewhat more complex, this procedure yielded consistently better mass measurements than the first approach.

A third procedure consisted in using the Lagrange interpolation

$$y = y_1 \left[\frac{(x - x_2)}{(x_1 - x_2)} \times \frac{(x - x_3)}{(x_1 - x_3)} \times \frac{(x - x_4)}{(x_1 - x_4)} \times \frac{(x - x_5)}{(x_1 - x_5)} \times \cdots \right] +$$
$$y_2 \left[\frac{(x - x_1)}{(x_2 - x_1)} \times \frac{(x - x_3)}{(x_2 - x_3)} \times \frac{(x - x_4)}{(x_2 - x_4)} \times \frac{(x - x_5)}{(x_2 - x_5)} \times \cdots \right] + \cdots \quad (3)$$

This interpolation yields intermediate y values (masses) corresponding to known x values (occurrence times), provided that there is available a set of known y and x values. These are furnished by the reference peaks.

Mass measurements obtained using the three approaches discussed above on the same set of experimental data are shown in Table II.

The data in Table II reflect 670 measurements, including those on peaks as small as 10 mV. Although the data obtained with the extrapolated τ values are the best, the Lagrange method is used routinely, because it is almost as accurate and requires less computer time than the former.

This mass measurement accuracy is sufficient for high voltage spectra. In this case formulas are obtained by an algorithm similar to those described in the literature (3, 4), based on the value of the fractional mass. The number and type of heteroatoms (up to 12 elements and 6 atoms for each element) to be considered in the calculation is an optional input. However, as mentioned above, low-voltage analyses require the correct and unequivocal recognition of all peaks in the spectrum. In order to achieve—or at least to approach—this objective, an additional step is used in the program for this type of analysis. This approach is based on the observed fact that mass measurement errors are minimum for peaks near reference masses, and, in general, are systematic. We are therefore using each identified sample peak as the reference standard for the next sample peak compensating thus for systematic errors. The logic is summarized below:

1. Mass measurement is carried out with the Lagrange interpolation on all peaks.
2. Formulas are calculated with conventional algorithms only for peaks in the immediate neighborhood of the external standard peaks. Heteroatoms considered are O and S. The error limit used is ±3.9 mmu.
3. Formulas containing S atoms are converted to corresponding hydrocarbons using the equivalence S = 3C—4H. This step is required to eliminate equivocal identifications as the mass difference for the doublet SH_4—C_3 is only 3.4 mmu, and therefore below the error limit used. Aromatic hydrocarbons and aromatic sulfur compounds differing by this

Table II. Mass Measurement Accuracy

Method	Experimental τ	Extrapolated τ	Lagrange
Average absolute error, mmu	2.09	1.26	1.54
Percent of errors			
between 0–1 mmu	39	56	51
1–2 mmu	27	24	20
2–3 mmu	16	13	18
>3 mmu	18	7	11

Table III. Comparison of Lagrange and Lagrange + Interpolation Methods for Correct Formula Recognition

		Formulas recognized correctly	
Sample	No. of peaks	Lagrange method	Lagrange + interpolation method
1	396	343	388
2	421	331	414
3	446	331	415
4	331	231	307
5	298	214	281

doublet are separated in the subsequent program developed for quantitative analysis (1).

4. The remaining formulas are calculated by linear interpolation, equating the measured mass differences with differential formulas within multiplets and between adjacent clusters of multiplets. For example, a mass difference of 2.016 ± 0.010 is equated to a formula difference of 2H. Tables considering all possible mass differences between CH, CHO, and CHO_2 formulas are stored in the computer.

5. The linear interpolation procedure starts anew at each external standard. In addition, provisions are made to attempt to calculate formulas in the conventional manner in case the linear interpolation fails. These safeguards prevent error accumulation.

This approach eliminates most systematic errors, and yields correct and unique formulas, although a few random errors are still present. The number of errors is, of course, a function of the stability of the instrument and varies in day-to-day operation.

The use of this interpolation greatly increases the number of correctly identified formulas, as shown in Table III.

The interpolation is applicable only to complex petroleum fractions containing a large number of relatively simple molecules, prevalently containing only C, H, O, and S atoms. Components containing N atoms are at present determined only indirectly using both the mass measurements and the presence of residuals after isotope correction (7).

In addition to intensities, mass measurements, formulas, and absolute errors, the program also determines the appropriate homologous series for each component belonging to the 58 compound types considered at the present for quantitative analysis. A partial output is shown in Table IV.

This program is capable of coping with very complex systems as illustrated by the data obtained on the multiplets shown in Table V (7).

(7) Thomas Aczel, J. Q. Foster, and J. H. Karchmer, 157th Meeting of the American Chemical Society, Minneapolis, Minn., April 13–18, 1969, Preprints of the Division of Fuel Chemistry, Volume 13, No. 1, p 8.

Table IV. Computer Identification of Formula and Series
MS9 High Resolution Mass Spectrometer Run

Intensity	Meas M/E	MMU Error	C12	Hyd	O16	S32	Series
0.0173	280.129	3.7	22	16			C(N)H(2N-28)
0.0256	280.180	−2.8	20	24	1		C(N)H(2N-16)O
0.4718	280.218	−1.1	21	28			C(N)H(2N-14)
0.1329	281.221	* ISOTOPE					
0.0233	282.142	1.3	22	18			C(N)H(2N-26)
0.5927	282.233	−1.3	21	30			C(N)H(2N-12)
0.1427	283.239	* ISOTOPE					
0.0323	284.156	−1.0	22	20			C(N)H(2N-24)
0.5247	284.251	0.3	21	32			C(N)H(2N-10)
0.0063	285.156	* ISOTOPE					
0.1440	285.256	* ISOTOPE					
0.2298	285.984	REFERENCE					
0.1374	286.173	1.0	22	22	0	0	
0.1374	286.173	−2.2	19	26	0	1	C(N)H(2N-22)
0.7991	286.267	0.7	21	34	0	0	
0.7991	286.267	−2.7	18	38	0	1	C(N)H(2N-8)
0.0510	287.177	* ISOTOPE					
0.1449	287.270	* ISOTOPE					
0.2091	288.189	1.2	22	24			C(N)H(2N-20)
0.8082	288.282	0.0	21	36			C(N)H(2N-6)
0.0314	289.194	* ISOTOPE					
0.1359	289.285	* ISOTOPE					
0.0107	290.167	0.4	21	22	1		C(N)H(2N-20)O
0.4311	290.205	1.4	21	26			C(N)H(2N-18)
0.0061	290.282	* COMPN NOT FOUND					
0.0871	291.206	* ISOTOPE					
0.0530	292.185	2.4	21	24	1		C(N)H(2N-18)O
0.4246	292.220	1.4	22	28			C(N)H(2N-16)
0.0922	293.225	* ISOTOPE					
0.0228	294.145	3.7	23	18			C(N)H(2N-28)
0.0195	294.197	−1.8	21	26	1		C(N)H(2N-16)O
0.4273	294.236	1.3	22	30			C(N)H(2N-14)
0.1374	295.235	* ISOTOPE					
0.0119	296.164	7.3	23	20			C(N)H(2N-26)
0.4631	296.250	−0.2	22	32			C(N)H(2N-12)
0.1036	297.255	* ISOTOPE					
0.0099	298.176	3.6	23	22			C(N)H(2N-24)
0.5838	298.267	1.3	22	34			C(N)H(2N-10)
0.1562	299.273	* ISOTOPE					
0.1246	300.187	−1.3	23	24			C(N)H(2N-22)
0.8352	300.282	0.8	22	36			C(N)H(2N-8)
0.0138	301.193	* ISOTOPE					
0.1932	301.287	* ISOTOPE					
0.1436	302.204	0.7	23	26			C(N)H(2N-20)
0.7080	302.296	−1.2	22	38			C(N)H(2N-6)
0.0339	303.207	* ISOTOPE					
0.1890	303.302	* ISOTOPE					
0.0053	304.183	0.2	22	24	1		C(N)H(2N-20)O
0.3396	304.219	0.2	23	28			C(N)H(2N-18)
0.0472	305.222	* ISOTOPE					
0.2246	305.830	REFERENCE					
0.0252	306.196	−2.0	22	26	1	0	C(N)H(2N-18)O
0.3210	306.233	−1.5	23	30	0	0	C(N)H(2N-16)
0.0785	307.238	* ISOTOPE					

The identified components are then sorted according to holomogous series and molecular weight. This summary information is both printed and punched on cards at the end of each computer run, in a format suitable for the subsequent quantitative analysis. An example of this output in shown below.

AROMDL	0401	−16.00	124.00	Blank	Blank	13.85	73.72	356.27
AROMDL	0408	911.38	1195.00	1069.59	486.32	169.60	18.31	23.70
AROMDL	0415	12.62	6.93	10.31	13.24	20.78	Blank	42.94
AROMDL	0422	40.48	42.78	42.78	58.79	20.62	Blank	Blank
AROMDL	0451	−18.00	122.00	Blank	Blank	Blank	Blank	427.99
AROMDL	0458	2790.32	5674.52	4835.16	2447.74	715.01	304.87	55.56
AROMDL	0465	58.02	11.70	4.77	13.70	31.86	35.00	37.09
AROMDL	0472	31.40	43.40	18.62	6.62	6.16	Blank	Blank

Table V. Computer Identification of Multiplets

Intensity, % ΣI	Measured mass	Error, mmu	Formula
0.0306	150.1421	1.3	$C_{11}H_{18}$
0.0830	150.1269	3.1	$C_9{}^{13}CH_{15}N$
0.8373	150.1034	−1.1	$C_{10}H_{14}O$
0.9736	150.0682	0.1	$C_9H_{10}O_2$
0.1916	150.0490	−1.1	$C_9H_{10}S$
0.1667	150.0133	−0.5	C_8H_6SO
0.0563	149.9631	...	Reference ($C_6H_4{}^{37}Cl_2$)
0.0208	302.0695	−7.0	$C_{20}H_{14}SO$
0.8592	302.1125	3.0	$C_{24}H_{14}$
0.0074	302.1684	1.4	$C_{22}H_{22}O$
2.0314	302.2021	−1.3	$C_{23}H_{26}$
0.0248	302.2557	−5.3	$C_{21}H_{34}O$
0.5962	302.2956	−1.7	$C_{22}H_{38}$

Table VI. High Resolution–Low Voltage Analysis of a Heavy Coker Gas Oil

Partial Compound Type–Carbon Number Distribution, Weight Percent on Aromatics

Carbon number	C_nH_{2n-12}	C_nH_{2n-14}	C_nH_{2n-16}	C_nH_{2n-18}	C_nH_{2n-20}
1–9	0.0	0.0	0.0	0.0	0.0
10	0.172	0.0	0.0	0.0	0.0
11	1.333	0.0	0.0	0.0	0.0
12	2.991	0.044	0.020	0.0	0.0
13	3.366	0.744	0.114	0.0	0.0
14	2.456	1.553	0.593	0.519	0.0
15	1.115	1.962	1.634	3.649	0.241
16	0.406	0.874	2.298	7.961	0.538
17	0.250	0.851	2.195	7.245	1.504
18	0.177	0.272	1.061	3.901	2.553
19	0.164	0.112	0.392	1.208	1.733
20	0.134	0.077	0.045	0.544	1.207
21	0.045	0.024	0.061	0.104	0.292
22	0.098	0.031	0.034	0.115	0.150
23	0.025	0.013	0.020	0.024	0.072
24	0.0	0.0	0.030	0.010	0.0
25	0.046	0.018	0.041	0.031	0.0
26	0.027	0.165	0.067	0.075	0.064
27	0.253	0.212	0.159	0.0	0.073
28	0.192	0.165	0.149	0.094	0.099
29	0.185	0.146	0.146	0.083	0.023
30	0.099	0.220	0.160	0.119	0.063
31	0.090	0.111	0.227	0.053	0.030
32	0.056	0.126	0.082	0.019	0.0
33	0.0	0.0	0.0	0.019	0.0
34	0.0	0.0	0.0	0.0	0.0
35–50	0.0	0.0	0.0	0.0	0.0
Totals	13.680	7.719	9.527	25.772	8.645

Each compound type is alloted an array of 50 positions, identified by progressive AROMDL numbers. The first two positions in each array list the Z number and the nuclear molecular weight of each type. The successive positions contain the intensities of each member of the series expressed in percent of total ionization times one thousand, in order of increasing molecular weight. Blank positions indicate that the computer did not find a peak at the corresponding mass number. If this is due to an error in mass measurement, the series is completed by posting the missing intensity value through human intervention. In most cases, of course, blanks are due to the fact that no peaks were observed in the spectrum at the corresponding positions. This output format provides one with a rapid look at the spectrum and the opportunity to correct random mass measurement errors prior to the final quantitative analysis.

Table VII. High Resolution–Low Voltage Analysis of a Heavy Coker Gas Oil

Summarized Data on Compound Type Distribution

Compound type	Wt % on aromatics	Wt % on sample	Av MW	Av. carbon no.	Av. carbons in side-chains
C_nH_{2n}	0.126	0.071	182.00	13.00	7.00
C_nH_{2n-2}	0.251	0.141	146.16	10.58	4.58
C_nH_{2n-4}	0.238	0.133	178.60	13.04	7.04
C_nH_{2n-6}	1.684	0.942	179.03	13.22	7.22
C_nH_{2n-8}	2.383	1.333	176.09	13.15	4.15
C_nH_{2n-10}	1.440	0.805	205.86	15.42	6.42
C_nH_{2n-12}	13.680	7.651	178.55	13.61	3.61
C_nH_{2n-14}	7.719	4.317	212.42	16.17	4.17
C_nH_{2n-16}	9.527	5.328	222.34	17.02	5.02
C_nH_{2n-18}	25.772	14.414	216.06	16.72	2.72
C_nH_{2n-20}	8.645	4.835	238.31	18.45	3.45
C_nH_{2n-22}	9.591	5.364	239.94	18.71	2.71
C_nH_{2n-24}	3.095	1.731	260.29	20.31	2.31
C_nH_{2n-26}	2.008	1.123	261.11	20.51	1.51
C_nH_{2n-28}	1.236	0.691	282.92	22.21	2.21
C_nH_{2n-30}	0.302	0.169	306.68	24.05	2.05
C_nH_{2n-32}	0.565	0.316	295.25	23.38	1.38
C_nH_{2n-34}	0.152	0.085	322.21	25.44	1.44
C_nH_{2n-36}	0.055	0.031	321.08	25.51	1.51
C_nH_{2n-38}	0.067	0.038	340.17	27.01	1.01
C_nH_{2n-40}	0.0	0.0	0.0	0.0	0.0
C_nH_{2n-42}	0.003	0.002	350.00	28.00	0.0
C_nH_{2n-44}	0.0	0.0	0.0	0.0	0.0
$C_nH_{2n-2}S$	0.0	0.0	0.0	0.0	0.0
$C_nH_{2n-4}S$	0.022	0.012	126.00	7.00	3.00
$C_nH_{2n-6}S$	0.0	0.0	0.0	0.0	0.0
$C_nH_{2n-8}S$	0.020	0.011	164.00	10.00	3.00
$C_nH_{2n-10}S$	0.502	0.281	167.62	10.40	2.40
$C_nH_{2n-12}S$	0.0	0.0	0.0	0.0	0.0
$C_nH_{2n-14}S$	0.062	0.034	202.78	13.20	2.20
$C_nH_{2n-16}S$	5.713	3.195	214.32	14.17	2.17
$C_nH_{2n-18}S$	0.173	0.097	233.14	15.65	1.65
$C_nH_{2n-20}S$	0.449	0.251	246.04	16.72	2.72
$C_nH_{2n-22}S$	0.761	0.426	253.61	17.40	1.40
$C_nH_{2n-24}S$	0.047	0.027	268.07	18.58	1.58
$C_nH_{2n-26}S$	0.019	0.011	276.98	19.36	1.36
$C_nH_{2n-28}S$	0.004	0.002	284.00	20.00	0.0
$C_nH_{2n-30}S$	0.018	0.010	312.90	22.21	1.21
$C_nH_{2n-32}S$	0.0	0.0	0.0	0.0	0.0
$C_nH_{2n-34}S$	0.0	0.0	0.0	0.0	0.0
$C_nH_{2n-2}O$	0.0	0.0	0.0	0.0	0.0
$C_nH_{2n-4}O$	0.0	0.0	0.0	0.0	0.0
$C_nH_{2n-6}O$	0.139	0.078	195.17	13.23	6.23
$C_nH_{2n-8}O$	0.0	0.0	0.0	0.0	0.0
$C_nH_{2n-10}O$	0.010	0.005	174.00	12.00	4.00
$C_nH_{2n-12}O$	0.021	0.012	396.00	28.00	17.00
$C_nH_{2n-14}O$	0.081	0.046	204.81	14.49	3.49
$C_nH_{2n-16}O$	2.229	1.247	213.14	15.22	3.22
$C_nH_{2n-18}O$	0.236	0.132	233.15	16.80	2.80
$C_nH_{2n-20}O$	0.040	0.022	231.03	16.79	2.79
$C_nH_{2n-22}O$	0.547	0.306	255.82	18.70	2.70
$C_nH_{2n-24}O$	0.258	0.144	275.44	20.25	3.25
$C_nH_{2n-26}O$	0.065	0.036	271.13	20.08	2.08
$C_nH_{2n-28}O$	0.016	0.009	308.91	22.92	2.92
$C_nH_{2n-30}O$	0.018	0.010	342.49	25.46	4.46
$C_nH_{2n-32}O$	0.007	0.004	334.00	25.00	3.00
$C_nH_{2n-34}O$	0.0	0.0	0.0	0.0	0.0
$C_nH_{2n-36}O$	0.006	0.003	330.00	25.00	1.00
Totals	100.003	55.931			

In the above example, the first peak in the C_nH_{2n-18} series was observed in position 5, that is, at m/e 178, and peaks were observed at each fourteen mass units through m/e 444, although the intensity of the peak at m/e 360 had to be posted manually. The nuclear molecular weight given, 122, is the molecular weight of the first member in the $C_nH_{2n-8}S$ series,

Table VIII. High Resolution–Low Voltage Analysis of a Heavy Coker Gas Oil
Carbon Number Distribution Summary

Carbon number	Wt. % on aromatics	Wt. % on sample	Carbon number	Wt. % on aromatics	Wt. % on sample
1–6	0.0	0.0	21	2.936	1.642
7	0.039	0.022	22	1.852	1.036
8	0.187	0.105	23	1.015	0.568
9	0.321	0.179	24	0.779	0.436
10	0.685	0.383	25	0.503	0.281
11	2.466	1.379	26	0.600	0.335
12	4.181	2.338	27	0.824	0.461
13	6.175	3.454	28	1.012	0.566
14	8.450	4.726	29	0.713	0.399
15	12.049	6.739	30	0.837	0.468
16	13.551	7.579	31	0.583	0.326
17	14.696	8.220	32	0.361	0.202
18	11.937	6.676	33	0.019	0.010
19	8.139	4.552	34–50	0.0	0.0
20	5.094	2.849	Totals	100.003	55.931

Table IX. Summary Table
Miscellaneous Averages and Summaries on Aromatics

Elemental analysis by MS, wt %		Characteristic averages on sample	
Atomic carbon	91.11	Molecular wt	218.124
Atomic hydrogen	7.48	Carbon no.	17.054
Atomic sulfur	1.16	Z number (CNH2N-Z)	17.086
Atomic oxygen	0.26	C atoms in sidechains	3.342
Atomic H/C ratio	0.978		

Distribution of Aromatic Rings[a]

	Hydrocarbons	Sulfur comp.	Oxygen comp.	Totals
Nonaromatics	0.615	0.042	0.0	0.657
1 Ring aroms	5.507	0.564	0.149	6.220
2 Ring aroms	30.925	6.335	2.332	39.591
3 Ring aroms	34.418	0.808	0.276	35.502
4 Ring aroms	14.695	0.041	0.870	15.606
5 Ring aroms	1.538	0.0	0.034	1.572
6 Ring aroms	0.772		0.013	0.785
7+ Ring aroms	0.070			0.070
Totals	88.539	7.790	3.674	100.003

Miscellaneous Averages and Summaries on Sample
Distribution of Aromatic Rings[a]

	Hydrocarbons	Sulfur comp.	Oxygen comp.	Totals
Nonaromatics	0.344	0.023	0.0	0.368
1 Ring aroms	3.080	0.315	0.083	3.479
2 Ring aroms	17.296	3.543	1.304	22.143
3 Ring aroms	19.250	0.452	0.154	19.856
4 Ring aroms	8.219	0.023	0.487	8.728
5 Ring aroms	0.860	0.0	0.019	0.879
6 Ring aroms	0.432		0.007	0.439
7+ Ring aroms	0.039			0.039
Totals	49.520	4.357	2.055	55.932

[a] Naphthenic and heterocyclic rings are not considered as aromatics; for example, tetralins, benzothiophenes, and octahydroanthracenes are listed as 1 ring aromatics.

which at this point in the program is not yet separated from the interfering C_nH_{2n-18} series, starting at m/e 178. The absence of peaks from m/e 122 to m/e 164 indicates this sample did not contain the first four members of the sulfur type.

Procedures for the quantitative analysis have been discussed previously (1, 2). The method at present handles up to 58 compound types and up to 50 homologs for each type. These 58 types include compounds with general formulas C_nH_{2n} through C_nH_{2n-44}, $C_nH_{2n-2}S$ through $C_nH_{2n-34}S$, and $C_nH_{2n-2}O$ through $C_nH_{2n-36}O$. The average number of components per sample is 300 to 500, although we have analyzed samples with up to 1500 components.

Several output formats have been devised to handle and summarize the detailed information thus obtained. Quantitative parameters given include the weight percent of each component, normalized on a sample basis, and, if required, to the original sample prior to silica gel percolation, or to the total charge if the sample contained a nonvolatile portion. A portion of this output is shown in Table VI. Summarized information includes the weight percent of each compound type (normalized, if required), its average molecular weight, carbon number, and the average number of carbon atoms in its side chains (Table VII). A molecular distillation type analysis is given by listing the fraction of sample at each carbon number regardless of type (Table VIII). A third summary table lists the average molecular weight, carbon number, Z number, carbon atoms in sidechains for the total sample, weight percent C, H, S, O, and atomic H to C ratio as calculated from MS data, and a distribution of the compound types according to the number of aromatic rings and the elements contained (Table IX). This last information can also be normalized. In addition, user oriented computer programs are available to merge the data from the aromatics analysis with those obtained for saturates, yielding conventional parameters such as carbon in aromatic rings, naphthenic rings, or in sidechains and specific correlations related to refinery or pilot plant processes.

These summaries apply at the present only to the 58 types included in the routine procedure; semiquantitative data on additional compound types, such as N compounds or compounds containing more than one heteroatom, can be obtained using manual computations.

CONCLUSION

The methods described have been used for the analysis of approximately 100 complex samples per month during the past 24 months, involving, on the average, the quantitative determination of 300 components in each sample. The method is applicable to all materials derived from petroleum boiling up to 1100 °F, regardless of origin, prior treatment, or width of boiling range interval. No previous separation is essential, and nonvolatile residua can be weighed and taken into consideration. Samples analyzed included petroleum streams, crude oils, and rock extracts, and more recently, coal liquefaction products (7). More than 150 compound types containing CH, CHO, CHO_2, CHO_3, CHS, CHS_2, CHSO, $CHSO_2$, $CHSO_3$, CHN, CHNS, and CHNO groups were identified in these samples. Analysis of a sample requires about three hours including instrument time, computer time, and human examination and interpretation of the various computer outputs. The actual computer time expended for data logging on the IBM 1802 varies from 2 to 16 minutes according to the scan rate. The CPU time expended for all successive calculations, including peak recognition, area measurement, formula calculation, and complete quantitative analysis is below 3 minutes on an IBM 360/50 computer. Elapsed time is usually two days. Unmechanized handling of the same type of sample would require about 4 to 8 hours for data acquisition, and a minimum of 24 hours for the computation of a quantitative analysis including only the compound type carbon number distribution. The calculation of the various summary tables would be practically impossible.

ACKNOWLEDGMENT

We thank G. R. Taylor and J. L. Taylor who obtained most of the experimental data, and M. S. B. Munson and H. E. Lumpkin for helpful suggestions in the course of this work.

RECEIVED for review September 22, 1969. Accepted December 19, 1969.

32

Copyright © 1971 by the American Chemical Society
Reprinted from *Anal. Chem.* 43:1790–1795 (1971)

Computer Methods in Analytical Mass Spectrometry

Development of Programs for Analysis of Low Resolution Mass Spectra

L. R. Crawford
Division of Chemical Physics C.S.I.R.O., Melbourne, Victoria 3000, Australia

J. D. Morrison[1]
Division of Physical Chemistry, La Trobe University, Bundoora, Victoria 3083, Australia

A computer program which analyzes low resolution mass spectra of organic compounds of molecular weight less than 200 is described. The program operates *ab initio*, that is, it does not require a reference library of mass spectra, except in its original compilation. The unknown mass spectrum is interrogated to yield successively the molecular mass if possible, the presence of functional groups, groups adjacent to the functional group, and finally the molecular skeleton. The conclusions are repeatedly checked for consistency. As details of the structure emerge, they are stored in a structure matrix. As output the program produces a conventional diagram of the molecular structure. Identification takes less than ten seconds. Obviously the program is not always successful but it does demonstrate the feasibility of carrying out in a general way the entire process of structure determination.

THE COMBINATION of mass spectrometry and gas chromatography can produce hundreds of mass spectra a day, and a trained chemist may take two or three weeks on the results of a single run to separate the common or simpler compounds from those he wants to investigate more fully.

Computer search of a library can be used and has been shown to be a very valuable technique (*1*), but the compounds with which the user is concerned may not be in the library; also such a library needs constant attention to keep it up to date and down to a reasonable size. A very desirable addition is a computer program which embodies all the rules for determining a structure from a mass spectrum *ab initio*, yet is capable of being used in a computer of modest size.

How far one can go in programming a computer to duplicate the performance of an organic mass spectrometrist is of course a very interesting question in itself, and some aspects of this problem have already been discussed. Computer routines have been developed by several authors (*2–5*) for interpreting the mass spectra of specific classes of compounds. These routines are capable in some degree of determining the chemical class of an unknown (*6, 7*). A detailed account of some of the problems of this approach has been given by McLafferty and his colleagues (*8*). In attempting to write a general purpose program, it is essential to have a scratch pad on which structural information can be stored and manipulated as required. A structural code was devised which gave a framework within which this could be done (*9*). This structure code, consisting of a nearest neighbor table, had the advantage that it at all times defined certain limitations in building up a structure. After each addition to the table a check could be made as to whether the structure was fully, although not necessarily correctly, specified.

This paper reports on an attempt to combine the various separate routines described above, so that the computer itself, on the basis of the mass spectrum alone, decides on the molecular mass, the molecular class, calls in the sub-routine available for that particular class, gathers the information together, checks it and any other information it can acquire for consistency, then draws out the structure. Failing this, it gives informative printout. In its present form the program can deal with molecules containing only the elements C, H, O, and N.

DESCRIPTION OF THE METHOD

The program which has been developed is an attempt to systematize the approach of an organic mass spectrometrist. As such, it differs in some respects from the more systematic approach of Lederberg and his colleagues (*10, 11*). In the latter, at a relatively early stage in the analysis, the computer generates all possible structures, then proceeds to eliminate these systematically on the basis of the mass spectral information. This approach could be expected to be very successful with small structures, but the computing time required should increase rapidly with the molecular weight. [This has been modified by these authors in later work (*12, 13*)

---

[1] Present address, Department of Chemistry, University of Utah, Salt Lake City, Utah 84112

(1) L. R. Crawford and J. D. Morrison, ANAL. CHEM., **40**, 1464 (1968).
(2) A. M. Duffield, A. V. Robertson, C. Djerassi, B. G. Buchanan, G. L. Sutherland, E. A. Feigenbaum, and J. Lederberg, *J. Amer. Chem. Soc.*, **91**, 2977 (1969).
(3) G. Schroll, A. M. Duffield, C. Djerassi, B. G. Buchanan, G. L. Sutherland, E. A. Feigenbaum, and J. Lederberg, *ibid.*, p 7440.
(4) M. Barber, P. Powers, P. Wallington, M. J. Wolstenholme, *Nature*, **212**, 784 (1966).
(5) K. Biemann, C. Cone, B. R. Webster, and G. P. Arsenault, *J. Amer. Chem. Soc.*, **88**, 5598 (1966).
(6) B. Pettersson and R. Ryhage, ANAL. CHEM., **39**, 790 (1967).
(7) L. R. Crawford and J. D. Morrison, *ibid.*, **40**, 1469 (1968).
(8) R. Venkataraghavan, F. W. McLafferty, and G. E. Van Lear, *Org. Mass Spectrom.*, **2**, 1 (1969).
(9) L. R. Crawford and J. D. Morrison, ANAL. CHEM., **41**, 994 (1969).
(10) J. Lederberg, G. L. Sutherland, B. G. Buchanan, E. A. Feigenbaum, A. V. Robertson, A. M. Duffield, and C. Djerassi, *J. Amer. Chem. Soc.*, **91**, 2973 (1969).
(11) J. Lederberg and M. Wightman, ANAL. CHEM., **36**, 2365 (1964).
(12) A. Buchs, A. M. Duffield, G. Schroll, C. Djerassi, A. B. Delfine, B. G. Buchanan, G. L. Sutherland, E. A. Feigenbaum, and J. Lederberg, *J. Amer. Chem. Soc.*, **92**, 6831 (1970).
(13) A. Buchs, A. B. Delfine, A. M. Duffield, C. Djerassi, B. G. Buchanan, E. A. Feigenbaum, and J. Lederberg, *Helv. Chim. Acta*, **53**, 1394 (1970).

but still appears to play a very significant part in their method of structure determination.] The approach of the present method is very close to that of the human chemist, in that it is much less systematic, but may be better suited to larger structures.

It is desirable that any such program should be capable of continual modification and improvement. Whenever an incorrect identification takes place, it should be relatively simple to modify the weighting factors given to various logical decisions, in order to rectify this error, and hopefully avoid it in future runs.

To obtain maximum flexibility in this way, the program was written in FORTRAN IV in the form of a relatively short main control program, which is able to call on a series of minor programs. The whole program is of approximately 26,000 words, small enough to allow it to be used on a small computer by chaining methods. The initial development of the program was carried out on a Control Data type CDC3200 computer. Later developments have run in a segmented form in a Digital Equipment type PDP9.

The main program reads in the mass spectrum, and preprocesses it, then calls into core the first sub-program. This returns to the main program information such as the molecular weight and main chemical class of the compound. On the basis of this information, the main program may call into core a second overlaying sub-program to analyze mass spectra belonging to a specific class of molecule. If this sub-program fails to identify the compound satisfactorily, or the main program decides otherwise, a third overlaying sub-program which interrogates other aspects of the mass spectrum in an attempt to deduce the structure may be called. The second and third subprograms leave the results of this interrogation in the form of a nearest neighbor table accessible by the main program, which may then call into core the final sub-program to draw out the chemical structure.

Main Program. The mass spectrum, recorded as integral mass numbers and peak intensities on cards or magnetic tape, is read and the intensities are normalized. A low resolution largest peak plot is produced and the group analysis routines are called into core. A table of probabilities of the compound belonging to any one chemical group and the molecular weight is returned. If the second most probable group has probability less than 10%, then it is decided that the compound has only one functional group, and one of the sub-programs giving a deductive routine for that specific group is called into core. If the routine fails to analyze the compound, or the compound appears to contain more than one functional group, the general purpose sub-program is called. If any one of these succeeds, the main program calls the structure drawing sub-program, and on return from this is ready to process the next mass spectrum. The generalized flow chart for the whole program is given in Figure 1.

Group Analysis Routines. The control routine calls subroutines which form tables of probability of the compound belonging to the twelve chosen groups: Aromatic, Ester, Ether, Acid, Ketone, Aldehyde, Alkene, Alkane, Alcohol, Cycloalkane, Diene, Amine. These probability tables are based on the four largest peaks in the spectrum, the four most significant peaks, fourteen peak condensation average spectra, and group mass spectral hypersphere coordinates (7). The control routine weights the probability values and sums them. A somewhat similar approach to this has been employed by Pettersson and Ryhage (6).

Molecular Weight Routine. The highest value of m/e at

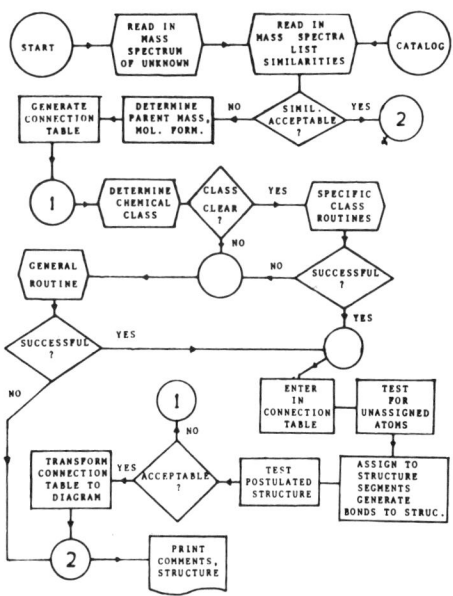

Figure 1. Generalized flowchart of program, showing relation of main program and routines

which a significant peak (significant = background +20%) is recorded, H, is taken initially to be an upper limit to the molecular weight. It is most often an isotope peak. The computer compares the peak height at this highest mass with the peak at one mass unit lower, and checks whether P(H) could be an isotope peak of P(H − 1). If the answer is positive, P(H − 1) is compared with P(H − 2), and so on until a negative answer is returned. This process allows the mass M of the molecular ion to be determined in many cases with fair certainty. A number of consistency checks are then applied. First, the characteristics of the peak believed to be the parent ion are checked for consistency with the molecular class. A second test is for the presence of significant peaks at masses M-3 and M-13, where a significant peak is considered to be one greater than 15% of the assumed parent peak, or greater than the sum of the peaks at ±1 mass unit from it. Another test, based on the greater stability of the even-electron ions, is that if the molecular mass is even, the sum of the odd mass fragment intensities is greater than that of the even and vice versa. If evidence suggests that the molecular peak is not observed, printout occurs.

Empirical Formula Routine. If high resolution mass spectra are available, the previous routine is much less necessary (14). However, it is possible to write a series of restrictive relations between the integral molecular mass and the possible atomic formulas, and when these are combined with other internal evidence in the mass spectrum, the possible atomic formulas are limited to a surprising degree.

The value of M sets an absolute upper limit to the number of C atoms, $q_c \geqslant \dfrac{M}{12}$ (integer division) in the molecule. It

(14) K. Bieman and P. V. Fennessey, *Chimica*, **21** (6), 226 (1967)

also of course sets upper limits for every other kind of atom, but these limits are less useful. The relative heights of P(M) and P(M + 1) give a less precise lower limit to q_C, and may indicate the approximate number of oxygen and nitrogen atoms q_O and q_N.

$$q_C \doteq \frac{P(M + 1)}{P(M)} \times 100 \quad (1)$$

$$q_O + q_N \doteq \frac{M}{13} - q_C \quad (2)$$

In the group identification, alcohols and esters, alkanes and ketones, etc., are frequently confused so a list of equivalent groups is referred to and the redundant classes are removed from the list of the three most probable classes.

In this present program, the three most probable class identifications are checked against the molecular weight in the following way.

On dividing the molecular weight by 14, the remainder is:

$$S_{obs} = 2 \times (q_O - R + 1) + q_N \quad (3)$$

where R is the number of rings and double bonds present. A list of q_C, q_O, and q_N and R values (with q_C a minimum) for all functional groups is maintained. Depending on what functional groups have been identified, an S_g value can be calculated using this formula, e.g.

	q_C	q_O	q_N	R	S_g
Alkane	1	0	0	0	2
Alkene	1	0	0	1	0
Ketone	3	1	0	1	2
Aldehyde	1	1	0	1	2
Cycloalkane	3	0	0	1	0
Alcohol	1	1	0	0	4

If several groups are believed to be present, the S_g value will be the sum of the separate values for each.

If the S_{obs} value obtained from the molecular weight is equal to this, it implies that either q_O, q_N, and R values for the molecule are the same as those for the functional group (or combination of groups) or excess q_O or q_N in the group minimum formula must be balanced by unsaturation.

If ($S_{obs} - S_g$) is zero, the combination is acceptable, and it is stored. If S_{obs} is greater than S_g and no hetero atoms are expected, then the combination is rejected. This information is then used to calculate a list of possible molecular formulas for the compound.

The maximum number of hydrogens in the formula is calculated (11) and then q_C is decreased from its maximum value to its minimum as restricted by P(M + 1)/P(M) in steps of one, and q_H is decreased in steps of two to zero, being reset to its maximum at each change in q_C. q_C is increased by one from zero to the difference between the minimum and maximum q_C. q_N is calculated from:

$$q_N = (M - 12q_C - 16q_O - q_H)/14 \quad (4)$$

and the unsaturations from:

$$R = q_C + 1 + (q_N - q_H)/2 \quad (5)$$

If q_N is neither zero nor integer, or twice $(R - 2)$ is greater than q_C, or R is not integer, or any value of q_C, q_O, or R is less than any corresponding value in the list of possible group combinations, the formula is rejected.

Special Chemical Class Routines. Many classes of molecule have fairly well defined breakup patterns, often involving the production of rearrangement ions. The control program selects the appropriate specific deductive routine for each chemical class if available, and calls it into core. One of these routines for alkyl benzenes, was based with very little modification on that described by S. Meyerson (15). Others for alkanes and for aliphatic esters were based on the descriptions given by Pettersson and Ryhage (6, 16). Others for amines, alcohols, and ethers (17), and aldehydes, ketones, acids, and esters (18), have been especially written. These routines (to the best of their ability) return the structure in the form of a nearest neighbor table, as well as informative printouts, and a true/false value for successful structure determination.

General Interrogation Routine. Satisfactory specific deductive routines have not yet been written for some chemical classes. Also, it has not yet proved possible to organize in such a systematic form much of the empirical information already known about breakup patterns.

In interrogating a mass spectrum, two methods may be employed. In the first, the serial method used in the specific rearrangement routines, the interrogation consists of a sequence of questions each of which depends on the answer to the previous question.

In the second, the parallel method, a number of quite independent questions are asked, the results are weighted and summed. While apparently less efficient, this second method surprisingly appears to be at least as successful, and is sometimes more so than the first.

When a mass spectrometrist has failed to deduce a structure by all the systematic methods, he is reduced to inspired guesswork. The present routine fulfils this role to a certain extent. It is based primarily on the use of diagnostic mass peaks in a way similar to that employed by McLafferty (19).

A list of 64 diagnostic masses is held in store. Each diagnostic mass has associated with it a table of peak characteristics, and structure addresses. The peak characteristics are a flag to indicate if the fragment to be considered is the ion, or the neutral part being formed at the same time, and whether the observed ion has the expected intensity, e.g., base peak, within a typical range, less than 4% of base peak, etc.

The basic structural storage is a code matrix which is decoded by appropriate subroutines into an atom connection matrix describing the structure. When referring to structures in the interrogation routine, the address of the basic structural code matrix in store is used.

The mass spectrum is scanned, and each mass number is checked for diagnostic masses at the mass modulo 14 plus an integral number of 14 mass units up to the mass number. The molecular mass minus the mass number is similarly checked. The characteristics of each observed peak are compared with those associated with each structure. If they do not agree, or if the formula of the structure is not a sub-set of the parent formula, then it is ignored. A list of identified structures is created and maintained and if a new structure is found it is added to the list. If the structure has

---

(15) S. Meyerson, *Appl. Spectrosc.*, **9**, 120 (1955).
(16) B. Pettersson and R. Ryhage, *Ark. Kemi*, **26** (25), 293 (1967).
(17) J. F. O'Brien and J. D. Morrison, *Org. Mass Spectrom.*, in press.
(18) J. D. Morrison, J. F. Smith, and J. Taranto, *ibid.*, in press.
(19) F. W. McLafferty, "Mass Spectral Correlations," American Chemical Society, Washington, D. C., 1963.

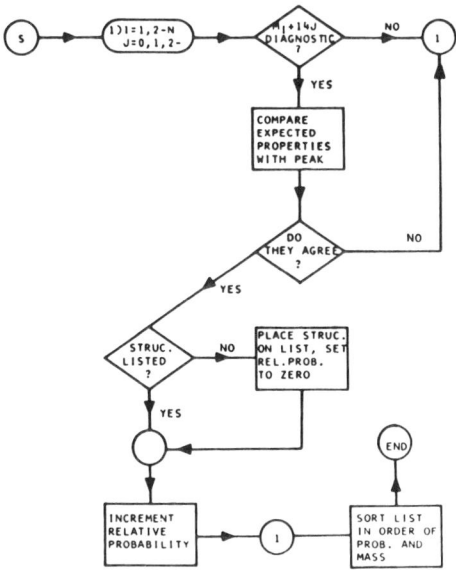

Figure 2. Flowchart of diagnostic peak interrogation part 1

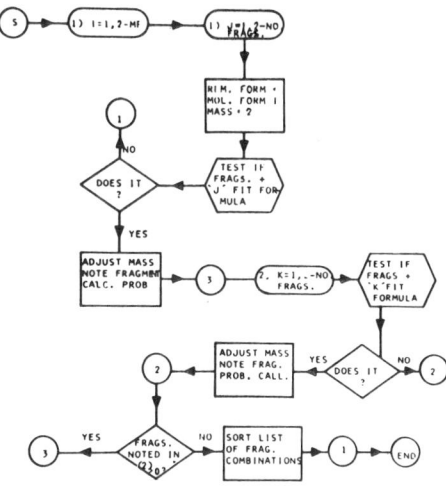

Figure 3. Flowchart of diagnostic peak interrogation part 2

been noted previously, then a frequency of occurrence is recorded in descending order of fragment mass, see Figure 2.

The resulting list of structures is then used to compile a table of likely fragment combinations. Each molecular formula is considered in turn. The list of structures is searched for one which is a sub-set of the formula, then searched again for one which is a sub-set of the remaining formula and so on until no more can be found. A probability value is calculated from the number of atoms remaining in the structure, the probabilities of the contributing structures and the probability of the molecular formula. This is repeated for each molecular formula, using different structures as the starting fragments. The resulting tables of formulas and structural combinations are sorted in order of probability value (Figure 3).

This whole process may be summed up briefly by saying

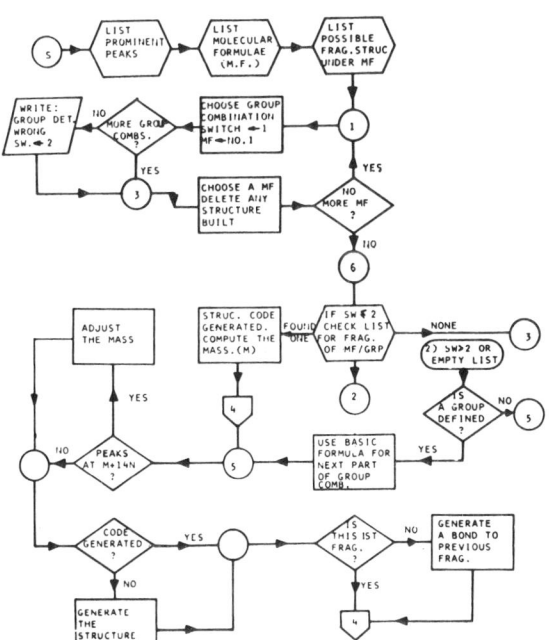

Figure 4. Flowchart of general interrogative routine part 1

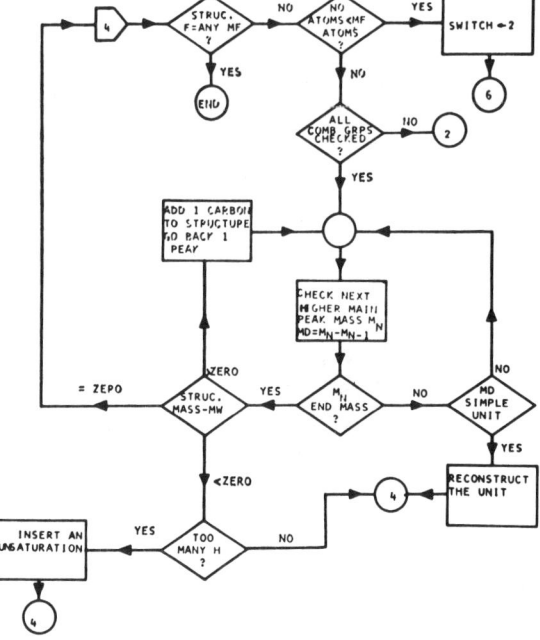

Figure 5. Flowchart of general interrogative routine part 2

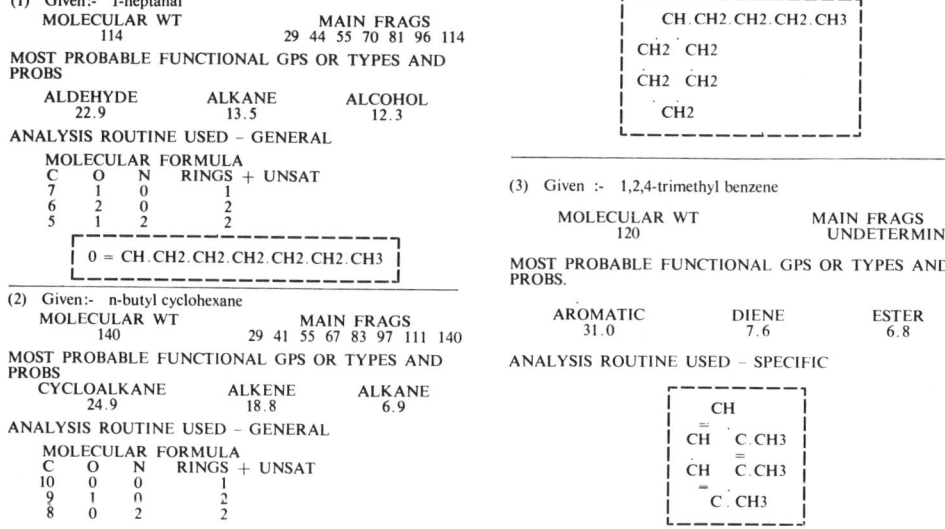

Figure 6. Examples of computer output for:
(1) 1-Heptanal (2) n-Butyl cyclohexane (3) 1,2,4-Trimethyl benzene

that structural information about the compound is accumulated by locating diagnostic peaks and remembering the structures indicated. The most probable structures are fitted to the molecular formulas to produce a table of formulas and probable sub-structures.

If no structure has been found, the program then proceeds with the following. The largest peak in each cluster of peaks is selected and stored separately. The mass of the minimum empirical formula for the first probable group combination is summed and n carbons are added, n varying from zero upward. If the resulting mass is one of the large peaks, then a nearest neighbor list is created for the fragment.

Table I. List of Substances Used for Test

Triethylamine	Cyclopentadiene
n-Butylamine	Methylcyclopentadiene
sec-Butylamine	n-Pentane
Diethylamine	2-Methylbutane
2-Ethylhexylamine	2,2 Diethylpropane
6-Methylquinoline	2-Phenyloctane
3-Methylpyridine	n-Hexane
Aniline	3-Methylhexane
Indole	2,2,3,3,5,6,6-Heptamethylheptane
Cyclohexylamine	Cyclobutane
1,2-Ethanediamine	Methylcyclohexane
2-Butanol	Cyclopropane
1-Butanol	n-Butylcyclohexane
1-Hexanol	Cyclooctane
2-5-Dimethyl-1-hexanol	Toluene
9-Heptadecanol	1,2,4-Trimethylbenzene
p-Methylcyclohexanol	Naphthalene
Furan	Biphenyl
Methyl isopropyl ether	1-Methylnaphthalene
Methyl ethyl ether	Styrene
Methyl isobutyl ether	Pyrene
Ethyl sec-butyl ether	Methyl butanoate
Di-ethoxyethane	Methyl methacrylate
1,2-Epoxypropane	Isobutyl propanoate
2-Methoxyethyl ethenyl ether	Ethyl acetoacetate
Ethyl nitrate	n-Butanoic acid
cis-2-Butene	2-Ethylbutanoic acid
Methylcyclopentene	Propanenitrile
1-Butene	2-Butanone
cis-2-Pentene	2,4-Dimethyl-3-pentanone
1-Pentene	6-Dodecanone
2-Methyl-2-pentene	Propyl benzyl ketone
4,4-Dimethyl-1-pentene	Methyl-cyclohexylketone
1-4-Pentadiene	Butanol
2,3-Pentadiene	2-Ethylhexanal
2-Methyl-1,3-pentadiene	Tetradecanal
Cyclohexene	1-Octyne
Cyclooctatetraene	2-Methylthiophene

Table II. Comparison of Performance of Humans and Computer in Interpreting 76 Mass Spectra

	Human	Computer
Parent mass correct	59	62
Correct formula in highest rating	48	53
One functional group correctly in highest rating	47	62
Correct structure or near isomer[a]	23	22
Correct or nearly[b] correct structure	50	45

[a] For example, error in position of double bonds in diene or of substituent methyl groups.
[b] For example, N,N-dimethyl n-butylamine instead of triethylamine, cycloheptane instead of methyl cyclohexane.

If it is not, then the next probable set of groups is attempted. If there are no more combinations, then it is assumed that the group identification was wrong, and this part is repeated. Common fragments, 14, 16, 18, 29, 41, 43 are looked for, and if they are present, the neighbor list is generated. At present no firm criteria for choosing an atom to which to attach this new fragment have been established, so it is attached to the first atom with a free valency. Each of the large peaks is examined to see if it resulted from the loss of a common fragment and each one found is attached, more or less arbitrarily at present, to the structure. If the fragment is not due to C, O, N, tert-butyl, loss of water, or CO_2, then an informative statement is printed, and it is assumed to be straight chain. The molecular formula and mass are checked against the generated molecular formula, and if an error occurs, the process begins again with the next probable group identification or molecular formula, see Figures 4 and 5.

It would be desirable to check the structure created against the mass spectrum, and routines capable of doing this are being tested, but this is not done in the present case.

Structure Drawing Routine. The nearest neighbor table is copied and the copy reduced to a polyvalent atom skeleton. Sidechains and atom substitution points are removed and listed in a structure component matrix. Rings and sections of cyclic structures are then listed in the matrix.

These components are then arranged on a previously cleared two-dimensional array and the sidechains, bonds, and hydrogen atoms added. The array is converted to character form and printed.

Results. In Figure 6 are given several examples of the output from the program. To test the operation of this program and its generality, and to compare its performance with human chemists, the low resolution mass spectra of 76 substances were processed by it. These spectra were also presented to 20 third- and fourth-year chemistry students who had completed a course of 16 lectures on organic mass spectrometry. The computer had to produce its answer for each unknown in six seconds, the humans were allowed as long as they chose. The mean time taken per substance by the humans was of the order of 10–15 minutes. The substances are listed in Table I. The performances are compared in Table II.

CONCLUSION

This work shows that it is possible to write a computer program which closely follows the human approach to deducing a chemical structure for a mass spectrum *ab initio*. In its present stage of development, the program is limited to very simple structures, usually containing not more than one structural group. Nevertheless its capability is of the same order as that of an undergraduate student. Its success is encouraging, all the more so since its approach is essentially an empirical one.

RECEIVED for review October 3, 1968. Resubmitted November 23, 1970. Accepted July 2, 1971.

Applications of Artificial Intelligence for Chemical Inference. IV.[1] Saturated Amines Diagnosed by Their Low Resolution Mass Spectra and Nuclear Magnetic Resonance Spectra[2]

Armand Buchs,[3] A. M. Duffield, Gustav Schroll,[4] Carl Djerassi,*
Allan B. Delfino, B. G. Buchanan, G. L. Sutherland, E. A. Feigenbaum,
and J. Lederberg

Contribution from the Departments of Chemistry, Computer Science, and Genetics, Stanford University, Stanford, California 94305. Received March 11, 1970

Previous publications[1,5] have described the results of computer interpretation of the low resolution mass spectra of aliphatic ketones and ethers. In the case of ethers a program was added to utilize nmr data (if available). The heart of the computer program (called Heuristic DENDRAL) was the DENDRAL algorithm which constructs complete and irredundant lists of aliphatic molecules or radicals, in a linear notation, corresponding to any desired empirical formula. Our general approach to the computer interpretation of mass spectra begins with the domain of all possible structures which might *a priori* fit the experimental data. In order to expand the challenge to more complex situations we decided to approach the general solution of the mass spectra of saturated amines, since for any given number of carbon atoms, the number of possible saturated amines is considerably larger than for aliphatic ketones or ethers.[6] It should be emphasized that our purpose has been to demonstrate the feasibility of the Heuristic DENDRAL approach solely to those classes of compounds that offer new problems rather than to one functional group after another.

The basic approach to the problem of interpreting low resolution mass spectra, with the aid of nmr data if desired, is described in our earlier publication[1] dealing with saturated ethers and can be summarized in the following paragraph.

Heuristic DENDRAL is divided into three main subprograms called PRELIMINARY INFERENCE MAKER, STRUCTURE GENERATOR, and PREDICTOR. The first part finds which particular structural features are consistent with the mass spectral data and the elementary composition of the compound studied. Its output is then sent to the STRUCTURE GENERATOR which builds an irredundant and complete list of structures compatible with the information supplied by the PRELIMINARY INFERENCE MAKER and the constraints imposed by BADLIST.[1,5] Each generated structure is then given as input to the PREDICTOR which predicts significant peaks of its mass spectrum. The program either rejects the candidate or accepts it depending upon the fit of the predicted spectrum with the experimental one. Finally the accepted candidates are ranked from the most to the least plausible.[7]

The PRELIMINARY INFERENCE MAKER has now been improved by incorporating much more mass spectrometric theory about fragmentation mechanisms, and by using nmr data at an early stage. This paper will now describe how this program infers plausible substructures from mass spectra and nmr data of saturated amines. As will be shown in this paper, the efficiency achieved in the PRELIMINARY INFERENCE MAKER with this class of compounds leads to results which are in most cases, even for large molecules, precise enough such that the two other phases of Heuristic DENDRAL (STRUCTURE GENERATOR and PREDICTOR) need not be used. This represents a somewhat different application of Heuristic DENDRAL than the one which was used for

---

* To whom correspondence should be addressed.
(1) Part III: G. Schroll, A. M. Duffield, C. Djerassi, B. G. Buchanan, G. L. Sutherland, E. A. Feigenbaum, and J. Lederberg, *J. Amer. Chem. Soc.*, **91**, 7440 (1969).
(2) Financial assistance from the Advanced Research Projects Agency (Contract SD-183), the National Aeronautics and Space Administration (Grant NGR-05-020-004), and the National Institutes of Health (Grants GM 11309 and AM 04257) is gratefully acknowledged.
(3) On leave of absence from the University of Geneva.
(4) Recipient of a Fulbright travel award.
(5) A. M. Duffield, A. V. Robertson, C. Djerassi, B. G. Buchanan, G. L. Sutherland, E. A. Feigenbaum, and J. Lederberg, *J. Amer. Chem. Soc.*, **91**, 2977 (1969).
(6) J. Lederberg, G. L. Sutherland, B. G. Buchanan, E. A. Feigenbaum, A. V. Robertson, A. M. Duffield, and C. Djerassi, *ibid.*, **91**, 2973 (1969).

(7) An nmr PREDICTOR is also available to the user at the very end of the process. It takes as input the structures ranked by the mass spectrum PREDICTOR and the experimental nmr spectrum, provided it was completely interpreted. It also either rejects or accepts candidates in a ranked order depending on how well the predicted nmr spectrum fits the experimental data.

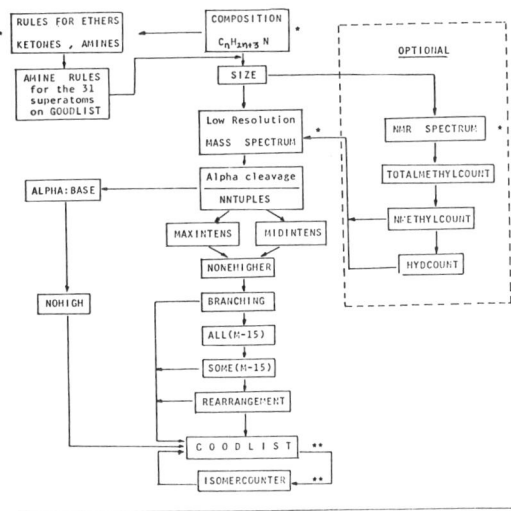

Figure 1. Sequence of the decision processes during inference phase.

ethers[1] and ketones.[5] *It should be emphasized that the approach employed previously with ketones*[5] (no complete list of superatoms and no nmr data) *or ethers*[1] (no complete list of superatoms and nmr data introduced at the end rather than the beginning of the pruning process) *would not have been sufficient to handle the problem posed by saturated amines.*

The decision processes invoked by Heuristic DENDRAL in the interpretation of amine mass spectra (supplemented by, but not dependent upon, the availability of nmr data) are schematically represented in Figure 1. If the composition of the unknown agrees with $C_nH_{2n+3}N$, the program extracts (decision "AMINE RULES," Figure 1) from memory the rules pertinent to amine mass (and nmr) spectra.

In order to approach the solution of an unknown low resolution mass spectrum of any saturated amine, it was necessary to define a complete set of possible amine subgraphs (*i.e.*, superatoms)[8] which could be inferred from the data. This is conveniently accomplished by using a combination of the four symbols T, S, P, and M for the superatom names. In this convention the number of symbols in a name refers to the number of α-carbon atoms bound to nitrogen, from one for primary amines to three for tertiary amines. The symbols themselves give the number of free valences on each α-carbon atom: P for one, S for two, and T for three free valences (see **1**, **2**, and **3**). The canonical order of the symbols is T > S > P > M. With this imposed order, PMM is the triplet chosen to represent the subgraph **4** instead of the equivalent but non-canonical names MMP and MPM. Using this nomenclature 31 superatoms[9] can be constructed from all possible combinations of the four symbols T, S, P, and M. Some additional examples of the use of this shorthand structure representation are listed below.

H_2N-CH_2- H_2N-CH H_2N-C-
1, P   **2**, S   **3**, T

$(CH_3)_2-N-CH_2-$ $-CH_2-NH-CH_2-$ CH_3-N-C-
                                      $-CH$
4, PMM **5**, PP **6**, TSM

$-CH-N-CH-$ $-CH_2-N-CH_2-$ $-CH_2-NH-CH-$
CH_2- CH_2-
7, SSP   **8**, PPP   **9**, SP

This notational scheme of representing the partial structure of saturated amines offers two major advantages. First, it is completely exhaustive. Thus, any saturated amine contains a subgraph which must belong to *one and only one* of the 31 superatoms. The second advantage is the ease of translation from a partial chemical structure and *vice versa*. The name of a superatom contains all the information needed for writing rules and conditions which will have to be satisfied by the data if the superatom is going to be an acceptable candidate (weight, free valences, rearrangement possibilities, etc).

In applying the proper processes to validate a particular superatom, the PRELIMINARY INFERENCE MAKER program is controlled by a table. To change the action of the program one need only change a table of superatom names and associated spectral features of molecules containing these superatoms. This way of driving the action of the program enables the user to change his mind about the properties he would expect to find in the mass spectra of compounds of any superatom class. Thus, before testing each superatom the program inquires in the property table about which processes should be used.

All 31 superatoms are initially found on a list, called GOODLIST, and each superatom is then tested for consistency with the experimental data, and either kept on GOODLIST or removed from it according to the results of the various tests. The first test is depicted in Figure 1 as "SIZE." The program compares the carbon content of the elementary composition with the number of carbon atoms required to construct the smallest molecule from the superatom by addition of only methyl groups to the free valences. If a superatom requires more carbon than is available to build the smallest molecule, that superatom is discarded from further consideration at this very early stage.

In our previous publication[1] concerning the ability of Heuristic DENDRAL to interpret low resolution mass spectra of saturated ethers, fully interpreted nmr data, if available, were used at the very end of the pruning process. This was done to test the validity of each candidate accepted by the mass spectrometric part of the program. With saturated amines, however, it was found desirable to introduce nmr data (if available) at the same time as the mass spectrum, *viz.*, in the PRE-

---

(8) As described in previous publications[1,6] a superatom is defined as a structural subunit having at least one free valence. In the present context only carbon atoms can be attached to the free valence(s).

(9) In addition to these 31 superatoms there exist three more molecules, M, MM, and MMM, which translate to methylamine, dimethylamine, and trimethylamine, respectively.

LIMINARY INFERENCE MAKER program. Thus, as soon as the program has normalized the amplitudes of the peaks in the mass spectrum and removed any improbable mass points (*e.g.*, M − 4 through M − 14) it accepts the nmr data. It should be stressed that the program does not require nmr input, such that if these data are unavailable, the program bypasses the nmr subroutine and proceeds directly to examine the mass spectrum. Neither does the program require completely interpreted nmr data; it is able to accept partial information from an nmr spectrum.

The nmr spectrum can be supplied to the program in different ways; if it is easily interpreted (no doubts about the multiplicity of the different carbon-bound methyl signals and an integration curve available) the nmr spectrum is given as a list of sublists in which each sublist is composed of three elements: the chemical shift (δ),[10] the number of protons responsible for the signal (n), and one of the following symbols S, D, T, Q, or M to denote the multiplicity of the signal (singlet, doublet, triplet, quartet, or multiplet, respectively). The supplied nmr data then have the form

$$((\delta_1\ n_1\ mult_1) \cdots (\delta_i\ n_i\ mult_i) \cdots (\delta_n\ n_n\ mult_n))$$

If the multiplicity of the methyl signals in the nmr spectrum is not readily apparent, as for example in the spectrum of *N*-methyl-*N*-propyl-*n*-hexylamine (**10**), the first sublist will contain the total height of the integration curve *vs.* the height at $\delta = 1.1$ ppm (Figure 2, run 1). The program will know that it has to calculate the number of carbon-bound methyl radicals by using the integral information and the number of hydrogen atoms in the compound (found by the program from the given empirical composition).

$$\underset{\underset{\textbf{10}}{}}{\text{CH}_3\text{CH}_2\text{CH}_2-\text{N}-\text{CH}_2\text{CH}_2\text{CH}_2\text{CH}_2\text{CH}_2\text{CH}_3}\atop{|\atop \text{CH}_3}$$

Sometimes, when no integration curve is available and the spectrum is not first order, some information can still be extracted easily from the nmr spectrum. Thus the presence of a sharp singlet in the *N*-methyl region will be used by the program and it will know that no significant information could be extracted from the *C*-methyl region of the nmr spectrum.

The program first examines the nmr spectrum to determine what kind of information was supplied and to decide how it should count the number of *C*-methyl groups. Thus the program looks first for signals at $\delta < 1.2$ ppm (either singlets, doublets, or triplets) and ensures that they originate from a number of protons exactly divisable by three. Otherwise the calculation is performed by using the two values from the integration curve and the total number of hydrogens in the compound.[11] Should neither of these two quantities be supplied, the program answers "NO INFORMATION" and no negative decision is made concerning the minimum number of methyl radicals needed to keep on GOODLIST the superatom under test.

---

(10) The chemical shifts are standardized against $\delta = 0$ ppm for tetramethylsilane.
(11) If A = total height, B = height in *C*-methyl region, and N = number of hydrogens in the elementary composition, the integer value of the relation $(((B/A) \times N)/3)$ is returned.

Figure 2

```
C10H23N   *PPM*    N-methyl-n-propyl-n-hexylamine

ADJUSTED SPECTRUM = ((41 . 25)(42 . 28)(43 . 38)(44 . 63)(55 . 5)
(56 . 3)(57 . 9)(58 . 34)(70 . 4)(72 . 3)(84 . 3)(86 . 100)
(87 . 4)(128 . 28)(129 . 1)(157 . 5))

Run 1
WAS A NMR SPECTRUM AVAILABLE                              ? YES
COULD ALL SIGNALS BE INTERPRETED                          ? NO
NMR SPECTRUM                           = ((201 54)(2.3  4  T)(2.2  3  S))
NUMBER OF CARBON-BOUND METHYLS                            = 2
NUMBER OF NITROGEN-BOUND METHYLS                          = 1
TOTAL NUMBER OF METHYLS                                   = 3
MINIMUM NUMBER OF ALPHACARBON BOUND HYDROGENS = NO VALID INFORMATION

GOODLIST = PPM

MASSES OF ATTACHED RADICALS :

PPM     (71, 29)                                 1 ISOMER.

TOTAL NUMBER OF ISOMERS = 1


Run 2
WAS A NMR SPECTRUM AVAILABLE                              ? YES
COULD ALL SIGNALS BE INTERPRETED                          ? NO
NMR SPECTRUM                           = ((2.2  ?  S))
NUMBER OF CARBON-BOUND METHYLS                            = NO INFORMATION
NUMBER OF NITROGEN-BOUND METHYLS                          = 1 OR 2
TOTAL NUMBER OF METHYLS                                   = NO INFORMATION
MINIMUM NUMBER OF ALPHACARBON BOUND HYDROGENS = NO VALID INFORMATION

GOODLIST = SSM   TPM   PPM   SMM   TM

MASSES OF ATTACHED RADICALS :

SSM     ((15, 29)(29, 29))                       1 ISOMER.
TPM     ((15, 15, 29)(29))                       1 ISOMER.
PPM     (71, 29)                                 8 ISOMERS.
SMM     (71, 29)                                 8 ISOMERS.
TM      (71, 29, 15)                             8 ISOMERS.

TOTAL NUMBER OF ISOMERS = 26
```

Figure 2. PRELIMINARY INFERENCE MAKER outputs with *N*-methyl-*n*-propyl-*n*-hexylamine (**10**) as an unknown. Use of partially interpreted nmr spectrum.

The nmr program then counts the number of nitrogen-bound methyl groups by searching for a singlet signal at $\delta = 2$–4 ppm. If the number of hydrogens responsible for the signal is exactly divisable by three, the program considers it as a signal due to nitrogen-bound methyl group(s) only after having performed some validation tests in order to ensure, for example, that the signal does not originate from 3 (or 6) hydrogens on the α-carbon atoms, all these carbon atoms being tertiary. Even if no integration curve is available, a singlet in the *N*-methyl region still means that at least one such methyl is present. In this case a question mark printed in place of the number of hydrogens causes the program to answer that there are either one or two nitrogen-bound methyl groups. The decision of keeping or rejecting a superatom will then be made on the basis of these two possibilities (see Figure 2). With our example (*N*-methyl-*n*-propyl-*n*-hexylamine) the nmr spectrum was supplied with or without an integral curve (see runs 1 and 2 of Figure 2). In the case of run 1 the program finds the correct number of *N*-methyl groups. However, when no integral curve is supplied (run 2), the presence of a sharp singlet at $\delta = 2.2$ ppm is a clear indication of the presence of one or two *N*-methyl groups. The total number of methyl groups, provided both the number of *C*-methyls and *N*-methyls had defined values, is remembered under the name TOTALMETHYLCOUNT,[12]

Figure 3

Test*	Superatoms eliminated	Why?
SIZE	TTT TTS	There are only 10 carbon atoms in the compound.**
TOTALMETHYLCOUNT	TTP TTM TSS TSP TSM TPP TPM TMM SSS SSP SSM SPP SPM SMM TT TS SS TP TM	They require more methyl groups than the three which are found by the NMR subroutine.
NMETHYLCOUNT	PPP SP PP T S P PMM	A N-methyl signal is in the NMR spectrum. Only one of the methyl groups is an N-methyl.
ALPHA:BASE	PM	m/e 44 is not the base peak.
MAXINTENS	SM	No valid ntuple can be found by using the allowed α-fission peaks for SM (72, 86, 100, 114, 128 and 142) and 157 as molecular weight. m/e 100 and 114 are not in the mass spectrum, and m/e 142 cannot be used due to insufficient number of carbon atoms. The only ntuple built is (72, 128) but the sum of the intensities of these two ions is less than 70%.***

* The names of the tests refer to Figure 1.
** N-methyl-n-propyl-n-hexylamine.
*** Rearrangement for molecular ions having SM as subunit is not a favored process.

Figure 3. Effect of the different tests upon the pruning of GOODLIST with N-methyl-n-propyl-n-hexylamine (see run 1 in Figure 2 for supplied data).

and the number of N-methyl groups under the name NMETHYLCOUNT.

Finally, the program counts the number of α-carbon bound protons, provided no nitrogen methyl signals were found. It searches the nmr spectrum for signals at $\delta > 2.2$ ppm having any multiplicity, and stores the value under the name HYDCOUNT. In our example (Figure 2), as an N-methyl group is already identified, the program does not search for signals originating from α-carbon hydrogens.

Having exhausted its survey of the nmr spectrum, the inference program commences its examination of the 31 superatoms on GOODLIST.

As the first necessary condition, each superatom has a number $(m_1 - 1)$ where m_1 represents the minimum number of methyl groups (1 for P, 2 for S, 4 for SS, up to 9 for TTT) which must be validated by the nmr subroutine for the superatom under test to remain on GOODLIST. The superatom passes this test (decision "TOTALMETHYLCOUNT," Figure 1) only if the number of methyl groups which the program finds exceeds $m_1 - 1$; otherwise it is deleted from GOODLIST at this very early stage, and will not be tested further. As shown in Figure 3 for N-methyl-n-propyl-n-hexylamine, 19 superatoms out of the remaining 29 (two were already eliminated by the test referred to as "SIZE" in Figure 1) are removed from GOODLIST by this test.

A second necessary condition, also related to the structure of each of the 31 superatoms, requires that the number m_2 of N-methyl groups found from the nmr spectrum must be the same as the number of M's in the superatom name. Any superatom requiring a number

(12) The program stores parameter values under identifiable names such as this for later use.

of nitrogen methyl groups different from that found by the program is deleted from GOODLIST. If the value of NMETHYLCOUNT is 1, for example, only superatoms with one M in their name are tested further. Should the value of NMETHYLCOUNT be partially undefined (1 or 2), only superatoms with 1 or 2 M's in the name would pass the test, and should it be totally undefined (NO INFORMATION), all superatoms would pass this test (decision "NMETHYLCOUNT," Figure 1).

If a superatom passes both these tests and has no M in its name a final nmr test is applied. For each superatom without M, a value m_3 related to its structure (2 for P, 3 for SP, etc.) represents the maximum number of α-carbon hydrogens which can be found in the spectrum in order for the superatom to be accepted for further consideration. If the value of HYDCOUNT is smaller than that of m_3, the superatom passes the test; otherwise it is removed from GOODLIST. It should be pointed out that each superatom requires an exact number of α-carbon hydrogens, but this number rapidly becomes small compared to the total number of hydrogens in the empirical formula when the size of the molecule increases. Therefore in order to avoid errors from marginal integration curves it was found safer to demand that the number found be just smaller than, or equal to, m_3.

If any of the values of TOTALMETHYLCOUNT, NMETHYLCOUNT, or HYDCOUNT are not defined, the corresponding test is passed successfully by default.

Following the nmr search the program then confronts the mass spectrum. The first condition programmed into the mass spectrometry section of the inference program relates to the well-documented[13,14] propensity of aliphatic amines to undergo α cleavage (see $11 \rightarrow a + b$).

For superatoms with only one free valence the first condition (decision "ALPHA:BASE," Figure 1) is that the only α-fission peak must be the base peak (respectively, 30 ($CH_2=N^+H_2$), 44 ($CH_2=N^+H—CH_3$), and 58 ($CH_2=N^+—(CH_3)_2$) for P, PM, and PMM), and the second condition (decision "NOHIGH," Figure 1) is that there should be no peaks with an intensity greater than 10% above the mass of one-half the molecular weight, provided m/e 30, 44, or 58 is not already above this limit, a fact which occurs for small molecules. This condition takes into account the possibility of β to ϵ cleavage and the rather improbable fact of finding intense peaks from cleavage occurring further away from the nitrogen atom. Since in the mass spectrum of N-methyl-n-propyl-n-hexylamine (10)

(13) H. Budzikiewicz, C. Djerassi, and D. H. Williams, "Mass Spectrometry of Organic Compounds," Holden-Day, San Francisco, Calif., 1967, pp 297–303.
(14) R. S. Gohlke and F. W. McLafferty, Anal. Chem., 34, 1281 (1962).

m/e 44 is not the base peak (see spectrum tabulated in Figure 2), the superatom PM does not pass this test (see Figure 3).

For any other superatom to be accepted there is a definite number of α-cleavage fragments which must be located in the experimental data (decision "NTUPLES," Figure 1). The lower mass limit of the α-cleavage peaks searched for is equal to the mass of the superatom in question (referred to as "overweight") added to $((n - 1) \times 15)$ where n is the number of free valences in that superatom. For example the superatom PPP (8) has $n = 3$ and a mass of 56 amu, so the search will begin at mass $((2 \times 15) + 56)$, i.e., mass 86. The program, to validate the presence of a PPP subgraph, must then search the mass spectrum in quest of sets of three peaks, the sum of the peaks in each set being equal to $[((n - 1) \times \text{mol wt}) + 56]$. If at least one of these sets (called ntuples)[15] is found, the α-cleavage condition is partially satisfied for that superatom. In the case of N-methyl-n-propyl-n-hexylamine (10), only SM and PPM remain as valid superatom candidates at this stage of the pruning process. Figure 3 shows why SM is eliminated by this test. For the superatom PPM the program finds (86, 128) as a valid ntuple, which translates to (M − C_5H_{11}, M − C_2H_5).

α cleavage is a major fragmentation mechanism only in those amines which cannot undergo a favored rearrangement process. This is true of amines having as subunit certain superatoms which bear this property in their α-carbon(s) and in their name. For these superatoms an ntuple will be kept only if the sum of the intensities of the α-fission peaks exceeds an empirically determined value of 70%, and only the ntuple with the highest sum of intensities[16] will be accepted (decision "MAXINTENS," Figure 1). With our example 10 the ntuple (86, 128) which is found for PPM has a sum of intensities in excess of 70% (see spectrum tabulated in Figure 2) and therefore passes the test.

With secondary and tertiary α-mono- or α-disubstituted amines (represented by superatoms with a name containing more than two symbols disregarding the M's, and with at least one S or T) α cleavage still remains a favored process but the major ion arises from a well-known rearrangement mechanism (see 11 → a + b → c + d). The ntuples are therefore tested against a less stringent condition for the total intensity of the α-fission peaks (decision "MIDINTENS," Figure 1). For these superatoms, all the ntuples with an intensity sum greater than 30% are kept for further test. If all ntuples for a superatom have been eliminated it is removed from GOODLIST.

Surviving ntuples are then submitted to a further test (decision "NONEHIGHER," Figure 1). The program requires that for each ntuple there be no intense peak at a mass higher than that of the ion with greatest m/e present in the ntuple. This rule protects the program from incorrectly identifying the highest mass α-cleavage ion, since this latter ion would be expected to have the greatest intensity of any peak found between itself and the $(M - 1)^+$ ion.[17] In the mass spectrum of 10 (tabulated in Figure 2) used to illustrate stepwise the pruning process, no peak with m/e greater than the mass of the α-fission ion $(M - C_2H_5)^+$ has an intensity above 10%. The correct superatom PPM with its associated ntuple (86, 128) is therefore not removed from GOODLIST by this test.

It is well known in mass spectrometry that at 70 eV[18] the larger alkyl group is preferentially expelled in an α-cleavage fragmentation. The program (decision "BRANCHING," Figure 1) exploits this concept to check whether it has correctly identified the ions resulting from α cleavage. In any ntuple the lowest mass value should have an intensity in excess of the next heavier α-fission fragment provided the difference in number of carbon atoms between the molecular ion and the heavier α-fission ion is less than three. Should the difference in mass between the molecular ion and the high-mass α-cleavage ion be larger than the mass of a C_2H_5 group then the program requires that the intensity of the low mass ion be in excess of $(0.5 + 0.1 \times \Delta C)$ of the intensity of the higher α-fission ion.[19] This process takes into consideration (by reducing the stringency of the condition and by even allowing the intensity of the high mass ion to be greater than that of the low mass ion) the known[13] ease of elimination in α-fission of a tertiary radical over a secondary and a secondary over a primary. In an ntuple, each α-cleavage ion is successively compared to the ion adjacent to it. The intensity of any α-cleavage ion is always normalized against the number of its occurrences on probability grounds. In the case of N-methyl-n-propyl-n-hexylamine (10), the difference between m/e 128 which is the high mass α-cleavage fragment in the ntuple (86, 128) and the molecular weight (m/e 157) represents a C_2H_5· radical. Since branching is impossible in an ethyl radical, the program requires that the intensity of the ion at mass 86 should be greater than that of the ion at mass 128 (see spectrum tabulated in Figure 2). However, if one considers a general molecule 12, of molecular weight 185, the correct

$$C_4H_9—CH_2—NH—CH_2—C_6H_{13}$$
12

ntuple (100, 128) for the superatom PP (5), i.e., (M − C_6H_{13}, M − C_4H_9), would be accepted even if the intensity of ion m/e 100 is less than that of ion m/e 128. The C_4H_9· group could be tertiary, secondary, or primary. In this case the program would allow the

---

(15) The term "ntuple" is used to refer to any set of possible α-cleavage peaks for a superatom in the context of a particular mass spectrum. For example, for the superatom SSP (7) and an amine having a molecular weight of 171 ($C_{11}H_{25}N$), the following sets would be built as *a priori* valid ntuples: $n = 5$; overweight = 54; sum of peaks = $(4 \times 171) + 54 = 738$; m/e of lowest α-fission peak possible = $(4 \times 15) + 54 = 114$; possible ntuples: (114, 156, 156, 156, 156), (128, 142, 156, 156, 156), and (142, 142, 142, 156, 156) (m/e 114 corresponds to M − C_4H_9, m/e 128 to M − C_3H_7, m/e 142 to M − C_2H_5, and m/e 156 to M − CH_3).

(16) All intensity values refer to relative abundances with intensity of the base peak = 100%. All threshold values were chosen on theoretical trends and corrected so they never eliminate the correct superatom but still give the maximum pruning effect.

(17) The $(M - 1)^+$ ion is not considered as an α-fission peak; the $(M - 15)^+$ ion, even if not present in the spectrum, is allowed to be used for building ntuples, provided its mass in conjunction with the masses of the other peaks in the ntuple satisfies the equation used for this purpose.

(18) This is not the case, however, at low ionizing voltage. See C. A. Brown, A. M. Duffield, and C. Djerassi, *Org. Mass. Spectrosc.*, **2**, 625 (1969).

(19) If this difference is C_3 or greater the possibility exists that the α-cleavage ion of higher mass can result from expulsion of a secondary or even tertiary radical. If this is possible, the program must weaken this condition such that a lower mass ion can be less intense than the higher mass ion. However, if the difference of mass between the highest mass α-cleavage ion found and the molecular weight is less than the mass of a C_3 unit, no possibility of branching exists. The expression $(0.5 + 0.1 \times \Delta C)$ was arrived at empirically.

```
C8H19N  *SP*      Ethyl-1,3-dimethylbutylamine

ADJUSTED SPECTRUM = ((15 . 4)(16 . 1)(17 . 1)(18 . 5)(27 . 1G)
(28 . 6)(29 . 9)(30 . 8)(31 . 1)(41 . 9)(42 . 7)(43 . 9)(44 . 36)
(45 . 3)(46 . 1)(55 . 1)(56 . 5)(57 . 1)(58 . 6)(59 . 1)(69 . 1)
(70 . 1)(71 . 1)(72 . 100)(73 . 1)(83 . 1)(84 . 1)(85 . 1)
(86 . 1)(87 . 1)(98 . 1)(112 . 1)(114 . 6)(128 . 1)(129 . 1))

WAS A NMR SPECTRUM AVAILABLE                              ? NO

GOODLIST =  PPM     SMM     TM      SP

MASSES OF ATTACHED RADICALS :

PPM     (57, 15)                                    4 ISOMERS.
SMM     (57, 15)                                    4 ISOMERS.
SP      ((15)(15, 57))                              4 ISOMERS.
TM      (57, 15, 15)                                4 ISOMERS.

TOTAL NUMBER OF ISOMERS = 16
```

Figure 4. PRELIMINARY INFERENCE MAKER output with ethyl-1,4-dimethylbutylamine (**14**) as an unknown.

intensity of the ion at mass 100 to be less than that of the ion at mass 128. It would require that the abundance of m/e 100 must be at least equal to (0.5 + 0.1 × 2), i.e., 0.7 times the abundance of m/e 128.

If any superatom still exists as a viable candidate to explain the experimental data and should its ntuple consist entirely of $(M - 15)^+$ α-fission ions, then it is subjected to another decision process (decision "ALL $(M - 15)$," Figure 1). The intensity of the $(M - 15)^+$ ion should in this condition be equal to or greater than $(1 - 1/n) \times 50$, where n is the number of times the mass of the $(M - 15)^+$ ion is present in the ntuple or number of possible α cleavages leading to an $(M - 15)^+$ ion. For example, in the case of diisopropylamine (**13**) the correct superatom SS would only be

$(CH_3)_2$—CH—NH—CH—$(CH_3)_2$
13

accepted if the $(M - 15)^+$ α-cleavage ion has an intensity greater than $(1 - 1/4) \times 50$, i.e., greater than 38% relative abundance. If besides some $(M - 15)^+$ α fissions an ntuple contains at least one ion of lower mass (loss of a group having more than one carbon atom) the condition about the intensity of the $(M - 15)^+$ ion is much less demanding (decision "SOME $(M - 15)$," Figure 1). This intensity needs only to be equal to or greater than the number of $(M - 15)$ α fissions minus one. This allows the $(M - 15)^+$ ion to be absent from the spectrum if only one α cleavage can lead to it.

After this test, superatoms with no rearrangement possibility or for which rearrangement can give rise to ions of only moderate intensity, and which still have at least one surviving ntuple, are accepted as plausible candidates. The masses of the alkyl fragments which should be attached to the free valences are then calculated by simply subtracting the mass of each α-fission peak in the ntuple from the molecular weight. These masses we refer to as "partition" because they are the masses of the alkyl groups which must be attached to the free valences of the superatom. The superatom with its list of partition is then examined by a subroutine which calculates the number of isomers compatible with both the structure of the superatom and the masses of the alkyl groups which have to be attached to this superatom (decision "ISOMERCOUNTER," Figure 1). With our example **10**, the correctly identified superatom PPM with its ntuple (86, 128) is definitely accepted at this stage of the pruning process. The program subtracts then both masses 86 and 128 from the molecular weight 157, translating the ntuple (86, 128) to the list of partition (29, 71). When no integral curve is supplied for the nmr spectrum, eight isomers can be constructed from the superatom PPM with (29, 71) as list of partition (see run 2 in Figure 2). However, when the nmr spectrum is supplemented by an integral curve, only one isomer is compatible with the information given by the PRELIMINARY INFERENCE MAKER program (see run 1 in Figure 2).

Superatoms for which rearrangement is a major process are further tested for the presence of at least one intense ion arising from rearrangement, with m/e in accordance with the structure of the superatom and its set of partition(s). The intensity of the rearrangement peak should be greater than 30% if the superatom is to be kept on GOODLIST. The subroutine which handles the rearrangement mechanism is programmed so that it takes into account all rearrangement possibilities for the superatom and the partitions under test. This allows the program to assign the correct alkyl fragments to each different α carbon. For example, in the case of ethyl-1,3-dimethylbutylamine (**14**), a molecule with an SP subgraph (**8**), the correctly inferred superatom will be assigned the following list of partition: "(15, 15, 57)." The alkyl groups which have to be attached to the three free valences of the SP superatom are two methyls and a C_4H_9 radical. This can be done in two ways (structures **15** and **16**).

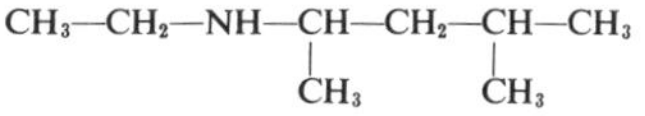

CH_3—CH_2—NH—CH—CH_2—CH—CH_3
   | |
  CH_3 CH_3
14

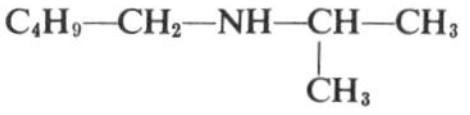

C_4H_9—CH_2—NH—CH—CH_3
                   |
                  CH_3
15

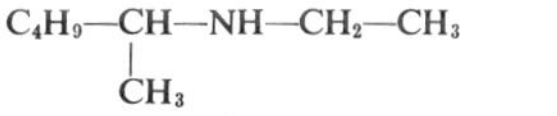

C_4H_9—CH—NH—CH_2—CH_3
       |
      CH_3
16

With both structures **15** and **16** rearrangement ions are expected at m/e 30 (CH_2=N^+H_2) and 44 (CH_3—N^+H=CH_2). But, if as generally accepted, the most favored rearrangement is the one where the C–N bond is broken and a hydrogen transferred to the nitrogen atom after expulsion by α cleavage of the heavier substituent, one can postulate that for the second structure (**16**), which is the correct one, the rearrangement leading to m/e 44 should be favored. With the first structure (**14**) the ion of m/e 30 would be expected to give a more intense signal than the ion of m/e 44. The intensities of m/e 30 and 44 in the mass spectrum of **14** are, respectively, 8 and 36%. So the program chooses the second structure and, instead of simply giving "SP (15, 15, 57)" as superatom and partition, it assigns the correct distribution of the alkyl fragments between the two different α-carbon atoms by answering "SP ((15, 57) (15))." Figure 4 shows the output of the PRELIMINARY INFERENCE MAKER program with ethyl-1,3-

Table I. Curtailment of the Search Space by Heuristic DENDRAL

	No. of amine isomers	Mass spectra alone		Mass spectra + nmr spectra	
Amine		No. of superatoms on GOODLIST	No. of possible isomers	No. of superatoms on GOODLIST	No. of possible isomers
3-Hexyl	39	1	2	1	1
1,3-Dimethylbutyl	39	2	8	1	2
2,2-Dimethyl-3-butyl	39	2	8	1	1
N-Methylethyl-n-propyl	39	8	15	2	2
N,N-Dimethyl-2-butyl	39	6	6	1	1
2-Heptyl	89	2	16	1	1
n-Propyl-n-butyl	89	5	10	1	1
1,3-Dimethylpentyl	89	2	16	1	4
1,5-Dimethylhexyl	211	2	34	1	9
N,N-Dimethyl-3-hexyl	211	3	6	1	1
N-Methyl-n-butylisopropyl	211	10	31	1	1
Diisopropylethyl	211	6	18	1	1
N-Methyl-n-propyl-n-butyl	211	5	24	1	1
n-Propyl-n-hexyl	507	7	42	1	1
3,3,5-Trimethylhexyl	507	1	89	a	
N-Methylisopropyl-n-amyl	507	5	20		
N-Methyl-n-propyl-n-hexyl	1,238	10	46	1	1
N,N-Dimethyl-3-octyl	1,238	8	36	1	1
n-Butyl-n-hexyl	1,238	3	48	1	1
N,N-Dimethyl-2-ethylhexyl	1,238	4	156		
n-Amyl-n-hexyl	3,057	9	112		
Tri-n-butyl	7,639	2	8		
Di-n-heptyl	48,865	6	510		
Triisoamyl	124,906	2	40	1	9
N-Methyl-8-hexadecyl	321,198	1	3,471		
n-Octyl-n-nonyl	830,219	2	6,942	1	1
N,N-Dimethyl-8-hexadecyl	12,156,010	4	14,418	1	1
Tri-n-hexyl	12,156,010	2	240	1	1
Tri-n-heptyl	38,649,142	2	1,938		

[a] Blanks in columns 5 and 6 indicate that no nmr spectrum was available.

dimethylbutylamine (14) as an unknown when only the mass spectrum is supplied.

The rearrangement process is the last test (decision "REARRANGEMENT," Figure 1) for those superatoms for which rearrangement is a favored mechanism. Each accepted candidate along with its correctly assigned list of partition(s) is then sent to the isomer counter subroutine which calculates the number of compatible isomers by taking into account the assignment of the different alkyl groups to specific positions.

Two different outputs for N-methyl-n-propyl-n-hexylamine (10) are reported in Figure 2. In the first case (run 1) the supplied nmr spectrum contains enough information to allow the program to calculate the number of C-methyl groups; the user made no decision about the multiplicity of the C-methyl signals. As shown, only the correct superatom remains on GOODLIST. The search space is curtailed from 1238 possible isomers for $C_{10}H_{23}N$ (see Table I) to one structure.

Then (Figure 2, run 2) it was assumed that no integration curve was available. Clearly, the only straightforward information from the nmr spectrum is a sharp singlet at $\delta = 2.2$ ppm. The output reflects this fact by showing additional superatoms on GOODLIST, which could not be eliminated for their required number of methyl groups, like TPM, TM, and SSM, or on the basis of the number of N-methyl groups needed, like SMM. The information is nevertheless sufficient to eliminate all superatoms without M's in their name.

It is interesting to note that with the aid of nmr data GOODLIST is pruned mainly by the nmr tests (decisions "TOTALMETHYLCOUNT," "NMETHYLCOUNT," and "HYD-COUNT," Figure 1). Figure 3 shows that already after the second nmr test (decision "NMETHYLCOUNT," Figure 1) only the three superatoms PM, SM, and PMM remain on GOODLIST in the case of run 1 (Figure 2). *Inserting the nmr tests at the beginning of the process is therefore very efficient in saving time.*

Even for rather large molecules and without the aid of nmr data, the PRELIMINARY INFERENCE MAKER program is able to curtail the search space in quite an impressive way. As can be seen from the outputs reported in Figure 5, only the correct superatom is inferred from the mass spectrum of N-methyl-8-hexadecylamine. The search space is reduced by a factor of 92 (see Table I). This factor is even larger in the case of tri-n-heptylamine; from 38,649,142 a priori possible amine isomers for $C_{21}H_{45}N$, the isomer counter subroutine finds 1938 structures compatible with the output of the inference phase, i.e., a reduction factor of nearly 20,000.[20]

The program has been successfully tested with 93 amines;[21] for 37 of them nmr data were available. With this set of examples the program always selects the correct answer in the final output.

Some results are reported in Table I. In every case when an nmr spectrum is used only the correct superatom is found on GOODLIST and with the exception of five cases out of 37, only one structure is compatible with the output. The five exceptions all include the correct compound.

(20) For both the above mentioned compounds the problem would have been completely solved with the use of nmr data to supplement the mass spectrum.
(21) Sixty-six mass spectra were taken from McLafferty's excellent publication on amines.[14]

Figure 5

```
C16H35N    *S*    8-hexadecylamine.

ADJUSTED SPECTRUM = ((41 . 23)(42 . 7)(43 . 31)(44 . 7)(53 . 2)
(54 . 2)(55 . 14)(56 . 24)(57 . 10)(58 . 1)(67 . 2)(68 . 1)
(69 . 9)(70 . 4)(71 . 1)(81 . 1)(82 . 1)(83 . 3)(84 . 2)(85 . 1)
(97 . 2)(98 . 2)(101 . 1)(102 . 1)(126 . 1)(128 . 100)(129 . 3)
(130 . 1)(140 . 1)(141 . 94)(142 . 4)(143 . 1)(155 . 1)(157 . 1)
(173 . 1)(240 . 1)(241 . 1))

WAS A NMR SPECTRUM AVAILABLE                    ? NO

GOODLIST = S

MASSES OF ATTACHED RADICALS :

S     (113, 99)                         3471 ISOMERS.

TOTAL NUMBER OF ISOMERS = 3471


C21H45N    *PPP*    Tri-n-heptylamine.

ADJUSTED SPECTRUM = ((15 . 1)(16 . 1)(17 . 1)(18 . 2)(27 . 1)
(28 . 5)(29 . 3)(30 . 8)(31 . 1)(41 . 5)(42 . 2)(43 . 6)(44 . 9)
(45 . 1)(55 . 3)(56 . 2)(57 . 6)(58 . 4)(59 . 2)(60 . 1)(69 . 1)
(70 . 1)(71 . 1)(72 . 1)(73 . 1)(83 . 1)(84 . 2)(85 . 1)(86 . 1)
(87 . 1)(98 . 3)(99 . 1)(100 . 1)(101 . 1)(112 . 2)(113 . 1)
(114 . 1)(115 . 1)(126 . 1)(127 . 1)(128 . 7)(129 . 1)(140 . 2)
(141 . 1)(142 . 1)(143 . 1)(154 . 1)(155 . 1)(156 . 1)(163 . 1)
(169 . 1)(170 . 1)(183 . 1)(184 . 1)(185 . 1)(197 . 1)(198 . 1)
(212 . 1)(213 . 1)(225 . 1)(226 . 100)(227 . 3)(240 . 1)(253 . 1)
(254 . 1)(268 . 1)(269 . 1)(282 . 1)(283 . 1)(296 . 1)(309 . 1)
(310 . 2)(311 . 2))

WAS A NMR SPECTRUM AVAILABLE                    ? NO

GOODLIST = PPP    TMM

MASSES OF ATTACHED RADICALS :

PPP   (85, 85, 35)                      969 ISOMERS.
TMM   (85, 85, 85)                      969 ISOMERS.

TOTAL NUMBER OF ISOMERS = 1938
```

Figure 5. PRELIMINARY INFERENCE MAKER outputs for N-methyl-8-hexadecyl- and tri-n-heptylamines when only the mass spectra were supplied.

It can be concluded that with the aid of nmr data only the first subprogram of Heuristic DENDRAL needs to be used. In fact, the PREDICTOR program is unable at the present stage to choose the correct molecule among candidates containing the same superatom as a subunit along with the same weight for the radicals attached to the α-carbon(s). This is not a surprising fact; mass spectrometry just does not give much information about structural features located too far away from the charge center. When no nmr data are used the search space is still greatly reduced as can be seen from the results recorded in Table I.

Clearly, more information could easily be extracted from the nmr spectra by the PRELIMINARY INFERENCE MAKER program. It would be possible to use the multiplicity of the C-methyl signals and not use the nmr spectrum only as a methyl counter (and occasionally as an α-carbon hydrogen counter). This would nevertheless require that the STRUCTURE GENERATOR program be able to accept overlapping information, a fact it cannot handle for the time being.

The results which have so far been obtained are sufficiently promising so as to stimulate further research on more general and complex problems. It remains to be seen, however, whether the heuristics can be made sufficiently precise for other types of organic molecules such that the present degree of efficiency obtained with amines can be maintained.

Experimental Section

The computer program described here is part of the complete Heuristic DENDRAL program. It runs on the IBM 360/67 computer at the Stanford Computation Center using the LISP programming language. Without nmr data, the computer needs 4.26 min to interpret 93 mass spectra. When nmr data are used the process is about 30% faster. Mass spectra which had not been reported in the literature were recorded in our laboratory,[22] some with a Varian MAT CH-4 mass spectrometer, others with an AEI MS-9 mass spectrometer. The majority (26 out of 37) of the nmr spectra used was recorded in our laboratory by Dr. L. J. Durham of Stanford University.

(22) We thank Mr. R. G. Ross and Mr. R. T. Conover for recording the mass spectra.

34

Copyright © 1972 by the Swiss Chemical Society, Basel

Reprinted from *Helv. Chim. Acta* **55**:2017-2029 (1972)

Heuristic Programming as an Ion Generator in Mass Spectrometry
I. Generation of Primary Ions with Charge Localization

by **Allan B. Delfino** and **Armand Buchs**

Department of Physical Chemistry,
University of Geneva, 1211 Geneva 4, Switzerland.

(3 V 72)

Résumé. Un programme connu sous le nom de *ION GENERATOR* a été élaboré. Pour le moment ce programme est capable de créer, à partir de n'importe quelle molécule organique, les ions primaires résultant de la fragmentation de l'ion moléculaire et de proposer des mécanismes de fragmentation pour expliquer la formation des ions.

In this paper we present a heuristic program we have devised to simulate the formation of ions in the ion source of a mass spectrometer. Our program, the *ION GENERATOR*, acts in much the same way as the chemist who is trying to rationalize ion formations with the help of paper and pencil; it is based on the well known method of electron book-keeping. The program is expandable to include in the future a great proportion of the mechanistic steps delineated for example in the textbook by *Djerassi et al.* [1]. With such a program one can expect to generate, from any molecule, a number of important ions present in a mass spectrum. The advantages of using a computer to look for fragmentation mechanisms are manyfold, but we would like to name two which we consider to be of primary importance. Since the electron book-keeping method is tedious, the chemist usually stops when he has realized a plausible rationalization of the fragmentation. But the computer will realize an *exhaustive* and *irredundant* set of plausible rationalizations that are consonant with the structures of the formed ions and that are not chemically unsound. Hence, the chemist who utilises the program could then design, on the basis of the proposed fragmentation paths, labelling experiments to determine which of the sets present indeed the best rationalizations, or which fragmentation modes are occuring simultaneously.

The second reason for interest in this kind of research is the fact that the methods of electron book-keeping were handed down by examples, after presenting of a few rules. If one wants a computer to handle electron book-keeping, a much more rigorous presentation must be worked out. One is compelled to generalize, as much as can be done, each and every step of a fragmentation process. Besides the research described here, another area of research where heuristic search techniques are applied to mass spectrometry has recently been published by *Buchanan et al.* [2].

In order to generate molecular ions from a molecule and to fragment them to form all plausible primary ions, the *ION GENERATOR* program must know the motivations and methods of altering structures, which we call primitive operations. By applying these primitive operations in all plausible sequences to a molecule, or to an ion, until the program finds a new ionic species different in mass from the parent species, it builds sets of primitive operations which represent fragmentation mechanisms and which yield ions of various masses.

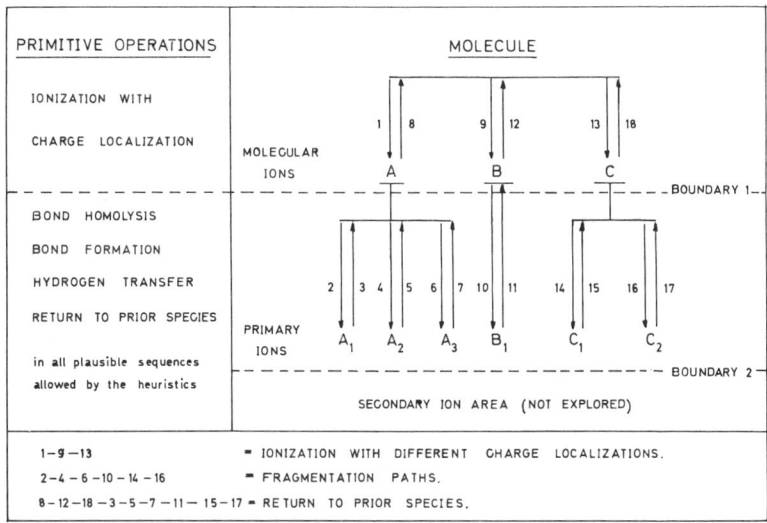

Figure 1. *Sequence of operations to ionize and fragment molecules.*

The primitive operations the program currently contains are the following:
- Ionization with charge localization,
- Bond homolysis β to a radical site,
- Bond formation between two adjacent radical sites,
- Transfer of a hydrogen atom to a radical site *via* cyclic transition states of various sizes.

The moves the program performs to get from a molecule to the primary ions, i.e. only to those ions which have the molecular ions as direct precursors, are illustrated in Fig. 1. As is shown in this Figure, for the time being, the program stops its work at boundary 2, after it has formed all plausible primary ions[1]). A particular mechanism will then be substantiated if the program finds in the actual mass spectrum a signal due to the ion generated through the mechanism.

The way the program applies the primitive operations will be illustrated with an example, the ionization and fragmentation of 5-diethylamino-pentan-2-one (**1**).

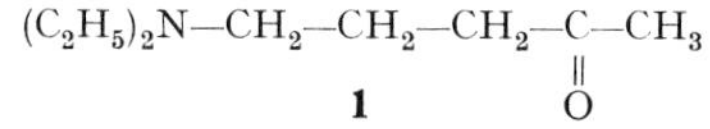

$$(C_2H_5)_2N-CH_2-CH_2-CH_2-C-CH_3$$
$$\underset{\mathbf{1}}{}\underset{O}{\overset{\|}{}}$$

Input of the data to the program. – The data submitted to the ION GENERATOR are a structure and its low resolution mass spectrum. The mass spectrum is given to the program as pairs of mass-intensity integer values. The first card is a title card which serves to identify the compound. The mass-intensity cards which follow the title card are punched with pairs of integers; four columns are reserved for the mass, followed by four columns for the corresponding intensity.

The input of the structure is explained by using the example 5-diethylamino-pentan-2-one (**1**). The atoms are first numbered in an *arbitrary* way. To make the input coding as short as possible, the longest chain of singly bonded atoms should be searched for and the atoms of that chain should be numbered in sequence. A convenient way to number the atoms of structure **1** is shown below (**2**).

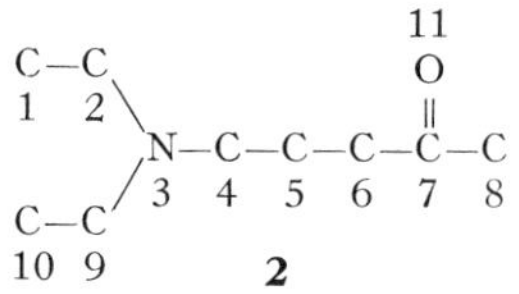

```
                           11
      C-C                   O
      1 2\                  ||
          N-C-C-C-C-C
         /3 4 5 6 7 8
      C-C
      10 9         2
```

All the atoms found in the longest chain, i.e. atom No. 1 through atom No. 8 are considered as a group of atoms. The first structure input card, which follows the last spectrum input card, is the *ATOM CARD*, which specifies what kind of atom each number represents. The chemical symbols of the atoms are punched in the columns corresponding to the assigned numbers. For structure **1** numbered as shown in **2**, the *ATOM CARD* would be as follows (**3**).

Column	1	2	3	4	5	6	7	8	9	10	11
Symbol	C	C	N	C	C	C	C	C	C	C	O

3

The program accepts up to fifty atoms not counting the hydrogen atoms.

BOND CARDS follow the *ATOM CARD*. They describe how the atoms are connected in the structure. There are no restrictions about representing structures with rings. The program is equipped to handle any structure which can be found in organic chemistry. The first column

---

[1]) The program can generate any depth in the ionic area, but lack of heuristics beyond boundary 2 (Fig. 1) prevents excursion in the ionic area retaining plausibility.

of a *BOND CARD* is used to differentiate between connections through a group of atoms and connections between two atoms only; the digit '1' is punched in that column for a group and the digit '0' for connections between two atoms only. The next two sets of three columns are used for the numbers assigned to the first and to the last atom of the chain. The eight and last column transmits the information about the bond order, i.e. 1, 2 or 3 for single, double and triple bonds, respectively.

With the numbering shown in **2** the *BOND CARDS* to input the structure of 5-diethylamino-pentan-2-one are the following (**4**):

BOND CARD No	Columns							
	1	2	3	4	5	6	7	8
1	1			1			8	1
2	0			9			3	1
3	0			9		1	0	1
4	0			7		1	1	2

4

These four *BOND CARDS* transmit the following information:

CARD No. 1: From atom No. 1 to atom No. 8 there is a group of atoms where each atom is connected to the following one in number by a single bond.
CARD No. 2: Atom No. 9 is connected to atom No. 3 by a single bond.
CARD No. 3: Atom No. 9 is connected to atom No. 10 by a single bond.
CARD No. 4: Atom No. 7 is connected to atom No. 11 by a double bond.

These cards can be given to the program in any sequence. The number of *BOND CARDS* needed to input a structure depends on how the atoms are numbered. The number of cards can always be minimized by choosing the appropriate numbering sequence.

Notational scheme for the output of chemical structures. – To present the structures of the ions and of the neutral species formed as a result of applying the primitive operations to a molecule or to an ion, a notation similar to *Lederberg*'s *DENDRAL* notation is used [3]. The differences between the linear notation used here and the *DENDRAL* notation are that our notation shows the hydrogen atoms of the structure and also indicates on which atoms radical sites and positive charges are found. Moreover, the notation used here has no canonical order. The *DENDRAL* notation has recently been nicely explained by *Buchanan et al.* in a review paper [4].

The two general rules of the notation used by the *ION GENERATOR* are:

Rule 1: An atom is printed along with its hydrogen atoms, followed by its free bonds.
Rule 2: The radical(s) attached to the free bond(s) of an atom are printed in the same order as were the bond(s) according to rule 1.

The following examples illustrate both rules.

1. CH2---CH3OH CH_3-CH_2-OH

2. NH2-CH---CH2-CH2-OHCH3 $H_2N-CH-CH_2-CH_2-OH$
                                    $|$
                                    CH_3

As an example of a ionic species one can consider the following structure (**5**):

$$CH_3-CH-CH-\overset{\overset{+\bullet}{O}}{\underset{\|}{C}}-CH_3$$
$$\underset{CH_3}{|}\underset{CH_2}{|}$$
$$\underset{CH_3}{|}$$

5

The linear notation for this structure can be written as (**6**).

$$CH3-CH---CH3CH---CH2-CH3C=-O(+.)CH3$$
$$\mathbf{6}$$

Ionization and charge localization. – In mass spectrometry, rationalization of bond cleavages by the electron book-keeping method assumes that in a molecular ion the charge is preferentially located at favorable sites, just prior to fragmentation. It is not unreasonable to think about the primary ions as originating from different molecular ions. These molecular ions could differ with respect to the charge locus, to their relative stability, linked to their tendency to fragment before they can rearrange. Functional groups containing heteroatoms generally provide the best loci for the positive charge. Carbon-carbon multiple bonds also play an important role while carbon-carbon single bonds and carbon-hydrogen bonds are much less effective. The program has been designed to know that the most likely occurring molecular ions are those where the active site arises by removal of a non-bonding electron, followed by those in which the expelled electron is a Π type electron. It also knows which heteroatoms are more effective than others to stabilize the positive charge. For each kind of electron type and for each environment a unique ionization threshold value has been semi-empirically chosen. Table 1 shows threshold values for a number of struc-

Table 1. *Threshold values for ionization.*

Substructure	$-N\langle$	$-NH$	$-NH_2$	$\rangle C=S$	$\rangle C=O$	$-S-$	$-O-$	$-SH$
Ionization threshold value	1000	950	900	850	800	750	700	650
Substructure		$-OH$	$-C=C-$	$\rangle C=C\langle$	$-\overset{|}{\underset{|}{C}}-$	$-CH-$	$-CH_2-$	$H-\overset{|}{\underset{H}{C}}-H$
Ionization threshold value		600	550	500	450	400	350	300

tural environments. These values, which range arbitrarily from 300 to 1000, allow the user of the program to select, by means of an option card, the kind(s) of ionization that he wants the structure to undergo. For example, if one wants ionization to take place only on the heteroatoms present in a molecule, one will select a range extending from 600 to 1000. In a difunctional molecule, it is possible to generate the ions arising from charge localization on the two heteroatoms separately. The program would have to be run twice, each time with the appropriate range for the ionization threshold values. For any chosen range, including the whole range, i.e. 300 to 1000, the *ION GENERATOR* builds the molecular ions and fragments them in order of decreasing threshold values.

Bond homolysis and bond formation. – The only motivation currently known to the program to homolyze a particular bond is the availability of a radical site in a β position. This allows, afterwards, the formation of a new bond either in the ionic or in the neutral species. Another requirement a particular bond must satisfy in order to be fissionable is that it must not have been formed by the *ION GENERATOR* in a previous operation of the ongoing fragmentation mode. Regarding C–H bond cleavages, the program considers all such bonds as equivalent when they originate from the same atom. Only one of the equivalent bonds will be homolyzed. Each time an ion has been formed, the program searches the mass spectrum for the relative abundance of the peak corresponding to the ion and prints it out, along with the structures of the ionic and neutral (if any) species. Even homolytic cleavages which do not change the mass of an ionic species are performed by the program. This affords a way to create a new radical site which can then play the trigger role for the next step of the fragmentation mode. However, homolysis performed on a double bond is not allowed to be the last step of a sequence of primitive operations; a fragmentation mechanism always ends with an ionic species having a mass which is smaller than the mass of the precursor ion.

Transfer of a hydrogen atom to a radical site via *cyclic transition states of various sizes.* – The tendency of a radical site to form a new bond can also be satisfied by the transfer of a hydrogen atom to the radical site. This creates a new radical site which is then able either to induce β-cleavage or to trigger a new hydrogen transfer. The sizes of the cyclic transition states which are presently allowed are from four- to eight-membered rings. Some loose restrictions are placed on the sizes which are allowed. Thus, for example, a hydrogen atom cannot be abstracted from a carbon atom bearing a double bond, or from an atom on which a hydrogen atom has already been transferred during a previous operation of the ongoing mechanism. Situations leading to blatant strain are also avoided. Moreover, the *ION GENERATOR* allows only two transition states to occur in any sequence of primitive operations. This restriction is based on the fact that extensive rearrangement before fragmentation, leading to drastic scrambling of the hydrogen atoms, is contrary to the ability of mass spectrometry to elucidate structures.

Mechanisms proposed by the ION GENERATOR for the formation of primary ions from 5-diethylaminopentan-2-one (**1**). – The ionization range which was selected extended from 750 to 1000. From Table 1 it can be seen that ionization then occurs only on the two heteroatoms. Two molecular ions were thus generated, one by lone-pair ionization on the nitrogen atom (atom No. 3 according to the numbering shown in **2**) and the other by lone-pair ionization on the oxygen atom (atom No. 11).

The program returns two kinds of outputs. On the main output every step of the various processes leading to the formation of an ion is shown in detail, with the structures of both the ionic and neutral (if any) species printed in the linear notation. A second output shows what occurred in a shorthand notation. The structures of the two molecular ions as they are printed are shown below with their respective translations (**7** and **8**).

Molecular ion No. 1

CH3–CH2–N(+.)– –CH2–CH2–CH2–C– – CH3OCH2–CH3   i.e.

$$\begin{array}{c} CH_3-CH_2 \\ \diagdown \\ CH_3-CH_2 \diagup \end{array} \overset{+\,\cdot}{N} - CH_2-CH_2-CH_2 - \overset{\overset{O}{\|}}{C} - CH_3$$

7

The abbreviated notation for that process is printed as follows:

ION 3   IMASS 157   INT 5,1%   NMASS 0,   which means:

'ionization affects atom No. 3, mass of the ionic species is 157, relative abundance of the peak found in the actual mass spectrum is 5,1%, mass of the neutral fragment lost is zero'.

Molecular ion No. 2

$$N----CH2-CH3CH2-CH2-CH2-C=-O(+.)CH3CH2-CH3 \quad \text{i.e.}$$

$$\begin{array}{c} CH_3-CH_2 \\ CH_3-CH_2 \end{array} \!\!\!\! N-CH_2-CH_2-CH_2-\overset{\overset{+\cdot}{O}}{\underset{\|}{C}}-CH_3$$

8

The abbreviated notation is:

ION 11   IMASS 157   INT 5,1%   NMASS 0

From these two molecular ions the program generated altogether 16 ions of different masses by various mechanisms. The list of the ions which were generated is

Table 2. *Number of mechanisms proposed for the formation of primary ions from 5-diethylaminopentan-2-one.*

	Ionization on nitrogen				Ionization on oxygen				Intensity[a] %
	Mechanisms involving			Total number of mechanisms	Mechanisms involving			Total number of mechanisms	
	0	1	2		0	1	2		
	H transfers				H transfers				
m/e									
43					1	0	2	3	52,8
44					0	0	1	1	5,1
58					0	1	1	2	26,4
71					0	0	2	2	3,6
72	0	0	3	3					3,4
73	0	1	1	2					1,0
85					0	0	1	1	30,5
86	1	0	0	1	0	1	2	3	100,0
87	0	1	1	2					11,0
100	0	0	3	3					0,8
114	0	1	1	2					0,6
115	0	1	0	1					0,1
128	0	0	3	3	0	0	1	1	0,2
129	0	1	0	1					0,1
142	1	0	0	1	1	0	0	1	3,6
156	2	0	0	2					0,8
157				1				1	5,1
				22				15	

[a] Intensity values refer to relative abundances with intensity of the base peak = 100%.

shown in Table 2 along with the number of non-equivalent mechanisms proposed for the formation of each ion. In the structure of **1**, carbon atoms No. 1 and No. 2 are respectively equivalent to carbon atoms No. 10 and No. 9. Equivalent mechanisms involving these carbon atoms have been counted once only. Moreover, when transfer of hydrogen atoms occurs, only those mechanisms which could be, if desired, subjected to direct experimental proof by means of labelling with deuterium, have been retained in Table 2. Mechanisms in which all the rearrangements of hydrogen atoms are internal to the product ion, or to the expelled neutral fragment, have been rejected manually.

Table 3. *Mechanisms proposed for the fragmentation of 5-diethylamino-pentan-2-one down to primary ions*

m/e	Mechanisms proposed					Order
43	ION 11	HOM 6,7	MKB 7,11			25
	ION 11	6MRTH4	6MRTH8	HOM 6,7	MKB 7,11	29
	ION 11	8MRTH9	8MRTH8	HOM 6,7	MKB 7,8	37
44	ION 11	8MRTH9	5MRTH5	HOM 6,7		34
58	ION 11	6MRTH4	HOM 5,6			26
	ION 11	8MRTH9	4MRTH4	HOM 5,6		31
71	ION 11	6MRTH4	4MRTH6	HOM 4,5	MKB 5,6	27
	ION 11	8MRTH9	6MRTH6	HOM 4,5	MKB 5,6	36
72	ION 3	4MRTH5	5MRTH9	HOM 3,4	MKB 3,9	15
	ION 3	5MRTH6	6MRTH9	HOM 3,4	MKB 3,9	19
	ION 3	7MRTH8	8MRTH9	HOM 3,4	MKB 3,9	22
73	ION 3	4MRTH10	6MRTH5	HOM 3,4		10
	ION 3	4MRTH5	HOM 3,4			13
85	ION 11	8MRTH9	5MRTH5	HOM 3,4	MKB 5,4	35
86	ION 3	HOM 4,5	MKB 3,4			4
	ION 11	6MRTH4	4MRTH9	HOM 3,4		28
	ION 11	8MRTH9	HOM 3,4			30
	ION 11	8MRTH9	4MRTH2	HOM 3,4		33
87	ION 3	5MRTH6	HOM 4,5			17
	ION 3	4MRTH10	7MRTH6	HOM 4,5		12
100	ION 3	4MRTH10	5MRTH4	HOM 5,6	MKB 4,5	8
	ION 3	5MRTH6	4MRTH4	HOM 5,6	MKB 4,5	18
	ION 3	7MRTH8	6MRTH4	HOM 5,6	MKB 4,5	21
114	ION 3	4MRTH5	HOM 6,7	MKB 5,6		14
	ION 3	4MRTH10	6MRTH5	HOM 6,7	MKB 5,6	11
115	ION 3	7MRTH8	HOM 6,7			20
128	ION 3	4MRTH10	5MRTH4	HOM 3,9	MKB 3,4	7
	ION 3	4MRTH10	5MRTH2	HOM 3,9	MKB 2,3	9
	ION 3	4MRTH1	5MRTH4	HOM 2,3	MKB 3,4	16
	ION 11	8MRTH9	4MRTH4	HOM 3,9	MKB 3,4	32
129	ION 3	4MRTH10	HOM 3,9			6
142	ION 3	HOM 1,2	MKB 2,3			2
	ION 11	HOM 7,8	MKB 7,11			24
156	ION 3	HOM 2,2'	MKB 2,3			3
	ION 3	HOM 4,4'	MKB 3,4			5
157 (M$\dot{+}$)	ION 3					1
	ION 11					23

As can be seen from Table 2, the program proposed 35 fragmentation mechanisms for the formation of the 16 aforementioned ions. Some of the ions, those with m/e 43, 44, 58, 71 and 85, originate, according to the *ION GENERATOR*, only from the molecular ion No. 2. Others, with m/e 72, 73, 87, 100, 114, 115, 129 and 156, have been found only through mechanisms starting with the molecular ion No. 1. Finally, those ions with m/e 86, 128 and 142 have been explained by the program starting with both molecular ions. Table 2 also shows, for each ion, how many mechanisms involved no hydrogen transfer before cleavage, those which involved the transfer of one hydrogen atom and those which involved the transfer of two hydrogen atoms before bond homolysis occurred to yield the ion.

All of the 35 proposed mechanisms are shown, in shorthand notation, in Table 3 along with the order in which the program presented them. For every mechanism listed in Table 3, a labelling experiment could be designed to substantiate or reject the proposed hydrogen transfer[2]).

Some of the mechanisms proposed by the *ION GENERATOR* will now be illustrated with examples.

α-cleavage next to the nitrogen atom. –

a) m/e 142

   ION 3   HOM 1,2   MKB 2,3   IMASS 142   INT 3,6%   NMASS 15

i.e.: 'ionization on atom No. 3, homolysis between atom No. 1 and atom No. 2, bond formation between atom No. 2 and atom No. 3'. The information concerning the masses of the ionic and neutral species, and the intensity of the corresponding peak found in the mass spectrum are self explanatory.

The shorthand notation shown above corresponds to the mechanism outlined in Scheme 1. The detailed output shows the following structures:

Scheme 1

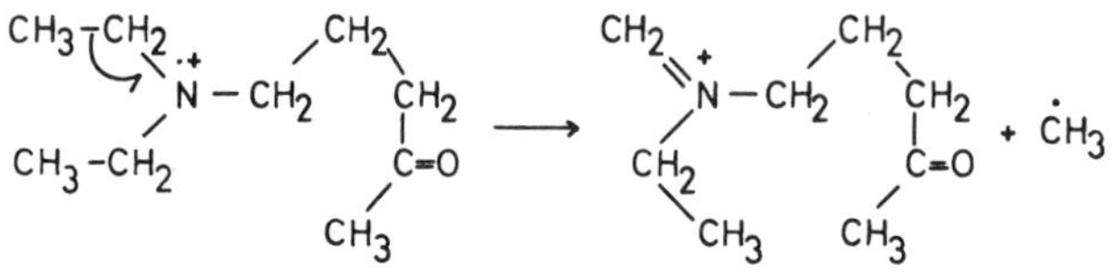

m/e 142

IONIC SPECIES: $N(+) = --CH2CH2-CH2-CH2-C- = CH3OCH2-CH3$
NEUTRAL SPECIES: $C(.)H3$

An equivalent mechanism is proposed for the formation of m/e 142 by α-cleavage. It involves loss of carbon atom No. 10 instead of carbon atom No. 1. As was mentioned before, whenever two mechanisms proposed by the program are equivalent, only one of them is retained in the list given in Table 3.

b) m/e 86

   ION 3   HOM 4,5   MKB 3,4   IMASS 86   INT 100%   NMASS 71

---

[2]) For some of the mechanisms with double hydrogen transfer, labelling experiments could only verify one of the hydrogen transfers.

This mechanism is shown in Scheme 2. The following structures are shown in the output:

Scheme 2

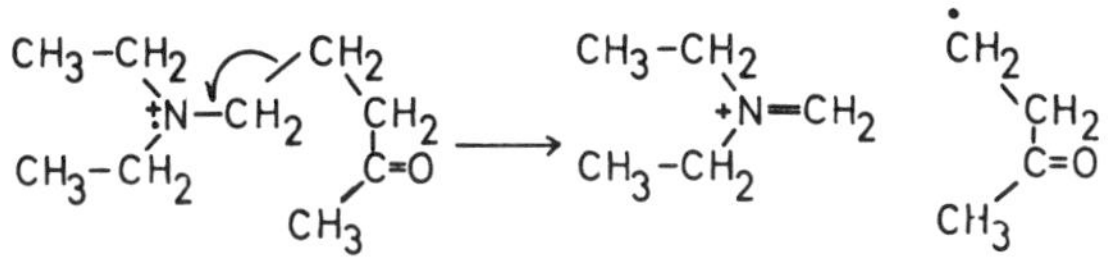

CH₃–CH₂ ... CH₂ ... CH₃–CH₂ ... CH₂
  \•+ /   \   \+   /
   N–CH₂ CH₂ →   N=CH₂   CH₂
  / | / |
CH₃–CH₂   C=O   CH₃–CH₂   C=O
          | |
          CH₃   CH₃
                                   m/e 86

IONIC SPECIES: N(+)–=–CH2–CH3CH2CH2–CH3
NEUTRAL SPECIES: C(.)H2–CH2–C=–OCH3

Cleavage of molecular ion No. 1 with transfer of hydrogen atoms. – An example in which two consecutive transfers of hydrogen atoms to a radical site occur before a

Scheme 3

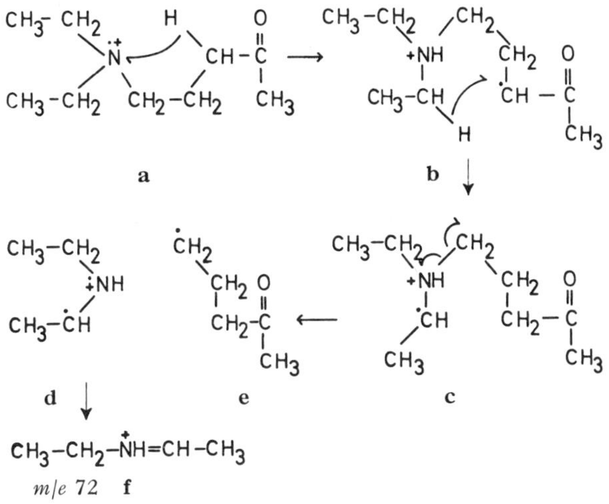

CH₃–CH₂   H   O   CH₃–CH₂   CH₂
  \••+   /   ‖   \+   \
   N   CH–C   →   NH   CH₂ O
  / | | | | ‖
CH₃–CH₂ CH₂–CH₂ CH₃   CH₃–CH   CH–C
                             \H |
                                        CH₃
     a   **b** ↓

CH₃–CH₂   CH₂   CH₃–CH₂   CH₂
  \•+   \   \+   \
   NH   CH₂ O   NH   CH₂ O
  / | ‖ | | ‖
CH₃–CH   CH₂–C   ← CH   CH₂–C
              | | |
              CH₃   CH₃   CH₃
     d ↓   **e**   **c**

CH₃–CH₂–ṄH=CH–CH₃
m/e 72   **f**

N(+.)––––CH2–CH3CH2–CH2–CH2–C=–OCH3CH2–CH3
                   a ↓

N(+)H––––CH2–CH3CH2–CH2–C(.)H–C=–OCH3CH2–CH3
                   b ↓

N(+)H––––CH2–CH3CH2–CH2–CH2–C–=CH3OC(.)H–CH3
                   c ↓

       N(+.)H––C(.)H–CH3CH2–CH3
                 d
     ↙   ↘

NEUTRAL SPECIES   *IONIC SPECIES*
C(.)H2–CH2–CH2–C–=CH3O   N(+)H=–CH–CH3CH2–CH3
     e   **f**

neutral fragment is expelled is afforded by the mechanism proposed for the formation of *m/e* 72.

ION 3 5MRTH6 6MRTH9 HOM 3,4 MKB 3,9 IMASS 72 INT 3,4% NMASS 85

i.e. 'ionization on atom No. 3, transfer of a hydrogen atom from atom No. 6 *via* a five-membered transition state, transfer of a hydrogen atom from atom No. 9 *via* a six-membered transition state, bond homolysis between atom No. 3 and atom No. 4, bond formation between atom No. 3 and atom No. 9'.

The corresponding conventional way of presenting that sequence of operations is illustrated in Scheme 3. For that sequence the structures depicted in Scheme 3 (**a** to **f**) are printed by the program as follows.

Fragmentation starting with molecular ion No. 2. – The two α-cleavages next to the carbonyl function are the only mechanisms without hydrogen transfer proposed by the *ION GENERATOR* for the fragmentation of molecular ion No. 2.

a) *m/e* 43

ION 11 HOM 6,7 MKB 7,11 IMASS 43 INT 52,8% NMASS 114

IONIC SPECIES: C = −O(+)CH3
NEUTRAL SPECIES: C(.)H2–CH2–CH2–N−−−CH2–CH3CH2–CH3

This corresponds to the mechanism depicted in Scheme 4.

Scheme 4

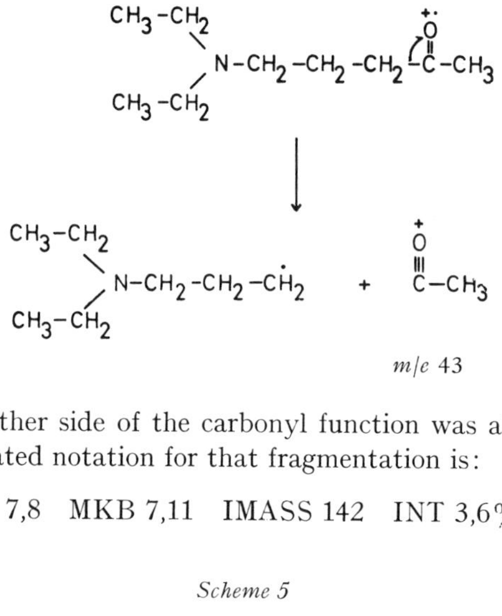

$$\begin{array}{c}
CH_3\text{-}CH_2 \\
\phantom{CH_3\text{-}}\diagdown \\
\phantom{CH_3\text{-}CH_2}N\text{-}CH_2\text{-}CH_2\text{-}CH_2\text{-}\overset{+\cdot}{\underset{\|}{C}}\text{-}CH_3 \\
\phantom{CH_3\text{-}}\diagup \\
CH_3\text{-}CH_2
\end{array}$$

$$\downarrow$$

$$\begin{array}{c}
CH_3\text{-}CH_2 \\
\phantom{CH_3\text{-}}\diagdown \\
\phantom{CH_3\text{-}CH_2}N\text{-}CH_2\text{-}CH_2\text{-}\overset{\cdot}{C}H_2 \quad + \quad \overset{+}{\underset{\|}{\overset{O}{C}}}\text{-}CH_3 \\
\phantom{CH_3\text{-}}\diagup \\
CH_3\text{-}CH_2
\end{array}$$

m/e 43

Cleavage on the other side of the carbonyl function was also proposed to yield *m/e* 142. The abbreviated notation for that fragmentation is:

ION 11 HOM 7,8 MKB 7,11 IMASS 142 INT 3,6% NMASS 15

b) *m/e* 58

Scheme 5

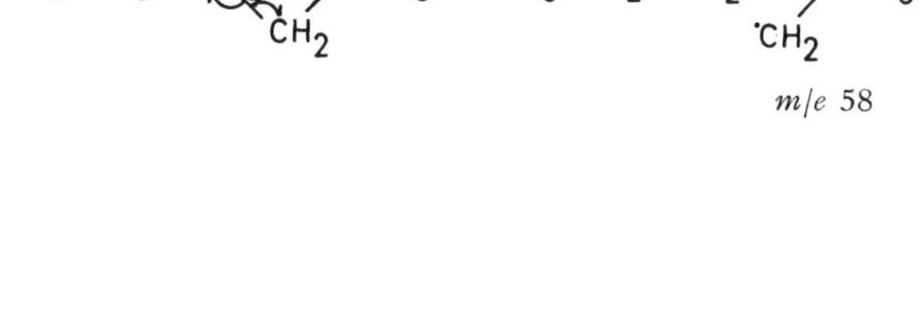

$$\begin{array}{cc}
CH_3\text{-}CH_2 \quad\quad H & CH_3\text{-}CH_2 \\
\phantom{CH_3\text{-}}\diagdown\diagup & \phantom{CH_3\text{-}}\diagdown \\
\phantom{CH_3\text{-}CH_2}N\text{-}CH \quad O\overset{+\cdot}{} & \phantom{CH_3\text{-}CH_2}N\text{-}CH \quad \overset{+}{O}H \\
\phantom{CH_3\text{-}}\diagup \quad|\| & \phantom{CH_3\text{-}}\diagup \quad\|+\| \\
CH_3\text{-}CH_2 \quad CH_2C\text{-}CH_3 \rightarrow CH_3\text{-}CH_2 \quad CH_2 \quad\quad C\text{-}CH_3 \\
\phantom{CH_3\text{-}CH_2\quad}\diagdown\diagup & \phantom{CH_3\text{-}CH_2 \quad}\diagup \\
\phantom{CH_3\text{-}CH_2\quad aa}CH_2 & \phantom{CH_3\text{-}CH_2\quad}CH_2
\end{array}$$

m/e 58

The *McLafferty* rearrangement shown in Scheme 5, which leads to m/e 58, was proposed in the following way:

ION 11   6MRTH4   HOM 5,6   IMASS 58   INT 26,4%   NMASS 99

IONIC SPECIES: O(+)H = C--CH3C(.)H2
NEUTRAL SPECIES: CH2 = CH–N--CH2–CH3CH2–CH3

As can be seen from Table 3, the program proposed only one mechanism for the formation of m/e 85. The fragmentation was visualized as starting from the molecular

Scheme 6

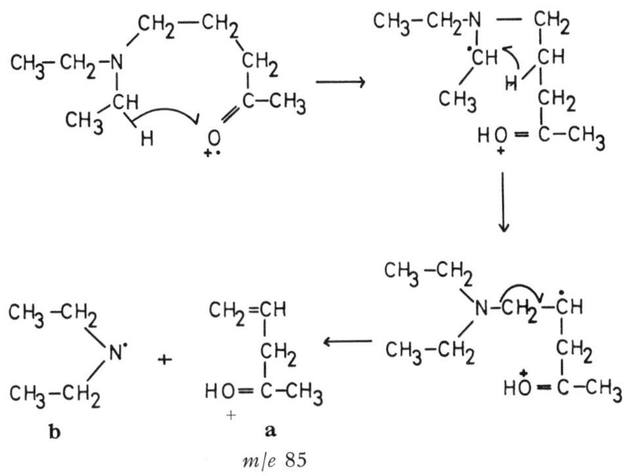

ion No. 2 to yield ion **a** shown in Scheme 6. The abbreviated notation corresponding to the mechanism illustrated in Scheme 6 and the structures of the ionic and neutral species are:

ION 11 8MRTH9 5MRTH5 HOM 3,4 MKB 4,5 IMASS 85 INT 30,7% NMASS 72

IONIC SPECIES: O(+)H = C--CH3CH2–CH = CH2
                                    (structure **a** in Scheme 6)

NEUTRAL SPECIES: N(.)--CH2–CH3CH2–CH3
                                    (structure **b** in Scheme 6)

A high-resolution measurement showed the signal at m/e 85 to be actually a singulet, due to an ion with the elemental composition C_5H_9O.

Table 3 shows also the order in which the ions were generated. The first ion to be created was, as has already been mentioned, the molecular ion No. 1. After that, the program generated all the ions arising from that form of the molecular ion, before building the other form of the molecular ion. Mechanisms which yield the same ion, from the same molecular ion, do not necessarily follow each other as is apparent from Table 3. This is due to the several points of view taken by the program in order to generate *all* plausible mechanisms.

By comparing the actual mass spectrum (Figure 2) of 5-diethylamino-pentan-2-one with the list of the ions shown in Table 2, it can be concluded that most of the important ions were generated. Although 22 of the 35 proposed mechanisms involve a double hydrogen transfer, they are plausible mechanisms and cannot be rejected *a priori*, without disproving them by means of deuterium labelling experiments. We are convinced that the *ION GENERATOR* has the potentiality to discover new mechanisms which have been missed in past studies of mass spectral fragmentation. The next step will be to implement more sophisticated primitive operations and to allow the program to explore any depth in the ionic area.

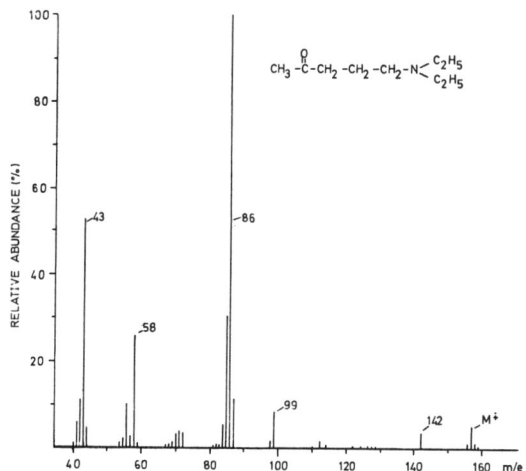

Figure 2: *Mass spectrum of 5-diethylamino-pentan-2-one (70 eV)*.

Experimental Part

The computer program was developed for the CDC 3800 computer at the Computer Center of the University of Geneva. It is written in the FORTRAN programming language. To fragment the structure of 5-diethylaminopentan-2-one down to primary ions the computer needed approximately 3 minutes. The low-resolution mass spectrum of **1** was recorded at 70 eV with a *Varian* MAT CH-4 mass spectrometer. High-resolution measurements were performed with a *Varian* MAT SM-1 instrument.

Support of part of this work by a grant (No 2.281.70) of the *Fonds National Suisse de la Recherche Scientifique* is gratefully acknowledged. We also thank Prof. *B. Levrat* for financial assistance to one of us (A.D.) and Mrs. *F. Kloeti* for her technical assistance with the mass spectrometer and for the drawing of the figures.

BIBLIOGRAPHY

[1] H. Budzikiewicz, C. Djerassi & D. H. Williams, 'Mass Spectrometry of Organic Compounds', Holden-Day, San Francisco 1967.
[2] B. G. Buchanan, E. A. Feigenbaum & J. Lederberg, MEMO AIM-145, Report No. CS-221, Stanford Artificial Intelligence Project, 1971.
[3] J. Lederberg, 'Topology of Molecules', in The Mathematical Sciences', M.I.T. Press, Cambridge, Mass. 1969, pp. 37–51.
[4] B. G. Buchanan, A. M. Duffield & A. V. Robertson, in 'Mass Spectrometry. Techniques and Applications'. Ed. G.W.A. Milne, Wiley-Interscience, New-York 1971, pp. 121–178.

Computer-assisted Analysis of Complex Synthetic Problems

By E. J. Corey
DEPARTMENT OF CHEMISTRY, HARVARD UNIVERSITY, CAMBRIDGE,
MASSACHUSETTS 02138, U.S.A.

1 Introduction

For the past several years a programme of research has been carried out at Harvard, the major objective of which has been the application of digital computers to assist in the derivation of synthetic routes to complex molecules. A review of this work will be presented here.

A complex synthetic problem for purposes of this discussion can be defined by the following criteria:

1. A solution cannot be obtained simply by analogy with previously solved problems.
2. The starting point(s) or material(s) for the synthesis are not directly apparent.
3. Many possible pathways of synthesis must be examined in order to ensure the selection of the simplest and most useful approach.
4. The structure to be synthesised is itself complex—not merely because of its size but also in terms of the presence of complicating structural features such as functional groups or reactive centres, rings, chiral or geometric stereocentres, destabilising interactions, *etc.*

With this definition it becomes apparent to a synthetic chemist that a project to develop a general problem-solving procedure for use by a computer must be regarded as long range in character. The task is too large to be accomplished in a five- or ten-year period and indeed is unlikely to be complete, in a final sense, in the foreseeable future. As with a 'theory' for complex situations, any general procedure for complex problem solving will be subject to further improvement and hence to an evolutionary course of development. At the outset of the studies described herein, it was by no means certain that meaningful progress could be made even in the development of simple problem-solving techniques and, indeed, there will doubtless continue to be a large group of chemists who take a sceptical view of the whole enterprise. Such scepticism may reasonably be based on the almost incredible diversity of organic structures, the complexities of stereochemistry, and the very large variety and number of chemical processes available to the synthetic chemist (not to mention the remarkably complicated limits on the scope of each process). Further, the need for compromise or 'trade-off' between 'generality' and 'power' in general problem-solving procedures[1] must be borne in mind. On the other hand, even if the effort to devise an effective problem-solving computer program were to fail utterly, a deeper comprehension of the strategies, principles, and elements of chemical synthesis would be gained, the classification and organisation of basic chemical data according to the requirements of synthesis would be advanced, and new and more powerful methods of teaching chemical synthesis and solving synthetic problems would result. These advances in the understanding and codification of an important area of chemistry can be regarded as a goal of *fundamental* synthetic research, the attainment of which is destined to yield results of considerable value.

Two computers have been used in the studies on machine-assisted synthetic analysis at Harvard, a PDP-1, vintage *ca.* 1960, and a modern PDP-10 (both Digital Equipment Corp.). The newer machine has the capacity to handle properly the very large program which is evolving. This program will be transferable to other machines, since it is being written in the most commonly used higher level language, FORTRAN IV, and designed so as to minimise hardware dependence. In contrast, the older machine has quite limited memory resources and must be programmed in a specialised assembler language (DECAL) which is unique to it. The program currently being used with the PDP-1, designated LHASA-1 (Logic and Heuristics Applied to Synthetic Analysis), lacks a stereochemical capability and is incomplete with regard to chemical program modules. However, it is very useful in the development and testing of new ideas and program modules which will be used eventually in LHASA-10, the PDP-10 program which is expected to become operational in the mid-to-later 1970s. The aspects of computer-assisted synthetic analysis discussed herein, unless otherwise indicated, have been implemented in the LHASA-1 and/or LHASA-10 programs.

An outline of the general approach which has guided the initial phase of program development has been presented previously.[2] This paper can also serve as an introduction to the present Review which will be oriented mainly towards points of chemical interest rather than programming details or computational aspects. In connection with the latter there are a number of printed works[3] which may be used as reference texts in connection with this Review or for the purpose

---

[1] G. W. Ernst and A. Newell, 'GPS: A Case Study in Generality and Problem Solving', Academic Press, New York, 1969.
[2] E. J. Corey and W. T. Wipke, *Science*, 1969, **166**, 178.
[3] The following references are listed approximately in order of increasing sophistication. (*a*) A. I. Forsyth, T. A. Keenan, E. I. Organick, and W. Sternberg, 'Computer Science, a First Course', J. Wiley and Sons, New York, 1969; (*b*) A. Ralston, 'Introduction to Programming and Computer Science', McGraw-Hill, New York, 1971; (*c*) I. Flores, 'Computer Programming', Prentice-Hall, Englewood Cliffs, N.J., 1966; (*d*) A. T. Berztiss, 'Data Structures, Theory and Practice', Academic Press, New York, 1971; (*e*) D. G. Hays, 'Introduction to Computational Linguistics', Elsevier, New York, 1967; (*f*) P. Wegner, 'Programming Languages, Information Structures and Machine Organization', McGraw-Hill, New York, 1968; (*g*) 'Computers and Thought', ed. E. A. Feigenbaum and J. Feldman, McGraw-Hill, New York, 1963; (*h*) 'Semantic Information Processing', ed. M. Minsky, MIT Press, Cambridge, Mass., 1968.

of gaining a general background in the computational and data processing techniques which are fundamental to this sort of computer application.

One approach[2,4] to the derivation of synthetic pathways to some target structure involves the generation of a set of intermediates which can be converted into that structure by one synthetic step and the iteration of this procedure for each intermediate so generated until a 'tree' of intermediates is developed (Figure 1). This technique forms the basis for the computer programs currently being

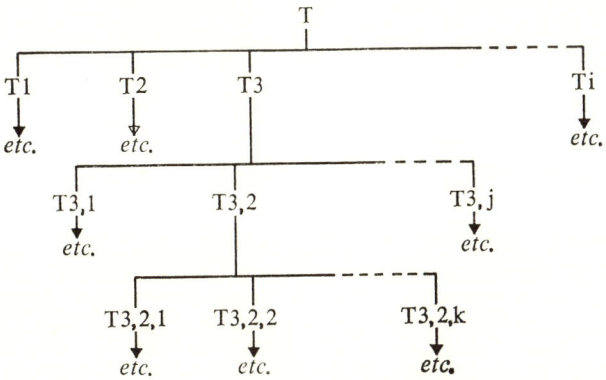

Figure 1 *Synthetic analysis of target T generates a 'tree' of intermediate precursor structures*

used and extended. It involves analytical processes which depend heavily upon the structural features of *reaction products* (as contrasted with starting materials) and the consideration of molecular changes in the *retro-synthetic* sense. In order to avoid confusion, two special terms and a special graphic have been employed to provide a distinction between the nomenclature appropriate to these analyses and that which is conventionally applied to synthesis in the direction of laboratory execution. The terms *antithetic* and *transform* and the 'double-arrow' graphic will be used strictly as is indicated in Scheme 1. It is noteworthy that this usage of the term *transform* has a parallel to the mathematical meaning (a *function*

1. Direction of laboratory execution is '*synthetic*'
2. Represented as
   (s)
   ⟶
3. Process is called a '*reaction*'

vs.

1. Direction of computer analysis is '*antithetic*'
2. Represented as
   (a)
   ⟹
3. Process is called a '*transform*'

Scheme 1 *Terminology for chemical structural changes in either of two directions*

[4] E. J. Corey, *Pure Appl. Chem.*, 1967, **14**, 19.

operating on an *argument* to produce a *result*) in the sense that a *transform* operates on a chemical structure to produce a different (transformed) structure.

By way of introduction a précis of the key features of current programs is given in Scheme 2. Communication between man and machine is accomplished graphically by a method first developed for this project in 1967.[2,5] The chemist 'inputs' a structure by drawing a standard two-dimensional structural diagram

1. Graphical Man-Machine Communication
    Structural input: electrostatic (Rand) tablet and stylus.
    Structural output: CRT displays (2), plotter (hard copy).
2. Emphasis on Interactive Relationship between Man and Machine
3. Tabular Internal Representation of Structure
    Atom and bond connection tables (input structure).
    Structure information blocks (each intermediate).
4. Machine-Oriented Perception of Synthetically Significant Structural Features
    Functional groups, rings, stereorelationships, *etc.*
5. Automatic Generation of Synthetic Intermediates
    Direction of analysis: antithetic.
    'Tree' collection of synthetic intermediates with target molecule as parent.
    Target-oriented data files, process (transform) selection and evaluation.
6. Multiple Problem-Solving Strategies
    Heuristic rules; goal and subgoal generation; strategic bond disconnections.

Scheme 2 *A program for computer-assisted synthetic analysis—LHASA (Logic and Heuristics Applied to Synthetic Analysis)*

using an electrostatic tablet (Rand tablet) capable of sensing positions on a 2^{10} by 2^{10} point grid and a pen equipped with a switch which closes and allows communication with the tablet when pressed down. The pen leaves no visible trace on the tablet but creates a display of the structure being drawn (and a tracking cross which locates the pen) on a cathode ray tube. All structural information output from the computer is also displayed on a cathode ray tube as a conventional formula drawing. Further, output structures can be provided as hard copy on paper using a commercial graphics plotter. In this form easy and rapid communication occurs in the natural pictorial language of the chemist in a way which requires neither training nor special skills. Because of the availability of such ready man–machine communication and the desirability of allowing the chemist to influence and direct the analysis of a problem to whatever extent he judges appropriate, the programs for synthetic analysis are designed to be highly interactive. The chemist at each stage has the option to specify, modify, or channel the flow of analysis.

[5] E. J. Corey, W. T. Wipke, R. D. Cramer, and W. J. Howe, *J. Amer. Chem. Soc.*, in press.

As is indicated in Scheme 2, the internal representation of chemical structure within the computer involves connection tables for the atoms and bonds within the structure (together with x, y co-ordinates for graphical display). Structure information blocks which contain data on *changes* in atom and bond tables may also be used to provide information on any 'offspring' structures which result from structural manipulation of an input or target structure. From the atom and bond tables information is extracted which is needed by the program. This process, which is designated 'perception', makes available data on synthetically significant structural features such as functional groups, rings, *etc*. The program provides for automatic selection of transforms, their evaluation, and their application to generate a 'tree' of synthetic intermediates. Transform selection may be guided by the chemist or by a number of strategies which are being added to the program. These strategies, which are allowed to operate independently of one another and which vary greatly in nature, parallel to a certain extent those used by the expert chemist.

Before proceeding further in the discussion of machine program(s) for synthetic analysis, it is instructive to review the most common approach of the chemist to problem solving, and this is outlined in Scheme 3. This approach could be simulated by two computer programs being executed simultaneously and intercommunicating (for example, in a time-sharing environment). One program would operate in the antithetic direction and the other in the synthetic direction. The latter would require as input the target structure or the latest level of intermediates on the antithetic tree and one or more starting structures which would be specified by the chemist or derived by another program. The synthetic and antithetic programs would have to communicate to one another

1. Grow tree in antithetic (retrosynthetic) direction (τ_a) from target (T).
2. At some point associate one or more structures (I_n) at lowest levels of τ_a with possible available starting structures (I_o).
3. Grow tree in synthetic direction (τ_s) from I_o *toward* I_n *as a goal*.
4. Alternately extend τ_a and τ_s using latest intermediates as strategy-providing subgoals (*match phase*).
5. Examine intermediates on a linear path from I_o to T in synthetic direction to *optimise ordering of sequence of I's* and to optimise application of *control elements*, *i.e.*, activating, deactivating, or stereo-correcting operations.

Scheme 3 *Typical course of problem analysis by a synthetic chemist*

the latest structures generated for their respective trees. The latest intermediates of one program would serve as goals which guide the operation of the other. The realisation of such a scheme is regarded as a major long-term objective of the Harvard program.

2 Internal Representation of Chemical Structure

The utility of connection tables for both bonds and atoms to provide an internal representation of a target structure for synthetic analysis has already been discussed, and a particular arrangement of these tables has been described. Recently, an improved version of these tables has been devised[6] for LHASA-10 which is illustrated in Table 1 for the specific case of 1,1-dimethylcyclopropane. The atom part of the table (one computer word of 32 bits for each entry) contains for each atom an atom number (sequence of atom input from Rand tablet), the number of attachments, charge, the valence of atom, the atom type (C, N, etc.), and a pointer to (relative address of) the first bond entry for this atom. The x, y co-ordinates of the atom in the external representation are also stored in a

Table 1 *Sample table for* 1,1-*dimethylcyclopropane*

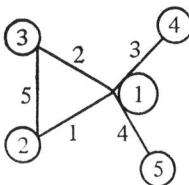

ATNO	NATCH	CHARGE	VALENCE	TYPE	POINTER*
1	4	1	4	4	350
2	2	1	4	4	349
3	2	1	4	4	347
4	1	1	4	4	345
5	1	1	4	4	343

ATNO	BNDNO	BNDTYP	POINTER**	LOCATION
2	5	1	0	341
3	5	1	0	342
1	4	1	0	343
5	4	1	0	344
1	3	1	0	345
4	3	1	344	346
1	2	1	341	347
3	2	1	346	348
1	1	1	342	349
2	1	1	348	350

*First bond entry for this atom. **Bond entry for next attached atom (in number sequence).

second computer word. In the bond part of the connection table, an entry is made twice for each bond between atoms (once for each atom). Each entry (one 32-bit word) for a bond contains the sequence numbers for the attached atoms, the bond number (according to the sequence with which bonds were input from the Rand tablet), the bond type (single, double, etc.), and a pointer to (address of) the bond entry for the next attached atom (according to number sequence). Information on up to 64 explicit atoms can be accommodated in

⁶ E. J. Corey and D. A. Pensak, to be published.

LHASA-10. Because of the storage of atom and bond data according to sequence, information on each is available without searching. Although information for a structure which has been input is in the form of a connection table, offspring of the parent structure which are generated by LHASA are internally represented by the *physical differences* between the table of the parent and the table that would correspond to the offspring. This approach is very economical in terms of memory. Generation of an offspring structure involves only arithmetic replacement operations, and this constitutes another advantage of the LHASA-10 system over its predecessor.[2]

3 Perception

One of the most challenging aspects of developing a program for synthetic analysis by machine is the gathering and storage of the synthetically significant structural information which is required for the parts of the program which select, evaluate, and apply chemistry so as to generate the tree of synthetic intermediates.[2,7] This function, which is essentially a form of perception, is performed by a separate module of the program. The techniques used for machine perception are on the whole very different from those used by a chemist. They are highly formalised in a way which is efficient with regard to machine memory and execution time, and they are applied systematically.

Among the large variety of data generated by the perception module are certain types which are obtained and stored in binary set form. Set information is obtained for atoms or bonds which possess a particular property. Examples of simple atom sets are: BOND1SET (atoms to which at least one single bond is attached), NITROGEN (set of all N atoms), HETERO (set of all N, O, S, P atoms), and RINGSET (set of all atoms in rings). Examples of bond sets are: BOND1 (single bonds), RING5 (bonds in a five-membered ring), RESON (bonds in an aromatic ring), CJBD (multiple bonds in conjugation). Starting with the most basic sets, increasingly complex sets can be constructed by standard set operations. Sets can be manipulated and combined by computer with great facility using basic instructions such as the logical AND^{8a} or the inclusive $OR.^{8b}$ In this way sets are obtained which deal with many types of structural features including, for example, those having to do with molecular topology, sites of reactivity, electronic properties, and structural redundancy.[7] Table 2 shows some simple binary sets which are derived by the perception module for a particular structure and indicates the way in which data are 'bit' coded and stored in memory. In LHASA-1 two 18-bit computer words (locations) are used to accommodate each set (up to 36 atoms). The leftmost bit in word 1 corresponds to the first atom (or bond) according to input sequence, and each bit to the right thereafter corresponds to the next atom (or bond) in the sequence.

---

[7] E. J. Corey, W. T. Wipke, R. D. Cramer and W. J. Howe, *J. Amer. Chem. Soc.*, in press.
[8] (*a*) The result of the *AND* operation between two sets is the intersection of the sets; (*b*) the result of the inclusive *OR* operation on two sets is the union of the sets; (*c*) the exclusive *OR* of two binary sets returns a set whose non-zero elements are present in one but not both of the original sets.

Table 2 *Some sample sets*

SETNAME	WORD1						WORD2
BOND1SET	111	111	111	001	010	000	zero
OXYGEN	000	000	010	110	100	000	zero
BOND2SET (atoms)	000	011	101	111	100	000	zero
BOND2 (bonds)*	000	010	000	011	010	000	zero
RINGSET	111	111	111	000	000	000	zero
JUNCTSET	011	000	000	000	000	000	zero
ALLYLIC	111	100	010	000	010	000	zero

* Bonds also are numbered in the order of drawing.
Illustrations of set operations:
  Carbonyl oxygens: BOND2SET *AND* OXYGEN.
  Atoms defining any double bond in a ring: RINGSET *AND* BOND2SET atoms.

The binary digits 0 or 1 indicate, respectively, that the corresponding atom (or bond) is not or is a member of the set in question.

The perception of functional groups has been accomplished by a variety of schemes, and that used by LHASA-1 has previously been outlined.[2,7] In the approach[6] which has been implemented in LHASA-10, all non C—C bonds are encoded (assigned numerical names) and a search is conducted in the order: triple bonds, double bonds (C=O first), then single bonds. The 'recogniser' program then reads a table which contains information on 64 functional groups and which is suitable for binary search. Starting with the bond which determines entry into the table, the group of contiguous bond(s) in the structure is matched against the table. Success or fail pointers then reference the address of the next appropriate table entry. Scheme 4 illustrates this operation diagrammatically for a small portion of the table. The functional groups thus found are stored in list form,[9] each along with the group name, level of reactivity (normal for the group, above- or below-normal), and the atom to which the functional group is attached (group origin).

[9] A list (singly linked) is a collection of elements in memory each of which is composed of two contiguous fields (storage locations). The first field contains datum or a pointer to (address of) a sublist, and the second contains a pointer to the next element on the list. As a data structure the list has the advantage of representing relationships between data as well as storing the data. In addition, the elements in the list need not be stored in contiguous memory locations. See J. M. Foster, 'List Processing', Elsevier, London, 1967, and ref. 3d.

Binary tree search by recognizer:

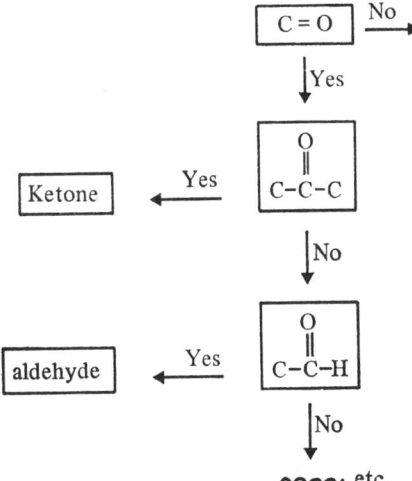

Scheme 4 *Functional group recognition*

A procedure for the perception of rings has been devised which selects the subset of synthetically significant rings ('synthetic subset') from an *n*-cyclic structure with high computing efficiency even for complex networks.[10] This has now been implemented in both LHASA-10 and LHASA-1 (replacing an algorithm described earlier[2]). Since ring-closure reactions depend on the size of the smallest ring containing the newly formed bond, 'envelope' or 'peripheral' rings must not be present in the synthetic subset of rings. The elimination of peripheral rings can be accomplished by the use of the exclusive *OR* operation on pairs of rings. The algorithm[10] for ring perception which is based on this elimination is summarised in Scheme 6. Definitions for some of the graph theoretic terms used in this summary appear in Scheme 5. The 'synthetic subset' is defined as the set of all minimum spanning rings plus any rings of size ≤ 6.

1. *Ring sum* = logical exclusive *OR* of rings ($\oplus$).

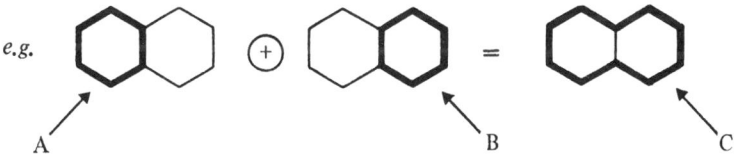

C contains bonds in A or B but *not both*, $A \oplus B = C$

[10] E. J. Corey and G. A. Petersson, *J. Amer. Chem. Soc.*, in press.

2. *Spanning tree* = acyclic molecular graph corresponding to same cyclic graph.

e.g. for [square ABCD] , spanning tree = [A–B, A–D, D–C path]

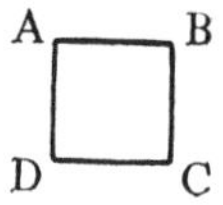

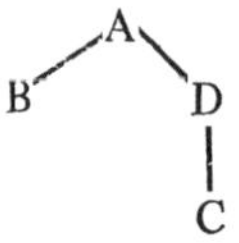

3. *Ring-closure bond* = bonds required for conversion of a spanning tree to corresponding cyclic graph.
   e.g. for above example, bond BC.
4. *Fundamental ring* = spanning tree + a ring-closure bond.

Scheme 5 *Perception of rings in polycyclic molecules. Some definitions*

Eliminate acyclic appendage atoms (successively eliminate atoms of connectivity 1); if cyclic order > 0:
1. *Grow a spanning tree. Find fundamental rings (FR)*:
   Encountering an atom already in the spanning tree indicates FR.
2. *Remove envelope rings to form reduced basis*:
   For each triplet of rings, R_i, R_j, $R_i \oplus R_j$, retain two smallest rings.
3. *For each ring-closure bond, b_c, find rings containing b_c not larger than FR or largest reduced basis ring:*
   Grow a tree from each end of b_c; a common atom in the two trees indicates a ring; iterate until the smaller of (a) fundamental or (b) the largest reduced basis ring is found.
4. *Order these rings by size and store as bond sets* (U FR = {ring bonds}).
5. *Select smallest ring* not in {MSR} with bonds not in U MSR and place them in {MSR}; iterate until U MSR = U FR.

Scheme 6 *Ring perception algorithm*

An indication of the efficiency of the ring perception technique may be found in the performance of LHASA-10 for the case of dodecahedrane, an undecacyclic structure with a total of 1168 possible rings. The synthetic subset which consists of 12 rings (all the five-membered rings) can be found in 0·264 s of PDP-10 time, and only 12 rings need to be grown. The various rings in a synthetic subset are stored in list form,[9] and the atoms and bonds in each ring are stored as sublists of the rings list. This is illustrated by Figure 2 for the specific example of bicyclo[2,1,0]pentane.

Other types of structural information are perceived by the program in considerable number. These include (i) interconnecting paths (*e.g.* between pairs of structural features such as functional groups or asymmetric centres), (ii) appendages on rings or functional group origins, (iii) relative levels of reactivity of each type of functional group in terms of high, low, or normal steric accessibility, electrophilicity, or nucleophilicity, (iv) sensitivity of functional groups to

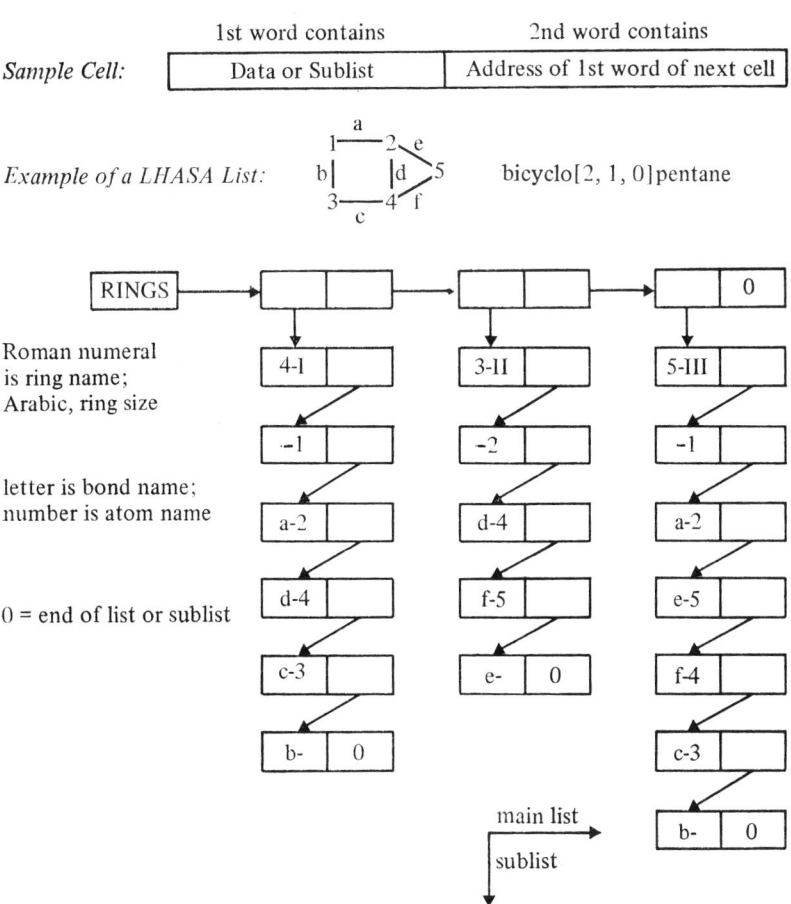

Figure 2 *Linked lists*

various reagents (*e.g.* oxidising, reducing, acid, or base), (v) especially strategic bond disconnections, (vi) aromatic ring systems, (vii) dihydroaromatic ring systems, (viii) asymmetric centres, and (ix) stereorelationships between groups on asymmetric centres. A procedure for the perception of stereochemical features and stereorelationships has already been developed for LHASA,[11] and this will serve as a basis for a stereochemical capability in synthetic analysis.

In the presently existing programs the perception of such fundamentally important structural features as are described above occurs prior to transform selection and, indeed, provides the basic information required for transform selection. However, additional perception of a much more varied and context-

[11] E. J. Corey and W. J. Howe, to be published.

dependent sort is needed for evaluation of the suitability of the various theoretically useful transforms. These perceptual processes are carried out by the program at the later stage of transform evaluation, as will be described in the next section. Machine perception therefore plays a key role in strategy selection, transform selection, and transform evaluation.

4 Organisation and Utilisation of Chemical Data

The process of generating an antithetic tree depends upon the recognition of key structural features of a target (or 'parent') structure which signal the applicability of certain transforms to the manipulation of the target structure. Once identified, each of these transforms can then be utilised to derive the structure of the corresponding precursor. The flow of events may thus be regarded as:

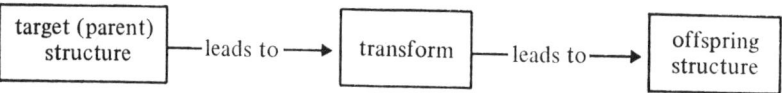

The identification of applicable transforms is made on the basis of the target structure and is independent of offspring structure.[12] Clearly then, it is both possible and useful to classify transforms according to the nature of the critical structural features ('synthons')[4] of a target molecule to which the transforms may be keyed. Several important classes of transforms are illustrated by the following entries:[13]

1. Transforms which create two functional groups in the synthetic product (group pair transforms).

$$\underset{R\ R}{\overset{O\ R\ OH}{\underset{|\ |}{R-C-C-C-R}}} \;\overset{\text{synthetic (s)}}{\underset{\text{antithetic (a)}}{\rightleftarrows}}\; \underset{R}{\overset{O\ R}{R-C-C-H}} \quad \underset{R}{\overset{O}{C-R}}$$

(aldol transform)

[12] As is indicated in a later section, however, the *evaluation* of the merit of a particular transform in a specific situation is dependent on *both* parent and offspring structures. Nonetheless, since for a particular transform the offspring structure is determined by that of the parent, it is possible to express such an evaluation solely in terms of parent structure.

[13] For a discussion of 'transform'-based data tables and the method by which these are used for the automatic generation of synthetic intermediates by LHASA, see E. J. Corey, R. D. Cramer, and W. J. Howe, *J. Amer. Chem. Soc.*, in press.

2. Transforms which change one functional group and which also modify the structural skeleton in the synthetic product (single group transforms).

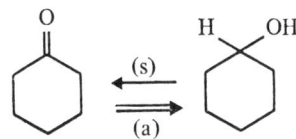

3. Single group transforms which modify only a functional group (functional group interchange or FGI).

4. Transforms which add a functional group in the antithetic direction (FGA).

or

or

5. Transforms which form or modify some particular type of ring (ring or cycle transforms).

6. General pair transforms (synthetically significant pairs other than FG pairs).
Examples: FG + appendage, FG + ring fusion, FG + stereocentre.

(C=O + β-R)

(R_{axial} + OH_{axial})

(C = O + *trans*-decalin)

7. Stereochemical transforms.

Each class of transforms may be subdivided in various ways for purposes of convenient organisation, table searching, or use by a computer program. For example, the functional group pair transform class may be segmented according to the number of atoms in the path which connects the two functional groups. The 'synthon' for a specific group pair transform, which consists of the group pair and the interconnecting path, likewise can be placed into an appropriate subdivision of the 'two-group synthon class' on the basis of path length. Ring transforms can be subdivided according to ring type—alicyclic, heterocyclic, aromatic, dihydroaromatic, for example—and also ring size. Clearly there are a very large number of important subdivisions in the ring transform class even with only these two distinguishing criteria. Despite the evident proliferation of subdivisions, it is useful to make a number of further distinctions between individual transforms based on other important properties. For example, a stereospecific transform which requires a particular stereochemical arrangement within the target structure should be differentiated from a non-stereospecific transform or even a stereospecific transform for a diastereomeric arrangement. In general, distinctions between transforms may be based on the structural *changes* they effect as well as on the basis of synthon type. Transforms may result in a change of:

(i) molecular skeleton (disconnection, connection, or rearrangement)
(ii) functional groups (addition, removal, interchange)
(iii) stereocentres or stereorelationships (addition, inversion, removal).

The separation of single group transforms into disconnective and FGI classes in the manner indicated above is the result of the consideration of both synthon type and structural change.

On an even more general basis it is important to note that certain transforms simplify molecular structure (in the antithetic direction), whereas others either do not, or actually cause an increase in structural complexity. Transforms of the last two types are obviously useful if their operation results in the generation of structures which then are susceptible to the operation of simplifying transforms. Transforms which simplify molecular structure vary with regard to the degree of simplification which their application produces; those whose operation results in major simplification are clearly more powerful than those which yield only a small decrease in molecular complexity. The Diels–Alder transform is one of the most powerful of all, since its application can result simultaneously in (i) a decrease in the number of rings, (ii) a decrease in the number of asymmetric centres, (iii) disconnection of molecular skeleton to generate two fragments, and (iv) simplification of functionality. (An equivalent statement can, of course, be made concerning the effectiveness of the Diels–Alder *reaction* in *increasing* molecular complexity in the *synthetic direction*.) Pair transforms which disconnect molecular skeleton or remove functionality and/or stereorelationships also are of considerable power, though in general the ring transform group may be regarded as the most powerful. It is evident that one strategy which should be useful in a computer program for synthetic analysis is that of trying to apply the most powerful transforms even though direct application may not be possible for a particular target structure. When the target structure lacks one or more of the features required for the application of a major simplifying transform, the strategy is to define the direct application of that transform as a goal toward which a number of other steps (*subgoals*) may be tried. These steps will in general involve less powerful or even non-simplifying transforms. Success of the strategy requires the generation from a target structure of a series of intermediate 'subgoal' structures leading finally to a 'goal' structure which allows the effective operation of the major or simplifying transform. This approach is treated in somewhat more detail in the section on strategies which follows. It is a strategy which chemists frequently are influenced by in some measure even though they may not have articulated a completely systematic and formal strategic technique.

The basic organisation of the chemical data tables in LHASA has been formed about the 'transform' as a key element. For *each class* of transform there is a 'data table' which contains an entry for *each transform* within the class. Each table entry contains the following types of information: (i) transform name, (ii) characteristic synthon, (iii) an intrinsic (target-independent) numerical rating of the transform (to be used in transform evaluation), (iv) the bonds within the synthon which are made or broken, (v) a set of conditional statements ('qualifiers') which cause the basic rating to be increased or decreased by certain amounts if certain structural features are present in the particular target structure. These conditional statements, which reflect what is known about the 'scope' of a given transform (or the corresponding reaction), allow the derivation of a rating for a transform as applied to a particular target. This rating is essentially a measure of the probability that the synthetic process corresponding to the transform is a realisable operation.

Within LHASA there exist packages of chemical information which might be termed 'chemistry units', each of which consists of three components.[13]

1. A data table of the type described above which refers to one *class* of transform.
2. A transform-choosing program which matches the features of a target (or 'current target') structure against the data table, evaluates all transforms for which there is a match, and stores all transforms passing evaluation with a rating above a pre-set cut-off value.
3. A transform-executing program which executes stored transforms one-by-one to generate new intermediates in the synthetic tree.

The chemistry unit for the two-group class of transform will now be described to illustrate in somewhat greater detail the operation of that part of LHASA which actually is concerned with the manipulation of chemical structures and the generation of an antithetically directed tree of intermediates.[13] At present there are about 125 entries in the two-group unit in LHASA-1. For each of the $n(n - 1)/2$ pairs of functional groups in a molecule of n functional groups, the interconnecting path(s) are determined and the occurrence(s) of full matches with the table entries for each are recorded. Further, for certain of the more powerful transforms in the two-group class, part-matches (one functional group and path, but not the other functional group) are also determined for use later in connection with the application of functional group interchange (FGI) transforms as a subgoal of group-pair transform application. The transform(s) corresponding to full matches are then evaluated from the 'qualifiers' in the data table to derive a rating for each transform.

The operation of the LHASA scheme for the evaluation of suitability of a transform within the context of a particular target structure is best explained by the use of an example. The aldol transform, being both suitable for this purpose and relatively important within the class, is chosen for the illustration. A very brief summary of the kinds of information found in the table entry for the aldol transform is presented in Scheme 7.[14] As indicated in item 1, the aldol transform is assigned an intrinsic rating of 70 (relative to a cut-off value of -50) and is

Brief summary of table entry:

1. ①②
    HO-C-C-W $\Longrightarrow$ O=C+H-C-W

    HO at atom①, W (an electron-withdrawing group) at atom②

    2-atom path, bond 1 broken, initial rating 70, try FGI (subgoal flag).

2. *'Standard' qualifiers*—statements modifying initial rating according to target structure—combined by inclusive *OR*.

    . . *addt 30* / *if* / *grp2*   / . . *is nitro*
    action   optype   modifier phrase

3. *Control phrases* permit qualifiers to interact by modes such as logical *AND*, exclusive *OR*.

4. *Condition statements* pertain to reaction conditions.

Scheme 7 *Table-driven rating of an aldol transform*

designated as a goal in case of a part match *via* the FGI subgoal. The example of a standard qualifier which is given in item 2 instructs the computer to increase the rating by 30 if group 2, the electron-withdrawing (W) group, is nitro. This and other statements in the data tables are written in a 'chemist-oriented' higher level language which is translated by a separate program (compiler) into machine language. The aldol entry in LHASA-1 contains *ca.* 40 standard qualifiers which raise or lower the transform rating by certain amounts. A representative collection of structural features in the target structure which raise or lower the rating is given in Scheme 8. The complete listing of the aldol table entry, which is presented elsewhere,[13] should be consulted for additional detail, including the use of control phrases to permit qualifiers to depend upon certain other qualifiers

---

[14] This table for the 'aldol transform' corresponds to what would be appropriate for the 'aldol reaction' in the synthetic direction. For the 'retro-aldol transform', which may be defined as the transform corresponding to the 'retro-aldol reaction', an *entirely* different table is required. Also it should be noted that although the cyclic version of the retro-aldol transform is a two-group transform, the non-cyclic version is a one-group transform (connective).

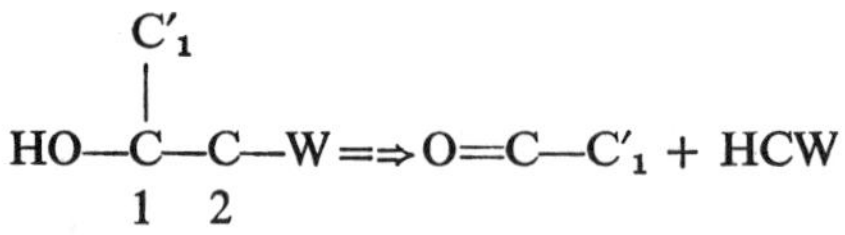

$$\underset{12}{\text{HO}-\text{C}-\text{C}-\text{W}} \Longrightarrow \text{O}=\text{C}-\text{C}'_1 + \text{HCW}$$
(with C'_1 substituent on C_1)

Rating decreased by: Hal, O, or S β to W; W = CONHR, CN, COOR; C=C— on C_1; for each Alk at C_2; W = CH=CHW; presence acid- or base-sensitive groups elsewhere.

Rating increased by: additional W at C_2; W = NO_2; no hydrogens on C'_1.

If C_1—C_2 in ring: rating is decreased if ring size other than 5 or 6 and increased for 5 or 6; rating is decreased if W or Alk groups exist at positions (*e.g.* C'_1) which would favour aldol cyclisation to a structure other than target.

Scheme 8 *Rating of an aldol transform*

(*i.e.* a 'nesting' of qualifiers). Condition statements play the useful role of allowing the identification of interfering groups remote to the reaction site. The target structures (1)—(4) which are candidates for the aldol transform, since they each possess the required synthon, can be used to exemplify the rating performance of the present version of the aldol table entry. The current LHASA-1 ratings of structures (1)—(4) are, respectively, $+300$, fail, -30, and fail (cut-off = -50).

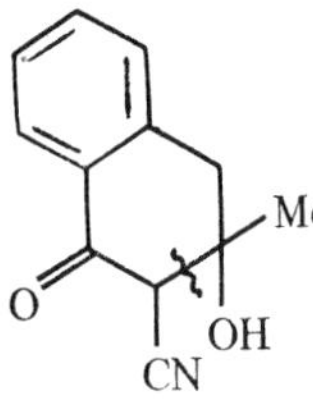

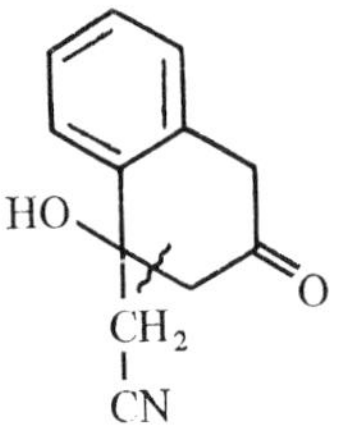

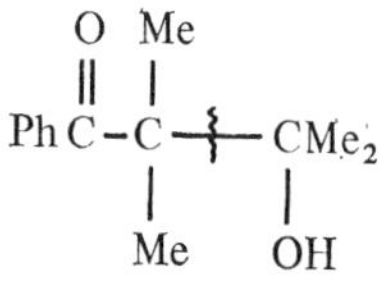

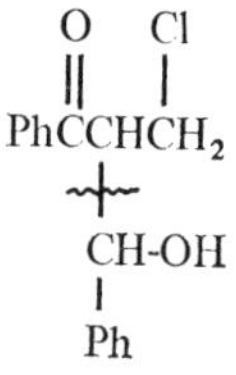

The ratings are used to circumvent the generation of intermediates corresponding to naïve or highly dubious synthetic processes and also to order the output of structures on a given level of the tree of synthetic intermediates. The intermediates for each level are displayed in order of decreasing rating by machine. The chemist can cause the computer to make any specific further deletions which he deems appropriate by use of the Rand tablet input; further, he can alter the cut-off value of the ratings.

The group-pair transforms are frequently highly effective in the generation of synthetic pathways to a target structure. Many of the published syntheses of alkaloids, for example, can be reproduced by machine using almost exclusively group-pair transforms. The pathways of synthesis shown in Schemes 9 and 10 were derived by computer solely through application of the group-pair chemistry unit.

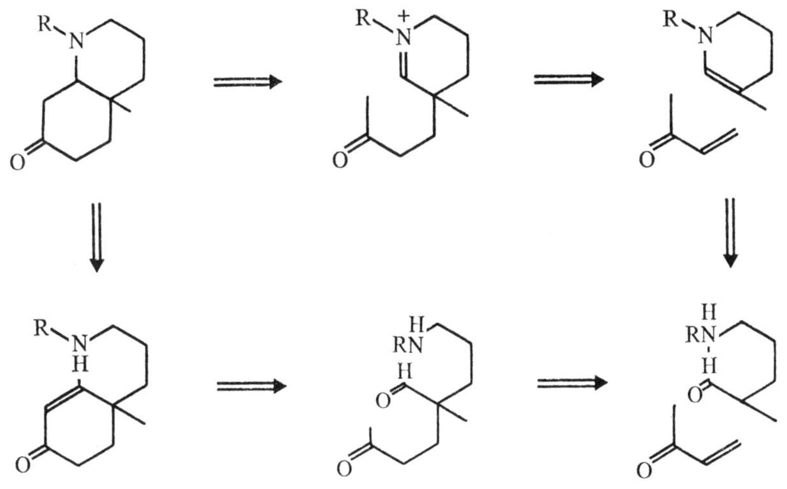

Scheme 9

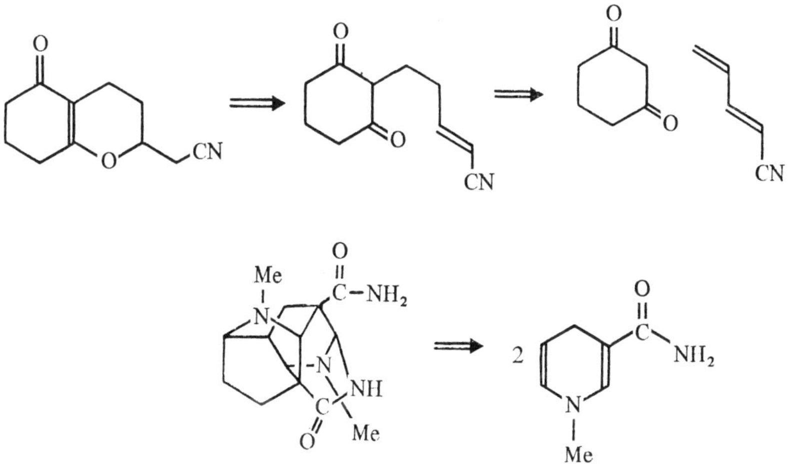

Scheme 10

When a part match occurs between a specific functional group-pair-path combination in some target structure and a group-pair table entry which corresponds to an important simplifying transform, a request is made to a functional group interchange (FGI) chemistry unit to ascertain whether there exists an FGI transform which can convert the non-matching group into that required for match to the pair table entry. An FGI transform selecting program scans the FGI data table to find whether the required transform exists and is applicable to the specific target structure. If this subgoal is achieved, FGI transform execution then occurs to generate a new intermediate which is further modified by the action of the pair transform that generated the FGI request initially. Two examples of sequential FGI and group transform application as executed by LHASA-1 are shown in Scheme 11.

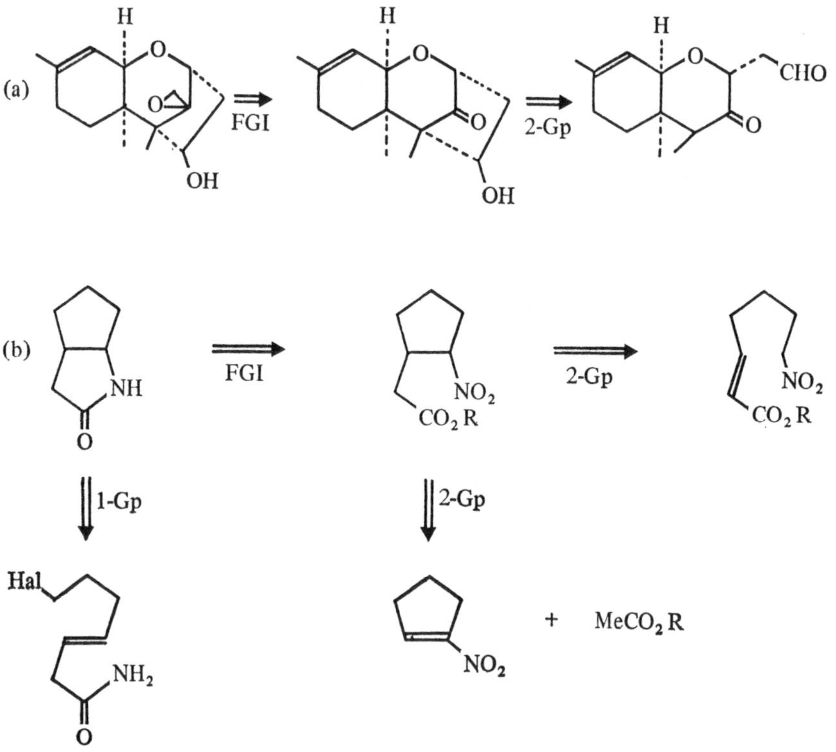

Scheme 11 *Application of FGI transforms as a subgoal of pair transforms*

The FGI and functional group addition (FGA) transforms indicated above are applied only as subgoals which allow the utilisation of a simplifying transform, for example, of the pair or ring type. Disconnective single functional group transforms are more versatile. Although they may be used to generate subgoals to satisfy some defined goal, they are also allowed by LHASA to operate *directly* under certain circumstances including cases where the target structure has less than four functional groups or cases where a single group transform effects 'strategic bond disconnection' (*see later*). When the opportunities for the exercise of single group transforms are especially numerous, their use must be controlled by one or more strategies.

The applicability of ring transforms to a particular target structure is normally quite limited (even more so than for group-pair transforms), and there is no difficulty in selecting and executing such transforms where direct matching techniques suffice. However, it is often the case that some ring transform can be applied successfully only after a number of other transform types are utilised to pave the way for a direct match. Some techniques for accomplishing the analysis required for this approach are discussed in the following section on strategies. The use of *general* pair transforms has obvious utility, since it broadens the range of synthons which can be matched against an organised data table and since its application can be accomplished by the same techniques which are used for the group-pair class of transform.

5 Strategy Selection

Probably the most fascinating and exciting area of the theory of synthetic analysis is that which concerns the strategies of synthetic chemistry and their effective use in problem solving or, perhaps more aptly, problem simplification.[4] The creative challenge, the formidable intellectual barriers, and the satisfaction which are associated with the design of a fine synthetic plan are all rooted in the process of devising a good strategy. Although only a brief discussion of this subject can be presented here, a more comprehensive and detailed treatment is planned for a future publication.

Synthetic organic chemistry makes use of a considerable number of different strategies, and it is no mean task to select one which is singularly appropriate to a problem. It is even more difficult to invent a basically new and original strategy in response to a refractory and complex problem. It is not unexpected therefore that the formulation and selection of strategies for a wide range of synthetic problems is the most formidable task in the development of a sophisticated computer program for synthetic analysis. The process of strategy selection and its relationship to synthetic problem solving are outlined in simplified form in Scheme 12. Analysis must start with perception, which in the initial stage is

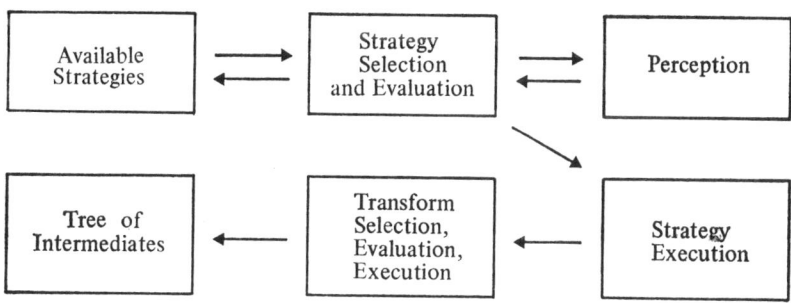

Scheme 12 *Synthetic problem solving*

systematic and also limited to certain basic features of the target molecule, and with explicit information on the various available strategies. If strategy selection is not to be perception-limited, there must be an additional type of perception based upon (and driven by) the more specialised data required for selection of individual strategies. This situation involving target-driven and strategy-driven perception (effectively a recursive type of perception) is clearly similar to the analog for transform selection and evaluation discussed earlier, although more complex information structures are involved. It is also true that the information content of individual strategies and their evaluation are more complex than is the case for transforms and, of course, our basic understanding of synthetic strategies in generalised form is relatively primitive. The analogy between 'transform' and 'strategy' is, however, useful in helping us to scrutinise the latter, even though it straddles two different levels[4] of problem solving. Like 'transforms', strategies can be allowed to operate independently of one another on a particular target structure; further, the more strategies which are available, the greater will be the power of a general problem-solving procedure. A rational and systematic classification and definition of strategies is fully as essential as in the case of transforms. Finally, just as there is a variation among transforms in their 'power' to reduce molecular complexity, there is an analogous variation of power among strategies. No strategy is universal, some strategies will only rarely be useful, and some strategies are helpful specifically because they remove the obstacles to the application of other strategies. Furthermore, the application of various strategies at more than one level of a synthetic tree means that along one or more vertical pathways in the tree there will be a *de facto* 'nesting' of strategies with those of the upper part of the path being of greater influence and consequence than those applied below.

Many of the most useful synthetic strategies fall into four categories which can be summarised as:

1. Strategies based on particular structural characteristics of a target molecule.
2. Strategies based on the selection of certain key transforms, or more generally, certain chemical information, the application of which becomes a goal.
3. Strategies based on the matched development of 'antithetic' and 'synthetic' trees, or certain assumed starting materials.
4. Strategies based on certain special (but external) circumstances connected with the problem (for example, the desirability of synthesising two or more related compounds *via* a common route or common intermediate).

The last two categories of strategy will not be considered here, since they are outside the scope of the LHASA programs. In the first category are included strategies which deal with the selection of certain structural features for modification because of their reactivity. For example, the presence in a target of one or more functional groups which would be highly sensitive to acids or bases signals the strategy of removal or modification of that group by the application of an appropriate transform, since it would clearly be difficult to carry out a multistep synthesis with such a group present. The existence within a target structure of certain functional groups which are very easily interconvertible with others indicates the desirability of considering several 'close relatives' of the target structure (for instance: cyclic ketal *and* ketone; or carboxylic acid *and* carboxylic ester; or lactone *and* hydroxy-acid; or $\alpha\beta$- and $\beta\gamma$-enones). In molecules containing a large number of functional groups (especially of the same type, or clustered on contiguous or nearby atoms), the application of transforms leading to reduction in the number of functional groups (for instance, transforms which generate unsaturated or aromatic units by elimination or connective processes) may be heuristically effective in uncovering especially simple synthetic routes. Functional groups which interfere with the operation of important transforms (for example, by their presence within a synthon) and which are detectable by recursive perception are especially deserving subjects for this strategy.

Tricyclic or higher polycyclic ring systems, especially of the bridged ring type, provide an opportunity to apply a network-oriented strategy which also comes under the first category. This strategy depends upon the possibility of identifying certain bond disconnections which are strategic in the sense that they lead to especially simple or accessible ring systems of lower cyclic order. A relatively simple but nonetheless useful algorithm for identifying such strategic bond disconnections for polycarbocyclic systems which has been implemented in LHASA-1[15] is outlined in Scheme 13.[16] This procedure leads to the generation of intermediates in which the following structural features are minimised (m) or avoided (a): (i) appendages (m), (ii) appendages carrying asymmetric centres (a), (iii) rings having greater than six members (m), and (iv) bridged rings (m). Rule 5, which is the only stereochemical provision in the algorithm outlined in Scheme 13, is itself a very powerful strategic guide that is both easy to apply and extremely useful from a chemist's point of view. Surprisingly, this rule and the others given in Scheme 13 have not previously been formulated. An example of

A strategic C—C bond must
1. be *endo* to a 5-, 6-, or 7-membered ring.
2. be *exo* to a ring larger than 3.
3. be a *perimeter bond*. (The *set of perimeter bonds* is obtained from all pairs of minimum spanning rings by taking the ring sums (logical exclusive *OR*, *XOR*, $R_i \oplus R_j$), or $R_i \cup R_j$ if R_i and R_j are both smaller than $R_i \oplus R_j$ but $R_i \oplus R_j$ is not larger than 6.)
4. be *endo* to a ring of maximum bridging [*i.e.* ring(s) bridged to max. no. of other rings].
5. not leave stereocentres on side-chains after cleavage.
6. minimise the cyclic order of the largest resulting substructure.

Scheme 13 *Rules for identification of strategic bond disconnections for polycyclic structures*

[15] E. J. Corey and G. A. Petersson, to be published.
[16] For a forerunner of historical interest see E. J. Corey, M. Ohno, P. A. Vatakencherry, and R. B. Mitra, *J. Amer. Chem. Soc.*, 1964, **86**, 478.

the application of the algorithm to the tricyclic molecule sativene[17] is given in Table 3. With a few minor additions this algorithm for strategic bond disconnections can be extended to cover polycyclic systems containing the heteroatoms O, N, and S.

Table 3 *Sample analysis for strategic bond disconnections*

Rule*	Bond No.											
	1	2	3	4	5	6	7	8	9	10	11	12
1	×	×	×	×	×	×	×	×	×	×	×	×
2	×	×	×	×	×			×	×		×	×
3	×	×	×	×	×	×	×		×	×	×	
4		×	×	×					×			×
Strat. Bonds	×	×	×									

* As in Scheme 13.

At this point mention should be made of a most interesting interactive feature which has been incorporated into LHASA-1. Through the Rand tablet and pen device for graphical input together with a graphically displayed control switch accessible on the tablet, the chemist can specify one or more bonds in a target structure as 'strategic' for disconnection. This simple device gives the chemist a powerful and unusual tool for guiding (and experimenting with) the course of synthetic analysis. It can even be used in effect to specify which part of a target structure is the starting point for the synthesis (by the designation of all other bonds as strategic).

The automatic or interactive identification of one or more strategic bond disconnections has an important part in other target-oriented strategies which have been devised. The designation of some bond disconnections as strategic can serve as a guide to enable the selection of one or more single functional group transforms or general pair transforms which break that bond. One of the schemes for synthesis of longifolene as generated by LHASA-1 using this guidance of single group transform selection is illustrated in Scheme 14. If reducible to practice, this would constitute a very simple route to this interesting sesquiterpene

[17] For a simple synthesis of sativene according to the guidance of this strategy, see J. E. McMurry, *J. Amer. Chem. Soc.*, 1968, **90**, 6821.

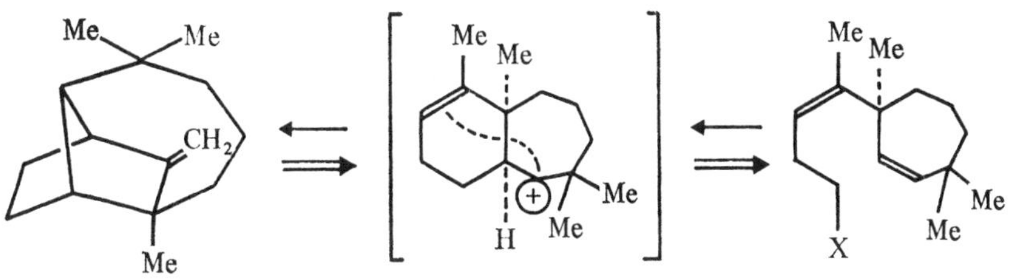

Scheme 14 *A hypothetical synthetic route to longifolene*

which has already been synthesised by another route[16] which also happens to follow the 'strategic bond' strategy.

Clearly, a designated strategic bond can also be used to direct the *introduction* of a functional group, *i.e.* FGA transform selection, if that bond is not disconnectable by other types of transforms, for example, pair or single group transforms. A less obvious technique based on strategic bond disconnections permits the effective use of double functional group interchange as a subgoal of important group-pair transforms even when there is *no* match between a group-pair in a target structure and a group-pair table entry. The algorithm for accomplishing this double FGI operation for the aldol transform is given in Scheme 15.

1. Strategic bond on path between functional groups G_1 and G_2 in 5- or 6-membered ring?

2. If yes, $G_1 \xrightarrow{FGI}$ OH attached to C(1), where C(1)-C(2) = strategic bond?

3. If yes to (2), $G_2 \xrightarrow{FGI}$ W attached to C(2)?

4. If no to (2), $G_2 \xrightarrow{FGI}$ OH attached to C(2)?

5. If yes to (4), $G_1 \xrightarrow{FGI}$ W attached to C(1)?

Scheme 15 *Double FGI as a subgoal of the aldol transform using an identified strategic bond disconnection*

A specific application of this strategy which leads to an already demonstrated synthesis of the plant toxin helminthosporal is outlined in Scheme 16.[18]

Another set of target-oriented strategies, which in a certain sense represent an opposite of strategic bond disconnections, are those which lead to the goal of applying to a target a connective transform which introduces a bond or a bridge between two of the atoms in the structure. These 'connective' strategies depend for their use on the presence within the target of units such as: (i) asymmetric centres on a functionalised appendage or chain, (ii) a functionalised appendage at an asymmetric centre in a five-, seven-, or higher-membered ring, (iii) two appendages (functionalised or non-functionalised) in proximity and involved in

---

[18] The total synthesis of helminthosporal has been described by E. J. Corey and S. Nozoe, *J. Amer. Chem. Soc.*, 1963, **85**, 3527.

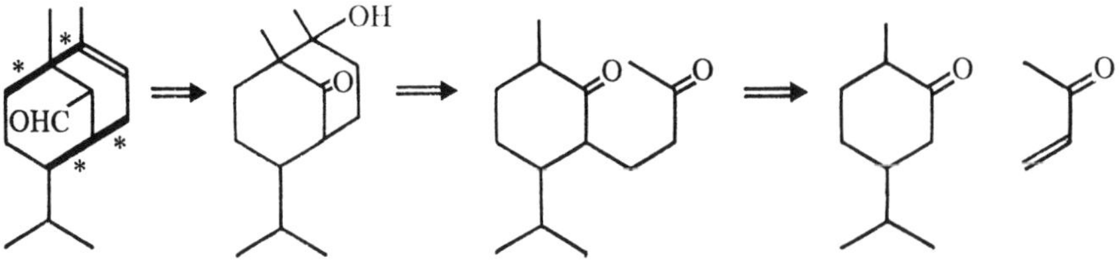

*Strategic bond disconnection

Scheme 16 *Application of double FGI and group-pair transforms to a synthetic scheme for helminthosporal*

non-bonded repulsion, and (iv) a ring of medium size, especially of 8, 9, or 10 members but in certain circumstances also of 7 or 11 members. The major goal of these connective strategies is the generation of six-membered cycles. It will be seen that condition (i) for the application of a strategy of connection is the opposite of rule 5 of Scheme 13 which refers to a disconnective operation. A case which illustrates how effective the connective strategy in situation (i) can be is that of the acyclic ketoamido-diester (5). An outstandingly clever application of connection under condition (ii) is found in the recently outlined synthetic approach to vitamin B-12.[19] A fine use of the connective strategy under circumstances of type (iii) is seen in the synthesis of o-di-t-butylbenzene.[20] The strategy implied in the application of oxidative C–C cleavage and fragmentation reactions to the synthesis of medium ring compounds falls in the category covered by condition (iv) as, for example in the synthesis of caryophyllene.[21]

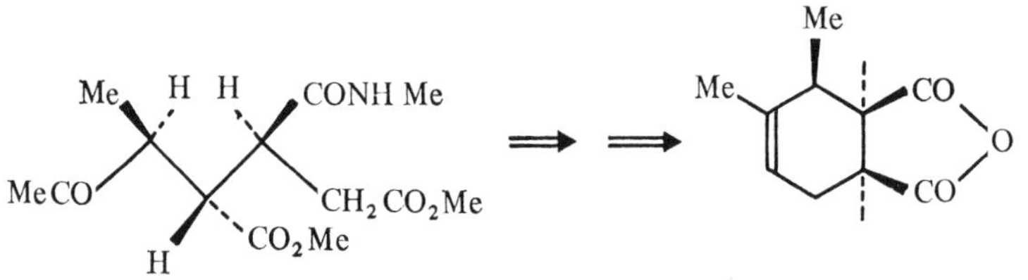

Several of the most important target-based strategies are stereochemical in nature. Among the most interesting of these is a strategy being developed for inclusion in LHASA for restricting the order of removal of asymmetric centres from a target structure which contains three or more such centres. Basically the strategy depends upon the perception of stereorelationships between groups

[19] R. B. Woodward, *Pure Appl. Chem.*, 1968, **17**, 519. An especially convenient flow chart of this synthesis appears in the excellent compilation of N. Anand, J. S. Bindra, and R. Ranganathan, 'Art in Organic Synthesis', Holden-Day, San Francisco, 1970.
[20] L. R. C. Barclay, C. E. Milligan, and N. D. Hall, *Canad. J. Chem.*, 1962, **40**, 1664.
[21] E. J. Corey, R. B. Mitra, and H. Uda, *J. Amer. Chem. Soc.*, 1964, **86**, 485.

on asymmetric centres along a stereopath which connects the centres. An algorithm for this strategy is currently under test which directs the preferential removal of terminal or peripheral stereocentres on the stereopath.

A final comment with regard to target-based strategies concerns what might be called *opportunistic* strategies. These are applied whenever some particular structural feature occurs in the target molecule. For example, the group-pair transforms can be utilised opportunistically with considerable effectiveness. The occurrence of group-pair matches is usually sufficiently limited so that the intermediates so generated are not excessively numerous.

At this point it is appropriate to consider the transform-oriented strategies, a class which appears to be of major significance. The algorithm outlined in Scheme 15 for the application of double functional group interchange followed by a disconnective aldol transform in a sense deals with a transform-oriented strategy, although in this case it is one which is keyed by a target-oriented strategy (identification of a strategic bond disconnection). In fact, all transform-oriented strategies depend upon some initial perception of the target structure which serves the function of preselecting one or more transform-oriented strategies for trial. By way of illustration we may consider the preselection and application of transform-oriented strategies for a structure containing a non-aromatic six-membered carbocyclic ring. For this structural unit there are at least four powerful ring transforms which are of sufficient importance to justify trial application even if this would entail a search procedure of some length and complexity which may fail. These are the Diels–Alder, Robinson annulation, Birch reduction, and cation–olefin cyclisation transforms. Each transform which has been identified by preselection is examined separately. In the case of the Diels–Alder transform, an appropriate subclass of the search strategy for this transform is entered according to whether the six-membered ring in question is an isolated ring, part of a fused ring system, part of a bridge ring system, or part of a spiro ring system. Next a rating is derived for the search strategy on the basis of the particular substitution, functionality, and stereochemistry about the six-membered ring in question. This rating determines the maximum depth of search (and number of intermediates) which will be allowed in the attempt to apply the Diels–Alder transform. The search for subgoals follows a binary decision pattern and is table driven.[22] Typically, an entry in the search tree poses a question (*e.g.* endocyclic C=C present?) which leads to one follow-up question if the answer is yes or a different follow-up question if the answer is no, and the process is continued. The elements in the binary search tree refer either to whether some structural feature is present on the six-membered ring in question or whether some transform is applicable.[23] A pathway leading to a success point corresponds to a sequence of transforms which, if applicable, would produce from the target structure an intermediate containing the essential features

[22] E. J. Corey, D. A. Pensak, W. J. Howe, and R. D. Cramer, to be published.
[23] Certain parts of the binary decision tree which occur multiply can be handled as subroutines. For example, the subroutine 'epimerise' contains all the necessary decision elements to test for and attempt epimerisation at some centre on the six-membered ring.

required for the operation of the Diels–Alder transform. The accumulated list of transforms on that pathway are then selected for successive application and evaluation. If none of these transforms fail, the goal has been reached of matching the original target to the Diels–Alder transform, which is then applied and evaluated in the normal way.

The above example of a transform-oriented strategy illustrates the current approach of the project at Harvard to one of the most complex and difficult areas of synthetic problem solving. It also provides some grounds for an optimistic view of the eventual possibilities of computer-assisted synthetic analysis as an aid to the chemist and a guide to those who would teach and/or learn this fascinating branch of science. In the final analysis the effectiveness of a general problem-solving computer program must be judged by its performance on a range of specific problems together with the extent to which it includes important chemical information on strategies and transforms. On this basis the performance of LHASA-1 (which is now being equipped with a stereochemical capability) is decidedly encouraging. Among those chemists for whom the program has been demonstrated there is essentially complete agreement on this point and an enthusiasm for the excitement and liveliness which result from the interactiveness and graphical communication which are basic to the functioning system.

Acknowledgment. A major part of the credit for the progress which is outlined in this lecture belongs to a group of unusually able and dedicated research students who have worked on the LHASA and OCSS programs at Harvard. These individuals (in the order of joining the project) are: W. Todd Wipke, W. Jeffrey Howe, Richard D. Cramer, David A. Pensak, Donald E. Barth, and George A. Petersson.

Financial assistance from the National Institutes of Health and the National Science Foundation is also gratefully acknowledged. The Advanced Research Projects Agency provided support in the form of grants to the Harvard Centre for Research in Computer Science and the PDP-1—PDP-10 facility.

AUTHOR CITATION INDEX

Abbey, K. M., 194
Abe, H., 280
Abel, R., 178
Abramson, F. P., 121
Abrahamsson, S., 280, 291
Achenbach, M., 121
Ackerman, J. L., 6, 126
Aczel, T., 321, 323
Adam, H. H., 216
Adams, R. N., 216
Ader, R. E., 6, 126
Ahlberg, J. H., 216
Alberts, G. S., 215
Aldous, K. M., 50
Alkins, J. R., 151
Allen, L. C., 240
Allerhand, A., 141
Alon, P., 240
Althaus, J. R., 288
Altman, D. E., 151
Altman, L., 50
American Instrument Co., 128
American Petroleum Institute Research, 282
Amodei, J. J., 178
Amy, J. W., 154, 158, 161, 259
Anand, N., 380
Anderson, D. H., 267, 290
Anderson, L. W., 151
Anderson, P. J., 121
Anderson, R. F., 178
Anderson, W. A., 137, 224
Andrieux, C. P., 216
Angerstein-Kozlowska, M., 6, 7, 176
Anson, F. C., 178, 216
Argaver, R., 128
Arsenault, G. P., 155, 328
Ash, J., 266
Ash, K. C., 255
Atkinson, T. V., 104
Averbach, B. L., 239
Axelson, M., 121

Bacchin, R., 121
Bacon, V. A., 279
Bailey, C. A., 41
Bankston, D. C., 251
Barber, M., 154, 328
Barclay, L. R. C., 380
Bard, A. J., 215, 216
Barradas, R. G., 7, 176
Bartholdi, E., 238
Barton, D. E., 277
Bassler, G., 256
Baumann, F., 80
Becker, E. D., 141, 238
Bennett, T., 50
Benson, F. C., 7, 176
Berman, S., 16
Bernstein, H. J., 238
Bertsch, W., 121
Berztiss, A. T., 356
Beynon, J. H., 153
Bieber, L., 134
Biemann, K., 79, 112, 121, 153, 155, 159, 161, 271, 279, 280, 286, 288, 291, 321, 328, 329
Biller, J. E., 112, 288
Bindra, J. S., 380
Black, H. S., 137
Blaisdell, B. E., 121
Blesch, J., 80
Blount, H. N., 216
Board, R. D., 154, 158
Bolton, H. C., 239
Bommer, P., 271, 321
Bond, A. M., 176, 177, 194, 216
Booman, G. L., 80, 181, 216
Bowen, E. J., 132
Bowen, H. C., 157
Bracewell, R., 225, 240
Brand, L., 128
Braun, W., 306
Brazie, T. L., 290
Bregman, J. D., 240

Author Citation Index

Brickstock, A., 280
Bridges, J. C., 279
Brigham, E. O., 194, 238
Bristow, Q., 58
Britton, W., 216
Brown, A. C., 80
Brown, C., 267
Brown, E. R., 80, 180, 181
Brown, H., 112, 311
Brown, R. D., 158
Bryant, M. F., 152
Buchanan, B. G., 112, 155, 280, 328, 334, 354
Buchs, A., 112, 280, 328
Buck, R. P., 216
Budzikiewicz, H., 337, 354
Burger, H. C., 239
Burke, M. F., 91, 94
Burlingame, A. L., 121, 157, 161, 256, 271, 321
Bushor, W. E., 178
Butterworth, J., 231
Byron, P., 128

Caldecourt, V. J., 251
Capelle, G., 151
Carhart, R. E., 112, 311
Carpenter, R. O'B., 251
Carver, R. D., 41
Casey, R. G., 320
Cehelnik, E. D., 152
Chalmers, R. A., 121
Chang, R. A., 121
Chesler, S. N., 58
Chilcote, D. C., 112
Childs, C. W., 270
Childs, W. V., 215
Chrisman, R. W., 127
Christie, J. H., 215, 216
Claveau, J. C., 121
Clayton, E., 157
Clerc, J. T., 79
Clyburn, S. A., 255
Coburn, T. C., 279
Cochran, D. W., 141
Codding, E. G., 127
Cole, H., 16
Coleman, G. N., 251
Coleman, I., 230, 231
Comisarow, M. B., 238
Cone, C., 155, 328
Conrad, R. M., 127
Conte, S. D., 112
Conway, B. E., 6, 7, 176
Cook, H. D., 251
Cook, T., 50
Cooley, J. W., 6, 141, 240

Cooper, J. W., 222, 238
Corey, E. J., 356, 357, 358, 361, 363, 366, 377, 379, 380
Coutant, J. E., 256, 257
Covert, G. L., 267, 290
Cox, R. C., 128
Cox, R. E., 121
Cramer, R. D., 358, 361, 366
Crawford, L. R., 270, 271, 280, 328
Crawhall, J. C., 121
Creason, S. C., 181, 194
Crepeau, R. M., 127
Crisler, R. O., 80
Cronholm, T., 121
Crouch, S. R., 58, 255
Culp, R. A., 91
Curry, A. S., 267
Curstedt, T., 121

Dagnall, R. M., 128
Davidson, D. E., 306
Davis, S., 50
Dawson, J. B., 222
Dayringer, H. E., 121
DeAngelis, T. P., 104
de Boor, C., 112
Decius, J. C., 239
DeFord, D. D., 80, 180
de Galan, L., 239, 245
Delahay, P., 216
Delany, E. B., 256
deLevie, R., 194
Delfino, A. B., 112, 280, 328
Delmastro, J. R., 216
Deltour, J., 222, 243
Deming, S. N., 216
de Mul, F. F. M., 240
Dendramis, N., 121
Denton, M. S., 104
Desiderio, D. M., 159, 271, 321
Dessy, R. E., 50, 78, 238
DeTemple, T. A., 151
Digital Equipment Corp., 86, 94, 182
Dillard, J. W., 216
Djerassi, C., 112, 155, 280, 311, 328, 334, 337, 354
Donaghue, P. F., 288
Dowden, B. F., 41, 80
Drake, K. F., 194
Dromey, R. G., 112, 121
Drushel, H. V., 128
Dryden, P. C., 152
DuBois, E., 251
Duffield, A. M., 112, 121, 155, 280, 328, 334, 354
Dulaney, J., 12

Dupzyk, R. J., 41

Eagleston, J. F., 92
Eakin, D. R., 306
Eastman, S. W., 135
Eaton, H. E., 12, 127
Edelstein, S. J., 127
Ehlers, V. J., 151
Eldjarn, L., 121
Elliott, R. M., 154
Ellis, D. J., 222
English, J. C., 127
Enke, C. G., 127, 255
Ergun, S., 239
Erley, D. S., 267, 290
Ernst, G. W., 356
Ernst, K., 50
Ernst, R. R., 137, 141, 224, 238
Ettinger, D. G., 255
Evans, D. A., 288
Evans, D. H., 216

Faggin, F., 50
Fales, H. M., 266, 306
Farrar, T. C., 141, 238
Fausett, D. W., 126
Feigenbaum, E. A., 112, 155, 271, 280, 320, 328, 334, 354, 356
Fein-Marquart Associates, Inc., 306
Feldberg, S. W., 215, 216
Feldman, A. P., 306
Feldman, J., 320, 356
Feldmann, R. J., 266, 306
Felkel, H. L., Jr., 127
Feller, W., 276
Fennessey, P. V., 280, 329
Ferretti, J. A., 141
Fitch, W. L., 121
Fitzsimmons, W. A., 151
Flanagan, J. B., 216
Fleet, B., 177
Fleischmann, M., 216
Fletcher, S., 7, 176
Flores, I., 356
Flynn, G. J., 80
Förster, H.-J., 121, 288
Forsyth, A. I., 356
Foster, J. M., 362
Foster, J. Q., 323
Foster, K. L., 41
Frank, P., 251
Franklin, M. L., 251, 255
Frazer, J. W., 50, 153
Frew, N. M., 320
Frohman-Bentchkowsky, D., 58
Froix, M. F., 6, 126

Fry, A. J., 216
Fu, K. S., 320
Fugmann, R., 306

Gallaway, W. S., 88
Gates, S. C., 112, 121
Gayles, J., 41
Gebbie, H. A., 223
Geller, M., 151
George A. Philbrick Researches, Inc., 194
Giddings, J. C., 81, 94, 95, 112
Giessner, B. G., 256
Gilbert, T. W., 104
Gill, J. M., 50
Gladney, H. M., 41, 80, 240
Gladstone, B., 58
Glarum, S. H., 240
Glover, D. E., 194
Goedde, A. O., 6, 126
Goedert, M., 50, 58
Gohlke, R. S., 271, 282, 337
Golay, M. J. E., 58, 88, 94, 222, 243
Goldman, A., 240
Goldman, S., 255, 277
Gordon, W. A., 251
Grabaric, B. S., 177
Grant, G. C., 215
Grant, P. M., 41
Greene, B. N., 157, 161, 321
Griffiths, J. E., 255
Griffiths, P. R., 238
Grigsby, R. D., 280
Grotch, S. L., 121, 274, 278, 283, 291
Grove, W. M., 178
Grubner, O., 81, 94
Grushka, E., 81, 94, 112, 239
Guiochon, G., 58, 78
Gutnecht, W. F., 91, 153, 178

Häggström, G., 280
Hahn-Weinheimer, P., 251
Hall, N. D., 380
Halliday, J. S., 256
Hallman, P. S., 270
Halpern, B., 279
Hambidge, K. M., 251
Hamliton, R. G., 216
Hamming, M. C., 280
Hamming, R. W., 216
Hanck, K. W., 216
Hancock, H. A., 80
Hannon, D. M., 41
Hansch, T. W., 151
Hardy, A. C., 240
Harrar, J. E., 178
Harris, W. E., 58

Author Citation Index

Harrison, J. H., 215
Harten, J., 121
Hawley, M. D., 216
Hayes, J. W., 194
Hays, D. G., 356
Hazard, G. F., 306
Hedfjall, B., 161
Heineman, W. R., 104
Heller, S. R., 266, 306
Helz, A. W., 16
Herlicska, E., 80
Herman, H. B., 215
Hermans, J. J., 129
Hertel, R. H., 121
Hertz, H. S., 288, 291
Hettinger, J. D., 50
Hewitt, J. W., 135
Hieftje, G. M., 58
Hignite, C. E., 288
Hill, H., 306
Hirschfeld, T., 222
Hites, R. A., 112, 121, 161, 279, 286, 291
Holland, J. F., 112, 121, 134, 136, 279
Holt, R. M., 50, 58
Holub, K., 216
Honzik, W., 41
Hopkins, J. W., 222, 239
Horlick, G., 127, 223, 224, 238, 240, 251, 255
Horne, D. E., 41
Horning, E. C., 121
Horning, M. G., 121
Howe, W. J., 358, 361, 366
Howerton, H. K., 128, 129
Hsu, H. D., 239
Hubbard, J. R., 50
Hudson, J. B., 128
Hulbert, M. H., 216
Hung, H. L., 181
Hutterer, F., 121
Hyde, E., 266, 306

Ingle, J. D., Jr., 255
Intel Corp., 58
Irving, P., 160, 261
Isenhour, T. L., 271, 280, 320

Jackson, K. W., 50
Jacobus, D. P., 306
Jakob, F., 178
James, G. E., 91
Jansson, P-A., 161
Jeftic, L., 216
Jellum, E., 121
Jenkins, F. A., 251
Johnson, B., 41
Johnson, B. H., 321
Johnson, W. F., 121

Jones, A. F., 240
Jones, D. O., 91, 153, 178, 238
Jones, D. T. L., 88
Jones, G. E., 153
Jones, G., Jr., 121
Jones, K., 80
Jones, R., 255
Jones, R. N., 222, 227, 239
Joslin, T. A., 216
Jurs, P. C., 271, 280, 320
Juvet, R. S., Jr., 50, 58

Kaikara, M., 129
Kaiser, H., 16, 267, 278
Karchmer, J. H., 323
Karohl, J. G., 80
Keenan, T. A., 356
Keller, H. E., 178
Kelley, J. A., 121
Kelly, P. C., 58, 251
Kelly, W., 280
Kendrick, E., 270
Kenyon, W. C., 267
Keszthelyi, C. P., 215
Khidir, A. L., 239
Kieselbach, R., 87
Kimble, B. J., 121
Kirmse, D. W., 240
Kissel, G., 216
Kleir, M. J., 222
Klimowski, R. J., 158, 161, 256
Klinger, J., 6, 7, 176
Klopfenstein, C. E., 41
Knock, B. A., 280, 291
Kobylarz, T., 225
Kojima, H., 216
Koniver, D. A., 266, 306
Koopman, R. I., 215
Koryta, J., 216
Koutecky, J., 216
Kovats, E., 121
Kowalski, B. R., 7, 271, 280, 320
Krichevsky, M. I., 279
Kucera, E., 81
Kuga, T., 41
Kutter, M., 79
Kuwana, T., 215, 216

Lamm, P., 121
Lancaster, G., 121
Lauer, G., 80, 153, 178
Lawler, J. E., 151
Lawrence, J., 178
Lawson, A. M., 121
Lederberg, J., 112, 155, 271, 279, 280, 328, 334, 354
Ledley, R. S., 112

Lee, J., 128
Lee, R., 50
Leemans, F. A. J. M., 279
Lefkovitz, D., 306
Leiter, D. P., 270
Lemas, M. R., 58
Lepley, A. R., 6, 126
Levine, M. D., 320
Levine, S. P., 112, 121
Levinson, S., 129
Levinthal, E. C., 279
Lewis, D. R., 50
Leys, J. A., 251
Li, C., 216
Lichte, F. E., 251
Lichtenstein, I., 80
Lineberger, W. C., 151
Lipsett, P. R., 128
Lipsky, S. R., 157, 161, 321
Lochmüller, C. H., 91
Lorenz, L. J., 91
Lovell, J., 50
Low, M. J. D., 16
Luyten, J. A., 121
Lytle, F. E., 290

McCloskey, J. A., 279
McCord, T. G., 180, 181
McCullough, R. D., 80
MacDonald, H. D., Jr., 178
McDougal, A. O., 80
McFadden, W. H., 112, 279
McFarland, J., 50
Mackay, I. S., 238
McLafferty, F. W., 121, 153, 154, 155, 158, 159, 160, 161, 256, 257, 259, 261, 280, 291, 328, 330, 337
MacLaughlin, P. E., 231
McMurray, W. J., 157, 161, 321
McMurry, J. E., 378
Macnaughtan, D., 91
McPherron, R. V., 121
McWilliam, I. G., 239
Major, H. W., 159
Malmstadt, H. V., 251, 255
Maloy, J. T., 215
Mamer, O., 121
Mandelbaum, A., 280
Marcoux, L., 215, 216
Marde, Y., 161
Margoshes, M., 16
Mark, H. B., Jr., 104, 178
Markey, S. P., 112, 121
Markl, G., 251
Marshall, A. G., 238
Marshall, R. C., 80
Marson, S., 6

Mastragostino, M., 216
Matsuda, K., 194
Mattson, C. S., 311
Mattson, J. S., 178, 222, 311
Mead, T. E., 154
Melhuish, W. H., 128
Melkersson, S., 161
Mendel, J. M., 320
Menis, O., 251
Mertz, L. N., 230
Metcalf, P., 151
Meyerson, S., 330
Mielenz, K. D., 152
Milano, M. J., 50
Miller, L. A., 50
Miller, T. L., 104
Milligan, C. E., 380
Milne, G. W. A., 266, 306
Milne, M., 306
Minsky, M., 298, 320, 356
Misell, D. L., 240
Mitchell, A. C. G., 239
Mitchell, D. G., 50
Mitra, R. B., 377, 380
Mohilner, D. M., 178
Moler, F., 78
Mooney, R. W., 128
Moore, S. E., 255
Moreland, A. K., 91
Morgan, H. L., 270, 306
Morgan, S. L., 216
Morrey, J. R., 239
Morris, M. D., 12
Morrison, J. D., 270, 271, 280, 328, 330
Morton, G. A., 255
Moss, B. C., 231
Mowery, R. A., Jr., 58
Mueller, K. A., 91
Munk, M. E., 311
Murphy, J. A., 50
Murphy, M. K., 255
Murphy, R. C., 288
Myers, M. N., 81, 94, 112

Nadjo, L., 216
Nagy, G., 320
Nakajima, Y., 159
Nakamura, J. K., 255
Nau, H., 79, 121
Neff, B. L., 6, 126
Nelson, R. F., 216
Newell, A., 356
Nicholson, R. S., 216
Niemczyk, T. M., 255
Nilson, E. N., 216
Nilsson, N. J., 320
Noda, T., 159

Author Citation Index

Norris, J. A., 251
Nozoe, S., 379
Nunn, G., 78

Oberholtzer, J. E., 81, 91, 94
O'Brien, J. F., 330
O'Brien, T. J. P., 216
Ochiai, S., 280
O'Halloran, R. J., 216, 222
O'Haver, T. C., 12
Ohnesorge, W. E., 135
Ohno, M., 377
Okamoto, J., 159
Oliver, C. J., 255
Olmstead, M. L., 216
Olsen, R. W., 121, 157, 161, 256, 321
Organick, E. I., 356
Osteryoung, R. A., 80, 153, 178, 216
Ouki, T., 280
Overton, M. W., 194

Palumbo, D. T., 128
Pao, Y.-H., 255
Papert, S., 298, 320
Pardue, H. L., 50, 58, 91, 127, 153
Parker, C. A., 128, 129
Parker, R. A., 58
Passwater, R. A., 135
Pauling, L., 121
Pawle, G. B., 238
Pence, D. T., 216
Pennington, R. H., 162
Pereira, W. E., 112
Perone, S. P., 50, 91, 92, 153, 177, 178, 238, 320
Perrin, D. D., 270
Perry, J. A., 151
Persinger, H. E., 80
Peterson, E. W., 251
Peterson, G. V., 50
Petersson, G. A., 363
Pettersson, B., 271, 280, 328, 330
Peysna, G. M., 121
Phillips, D., 151
Piepmeier, E. H., 255
Pike, E. R., 255
Pilla, A. A., 194
Pitt, W. W., Jr., 112, 121
Pletcher, D., 216
Polcyn, D. S., 216
Poole, J. S., 50
Pople, J. A., 238
Power, R., 306
Powers, P., 328
Pratt, S. S., 128
Preti, G., 288

Price, J. M., 129
Prokof'ev, V. K., 227
Pruett, H. D., 255

Quantum Science Corp., 50

Rains, T. C., 251
Raisku, S. M., 227
Ralston, A., 356
Ramaley, L., 91
Ramsay, D. A., 222, 239
Ranganathan, R., 380
Ray, B. D., 279
Raznikov, V. V., 271, 280
Read, J. F., 267
Rees, W. T., 128, 129
Reilley, C. N., 216, 271, 280, 320
Reilly, C. A., 320
Reimendal, R., 121
Reinhard, M., 79
Reinhold, V. N., 288
Reynolds, D., 78
Reynolds, W. E., 112, 279
Rice, R., 50
Richards, J. A., 216
Richman, S., 306
Ridgway, T. H., 215, 216
Ridley, R. G., 280, 291
Rieman, T. A., 127
Rigdon, L. P., 178
Rindfleisch, T. C., 112, 121
Ritz, G. P., 12
Robben, F., 255
Robertson, A. V., 155, 280, 328, 334, 354
Robinson, A. B., 121
Roboz, J., 121, 256, 257
Rogers, L. B., 81, 91
Roitman, E., 121
Rolfe, J., 255
Rosen, C. A., 320
Ross, D. T., 81
Rowlett, R. J., 306
Ruhig, A., 121
Rusakowica, R., 129
Rutten, G. A. F. M., 121
Ruzic, I., 216
Ryan, M. D., 216
Ryhage, R., 161, 271, 280, 288, 328, 330

Sakamoto, M., 280
Salaita, G. N., 41
Saliger, H. H., 128
Sandifer, J. R., 216
Sarkozi, L., 121
Sasaki, S-I., 280
Saucedo, R., 194

Saunderson, J. L., 251
Savenant, J. M., 216
Savitzky, A., 58, 94, 222, 243
Schafer, J. A., 306
Schenck, H., 151
Schettler, P. D., 81, 94
Schiring, E. E., 194
Schneider, W. G., 238
Schrijver, H., 240
Schroeder, F., 134
Schroll, G., 280, 328, 334
Schultz, G. W., 50
Schultz, M. H., 216
Schwall, R. J., 176, 194
Schwarz, S. E., 255
Schwarz, W. M., 215
Schwarzenbach, R., 79
Scott, C. D., 112, 121
Scriver, C. R., 121
Seager, S. L., 95
Sebestyen, G. S., 320
Senn, M., 155
Shain, I., 215, 216
Shank, J. T., 80
Shannon, T. W., 154
Shapiro, K. S., 239
Shelley, C. A., 311
Shields, D. J., 157
Shima, M., 50
Shoesmith, D. W., 215
Shulman, G. P., 270
Shuman, M. S., 216
Siena, W. R., 50
Silverstein, M., 256
Simmonds, P. G., 270
Sindmack, G. K., 136
Sjövall, J. B., 121
Skogerboe, R. K., 251
Slavin, M., 251
Slavin, W., 128
Sluyters, J. H., 194
Sluyters-Rehbach, M., 194
Smith, D. E., 80, 176, 180, 181, 194, 216, 222
Smith, D. H., 112, 121, 157, 161, 256, 271, 311, 321
Smith, E. G., 266
Smith, G. D., 216
Smith, I. C., 280, 291
Smith, J. F., 330
Smith, R., 128
Snelleman, W., 251
Snelling, C. R., Jr., 311
Solomonoff, R. J., 320
Sommers, A. L., 128
Sondheimer, F., 216
Songco, D. C., 6, 126

Spencer, F. W., 311
Spenser, M. J., 311
Stainier, H. M., 157
Steed, E., 279
Stefik, M. J., 121
Steiner, J., 222, 243
Steinfeld, J. I., 151
Stembridge, C. H., 270
Stenhagen, E., 280, 288, 291
Stephens, F. B., 178
Sternberg, J. C., 81, 88
Sternberg, W., 356
Sterner, J., 251
Stewart, G. H., 95
Stiehl, G. L., 216
Stillwell, R. N., 78, 112
Stillwell, W. G., 78
Stobaugh, R. E., 270
Stokke, O., 121
Strojek, J. W., 216
Stuart, J. D., 12, 127
Stuki, L. R., 95
Summons, R. E., 112
Sutherland, G. L., 155, 280, 328, 334
Swalen, J. D., 41, 80
Swanson, R., 238
Swarg, G., 78
Sweeley, C. C., 112, 121, 279
Swingle, R. S., 91
Sybrandt, L. B., 178, 320
Szöke, J., 239

Tal'roze, V. L., 271, 280
Tamamushi, R., 194
Tantsyrev, G. D., 280
Tao, F., 80
Taranto, J., 330
Tate, F. A., 306
Teale, F. W. J., 132
Teets, R. E., 136
Termonia, Y., 222, 243
Testa, A. C., 129
Thoennes, D. J., 238
Thomas, J. W., 194
Thomas, R. A., 41
Thompson, G., 251
Thompson, J. A., 121
Thompson, R. H., 121
Thurman, R. G., 91, 94
Timnick, A., 136
Titus, C. A., 78
Titus, J. T., 50, 238
Tjoa, S., 121
Tobias, R. S., 127
Tou, J. T., 320
Tsuyama, H., 159

Author Citation Index

Tucker, R. B., 279
Tukey, J. W., 6, 141, 240
Tull, R. G., 255
Tunnicliff, D. D., 161, 259, 321
Turner, G. K., 128

Uda, H., 380
Urban, W. G., 112, 121

Valcher, S., 216
van Cittert, P. H., 239
Van Duyne, R. P., 215, 216
van Hulst, H. C., 239
Van Lear, G. E., 155, 160, 280, 328
Vatakencherry, P. A., 377
Vaupel, W., 306
Vavilov, S., 132
Veillon, C., 255
Venkataraghavan, R., 121, 154, 155, 158, 160,
    161, 222, 239, 256, 259, 261, 280, 328
Vianello, E., 216
Villivock, R. D., 121
Vink, H., 81
Von Neumann, J., 320

Wadsworth, P. A., 161, 259, 321
Wagenaar, H. C., 245
Wallen, D. J., 12
Waller, G., 256
Wallington, P., 328
Walls, F. C., 121, 157, 256
Walsh, J. L., 216
Walthall, F. G., 16
Wampler, R. H., 166
Wangen, L. E., 320
Warner, C. G., 154
Warren, B. E., 239
Watanabe, S., 320
Watts, R. W. E., 121
Waugh, J. S., 6, 126, 141
Webb, J. P., 151
Weber, G., 132
Weber, J., 216
Weber, J. H., 126
Webster, B. R., 155, 328
Weekley, B. E., 251
Wegner, P., 356
Weisbecker, J., 50
Weiss, C. D., 50
Weissberger, A. J., 50
Wenke, D. C., 255

West, T. S., 128
Westerberg, A. W., 80, 153, 240
White, C. E., 128
White, H. E., 251
Whitson, P. E., 216
Wiener, H., 58
Wightman, M., 328
Wikstrom, S., 161
Wilcox, R. H., 320
Wilkins, C. L., 238
Williams, D. H., 337, 354
Williams, D. J., 6, 126
Williams, R. S., 238
Willis, B. E., 58
Wilson, D., 41
Wilson, G. S., 91, 216
Wilson, J. M., 216
Wilson, R. M., 104
Winefordner, J. D., 239
Winograd, N., 216
Wipke, W. T., 6, 266, 356, 358, 361
Wise, S. A., 50
Witholt, B., 128
Wolfe, W. L., 227
Wolstenholme, M. J., 328
Wong, H. N. C., 216
Wood, H., 104
Wood, W. I., 279
Woodruff, H. B., 311
Woodward, B. W., 151
Woodward, M., 238
Woodward, R. B., 380
Wopschall, R. H., 216
Wright, D. E., 280, 291

Yacynych, A. M., 104
Yamazaki, H., 81
Yeager, W. J., 121
Yee, K. W., 251
Yguerabilde, J., 130
Young, A. T., 255
Young, D. S., 121
Young, F. M., 240
Young, L., 50
Young, N. D., 112, 121

Zaidel', A. N., 227
Zemansky, M. W., 239
Zemany, P. D., 279
Zipper, J. J., 177
Zitter, R. N., 255
Zlatkis, A., 121

SUBJECT INDEX

Algorithm
  for averaging scans in mass spectrometry, 166
  for chromatography, 47
  for comparison of mass spectra, 264
  for data acquisition in mass spectrometry, 165
  for on-line calibration, 163–164
  for ring perception, 363–364
  for searching mass spectra, 279–280, 294–295
Aliasing, 225, 235
Analog-to-digital (A/D) converters, 2, 54, 85–86, 92, 157, 164, 187, 232
  resolution of, 230–231
    random noise, effect on, 231
Apodization, 15
  assembly language, 45
Automation, 11, 18
  adaptive experimentation, 19
  atomic absorption, 126
  control, 18, 62
  hardware, 20
  interactive experimentation, 18
  in mass spectrometry, 125
  in nuclear magnetic resonance, 125
  on-line computation, 18
  of optical spectrometer, 49
  of photoacoustic spectrometer, 126
  of Raman difference spectrometer, 126
  software, 21
  of spectrometer, 48, 124
    data processing, 48
    instrument control, 48
  of vidicon spectrometer, 126

Background correction, 246–251
  approaches to, 246–247
  characteristics, 247
  characterization of, 248
  different methods, effect of, 250
  dynamic, 220
  schematic of, 248
  system for, 247–248
Bandwidth, 218, 225–226
Bias, dc, in data acquisition, 233

Chemical data, organization of, for computer synthesis, 366–375
Chemical structures, computer representation of, 300, 359–361
Chemometrics, 5
Chromatographic data, precision of, 98
Chromatographs, multiple, problems with automation, 47
Chromatography, algorithms for, 47
Computer programs
  DENDRAL, 309, 334–341, 345
  LHASA, 356–382
  MSSMET, mass spectral metabolite program, 114–116
Connection table, 360
Convergence rate, of decision makers, 317
Cross assemblers, 45
Cyclic voltammetry
  sampling technique for, 178–185
  schematic for, 179

Data acquisition, 18, 83–86
  bases, for substructure searches, 301
  collection, computer control of, 221
  compression, 265
  dc bias in, 223
  display, in liquid chromatography, 100–104
  flow chart, for mass spectrometry, 165
  in liquid chromatography, 100–104
  in mass spectrometry, 321–327
  with microcomputers, 59–73
  sampling, of transient waveforms, 178–185
  timing in, 234
Data enhancement, 218–222
Data handling, errors in, 219, 232–238

Subject Index

Data logger, 83
Data processing, in electrochemistry, 175
Deconvolution
  advantages, 245
  applications, 244–245
  broadening processes, 239
  evaluation of, 242–243
  in Faradaic admittance measurements, 190–192
  Fourier transforms in, 219, 240
  of Gaussian peak, 240–242
  methods of, 239–246
  noise, influence of, 243
  pseudo-, 239–240
  real time, 219, 239–245
DENDRAL, computer program, 309, 334–341, 345
Detector time constant, in gas chromatography, 87
Differential pulse polarography, microprocessor controlled, 176
Digital control, in gas chromatography, 83–86
Digital integrator, 46
  general features, 47
Digital logic system, for gas chromatography, 80–90
Digital-to-analog (D/A) converter, 54
Discriminant function analysis, 311
Distributed systems, microcomputers in, 48–49

Economics, and computerized instruments, 15
Editors, 45–46
Electrochemical mechanisms, analysis of, 195–216
  accuracy of, 199–202
  experimental data, analysis of, 202–215
  reliability of, 199–202
  simulation, finite difference, 198–199
  working curves, 197–202
Ensemble averaging, 259
Exclusive OR, 295

Feedback control, 48, 157, 159
Fellgett's advantage, 15
Fluorescence, 128–136, 150–151
  spectrometer, computer controlled, 128–136
    absorption correction, 135–136
    computer program, 130–132
    flow chart, 130
    instrumental variables, correction for, 128
    mathematical corrections, 129
    optical system, 129–130
    output, 131
    performance, 132–133
Fourier transform (FT), 14, 240
  data
    errors in processing of, 236–237
    handling in, 234–238
  for deconvolution, 219, 240
  fast, 174–175, 186
  infrared spectrometer, 15
  for interpolation of sampled data, 221–222
  in nuclear magnetic resonance
    basic configuration, 138
    data acquisition, 138–141
    instrumentation, 138–140
    interactive phase correction, 141–142
    memory overflow, 233
    resolution in, 235–236
    sensitivity enhancement, 142
    software, 140–141
    system, 137–143
  resolution, in FT-ICR and FT-IR, 236
  spectrometer, 223
  spectroscopy, errors in, 232–238
Free induction decay signal, 137–138, 224
Frequency domain analysis, 174, 186, 190–193

Gas chromatograph
  automation of, 48, 83–86
  computer controlled, 91–99
    block diagram, 92
    hardware, 92–94
    interface, 92–94
    output control functions, 94
    signal conditioning, 93
    software, 94–97
    system design, 91–92
Gas chromatography (GC)-mass spectrometry, 3, 77–78, 105–122, 264, 279
  automated analysis, 113–121
  detection of elutants, 107–108, 116
  library for, 116
  mass fragmentograms, 106
  mass spectra, extraction from background, 105–112
  peak shapes, 108
  profile analysis, problems with, 114
  quantitation, 116–120
  retention index, 116–117
  reverse library search, 118–119
  saturated peaks, reconstruction of, 109–110
  spectral intensities, estimation of, 108–109

unresolved elutants, 109, 110–112

Hadamard transform spectroscopy, 221
Hall effect probe, 161
Hash coding, 291
Heuristics, 356
Hewlett Packard 2114, 2115, 32

IBM 1800, 21, 26, 29, 32, 137, 284
Inclusive OR, 361
Information theory, 224–225, 265, 270, 291
Infrared
   Fourier transform spectrometry, 14
   high resolution, 15
   spectra, searching of, 267–269
Instrumental techniques, information available from, 153
Intel 8080, 147–148
Intelligent instruments, 47
Intel MCS 8008, 60, 72
Intel SIM 8-01, 52–53
Interferograms, 15
   sampling and digitizing, 223–231
Interpolation, of surfaces, 202–211
Interpretation, 4
   of C^{13}-NMR, 310
   of electrochemical data, 195–216
   of experimental data, 312–320
   of mass spectra, 309–310, 316–341
Interrupt, 53
   timing diagram for data handling, 54
Inverted files, 264–269
   characteristics of, 269
   computerized searching of, 267–269
   data in, 268
   flow chart, of search program, 269
   sifting factor in, 268
   small computer aspects of, 268–269
Ion-selective electrodes, computer control of, 176

Jacquinot's advantage, 15

Large-scale integration (LSI), 42
Learning machines, 312–320
   features of, 316–319
LHASA, computer program, 356–382
Logical AND, 295, 361

Mass spectra (MS)
   abbreviation of, 281
   computer techniques in analysis, 321–327
   data reduction and display, 155
   identification of, 279–288
      of formula and series, 324
      of multiplets, 325
   information content of, 277–278, 294
   interpretation of, 309, 316–341
      comparison of humans and computer, 333
      diagnostic peak flowchart, 331
      empirical formula, 329–330
      method, 328–333
      molecular weight, 329
   mass measurement, 154, 260, 322–323
   matching of, 270–278
      statistics of, 273–277
   one-bit encoding, 272–273
   on-line computers in, 153–160
   photoplate recorded, 154–156
   presearch of, 282–283
   real-time computing system, 156–157
      closed-loop feedback control in, 159
      design factors, 156
      hardware characteristics, 158
      interface, components of, 157
      software, characteristics of, 158–159
      specifications, of computer, 157
      system operation, 158
   reference file, 281–282
   resolving power, 154–155
   similarity index, 283–286
   simulation of, 342–354
      examples of, 350–354
   spectrum representation and coding, 292–293
Mass storage, 5
Michelson interferometer, 14
Microcomputer, 42
   applications, 46
      in GC/MS, 57
      in instrumental noise evaluation, 58
      on-line filtering of analog chromatographic signals, 55–57
      scanning dye laser, 144–152
      system for instrumental, 43
   in data acquisition, 59–73
      components, 63–65
      design considerations, 65
      features, 61–63
      philosophy of, 61–63
      system organization, 61
   design of, 43
      procedures, 68–70
      steps in, 51–55
   development of, recommended procedures, 43
   distributed systems, 48–49
   general purpose, 51–58
   hardware considerations, 44
   schematic of, 53

Subject Index

software for, 72–73
  basic tools, 45
  considerations, 45
  implementation steps, 45
Microprocessors, 2, 11–12, 42–73
  characteristics, 44
  in chemical instrumentation, 42–50
  elements of, 42
Moving average, 55–56
MSSMET, mass spectral metabolite program, 114–116
Multiplex advantage, 15

NIA-EPA substructure search system, 300–306
Noise
  coherent, 233
  in deconvolution, 243
  errors due to, in mass spectrometry, 169–171
  in mass spectra, 278
Nyquist frequency, 234
Nuclear magnetic resonance (NMR). See also Fourier transform
  interpretation of spectra, 310, 334, 336–337
  pulsed, 15

On-line computer, for mass spectrometry, 161–172
  calibration algorithm, 163–164
  configuration, 161
  data reduction, 164
  interfaces, 162
  scan algorithm, 162
Operational amplifiers, 1, 2, 92–93, 190

Pattern recognition, 312–320
  decision development in, 314–315
  feature selection in, 313–314
  prediction, 317–318
  reliability, 317
  stages of, 313–316
PDP-1, 356
PDP-7, 29, 30
PDP-8, 31, 33, 35, 130, 158, 161, 180, 254, 258
PDP-10, 21, 27, 356
PDP-11/45, 110
Peak detection, errors in, 237
Phase correction, 15
Photon-counting spectrometer, 252–255
  advantages of, 252–253
Plotting, errors in, 237
Precision, effect of measurement time on, 249

Problem solving, approach of synthetic chemist, 359
PROM, 43

Radiofrequency interference, shielding digital electronics from, 145
RAM, 42
Rand tablet, 358, 378
Rapid scanning spectrometry, in liquid chromatography, 100–104
Raytheon-704, 100
Real time
  clock, 83
  monitor, advantage of, 38
Resolution
  in FT-ICR, 236
  in FT-NMR, 235
  in mass spectrometry, 259–260
ROM, 42

Sampling
  bad data, effect of, 230
  errors, simulation of, 228–231
  extra and missed points, effect of, 228–230
  interval, effect of, 226–228
  rate, effect of, 224
  technique for transient waveforms, 174, 178–185
    components for, 180–182
    program, 182–184
    signal conditioning, 181–182
  theorem, 86
  valve, for gas chromatography, 82, 88–90
Scanning dye laser, microprocessor controlled, 144–152
  control circuit, dye cell carriage, 148
  detection of equipment problems, 149–150
  grating drive and control, 147
  instrument design, 144–150
  interface schematic, 149
  microprocessor control, 147–149
  programs for, 149
  RFI shielding, 45
  trigger circuit, for laser, 145
Searches
  fragments, 303–304
  of mass spectra, 279–288
    algorithm for, 294–295
    coding logic, 292–294
    methods for, 290–292
    programs, 295–296
    reduction of dimensionality, 293–294
    system for, 290–299
    techniques for, 280

match coefficient, 115–117
molecular formula, 303
retention index in, 116–119
reverse library search, 118–119
rings, 304
structural features, 305–306
for structures, 302
for substructures, 302
steps in, 302–305
system, for interactive, 300–306
Signal
averaging, 232
dependent scanning, 253
enhancement, 220
in mass spectroscopy, 256–261
to noise ratio, 15, 220, 232, 253, 259
factors in photon counting, 253–254
Simplex, 202, 208–210
interpolation, 208–215
Simulation
of double potential step current, charge and absorbance responses, 195–216
of errors in data, 219
of formation of ions in mass spectrometry, 342–354
of passivating film on electrode, 175
of resolution in A/D converters, 231
of sampling errors, 228–231
of surface reactions, kinetic behavior of, 175
Software, 11, 50, 72–73
for FT-NMR, 140–141
for gas chromatography, 47, 94–95
Spline fit interpolation, 167
Spline functions, 203–205

Statistical moments, 80, 87–88
Stepping circuits, 257
Structural information
storage of, 361
types of, 364
Superatoms, in interpretation of mass spectra, 335
SYDAGES, synchronous data generation and sampling system, 186
instrumentation, 187–190
programs, 190
Synthesis
analysis of, 355–382
approach to, 357
criteria for, 355
functional groups, perception of, 362–363
strategy selection, 375–382

Threshold logic unit (TLU), 315
Throughput advantage, 15
Time-shared computer, 21, 25, 28, 156
configuration of, 27
hierarchical, 39
philosophy of, 21–22
real time system, characteristics and disadvantages, 37–38
services of, 66–68
Timing jitter, effect on data acquisition, 234
Transforms, for computer synthesis, 366–375

Waveform, properties of pseudo-random, 186

About the Editors

THOMAS L. ISENHOUR is a professor of chemistry at the University of North Carolina at Chapel Hill where he teaches general and analytical chemistry. He received the B.S. degree at the University of North Carolina in 1961 and the Ph.D degree in 1965 at Cornell University. Professor Isenhour held the I. M. Kolthoff Senior Fellowship in Anayltical Chemistry at the Hebrew University in Jerusalem from October through December 1980. He has authored or coauthored eight books and over 90 technical papers.

JOSEPH B. JUSTICE, Jr., is an assistant professor of chemistry at Emory University. He received the B.S. degree from Rutgers University, New Brunswick, New Jersey, and the Ph.D. degree in 1974 from the University of North Carolina. His research interests include the analytical chemistry of the nervous system and microcomputer-controlled instrumentation for *in vivo* measurements. He is a member of Sigma Xi and has authored or coauthored over 20 technical papers.